ANNALS *of* THE NEW YORK ACADEMY OF SCIENCES

EDITOR-IN-CHIEF
Douglas Braaten

ASSOCIATE EDITOR
Rebecca E. Cooney

PROJECT MANAGER
Steven E. Bohall

EDITORIAL ADMINISTRATOR
Daniel J. Becker

Artwork and design by Ash Ayman Shairzay

The New York Academy of Sciences
7 World Trade Center
250 Greenwich Street, 40th Floor
New York, NY 10007-2157

annals@nyas.org
www.nyas.org/annals

Published by Blackwell Publishing
On behalf of the New York Academy of Sciences

Boston, Massachusetts
2011

ANNALS *of* THE NEW YORK ACADEMY OF SCIENCES

VOLUME 1244

ISSUE

Responding to Climate Change in New York State

The ClimAID Integrated Assessment for Effective Climate Change Adaptation Final Report

Climate change is already beginning to affect New York State, and these impacts are projected to grow. At the same time, the state has the ability to develop adaptation strategies to prepare for and respond to climate risks now and in the future. The ClimAID assessment provides information on climate change impacts and adaptation for eight sectors in New York State: water resources, coastal zones, ecosystems, agriculture, energy, transportation, telecommunications, and public health. Observed climate trends and future climate projections were developed for seven regions across the state. Within each of the sectors, climate risks, vulnerabilities, and adaptation strategies are identified. Integrating themes across all of the sectors are equity and environmental justice and economics. Case studies are used to examine specific vulnerabilities and potential adaptation strategies in each of the eight sectors. These case studies also illustrate the linkages among climate vulnerabilities, risks, and adaptation, and demonstrate specific monitoring needs. Stakeholder participation was critical to the ClimAID assessment process to ensure relevance to decision makers across the state.

Responding to Climate Change in New York State: The ClimAID Integrated Assessment for Effective Climate Change Adaptation in New York State

Final Report

Editors
Cynthia Rosenzweig, William Solecki, Arthur DeGaetano, Megan O'Grady, Susan Hassol, and Paul Grabhorn

Team Leaders
Frank Buonaiuto
Stephen A. Hammer
Radley Horton
Klaus Jacob
Patrick L. Kinney
Robin Leichenko
Andrew McDonald
Lesley Patrick
Rebecca Schneider
David W. Wolfe

Layout and Graphics
Joshua Weybright, Paul Grabhorn

Prepared for
The New York State Energy Research and Development Authority
Albany, NY
Amanda Stevens, Project Manager
Mark Watson, Program Manager
Full report may be found at www.nyserda.ny.gov

NYSERDA
Report 11-18

NYSERDA 10851

---

doi:10.1111/j.1749-6632.2011.06331.x

ClimAID Leadership Team

Cynthia Rosenzweig (PI), NASA Goddard Institute for Space Studies and Columbia University
William Solecki (PI), City University of New York, CUNY Institute for Sustainable Cities (CUNY CISC)
Arthur DeGaetano (PI), Northeast Regional Climate Center, Department of Earth and Atmospheric Science, Cornell University
Amanda Stevens (Project Manager), New York State Energy Research and Development Authority (NYSERDA)
Mark Watson (Program Manager, Environmental Research), New York State Energy Research and Development Authority (NYSERDA)
Megan O'Grady (Project Manager), Columbia University
Lesley Patrick (Project Manager), City University of New York
Susan Hassol (Science Writer), Climate Communication, LLC
Paul Grabhorn (Graphic Designer), Grabhorn Studio, Inc.
Josh Weybright (Graphic Designer), Graphic Sky, Inc.

ClimAID Teams

Climate

Radley Horton (Lead), Columbia University
Daniel Bader, Columbia University
Arthur DeGaetano, Cornell University
Cynthia Rosenzweig, Columbia University
Lee Tryhorn, Cornell University
Richard Goldberg, Columbia University

Adaptation and Vulnerability

William Solecki, City University of New York
Lee Tryhorn, Cornell University
Arthur DeGaetano, Cornell University

Equity and Environmental Justice

Robin Leichenko (Lead), Rutgers University
Peter Vancura, Rutgers University
Adelle Thomas, Rutgers University

Economics

Yehuda Klein (Lead), City University of New York
Robin Leichenko, Rutgers University
David C. Major, Columbia University
Marta Panero, New York University

Water Resources

Rebecca Schneider (Sector Lead), Cornell University
Andrew McDonald (Sector Lead), New York State Water Resources Institute
Stephen Shaw, Cornell University
Susan Riha, Cornell University
Lee Tryhorn, Cornell University
Allan Frei, City University of New York
Burrell Montz, East Carolina University

Coastal Zones

Frank Buonaiuto (Sector Lead), City University of New York
Lesley Patrick (Sector Lead), City University of New York
Ellen Hartig, New York City Department of Parks and Recreation
Vivien Gornitz, Columbia University
Jery Stedinger, Cornell University
Jay Tanski, Cornell University
John Waldman, City University of New York

Ecosystems

David W. Wolfe (Sector Lead), Cornell University
Jonathan Comstock, Cornell University
Holly Menninger, Cornell University
David Weinstein, Cornell University
Kristi Sullivan, Cornell University
Cliff Kraft, Cornell University
Brian Chabot, Cornell University
Paul Curtis, Cornell University

Agriculture

David W. Wolfe (Sector Lead), Cornell University
Jonathan Comstock, Cornell University
Alan Lakso, Cornell University
Larry Chase, Cornell University
William Fry, Cornell University
Curt Petzoldt, Cornell University

Energy

Stephen A. Hammer (Sector Lead), Massachusetts Institute of Technology, formerly with Columbia University
Lily Parshall, formerly with Columbia University

Transportation

Klaus Jacob (Sector Lead), Columbia University
George Deodatis, Columbia University
John Atlas, formerly with Columbia University
Morgan Whitcomb, Columbia University
Madeleine Lopeman, Columbia University
Olga Markogiannaki, Columbia University
Zackary Kennett, Columbia University
Aurelie Morla, Columbia University

Telecommunications

Klaus Jacob (Sector Lead), Columbia University
Nicholas Maxemchuk, Columbia University
George Deodatis, Columbia University
Aurelie Morla, Columbia University
Ellen Schlossberg, Columbia University
Imin Paung, Columbia University
Madeleine Lopeman, Columbia University

Public Health

Patrick L. Kinney (Sector Lead), Columbia University
Perry Sheffield, Mount Sinai School of Medicine
Richard S. Ostfeld, Cary Institute of Ecosystem Studies
Jessie L. Carr, Columbia University

NOTICE

ABSTRACT

Climate change is already beginning to affect New York State, and these impacts are projected to grow. At the same time, the state has the ability to develop adaptation strategies to prepare for and respond to climate risks now and in the future. The ClimAID assessment provides information on climate change impacts and adaptation for eight sectors in New York State: water resources, coastal zones, ecosystems, agriculture, energy, transportation, telecommunications, and public health. Observed climate trends and future climate projections were developed for seven regions across the state. Within each of the sectors, climate risks, vulnerabilities, and adaptation strategies are identified. Integrating themes across all of the sectors are equity and environmental justice and economics. Case studies are used to examine specific vulnerabilities and potential adaptation strategies in each of the eight sectors. These case studies also illustrate the linkages among climate vulnerabilities, risks, and adaptation, and demonstrate specific monitoring needs. Stakeholder participation was critical to the ClimAID assessment process to ensure relevance to decision makers across the state.

KEYWORDS

Climate change; adaptation; impacts; vulnerability; climate risk; sector impacts

ACKNOWLEDGMENTS

It has been an honor to work with the New York State Energy Research and Development Authority (NYSERDA) and the stakeholders of New York State to assess vulnerability and adaptation capacity related to climate change. At NYSERDA, we thank Janet Joseph, Vice President for Technology and Strategic Planning, Mark Watson, Program Manager, and especially Amanda Stevens, the ClimAID Project Manager who guided the assessment on a daily basis. Alan Belensz, formerly of the Climate Change Office of the NYS Department of Environmental Conservation, was a source of support throughout the ClimAID process. They are all exemplary civil servants committed to developing effective ways for the State to confront climate change challenges.

This report is the product of the work of the dedicated sector and integrating theme leaders and the team members of ClimAID. We express our sincere thanks to each of them for their contributions.

We thank Megan O'Grady and Lesley Patrick for their great work as the ClimAID Project Managers, without whom ClimAID could not have completed its tasks in such a comprehensive way. We are grateful to Daniel Bader for his technical climate expertise and for coordinating the finalization of the report. At the Goddard Institute for Space Studies and the Columbia Center for Climate Systems Research, we also thank Richard Goldberg and José Mendoza for their technical and graphics contributions. At the CUNY Institute for Sustainable Cities, we thank Michael Brady, Kristen Grady, Bridget Ripley, Andrew Maroko, and Andrew Lynch for technical expertise as well. The technical assistance provided by Brian Belcher at Cornell University is also appreciated.

It has been a tremendous experience to work with Susan Hassol and Paul Grabhorn, both national leaders in communicating the science of climate change to a broad array of citizens.

We thank the expert reviewers of the ClimAID assessment, without whom the independent provision of sound science for climate change adaptation cannot proceed.

Finally, we thank the many stakeholders who provided inputs and feedback to the ClimAID assessment. They generously shared their knowledge so that New York State can respond effectively to changing climate conditions.

Cynthia Rosenzweig, NASA GISS and Columbia University
William Solecki, Hunter College at City University of New York
Arthur DeGaetano, Cornell University

Integrated Assessment for Effective Climate Change Adaptation in New York State

Contents

Introduction....1
Chapter 1. Climate Risks....15
Chapter 2. Vulnerability and Adaptation....49
Chapter 3. Equity and Economics....61
Chapter 4. Water Resources....79
Chapter 5. Coastal Zones....121
Chapter 6. Ecosystems....163
Chapter 7. Agriculture....217
Chapter 8. Energy....255
Chapter 9. Transportation....299
Chapter 10. Telecommunications....363
Chapter 11. Public Health....397
Chapter 12. Conclusions and Recommendations....439
Annex I. Expert Reviewers for the ClimAID Assessment
Annex II. New York State Adaptation Guidebook
Annex III. An Economic Analysis of Climate Change Impacts and Adaptations in New York State

ClimAID: Integrated Assessment for Effective Climate Change Adaptation Strategies in New York State

Authors: Cynthia Rosenzweig, William Solecki, and Arthur DeGaetano

Contents

Introduction 2
Climate Change in New York State 3
ClimAID Team, Meetings, and Reviews 3
Stakeholder Interactions 4
Integrating Themes 4
    Climate 5
    Vulnerability 6
    Adaptation 6
    Equity and Environmental Justice 7
    Economics 8
Sectors 8
    Water Resources 8
    Coastal Zones 9
    Ecosystems 9
    Agriculture 10
    Energy 10
    Transportation 10
    Telecommunications 11
    Public Health 11
Case Studies 11
Assessment Outcomes 12
Appendix A. Project Advisory Committee Members 12
Appendix B. Stakeholder Organizations Engaged in ClimAID Assessment 13

Introduction

New York State is already experiencing impacts as a result of climate change, and impacts are projected to increase with further warming. At the same time, the state has great adaptive capacity to address them. From the Great Lakes to Long Island Sound, from the Adirondacks to the Susquehanna Valley, climate change will affect the people and resources of New York State. Risks associated with climate change include greater incidence of heat stress caused by more frequent and intense heat waves; greater incidence of heavy rainfall events affecting food production, natural ecosystems, and water resources; and sea level rise leading to increased flooding in coastal areas. Climate change may exacerbate existing stresses on the people and activities of New York State and, in some cases, might provide opportunities such as enhancement of its water resources and agricultural potential. The goals of the Integrated Assessment for Effective Climate Change Adaptation Strategies in New York State (ClimAID) are to provide New York State decision-makers with cutting-edge information on its vulnerability to, as well as its ability to derive benefits from, climate change and to facilitate the development of adaptation strategies informed by both local experience and scientific knowledge. Further aims of ClimAID are to highlight areas related to climate change and New York State that warrant additional research and to identify data gaps and monitoring needs in order to help guide future efforts.

Initiated in 2008, ClimAID is funded by the New York State Energy Research and Development Authority (NYSERDA) as part of its Environmental Monitoring, Evaluation, and Protection Program (EMEP). The assessment proceeds from the acknowledgement that the unique combination of natural resources, ecosystems, economic activities, and human population of New York State will not be immune to the impacts of climate change, and that local communities and the State as a whole, therefore, need to plan for and adapt to these effects. Climate change poses special challenges for New York State decision-makers related to the uncertainties inherent in future climate projections and the complex linkages among climate change, physical and biological systems, and socioeconomic sectors.

Working interactively with stakeholders, the ClimAID team focused on five integrating themes across a broad range of key sectors (**Figure 1**). The five integrating themes are climate, vulnerability, adaptation, equity and environmental justice, and economic costs associated with climate change impacts and adaptive measures, as well as the benefits of avoiding impacts. The five integrating themes were selected based on discussions with NYSERDA and sector stakeholders about factors of key relevance for responding to a changing climate in the state. The eight sectors are Water Resources, Coastal Zones, Ecosystems, Agriculture, Energy, Transportation, Telecommunications, and Public Health.

Climate
Vulnerability
Adaptation
Equity & Environmental Justice
Economics
Water Resources
Coastal Zones
Ecosystems
Agriculture
Energy
Transportation
Communication
Public Health

Figure 1 ClimAID integrating themes and sectors, illustrating the interwoven fabric of climate change assessment

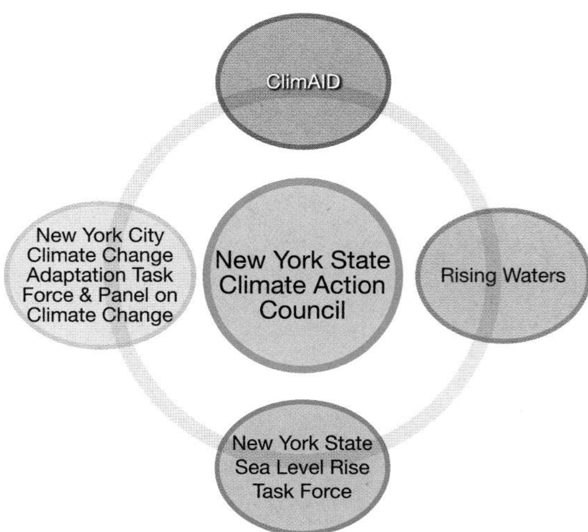

Figure 2 Interactions of ClimAID assessment with other climate change adaptation initiatives in New York State

Progress in reducing vulnerability and building adaptive capacity to respond to climate change depends on integrating the best available local and scientific knowledge with lessons learned from previous and current efforts. To that end, ClimAID built on key findings from prior assessments, and coordinated with concurrent research and policy initiatives on climate change adaptation, such as the Rising Waters project (a regional planning effort for the Hudson Valley convened by The Nature Conservancy and partners),[1] the New York State Sea Level Rise Task Force convened by the New York State Legislature,[2] the New York City Climate Change Adaptation Task Force led by the Mayor's Office of Long-Term Planning and Sustainability of New York City,[3] and the New York City Panel on Climate Change (NPCC) convened by Mayor Bloomberg of New York City[4] (**Figure 2**). The New York State Climate Action Council is preparing mitigation and adaptation policy recommendations for the Governor's Office.[5]

Climate Change in New York State

Climate change is already affecting and will continue to affect a broad set of activities across New York State. Its geographical and socioeconomic diversity means that New York State will experience a wide range of effects. There will be opportunities to explore new varieties, new crops, and new markets associated with higher temperatures and longer growing seasons. New York's relative wealth of water resources, if properly managed, can contribute to resilience and new economic opportunities. On the other hand, higher temperatures and increased heat waves have the potential to increase fatigue of materials in the water, energy, transportation, and telecommunications sectors; affect drinking water supply; cause a greater frequency of summer heat stress on plants and animals; alter pest populations and habits; affect the distribution of key crops such as apples, grapes, cabbage, and potatoes; cause reductions in dairy milk production; increase energy demand; and lead to more heat-related deaths and declines in air quality. Projected higher average annual precipitation and frequency of heavy precipitation events could also potentially increase the risks of several problems, including flash floods in urban areas and hilly regions; higher pollutant levels in water supplies; inundation of wastewater treatment plants and other vulnerable development in floodplains; saturated coastal lands and wetland habitats; flooded key rail lines, roadways, and transportation hubs; and travel delays. Sea level rise will increase risk of storm surge-related flooding, enhance vulnerability of energy facilities located in coastal areas, and threaten transportation and telecommunications facilities.

Across the varied geography of New York State, many individuals, households, communities, and firms are at risk of experiencing climate change impacts. Some will be especially vulnerable to specific impacts due to their location and lack of resources.

ClimAID Team, Meetings, and Reviews

Because New York is large and diverse, special emphasis in ClimAID was placed on integration and coordination so that climate change impacts and potential responses could be addressed coherently across the geographic regions and the multidimensional sectors of the state.

The ClimAID team was made up of university and research scientists who are specialists in climate change science, impacts, and adaptation. Researchers came primarily, but not exclusively, from Columbia University, Cornell University, and Hunter College of the City University of New York (See front matter for list of ClimAID team members). The team was organized into groups that addressed the five integrating themes and the eight sectors.

Approximately every six months over the period November 2008 to June 2010, ClimAID team members gathered from around the state for face-to-face meetings. The kickoff meeting was held in Albany in the fall of 2008 to present the scope of work, identify sectors and stakeholders, and set priorities. The second meeting was held early in 2009 at Cornell University and focused on initial findings, overall emerging messaging, and identifying common themes. The third meeting was again held in Albany in the fall of 2009 and provided a further update of the findings. The final ClimAID meeting was held at Hunter College CUNY to discuss the major conclusions. The team also held regular teleconferences, approximately every two weeks in the beginning and then once a month. The integrating theme groups interacted directly with each of the sector groups throughout the process.

A Project Advisory Committee, convened by NYSERDA, was made up of experts from the sectors covered by ClimAID (Appendix A). The committee met approximately every six months to review draft materials and to advise on the overall scope and direction of the assessment.

Besides the Project Advisory Committee reviews, external reviews were conducted for the sector chapters in the early summer 2009 and early in 2010. Each chapter was reviewed by multiple outside experts in relevant fields (see Annex I).

Stakeholder Interactions

To ensure that the information provided by ClimAID was relevant to the climate-related decisions made by practitioners, stakeholder interactions were a key part of the process (**Figure 3**). Working with NYSERDA and the Project Advisory Committee, the sector leaders identified relevant stakeholders from the public sphere (e.g., state and local agencies), nonprofit organizations (e.g., non-governmental community and environmental groups), private-sector entities (e.g., businesses), and academic institutions for each of the sectors, and organized the stakeholder interactions. (For a list of stakeholder organizations by sector, see Appendix B and the sector chapters).

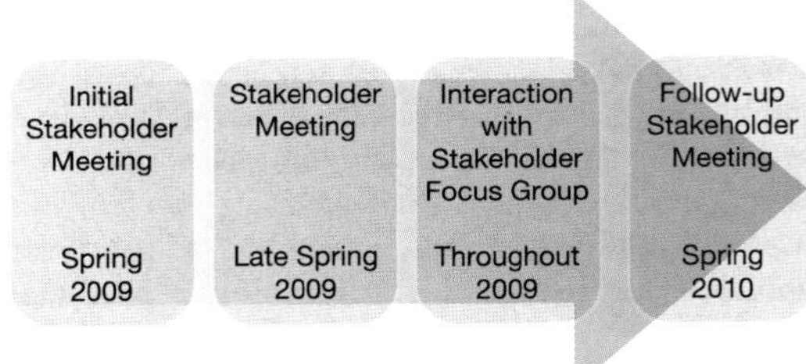

Figure 3 ClimAID sector stakeholder process

While each sector developed its own stakeholder process, sectors generally convened sector-specific stakeholder meetings in the first four-month period of ClimAID; developed, administered, and analyzed a survey to a wider group of sector stakeholders; formed a focus group of key stakeholders for ongoing discussion and advice throughout the assessment; and held a final stakeholder meeting in the third four-month period to present preliminary results and get feedback on draft conclusions and recommendations. See sector chapters for fuller descriptions of stakeholder interactions.

Integrating Themes

ClimAID developed five key themes to integrate across the eight sectors: climate, vulnerability, adaptation, equity and environmental justice, and economics.

Sector	Case Study Title
Water Resources	Susquehanna River Flooding, June 2006* Orange County Water Supply Planning
Coastal Zones	1-in-100-Year Flood and Environmental Justice* Modeling Climate Change Impacts in the Hudson River Estuary Salt Marsh Change at New York City Parks and Implications of Accelerated Sea Level Rise
Ecosystems	Hemlock—Cascading Effects of Climate Change on Wildlife and Habitat Creative Approaches to Monitoring and Adaptive Management—New York's Invasive Species Program as a Model Maple Syrup Industry—Adaptation to Climate Change Impacts Brook Trout—Reduction in Habitat Due to Warming Summers*
Agriculture	Frost Damage on Grapes Potato Late Blight Drought Dairy Heat Stress*
Energy	Impact of Climate Change on New York State Hydropower Climate Change-Induced Heat Wave in New York City*
Transportation	Future Coastal Storm Impacts on Transportation in the New York Metropolitan Region*
Telecommunications	Winter Storm in Central, Western, and Northern New York*
Public Health	Heat-related Mortality Among People Age 65 and Older* Ozone and Respiratory Diseases Extreme Storm and Precipitation Events West Nile Virus

*In-depth case study including economic and environmental justice analysis

Table 1 Case studies by sector

A group of ClimAID scientists focused on each of these themes, working with the sectors to ensure broad coordination across the assessment. Case studies of specific impacts and/or locations were selected for each of the sectors that provided special analysis of the five themes (**Table 1**). See chapters 1, 2, and 3 for detailed descriptions of the concepts and methods used in the integrating themes.

Climate

The Climate group analyzed both past and future climate in New York State (see "Climate," Chapter 1). Climate observations from across New York State obtained from the NOAA Northeast Regional Climate Center were used to analyze trends in key variables and thus to answer the question, "Is climate changing in New York State?" The Climate group also developed a set of climate change scenarios for New York State to facilitate the assessment of potential impacts under future conditions. The group assessed the degree to which current-generation climate models are able to replicate observed climate and climate trends over the past several decades in New York State and analyzed the relevant results for New York State of regional climate models and statistical downscaling. As part of this effort, results were analyzed from the North American Regional Climate Change Assessment Program (NARCCAP), which is conducting a coordinated set of current and future regional climate model simulations. The Climate group assessed global and regional climate model simulations on the basis of the availability of climate change simulations, spatial and temporal resolution, selection of climate variables, and accuracy in representing New York State climate.

The Climate group developed a set of climate change projections for New York State as a whole and for seven climate regions within the state based on 16 global

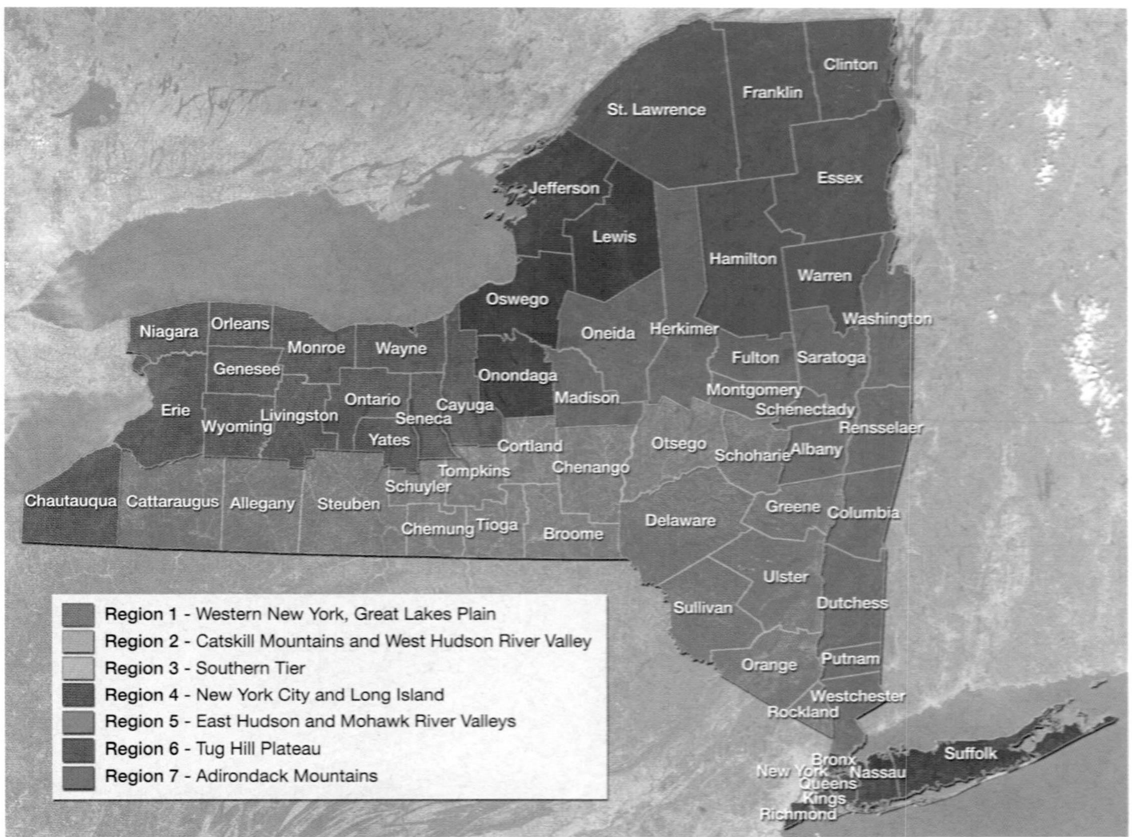

Figure 4 Climate regions of New York State

climate models and three emission scenarios (**Figure 4**). The outcomes are presented as model-based probabilities or "uncertainty envelopes"[6] around the projections. Because extreme climate events are associated with the greatest risks, the Climate group also developed model-based probabilities of the future occurrence of extreme events (e.g., heat waves, droughts, and floods) in New York State. The Climate group associated confidence levels with the projections, using a rating based on that employed by the Intergovernmental Panel on Climate Change Fourth Assessment Report (AR4) (IPCC, 2007). These ratings (e.g., "extremely likely," "very likely") are based on the correspondence between observations and climate model projections, agreement among climate models, and expert judgment.

For sea level rise, the Climate group developed a set of projections for the coastal area of New York State and conducted a historical analysis of current and historical storm damage on its infrastructure systems. The sea level rise projections were based on global climate models and methods with the addition of a "rapid ice-melt scenario" that takes into account current accelerated rates of ice melt in Greenland and Antarctica and documented rates of melting in paleoclimate records. The Climate group also analyzed tide gauge records, historical storms, future climate model simulations, and storm surge model simulations to assess the potential for changes in the spatial and temporal distribution of coastal storms (hurricanes and nor'easters) for the state.

Working with each of the sectors, the Climate group defined specific climate hazards as identified by stakeholders and produced tailored products for use in the sector assessments. Based on the vulnerabilities identified for each sector, the Climate group identified and developed a set of climate hazard indicators that could be monitored to track current climate trends and variability, and enable comparisons to historical data and future scenarios. These climate hazard indicators will help inform the development of appropriate adaptation programs and policies.

Throughout the assessment, the Climate group presented uncertainties surrounding climate modeling in general and in downscaling to regional levels within the state in particular, highlighting data gaps and monitoring needs, and indicating areas that require further research related to climate hazard and risk analysis, modeling uncertainty, and downscaling in New York State.

Vulnerability

The ClimAID Vulnerability group identified both near-term and longer-term climate vulnerabilities for New York State (see "Vulnerability and Adaptation," Chapter 2). To ensure that the assessment was aimed at the most pressing near-term climate impacts, the sectors worked with stakeholders to target vulnerabilities that currently affect the state, such as heat waves, floods, droughts, and coastal and inland flooding. The assessment also identified future climate vulnerabilities in order to provide information that will enable the state to take early action to reduce the possibility of catastrophic or large-scale climate impacts and/or to take full advantage of climate change opportunities.

Based on published literature and salience for New York State stakeholders, the Vulnerability group brought forward a range of criteria for identifying the climate change vulnerabilities in each sector. Key vulnerability criteria related to climate impacts in New York State include the following:

- magnitude
- timing (e.g., seasonality)
- persistence and reversibility
- likelihood (based on estimates of uncertainty)
- distributional aspects within a region or among socioeconomic groups
- importance of the at-risk systems
- thresholds or trigger points that could exacerbate the change

Through stakeholder interactions, analysis of previous studies, and case studies, the ClimAID assessment then evaluated how these and related criteria affect potential vulnerabilities in each sector. In communication with stakeholders, ClimAID then compiled a set of potential vulnerabilities related to climate change, highlighting those with higher impacts for New York State.

Adaptation

The ClimAID Adaptation group identified, developed, and assessed adaptation strategies for the eight sectors included in the assessment (see Chapter 2,

"Vulnerability and Adaptation"). Developing climate change adaptation strategies requires input from a breadth of academic disciplines as well as stakeholder experience to ensure that recommendations are both scientifically valid and practically sound. For each sector and in direct response to the climate hazards, key vulnerabilities, and stakeholder priorities identified in the assessment, the Adaptation group worked with the sector teams and stakeholders first to catalog existing adaptation practices already in place and then to develop and assess potential adaptation strategies to the expected climate impacts. The Adaptation group used empirical, quantitative, and qualitative methods, taking into account the relevant issues in each sector. To further expand the range of options, the group also conducted a benchmark study examining adaptation strategies such as health-alert systems already being implemented in other regions of the United States and the world.

The Adaptation group categorized existing and potential adaptations with respect to various mechanisms (see **Table 2**).

The potential for synergistic or unintended consequences of adaptation strategies was considered as part of the assessment process.

The Adaptation group also identified research gaps, data requirements, and monitoring needs in the area of climate change adaptation to help guide future research efforts in New York State. Of particular interest was the identification of existing linkages between climate science and existing and potential policy adjustments, as well as opportunities to enhance these science-policy linkages through the identification of co-benefits and other conditions. Co-benefits are positive effects that adaptation actions can have on mitigating climate change (e.g., reduction of greenhouse gas emissions) or on improving other aspects of the lives of New York State citizens. An example of a mitigation co-benefit is the establishment of green roofs that keep residents cooler while reducing the use of air conditioners, thereby reducing fossil fuel emissions at power plants. An example of a co-benefit with other aspects is the upgrading of combined sewer and stormwater systems to reduce current pollution, while helping to prepare for future climate change impacts.

Adaptation Mechanism	Definitions
Type	Behavior, management/operations, infrastructure/physical component, risk-sharing, and policy (including institutional and legal)
Administrative group	Private vs. public; governance scale – local/municipal, county, state, national
Level of effort	Incremental action, paradigm shift
Timing	Years to implementation, speed of implementation (near-term/long-term)
Scale	Widespread, clustered, isolated/unique

Table 2 Adaptation mechanisms and definitions

Equity and Environmental Justice

The equity and environmental justice component of ClimAID involves three types of parallel efforts: 1) development of equity and environmental justice assessments for each sector, based on a review of background literature in these areas, 2) participation in sector case studies, and 3) attention to participation of a broad range of groups or representatives in the sector meetings with stakeholders (see "Equity and Economics," Chapter 3).

For the selected case studies, the Equity and Environmental Justice group explored critical environmental justice issues with respect to intensity and extent of impacts and vulnerability and the potential for unintended consequences of adaptation. Both distributional and procedural aspects of equity and environmental justice were included in the assessment. In terms of distribution of climate hazards, there is an emphasis on identification of situations where particular groups may be systematically disadvantaged, either in terms of differences in vulnerability or capacity to adapt to climate change or in terms of the impacts of policies surrounding adaptation. Concerning distributional issues, the analysis is focused on inequalities in vulnerability to climate change, capacity to adapt to climate change, and effects of adaptation policies.

In terms of procedural elements, key considerations include incorporation of equity issues in adaptation discussions and policies, mechanisms for participation among a broad range of societal groups in future adaptation planning and policy efforts, and incorporation of equity and environmental justice stakeholders (such as associations of elderly, disabled, and health-compromised—for example, those suffering from asthma—people; low-income groups; farm workers; and small business owners) in climate vulnerability and adaptation assessments.

These distributional and procedural definitions of equity and environmental justice are used throughout ClimAID. The broader aims include consideration of potential inequalities associated with climate change along both traditional lines that have been identified within the environmental justice literature (e.g., underprivileged, minority groups), as well as along new lines that may emerge under an altered climatic regime (e.g., different-sized firms) or may result from the implementation of adaptation policies and plans.

Economics

The Economics group broadly surveyed the economic value of each of the eight ClimAID sectors as well as the potential damages and adaptation costs associated with climate change impacts. They also carried out a detailed economic assessment for a selected case study in each of the sectors regarding the monetary costs of climate change impacts and adaptation (see Chapter 3, "Equity and Economics"). The goal was to evaluate the economic costs associated with the impacts of and adaptations to climate change that are likely to affect the different sectors of the New York State economy. Where possible, variations in costs were calculated across time and space. Measures of economic impact reviewed in the analysis included human welfare losses incurred due to healthcare costs, lost income and wage differentials, and productivity and consumer losses.

The economic analysis builds on the impacts and adaptation information in each of the sectors as well as economic data from New York State and analyses of the costs of impacts and adaptation strategies elsewhere. Methods included interviews, risk-based assessment of key impacts of climate change on sectors, and the framework of cost-benefit analysis (recognizing its significant limitations in evaluating adaptation to climate change) to provide an overview of the costs of impacts and adaptation strategies.

For a selected case study in each sector, the Economics group worked with the sector groups and stakeholders to create a "short list" of potential impacts and adaptation strategies. The Economics group then ranked these impacts and strategies based on the potential costs and benefits (as avoided impacts from the vulnerability assessment) associated with each. Cost and benefit estimates were derived from standard pricing protocols and discount rate measures. For the selected case studies, the Economics group conducted economic analyses for different time horizons depending on the sector adaptations under consideration: short-term (i.e., actions within the next five years), medium-term (i.e., actions within the next five to 15 years), and long-term (i.e., actions to be taken beyond 15 years) responses. The Economics group, when possible, included the expected lifetimes (e.g., for capital-intensive infrastructure), amortization times (often linked to bonds that public entities use to finance public projects), and discount rates.

Sectors

ClimAID assessed how Water Resources, Coastal Zones, Ecosystems, Agriculture, Energy, Transportation, Telecommunications, and Public Health in New York State are currently affected by climate, how they may be affected by future climate change, and how they may adapt. Stakeholders provided key information about climate-related decisions for each sector. Each sector chapter begins with a description of the sector followed by sections that present the five integrating themes of climate hazards, vulnerabilities, adaptation, equity and environmental justice, and economics specific to the sector. A key focus of each sector chapter is a highlighted case study with in-depth analysis of potential adaptation strategies for a major climate hazard related to the sector. Case studies of other key climate vulnerabilities and adaptation methods are included in the sector chapters as well. An Appendix to each sector chapter describes how stakeholders were engaged in the assessment.

Water Resources

Water resources are dependent on multiple interacting climate factors, including air temperature and the timing and quantity of snow, rainfall, and evaporation. The Water Resources sector emphasizes in Chapter 4 that water resources in New York State are already subject to numerous human-induced stresses and these pressures are likely to increase over the coming decades.

Potential vulnerabilities for water resources and related infrastructure described in Chapter 4 include flooding, increase in duration and/or frequency of dry periods affecting drinking water supplies in systems with low storage relative to demand, changes in demand for commercial and agricultural water related in part to climate-related factors, and declines in water quality due to higher water temperatures and decreased stream flows in summer. There may be enhanced opportunities for New York State as a potentially water-rich area in future climate conditions.

Examples of adaptation strategies for water resources detailed by the Water Resources sector for floods include 1) development of cost-effective stormwater-management infrastructure that enhances natural hydrologic processes (infiltration into soils, recharging groundwater, evaporation) and slows the movement of stormwater instead of rapidly conveying it to waterbodies, and 2) consideration of phased withdrawal of infrastructure from high-risk, flood-prone areas. For water supplies, adaptation strategies include establishment of guidelines for systematic management of water supplies under drought and implementation of an automatic gauging and reporting network to provide improved early-warning systems for supply shortages. For non-potable water supplies, the chapter suggests mechanisms for better coordination of water use in shared water bodies, the development of a public online system for tracing water usage across the state, establishment of minimum flow requirements for water withdrawals, and the preparation of a statewide water plan.

In regard to water quality, adaptation strategies evaluated in the chapter include design modifications to insure that regulatory requirements are met under current and future climate conditions; research and monitoring are needed to understand impacts of low-flows and higher temperatures on water quality, and potential changes to nutrient, sediment, and pathogen pollution in a changing climate.

Coastal Zones

The Coastal Zones sector in the ClimAID assessment focused on the regions close to the ocean, rather than on the coastal areas of the Great Lakes. Climate change vulnerabilities related to the Great Lakes are discussed in Climate Risks (Chapter 1), Water Resources (Chapter 4), Ecosystems (Chapter 6), Energy (Chapter 8), Transportation (Chapter 9), and Telecommunications (Chapter 10). Climate hazards related to coastal zones encompass the distinct but related factors of sea level rise, coastal storms, increasing coastal water temperatures, and changes in precipitation patterns. These hazards are likely to occur in combination. As highlighted in Chapter 5, coastal zones in New York State are already stressed by high levels of development, which tend to reduce groundwater recharge and degrade water quality in the region.

Potential vulnerabilities for coastal zones described in the Coastal Zones sector in Chapter 5 include more frequent coastal flooding over larger areas during storms, increased shoreline erosion leading to alteration of the coastline, changes in the location of the salt front in the Hudson River estuary, loss of coastal wetlands, and changes in fish and shellfish populations.

Examples of adaptation strategies for coastal zones detailed in Chapter 5 include incorporating climate change and sea level rise information into land-use planning (for instance, setback zones requiring that new coastal development be a minimum distance from the shore). Other adaptation strategies include preparation of a detailed inventory of shoreline assets located in at-risk areas; acquiring of open coastal land for storm protection, recreation, and ecosystems; development of design criteria for new infrastructure; design of retrofit and/or relocation options for existing infrastructure that are more flexible to changing conditions and periodic reassessment; creation of a dynamic framework for updating policy guidelines given the "moving target" of climate change; and establishment of a network of stakeholders and volunteers to assist in monitoring for sea level rise response and coordinating outreach and education efforts. Key tasks for climate change adaptation in the coastal zones are to monitor coastal hazard zones over time in order to determine optimal timing of adaptation measures and to coordinate efforts across the state.

Ecosystems

Climate hazards of particular relevance as detailed by the Ecosystems sector are warmer winter temperatures, increased frequency of summer heat stress, increased frequency of heavy rainfall events, and increased frequency of late summer droughts.

Chapter 6 characterizes the potential vulnerabilities of natural ecosystems to climate changes as the loss of spruce/fir forests in the Adirondacks and major shifts in tree species composition across the state, the loss of hemlock stands as a result of the wooly adelgid insect pest expanding its range northward, the effects on coldwater fish with repercussions for sport fishing, and the impacts on ski and snowmobile businesses.

Examples of adaptation strategies presented in Chapter 6 for ecosystem management include creation of migration corridors, reduction of human impacts in particularly vulnerable areas, and creation of more protected areas.

Agriculture

The Agriculture sector describes the climate change hazards of particular relevance in Chapter 7. These include warmer summer temperatures and longer growing seasons, increased frequency of summer heat stress and warmer winters, reduced snow cover, increased frequency of late-summer droughts, and increased frequency of heavy rainfall events.

Chapter 7 characterizes the potential vulnerabilities for agriculture as increased insects and diseases, heightened weed pressure, and the effects of excess water and drought. Key agricultural industries in the state that may be affected include dairy and livestock (via heat stress effects on productivity and changes in feed availability and prices), and poor spring bloom and yields of apples and other temperate fruit crops because of inadequate winter chill hours.

Examples of on-farm adaptation strategies described in the chapter for dairy and livestock industries include diet and feeding management; use of fans, sprinklers, and other cooling systems; and enhancement of cooling capacity in housing facilities. For crops, on-farm adaptations include shifting planting dates; diversification of crop varieties and crops; chemical and non-chemical control of insects, disease, and weeds; expanded irrigation capacity and other capital investments; and freeze and frost protection for perennial fruit crops. Chapter 7 explores adaptation strategies beyond the farm through such mechanisms as information delivery/extension systems, locally available design and planning assistance, disaster-risk management and insurance, financial assistance, and policy and regulatory decisions.

Energy

Climate hazards of particular relevance described by the Energy sector in Chapter 8 include anticipated changes in heating and cooling degree days; changes in hydrology affecting hydropower potential, including increased flooding along the coasts and in rivers and declines in streamflow; higher water temperatures; ice and snow storms; and wind. Based on the climate change projections, the Energy chapter evaluated the implications of changing loads on energy system operations in different parts of the state.

The energy sector explored climate vulnerabilities for electricity generation, transmission, and distribution. Potential vulnerabilities for energy supply include impacts on thermoelectric power generation and power distribution, impacts on natural gas distribution infrastructure, and impacts on renewable power generation. Potential vulnerabilities for energy demand include changes in total demand, seasonal variability, and peak demand.

Examples of adaptation strategies for the Energy sector described in Chapter 8 include changes in power dispatch rules to de-emphasize the use of vulnerable system assets; establishment of larger incentives to promote energy efficiency in order to reduce energy demand during extreme heat events and associated peak load demands; strategies to promote the more rapid deployment of distributed generation technologies (including solar, on-site combined heat and power technology, etc.) to both reduce demand on the grid and reduce site-specific system vulnerabilities; construction of additional power generation capacity to offset anticipated periodic losses in hydropower availability; changes in flood protection land-use practices to site power generation capacity in areas less vulnerable to flooding or extreme weather events; and requirements that utilities begin upgrading their transmission and distribution systems to prepare for demand growth associated with changing temperature levels around the state.

Transportation

Climate hazards of particular relevance to transportation include warmer summer and winter temperatures, increased precipitation, decreased snowfall, sea level rise, increased likelihood of heat waves, increased likelihood of coastal and inland floods, changes in extreme events including hurricanes and nor'easters, and potential changes in wind speed and patterns and associated changes in wave climate.

Examples of potential vulnerabilities for transportation from Chapter 9 include increased stress on materials from increased temperatures and precipitation, increased coastal flooding risks due to sea level rise and storm surges, increased inland flooding from more intense precipitation events (especially in urban and hilly areas), and potential impacts of saltwater intrusion on coastal infrastructure.

Examples of adaptation strategies for the Transportation sector described in Chapter 9 relate to coastal hazards, heat hazards, precipitation hazards, and winter storms including snow and ice. Strategies explored include raising the level of new critical infrastructure and essential service sites; including climate change adaptation knowledge when retrofitting older infrastructure; switching to more durable materials; changing land-use planning mechanisms; and creating increased resilience through flexible adaptation pathways in operations, management, and policy decisions.

Telecommunications

Climate hazards related to telecommunications described in Chapter 10 are extreme temperatures, heat waves, and intense precipitation; sea level rise, coastal floods, and storms; and ice storms. The primary vulnerability brought forward is communications outages caused by these climate hazards.

Adaptation strategies presented by the Telecommunications sector in Chapter 10 include tree trimming to avoid damage to existing wires, switching from aboveground to belowground infrastructure, the use of fiber optics rather than wires, and wireless systems instead of land lines. A general adaptive principle in the sector is to increase redundancy via the use of generators and backup solar-powered battery banks. Other useful adaptation strategies include relocation of central offices away from floodplains and diversification of telecommunications media, e.g., the continued development of high-speed broadband and wireless services throughout the state.

Public Health

Climate hazards described by the Public Health sector in Chapter 11 include increasing temperature (especially heatwaves), extreme precipitation and flooding events, and changing patterns of monthly temperatures and precipitation.

Vulnerabilities for public health detailed in Chapter 11 include illness and death associated with more frequent and severe heat waves. Cold-related death is projected to decrease, although increases in heat-related death are projected to outweigh reductions in cold-related death. Vulnerabilities related to climate change also include illness and death associated with ozone and fine-particle air pollution, asthma and other respiratory diseases including allergies associated with altered pollen and mold seasons, cardiovascular disease, and infectious diseases. Climate plays a strong role in the emergence and/or changing distributions of vector-borne diseases, such as those spread by mosquitoes and ticks.

Examples of adaptation strategies for public health from Chapter 11 include integrating specific information about climate-related vulnerabilities into ongoing programs of public health surveillance, prevention, and response, rather than developing new programs to deal with unique challenges; developing scenarios that integrate climate forecasts into planning around heat emergencies and heat-warning systems; and integrating climate forecasts into ongoing planning for air quality.

Case Studies

Case studies were done for each of the ClimAID sectors (**Table 1**) and are found at the ends of the chapters. Some of these served in particular as a crosscutting element across the sectors and as a way to highlight concrete climate change adaptation challenges and opportunities across the state. For these in-depth case studies, the ClimAID sectors identified, with input from stakeholders, high-priority vulnerabilities in each sector. The in-depth case studies targeted those areas, communities, subpopulations, or sub-sectors that experience frequent climate impacts under current climate conditions. The aim was to identify useful climate adaptation strategies, taking into account uncertainties in future climate projections. Through the case studies, the ClimAID team identified and illustrated the linkages connecting climate hazards, vulnerabilities, adaptation strategies, equity and environmental justice, and economics. Specific monitoring needs were identified as well.

Assessment Outcomes

The ClimAID assessment identified key climate change vulnerabilities and presented potential adaptation strategies for eight sectors in New York State. The assessment developed a coordinated set of climate change scenarios for the state as a whole and for seven regions within the state. This information contributed to the New York State Climate Action Council (CAC) process through the creation of a generalized set of adaptation guidelines and sector templates (see **Table 3** and Annexes I and II for a description of the CAC adaptation process and the ClimAID Climate Change Adaptation Guidebook and sector templates). Further ClimAID economic analyses that contributed to the CAC process are included in Annex III of this report.

The generalized set of guidelines described in the guidebook could be used by practitioners around the state to develop flexible yet prioritized responses to the risks of climate change. The guidebook provides a stakeholder guide to climate change adaptation, including a series of steps that can help to guide the process of considering how to assess vulnerabilities and establish adaptation plans within an organization. These were developed and tested as part of the activities of the Climate Action Council Adaptation Technical Working Group. Such decision-support tools aid development of science-based adaptation strategies and describe a coordinated approach to the development of effective adaptations among and across sectors. In some cases, climate change might provide opportunities to the state; these were brought forward as well.

Another ClimAID outcome is the identification of information gaps and research needs developed in conjunction with stakeholders and decision-makers. The ClimAID data, climate change projections, reports, and findings are publicly available on the Internet.

A key lesson of the ClimAID assessment is that such a coordinated approach is useful in dealing with the challenges and opportunities inherent in climate change and the complexities of integrating adaptation into the myriad of New York State activities.

Appendix A. Project Advisory Committee Members

Name	Affiliation
Jim Austin	NYS Department of Public Service
Alan Belensz	NYS Department of Environmental Conservation
Adam Freed	New York City Office of Long-Term Planning and Sustainability
John Kahabka	New York Power Authority
Naresh Kumar	Electric Power Research Institute
Jason Lynch	U.S. Environmental Protection Agency, Clean Air Markets
Lisa Moore	Environmental Defense Fund
Christina Palmero	NYS Department of Public Service
Barry Pendergrass	NYS Department of State
Ron Rausch	NYS Department of Agriculture and Markets
Patricia Reixinger	NYS Department of Environmental Conservation
Victoria Simon	New York Power Authority
James Wolf	Consultant
John Zamurs	NYS Department of Transportation

Review and assess existing climate stress conditions. Clarify goals and identify intersections of climate trends and changes with respect to achieving goals.
Compare with climate trends and change projections for New York State. How will these affect particular sectors?
Characterize adaptation strategies for operations and management, infrastructure, and policies for thresholds and ranges of key climate variables (e.g., temperature, sea level, storm surge, and precipitation) needed to maintain resilience.
Evaluate potential adaptation strategies (cost/benefit, environmental impacts) for management, infrastructure, and policy in the short, medium, and long terms; assess effectiveness and costs relative to benefits accrued; evaluate human capacity to respond or implement the strategies.
Prioritize flexible adaptation pathways (over decades).
Review climate trends and scenarios (at regular intervals).

Table 3 Climate change adaptation assessment guidelines (see ClimAID Climate Change Adaptation Guidebook, Annex II, for full description)

Appendix B. Stakeholder Organizations Engaged in ClimAID Assessment (see Sector Chapters for complete lists)

Water Resources

NYS Department of Environmental Conservation, NYS Department of Health, NYC Department of Environmental Protection, county and local water and sewer departments, U.S. Geological Survey, U.S. Army Corps of Engineers, Delaware River Basin Commission, NYS Soil and Water Conservation Committee, Cornell Cooperative Extension, Susquehanna River Basin Commission, Directors of Lake Associations, American Wildlife Conservation Foundation, American Public Works Association, Federation of NYS Solid Waste Association, NY Forest Owners Association, Ontario Dune Coalition, Director of State Wetland Managers and State Floodplain, municipal engineers, town planners, watershed council program managers, private engineering consultants.

Coastal Zones

NY District of U.S. Army Corps of Engineers, NYS Department of State, NYS Department of Transportation, NYS Emergency Management Office, NYC Department of Environmental Protection, Metropolitan Transportation Authority, The Nature Conservancy, the Port Authority of New York and New Jersey, NYC Office of Emergency Management, FEMA Region II, National Park Service, Stony Brook University, Suffolk and Nassau Counties, NYC Department of Parks and Recreation, DEC Hudson River Estuary Program and NYS Climate Change Office, New York Sea Grant Extension, NYC Climate Change Adaptation Task Force, NYC Panel on Climate Change, NYS Sea Level Rise Task Force.

Ecosystems

NYS Department of Environmental Conservation; U.S. Geological Survey; U.S. Fish and Wildlife Service; Cornell Cooperative Extension; The Nature Conservancy; National Wildlife Federation; Audubon New York; Wildlife Conservation Society; Adirondack Mountain Club; NY Forest Landowners Association; Empire State Forest Products Association; Olympic Regional Development Authority; land, fish, and wildlife managers; maple growers.

Agriculture

Cornell Cooperative Extension, Cornell Integrated Pest Management Program, NYS Department of Agriculture and Markets, U.S. Department of Agriculture, Finger Lakes Grape Growers Association, Sweet Corn Grower Association, crop consultants, individual farmer collaborators.

Energy

Electric Power Research Institute, New York State Energy Research and Development Authority, New York State Department of Public Service, New York State Independent System Operator, Alliance for Clean Energy New York, Cogentrix, Con Edison, Dynegy, FirstLight Power, Suez GDF, National Grid, NRG Energy, NY Power Authority, TransCanada, Ravenswood, USPowerGen, Environmental Energy Alliance of NY, AES, Long Island Power Authority, NYS Department of Environmental Conservation, New York City Mayor's Office of Long-Term Planning and Sustainability.

Transportation

NYS Department of Transportation, Metropolitan Transportation Authority, Port Authority of New York and New Jersey, Amtrak, U. S. Army Corps of Engineers, CSX Corporation, NYS Emergency Management Office, NJ TRANSIT.

Telecommunications

Verizon, AT&T, Sprint Nextel, T-Mobile, UPS, FedEx, Time-Warner Cable, Federal Communications Commission, Cablevision, CTANY, National Grid, New York City Mayor's Office, Department of Homeland Security, NYS Emergency Management Office, NYC Office of Emergency Management, NYS Department of Environmental Conservation, Sea Level Rise Task Force, NYS Department of Public Service, NYS Energy Research and Development Authority.

Public Health

NYS Department of Health; NYS Department of Environmental Conservation; NYC Department of Health and Mental Hygiene; U.S. Environmental Protection Agency Region II; U.S. Centers of Disease Control; National Oceanic and Atmospheric Administration; city, State, and federal governmental agencies in the areas of environment, health, planning, and emergency management; non-governmental environmental organizations; academic institutions with research interests in public health and climate change; environmental justice organizations; clinical health sector organizations.

[1] http://www.nature.org/wherewework/northamerica/states/ newyork/science/art23583.html
[2] http://www.dec.ny.gov/energy/45202.html
[3] http://www.nyc.gov/html/planyc2030/html/plan/climate_task-force.shtml
[4] http://www.nyas.org/Publications/Annals/Default.aspx
[5] http://nyclimatechange.us/InterimReport.cfm
[6] The model-based probabilities do not encompass the full range of potential climate changes, since future greenhouse gas emissions may be higher than the emissions scenarios used in their construction.

Chapter 1

Climate Risks

Authors: Radley Horton,[1,2] Daniel Bader,[2] Lee Tryhorn,[3] Art DeGaetano,[3] and Cynthia Rosenzweig[2,4]

[1] Integrating Theme Lead
[2] Columbia University Earth Institute Center for Climate Systems Research
[3] Cornell University Department of Earth and Atmospheric Sciences
[4] NASA Goddard Institute for Space Studies

Contents

Introduction......16
1.1 Climate Change in New York State......17
1.2 Observed Climate......18
1.2.1 Average Temperature and Precipitation......18
1.2.2 Sea Level Rise......19
1.2.3 Snowfall......19
1.2.4 Extreme Events......19
1.2.5 Historical Analysis......21
1.3 Climate Projections......23
1.3.1 Climate Model Validation......24
1.3.2 Projection Methods......27
1.3.3 Average Annual Changes......30
1.3.4 Changes in Extreme Events......33
1.4 Conclusions and Recommendations for Future Research......37
References......38
Appendix A. Uncertainty, Likelihoods, and Projection of Extreme Events......39
Appendix B. Indicators and Monitoring......42
Appendix C. Regional Climate Models......43
Appendix D. Statistical Downscaling in the ClimAID Assessment......47

Introduction

This chapter describes New York State's climate and the climate changes the state is likely to face during this century. The chapter contains: 1) an overview; 2) observed climate trends in means and extremes; 3) global climate model (GCM) validation, methods, and projections (based on long-term average changes, extreme events, and qualitative descriptions); and 4) conclusions and recommended areas for further research. To facilitate the linking of climate information to impacts in the eight ClimAID sectors, the state is divided into seven regions. Three appendices describe the projection methods, outline a proposed program for monitoring and indicators, and summarize the possible role of further downscaling climate model simulations for future assessments.

The climate hazards described in this chapter should be monitored and assessed on a regular basis. For planning purposes, the ClimAID projections focus on the 21st century. Although projections for the following centuries are characterized by even larger uncertainties and are beyond most current infrastructure planning horizons, they are briefly discussed in Appendix A because climate change is a multi-century concern.

Observed Climate Trends

- Annual temperatures have been rising throughout the state since the start of the 20th century. State-average temperatures have increased by approximately 0.6°F per decade since 1970, with winter warming exceeding 1.1°F per decade.
- Since 1900, there has been no discernable trend in annual precipitation, which is characterized by large interannual and interdecadal variability.
- Sea level along New York's coastline has risen by approximately 1 foot since 1900.
- Intense precipitation events (heavy downpours) have increased in recent decades.

Climate Projections

These are the key climate projections for mean changes and changes in extreme events.

Mean Changes

- Mean temperature increase is extremely likely this century. Climate models with a range of greenhouse gas emissions scenarios indicate that temperatures across New York State[1] may increase 1.5–3.0°F by the 2020s,[2] 3.0–5.5°F by the 2050s and 4.0–9.0°F by the 2080s.
- While most climate models project a small increase in annual precipitation, interannual and interdecadal variability are expected to continue to be larger than the trends associated with human activities. Projected precipitation increases are largest in winter, and small decreases may occur in late summer/early fall.
- Rising sea levels are extremely likely this century. Sea level rise projections for the coast and tidal Hudson River based on GCM methods are 1–5 inches by the 2020s, 5–12 inches by the 2050s, and 8–23 inches by the 2080s.
- There is a possibility that sea level rise may exceed projections based on GCM methods, if the melting of the Greenland and West Antarctic Ice Sheets continues to accelerate. A rapid ice melt scenario, based on observed rates of melting and paleoclimate records, yields sea level rise of 37–55 inches by the 2080s.

Changes in Extreme Events[3]

- Extreme heat events are very likely to increase and extreme cold events are very likely to decrease throughout New York State.
- Intense precipitation events are likely to increase. Short-duration warm season droughts will more likely than not become more common.
- Coastal flooding associated with sea level rise is very likely to increase.

A Note on Potential Changes in Climate Variability

Climate variability refers to temporal fluctuations about the mean at daily, seasonal, annual, and decadal timescales. The quantitative projection methods in ClimAID generally assume climate variability will remain unchanged as long-term average conditions shift. As a result of changing long-term averages alone, some types of extreme events are projected to become more frequent, longer, and intense (e.g., heat events), while events at the other extreme (e.g., cold events) are projected to decrease.

In the case of brief intense rain events (for which only qualitative projections can be provided), both the mean and variability are projected to increase, based on a combination of climate model simulations, theoretical understanding, and observed trends. Both heavy precipitation events and warm season droughts (which depend on several climate variables) are projected to become more frequent and intense during this century. Whether extreme multi-year droughts will become more frequent and intense than at present is a question that is not fully answerable today. Historical observations of large interannual precipitation variability suggest that extreme drought at a variety of timescales will continue to be a risk for the region during the 21st century.

1.1 Climate Change in New York State

Global average temperatures and sea levels have been increasing for the last century and have been accompanied by other changes in the Earth's climate. As these trends continue, climate change is increasingly being recognized as a major global concern. An international panel of leading climate scientists, the Intergovernmental Panel on Climate Change (IPCC), was formed in 1988 by the World Meteorological Organization and the United Nations Environment Programme to provide objective and up-to-date information regarding the changing climate. In its 2007 Fourth Assessment Report, the IPCC states that there is a greater than 90 percent chance that rising global average temperatures, observed since 1750, are primarily due to human activities. As had been predicted in the 1800s (Ramanathan and Vogelman, 1997; Charlson, 1998), the principal driver of climate change over the past century has been increasing levels of atmospheric greenhouse gases associated with fossil-fuel combustion, changing land-use practices, and other human activities. Atmospheric concentrations of the greenhouse gas carbon dioxide are now more than one-third higher than in pre-industrial times. Concentrations of other important greenhouse gases, including methane and nitrous oxide, have increased as well (Trenberth et al., 2007). Largely as a result of work done by the IPCC and the United Nations Framework Convention on Climate Change (UNFCCC), efforts to mitigate the severity of climate change by limiting levels of greenhouse gas emissions are under way globally.

Some impacts from climate change are inevitable, because warming attributed to greenhouse gas forcing mechanisms is already influencing other climate processes, some of which occur over a long period of time. Responses to climate change have grown beyond a focus on mitigation to include adaptation measures in an effort to minimize the current impacts of climate change and to prepare for unavoidable future impacts. Each ClimAID sector used the climate-hazard information described in this chapter to advance understanding of climate change impacts within the state, with the goal of helping to minimize the harmful consequences of climate change and leverage the benefits.

New York State was divided into seven regions for this assessment (**Figure 1.1**). The geographic regions are grouped together based on a variety of factors, including type of climate and ecosystems, watersheds, and dominant types of agricultural and economic activities. The broad geographical regions are: Western New York and the Great Lakes Plain, Catskill Mountains and the West Hudson River Valley, the Southern Tier, the coastal plain composed of the New York City metropolitan area and Long Island, the East Hudson and Mohawk River Valleys, the Tug Hill Plateau, and the Adirondack Mountains.

Climate analysis was conducted on data from 22 meteorological observing stations (**Figure 1.1**; **Table 1.1a**). These stations were selected based on a combination of factors, including length of record, relative absence of missing data and consistency of station observing procedure, and the need for an even

Figure 1.1 ClimAID climate regions. Circles represent meteorological stations used for the climate analysis

spatial distribution of stations throughout the regions and state.

Global climate model-based quantitative projections are provided within each region for:

- temperature,
- precipitation,
- sea level rise (coastal and Hudson Valley regions only), and
- extreme events.

The potential for changes in other variables is also described, although in a more qualitative manner because quantitative information for them is either unavailable or considered less reliable. These variables include:

- heat indices,
- frozen precipitation,
- lightning,
- intense precipitation of short duration, and
- storms (hurricanes, nor'easters, and associated wind events).

Station	Location	NYSERDA region	Data source	Length of coverage	Time-scale
Buffalo/Niagara International Airport	Buffalo	Region 1	COOP	1970–2008	Daily
Rochester International Airport	Rochester	Region 1	COOP	1970–2008	Daily
Geneva Research Farm	Geneva	Region 1	COOP	1970–2008	Daily
Fredonia	Fredonia	Region 1	COOP	1970–2008	Daily
Mohonk Lake	Mohonk Lake	Region 2	COOP	1970–2008	Daily
Port Jervis	Port Jervis	Region 2	COOP	1970–2008	Daily
Walton	Walton	Region 2	COOP	1970–2008	Daily
Binghamton Link Field	Binghamton	Region 3	COOP	1970–2008	Daily
Cooperstown	Cooperstown	Region 3	COOP	1970–2008	Daily
Elmira	Elmira	Region 3	COOP	1970–2008	Daily
Bridgehampton	Bridgehampton	Region 4	COOP	1970–2008	Daily
Central Park	New York	Region 4	COOP	1970–2008	Daily
Riverhead Research Farm	Riverhead	Region 4	COOP	1970–2008	Daily
Saratoga Springs 4 S	Saratoga Springs	Region 5	COOP	1970–2008	Daily
Yorktown Heights 1 W	Yorktown Heights	Region 5	COOP	1970–2008	Daily
Utica - Oneida Country Airport	Utica	Region 5	COOP	1970–2008	Daily
Hudson Correctional	Hudson	Region 5	COOP	1970–2008	Daily
Boonville 4 SSW	Boonville	Region 6	COOP	1970–2008	Daily
Watertown	Watertown	Region 6	COOP	1970–2008	Daily
Indian Lake 2 SW	Indian Lake	Region 7	COOP	1970–2008	Daily
Peru 2 WSW	Peru	Region 7	COOP	1970–2008	Daily
Wanakena Ranger School	Wankena	Region 7	COOP	1970–2008	Daily

Table 1.1a The 22 New York State stations used in regional baseline averages and extreme events

1.2 Observed Climate

This section describes New York State's mean climate, trends, and key extreme events since 1900. The climate and weather that New York State has experienced historically provides a context for assessing the climate changes for the rest of this century (Section 1.3.3 and Section 1.3.4).

1.2.1 Average Temperature and Precipitation

New York State's climate can be described as humid continental. The average annual temperature varies from about 40°F in the Adirondacks to about 55°F in the New York City metropolitan area (**Figure 1.2**). The wettest parts of the state—including parts of the Adirondacks and Catskills, the Tug Hill Plateau, and portions of the New York City metropolitan area—average approximately 50 inches of precipitation per year (**Figure 1.3**). Parts of western New York are relatively dry, averaging about 30 inches of precipitation per year. In all regions, precipitation is relatively consistent in all seasons, although droughts and floods are nevertheless not uncommon.

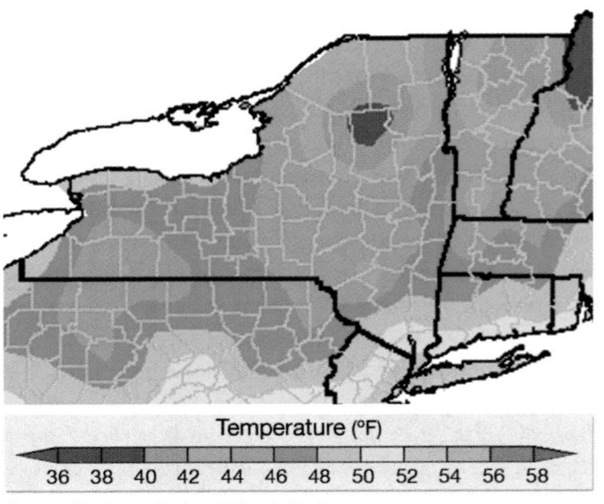

Source: Northeast Regional Climate Center

Figure 1.2 Normal average temperature in New York State

observed values at Indian Lake, but is representative of the region as a whole, which includes areas that receive more than 50 inches of precipitation per year.

Trends

Historical trend analysis is challenging for multiple reasons. First, over the historical period, the climate change signal from greenhouse gases was not as strong as it is expected to be during this century. Additionally, because the ocean and atmosphere in the climate models interact, the oceans in the models evolve independently from the real ocean through time. As a result, the global climate model historical simulations do not feature the same ocean temperatures and forcing that actually occurred at multi-year to decadal timescales. Thus, the role of natural variability relative to climate change in generating a trend in the models—or in the models relative to observations—cannot be easily assessed. Trends and statistical significance are therefore calculated independently for observations and models.

In Western New York, annual observed temperatures increased 0.2°F per decade over the 20th century. Only the fall trends were not significant at the 95 percent level. Modeled temperatures have warmed by 0.13°F per decade since 1900. The annual and seasonal model trends are all significant at the 99 percent level, with the greatest seasonal warming(0.17°F) present in winter. For the 1970–1999 period, the observed warming increased to 0.43°F per decade. No trends for the 1970–1999 observed period were significant. Over the same period, modeled annual warming was 0.34°F; both the modeled annual trend and the fall trend of 0.53°F per decade are significant at the 99 percent level.

The only significant trend in Rochester's observed average precipitation was for the fall season over the 20th century, at 0.20 inch per decade. The global climate models ensemble precipitation for the 20th century was significant annually and for all seasons but the summer. While the observed trends were not significant for the 1970–1999 period, the global climate model ensemble showed a significant increase in annual average precipitation.

For the Adirondack region (**Table 1.5**, Indian Lake station), the observed warming trend of 0.15°F per decade for the 1900s is well simulated by the global climate model hindcast of 0.14°F per decade. In the observations, approximately half of the warming is due to winter warming; in the global climate models, winter warming exceeds warming in other seasons, but each of the four modeled seasonal trends is similar and significant at the 99 percent level. Over the 1970–1999 period, the global climate model ensemble underestimates the observed annual temperature trend (0.34°F modeled versus 0.87°F observed per decade), although both trends are significant at the 99 percent level. While the observed warming during that time period is primarily in the winter, the global climate model ensemble warming is only significant at the 99 percent level in the summer and fall, when the warming trend in the model is also the largest.

1900–1999 Annual and Seasonal Temperature Trends (°F/decade)
Region 7 – Indian Lake***

100-year average temperature	17%	83%	ENS	Observed
December–February	0.05	0.29	0.16**	0.29*
March–May	0.00	0.26	0.12**	0.13
June–August	0.05	0.24	0.12**	0.05
September–November	0.07	0.21	0.15**	0.14*
Annual	0.04	0.27	0.14**	0.15**

1970–1999 Annual and Seasonal Temperature Trends (°F/decade)
Region 7 – Indian Lake

30-year average temperature	17%	83%	ENS	Observed
December–February	-0.48	0.84	0.16	2.02**
March–May	-0.36	0.67	0.22	0.70
June–August	0.17	0.56	0.40**	0.33
September–November	0.19	0.96	0.55**	0.48
Annual	0.1	0.59	0.34**	0.87**

1900–1999 Annual and Seasonal Precipitation Trends (inches/decade)
Region 7 – Indian Lake***

100-year average precipitation	17%	83%	ENS	Observed
December–February	-0.05	0.16	0.40*	-0.10
March–May	0.02	0.15	0.01	-0.01
June–August	-0.16	0.07	0.03**	-0.04
September–November	-0.02	0.12	0.06*	0.08
Annual	-0.01	0.41	0.14**	-0.06

1970–1999 Annual and Seasonal Precipitation Trends (inches/decade)
Region 7 – Indian Lake

30-year average precipitation	17%	83%	ENS	Observed
December–February	-0.15	0.42	0.15	-0.60
March–May	-0.35	0.29	-0.08	-0.24
June–August	-0.45	0.28	-0.02	-0.56
September–November	-0.22	0.40	0.13	-0.36
Annual	-0.44	0.80	0.10	-1.76

* Significant at the 95% level. ** Significant at the 99% level.
*** Observed data set came from Indian Lake, New York, 1901–2000.
Shown are the observed values for Indian Lake, the GCM ensemble average (ENS), and two points on the GCM distribution (17th and 83rd percentiles) representing the central range. Source: Columbia University Center for Climate Systems Research. Data are from WCRP and PCMDI

Table 1.5 Indian Lake validation

Indian Lake's observed average precipitation trends are not significant in any seasons for both the 1900s and 1970–1999 periods (**Table 1.5**). The same is true of the global climate ensemble for the 1970–1999 period; however for the 1900–1999 period, the ensemble shows statistically significant (99 percent) increases in precipitation both annually and during the summer.

In the coastal plain, the modeled annual temperature increases by 0.13°F per decade during the 1900s. This can be attributed to the 0.32°F per decade trend from 1970 through 1999. The observed 1970–1999 trend is greater at 0.67°F per decade. Observed per-decade temperature increases over the entire 1900s, however, are nearly triple that of the models, at 0.39°F. The 1900s model ensemble trend is similar in each season, while the 1970–1999 model ensemble shows the most temperature increase in the fall and summer. Observed temperature increases during the 1900s, by contrast, were largest in the winter and the smallest during the fall, though all seasons showed significant warming in all seasons. The entire observed warming trend during the past three decades can be attributed to winter warming.

The ensemble average model precipitation trend for the coastal plain is negligible over the 100-year record. The 1970–1999 30-year record shows a small increase of 0.18 inch per decade, due almost entirely to a small increase in winter precipitation. Nevertheless, in all four seasons, the central range of global climate models span from decreasing to increasing values. Over the 1970–1999 period, observed precipitation patterns show a small decrease in precipitation, which is due to decreases in summer and fall precipitation that outweigh increases in spring precipitation. This trend, however, is highly dependent on the selection of years, suggesting that 100-year trends for precipitation are more appropriate, given precipitation's high year-to-year and decade-to-decade variability in the region.

Validation Summary

While the global climate models are able to reproduce the state's climatology with limited biases, departures from observations over the hindcast period (due largely to spatial scale discontinuities between point data and GCM gridboxes)—are large enough to necessitate the use of climate change factors—future global climate model departures from global climate model baseline values—rather than direct model output. This finding provides a rationale for bias-correction such as the change factors or delta-method approach used for the ClimAID assessment (see section 1.3.3 for a description of this method).

The picture regarding trend validation is more complex. Ideally the global climate change factors from each model could be trained using historical trends, but this is not advisable for several reasons. While the 30-year modeled trends deviate from observations, these deviations do not necessarily indicate that global climate model sensitivity and regional response to

Climate Model Acronym	Institution	Atmospheric Resolution (latitude x longitude)	Oceanic Resolution (latitude x longitude)	References
BCCR	Bjerknes Center for Climate Research, Norway	1.9 x 1.9	0.5 to 1.5 x 1.5	Furevik et al., 2003
CCSM	National Center for Atmospheric Research, USA	1.4 x 1.4	0.3 to 1.0 x 1.0	Collins et al., 2006
CGCM	Canadian Center for Climate Modeling and Analysis, Canada	2.8 x 2.8	1.9 x 1.9	Flato 2005
CNRM	National Weather Research Center, METEO-FRANCE, France	2.8 x 2.8	0.5 to 2.0 x 2.0	Terray et al., 1998
CSIRO	CSIRO Atmospheric Research, Australia	1.9 x 1.9	0.8 x 1.9	Gordon et al., 2002
ECHAM5	Max Planck Institute for Meteorology, Germany	1.9 x 1.9	1.5 x 1.5	Jungclaus et al., 2005
ECHO-G	Meteorological Institute of the University of Bonn, Germany	3.75 x 3.75	0.5 to 2.8 x 2.8	Min et al., 2005
GFDL-CM2.0	Geophysical Fluid Dynamics Laboratory, USA	2.0 x 2.5	0.3 to 1.0 x 1.0	Delworth et al., 2006
GFDL-CM2.1	Geophysical Fluid Dynamics Laboratory, USA	2.0 x 2.5	0.3 to 1.0 x 1.0	Delworth et al., 2006
GISS	NASA Goddard Institute for Space Studies	4.0 x 5.0	4.0 x 5.0	Schmidt et al., 2006
INMCM	Institute for Numerical Mathematics, Russia	4.0 x 5.0	2.0 x 2.5	Volodin and Diansky, 2004
IPSL	Pierre Simon Laplace Institute, France	2.5 x 3.75	2.0 x 2.0	Marti, 2005
MIROC	Frontier Research Center for Global Change, Japan	2.8 x 2.8	0.5 to 1.4 x 1.4	K-1 Developers, 2004
MRI	Meteorological Research Institute, Japan	2.8 x 2.8	0.5 to 2.0 x 2.5	Yuikimoto and Noda, 2003
PCM	National Center for Atmospheric Research, USA	2.8 x 2.8	0.5 to 0.7 x 1.1	Washington et al., 2000
UKMO-HadCM3	Hadley Center for Climate Prediction, Met Office, UK	2.5 x 3.75	1.25 x 1.25	Johns et al., 2006

Table 1.6 Global climate models used in the ClimAID assessment

greenhouse gas forcing is incorrect in the models. For example, observed trends, especially for precipitation, also vary substantially based on the time period selected due to high year-to-year and decade-to-decade variability, which the models are not expected to experience concurrently with their freely evolving climate system. The fact that some important, regionally varying external forcings, including some aerosols, are not included in all the global climate models would be expected to further lead to departures from observations over the historical period. Finally, the models are missing local features that may have influenced the trends, including the urban heat island and precipitation island in those stations that are urban centers. In the New York metropolitan region, the heat island effect has been substantial (Rosenzweig et al., 2009; Gaffin et al., 2008). While these missing forcings may contribute to errors in the future, these errors are expected to become relatively less important as the warming role of increasing greenhouse gas concentrations becomes more and more dominant.

1.3.2 Projection Methods

For the ClimAID assessment, global climate models were used to develop a set of climate projections for New York State. Projections were made for changes in mean annual climate (Section 1.3.3) and extreme events (Section 1.3.4). Model-based probabilities for temperature, precipitation, sea level rise, and extreme events are created based on global climate model simulations (see **Table 1.6** for more information about the global climate models) and greenhouse gas emissions scenarios (IPCC, 2000) used in the IPCC Fourth Assessment Report (IPCC, 2007). This approach has been applied to many regions, including locally for New York City as part of the New York City Panel on Climate Change activities in support of New York City's Climate Change Adaptation Task Force (New York City Panel on Climate Change, 2010; Horton et al., 2010).

Emissions Scenarios

To produce future climate scenarios, global climate model simulations are driven with projected greenhouse gas emissions scenarios (**Figure 1.5**). Each emissions scenario represents a unique blend of demographic, social, economic, technological, and environmental assumptions (IPCC, 2000). The following three scenarios are used for this analysis:

A2: Relatively rapid population growth and limited sharing of technological change combine to produce high greenhouse gas levels by the end of this century, with emissions growing throughout the entire century.

A1B: Effects of economic growth are partially offset by introduction of new technologies and decreases in global population after 2050. This trajectory is associated with relatively rapid increases in greenhouse gas emissions and the highest overall carbon dioxide levels for the first half of this century, followed by a gradual decrease in emissions after 2050.

B1: This scenario combines the A1 population trajectory with societal changes tending to reduce greenhouse gas emissions growth. The net result is the lowest greenhouse gas emissions of the three scenarios, with emissions beginning to decrease by 2040.

Additional IPCC-based scenarios, such as the high-end A1FI scenario, yield moderately higher greenhouse gas concentrations (and therefore climate response) by the end of this century than the three scenarios indicated

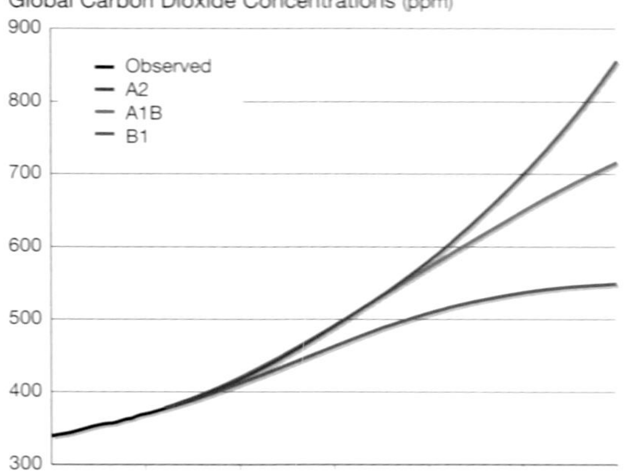

Based on IPCC emissions scenarios. Observed carbon dioxide concentrations through 2003 and future carbon dioxide concentrations in the A1B, A2, and B1 scenarios (2004 to 2100). Source: Columbia University Center for Climate Systems Research. Data are from WCRP and PCMDI

Figure 1.5 Future carbon dioxide concentrations used in the ClimAID assessment

above. High-end climate change scenarios along the lines of A1FI are discussed qualitatively, especially with regard to the rapid ice melt scenario. Such trajectories should continue to be monitored and reassessed over time. The A1FI scenario was not included in the model-based approach described here due to few available corresponding global climate model simulations.

Model-based Probability

The combination of 16 global climate models and three emissions scenarios produces a matrix with 48 scenarios for temperature and precipitation;[9] for each scenario time period and variable, the results constitute a model-based probability function. The results for the future time periods are compared to the model results for the 1970–1999 baseline period. Average temperature change projections for each month are calculated as the difference between each model's future simulation and the same model's baseline simulation, whereas average monthly precipitation is based on the ratio of a given model's future precipitation to the same model's baseline precipitation (expressed as a percentage change).[10] Sea level rise methods are more complex since sea level rise is not a direct output of most global climate models.

Sea Level Rise

The GCM-based methods used to project sea level rise for the coastal plain and Hudson River include both global components (global thermal expansion, or sea level rising as a result of increases in water temperature, and meltwater from glaciers, ice caps, and ice sheets) and local components (local land subsidence, i.e., sinking, and local water surface elevation).

Within the scientific community, there has been extensive discussion of the possibility that the GCM approach to sea level rise may substantially underestimate the range of possible increases. For this reason, an alternative rapid ice melt approach has been developed based on paleoclimate studies. Starting around 20,000 years ago, global sea level rose 394 feet; present-day sea level was reached about 8,000 to 7,000 years ago. The average rate of sea level rise during this 10,000 to 12,000-year period was 0.39–0.47 inch per year. This information is incorporated into the rapid ice melt scenario projections. More information on this method, including how it is integrated with the global climate model-based methods, can be found in Appendix A, "Rapid Ice Melt Sea Level Rise Scenario."

Extreme Events

Extremes of temperature and precipitation (with the exception of drought) tend to have their largest impacts at daily rather than monthly timescales. However, monthly output from climate models has more observational fidelity than daily output (Grotch and MacCracken, 1991), so a hybrid projection technique was employed for these events. The modeled mean changes in monthly temperature and precipitation for each of the 16 global climate models and three emissions scenarios were applied to each region's observed daily data from 1971 to 2000 to generate 48 time series of daily data.[11]

This is a simplified approach to projections of extreme events, since it does not allow for possible changes in variability through time. While changes in variability are generally highly uncertain (rendering the precise changes in extreme event frequency highly uncertain as well), changes in frequency associated with average monthly shifts alone are of sufficient magnitude to merit consideration by long-term planners as they develop adaptation strategies that prepare for extreme events.

Regional Projections

The projections for the seven regions of New York State are based on global climate model output from each model's single land-based model gridbox covering the center of each region. The precise coordinates of each model's gridboxes differ since each global climate model has a different spatial resolution. These resolutions range from as fine as about 75 by 100 miles to as coarse as about 250 by 275 miles, with an average resolution of approximately 160 by 190 miles. Changes in temperature (**Figure 1.6a**) and precipitation (**Figure 1.6b**) through time are region-specific (for example, 3°F degrees of warming by a given timeframe for a particular region). Neighboring regions, however, exhibit similar average changes in climate. This spatial similarity indicates that the average change results shown here are not very sensitive to how the region was defined geographically.

By applying the projected changes from the relevant gridbox to observed data, the projections become specific to the region. For example, although Rochester's projected change in temperature through time is similar to New York City's, the number of current and projected days per year with temperatures below 32°F degrees differs between the two locations because they have different baseline temperatures. Thus, the spatial variation in baseline climate is much larger than the spatial variation of projected climate changes.

Annual temperature change (°F)
4 4.5 5 5.5 6 6.5 7 7.5

Source: Columbia University Center for Climate Systems Research. Data are from WCRP and PCMDI

Figure 1.6a Projected change in annual temperature for the 2080s in the Northeast relative to the 1980s baseline period

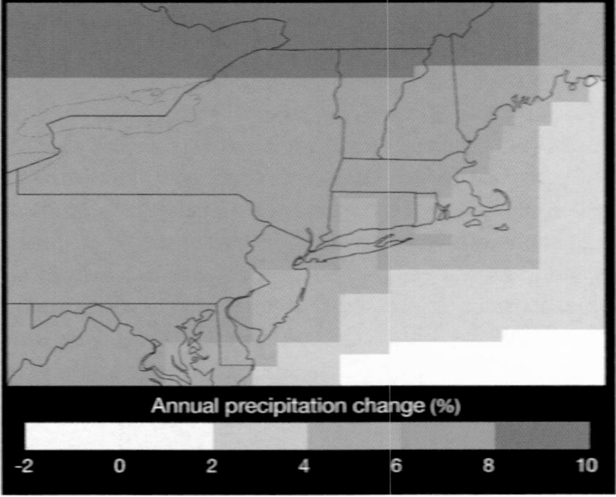

Source: Columbia University Center for Climate Systems Research. Data are from WCRP and PCMDI

Figure 1.6b Projected change in annual precipitation for the 2080s in the Northeast relative to the 1980s baseline period

		Baseline[1] 1971–2000	2020s	2050s	2080s
Region 1					
Stations used for Region 1 are Buffalo, Rochester, Geneva and Fredonia.	Air temperature[2]	48°F	+1.5 to 3.0°F	+3.0 to 5.5°F	+4.5 to 8.5°F
	Precipitation	37 in	0 to +5%	0 to +10%	0 to 15%
Region 2					
Stations used for Region 2 are Mohonk Lake, Port Jervis, and Walton.	Air temperature[2]	48°F	+1.5 to 3.0°F	+3.0 to 5.0°F	+4.0 to 8.0°F
	Precipitation	48 in	0 to +5%	0 to +10%	+5 to 10%
Region 3					
Stations used for Region 3 are Elmira, Cooperstown, and Binghamton.	Air temperature[2]	46°F	2.0 to 3.0°F	+3.5 to 5.5°F	+4.5 to 8.5°F
	Precipitation	38 in	0 to +5%	0 to +10%	+5 to 10%
Region 4					
Stations used for Region 4 are New York City (Central Park and LaGuardia Airport), Riverhead, and Bridgehampton.	Air temperature[2]	53°F	+1.5 to 3.0°F	+3.0 to 5.0°F	+4.0 to 7.5°F
	Precipitation	47 in	0 to +5%	0 to +10%	+5 to 10%
Region 5					
Stations used for Region 5 are Utica, Yorktown Heights, Saratoga Springs, and the Hudson Correctional Facility.	Air temperature[2]	50°F	+1.5 to 3.0°F	+3.0 to 5.5°F	+4.0 to 8.0°F
	Precipitation	51 in	0 to +5%	0 to +5%	+5 to 10%
Region 6					
Stations used for Region 6 are Boonville and Watertown.	Air temperature[2]	44°F	+1.5 to 3.0°F	+ 3.5 to 5.5°F	+4.5 to 9.0°F
	Precipitation	51 in	0 to +5%	0 to +10%	+5 to 15%
Region 7					
Stations used for Region 7 are Wanakena, Indian Lake, and Peru.	Air temperature[2]	42°F	+1.5 to 3.0°F	+3.0 to 5.5°F	+4.0 to 9.0°F
	Precipitation	39 in	0 to +5%	0 to +5%	+5 to 15%

[1] The baselines for each region are the average of the values across all the stations in the region.
[2] Shown is the central range (middle 67%) of values from model-based probabilities; temperature ranges are rounded to the nearest half-degree and precipitation to the nearest 5%.
Source: Columbia University Center for Climate Systems Research. Data are from USHCN and PCMDI

Table 1.7 Baseline climate and mean annual changes for the 7 ClimAID regions

Projections for extreme events use baseline climate and projected changes in temperature, precipitation, and sea level rise relative to the given baseline for the timeslices, which are defined by averaging all 22 stations within a given region (**Table 1.7**).

Timeslices

Although it is not possible to predict the temperature, precipitation, or sea level for a particular day, month, or even specific year due to fundamental uncertainties in the climate system, global climate models can project the likely range of changes over decadal to multi-decadal time periods. These projections, known as timeslices, are expressed relative to the given baseline period, 1970–1999 (2000–2004 for sea level rise). The timeslices are centered around a given decade. For example, the 2050s timeslice refers to the period from 2040–2069.[12] Thirty-year timeslices (10 years for sea level rise) are used to provide an indication of the climate normals for those decades. By averaging over this period, much of the random year-to-year variability—or noise—is cancelled out,[13] while the long-term influence of increasing greenhouse gases—or signal—remains (Guttman, 1989; WMO, 1989).

1.3.3 Average Annual Changes

Higher temperatures and sea level rise are extremely likely for New York State. For temperature and sea level rise, all simulations project continued increases over the century, with the entire central range of the projections indicating more rapid temperature and sea level rise than occurred during the last century. Although most projections indicate small increases in precipitation, some do not. Natural precipitation variability is large; thus, precipitation projections are less certain than temperature projections. There is a distinct possibility that precipitation will decrease over both 10-year and 30-year timescales. For all variables, the numerical projections for later in this century are less certain than those for earlier in the century (i.e., the ranges of outcomes become larger through time), due to uncertainties in the climate system and the differing possible pathways of the greenhouse gas emission scenarios.

Comparing observed data with projected changes for temperature and precipitation provides context with regard to how projected changes in the region compare to historical trends and long-term variability (**Figure 1.7**). To emphasize the climate signal and deemphasize the unpredictable year-to-year variability, a 10-year filter has been applied to the observed data and model output.

Temperature

Average annual temperatures are projected to increase across New York State by 1.5–3.0°F in the 2020s, 3.0–5.5°F in the 2050s, and 4.0–9.0°F in the 2080s (**Table 1.7; Figure 1.6a**). By the end of the century, the greatest warming may be in the northern parts of the state. The state's growing season could lengthen by about a month, with summers becoming more intense and winters milder. The climate models suggest that each season will experience a similar amount of warming relative to the baseline period.

Beginning in the 2030s, the emissions scenarios diverge, producing temperature patterns that are distinguishable from each other (**Figure 1.7**). This is because it takes several decades for the climate system to respond to changes in greenhouse gas concentrations. It also takes several decades for different emissions scenarios to produce large differences in greenhouse gas concentrations.

Precipitation

Regional precipitation across New York State may increase by approximately 0–5 percent by the 2020s, 0–10 percent by the 2050s, and 5–15 percent by the 2080s (**Table 1.7; Figure 1.6b**). By the end of the century, the greatest increases in precipitation may be in the northern parts of the state. While seasonal projections are less certain than annual results, much of this additional precipitation may occur during the winter months. During September and October, in contrast, total precipitation is slightly reduced in many climate models.

Precipitation is characterized by large historical variability, even with 10-year smoothing (**Figure 1.7**). Beginning in the 2040s, the climate models diverge, with the lower-emission B1 scenario producing smaller increases in precipitation than the high-emission A1B and the mid-emission A2 scenarios. However, even after the 2040s there are occasional periods where the B1 scenario projects more precipitation than that of A2. At no point in the century are the A2 and A1B scenario-based precipitation projections consistently distinguishable.

Sea Level Rise

Sea level is projected to rise along the coast and in the tidal Hudson by 1–5 inches in the 2020s, 5–12 inches in the 2050s, and 8–23 inches in the 2080s, using the

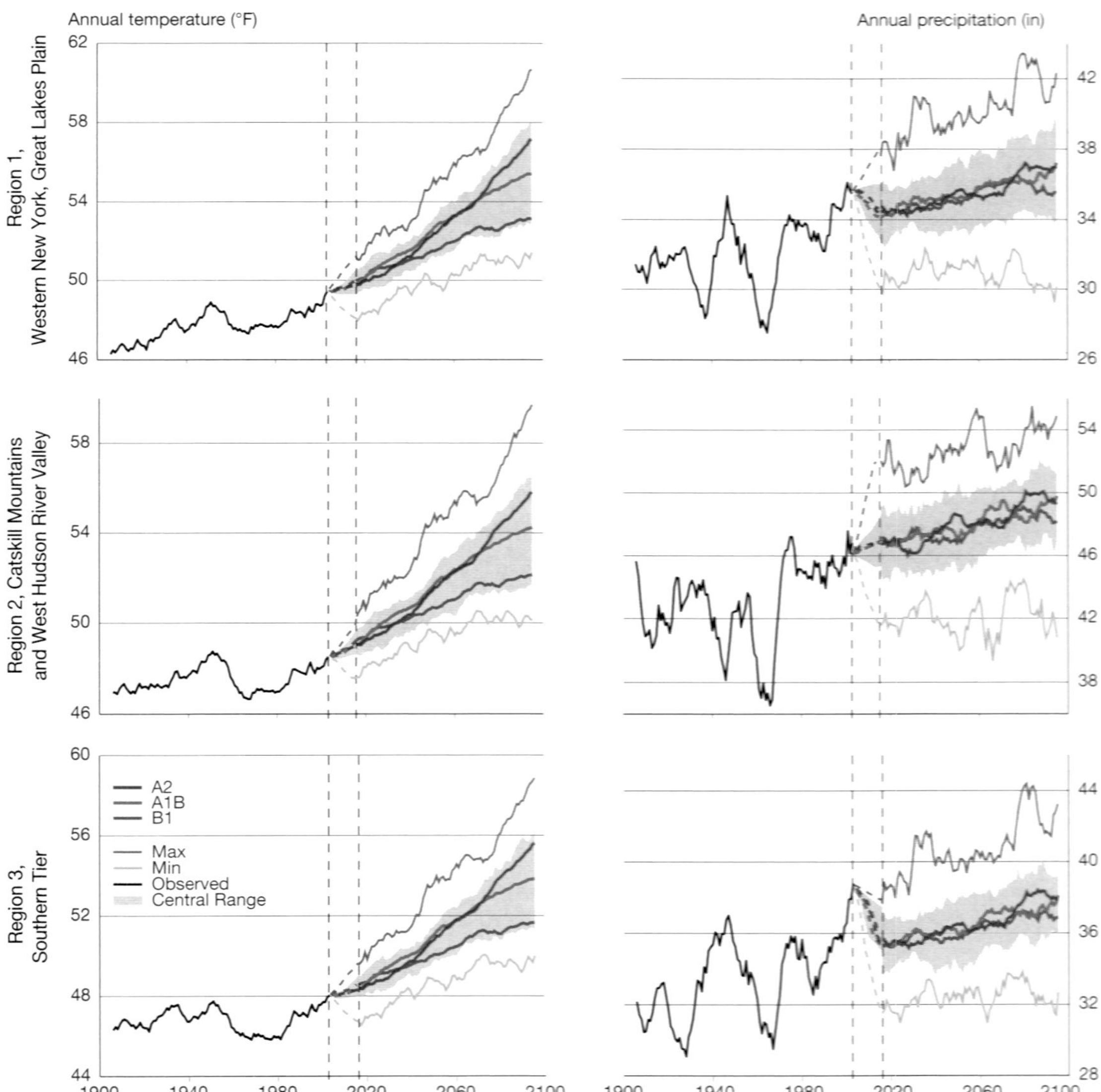

Observed (black line) and projected temperature (left) and precipitation (right). Projected model changes through time are applied to the observed historical data. The green, red, and blue lines show the average for each emissions scenario across the 16 global climate models. The shaded area indicates the central range. The bottom shows the minimum projection across the suite of simulations, and the top line shows the maximum projections. A 10-year filter has been applied to the observed data and model output. The dotted area between 2004 and 2015 represents the period that is not covered as a result of 10-year filter. Note different scales for temperature and precipitation.
Source: Columbia University Center for Climate Systems Research. Data are from USHCN, WCRP and PCMDI

Figure 1.7 Observed and projected temperature (left) and precipitation (right) for the ClimAID regions of New York State. Note that the y-axis is specific to each graph (continues on next page)

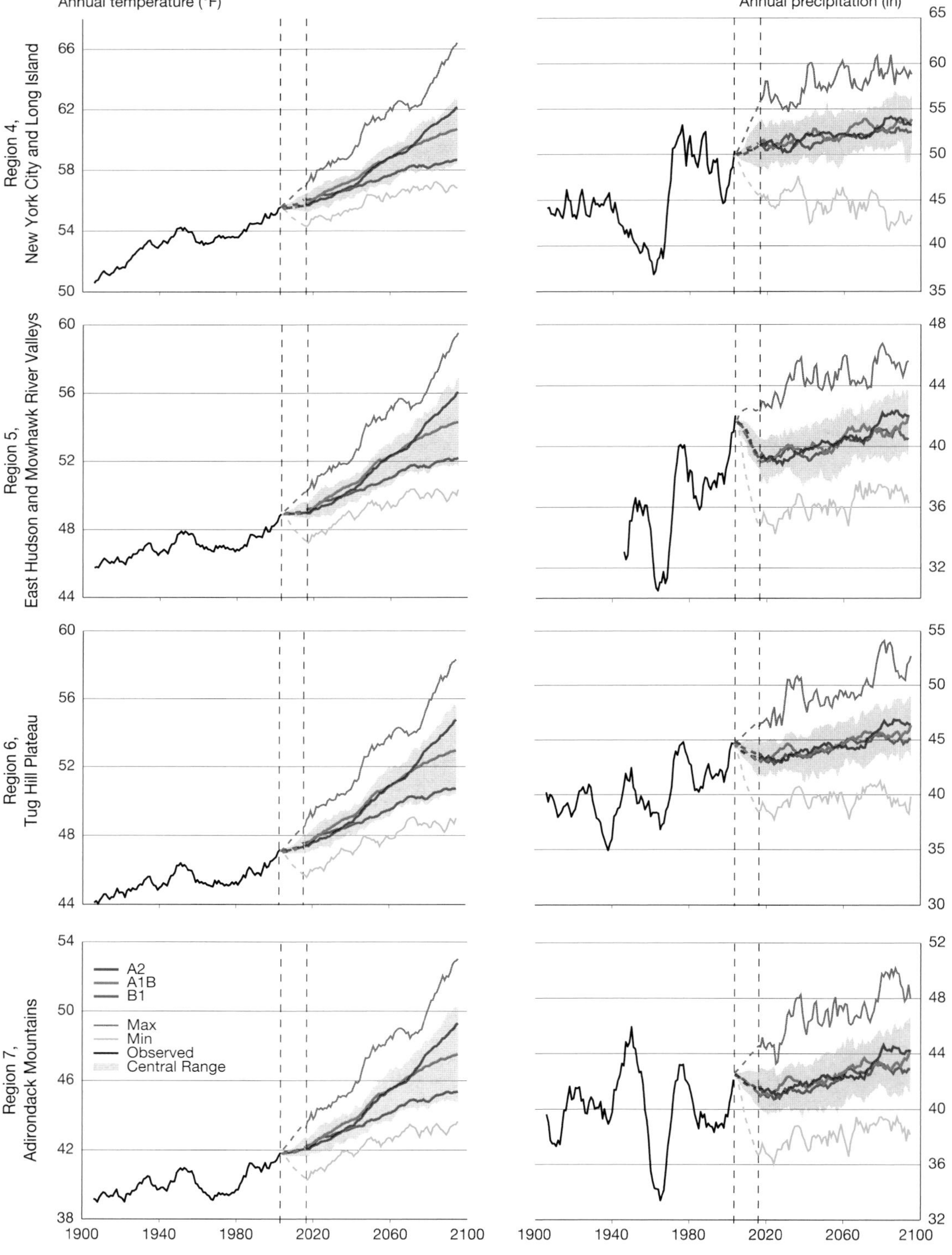

Annual temperature (°F)
Annual precipitation (in)
Region 4, New York City and Long Island
Region 5, East Hudson and Mowhawk River Valleys
Region 6, Tug Hill Plateau
Region 7, Adirondack Mountains
A2
A1B
B1
Max
Min
Observed
Central Range
1900
1940
1980
2020
2060
2100

GCM-based model projections (**Table 1.8**). Beginning in the 2050s, the low-emissions B1 scenario produces smaller increases in sea level than the higher-emissions A1B and A2 scenarios, and in the 2080s, the A2 scenario projects more sea level rise than A1B. The A2 scenario diverges from A1B approximately 10 years earlier for temperature than it does for sea level rise, in part reflecting the large response time of the ocean and ice sheets relative to the atmosphere.

The model-based sea level rise projections are characterized by greater uncertainty than the temperature projections, largely due to the possibility that future changes in polar ice sheets may accelerate melting beyond currently projected levels; this possible change is not captured by global climate models. This uncertainty is weighted toward the upper bound; that is, the probability that sea level rise will be lower than the GCM-based projection is very low, but the probability that sea level rise will exceed the GCM-based projection is higher.

The rapid ice melt sea level rise scenario addresses the possibility of the ice sheets melting more rapidly. This scenario is based on extrapolating the recent accelerating rates of ice melt from the Greenland and West Antarctic ice sheets and on paleoclimate studies that suggest sea level rise on the order of 0.39–0.47 inch per year may be possible. This scenario projects a sea level rise of 37 to 55 inches by the 2080s. The potential for rapid ice melt should be considered, in part, because of its potential for large consequences. It is also uncertain how rapid ice melt might indirectly influence sea level in the New York region through second-order effects, including gravitational, glacial isostatic adjustments, and rotational terms (e.g., Mitrovica et al., 2001, 2009).

Region 4: New York City and Long Island	2020s (inches)	2050s (inches)	2080s (inches)
GCM-based[1]	+2 to +5	+7 to +12	+12 to +23
Rapid ice-melt scenario[2]	~5 to +10	~19 to +29	~41 to +55
Region 5: East Hudson and Mohawk River Valleys	**2020s (inches)**	**2050s (inches)**	**2080s (inches)**
GCM-based[1]	+1 to +4	+5 to +9	+8 to +18
Rapid ice-melt scenario[2]	~4 to +9	~17 to +26	~37 to +50

[1] Shown is the central range (middle 67%) of values from global climate model-based probabilities rounded to the nearest inch.
[2] The rapid-ice melt scenario is based on acceleration of recent rates of ice melt in the Greenland and West Antarctic Ice sheets and paleoclimate studies.

Table 1.8 ClimAID Assessment sea level rise projections

To assess the risk of accelerated sea level rise over the coming years, scientific understanding as well as many key indicators should be monitored and reassessed on an ongoing basis (Appendix B).

1.3.4 Changes in Extreme Events

Despite their brief duration, extreme climate events can have large impacts, so they are a critical component of this climate change impact assessment. The frequencies of heat waves, cold events, intense precipitation, drought, and coastal flooding in the seven regions are projected to change in the coming decades, based on average global climate model shifts (**Table 1.9**). The average number of extreme events per year for the baseline period is shown, along with the middle 67 percent and full range of the model-based projections. Because the model-based probability does not represent the actual probability distribution, and shifts in extreme event distributions are not constrained to the types of average shifts described above, the relative magnitude of projected changes, rather than the actual projected number of events, should be emphasized.

Heat Waves and Cold Events

The total number of hot days in New York State is expected to increase as this century progresses. The frequency and duration of heat waves, defined as three or more consecutive days with maximum temperatures at or above 90°F, are also expected to increase (**Table 1.9**). In contrast, extreme cold events, defined both as the number of days per year with minimum temperature at or below 32°F, and those at or below 0°F, are expected to decrease. Some parts of each region, such as cold high-altitude zones, are likely to experience fewer heat events and more cold events in the future than regional averaging would suggest, because of the cold tendency in their baseline climates.

Intense Precipitation and Droughts

Although the increase in total annual precipitation is projected to be relatively small, larger increases are projected in the frequency, intensity, and duration of extreme precipitation events (defined as events with more than 1, 2, or 4 inches of rainfall) at daily timescales. The projection for New York State is

Table 1.9 Extreme events projections

Rochester (Region 1): Full range of changes in extreme events: minimum, (central range*), and maximum

	Extreme event	Baseline	2020s	2050s	2080s
Heat Waves & Cold Events	Number of days per year with maximum temperature exceeding				
	90°F	8	8 (10 to 17) 23	12 (17 to 30) 44	16 (22 to 52) 68
	95°F	0.8	0.9 (2 to 4) 6	2 (3 to 9) 17	3 (6 to 22) 38
	Number of heat waves per year[2]	0.8	0.9 (1 to 2) 3	2 (2 to 4) 6	2 (3 to 7) 8
	average duration	4	4 (4 to 4) 5	4 (4 to 5) 5	4 (4 to 5) 7
	Number of days per year with min. temp. at or below 32°F	133	99 (104 to 116) 126	76 (90 to 103) 108	53 (75 to 97) 106
Intense Precipitation	Number of days per year with rainfall exceeding:				
	1 inch	5	3 (4 to 5) 6	3 (4 to 6) 7	3 (4 to 6) 7
	2 inches	0.6	0.4 (0.5 to 0.7) 0.9	0.3 (0.5 to 0.8) 1	0.2 (0.5 to 1) 1

Port Jervis (Region 2): Full range of changes in extreme events: minimum, (central range*), and maximum

	Extreme event	Baseline	2020s	2050s	2080s
Heat Waves & Cold Events	Number of days per year with maximum temperature exceeding				
	90°F	12	13 (14 to 24) 34	16 (22 to 40) 53	21 (28 to 65) 75
	95°F	2	2 (2 to 5) 10	3 (5 to 12) 20	4 (7 to 28) 39
	Number of heat waves per year[2]	2	2 (2 to 3) 5	2 (3 to 5) 7	3 (4 to 9) 10
	average duration	4	4 (4 to 5) 5	5 (5 to5) 6	5 (5 to 6) 8
	Number of days per year with min. temp. at or below 32°F	138	101 (111 to 121) 128	70 (91 to 111) 115	57 (70 to 101) 112
Intense Precipitation	Number of days per year with rainfall exceeding:				
	1 inch	12	10 (11 to 13) 14	10 (12 to 14) 14	10 (12 to 14) 15
	2 inches	2	1 (2 to 2) 3	1 (2 to 3) 3	1 (2 to 3) 3

Elmira (Region 3): Full range of changes in extreme events: minimum, (central range*), and maximum

	Extreme event	Baseline	2020s	2050s	2080s
Heat Waves & Cold Events	Number of days per year with maximum temperature exceeding				
	90°F	10	11 (14 to 19) 25	15 (21 to 33) 45	19 (26 to 56) 70
	95°F	1	2 (2 to 4) 7	2 (4 to 10) 18	4 (7 to 24) 38
	Number of heat waves per year[2]	1	1 (2 to 3) 3	2 (3 to 4) 6	2 (3 to 8) 9
	average duration	4	4 (4 to 5) 5	4 (4 to 5) 5	4 (5 to 5) 7
	Number of days per year with min. temp. at or below 32°F	152	116 (122 to 124) 145	86 (106 to 122) 168	68 (87 to 114) 124
Intense Precipitation	Number of days per year with rainfall exceeding:				
	1 inch	6	5 (6 to 7) 8	5 (6 to 7) 8	5 (6 to 8) 10
	2 inches	0.6	0.5 (0.6 to 0.9) 1	0.5 (0.6 to 1) 1	0.4 (0.7 to 1) 2

New York City (Region 4): Full range of changes in extreme events: minimum, (central range*), and maximum

	Extreme event	Baseline	2020s	2050s	2080s
Heat Waves & Cold Events	Number of days per year with maximum temperature exceeding				
	90°F	19	20 (23 to 31) 42	24 (31 to 47) 58	31 (38 to 66) 80
	95°F	4	4 (6 to 9) 15	6 (9 to 18) 28	9 (12 to 32) 47
	Number of heat waves per year[2]	2	3 (3 to 4) 6	3 (4 to 6) 7	4 (5 to 8) 9
	average duration	4	4 (5 to 5) 5	5 (5 to 5) 6	5 (5 to 7) 8
	Number of days per year with min. temp. at or below 32°F	72	48 (53 to 62) 66	31 (45 to 54) 56	22 (36 to 49) 56
Intense Precipitation	Number of days per year with rainfall exceeding:				
	1 inch	14	11 (13 to 15) 16	11 (14 to 16) 16	11 (14 to 16) 17
	2 inches	3	2 (3 to 4) 5	3 (3 to 4) 5	2 (4 to 5) 5

Saratoga Springs (Region 5): Full range of changes in extreme events: minimum, (central range*), and maximum

	Extreme event	Baseline	2020s	2050s	2080s
Heat Waves & Cold Events	Number of days per year with maximum temperature exceeding				
	90°F	10	11 (14 to 20) 28	17 (20 to 35) 49	18 (26 to 60) 75
	95°F	1	1 (2 to 4) 7	3 (3 to 10) 18	3 (6 to 25) 42
	Number of heat waves per year[2]	2	2 (2 to 3) 4	3 (3 to 5) 7	3 (4 to 8) 9
	average duration	4	4 (4 to 5) 5	4 (4 to 5) 6	4 (5 to 6) 9
	Number of days per year with min. temp. at or below 32°F	134	121 (128 to 139) 147	92 (111 to 127) 135	78 (90 to 120) 131
Intense Precipitation	Number of days per year with rainfall exceeding:				
	1 inch	10	8 (10 to 11) 12	9 (10 to 11) 12	10 (10 to 12) 14
	2 inches	1	1 (1 to 2) 2	1 (1 to 2) 2	1 (1 to 2) 2

Watertown (Region 6): Full range of changes in extreme events: minimum, (central range*), and maximum					
	Extreme event	Baseline	2020s	2050s	2080s
Heat Waves & Cold Events	Number of days per year with maximum temperature exceeding				
	90°F	3	2 (4 to 7) 11	5 (8 to 17) 27	8 (12 to 36) 52
	95°F	0	0 (0.1 to 0.9) 2	0.2 (0.6 to 3) 7	0.8 (2 to 11) 23
	Number of heat waves per year[2]	0.2	0.2 (0.4 to 0.9) 1	0.6 (0.8 to 2) 4	0.6 (1 to 4) 6
	average duration	4	3 (4 to 4) 5	3 (4 to 4) 5	4 (4 to 5) 7
	Number of days per year with min. temp. at or below 32°F	147	114 (120 to 130) 140	93 (108 to 121) 126	78 (91 to 114) 122
Intense Precipitation	Number of days per year with rainfall exceeding:				
	1 inch	5	5 (6 to 8) 9	6 (6 to 8) 9	5 (7 to 10) 11
	2 inches	0.8	0.4 (0.6 to 0.9) 1	0.5 (0.6 to 1) 1	0.3 (0.6 to 1) 2

Indian Lake (Region 7): Full range of changes in extreme events: minimum, (central range*), and maximum					
	Extreme event	Baseline	2020s	2050s	2080s
Heat Waves & Cold Events	Number of days per year with maximum temperature exceeding				
	90°F	0.3	0.3 (0.5 to 1) 2	0.5 (1 to 5) 7	1 (2 to 13) 23
	95°F	0	0 (0 to 0.1) 0.2	0.1 (0.1 to 0.3) 0.6	0.1 (0.2 to 2) 6
	Number of heat waves per year[2]	0	0 (0 to 0.1) 0.2	0 (0.1 to 0.6) 0.7	0.1 (0.2 to 2) 3
	average duration	3	3 (3 to 3) 4	3 (3 to 4) 4	3 (4 to 4) 5
	Number of days per year with min. temp. at or below 32°F	193	155 (166 to 177) 184	125 (146 to 163) 173	108 (124 to 156) 166
Intense Precipitation	Number of days per year with rainfall exceeding:				
	1 inch	7	6 (7 to 8) 10	6 (7 to 9) 10	6 (7 to 10) 11
	2 inches	0.8	0.4 (0.7 to 1) 1	0.6 (0.7 to 1) 2	0.6 (0.8 to 1) 2

The values in parentheses in rows two through four indicate the central 67% range of the projected model-based changes to highlight where the various global climate model and emissions scenario projections agree. The minimum values of the projections are the first number in each cell and maximum values of the projections are last numbers in each cell.

* The central range refers to the middle 67% of values from model-based probabilities across the global climate models and greenhouse gas emissions scenarios.

1 Decimal places shown for values less than 1, although this does not indicate higher precision/certainty. The high precision and narrow range shown here are due to the fact that these results are model-based. Due to multiple uncertainties, actual values and ranges are not known to the level of precision shown in this table.

2 Defined as three or more consecutive days with maximum temperature exceeding 90°F.

3 NA indicates no occurrences per 100 years.

Source: Columbia University Center for Climate Systems Research. Data are from USHCN and PCMDI

consistent with global projections (Meehl et al., 2007) and with trends observed nationally (Karl and Knight, 1998; Kunkel et al., 2008).

Drought projections for this century reflect the competing influences of more total precipitation and more evaporation due to higher temperatures. By the end of this century, the number of droughts is likely to increase, as the effect of higher temperatures on evaporation is likely to outweigh the increase in precipitation, especially during the warm months. Drought projections, however, are marked by relatively large uncertainty. Drought in the Northeast has been associated with local and remote modes of multi-year ocean-atmosphere variability, including sea surface temperature anomalies in the North Atlantic (e.g., Namias, 1966; Bradbury et al., 2002) that are currently unpredictable and may change with climate change. Changes in the distribution of precipitation throughout the year and the timing of snowmelt could potentially make drought more frequent as well. The length of the snow season is very likely to decrease throughout North America (IPCC, 2007).

Coastal Floods and Storms

As sea levels rise, coastal flooding associated with storms will very likely increase in intensity, frequency, and duration. The changes in coastal flood intensity shown here are solely due to gradual changes in sea level through time. Any increase in the frequency or intensity of storms themselves would result in even more frequent large flood events. By the end of this century, sea level rise alone may contribute to a significant increase in large coastal floods; coastal flood levels that currently occur once per decade on average may occur once every one to three years.

Due to sea level rise alone, flooding at the level currently associated with the 100-year flood may occur about four times as often by the end of the century, based on the more conservative IPCC-based sea level rise scenario. The rapid ice melt scenario, should it occur, would lead to more frequent flood events. It should be noted that the more severe, current 100-year flood event is less well characterized than the less severe, current 10-year flood, due to the limited length of the historical record.

The relative flood vulnerability between locations is likely to remain similar in the future. Thus, portions of the state that currently experience lower flood heights than those described here (for reasons including coastal bathymetry and orientation of the coastline relative to storm trajectories) are likely to experience lower flood heights in the future than these projections indicate.

Uncertainties Related to Extreme Events

Because extreme events are by definition rare, they are characterized by higher uncertainty than the annual averages described previously. The climate risks described in each sector chapter in the ClimAID assessment reflect the combination of the climate hazard probability and the related impacts. The method used with GCM projections assumes that the distribution of the extreme events described quantitatively will remain the same, while average temperature, precipitation, and sea level rise change (**Table 1.9**). A change in the distribution of extreme events could have a large effect on these results.

The occurrence of extreme events in a given year will continue to be characterized by high variability; in some cases, the pattern of changes will only become evident after many years, or even decades, are averaged. For example, much of New York State's record of significant drought was a multiyear event that occurred four decades ago in the 1960s; no drought since that time in the state has approached it in severity. Generally speaking, changes in variability in future climate are considered very uncertain, although there are exceptions. For example, precipitation at daily timescales is likely to increase in variability since the warming atmosphere can hold more moisture (Emori and Brown, 2005; Cubasch et al., 2001; Meehl et al., 2005).

Other Extreme Events

Some of the extreme events that have a large impact throughout the state cannot be quantitatively projected into the future at local scales due to the high degree of uncertainty. Qualitative information for some of these factors is provided, including:

- heat indices, which combine temperature and humidity,
- frozen precipitation (snow, ice, and freezing rain),
- large-scale storms (tropical storms/hurricanes and nor'easters) and associated extreme wind,
- intense precipitation of short duration (less than one day), and
- lightning.

By the end of the century, heat indices (which combine temperature and humidity) are very likely to increase, both directly due to higher temperatures and because warmer air can hold more moisture. The combination of high temperatures and high moisture content in the air can produce severe effects by restricting the human body's ability to cool itself. The National Weather Service heat index definition is based on the combination of these two climate factors.

Seasonal ice cover has decreased on the Great Lakes at a rate of 8 percent per decade over the past 35 years; models suggest this will lead to increased lake-effect snow in the next couple of decades through greater moisture availability (Burnett et al., 2003). By mid-century, lake-effect snow will generally decrease as temperatures below freezing become less frequent (Kunkel et al., 2002).

Intense mid-latitude, cold-season storms, including nor'easters, are projected to become more frequent and take a more northerly track (Kunkel et al., 2008).

Intense hurricanes and associated extreme wind events may become more frequent (Bender et al., 2010) as sea surface temperatures rise in the areas where such storms form and strengthen (Meehl et al., 2007; Emanuel, 2008). However, other critical factors in the formation and intensity of these storms are not well known, including changes in wind shear, the vertical temperature gradient in the atmosphere, and patterns of variability such as the El Niño Southern Oscillation climate pattern and large-scale ocean circulation (for example, the meridional overturning circulation). As a result, there is the possibility that intense hurricanes and their extreme winds will not become more frequent or intense. It is also unknown whether the tracks or trajectories of hurricanes and intense hurricanes will change in the future. Thus, the impacts of future changes in hurricane behavior in the New York State coastal region are difficult to assess given current understanding.

Downpours, with intense precipitation occurring over a period of minutes or hours, are likely to increase in frequency and intensity as the state's climate warms. Thunderstorm and lightning projections are currently too uncertain to support even qualitative statements.[14]

1.4 Conclusions and Recommendations for Future Research

Climate change is extremely likely to bring higher temperatures to New York State, with slightly larger increases in the north of the state than along the coastal plain. Heat waves are very likely to become more frequent, intense, and longer in duration. Total annual precipitation will more likely than not increase; brief, intense rainstorms are likely to increase as well. Additionally, rising sea levels are extremely likely and are very likely to lead to more frequent and damaging flooding along the coastal plain and Hudson River related to coastal storm events in the future.

Climate hazards are likely to produce a range of impacts on the rural and urban fabric of New York State in the coming decades. The risk-management adaptation strategies described in this report will be useful in reducing these impacts in the future, but are also likely to produce benefits today, since they will help to lessen impacts of climate extremes that currently cause damages. However, given the scientific uncertainties in projecting future climate change, monitoring of climate and impacts indicators is critical so that flexible adaptation pathways for the region can be achieved.

Region-specific climate projections are only a starting point for impact and adaptation assessments. For some sectors, climate changes and their impacts in regions outside New York may rival the importance of local climate changes, by influencing, for example, migration, trade, ecosystems, and human health. Furthermore, some of the hazards described here (such as drought), are often regional phenomena with policy implications (such as water-sharing) that extend beyond state boundaries. Finally, since climate vulnerability depends on many factors in addition to climate (such as poverty and health), some adaptation strategies can be initiated in the absence of region-specific climate change projections.

Given the existing uncertainties regarding the timing and magnitude of climate change, monitoring and reassessment are critical components of any climate change adaptation plan. A dense network of sustained observations with resolutions that allow more accurate projections on a decade-to-decade basis will improve understanding of regional climate, extreme events, and long-term trends. Monitoring climate indicators can also play a critical role in refining future projections and reducing uncertainties. In order to successfully monitor future climate and climate impacts, specific indicators must be identified in advance. For example, to assess the significant risk of accelerated sea level rise and climate change for the coastal regions over the coming years, polar ice sheets and global sea level should be monitored. These uncertainties of timing and magnitude point to the need for flexible adaptation strategies that optimize outcomes by repeatedly revisiting climate, impacts, and adaptation science rather than committing to static adaptations. Frequent science updates will help to reduce these uncertainties.

Future projections can also be refined with greater use of regional climate models (see Appendix C for a description of regional climate models), which can capture changes in local processes as climate changes, such as the difference in magnitude of temperature increases on land versus that of the ocean. Advanced statistical downscaling techniques (see Appendix D) that allow projections at more localized levels than those described here may be of use as well; such techniques tend to be more effective when they use predictor variables that are well simulated by global climate models and that are policy relevant.

There is also a need for improved simulation of future climate variability at year-to-year and decade-to-decade scales, a need that may be met by future generations of climate models. Even the background rates of climate variation and extremes such as the 100-year drought and coastal flood will be better understood as a wide range of approaches, such as long-term tree-ring and sediment records, are increasingly used.

References

Bradbury, J.A., S.L. Dingman, B.D. Keim. 2002. "New England drought and relations with large scale atmospheric circulation patterns." *Journal of the American Water Resources Association* 38:1287–1299.

Bindoff, N.L., J. Willebrand, et al. 2007. "Observations: Oceanic Climate Change and Sea Level." In Solomon, S., D. Qin, M. Manning, et al. *Climate Change 2007: The Physical Science Basis. Contribution of Working Group I to the Fourth Assessment Report of the Intergovernmental Panel on Climate Change.* Cambridge University Press.

Brekke, L.D., M.D. Dettinger, E.P. Maurer, and M. Anderson. 2008. "Significance of model credibility in estimating climate projection distributions for regional hydroclimatological risk assessments." Climatic Change 89:371– 394. doi:10.1007/s10584-007-9388-3.

Burnett, A.W., M.E. Kirby, H.T. Mullins, and W.P. Patterson. 2003. "Increasing Great Lake-effect snowfall during the twentieth century: a regional response to global warming?" *Journal of Climate* 16(21): 3535-3542.

Charlson, Robert J. 1998. "Direct Climate Forcing by Anthropogenic Sulfate Aerosols: The Arrhenius Paradigm a Century Later." In *The Legacy of Svante Arrhenius. Understanding the Greenhouse Effect*, edited by Henning Rodhe and Robert Charlson, 59-71. Stockholm: Royal Swedish Academy of Sciences.

Christensen, J. H., B.C. Hewitson, et al. 2007. Regional Climate Projections. In Solomon, S., D. Qin, M. Manning, et al. *Climate Change 2007: The Physical Science Basis. Contribution of Working Group I to the Fourth Assessment Report of the Intergovernmental Panel on Climate Change.* Cambridge University Press.

Church, J.A., N.J. White, T. Aarup, W.S. Wilson, P. Woodworth, C.M. Domingues, J.R. Hunter, and K. Lambeck. 2008. "Understanding global sea levels: past, present, and future." *Sustainability Science* 3:9–22.

Cubasch, U., et al. 2001. "Projections of future climate change." In J.T. Houghton, ed., *Climate Change 2001: The Scientific Basis: Contribution of Working Group I to the Third Assessment Report of the Intergovernmental Panel on Climate Change.* Cambridge University Press.

Emanuel, K. 2008. "Hurricanes and global warming: results from downscaling IPCC AR4 Simulations." *Bulletin of the American Meteorological Society* 89:347–367.

Emori, S., and S. J. Brown. 2005. "Dynamic and thermodynamic changes in mean and extreme precipitation under changed climate." *Geophysical Research Letters* 32:L17706. doi:17710.11029/12005GL023272.

Groisman, P.Y., R.W. Knight, T.R. Karl, D.R. Easterling, B. Sun, and J.H. Lawrimore. 2004. "Contemporary changes of the hydrological cycle over the contiguous United States, trends derived from in situ observations." *Journal of Hydrometeorology* 5(1):64–85.

Grotch, S. L. and M.C. MacCracken. 1991. "The use of general circulation models to predict regional climatic change." *Journal of Climate* 4:286–303.

Guttman, N.B. 1989. "Statistical Descriptors of Climate." *Bulletin of the American Meteorological Society* 70(6): 602–607.

Hamilton, R.S., D. Zaff, and T. Niziol. 2007. "A catastrophic lake effect snow storm over Buffalo, NY October 12–13, 2006." Accessed June 29, 2010. ams.confex.com/ams/pdfpapers/124750.pdf

Hayhoe, K. 2007. "Past and future changes in climate and hydrological indicators in the US Northeast." *Climate Dynamics* 29:381–407.

Hayhoe, K. 2008. "Regional climate change projections for the Northeast USA." *Mitigation and Adaptation Strategies for Global Change* 13:425–436.

Hegerl, G.C., F.W. Zwiers, et al. 2007. "Understanding and Attributing Climate Change." In Solomon, S., D. Qin, M. Manning, et al. *Climate Change 2007: The Physical Science Basis. Contribution of Working Group I to the Fourth Assessment Report of the Intergovernmental Panel on Climate Change.* Cambridge University Press.

Horton, R., C. Rosenzweig, V. Gornitz, D. Bader and M. O'Grady. 2010. "Climate Risk Information." In C. Rosenzweig and W. Solecki, eds. *Climate Change Adaptation in New York City: Building a Risk Management Response.* New York Academy of Sciences.

Horton, R., and C. Rosenzweig. 2010. "Climate Observations and Projections." In C. Rosenzweig and W. Solecki, eds. *Climate Change Adaptation in New York City: Building a Risk Management Response.* New York Academy of Sciences.

Intergovernmental Panel on Climate Change. *Climate Change 2007: Impacts, Adaptation and Vulnerability, Contribution of Work Group II to the Fourth Assessment Report of the IPCC*, Cambridge, UK: Cambridge University Press, 2007.

Intergovernmental Panel on Climate Change. *Climate Change 2000: Special Report on Emissions Scenarios.* Geneva: IPCC, 2000.

Intergovernmental Panel on Climate Change. *Climate Change 2007: Synthesis Report.* Geneva: IPCC, 2008.

Intergovernmental Panel on Climate Change. *Climate Change 2007: The Physical Science Basis*, Contribution of Working Group I to the Fourth Assessment Report: Cambridge University Press, 2007.

Kalnay, E., M. Kanamitsu, R. Kistler, W. Collins, D. Deaven, L. Gandin, M. Iredell, S. Saha, G. White, J. Woollen, Y. Zhu, A. Leetmaa, B. Reynolds, M. Chelliah, W. Ebisuzaki, W. Higgins, J. Janowiak, K.C. Mo, C. Ropelewski, J. Wang, R. Jenne, and D. Joseph. 1996. The NCEP-NCAR 40-year reanalysis project. *Bulletin of the American Meteorological Society*, 77: 437–71

Karl, T.R. and Knight, R.W. 1998. "Secular trends of precipitation amount, frequency, and intensity in the United States." *Bulletin of the American Meteorological Society* 79:231.

Kunkel, K.E., P.D. Bromirski, H.E. Brooks, T. Cavazos, A.V. Douglas, D.R. Easterling, K.A. Emanuel, P.Ya. Groisman, G.J. Holland, T.R. Knutson, J.P. Kossin, P.D. Komar, D.H. Levinson, and R.L. Smith. 2008. "Observed changes in weather and climate extremes." In *Weather and Climate Extremes in a Changing Climate: Regions of Focus: North America, Hawaii, Caribbean, and U.S. Pacific Islands*, edited by Karl, T.R., G.A. Meehl, C.D. Miller, S.J. Hassol, A.M. Waple, and W.L. Murray, 35-80. Synthesis and Assessment Product 3.3. U.S. Climate Change Science Program, Washington, DC.

Meehl, G.A., J.M. Arblaster, and C. Tebaldi. 2005. "Understanding future patterns of increased precipiation intensity in climate model simulations." *Geophysical Research Letters* 32:L18719. doi:18710.11029/12005GL023680.

Meehl, G.A., T.F. Stocker, et al. 2007. "Global Climate Projections." In *Climate Change 2007: The Physical Science Basis.* Contribution of Working Group I to the Fourth Assessment Report of the Intergovernmental Panel on Climate Change. S. Solomon, D. Qin, M. Manninget al. Cambridge University Press: 94 pp.

Menne, M.J., C.N. Williams, and R.S. Vose. 2009. "The United States Historical Climatology Network Monthly Temperature Data - Version 2." *Bulletin of the American Meteorological Society* 90(7):993-1107.

Mitrovica, J.X., M.E. Tamisiea, J.L. Davis, and G.A. Milne. 2001. "Recent mass balance of polar ice sheets inferred from patterns of global sea-level change." *Nature* 409:1026–1029.

Mitrovica, J.X., N. Gomez, and P.U. Clark. 2009. "The sea-level fingerprint of West Antarctic collapse." *Science* 323:753.

Namias J. 1966. "Nature and possible causes of the northeastern United States Drought during 1962–1965." *Monthly Weather Review* 94(9):543–557.

New York City Panel on Climate Change. 2010. *Climate Change Adaptation in New York City: Building a Risk Management Response*. C. Rosenzweig & W. Solecki, Eds. Prepared for use by the New York City Climate Change Adaptation Task Force. New York, NY: Annals of the New York Academy of Sciences 1196.

New York State Climate Office. 2003. "The Climate of New York: Snowfall." Updated 1 July 2003. Accessed 29 June 2010. http://nysc.eas.cornell.edu/climate_of_ny.html

Palmer, W.C. 1965. "Meteorological drought." *Weather Bureau Research* Pap. No. 45, U.S. Dept. of Commerce, Washington, DC: 58 pp.

Peltier, W.R. and Fairbanks, R.G. 2006. "Global glacial ice volume and last glacial maximum duration from an extended Barbados sea level record." *Quaternary Science Review* 25:3322-3337.

Pope, V.D., M.L. Gallani, P.R. Rowntree, and R.A. Stratton. 2000. The impact of new physical parameterizations in the Hadley Centre climate model-HadCM3. *Climate Dynamics* 16:123–146.

Ramanathan, V., and Andrew M. Vogelman. 1997. "Greenhouse Effect, Atmospheric Solar Absorption and the Earth's Radiation Budget: From the Arrhenius-Langley Era to the1990s." *Ambio* 26:38-46.

Randall, D.A., R.A. Wood, et al. 2007. "Climate Models and Their Evaluation." In *Climate Change 2007: The Physical Science Basis.* Contribution of Working Group I to the Fourth Assessment Report of the Intergovernmental Panel on Climate Change. S. Solomon, D. Qin, M. Manninget al, Cambridge University Press: 59 pp.

Rosenzweig, C., W.D. Solecki, L. Parshall, et al. 2009. "Mitigating New York City's Heat Island: Integrating Stakeholder Perspectives and Scientific Evaluation." *Bulletin of the American Meteorological Society* 90:1297-1312.

Taylor, J., C. Rosenzweig, R. Horton, D.C. Major, and A. Seth. 2008. "Regional Climate Model Simulations" *Final Report to the New York City Department of Environmental Protection* CU02650201, 57 pgs.

Trenberth, K.E. et al. 2007. "Observations: Surface and Atmospheric Climate Change." In *Climate Change 2007: The Physical Science Basis.* Contribution of Working Group I to the Fourth Assessment Report of the Intergovernmental Panel on Climate Change. S. Solomon, D. Qin, M. Manninget al, Cambridge University.

Tryhorn, L. and A. DeGaetano. 2011a. In press, A comparison of techniques for downscaling extreme precipitation over the Northeastern United States. *International Journal of Climatology*, n/a. doi: 10.1002/joc.2208.

Tryhorn, L. A. and DeGaetano. 2011b. In review, A methodology for statistically downscaling snow cover over the Northeast United States. *Journal of Applied Meteorology and Climatology*.

WMO, 1989: Calculation of Monthly and Annual 30-Year Standard Normals. *WCDP928 No.10,WMO-TD/No.341*, World Meteorological Organization.

Wilby, R.L., C.W. Dawson, and E.M. Barrow. 2002. SDSM – a decision support tool for the assessment of regional climate change impacts, *Environmental Modelling and Software* 17: 147–159.

Appendix A. Uncertainty, Likelihoods, and Projection of Extreme Events

Uncertainty and Likelihoods

Climate projections are characterized by large uncertainties. At the global scale these uncertainties can be divided into two main categories:

- *Uncertainties in future greenhouse gas concentrations* and other climate drivers, which alter the global energy balance, such as aerosols and land-use changes; and
- *Uncertainties in how sensitive the climate system will be* to greenhouse gas concentrations and other climate drivers.

When planning adaptations for local and regional scales, uncertainties are further increased for two additional reasons:

- *Climate variability* (which is mostly unpredictable) can be especially large over small regions, partially masking more uniform effects of climate change; and
- *Changes in local physical processes* that operate at fine scales, such as land/sea breezes, are not captured by the global climate models used to make projections.

By providing projections that span a range of global climate models and greenhouse gas emissions scenarios, the global uncertainties may be reduced, but they cannot be fully eliminated. Averaging projections over 30-year timeslices and showing changes in climate through time, rather than absolute climate values, reduces the local- and regional-scale uncertainties, although it does not address the possibility that local processes may change with time.

The treatment of likelihood is similar to that developed and used by the IPCC. The six likelihood categories used here are as defined in the IPCC WG I Technical Summary (2007). The assignment of climate hazards to these categories is based on global climate simulations, published literature, and expert judgment.

Droughts

Droughts reflect a complex blend of climate and non-climate factors that operate at a number of timescales and are fundamentally different from other extreme events in that they are of longer duration. The drought timescale can last from a few months to multiple years. For this analysis, an intermediate timescale of 24 consecutive months was selected. In addition to precipitation, the other critical drought component is potential evaporation, which has a more complex relationship to drought. High temperatures, strong winds, clear skies, and low relative humidity all increase evaporative potential. Actual evaporation will generally be less than potential evaporation, however, since water is not always present for evaporation. For example, there will be little evaporation from dry soils, and as plants become water stressed under drought conditions, they become more effective at restricting their water loss to the atmosphere. Drought is also driven by water demand, so water-management decisions and policies can influence the frequency, intensity, and duration of droughts.

The Palmer Drought Severity Index (PDSI) uses temperature and precipitation to generate region-specific measures of drought and soil water excess. Because the calculation is strongly influenced by conditions in prior months, the PDSI is a good indicator of long-term phenomena like droughts. Potential limitations of the PDSI as used in this analysis include, but are not limited to, the exclusion of the water-demand component and the challenge of accurately capturing how potential evaporation changes with time. This analysis also does not consider water supplies stored on the ground as snow and ice.

The drought analysis conducted included two phases. First, the monthly PDSI was calculated for each observed data station from 1901 to 2000. Based on this calculation, the lowest consecutive 24 month-averaged PDSI value was defined as the 100-year drought. It should be noted that: 1) the drought record over the last 100 years can only provide a very rough estimate of the true 100-year drought; and 2) drought over a 24-month interval is only one possible definition.

In the second phase, the monthly changes in temperature and percentage changes in precipitation through time for each global climate model and emissions scenario were applied to the observed station data. The number of times that the 100-year, 24-month drought threshold (as defined in the paragraph above) was exceeded was then recalculated. Only events that did not overlap in time were counted.

Coastal Flood and Storm-related Extreme Events

The quantitative analyses of changes in coastal flooding are based on changes in sea level only, not in storm behavior. Projections were made by superimposing future changes in average sea level onto the historical dataset. The sea level rise projections are for the decade-to-decade averages of the 2020s, 2050s, and 2080s relative to the average sea level of the 2000–2004 base period. For coastal flooding, the critical thresholds were the 10-year, 100-year and 500-year flood events.

The 10-year event was defined using historical hourly tide data from the Battery. Forty years' worth of hourly sea level data were available from a period spanning 1960 to 2006 (nearest-neighbor interpolation was used to fill in missing data points for those years with little missing data). The Battery tide gauge was used to assess the frequency and duration of extreme coastal flood events. The raw tidal data are accessible from the NOAA website (http://tidesandcurrents.noaa.gov).

Average sea level was used as the reference datum. For the purposes of the storm analysis, additional calculations were made. First, data were de-trended (to remove the linear sea level trend) and normalized by dividing the data by the long-term average. This procedure gives water levels that include the influence of astronomical tides. To calculate surge levels, which more directly reflect the strength of the storm itself than do water levels, the difference between the actual flood level and the predicted level (the astronomical tide) was calculated. This approach allows assessment of the frequency and duration of extreme flood events. The ClimAID assessment defines the 10-year event as the storm surge thresholds corresponding to the fourth-largest surge over the 40-year period of tide data. Once the 10-year threshold was identified, the final procedure involved adding sea level rise projections for this century to the historical storm data as modified above to assess how frequently these flood levels would occur during this century.

Inasmuch as hourly data are unavailable from tide gauges prior to 1960, different methods were applied for estimating the 100-year and 500-year floods. The 100- and 500-year storms were analyzed using flood return interval curves (stage-frequency relationships) that provide a correlation between the water elevation by coastal storms versus the likelihood of occurrence. These curves include both surge and tidal components. An increase in sea level results in a higher flood height for a storm of a given return interval. The alternative approach taken here is to calculate the decrease in the return period for a given flood height with sea level rise (e.g., what will be the change in return period for the current 100-year flood if sea level rises 2 feet by 2080?). The 500-year estimate especially must be considered highly uncertain.

The surge data for the 100-year and 500-year storm calculations are based on data provided by the U.S. Army Corps of Engineers for the Metro East Coast Regional Assessment (MEC, 2001). In that study, the Army Corps used the USACE Waterways Experiment Station (WES) Implicit Flood Model (WIFM) developed in the 1980s as the hydrodynamic storm surge model. This time-dependent model includes sub-grid barriers and allows grid cells to become flooded during a simulation. The surge data were calculated relative to the National Geodetic Vertical Datum of 1929 (NGVD29) at high tide (thus a storm-flood level), excluding the effects of waves, for combined nor'easters and hurricanes. The flood height data were converted to the North American Vertical Datum of 1988 (NAVD88) by subtracting 0.338 meters (1.11 feet) from the flood heights given by the Army Corps. The conversion factors can be obtained from the National Geodetic Survey.

As research continues to advance, it may become possible to better estimate the surges associated with the 100-year and especially the 500-year historical storms, which are currently not well known.

High-end Scenarios and Longer-term Projections

This section describes 1) the possibility that climate changes in this century may deviate beyond the ranges projected by global climate models, 2) the rapid ice melt sea level rise scenario, and 3) potential climate change beyond this century.

There are several reasons why future climate changes may not fall within the model-based range projected for the ClimAID assessment. Actual greenhouse gas emissions may not fall within the envelope encompassed by the three emissions scenarios used here (A2, A1B, B1). This could be due either to changes in greenhouse gas concentrations directly related to changes in human activities or indirectly due to changes in the Earth's carbon and methane cycles brought on by a changing climate. The simulations used here all have known deficiencies regarding carbon cycle feedbacks, and some global climate models do not include volcanic forcings, for example.

Additionally, the climate's sensitivity to increasing greenhouse gases during this century may fall outside the range of the 16 climate models used. Possible types of climate changes exceeding model-based estimates that could have large impacts on the region include shifts in the average latitudes or tracks of moisture-laden storms traversing eastern North America and/or changes in ocean circulation in the North Atlantic.

Rapid Ice Melt Sea Level Rise Scenario

The rapid ice melt scenario addresses the possibility of more rapid sea level rise than the IPCC-based approach yields. The motivation to consider sea level rise exceeding IPCC-based estimates is based on several factors, including:

- recent accelerated ice melt in Greenland and West Antarctica, which may indicate the potential for high levels of sea level rise over multiple centuries if ice melt rates continue to accelerate;[15]
- paleoclimatic evidence of rapid sea level rise;
- the fact that not all sea level rise components are properly simulated by global climate models, increasing uncertainty about global climate model-based sea level rise projections; and
- the potentially large implications for a coastal city of more rapid sea level rise.

While not a significant direct cause of sea level rise, recent well-documented decreases in summer and fall Arctic sea-ice area and volume are also raising concern, since the decreases point to polar climate sensitivity higher than predicted by models. This could potentially modify atmospheric and oceanic conditions over a broader region, with implications for Greenland's ice sheet. For example, if warmer air were transported out of the Arctic to Greenland, Greenland's coastal and low-elevation glaciers might receive more moisture in the form of rain and less as snow.

Around 21,000 to 20,000 years ago, sea level began to rise from its low of about 394 feet below current levels. It approached present-day levels about 8,000 to 7,000 years ago (Peltier and Fairbanks, 2006; Fairbanks, 1989). Most of the rise was accomplished within a 12,000–10,000 year period; thus, the average rate of sea level rise over this period ranged between 0.39 and 0.47 inch per year. During shorter periods of more rapid rise, known as meltwater pulses, lasting several centuries, maximum rates of sea level rise ranged between 1.6 and 2.4 inches per year. These meltwater pulse sea level rise rates are considered too high to be matched during this century, since they occurred 1) after the ice sheets had already been undermined by thousands of years of forcing and 2) as abrupt intervals associated with singular events (e.g., ice dams breaking) at a time when total ice extent was much greater than today.

The rapid ice melt scenario assumes that glaciers and ice sheets melt at an average rate comparable to that of the last deglaciation (i.e., total ice melt increases linearly at 0.39 to 0.47 inch per year until 2100). However, the ice melt rate is more likely to be exponential. Thus, the average present-day ice melt rate of 0.04 inch per year (sum of observed mountain glacier melt [Bindoff et al., 2007] and ice sheets [Shepherd and Wingham, 2007]) during the 2000–2004 base period is assumed to increase to 0.39 to 0.47 inch per year (all ice melt) by 2100. An exponential curve is then fitted to three points: 2000, 2002 (midpoint of the 2000–2004 base period), and 2100. The other components—thermal expansion, local ocean dynamics, and subsidence—are added from the global climate model-based simulations and local information to this exponential meltwater estimates for the three timeslices. The rapid ice melt values combine the central range of the global climate model components and the range of estimates of rapid ice melt from the paleoclimate literature for multi-millennia timescales.

Longer-term Projections

Projections for the 22nd century are beyond most current infrastructure planning horizons. However, planning for some long-lived infrastructure, which hypothetically could include, for example, new aqueducts and subway lines, would justify considering the climate during the next century. Furthermore, many pieces of infrastructure intended only to have a useful lifespan within this century may remain operational beyond their planned lifetime. It is also possible that future projects aimed specifically at climate change adaptation might benefit during their planning stages from long-term climate guidance.

Because next century's climate is characterized by very high uncertainty, only qualitative projections are possible, especially at a local scale. Despite uncertainties, the large inertia of the climate system suggests that the current directional trends in two key climate variables, sea level rise and temperature, will probably continue into the next century (Solomon et al., 2009). Given the large inertia of the ice sheets on Greenland and West Antarctica, continued evidence during the next decade of acceleration of dynamically induced melting would greatly increase the probability that these ice sheets would contribute significantly to sea level rise in the next century, even if greenhouse gas concentrations, and perhaps even global temperatures, were to stabilize at some point during this century.

Appendix B. Indicators and Monitoring

Monitoring and reassessment are critical components of any climate change adaptation plan. Adaptation plans should account for changes in climate science, impacts, technological advancements, and adaptation strategies.

In order to successfully monitor future climate and climate impacts, specific indicators to be tracked must be identified in advance. These indicators are of two types. First, climate indicators, such as extreme precipitation, can provide an early indication of whether climate changes are occurring outside the projected range.[16] Given the large uncertainties in climate projections, monitoring of climate indicators can play a critical role in refining future projections and reducing uncertainties. Second, climate-related impact indicators provide a way to identify consequences of climate change as they emerge. For example, lower water quality may be a climate-related impact of extreme precipitation.

Regional climate indicators to monitor include, but are not limited to the following:[17]

Temperature-related
- average annual temperatures
- degree days in the hot and cold seasons
- temperature extremes
- coastal and inland water temperatures

Precipitation-related
- average annual precipitation
- extreme precipitation events
- droughts

Sea level rise and coastal flood-related
- average sea level
- high water levels
- extreme wind events

Additional larger-scale climate indicators should include:

- nor'easter frequency and intensity,
- tropical storms over the entire North Atlantic basin, as well as climatic conditions (including upper-ocean temperatures) that support tropical cyclones,
- variability patterns that influence the region, such as the North Atlantic Oscillation (large-scale ocean circulation patterns) and the El Niño Southern Oscillation climate pattern, and
- evidence of changes in the Earth's carbon cycle.

The possibility of rapid climate change in general and sea level rise in particular are two areas where the importance of monitoring and reassessment is well documented. Indicators of rapid ice melt to monitor could include, but should not be limited to:

- status of ice sheets,
- changes in sea-ice area and volume,
- global and regional sea level, and
- polar upper-ocean temperatures.

Climate variables cause certain climate-related impacts, which will also need to be monitored. These impacts include, but are not limited to:

- shoreline erosion,
- localized inland flooding,
- biological and chemical composition of waters, and
- changes in vegetation.

In addition to monitoring climate changes and their impacts, advances in scientific understanding, technology, and adaptation strategies should also be monitored. Technological advances, such as those in material science and engineering, could influence design and planning, and potentially result in cost savings. Monitoring adaptation plans in the region should be done both to determine if they are meeting their intended objectives and to discern any unforeseen consequences of the adaptation strategies. Some adaptation strategies will also have to be reassessed in the context of non-climate factors that are based on uncertain projections. For example, by monitoring trends in population, economic growth, and material costs, managers can tailor future climate change adaptation strategies to ensure they remain consistent with broader statewide objectives. Monitoring and reassessment of climate science, technology, and adaptation strategies will no doubt reveal additional indicators to track in the future.

Appendix C. Regional Climate Models

Additional downscaling methods have been employed in the ClimAID case studies including all or portions of New York State. These downscaling initiatives include both regional climate modeling and statistical downscaling (see Appendix D).

Regional climate models (RCMs) are similar to the models used for global modeling, except they run at higher spatial resolution and use different physics parameters for some processes such as convective precipitation (rain events accompanied by instability often associated with lightning, thunder, and heavy rain). Higher resolution improves the depiction of land and water surfaces as well as elevation. Because the domain is not global, information from outside the domain must be provided by a global climate model. Regional climate model simulations depend on high-quality global climate model boundary conditions; global climate model biases may thus be inherited by regional climate models. Additionally, regional climate models cannot provide feedbacks to the global climate models, so important observed local factors that impact the global scale may be missing from these experiments. Because regional climate model resolutions are generally no finer than three to four times the lateral resolution of the driving global climate models, more complex double-nesting (essentially running a high-resolution RCM inside a lower-resolution RCM) computations may also be needed to achieve policy-relevant resolutions, which leads to further uncertainty in the regional climate models. Even at such fine scales, there are uncertainties regarding how the parameters of subgrid-scale processes (such as convective rainfall) are defined. Furthermore, even the most high-resolution regional climate model simulations generally require some corrections for bias.

Because regional climate modeling is computationally demanding, historically only a limited number of short-duration simulations have been performed, potentially limiting their value for climate change assessment. For example, in New York State, the New York City Department of Environmental Protection and Columbia University funded short-duration regional climate model simulations using both the Pennsylvania State University/National Center for Atmospheric Research mesoscale model (MM5) and the International Center for Theoretical Physics Regional Climate Model (ReGCM3) (Taylor et al., 2008). While validation of these proof-of-concept studies demonstrated that regional climate models can simulate historical average climate, the applicability of these results was limited by the fact that the experiments were limited to single-year runs. To be useful for climate change assessment, simulations over multiple decades driven by a number of climate models are needed.

An advantage of regional climate modeling relative to statistical downscaling techniques is that regional climate models do not depend on the assumption that historical relationships between predictors (the information provided by the global climate models) and predictands (the local information needed for impact analysis, e.g., daily precipitation) will continue in the future. Because regional climate models are physics-based, they do not need to rely on the assumption that relationships will remain the same, which may not be valid as the climate moves further from its present state. For example, regional climate models may be able to provide reliable information about how changes in land/sea temperature gradients may modify coastal breezes in the future.

The North American Regional Climate Change Assessment Program (NARCCAP) is an ongoing project designed to address stakeholders' need for high-resolution climate projections. The program is a repository for multi-decade simulations, based on pairings of six regional climate models and four global climate models (**Table 1.10**). For validation purposes, all six regional climate models were also driven by a global climate model from 1980–2004 (the National Centers for Environmental Prediction/Department of Energy Atmospheric Model Intercomparison Project II (NCEP/DOE AMIP-II) Reanalysis) (**Table 1.10**). These reanalysis simulations represent the best estimate of observed conditions as simulated by a combination of observations and short-term global model simulation. Long-term climate change simulations over the northeastern United States are currently available from NARCCAP (http://www.narccap.ucar.edu/) for 2041 to 2070 for the A2 emissions scenario from two regional-climate-model/global-climate-model combinations, at an approximately 50-kilometer resolution. These combinations are the Canadian Centre for Climate Modeling and Analysis (CCCma) Coupled Global Climate Model (CGCM3) with the Canadian Regional Climate Model (CRCM) and the Geophysical Fluid Dynamics Laboratory (GFDL) 2.1 global climate model with the International Centre for Theoretical Physics regional climate model (RegCM3). These same two regional-climate-model/global-climate-model pairings have been hindcast for the 1970–1999 period based on coupled global climate model simulations.

Climate Model	Full Name	Modeling group
CRCM	Canadian Regional Climate Model	OURANOS / UQAM
ECPC	Experimental Climate Prediction Center Regional Spectral Model	University of California, San Diego / Scripps
HRM3	Hadley Regional Model 3 / Providing Regional Climates for Impact Studies	Hadley Centre
MM5I	MM5 – PSU/NCAR mesoscale model	Iowa State University
RCM3	Regional Climate Model version 3	University of California, Santa Cruz
WRFP	Weather Research and Forecast Model	Pacific Northwest National Lab

Driver GCM	Full Name
CCSM	Community Climate System Model
CGCM3	Third Generation Coupled Global Climate Model
GFDL	Geophysical Fluid Dynamics Laboratory GCM
HadCM3	Hadley Centre Coupled Model, version 3
NCEP	NCEP/DOE AMIP-II Reanalysis

Table 1.10 North American Regional Climate Change Assessment Program (NARCCAP) models

Regional Climate Model Validation

Because the Reanalysis product is the best estimate of the actual chronological order of the boundary conditions for the 1980–2004 period, the Reanalysis-driven simulations are used to estimate regional climate model biases and strengths. The RegCM3 and CRCM NCEP-driven simulations are compared here to the observed data for the Northeast from the University of Delaware (also available from NARCCAP/not shown here). Temperature and precipitation are evaluated for the winter and summer seasons.

The National Centers for Environmental Prediction (NCEP) Reanalysis simulation with RegCM3 has a cold bias in both winter and summer over New York State, indicating lower temperatures than the historical observations. The RegCM3 does not capture the observed pattern of increasing temperatures from west to east of the Great Lakes (**Figure 1.8**). This cold bias east of the Great Lakes is also present in the CRCM regional climate model in winter, but not in summer (not shown). In both winter and summer, cool biases are more prevalent than warm biases across the six regional climate models.

The NCEP-RegCM3 pairing captures eastern New York's tendency to receive more winter precipitation than the western part of the state. It also captures the precipitation maximum (the state's highest precipitation area) downwind of Lakes Ontario and Erie (**Figure 1.9**). However, winter precipitation is overestimated by approximately 1 millimeter per day in the RegCM3 model. The summer precipitation minimum in western New York is also simulated; like the winter, summer precipitation is also overestimated by approximately 1 millimeter per day. The NCEP/CRCM pairing does not produce the overestimated precipitation bias seen with RegCM3 over New York State (not shown). Across the entire six regional climate models, winter precipitation biases span from strongly underestimating to strongly overestimating precipitation, while summer precipitation biases tend towards overestimates.

In general, the RCM results vary significantly among models. The majority of models show cool biases over the region, and there is a tendency for summer precipitation to be overestimated.

Regional Climate Model Projections

By comparing projected climate change from a global climate model only to projected changes from a regional climate model forced by the same global climate model, the effects of higher resolution can be emphasized. Discussed here are winter and summer temperature and precipitation results from the two

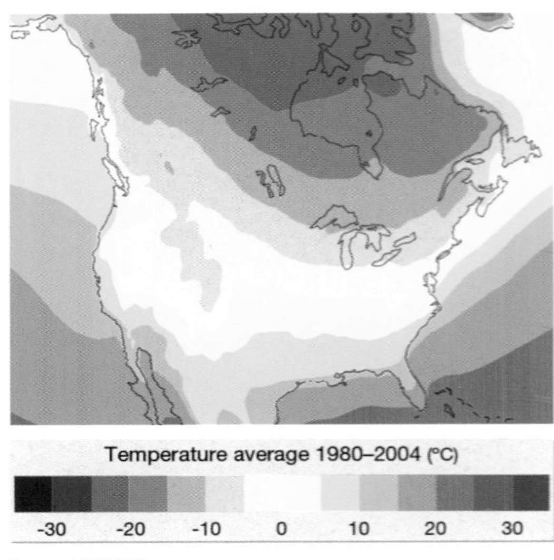

Source: NARCCAP

Figure 1.8 NCEP/RegCM3 winter (December, January, February) temperatures for 1980–2004

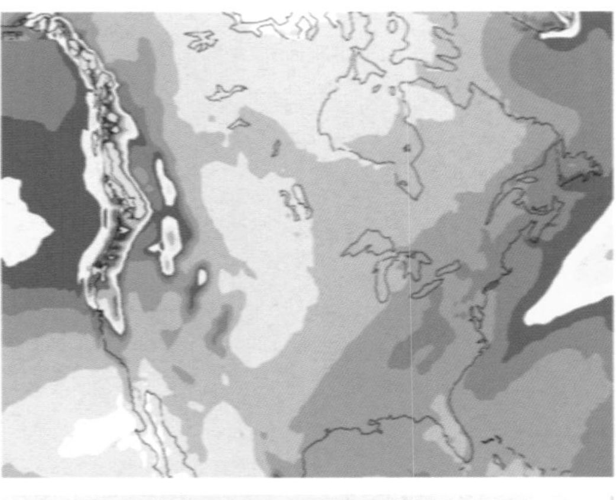

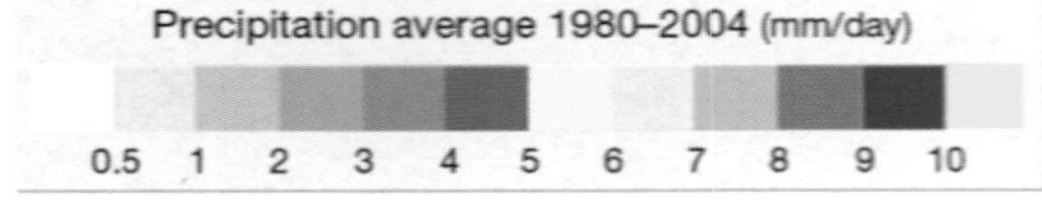

Source: NARCCAP

Figure 1.9 NCEP/RegCM3 winter (December, January, February) precipitation for 1980–2004

available global-climate-model/regional-climate-mode pairings described above.

Over northeast North America, the winter spatial pattern of warming in RegCM3 driven by the GFDL global climate model is quite different than the GFDL model warming pattern alone (**Figure 1.10**). Whereas GFDL features the characteristic pattern of greater warming moving north (not shown), the GFDL-RegCM3 pairing features a local minimum east of Hudson Bay. As a consequence, while both models indicate that southeastern New York will warm by approximately 5.4°F, the GFDL/RegCM3 produces less warming to the north than the GFDL global climate model. The CRCM regional climate model driven by CGCM3 over New York State produces a warming trend of 4.5–5.4°F by the 2050s relative to the base period and is also less than the CGCM3 global climate model's results (not shown).

In summer, GFDL global climate model warming over much of the central United States is 1.8–3.6°F higher than the paired GFDL/RegCM3 regional climate model warming over the same region. Both the GFDL global climate model and the GFDL/RegCM3 regional climate model simulations produce the greatest New York warming in the western portions of the state that are farthest from the coast, with the global climate model indicating slightly higher temperatures than the regional climate model in western New York (**Figure 1.11**). By contrast, for most of the United States including New York State, the CRCM regional climate model driven by the CGCM3 global climate model produces approximately 1.8°F more warming than the CGCM3 global climate model alone (**Figure 1.12**). The CRCM regional climate model indicates that summer temperatures over the state will increase by 5.4–7.2°F.

The GFDL global climate model produces large increases in winter precipitation—greater than 20 percent—in New York State, whereas the RegCM3 regional climate model driven by GFDL indicates a precipitation increase between 10 and 20 percent. Both the CGCM3 global climate model alone and the CGCM3/CRCM pairing indicate a 10–20 percent precipitation increase (not shown).

In summer the GFDL global climate model produces precipitation patterns that range from no change (0 percent) in southeastern New York to a greater than 10 percent decrease in precipitation in southwestern New York. Regional climate model precipitation changes have a fine spatial scale; precipitation increases by approximately 10 percent in much of the southern part of the state. The far west of the state shows precipitation decreases of approximately 10 percent. The CGCM3 global climate model produces slight decreases in precipitation ranging from 0 to 5 percent across the entire state (**Figure 1.13**). The CRCM regional climate model simulation driven by CGCM3

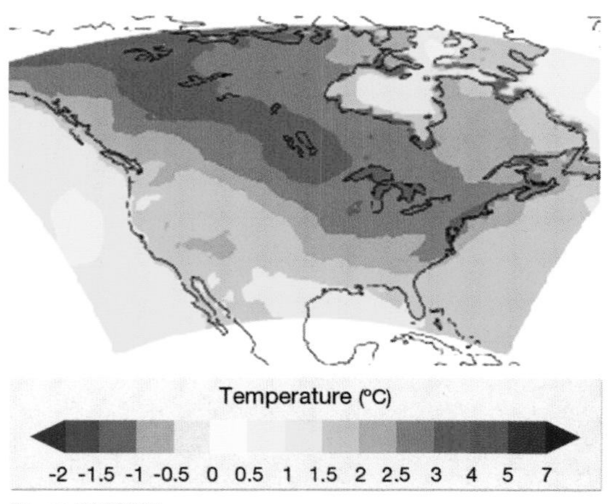

Source: NARCCAP

Figure 1.10 GFDL/RegCM3 modeled winter (December, January, February) temperature change for the A2 scenario in the 2050s

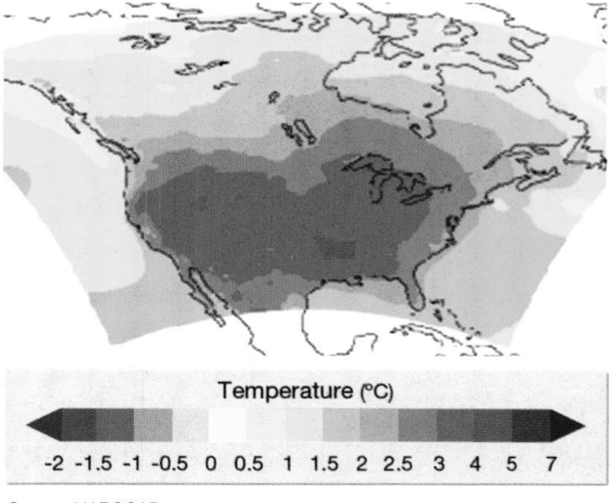

Source: NARCCAP

Figure 1.11 GFDL/RegCM3 modeled summer (June, July, August) temperature change for the A2 scenario in the 2050s

indicates even more drying throughout New York State, with precipitation decreases approaching 20 percent in New York's northern and western regions.

These two global climate model-regional climate model pairings demonstrate that a range of uncertainties persist in regional climate projections. Over New York State, the largest discrepancy is in summer precipitation.

Downscaling Extreme Events

Regional climate model simulations hold promise for the simulation of changes in climate extremes, since many extreme events occur at smaller spatial scales than global climate model gridboxes.

Regional climate model simulations have also been conducted for the ecosystems sector. Specifically, Weather Research and Forecasting (WRF) regional climate model sensitivity experiments were conducted at Cornell University on the effects of changing Great Lake and atmospheric temperatures on lake-effect snow (see Chapter 6, "Ecosystems").

Future work by the climate team will evaluate 3-hour outputs from NARCCAP, to assess how the climate model projections of extremes such as intense precipitation, heat waves, and cold events described in this chapter could be augmented by regional climate model output.

Appendix D. Statistical Downscaling in the ClimAID Assessment

An additional downscaling approach used in the ClimAID report to show potential changes in extremes to the end of the century is to utilize The Statistical DownScaling Model[18] (SDSM) Version 4.2 of Wilby et al. (2002, 1999). SDSM is described as a hybrid of a stochastic weather generator and regression-based methods. Large-scale circulation patterns and atmospheric moisture variables are used to linearly condition local-scale weather generator parameters (e.g., precipitation occurrence and intensity) for the predictand series. This approach is potentially better for estimating extremes, as it attempts to bridge the gap between dynamical and statistical downscaling.

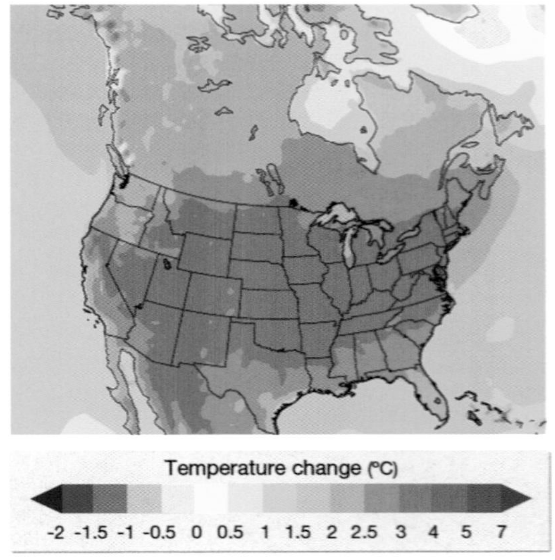

Source: NARCCAP

Figure 1.12 CGCM3/CRCM modeled summer (June, July, August) temperature change for the A2 scenario in the 2050s

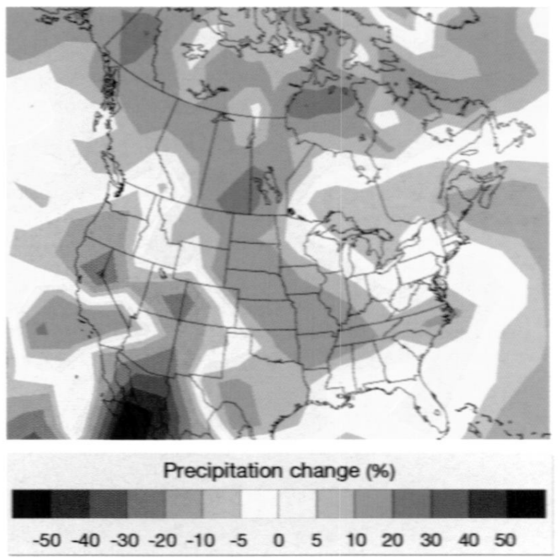

Source: NARCCAP

Figure 1.13 Summer precipitation change (June, July, August), from the CGCM3 model for the A2 scenario in the 2050s

Downscaling using SDSM in the ClimAID report was completed for extreme precipitation events (see Chapter 4, "Water Resources" and Chapter 7, "Agriculture") and winter snow cover (see Chapter 6, "Ecosystems"). In both cases, observed climate data were linked to large-scale predictor variables derived from the National Centers for Environmental Prediction (NCEP) reanalysis data set (Kalnay et al., 1996). For both projections in rainfall and snow cover, a dataset with an ensemble of 20 daily simulations was created using model output from the United Kingdom Meteorological Office Hadley Centre Climate Model version 3 (HadCM3; Pope et al., 2000).

For the precipitation events, the simulated daily data were used to construct extreme value series consisting of the annual maximum rainfall event for 30-year periods beginning in 1961. The first of these series included data from 1961–1990 and the last of these encompassed the 2071–2100 period. Additional statistical analysis was then conducted on these daily series (see Tryhorn and DeGaetano, 2011a). For snowfall, the two datasets were then combined by adding up the increases and decreases over time to give an estimate of the snow cover over the winter (Tryhorn and DeGaetano, 2011b).

[1] The range of temperature projections is the lowest and highest of values across the middle 67% of projections for all regions of New York State.

[2] The temperature and precipitation timeslices reflect a 30-year average centered around the given decade, i.e., the time period for the 2020s is from 2010–2039. For sea level rise, the timeslice represents a 10-year average.

[3] Probability of occurrence is defined as follows: Very likely (>90% probability of occurrence), Likely (>66% probability of occurrence), and More likely than not (>50% probability of occurrence).

[4] Preliminary analysis of those stations with lengthy records indicated that one station per region was generally sufficient to characterize each region's overall trends.

[5] The USHCN data are a selected group of stations that come from the COOP data set.

[6] Lower thresholds were used for the historical analysis than the projections, since warming is expected.

[7] A degree day is defined as the difference between the daily mean temperature and 65°F. Heating degree days occur when the daily mean temperature is below 65°F, while cooling degree days occur when the daily mean temperature is above 65°F.

[8] Changes in these additional factors are expected to have a smaller influence on climate change than increases in greenhouse gases during this century.

[9] Due to limited availability of model outputs, sea level rise projections are based on seven GCMs.

[10] The ratio approach is used for precipitation because it minimizes the impact of model biases in average baseline precipitation, which can be large for some models/months.

[11] Because they are rare, the drought and coastal storm projections were based on longer time periods.

[12] For sea level rise, the multidecadal approach is not necessary due to lower inter-annual variability; the 2050s timeslice for sea level (for example) therefore refers to the period from 2050–2059.

[13] The influence of interdecadal variability cannot be eliminated with 30-year timeslices, however. While longer timeslices would reduce the influence of interdecadal variability, it would be at the expense of information about the evolution of the climate change signal through time.

[14] Some research does suggest that lightning may become more frequent with warmer temperatures and more moisture in the atmosphere (Price and Rind, 1994, for example).

[15] Neither the Greenland nor West Antarctic ice sheet has yet to significantly contribute to global and regional sea level rise, but because potential sea level rise is large, should current melt patterns continue to accelerate, their status should be monitored.

[16] One potential pitfall of monitoring over short timescales, especially for small regions, is that it is easy to mistake natural variability for a long-term trend.

[17] Many of these indicators are already tracked to some degree by agencies within New York State.

[18] Available for download at http://www.sdsm.org.uk

Chapter 2

Vulnerability and Adaptation

Authors: William Solecki,[1,2] Lee Tryhorn,[3] Art DeGaetano,[1,3] and David Major[4]

[1] Integrating Theme Lead
[2] City University of New York Institute for Sustainable Cities
[3] Cornell University
[4] Columbia University Earth Institute Center for Climate Systems Research

Contents

Introduction....................50
2.1 Stakeholder Interactions....................51
2.2 Vulnerability....................52
 2.2.1 Vulnerability Concepts....................52
 2.2.2 Vulnerability Assessment Approaches....................53
 2.2.3 Vulnerability Measures and Metrics....................53
 2.2.4 Evaluating Vulnerability in ClimAID....................54
2.3 Adaptation....................55
 2.3.1 Adaptation Concepts....................55
 2.3.2 Adaptation Assessment Approaches....................56
 2.3.3 Assessing Adaptation in ClimAID....................57
2.4 Outcomes....................58
References....................59

Introduction

The objective of this chapter is to introduce the issues of vulnerability and adaptation as a framework for analysis of the potential impacts and adaptation responses to climate change in New York State. Within the ClimAID assessment, vulnerability and adaptation are key integrating themes and are examined directly by each of the sectors.

New York State is increasingly faced with a changing climate that is beyond the range of past experiences (See Chapter 1, "Climate Risks"). Determining the potential consequences of climate change and possible responses is a complex task, as the effects of changes in climate will vary over space, through time, and across social groups.

This chapter outlines definitions and concepts associated with climate change vulnerability and adaptation (**Box 2.1**; Schneider et al., 2007). It also provides background on approaches to vulnerability and adaptation assessments and the different factors that contribute to both in the context of New York State. Details of the approaches used in the ClimAID assessment, as well as a description of the stakeholder engagement undertaken, are provided. Toward the end of the chapter, guidelines for evaluation and prioritization of vulnerability and adaptation actions are introduced.

It is reasonable to expect that adaptation to climate change will not always be a smooth process nor will it always be optimal or ideal. Adaptation will be ongoing, with mid-course corrections in response to the evolving context. The goal of the ClimAID assessment is to provide information that will help the people of New York to better understand climate change in their own context and to decide on effective policies.

The objective of the ClimAID process was to define the vulnerability and adaptation potential within each of the eight sectors. Critical to the process was identifying the opportunities and challenges within each sector now and in the future. Because of the widely varying impact levels and adaptation possibilities, study of comparative vulnerability and adaptive capacity was not explicitly included in this assessment. Connections between the sectors (e.g., communication and energy, and ecosystems and agriculture) were made as part of the analytical process, but large-scale comparisons were deemed outside the scope of the study.

Vulnerability plays an essential role in determining the severity of climate change impacts. In ClimAID, *vulnerability* is defined as the degree to which systems are susceptible to, and unable to cope with, adverse impacts of climate change (Schneider et al., 2007).

A variety of approaches can help to reduce vulnerability to climate variability and extremes, including participatory planning processes, private initiatives, and specific government policies. Thus, there is an urgent need to understand the factors that affect the climate vulnerability of the state's residents, ecosystems, and economy. It is recognized, however, that efforts to reduce current vulnerability will not be sufficient to prevent all damages associated with climate change in the long term, and that the reduction of atmospheric greenhouse gas concentrations will be necessary as well.

Box 2.1 Definitions

Vulnerability
Vulnerability to climate change is the degree to which systems are susceptible to, and unable to cope with, adverse impacts of climate change.

Adaptation
Actions that reduce the level of physical, social, or economic impact of climate change and variability, or take advantage of new opportunities emerging from climate change.

Exposure
The degree to which elements of a climate-sensitive system are in direct contact with climate variables and/or may be affected by long-term changes in climate conditions or by changes in climate variability, including the magnitude and frequency of extreme events.

Sensitivity
The degree to which a system will respond to a change in climate, either beneficially or detrimentally.

Adaptive Capacity
The ability of a system to adjust to actual or expected climate stresses or to cope with the consequences.

Source: Derived by authors from Easterling et al. 2004; Schneider et al. 2007; Smit et al. 2001

Connected to the concept of vulnerability is the capacity and capability of a society to adjust its functioning to better respond to actual and projected climate changes. This condition is broadly defined as climate change *adaptation*. Adaptation, in this context, includes those strategies and policies that can make both human and natural systems better able to withstand the detrimental impacts of climatic changes, and also potentially take advantage of opportunities emerging with climate change. Adaptations can take place at the individual, household, community, organization, and institutional level, and are defined broadly in ClimAID as actions of stakeholders.

2.1 Stakeholder Interactions

Addressing vulnerability requires merging expert and decision-makers' knowledge to capture the complexity of the vulnerabilities that influence priorities, preferences, opportunities, and constraints (NRC 1996, 2005). Accordingly, a key component of this assessment was early and continuous participation from stakeholders in the identified sectors. Stakeholders were defined broadly as individuals or groups that have anything of value that may be affected by climate change or by the actions taken to manage climate vulnerability. Examples include owners as well as practitioners, such as policy-makers, communities, and natural resource managers.

The assessment began with stakeholder-driven identification of climate change vulnerabilities through both past experience and visualized (anticipated or predicted) damage. ClimAID took this approach because the stakeholders themselves are in the best position to understand their own challenges, to decide their own course of action, and to take responsibility for those decisions (Lynch and Brunner, 2007). This type of ongoing stakeholder engagement avoids the pitfall of researchers assigning their own values to an assessment. The many specific values that figure in the interests of stakeholders vary greatly across each scale and are subject to change. But typically the values include community, property, other tangible and intangible cultural artifacts, and the animate (living) and inanimate (nonliving) natural environment, in addition to minimizing the costs of protecting such things. Issues of equity—winners and losers—and more specific environmental justice questions were also critical to understanding the full character of the sector-specific vulnerabilities (see Chapter 3, "Equity and Economics").

Given this spatial and sector-specific variability, the format and scope of stakeholder interaction varied among the ClimAID sectors. Nonetheless, a general framework was followed by all sector teams that included the following:

1) An initial stakeholder meeting with presentations that described the ClimAID project, climate change, and likely types of impacts. At this meeting the researchers solicited input on the types of impacts and vulnerabilities likely to be faced by each stakeholder. This meeting focused on the identification of key climate vulnerabilities and associated climate variables for each sector.
2) Each sector developed a survey instrument and administered it either formally or semi-formally to elicit key sector vulnerabilities and potential adaptation strategies from a broader group of stakeholders across the state.
3) Focus groups were convened with key stakeholders for ongoing discussion and advice throughout the assessment. This entailed follow-up meetings and discussions to get feedback on the progress of the assessment and refine the analysis of sector-specific climate variables and vulnerabilities. These addressed vulnerabilities and climate variables and began a dialogue on adaptation alternatives and opportunities.
4) A final stakeholder meeting was conducted by each sector team to present the results of the assessment and to identify the steps required to act upon the findings.

Within individual sectors the form of this stakeholder process varied; these differences reflected the makeup of the stakeholder base for each sector. In the Energy sector, for instance, private industry comprised the majority of stakeholders, so stakeholder meetings tended to be one-on-one interviews with individual power generators. However, the Agriculture and Ecosystems sectors were a mix of government organizations, non-government organizations, citizens, and grower associations; broad workshops were followed by targeted focus-group sessions. Additional details on the sector-specific stakeholder engagements can be found in each sector chapter.

The stakeholders added vital insight about the range of risks and uncertainties they face and how they currently manage these challenges. Local experience was integrated with scientific knowledge from a variety of disciplines and used to identify key climate variables that were particularly relevant to each sector. The ClimAID Climate team then developed sector-specific "climate products" to guide scientific inquiry, such as the detailed analysis of flooding criteria in Chapter 4, "Water Resources." This decision-focused science led the assessment of vulnerability and the development of adaptation strategies to expand the range of informed choices for stakeholders.

2.2 Vulnerability

The concept of vulnerability is useful for organizing an investigation into the impacts of climate change on the human–environment system. This perspective is particularly pertinent because it is inclusive, and human and natural systems are viewed as intimately coupled.

2.2.1 Vulnerability Concepts

Any system's vulnerability to climate change is fundamentally determined by its exposure to shocks and stresses and its baseline sensitivity to those stresses (**Box 2.1**; Smit et al., 2001), concepts that are related to each other (see **Figure 2.1**). *Exposure* is the degree to which elements of a climate-sensitive system are in contact with climate and may be related to long-term changes in climate conditions or by changes in climate variability, including the magnitude and frequency of extreme events (Easterling et al., 2004). For example, as the population of New York State moves toward coastal areas, the state's exposure to sea level rise and coastal storms increases. *Sensitivity* refers to the degree to which a system will respond to a change in climate, either beneficially or detrimentally. For example, corn is more sensitive to hot and dry conditions and is less able to take advantage of higher carbon dioxide levels than wheat, making it more physiologically sensitive to climate change (Easterling et al., 2004).

Furthermore, any system's ability to cope with exposure and/or sensitivity depends on its level of *adaptive capacity*. Adaptive capacity describes the ability of a system to adjust to actual or expected climate stresses or to cope with the consequences. Capacity, however, does not ensure positive action or any action at all. Although New York State has considerable adaptive capacity, people and property have not always been protected from adverse impacts of climate variability and extreme weather events, such as winter ice storms and extended heat waves.

Exposure and sensitivity give information about the potential impacts of climate change, while adaptive capacity is a measure of the extent to which a sector or group can respond to those impacts. The significance of climate impact depends on both the climate change itself and the characteristics of the system exposed to it (Ausubel, 1991; Rayner and Malone, 1998). The characteristics of any system—both the physical properties of its environment as well as the socioeconomic context (Smit et al., 2001; Tol and Yohe, 2007)—determine its vulnerability. These elements are place- and system-specific and are similar to those that influence a system's adaptive capacity.

Human systems are distinguished from natural systems by their capacity to anticipate environmental changes and respond accordingly so as to best prepare for expected future conditions. The vulnerability of the people in New York State is largely determined by several key factors: behavioral norms that have been institutionalized through building codes, crop insurance, flood-management infrastructure, water systems, and a variety of other

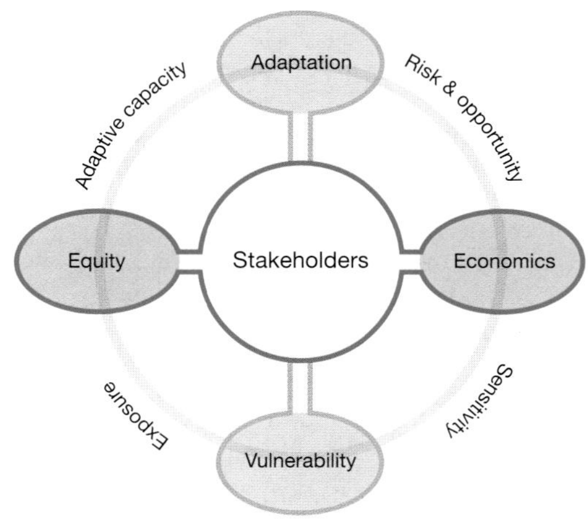

Figure 2.1 Vulnerability and adaptation

programs; socio-economic factors that affect access to technology, information, and institutions; geographic climate-sensitive health risks due to the proximity of natural resources, dependence on private wells for drinking water, and vulnerability to coastal surges or river flooding (Balbus and Malina, 2009); and biological sensitivity related to pre-existing medical conditions, such as the sensitivity of people with chronic heart conditions to heat-related illness (Balbus and Malina, 2009).

Natural systems are potentially more vulnerable to climate change than human systems because of their limited ability to adapt. Although biological systems have an inherent capacity to adapt to changes in environmental conditions, given the rapid rate of projected change, adaptive capacity is likely to be exceeded for many species (Easterling et al., 2004). Moreover, the vulnerability of ecosystems is increased by the effects of urbanization, pollution, invasion by exotic species, and fragmentation (or isolation) of habitats, all of which have already critically stressed ecosystems independent of climate change itself. An understanding of these components is essential for the formulation of effective climate policy.

2.2.2 Vulnerability Assessment Approaches

There are many different approaches to vulnerability assessment (Carter et al., 2007; Fussel, 2007; Polsky et al., 2007; Hahn, 2009). The main approaches are a risk-hazard approach (visualize future damages), a policy approach (visualize desired future), an adaptive capacity or resilience approach (assess current and future response capacity), and an integrated approach that combines aspects of these different approaches (e.g., the Center for Clean Air Policy Urban Leaders Adaptation Initiative). The risk approach is used for assessing the risk to a particular system that arises from exposure to hazards of a particular type and magnitude (e.g., Yohe, 1989; Preston et al., 2009). The policy approach is a goal- or problem-oriented approach in which analysis focuses on stakeholder-determined desired outcomes or solutions and analyzes the effectiveness of policies under climate change (e.g., Lynch et al., 2008; Tryhorn and Lynch, 2010). The adaptive capacity approach concentrates on the resources available, either actually or potentially, to cope with changes in the system (e.g., Vásquez-León et al., 2003; Brooks et al., 2005).

The ClimAID assessment uses an integrated approach that combines aspects of a risk-hazard approach and a policy approach. This approach aims to investigate vulnerability across a broad range of sectors and scales with a specific focus on the regions of New York State. Key interactions and feedbacks are represented through the use of climate scenarios (see Chapter 1, "Climate Risks") in combination with the assessment of the effects of biophysical and socio-economic stresses on society and ecosystems. The incorporation of climate change scenarios into these types of assessments is still relatively new, and few protocols (e.g., building codes and standards, flood-protection guidelines) have yet been established by practitioners and their governing bodies (e.g., engineering associations, insurance providers).

2.2.3 Vulnerability Measures and Metrics

There is a great diversity of methods and approaches for measuring vulnerability (Adger, 2006; Polsky et al., 2009). Because vulnerability reflects both social and physical aspects of systems, it is not easy to reduce to a single metric and is not easily quantifiable.

Specific variables do not measure vulnerability directly, so many assessments attempt to quantify vulnerability by using indicators as proxies. This is because focusing on purely physical or social variables may not capture the issues that make individuals or localities vulnerable to multiple stresses. Many assessments combine indicators to create a single numeric index (e.g., vulnerability to flooding and the Livelihood Vulnerability Index) (Speakman, 2008; Hahn et al., 2009). For example, the Human Development Index uses life expectancy, health, education, and standard of living as a measure of national well-being (UNDP, 2007). If this approach is used, variable and causal linkages between indicators (e.g., between standard of living and health) must be well established to ensure that the relationship is valid. The indicators that are chosen to represent vulnerability need to be sensitive to redistribution of risk within a vulnerable population or system (Adger, 2006).

While composite indices can provide valuable insight into current patterns of physical and socioeconomic vulnerability, they can also lead to a loss of information about how the different indicators contribute to vulnerability and are unable to incorporate changes in the larger national and global context. Patterns of vulnerability have become increasingly dynamic as the result of rapid, ongoing economic and institutional changes. The dynamic character of vulnerability means that it is particularly difficult to assess, as the factors that shape vulnerability—both the physical properties of a system and the socioeconomic context—are in a constant state of flux (Adger and Kelly, 1999; Thomas and Twyman, 2005). Under these circumstances, a flexible approach based on place-specific local variability within the broader state/federal policy guidelines and frameworks is suggested (Cutter and Finch, 2008). This requires replacing traditional indicators (e.g., share of drought-resistant crops, rainfall, per-capita staple food production, population density, infant mortality index) with dynamic indicators (e.g., change in access to credit, change in crop subsidy policies, change in national trade or investment policy stance, change in soil fertility, change in climate variability). (For more examples and explanation of the differences between traditional and dynamic indicators, see Leichenko and O'Brien, 2002.)

2.2.4 Evaluating Vulnerability in ClimAID

Throughout New York State, climate impacts and vulnerabilities vary widely by region and sector, as do the resources available to respond to climate change, necessitating regional solutions to adaptation rather than the proverbial one-size-fits-all approach. The ClimAID approach to assessment attempts to simplify the complex issues associated with climate change by dividing problems geographically and sectorally bringing into focus realities that are often discounted or overlooked in the development of the national- or state-level frameworks. Although detailed quantitative vulnerability studies were beyond the scope of this assessment, specific case studies for key vulnerabilities within each sector used a qualitative approach. A focus on key vulnerabilities is necessary to help policy-makers and stakeholders assess the level of risk, evaluate, and design pertinent response strategies.

The ClimAID assessment categorizes vulnerability through an evaluation framework and associated mapping activities (see Chapter 3, "Equity and Economics") across eight sectors and seven regions of the state. General conclusions and recommendations regarding vulnerability and potential vulnerability-reduction and adaptation strategies were then developed for each sector.

Within each sector chapter, vulnerabilities have been evaluated depending upon those systems or regions whose failure or alteration is likely to carry the most significant consequences. More details can be found in the sector chapters. In most instances, evaluation was qualitative, based on stakeholder input and the degree to which the relevant climate parameters were shown to change in the downscaled projections. A common set of criteria for evaluating vulnerabilities was used within each sector. The factors that were considered characterized anticipated impacts based on the "reasons of concern" developed by the IPCC Fourth Assessment Report (Schneider et al., 2007) (see **Box 2.2**).

The ClimAID assessment has not specifically identified vulnerability indices for New York State as a whole. Instead, each sector has worked individually to identify stakeholder characteristics that could potentially lead to climate vulnerability. The assessment uses different physical, socio-economic,

Box 2.2 Factors used to evaluate vulnerability in New York State

Magnitude (e.g., the area or number of people affected) and the intensity (e.g., the degree of damage caused)

Timing (is this impact expected to happen in the near term or in the distant future?)

Persistence (e.g., are previously rare events becoming more frequent?)

Reversibility (over the time scale of generations)

Likelihood (estimates of uncertainty)

Confidence in likelihood estimates

Distributional aspects within a region or among socio-economic groups

Importance of the at-risk systems—If the livelihoods of many people depend on the functioning of a system, this system may be regarded as more important than a similar system in an isolated area (e.g., a mountain snowpack system with large downstream use of the meltwater versus an equally large snowpack system with only a small population downstream using the meltwater)

Potential for adaptation (the ability of individuals, groups, societies, and nature to adapt to or ameliorate adverse impacts)

Thresholds or tipping/trigger points that could exacerbate change or initiate policy

Source: Schneider et al., 2007

and ecological indicators to measure vulnerability for different systems within the sectors. For example, the Coastal Zones sector uses coastal vulnerability index maps (Thieler and Hammer-Close, 2000; Gornitz et al., 2004) to illustrate the vulnerability of the New York State shoreline to sea level rise by considering a number of contributing geomorphological, geological, and oceanographic factors. The Water Resources sector has demonstrated that vulnerability to flooding in parts of New York State has often been related to socioeconomic factors. Similarly, the Public Health sector shows that those at higher risk for heat-related mortality are among the most vulnerable urban residents: elderly, the low-income populations, those with limited mobility and little social contact, those with pre-existing health conditions and belonging to certain racial/ethnic groups, and those lacking access to public facilities and public transportation or otherwise lacking air conditioning.

2.3 Adaptation

Adaptation to climate change focuses on actions that take place in response to a changing climate. Adaptation strategies do not directly include actions to reduce the magnitude of climate change, generally referred to as climate change mitigation, but instead present actions to lessen the impact of climate change or take advantage of changes caused by a shifting climate. In the context of the ClimAID project, two categories of adaptation strategies were examined, those that 1) reduce the level of physical, social, or economic impact of climate change and variability; or 2) take advantage of new opportunities emerging from climate change.

2.3.1 Adaptation Concepts

Adaptation strategies and actions have a direct connection to the risk and hazards management tradition. Individuals and organizations attempt to reduce their vulnerability and exposure to threats. Stakeholders and decision-makers within each ClimAID sector have developed extensive protocols to avert and manage hazards and to promote greater disaster-risk reduction. In many ways, adaptation to climate change fits into this tradition. How adaptation strategies are now being developed reflects, in turn, both historical risk management and the emerging understanding of the magnitude and significance of ongoing climate change. In this way, climate change represents either an increased manifestation of established hazards (e.g., possibly longer and more intense droughts) and/or new hazards (e.g., emergence of a new type of pathogen moving northward with climate change).

Potential adaptation strategies can be further defined within a range of elements, including economics, timing, and institutional organization. Economic issues include the costs and benefits of adaptation and the relative distribution of both (see Economics section in Chapter 3, "Equity and Economics" and additional economic analysis in Annex III). A critical issue is the overall cost-to-benefit ratio and how much economic advantage there is to taking a specified action. There are difficulties in calculating these costs due to the issues in determining the "social rate of time discount," that is, the rate used to compare the well-being of future generations to the well-being of those alive today. Potential opportunity costs also are important to determine, given what is understood about the rate of climate change and the sensitivities of the system in question. A primary question is whether the adaptation strategies take place in the short-term (less than 5 years), medium-term (5 to 15 years) or long-term (more than 15 years).

Crucial to the issue of timing is whether there are tipping points associated with dramatic shifts in the level of impacts and/or vulnerabilities and whether these tipping points become triggers for new policies and regulations. A tipping point can be defined as a moment in time when the operation of a system would move to a new phase as a result of changes in internal dynamics or a perceived need by associated managers. An example of a tipping point could be the occurrence of a major heat-mortality event such as occurred in Europe in 2003. Over 25,000 people, many of them elderly, died due to a heat wave that was five standard deviations away from normal (IPCC, 2007). This event triggered a massive public health adaptation response to heat waves in European countries that is in place today.

Another primary category of adaptation is the institutional organization of the entity responding to climate change. A key issue here is whether the stakeholder is administratively organized to collect and monitor climate change conditions and to incorporate this information into decision-making analysis on a regular and ongoing basis. These conditions are necessary for the development of adaptation strategies that enable flexible responses to evolving scientific understanding and uncertainty; that is, putting in place adaptations that can be adjusted or shifted over time (i.e., years or decades) as new information and evidence indicate the need for shifts in strategies and policies to better respond to emerging climate threats and opportunities.

2.3.2 Adaptation Assessment Approaches

Adaptation to climate change includes a wide diversity of issues and considerations that are important for assessing the context and need for adaptation strategies and their potential success. Broadly speaking, two primary sets of considerations for adaptation strategies can be defined during an assessment: 1) those associated with the entity implementing, proposing, and/or planning the adaptation; and 2) those associated with the character of the adaptation strategy itself, and its (potential) impact.

Within the scholarly literature on adaptation assessment, these two sets are further refined into several elements of the adaptation development, planning, and implementation process. These elements include focus on the type of entity from which the adaptation emerges, the character of the strategy (e.g., timing, extent, impact), and adaptation financing.

Public and Private Sectors as Agents of Change

In the first category, a key element focuses on whether the adaptation emerged from the private sector or from the public sector. A related consideration is whether the stakeholder is traditionally proactive or reactive with respect to decision-making, in general, and issues of risk and vulnerability, specifically. Some ClimAID sectors—especially public health and water resources—spend extensive time and resources preparing for crises and, in turn, could be seen as having heightened capacity to plan and respond to climate change. Additional adaptation strategies can be implemented during times of crisis, because these moments open a policy window during which an opportunity for administrative reform and change can occur.

Gradual vs. Transformative Change

Some stakeholders have pre-existing trigger points for regulatory and administrative action, such as those that are embedded in heat and drought advisories and alerts. These trigger points can become the administrative structure within which adaptation to climate change can be developed. Related to this point is the question of whether climate change adaptation can be implemented simply as an extension or adjustment of existing rules, guidelines, or regulation, or if it must be implemented as a more significant transition within the stakeholder organization or operation. For example, stakeholders in all of the sectors have climate- or weather-risk policies, some of which are more developed than others. (In the Transportation sector, this could vary from New York City Transit's flood-mitigation policies to rural municipalities' road salting and plowing schedules to deal with snowfall.) Another related consideration is the possibility that the adaptation can be derived as an extension of existing codes, standards, or practices, or it can require a more significant reorganization of the entities' management structure and agenda.

Technical vs. Non-technical Adaptations

Another key element is whether the adaptation is technical in nature (e.g., engineering modification, hard option) or non-technical in nature (e.g., non-structural, soft option), such as policy and/or regulatory change. A connected issue is whether the strategy involves a simple adjustment to how the climate hazard is managed or involves a larger, system-wide change. An example could include increased efforts to provide shoreline protection from increased flood frequency (structural) as opposed to a more dramatic staged retreat from the coast (non-structural). Other elements associated with the character of the adaptation strategies include the timing of adaptation and its consequence. For example, is there a trigger point for action when the likelihood of a negative impact becomes sufficiently great such that a stakeholder response becomes necessary? Critical related questions are: How is the trigger point defined, and who determines that the trigger has been reached? Underpinning these considerations are questions of uncertainty and system complexity that result from the fact that, at the sector level, the organization and structure of a system, in many situations, are not fully understood and the potential response to climate change remains only partially known.

Financial Elements

In regard to the character of the adaptation strategy itself, funding and expected benefit-cost ratio are two of the most important elements. The issue of liability is important as well because it directly relates the climate-hazard information to action. As information about climate change and its impacts becomes available, decision-makers are increasingly faced with the question of when and with what caveats to present this knowledge to the public. Will withholding information make them liable for potential future damages? Or will actively responding to the information result in liability issues if certain parties are more adversely affected as a result of the actions taken? For example, who will pay the costs of increased air conditioning? And who will pay for the costs associated with the loss of property use if sea level rise projections place additional property within the 100-year flood zone?

2.3.3 Assessing Adaptation in ClimAID

Within the ClimAID project, the investigators assessed adaptation strategies within New York State in a way that reflected the specific interests and information requirements of climate change stakeholders and decision-makers within the state. The assessment frame was distilled from the considerations and elements defined in Section 2.4.2 and translated into particular categories relevant to each ClimAID sector. The categorization procedure set the stage for the adaptation strategy evaluation process that followed.

To perform the adaptation assessment, ClimAID sector investigators inventoried a set of the sector stakeholders' present and planned adaptation strategies (the set does not include every possible adaptation strategy but highlights representative ones). As part of the analysis, each ClimAID sector team defined potential adaptation strategies that were identified by engaging in discussions and holding meetings with the stakeholders. The sector analyses focus both on those adaptations designed to limit exposure to increased climate risk as well as those that enhance the stakeholder's ability to take advantage of opportunities presented by climate change, such as a switch in crop choice or shifts in water availability (e.g., water shortages may occur in other parts of the country while water supplies may increase overall in New York; see "Agriculture" and "Water Resources" chapters).

Adaptation Categories

The adaptation strategies developed through the stakeholder process were first divided into categories: type, administrative group, level of effort, timing, and scale (**Box 2.3**). "Type" includes whether the strategies were focused on management and operations, infrastructural change, or policy adjustments. "Administrative Group" defines the strategies as either emerging from the public or private sectors and the level of government (e.g., local/municipal, county, state, national) to which they pertain. "Level of Effort" indicates whether the strategy represents an incremental action or a larger-scale paradigm shift. "Timing" highlights the period during which the adaptation strategy will be implemented—short-term (less than 5 years), medium-term (5 to 15 years), or long-term (more than 15 years)—as well as the speed of implementation and the presence of established or known tipping points and policy triggers. "Scale"

Box 2.3 Categories of adaptation strategies

Type
- Behavior
- Management/operations
- Infrastructural/physical component
- Risk-sharing
- Policy (including institutional and legal)

Administrative Group
- Public or private
- Local/municipal, county, state, national government

Level of Effort
- Incremental action
- Paradigm shift

Timing
- a) Period
  - Short-term (less than five years)
  - Medium-term (five to 15 years)
  - Long-term (more than 15 years)
- b) Abrupt Changes
  - Tipping points
  - Policy triggers

Scale
- Widespread
- Clustered
- Isolated/unique

includes the overall spatiality of the adaptation impacts, specifically cataloging if the adaptation strategy impact is widespread, clustered, or isolated/unique (e.g., impact associated with a specific site or location) throughout the state.

Adaptation Strategy Evaluation

Once adaptation strategies have been categorized, evaluating them is a critical yet complex task. Strategy evaluation can help stakeholders to determine an order to implement strategies and aid in developing a broader agency- or organization-wide adaptation plan. Criteria that can be used to help evaluate strategies include cost, feasibility, efficacy, timing, resiliency, impacts on environmental justice communities, robustness, and co-benefits/unintended consequences (Major and O'Grady, 2010). These are briefly described below:

- Cost—What will be the economic impact of the strategies, including an estimate of short-, medium-, and long-term benefits and costs?
- Feasibility—How feasible is the strategy for implementation both within an organization and from perspectives such as engineering, policy, legal, and insurance? Are there expected technological changes that would impact future feasibility?
- Efficacy—To what extent will the strategy, if successfully implemented, reduce the risk?
- Timing—When is the strategy to be implemented? What factors affect the implementation schedule?
- Resiliency—To what extent is the strategy, when implemented, able to withstand shocks or stresses—either physical or social (e.g., policy) in character?
- Impacts on environmental justice communities—Will strategy impacts be negative or positive for communities already stressed by environmental risk exposures?
- Robustness—Is there the potential to install equipment or upgrade infrastructure that is designed to withstand a range of climate hazards? Are there opportunities for flexible adaptation pathways, i.e., incremental management adjustments associated with the pre-determined objective of updating adaptation based on emerging science and management needs?
- Co-benefits/unintended consequences—Will any strategies have positive or negative impacts on another stakeholder or sector? Is there potential for cost sharing? Are there impacts on mitigation of greenhouse gases? Are there impacts on the environment or a vulnerable population?

Through meetings and discussions, sector leaders and stakeholders evaluated adaptation strategies via the criteria defined above. However, the quantification of benefits and costs was often confounded, particularly when sectors were represented by multiple stakeholders with diverse interests and values. This was particularly true in the Ecosystems sector, where the values of factors such as diversity of species and the preservation of natural areas are extremely difficult, if not impossible, to quantify. In many cases, net benefits to one group may be viewed as losses by a different stakeholder group; for example, warmer winters may benefit homeowners due to reduced heating costs and, at the same time, cause losses for the winter recreation industry. Finding the common interest under these circumstances is a complex task. Other topics to emphasize are the spatial and temporal character of the adaptation strategies and how easily modified they may be in response to a changing climate through time, i.e., do they contribute to the development of flexible adaptation pathways.

2.4 Outcomes

A major aim of the ClimAID assessment is to help New York State manage, rather than eliminate, uncertainties related to a changing climate. Drawing

Box 2.4 ClimAID vulnerability and adaptation assessment approach with links to the five integrating themes

1) Identify current and future climate hazards C
2) Conduct risk assessment inventory........................C, V, EEJ, E
3) Characterize risk of climate change........................C, V, EEJ, E
4) Develop initial adaptation strategies ....A
5) Identify opportunities for coordination....................................A
6) Link strategies to capital and rehabilitation cycles...............A, E
7) Prepare and implement adaptation plans............C, V, A, EEJ, E
8) Monitor and reassess vulnerability and adaptation ...C, V, A, EEJ,

C = Climate (Chapter 1);
V = Vulnerability (Chapter 2);
A = Adaptation (Chapter 2);
EEJ = Equity and Environmental Justice (Chapter 3);
E = Economics (Chapter 3)

Defining Equity and Environmental Justice within ClimAID

The maps below draw attention to spatial differences across New York State in the ability of communities to adapt to climate change. These differences, which are likely to influence climate change vulnerability and adaptation, emphasize issues of distributional equity. Within the climate change literature, distributional equity may be defined as the fair distribution of outcomes or impacts associated with climate change (Kasperson and Dow, 1991). However, an emphasis on environmental justice requires attention to and recognition of both distributional and procedural equity. Within the broader literature on environmental justice, distributional equity emphasizes securing benefits and amenities such as access to parks and greenspace to offset environmental burdens that specific communities face. Procedural or process equity entails an equitable approach to environmental decision-making (Lake, 1996). In the context of climate change, procedural equity may be defined as inclusion of representatives of all affected communities and groups in decisions about climate change adaptation, including emergency preparedness and emergency response. Efforts to achieve procedural equity include mechanisms to ensure participation of affected actors in policy decisions (O'Brien and Leichenko, 2010).

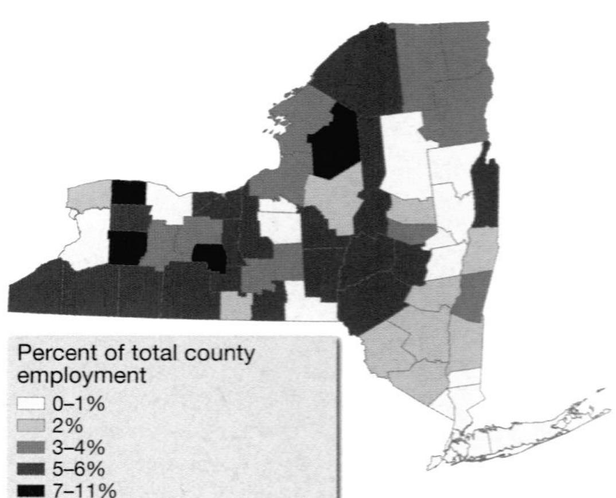

Source: U.S. Department of Commerce, Bureau of Economic Analysis: REIS Table CA25N

Figure 3.4 Employment in agriculture and forestry, fishing, and other related activities as a percentage of total county employment

In defining the equity and environmental justice element within the ClimAID assessment, the study draws insights from both distributional and procedural approaches used within the environmental justice and climate change literatures. In terms of distribution, there is an emphasis on identification of situations where particular groups may be systematically disadvantaged either in terms of differences in vulnerability or capacity to adapt to climate change or in terms of the impacts of policies surrounding adaptation. While the equity and environmental justice analysis for some sectors emphasizes commonly recognized groups within the environmental justice literature (including lower-income, minority, and Native American populations), there is also consideration of equity effects across other units of analysis, such as rural regions versus urban areas, small versus large firms, or small versus large cities, as appropriate for the type of analysis conducted for each of the sectors.

For all of the sectors, the analysis of distributional equity issues includes consideration of:

- Inequalities in vulnerability to climate change;
- Inequalities in the capacity to adapt to climate change;
- Inequalities in adaptation policy benefits; and
- Inequalities in the effects of the adaptation policies.

In terms of procedural equity elements, key considerations include the incorporation of equity issues in adaptation discussions and policies, the mechanisms for broad and meaningful participation in future adaptation planning and policy efforts, and the incorporation of input from the equity and environmental justice stakeholders in the ClimAID assessment.

These distributional and procedural definitions of equity and environmental justice are used in various components of the assessment. The broader aims include consideration of potential inequalities associated with climate change along traditional lines that have been identified within the environmental justice literature (e.g., underprivileged, minority groups), as well as along new lines that may emerge under an altered climatic regime (e.g., different-sized firms) or may result from the implementation of adaptation policies and plans.

3.1.2 Approach for Equity and Environmental Justice Assessment

The equity and environmental justice component of ClimAID involves three types of parallel efforts: 1) development of equity and environmental justice assessments for each sector, based on review of background literature in these areas; 2) development of integrated case studies; and 3) attention to input from environmental justice groups or representatives in the sector meetings with stakeholders. Descriptions of each of these elements are presented below.

General Assessment for Each Sector

The first task for the equity and environmental justice element entails the creation of assessments for each sector that identify the key, relevant equity and environmental justice issues based on review of past studies. (Elements of these sector-specific assessments, including references to the studies reviewed, are presented in the sector chapters.) Each of the sector assessments addresses the same general questions. The questions, which emphasize equity and environmental justice issues surrounding both vulnerability and adaptation within each sector, are as follows:

1) Are there preexisting socioeconomic or spatial inequalities that make certain regions, communities, or groups of individuals systematically more vulnerable to the impacts of climate change on the sector? What groups or areas are likely to shoulder a disproportionate share of the burden from these impacts? Potential differentiations by group include socioeconomic status, education, health/disability, race, age, gender, culture, or citizenship. Community differentiations include the extent of segregation, access to health care, unemployment, and poverty/wealth/assets.
2) Are there groups, communities, or regions that are less able to adapt to the impacts of climate change and, therefore, merit special attention during adaptation planning?
3) Within the range of adaptation strategies in each sector, which strategies are more likely than others to exacerbate underlying socioeconomic disparities? Could some strategies change social and environmental dynamics so as to create emergent or unintended disparities? Are there situations in which strengthening adaptive capacity in one area or for one group may, in turn, create, reinforce, or exacerbate maladaptation or vulnerability (either in an absolute or relative sense) in other groups or areas?

Each of the above questions was considered in a broad review of prior work on environmental justice and climate change. The questions also guided work on the integrated case studies, as discussed below. The sectoral assessments also touched on two additional questions related to the equity and environmental justice consequences of adaptation planning:

1) Are there certain groups, communities, or regions that may be systematically underrepresented during adaptation planning, unable to access or influence the process and procedures of decision-making, or otherwise disempowered, unable, or disinclined to consider adaptation when it is likely to be in their interest to do so?
2) When designing adaptation strategies, are there ways to insert mechanisms that encourage or ensure fair outcomes, whether preventive (e.g., avoiding and adjudicating disputes), corrective and compensatory (e.g., payments to an affected party to compensate for loss of access to a resource), or retributive (e.g., sanctions and penalties)?

Although a full assessment of these latter two questions was beyond the scope of the present study, raising these questions nonetheless represents an important starting point for incorporation of equity and environmental justice issues into future adaptation planning and policies in New York State.

Role in Integrated Case Studies

The equity and environmental justice component also entails participation in integrated case studies for each sector. (These case studies are presented in full in each of the sector chapters.) In some instances, these case studies explore impacts of past climate extremes, such as a past flood event or heat wave. Such cases serve as historical analogues, whereby the research teams may consider the equity and environmental justice consequences of a past climate event in order to extract lessons on how to reduce distributional inequalities in planning for future climate change. Other case studies project future climate change impacts on various sectors and industries and explore potential future equity issues

informed by a faith in the workings of markets coupled with a technological optimism. These premises assure society that as an exhaustible natural resource is depleted, its price will rise exponentially and demand will ultimately go to zero. During this inexorable process, society will make transitions to other technologies, and ultimately the use of the exhaustible resource will be displaced by an inexhaustible backstop technology (e.g., solar, wind, or hydrogen energy) that is expected to meet all our future needs(see Solow, 1974).

The strong sustainability criterion for sustainable development rejects the notion that natural and man-made capital are freely substitutable. The strong sustainability criterion requires that society preserves minimum quantities of natural capital stocks and ecosystem services, rather than allowing man-made capital to displace natural capital stocks over time. (For a full exposition of the ecological economic critique of neoclassical environmental economics, see Daly, 1997a; Solow, 1997; Stiglitz, 1997; and Daly 1997b.)

Another analysis clearly frames the implications of these alternative paradigms for climate change mitigation policy: A policy based on strong sustainability requires that a cap on greenhouse gas emissions be based on the assimilative capacity of the global ecosystem. A policy based on weak sustainability is based on the presumed tradeoffs between economic activities and the value of ecosystem services. For example, we might as a society accept a loss of biodiversity or an increase in coastal erosion if these costs are outweighed by other economic benefits.

In practical terms, cost-benefit analysis, which lies at the foundation of neoclassical economic policy analysis, acknowledges tradeoffs among human capital (e.g., labor), manufactured capital, and natural capital. This cost-benefit analysis must weigh the cost of adaptation and mitigation strategies against the costs associated with climate change.

The second point of contention in the climate change debate within economics is the tradeoff between the needs of the present and those of future generations. One report, the Brundtland Commission report, incorporates this tradeoff within the report's definition of sustainable development: "development that meets the needs of the present without compromising the ability of future generations to meet their own needs" (World Commission on Environment and Development, 1987). This debate has practical implications for climate change policy and, in particular, the cost-benefit analyses that economists use to critique alternative adaptation and mitigation strategies. In applying cost-benefit methods to the study of global climate change, it first must be acknowledged that: 1) the costs of increasing atmospheric concentrations of carbon dioxide (and other greenhouse gases) will be felt gradually over a number of decades; and 2) the benefits of reducing greenhouse gas emissions will be realized over a long time period.

One economic analyst who explicitly addresses the issue of intergenerational equity argues that we should treat present and future generations equally; thus he proposes that we should weight the benefits of mitigation and adaptation policies equally, whether they occur now or in the distant future (Solow, 1974).

A recent analysis reviews the ongoing debate between those who believe that the discount rate used in climate policy studies should reflect the real return on investments (the "descriptive" approach) and those who feel that it should be based on intergenerational equity concerns, i.e., the relative weight placed on the needs of current versus future generations (Dietz and Maddison, 2009). In the context of the ClimAID assessment, the intergenerational equity approach would justify increased investments in greenhouse gas adaptation and mitigation policies.

Finally, an economic analysis of global climate change policy must account for distributional equity. Although cost-benefit analysis essentially compares the total costs of climate change versus the total costs of adaptation and mitigation strategies, any policy intervention will have winners and losers. In practice, neoclassical economists have focused on economic efficiency and have tended to neglect issues of equity as being outside the realm of their analysis. The equity and environmental justice analyses in the ClimAID assessment, however, explicitly address the issue of distributional equity.

3.2.2 Methodological Foundations of Cost-benefit Analysis

The standard neoclassical economic criterion for a Pareto-optimal allocation of resources requires that there be no possible reallocation of resources that could make at least one person better off while making no individual(s) worse off (Bergson, 1938; Boulding, 1952; Tietenberg, 2000). As a rule, any policy change will have both winners and losers. Hence, the practical criterion for evaluating government policies and programs is whether the policies and programs result in a potential Pareto improvement, i.e., that, in principle, the winners should be able to compensate the losers. Consider, for example, a project to prevent riverine flood damage by selective retreat from the flood plain. If the projected benefits (avoided damage, ecosystem benefits) exceed the value of the property within the flood plain, the beneficiaries (society, in this instance) should pay the losers (property owners) to compensate them for their property losses, and create parklands or open space.

The social welfare maximization criterion does not require that any particular outcome will satisfy the ClimAID assessment's criterion for social justice (or distributional equity). Neoclassical welfare economics, however, enables the analysis to address the issue of equity within a market economy, as follows.

Economists often act as if the issues of economic efficiency and equity can be addressed independently. Economists prefer to leave issues of equity to the political process, through taxation and public finance. Once the government has addressed issues of equity through the budgetary process, the free market should be left to "work for itself" (Stiglitz, 1991). In practice, however, neoclassical environmental economists have addressed almost exclusively the issue of efficiency, arguing that actual compensation as it relates to equity issues is outside the realm of economics (Splash, 1993).

The basic tool of cost-benefit analysis in applied welfare economics is an implementation of this social welfare maximization criterion: If the benefits of a proposed policy change exceed their costs (i.e., if the benefit-cost ratio exceeds 1), then it would clearly be possible for the winners to compensate the losers.

It should be noted that an alternative framework for project evaluation has long been available that takes into account not only efficiency but also other objectives such as redistribution. (For an example of a classic study that takes this approach, see Dasgupta et al., 1972.) While this approach has many advantages, it is not used within the economics component, which is designed to study the efficiency costs and benefits of impacts and adaptations.

The choice of an appropriate social discount rate must accommodate the issues of economic efficiency, intergenerational equity, and the global nature of climate risk. As these issues remain unresolved, the ClimAID assessment performs sensitivity analyses with rates of 0 percent and 3 percent, reflecting lower rates of discount used in many climate studies, most notably in the Stern Report (Stern, 2007). Some analysts advocate higher rates, for example Nordhaus (2007a; 2007b). Stern (2009) argues that such higher rates are inappropriate for large-scale social decisions where the risks of inaction are to a significant extent unknown (and possibly very high), and the costs of present action are relatively low.

3.2.3 Economic Analysis in the ClimAID Assessment

For sectors whose goods or services are traded in organized markets, the ClimAID assessment relies on market data on observed input-output quantities and prices in order to directly estimate the social marginal benefits (i.e., the value to society of a small increase in the scale of an adaptation measure, such as an additional mile of shoreline protected from storm damage) and social marginal costs (i.e., the cost of that same adaptation measure, in this case the cost to protect an additional mile of coastline). With sufficient available data, social marginal benefits and social marginal costs can be estimated with conventional statistical economic modeling. When adequate data are not available, existing estimates from reliable sources are included in the analysis. In the case of sectors whose goods or services are not traded in markets, widely accepted techniques to represent values for such goods and services are used, as summarized below:

1) For sectors in which statistics on economic activity are available (e.g., energy, agriculture), a regression-based approach is followed. The relationship between demand and supply in a market are statistically estimated from price, quantity, and other data.
2) When market data are not available, the assessment follows a different approach, based on survey information and other sources. Various techniques are available to estimate the economic value of non-traded goods and services, including direct methods such as contingent valuation (surveys that provide a gauge of people's willingness

Chapter 4

Water Resources

Authors: Stephen Shaw,[2,7] Rebecca Schneider,[1,3] Andrew McDonald,[1,2,7] Susan Riha,[2,7] Lee Tryhorn,[2] Robin Leichenko,[4] Peter Vancura,[4] Allan Frei,[5] and Burrell Montz[6]

[1] Sector Lead
[2] Cornell University, Department of Earth and Atmospheric Sciences
[3] Cornell University, Department of Natural Resources
[4] Rutgers University, Department of Geography
[5] City University of New York Hunter College, Institute for Sustainable Cities
[6] East Carolina University, Department of Geography (formerly at Binghamton University)
[7] NYS Water Resources Institute

Contents

Introduction 80
4.1 Sector Description 80
4.1.1 Economic Value 80
4.1.2 Non-climate Stressors 80
4.2 Climate Hazards 81
4.2.1 Temperature 81
4.2.2 Precipitation 81
4.2.3 Sea Level Rise 82
4.2.4 Other Climate Factors 82
4.3 Vulnerabilities and Opportunities 83
4.3.1 Flooding 84
4.3.2 Drinking Water Supply 86
4.3.3 Water Availability for Non-potable Uses 92
4.3.4 Water Quality 94
4.4 Adaptation Strategies 97
4.4.1 Flood Adaptation Strategies 97
4.4.2 Drinking Water Supply Adaptation Strategies 98
4.4.3 Non-potable Water Supply Adaptation Strategies 99
4.4.4 Water Quality Adaptation Strategies 100
4.5 Equity and Environmental Justice Considerations 101
4.5.1 Local Management Capacity 101
4.5.2 Equity and Flooding 101
4.5.3 Equity and Water Supply 102
4.5.4 Equity and Water Quality 102
4.6 Conclusions 102
4.6.1 Key Existing and Future Climate Risks 103
4.6.2 Main Findings on Vulnerabilities and Opportunities 103
4.6.3 Adaptation Options 103
4.6.4 Knowledge Gaps 106
Case Study A. Susquehanna River Flooding, June 2006 107
Case Study B. Orange County Water Supply Planning 112
References 114
Appendix A. Stakeholder Interactions 117
Appendix B. New York State Flood Analysis 118

Introduction

This ClimAID chapter covers climate change vulnerabilities and possible adaptation strategies for four major water resource themes: 1) flooding in non-coastal regions, 2) drinking water supply, 3) water availability for non-potable uses (primarily agriculture and hydropower), and 4) water quality. Ensuring reliable water supplies, minimizing the disruptive and destructive impacts of flooding, and maintaining the recreational and aesthetic value of water bodies are fundamental needs, critical to the well-being of communities and businesses throughout New York State.

4.1 Sector Description

New York State has an abundance of water resources. Despite having only 0.3 percent of the world's population, the state is bordered by lakes containing almost 2 percent of the world's fresh surface water: Lake Erie, Lake Ontario, and Lake Champlain. It is home to the Finger Lakes in central New York, which are the largest of the state's 8,000 lakes as well as some of the largest inland water bodies in the United States. The state has several high-yielding groundwater aquifers, particularly those underlying Long Island. It has an average annual rainfall of almost 40 inches, readily supplying numerous small municipal reservoirs as well as the extensive New York City water supply system with surface water impoundments in the Catskill Mountains and the Croton watershed east of the Hudson River. The state contains the headwaters of three major river systems in the Northeast: the Hudson River, the Delaware River, and the Susquehanna River.

In 2000, New York State's 19 million residents consumed approximately 2,200 million gallons per day of fresh surface water and 890 million gallons per day of fresh groundwater for public water supply, irrigation, and industrial uses (Lumia and Linsey, 2005). Of this nearly 3,100 million gallons per day of consumption, only about 10 percent was for industrial and agricultural use.

4.1.1 Economic Value

There is no direct way to describe the economic value of water resources in the state. One could attempt to place an approximate market value on the water consumed. Treated water costs approximately \$3 per 1,000 gallons; given that New Yorkers consume around 3,100 million gallons per day, this works out to more than \$3 billion in revenue per year. Another way to look at economic value is in terms of infrastructure. An estimated value of this part of the sector can be gathered by considering that the New York City Department of Environmental Protection's capital program for 2010 through 2019 is just over \$14 billion (NYCMWFA, 2009, p. 24). Conversely, it is important to consider negative economic consequences associated with water. For instance, disaster assistance for a large flood event in the Susquehanna Basin in 2006 topped \$225 million; such a flood event typically occurs every few decades. Overall, a clean, reliable source of potable water is an essential underpinning of community stability and a necessity for numerous other economic activities.

4.1.2 Non-climate Stressors

There are several non-climate factors that will interact with possible changes in climate. First, much of the water resource infrastructure in New York is old and requires updating and rebuilding. Failing infrastructure can contribute to water pollution and reduce the reliability of treatment and distribution systems. Recent estimates suggest that over \$36 billion is needed to update wastewater treatment infrastructure in the next 20 years (NYSDEC, 2008a). Second, even without any possible changes in water supply with climate change, rapidly developing regions face increasing water demands. Continued increases in population are projected, particularly in the New York City metropolitan area. New York City alone anticipates an increase of about 0.7 million people by 2030 (NYCDEP, 2006). Newly developing suburban and exurban regions can face unique challenges. In many cases, older cities that built water infrastructure more than 100 years ago were able to develop water sources in undeveloped regions outside their immediate borders (e.g., New York City reservoirs, City of Troy Reservoir, Albany Reservoir). Newly developing communities face more extensive regulations, few undeveloped areas to claim for their use, and competition from other neighboring communities that may also have rapid growth.

4.2 Climate Hazards

This section focuses on the temperature, precipitation, sea level rise, and extreme event hazards of particular concern to the water resources sector.

4.2.1 Temperature

Increases in air temperature will lead to increases in water temperature. Up to a water temperature of approximately 77°F, water temperature directly increases with air temperature, with a proportionality constant of 0.6–0.8. For instance, an air temperature increase of 9°F would result in a water temperature increase of 5–7°F. Thus, increases in water temperature will be slightly less than increases in air temperature. Higher water temperatures will have direct impacts on certain elements of water quality such as oxygen content.

Additionally, increases in temperature are likely to decrease the fraction of precipitation falling as snow. This will lead to shifts in seasonal stream flows. Many observational and modeling studies suggest that late winter and early spring flows will increase and that spring snowmelt will occur earlier in the year (Hayhoe et al., 2007; Burns et al., 2007; Hodgkins et al., 2003; Neff et al., 2000). Thus, even if there is more annual streamflow, it may be distributed unevenly over the year, with lower flows in the late summer and autumn and higher flows in the late winter and spring. This shift in timing of flow magnitudes has already been observed in stream records.

Temperature will also have some impact on evaporation rates, either by extending growing seasons or by increasing the potential rate of water vapor transfer to the atmosphere from soils, vegetation, or open water. Although evaporation has some dependency on air temperature, the primary driver of evaporation in humid temperate regions such as New York State is the net amount of energy from sunlight plus the net amount of energy emitted from the Earth's own atmosphere (as demonstrated, for example by Brutsaert, 2006). Some models used to estimate evaporation (such as the Thornthwaite Equation) only consider temperature; we suggest that these temperature-based models may overestimate changes in future evaporation. Studies have long indicated the greater sensitivity of temperature-based evaporation equations to changes in temperature relative to more physically based equations for estimating evaporation (McKenney and Rosenberg, 1993). More accurate estimate of changes in evaporation will come from equations that consider the complete energy balance at the land surface (such as in the work by Hayhoe et al., 2007).

4.2.2 Precipitation

Precipitation feeds the hydrologic cycle. Changes in precipitation amounts and frequency can cause changes in stream and river discharges, lake levels, and groundwater levels. However, as discussed further in 4.2.4, hydrology is dependent on a number of interacting factors; changes in precipitation alone rarely explain likely changes to water resources. Based on historical observations, precipitation in New York State has been increasing both in total annual amount and in intensity. As noted in Chapter 1 ("Climate Risks"), annual average precipitation has been increasing by 0.37 inches per decade since 1900. In terms of intensity, increases in the frequency of heavy rainfall have been observed across much of the United States, with such upward trends strongest in the Northeast (DeGaetano, 2009; USGRCP, 2009). For instance, in New York State, the number of rainfall events each year with greater than 1 inch of precipitation in 24 hours has increased over time (**Figure 4.1**, black line).

As discussed in Chapter 1, climate models indicate that annual average precipitation in New York State will

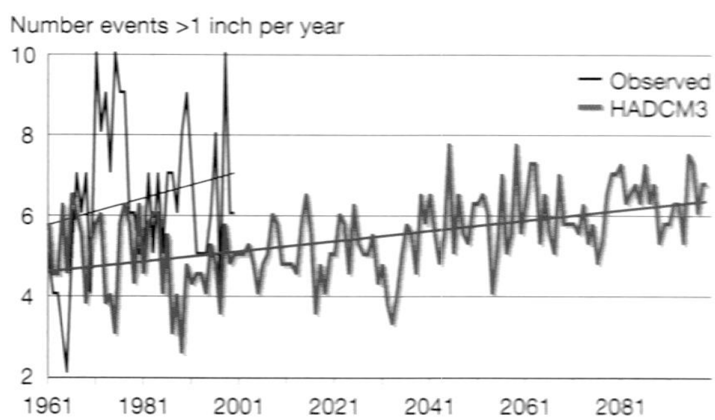

Note: UK Meteorological Office Hadley Centre Climate Model version 3 (HADCM3) projections adjusted to reflect regional climatology are shown in blue and observations are shown in black. These results are broadly consistent with those of the other 15 GCMs used by ClimAID.
Source: Tryhorn and DeGaetano, 2010

Figure 4.1 The observed and projected (by one global climate model) number of rainfall events exceeding one inch from 1960 to 2100, averaged over four stations in New York State

increase by 5 to 10 percent by 2080. Based on climate modeling, the frequency of heavy rainfall events is projected to increase as well. Applying a model (the Statistical Downscaling Model, or SDSM, Version 4.2) that relates large-scale circulation patterns and atmospheric moisture to local weather conditions, Tryhorn and DeGaetano (2010) simulated daily rainfalls from 1961 to 2100. By the end of this century, precipitation from storms that now occur on average every 100 years is projected to increase by 0.2 inch (**Figure 4.2**). With these increased event rainfall amounts, storms that now occur on average every 100 years are likely to become more frequent, recurring on average every 80 years by the end of the century. These trends, however, likely underestimate the future changes, given that the model upon which these predictions are based is underestimating current trends.

It is important to note that only recently have researchers started to investigate changes in the intensity of sub-daily precipitation events (Berg et al., 2009; Lenderink and Van Meijgarrd, 2008). The intensity of sub-daily rainfall (particularly in periods of less than an hour) is of particular relevance. It is usually these intense short events that exceed a landscape's ability to allow water to infiltrate. Particularly in urban areas or steep basins, these intense rainfall events can result in flooding. For example, 1 inch of steady rainfall spread evenly over a day would likely produce less surface runoff than 0.5 inch of rain in an intense 15-minute event. There is evidence from historical data and regional climate modeling to suggest that the intensity of sub-daily rainfall events will increase in a warming climate. For example, one study found that 1-hour rainfall amounts increased 7 percent for every degree Fahrenheit that the air temperature increased in the Netherlands (Lenderink and Van Meijgarrd, 2008). Similar analyses of sub-daily rainfall intensities have not yet been carried out for New York State.

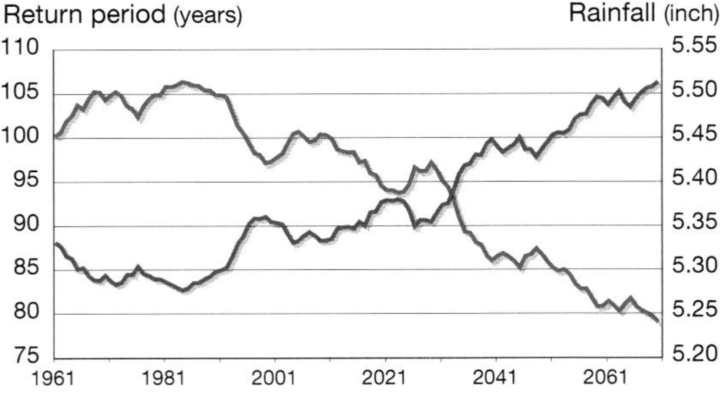

Note: The rainfall amount of the 100-year storm computed for each 30-year period beginning at the date in the graph (red) and the change in the return period associated with the amount of the 1961–1990 100-year storm (blue). The return period is the average interval of time between storm events of a given magnitude; a decreasing return period indicates that a given storm event occurs more often. These results from the HadCM3 model are broadly consistent with those of the other 15 GCMs used by ClimAID.
Source: Tryhorn and DeGaetano, 2010.

Figure 4.2 Projected rainfall and return period of the 100-year storm

If storm rainfall amounts increase in the future, the frequency of storm events could decrease (Trenberth et al., 2003; Hennessy et al., 1997), leading to longer periods with no rainfall. However, to date, an analysis of the time interval between historical storm events indicated no change in the Northeast despite increasing dry periods in other regions (Groisman and Knight, 2007).

4.2.3 Sea Level Rise

By the 2080s, sea levels could rise under rapid ice-melt as much as 55 inches (see Chapter 1, "Climate Risks"), with important implications for coastal storm flooding potential. With the additional water under the high-end scenario, the current 1-in-100-year flood could occur approximately an order of magnitude more frequently along the New York State coast (see Chapter 5, "Coastal Zones"). This shift will have ramifications for a broad set of coastal management processes including those for coastal water resources, groundwater protection from saltwater intrusion, and operation of wastewater treatment facilities.

4.2.4 Other Climate Factors

Changes in water resources rarely have a one-to-one link with a single climate factor. Numerous processes combine to determine the level of discharge seen in a stream or the amount of water available to a well. This can be illustrated by comparing typical monthly rainfall and stream discharge amounts for a stream in New York (see **Figure 4.3**). Rainfall is relatively even over most of the year, with the exception of lower amounts in the winter months. However, most evaporation and transpiration (water loss from plants) occurs between May and October when plants are active, making streamflows lower and soils dryer during summer and early fall. In addition, winter and spring streamflows may be further increased by the contribution of melting snow, causing streamflow to peak in the spring despite lower precipitation amounts. Thus, streamflows combine the effects of many interacting climate factors, as well as the water catchment's capacity for infiltration and storage.

A study of streamflows at 400 U.S. sites from 1941 to 1999 documented an increase in annual minimum and median daily streamflow beginning around 1970, particularly in the East. Notably, peak streamflow (i.e., floods) did not show a consistent increase in the studied streams (McCabe and Wolock, 2002). This is consistent with findings specific to New York. In the Catskill Mountain region, runoff increased from the 1950s to 2000s with an increase in annual warm-season streamflow (June to October) (Burns et al., 2007). Also, a study in Monroe County noted an increase in seven-day low-flows in rural streams from 1965 to 2005 (Coon, 2005).

There have also been efforts to project future streamflows in the northeastern United States. The basic approach in these studies is the same: 1) global climate models project future temperature and precipitation amounts for large-scale regions of the globe, 2) a downscaling procedure is used to adjust these projections for the climate conditions of the areas of interest, and 3) the downscaled climate data are incorporated into a hydrologic model that predicts streamflows and groundwater levels. Within each study, several scenarios comprising different emission levels, global climate models, downscaling techniques, and model parameterizations may be chosen, resulting in an average and range of possible outcomes. However, there is a growing recognition that a large number (on the order of thousands) of equally plausible scenarios could be used. For example, one study in the United Kingdom demonstrated that streamflow projections are most dependent on the choice of global climate model, but that each global climate model and hydrologic model can also be parameterized slightly differently to result in additional variation in possible outcomes (New et al., 2007). In brief, no one outcome based on a single scenario should be granted much weight. Instead, multiple models can be used, as in the ClimAID study, to suggest the direction and relative magnitude of possible changes in hydrology.

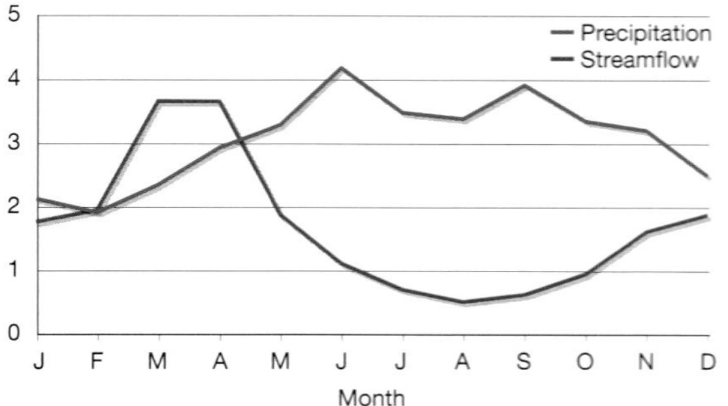

Figure 4.3 Average monthly streamflow discharge and precipitation for the Fall Creek watershed in Central New York

In general, nearly all modeling studies that have assessed water bodies in the northeastern United States have estimated that, on average, annual streamflow should change little. However, studies do differ in their estimates of the largest and smallest possible amounts of change, as would be expected given the widely different choices of modeling approaches. One study using nine different global climate models predicted an increase in annual streamflow of 9 to 18 percent for 2070–2099 in the New York State/Pennsylvania region (Hayhoe et al., 2007). Another study that used projections from the United Kingdom Meteorological Office Hadley Centre Climate Model version 2 (HADCM2) and the Canadian Centre for Climate Modeling and Analysis Coupled Global Climate Model (CGCM1) estimated a -4 to +24 percent change in annual streamflow in the Susquehanna River by 2099 (Neff et al., 2000). Using four different scenarios from the same models, a different study estimated changes in annual streamflow of -28 to +12 percent for the 2080s for the Cannonsville Basin in Delaware County (Frei, 2002). More recently, another study estimated little annual streamflow change for Moodna Creek in Orange County when using high-end and low-end estimates of climate change from 16 different climate models (Frei et al., 2009).

4.3 Vulnerabilities and Opportunities

This section gives an overview of the likely consequences of climate change on flooding, water supply, and water quality. It discusses the certainty of different outcomes and speculates on possible favorable opportunities that could arise with a changing climate.

4.3.1 Flooding

Non-coastal floods occur when rivers or streams overflow their channels, flooding the adjacent land or floodplain (coastal flood issues are discussed in Chapter 5, "Coastal Zones"). In areas prone to river and stream flooding, damage is contingent on the presence of humans and infrastructure. In New York State many original settlements were concentrated within the most viable transportation corridors, typically along rivers and their valleys, and much of the state's infrastructure reflects these early patterns of development. Many major roadways in Central New York lie within the Federal Emergency Management Agency's (FEMA) 100-year floodplain (**Figure 4.4**). Wastewater treatment plants are also at risk during floods (**Figure 4.5**). These plants are typically located at the lowest point in a landscape so that sewage can be conveyed by gravity.

Across the state, flooding continues to be an expensive and disruptive phenomenon. Record flooding in the Susquehanna Basin in 2006 required more than $225 million in disaster assistance, as reported by FEMA. Over a recent 12-year period, nine New York counties in the Southern Tier and Catskill regions experienced more than four FEMA-designated flood disasters; Delaware County had flood damage in 7 of the 12 years (**Figure 4.6**). However, the question of whether flooding will increase with climate change remains inconclusive.

Increases in total annual rainfall as well as higher rainfall intensities—both likely as a result of climate change—are often used as justification for predictions of an increased likelihood of flooding. For some parts of the country, this direct link between precipitation and flooding is likely to be the case. For example, a study looking at the relationship between precipitation and flooding at a national scale found that a 13.5-percent increase in overall annual precipitation could increase future flood damage by approximately 130 percent (Choi and Fisher, 2003). However, large-scale flood damage predominantly occurs in the Mississippi Basin and such a study at the national scale is probably more reflective of the central United States than New York.

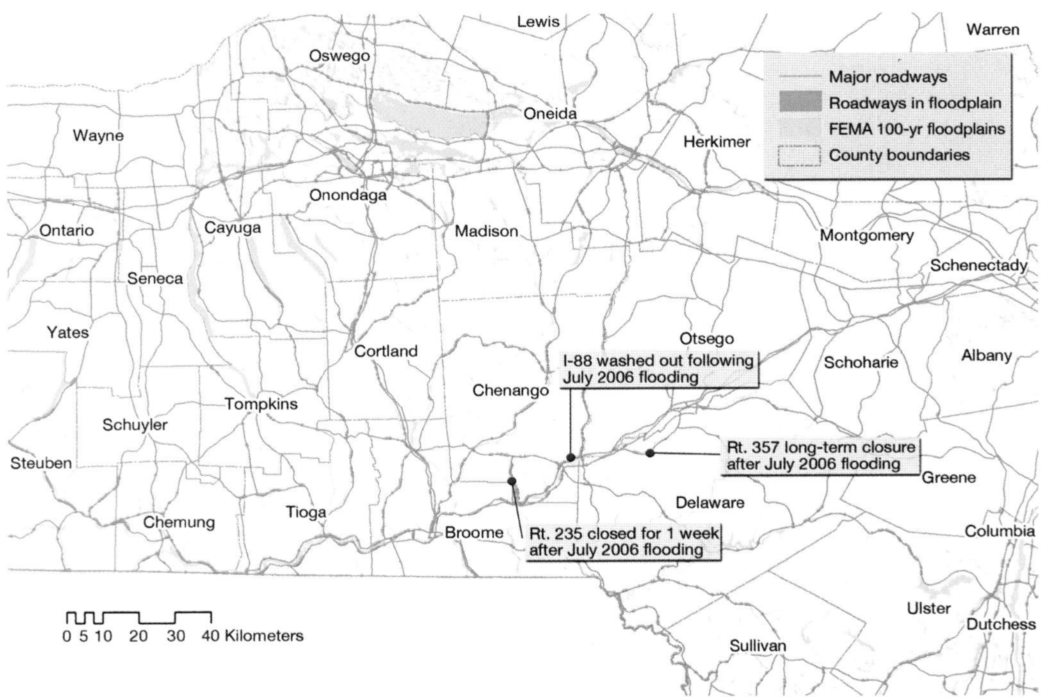

Note: Roads are highlighted that were temporarily closed after severe flooding along the Susquehanna River in July 2006. Sources: FEMA Q3 Flood Zone Data, Census 2000 Railroads, USGS NYS Transportation Coverage

Figure 4.4 Major roadways in proximity to FEMA-mapped 100-year floodplains

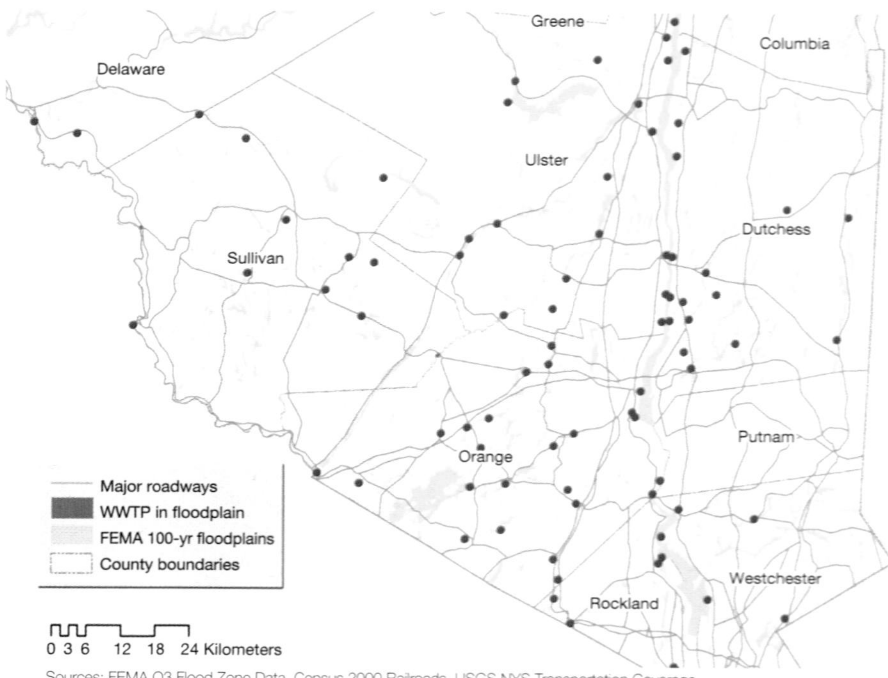

Figure 4.5 Wastewater treatment plants (WWTP) in proximity to floodplains in the Hudson Valley and Catskill Region

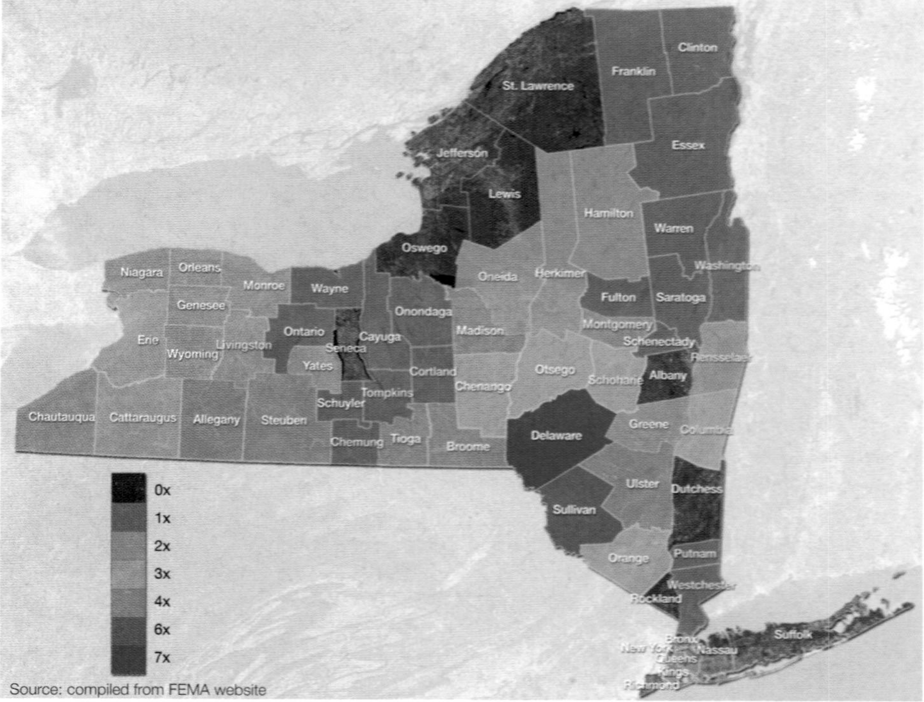

Figure 4.6 Number of FEMA-declared (Federal Emergency Management Agency) flood disasters in New York counties

To appropriately assess whether flooding in New York State may increase, it is important to focus on the dominant processes leading to large flows in streams and rivers in the New York region and to realize that studies carried out in other regions may not be applicable. For example, one study suggests that different flooding factors dominate in different regions: Damage in the Northeast is related to three-day heavy rainfalls, while damage in the central United States is related to the number of wet days across a season (Pielke and Downton, 2000).

To better understand flood processes in New York State, linkages among stream discharge, precipitation, and snowmelt were examined for three moderately sized watersheds in three different regions. The three watersheds generally reveal the same patterns in flooding. Most notably, less than 20 percent of the largest yearly stream flows correspond to the largest yearly rain events. Most large rainfall events occur between May and October when soils are dry and able to store rain. Instead of being associated with large rainfall amounts, many floods in New York State occur from snowmelt or moderate rainfall amounts on very wet soils. Given the number of interacting factors that affect flooding (snowmelt, precipitation, growing season length, soil wetness), it remains uncertain whether the magnitude of annual maximum flows will increase with climate change. Further details of this flood assessment are described in Appendix B.

There are some cases in which changes in flooding can be predicted with more certainty. More-frequent, larger-magnitude floods as a result of climate change are possible in areas in which flooding is directly linked to the intensity and amount of rainfall, such as in urban areas and steep basins. Urban areas tend to have impervious surfaces, reduced vegetative cover, and compacted soils that minimize the ability of soil to store water; thus, intense precipitation events can increase streamflows quickly. Similarly, small, steep basins, such as those found in the Southern Tier of New York State, rapidly collect water and have a limited capacity to lessen the impacts of rainfall, increasing the likelihood of flash floods following increases in rainfall intensity.

4.3.2 Drinking Water Supply

To assess the vulnerability of water supplies due to climate change, we assume that long-term average water supply will remain largely the same, but, consistent with the ClimAID climate projections, the duration and/or frequency of dry periods may increase. To compare vulnerability, water systems in the state have been classified based on the amount of time over which they can handle a temporary, but sizable, decrease in water supply.

For both surface water supplies and groundwater supplies, the systems are divided into three categories: 1) sensitive to short droughts (two months) and longer, 2) sensitive to moderate droughts (six months) and longer, and 3) relatively insensitive to any droughts. This provides a basic sense of the population and the characteristics of communities likely to be most vulnerable to the uncertain changes in water supply. This analysis was conducted for water systems that serve more than 3,000 people; thus, very small water systems are not directly represented.

Surface Water Supplies

Water supply systems that are relatively insensitive to droughts draw from a water source that greatly exceeds any potential demand. For instance, the City of Buffalo draws approximately 200 million gallons of water per day from Lake Erie. Lake Erie has a total volume of 128 trillion gallons, making Buffalo's daily withdrawal 1/10,000 of the total lake volume. Many other communities also fall into this category (**Table 4.1**). For instance, numerous small towns with demands less than 5 million gallons per day draw water from the Finger Lakes (containing nearly 1 trillion gallons). This analysis does not include possible emergency interconnections; the cities of Rochester and Syracuse as well as additional portions of the

Source	Population
Lake Erie & Niagara River	930,000
Lake Ontario	486,000
Hudson River*	122,000
Finger Lakes	115,000
Mohawk River	94,000
Susquehanna River	68,000
Chemung River	65,000
Other major rivers	53,000
Total	**1,933,000**

* Hudson River water withdrawn at Poughkeepsie could potentially be threatened by upstream movement of the salt front, as the division between saltwater and freshwater moves inland. Source: NYSDOH Public Water Supply Database

Table 4.1 Large New York water bodies and their dependent community populations

Onondaga County Water Authority's supply region could potentially use Lake Ontario water, adding upwards of 300,000 people within this category. Cumulative demands from communities outside New York State can also reduce Great Lakes' water availability. Nonetheless, on a relative basis, the water supplies within this category can generally be considered highly resilient to climate change. That is, although lake levels will likely be affected by climate change, lake volume is not expected to be altered enough to constitute a risk for these supplies.

Most water suppliers drawing from major river systems in New York State, such as the Niagara River, also fall under the category of relatively insensitive to any drought since the rivers' minimum flows greatly exceed the maximum likely demands given existing uses. Although municipalities drawing from large rivers, such as the Hudson, have been classified in this report as water supply systems with low sensitivity to drought, there could be circumstances in which this classification should change. On the Hudson, approximately 75,000 people rely on Hudson River withdrawals at Poughkeepsie. While this withdrawal of 10 million gallons per day is only a small fraction of total river flow, the intake is located far enough downriver that the saltwater/freshwater interface (salt front) could move above the City of Poughkeepsie's intake as a result of reduced freshwater inflows or sea level rise. Such a shift in the salt front would cause the supply to no longer meet regulatory standards for drinking water. Historical measurements during periods of low flow on the Hudson River give some indication of the possible movement of the salt front. During the 1960s drought, average freshwater flow as measured at Green Island was only 2,090 million gallons per day (the annual mean is 9,000 million gallons per day), and the salt front was observed at the Poughkeepsie intake (de Vries and Weiss, 2001).[1] See the "Coastal Zones" chapter in this report for additional information on salt fronts (Chapter 5).

Surface water supplies sensitive to short-drought periods are those served by what are called run-of-the-river systems (i.e., where water is pumped directly from the river) within a small drainage basin. A run-of-the-river system has either no storage reservoir or a very small storage reservoir. This design assumes that the minimum river flow always exceeds human demand, plus some required conservation flow necessary to sustain fish and other aquatic organisms and the needs of any downstream communities or other permitted users (e.g., industries) that rely on the same river.

Of primary concern are communities that use smaller rivers and streams. As a consequence of shifts in the timing of stream discharge (i.e., more discharge in winter time and less in summer time) and reduced frequency of summer rainfall, there may be new lows in streamflow. A brief period of very low water flows could greatly disrupt the habitability of a community relying on a small run-of-the-river supply. Only six water supplies in the state appear to rely solely on a single small stream, and the population served in most cases is below 5,000 people (Cornell University, Village of Warsaw, Village of Saugerties, Village of Herkimer, Village of Carthage, and the City of Hudson). The Cornell University supply is the largest such system and currently supplies about 20,000 people at an average demand of 2 cubic feet per second, with water being drawn from the 126-square-mile Fall Creek watershed in the Cayuga Lake basin. Based on the 84-year historical record for Fall Creek, the lowest recorded flow was 3.3 cubic feet per second in September 1999, slightly more than normal demand. It seems probable that a shift in streamflow timing could lead to periods when demand does exceed supply; however, the Cornell water system has proactively addressed this risk by recently completing an interconnection with a nearby municipality. Without secondary sources, these run-of-the-river systems with small drainage basins are considered to be at risk for occasionally running out of water under conditions of climate change.

Other surface-water systems in New York are those with a reservoir located on a stream or river and fall into the category of sensitive to moderate drought. A reservoir is constructed when the long-term average surface water supply is sufficient to meet long-term average demand but when short-term variations (from months to years, depending on the system) may lead to deficits between supply and demand. For example, runoff is generally much higher in the winter, while demand is much higher in the summer (for uses such as lawn watering, car washing, pools, commercial air chillers). Reservoirs are frequently constructed to store high winter and spring runoff for use in the summer and autumn. Inherently, reservoirs are only designed to extend supply over a dry period of a certain length. All reservoir systems will be stressed if there is a downward shift in their long-term average supply (although this appears unlikely based on the ClimAID and other climate projections for New York) or a large increase in demand associated with population influx, increased irrigation, or growth of water-dependent industries. Water systems with sufficient reservoir storage will face limited negative consequences even if short-term variations change, as long as the long-term mean inflows and demands remain similar to historical conditions.

There are approximately 40 reservoir systems in New York that each serve at least 3,300 people. These systems range in size from the New York City system, serving more than 9 million (NYCDEP, 2008; p. 34) and consisting of 580 billion gallons of storage, to a municipality such as the City of Mechanicville with a population of 5,000 and less than 100 million gallons of storage. An exact determination of each system's sensitivity to droughts of differing duration cannot be made without a thorough study of each system's infrastructure, demands, supplies, and operational procedures.

To provide a simplified picture of the vulnerability of reservoir systems across the state to droughts of varying durations, we calculated the number of days it would take for the maximum storage volume in a reservoir system to be depleted given historical rates of demand, adjusted down by 20 percent to account for conservation (**Table 4.2**). This approach notably ignores factors such as inflows during the dry period, required discharges to protect fisheries, and the fraction of the total storage volume that is not usable. The days of supply range from 99 to more than 1,000. Most systems with a supply of less than 200 days also have alternate sources that can provide at least a portion of daily demand. (Note: Days of supply is only intended as a simple metric to allow for comparison across systems of greatly varying size and should not be interpreted as a definitive measure of system resilience to drought.)

Municipality	Demand (million gallons/day)	Storage (million gallons)	Secondary Source	Days of Supply w/ No Inflow
Ithaca	3.3	261	Yes	99
Oneonta	1.5	140	Yes	117
Beacon	2.3	218	Yes	118
Ilion	1.97	225	Yes	143
Rome	9.5	1,419	No	187
Colonie	10.4	1,797	Yes	216
Plattsburg	2.3	457	Yes	248
Guilderland/Watervliet	7.3	1,700	Yes	291
Fredonia	1.4	335	No	299
Albany	18.5	13,500	No	912
Troy	14.4	12,912	No	1121

Note: Storage volume information was taken from a USGS inventory of large dams in New York and from a New York State Department of Health (1974) report

Table 4.2 Average daily demand, total storage, and approximate days of supply for a sample of reservoir systems in New York State

Reservoir systems provide a measurable quantity of stored water. With reasonable estimates on the timing of additional inflows, reservoirs can be conservatively operated by adjusting releases long before severe shortages occur. As an example, the Drought Management Plan for the New York City water supply system has three operational phases: drought watch, drought warning, and drought emergency. Different phases lead to different use restrictions. Phases are determined based on the probability that a major reservoir will fail to fill by the end of spring, which is typically when reservoirs reach their maximum water storage. For instance, a drought watch is declared when there is a 50-percent probability that a major reservoir will not be filled by June 1. However, while New York City has a formal protocol for monitoring supply sufficiency, most other water suppliers in New York State operate on an *ad hoc* basis. In conjunction with the New York State Department of Health, the New York State Department of Environmental Conservation tracks drought indicators across the state and issues drought declarations for regions outside of New York City. Individual water supplies may have more or less stored water per capita than suggested by these types of general drought declarations; in any case, it is up to individual water suppliers to decide how stressed their actual systems are and whether to implement appropriate management responses to these conditions. Water suppliers with limited technical resources or insufficient risk aversion in their operating procedures may fail to reduce releases to a point where an imminent shortage is unlikely.

New York City Water Supply System

The New York City water supply system supplies nearly half the state population with water. The system consists of 18 supply reservoirs located in three different drainage basins: the Delaware River, the Hudson River, and the Croton River. Located up to 125 miles north of New York City, the reservoirs are connected to the city by three aqueducts. In the last five decades (since the completion of the current upstate reservoir system in the early 1960s), drought

concentrations below current levels, although dissolved oxygen depletion would be limited to a shorter section of river.

Among the 100 water bodies in New York listed as having "impaired" water quality, 26 were noted as being impaired due to low dissolved oxygen (NYSDEC, 2008). Many were in the lower Hudson region or in urban areas of central New York. Most of the impaired water bodies were not associated with WWTP discharge but with non-point source loads, suggesting most of the thousands of WWTPs throughout New York do not currently have a strong impact on stream-dissolved oxygen under normal stream flow conditions. Presumably, an increase in temperature by only several degrees would not result in a dramatic increase in point-source-related dissolved oxygen depletion in waterways.

Impacts of Decreased Flows

Climate change will not only increase stream water temperatures but also potentially result in decreased stream flows, particularly during the summer when stream flow is already at its lowest for the year. At low-flow levels there is less dilution and the pollutant concentration is effectively higher. For water-quality-based SPDES effluent permits (issued in place of a general permit when the water body has an obvious impairment related to the pollutant for which a release permit is being sought), the in-stream concentration of the emitted pollutant is determined from the dilution capacity of the seven-day, ten-year return period of low-flow (which are based on data from the 1940s to 1975). Thus, decreases in low-flows may require reconsideration of these water-quality-based permits as well as a reconsideration of which facilities should still receive general SPDES permits. There are no direct means for estimating low-flows in streams across New York State under a changing climate. Existing regression models for predicting low-flows do include mean annual rainfall as a predictor (Ehlke and Reed, 1999), but the relationships behind these regression models were established from historical records (Eissler, 1979) and do not reflect possible changes in the frequency and size of summer storm events likely with climate change. More accurate predictions of low-flows will only be possible with a better understanding of how the temporal distribution of rainfall will likely change in the future.

4.4 Adaptation Strategies

A variety of adaptation strategies is possible for the water resources sector. Potential strategies span a range of temporal and spatial scales and system-level adjustments.

4.4.1 Flood Adaptation Strategies

As part of an ongoing effort to improve water quality, federal stormwater management regulations under the National Pollutant Discharge Elimination System stormwater program—applicable to both large and small communities—are in the process of being implemented. When retrofitting existing developments and designing new developments, a continued emphasis should be placed on encouraging cost-effective stormwater-management infrastructure that enhances natural hydrologic processes (infiltration into soils, recharging groundwater, evaporation) and slows the movement of stormwater instead of rapidly conveying it to water bodies.

Due to multiple interacting factors (snowmelt, rainfall amount, ability of soil to store moisture, evaporation rates), changes in flooding in large, rural-to-forested basins is uncertain. However, because of the steep slopes, convergent topography, and narrow valley bottoms, the Chemung, Susquehanna, and Delaware River basins have historically been subject to damaging floods. Consideration could be given to moving development out of floodplains as buildings, infrastructure, and flood-protection structures age and it becomes time to rebuild. This strategy of phased withdrawal from the highest-risk, flood-prone areas is currently recommended by the National Association of Floodplain and Stormwater Managers and was publicly endorsed by the New York State Department of Environmental Conservation commissioner at the 2008 Flood Summit.

In particular, wastewater treatment plants within floodplains may require a more thorough examination, even with a limited degree of change in flood risk. A brief interruption of operations during infrequent floods may be acceptable (high floodwater would dilute and rapidly transport discharge from the plant), but floods that routinely interrupt operations for an extended time pose a risk to public health as well as water body health. Many wastewater treatment plants are located in floodplains, since this often coincides with a topographic low point in a municipality and sewage can be conveyed to the plant by gravity. Relatively simple siting modifications or the raising of the facility by several feet may prevent severe inundation and entail little additional cost if incorporated at the time of construction. Since many aging wastewater treatment plants are in need of replacement, the possibility of moving a plant out of the floodplain should be considered when new facilities are designed.

4.4.2 Drinking Water Supply Adaptation Strategies

Many reservoir systems lack the type of formal operating rules that are useful for mitigating the risk of shortages. Accordingly, one adaptation strategy that would be useful for managing current and future climate risks would be to require public water suppliers to establish "rule curves" for water supply reservoirs and aquifers (e.g., a rule curve sets specific guidelines for reservoir releases given the amount of stored water at different times of the year). Currently, New York City is one of the few entities in the state that has a drought-response plan triggered by set water-related thresholds. For the rest of the state, the New York State Department of Environmental Conservation, in conjunction with the New York State Department of Health, determines when to make regional drought declarations. Ultimately, local municipalities are responsible for avoiding shortages based on their own judgment and operational practices. The development of rule curves would provide a systematic, unbiased protocol for managing water supplies under current and future drought.

Additionally, while there are numerous stream gauging stations throughout New York, there are few routine measurements of reservoir, aquifer, or lake levels. Consideration should be given to developing an automated gauging network or, at a minimum, a formal reporting network (e.g., routine manual measurements submitted to a central online clearinghouse) of water levels in public water supply reservoirs and aquifers. This would provide the basis for an improved, statewide early-warning system for recognizing supply shortages, while also establishing a long-term record for better understanding the link between the hydrology of specific watersheds and climate.

Nearly 4 million people in the state rely on homeowner or small public water systems. These individual homeowners or small water utilities may lack the expertise or resources to make proactive decisions prior to running out of water. However, due to the dependence of such small systems on localized conditions, it is unlikely these systems will fail simultaneously or that all systems in a given geographic region will run dry. Given that failures are likely to be small and localized, there should be sufficient resources to assist in developing temporary alternate water sources, either by trucking in water or by tapping a nearby store of surface water. The New York State Department of Health currently maintains a stockpile of equipment (mobile pumps, water tanks, filters, etc.) that can be used by municipalities to assist in supplementing critically low water supplies. Given the potential vulnerability of small water supply systems to climate change, the New York State Department of Health should consider updating and possibly enlarging its stockpile of drought emergency equipment.

In regions with large or growing populations, a possible adaptation could involve creating new water management commissions to oversee water allocations among multiple competing users. The Delaware and Susquehanna River basins already have such commissions. Basin-level commissions could be established for other major rivers in the state, in particular the Hudson and Mohawk Rivers, where population density and growth are the greatest. Other major rivers (the Genesee River, Black River, and Oswego River/Finger Lakes Region) have fewer users and already fall into the Great Lakes Basin, and so will be subject to some oversight due to the Great Lakes Compact (an agreement among eight U.S. states and two Canadian provinces that will ultimately require more extensive reporting of water use in regions of New York located in the basin). There is already an existing Hudson/Black River Regulating District, although it is tasked with an important, but relatively narrow, set of responsibilities. In the upper Hudson River basin, the Regulating District manages discharge from the Great Sacandaga Lake and Indian Lake (two tributaries of the upper Hudson) in order to reduce flood risks and increase summertime low-flows. This regulation only has a moderate impact on the lower Hudson, since the portion of the Hudson below Troy is dominated by tidal flows rather than freshwater inflows. A Hudson and Mohawk River commission could more broadly consider water allocations as well as potential water quality impacts within the entire basin.

Finally, across the state the threat posed by less certain water supplies can be most readily addressed by reducing consumption. In addition, measures to reduce consumption can help control possible increases in water demand due to higher temperatures that arise from increased landscape irrigation, opening of fire hydrants, and water use in commercial air conditioning. Comprehensive water conservation plans have already been developed for certain regions of the state (e.g., New York City) as well as other parts of the country; measures often include the use of low-flow showerheads, toilets, and washing machines; limited car washing and lawn watering; and increased use of rain barrels for gardens. New York City's efforts during the 1980s probably served to avoid drought emergencies during the 1990s. From a review of water use data for 40 larger municipalities in the state, per capita water consumption varies between 89 and 237 gallons per day with a median value of 148 gallons per day. In many cases, a large portion of water usage is related to nonessential uses such as landscape irrigation, swimming pools, and car washing. For example, in Rockland County, water demand rises to upwards of 37 million gallons per day in the summer from around 27 million gallons per day during the winter (Haverstraw Water Supply Project, 2009; DEIS, United Water NYS). Thus, if demands must be cut, water usage could be greatly reduced in some locales without directly affecting basic activities related to hygiene and sanitation.

Western states that experience frequent water shortages have experimented with pricing schemes to modify consumptive behavior. An important consideration is the sensitivity of water demand to the price of water. Studies consistently find that increasing the price of water leads to only small declines in consumption (Kenney et al., 2008). This has been attributed to people's general lack of knowledge of water rates, presumably because water bills are such a small percentage of their total annual expenses. A new technology that is currently being evaluated for its ability to raise awareness of water rates and to modify consumer behavior is the smart water meter. These meters allow different rates to be charged when overall system demand is higher. Dubuque, Iowa, will be one of the first cities in the United States with widespread implementation of these meters (*New York Times*, October 11, 2009, "To do more with less, governments go digital"). A pilot study of smart-meter users in Colorado found that homes with smart meters increased consumption, but they did so by using more water during less expensive periods (Kenney et al., 2008). This suggests that water consumers' behavior is malleable and that with the right pricing structure, overall water use could be reduced.

4.4.3 Non-potable Water Supply Adaptation Strategies

While average annual water supplies are likely to remain at current levels, there may be greater variability in flows throughout the year. Water users with sizable storage capacity (such as most public water supplies) are not likely to be significantly affected by temporary low-flow periods, but water users that depend on run-of-the-river withdrawals (agriculture, power plants, commercial users such as golf courses) may face periods of critical shortages. Thus, it may be important to implement measures to better coordinate water use on shared water bodies. Starting in 2010, the New York State Environmental Conservation Law Article 15 Title 33 will require all water withdrawals that exceed 0.1 million gallons per day to be reported to the New York Department of Environmental Conservation. This will result in more complete information on water usage, in part to comply with elements of the Great Lakes Compact. Consideration could be given to developing a publicly accessible, online system for tracking water usage of all users across the entire state. A complete inventory of water usage across the state will be critical to planning and conflict resolution if competing demands for water usage for drinking water supply, agriculture, energy, industry, export for bottled water, or other uses intensify with climate change. Allocation records are currently maintained by the Department of Environmental Conservation, the Susquehanna River Basin Commission, and the Delaware River Basin Commission; however, a more open system would enhance accountability and potentially allow for more input from additional stakeholders groups (e.g., for recreation and/or ecosystem conservation).

Currently, agricultural, industrial, and commercial users do not need permits to withdraw water in New York State outside the Delaware and Susquehanna River basins. Therefore, water withdrawals from rivers and streams throughout much of the state are not subject to minimum flow requirements. Given that droughts and, consequently, low-flow periods may become more frequent, establishing minimum flow requirements using biological criteria could help to better determine and permit the maximum amount of water that can be withdrawn from a water body during different times of the year. As mentioned in the drinking water supply subsection, basin-level commissions could be established in areas of the state outside of the Delaware River Basin Commission and Susquehanna River Basin Commission boundaries to implement water allocations. The commissions already have guidelines for determining acceptable withdrawals during low-flow periods, and other possible guidelines have recently been proposed in the generic environmental impact statement related to shale gas drilling in New York State.

Finally, more severe climate changes in other parts of the United States (relative to the Northeast) could shift population growth and water-intensive economic activities from western and southern states to eastern states, including New York. Despite relatively plentiful water resources in the state as a whole, the state's most densely populated regions have few additional water sources with which to meet increased demand. A statewide water plan could provide an overview of which areas of the state have excess existing capacity or, at a minimum, have the potential for the development of additional water resources. Specifically, a statewide water plan could provide guidance to commercial and industrial entities, as well as homebuilders and home buyers, on which communities are most likely and least likely to face water shortages, particularly with the additional stress of climate change.

4.4.4 Water Quality Adaptation Strategies

Nearly all major cities in the state have or are developing plans to address impacts from combined sewer overflows. However, overflows are difficult and expensive to eliminate. For example, the City of Rochester began planning its tunnel system in the early 1970s; implementation took more than 20 years and half a billion dollars (75 percent of the funds were provided by the federal government). Due to the cost and complexity of such infrastructure, recent plans for handling combined sewer overflows have often emphasized the removal or elimination of pollutants instead of the attenuation of runoff volumes. For instance, Buffalo has implemented systems to capture large, floatable debris and to disinfect the discharges to kill pathogens (Di Mascio et al., 2007). Syracuse is in the process of constructing facilities that store volume from smaller combined sewer overflow events and that disinfect and remove solids from larger-volume events (see Onondaga County Department of Water Environment Protection Documents: www.ongov.net/lake/index.htm).

In terms of meeting regulatory requirements, communities designing to allow only four overflow events per year may need to consider design modifications to ensure this standard can be met in the future, as indicated above (see section 4.3.4) by the USEPA study (2008). Communities planning on using mitigation measures that remove the majority of the pollutant load in the combined sewer overflow likely have systems that can be scaled up without great difficulty, because these systems treat flows, not volumes. Disinfection and primary filtering facilities would have to operate longer or more often during high-flow events, but would not necessarily have to increase in size.

Gaps remain in the scientific and regulatory communities' understanding of certain basic water quality issues related to climate change. There is a clear need to better understand the impact of low-flows and higher temperatures on the pollutant assimilative capacity of streams and rivers in the State. This entails better understanding of the in-stream chemistry at higher water temperatures (the fundamentals are well established but should be evaluated on actual streams) as well as improving means to predict low-flows on streams so that the most-vulnerable streams can be identified. Improving low-flow estimates would require better accounting of the changes in the temporal distribution of rainfall, and in understanding fundamental subsurface geologic characteristics that create differences in low-flows among streams. There has been recent work to develop methods to estimate low-flows on streams with a minimum of two measurements during periods of stream recession (Eng and Milly, 2007) that could be employed on streams in New York.

An additional potential source of deteriorating water quality in the future will is shift in land use motivated by climate-related factors (e.g., addition of new farm land for biofuels production, unconventional gas well drilling to replace coal). Some of this new development will likely be on marginal land with steep slopes or wet soils, traits that may increase the potential for pollutant generation. Thus there is a need for additional applied research to identify areas that create a disproportionate amount of pollution relative to their size. This research would also allow for more targeted implementation of management measures directed to the critical areas and processes rather than intervening with multiple management options across entire watersheds without regard to the primary pollution source (Garbrecht et al., 2007). For instance, farm-scale research in the Catskills has demonstrated that fencing streams to keep cows out can be as effective at managing water quality as more extensive and costly changes to soil and manure management (Easton et al., 2008).

Potential changes to nutrient, sediment, and pathogen pollution in a changing climate are difficult to predict due to multiple interacting processes and other drivers of change, including urbanization and agricultural intensification. Currently, there is sparse and infrequent pollutant sampling, which limits our ability to separate the impact of the climate and changing land use on water quality. As a starting point, frequent monitoring of primary nutrients, turbidity, and pathogen indicators on major rivers (Chemung, upper Susquehanna, and Delaware) would enable a clearer picture to emerge of the associations among climate factors, land use, and water quality in New York State at a large spatial scale. This effort could integrate with ongoing work to manage nutrient loads in the Chesapeake Bay Region.

4.5 Equity and Environmental Justice Considerations

The anticipated impacts of climate change on livelihoods and ecosystems will be distributed unevenly, depending both upon changes in physical parameters and institutional and socioeconomic conditions that influence local capacity to adapt. Efforts to manage the effects of climate change may create new patterns of winners and losers, further emphasizing the need to consider social equity in adaptation planning.

Equity and environmental justice issues are likely to arise with regard to local management capacity, adaptive flood management, and water supply and quality. Because water resources are closely coupled to other sectors, particularly coasts and agriculture, a number of equity issues overlap and are addressed in more depth elsewhere (see Chapters 3, 5, and 7).

4.5.1 Local Management Capacity

The capacity of local municipalities to manage water resources is a critical indicator of ability to adapt to climate change. This capacity varies widely across local governments (Gross, 2003), and whether new policies that consider climate change are enacted depends largely upon the importance placed on water resource management by local governments, access to technology and information, and the willingness of water resource managers. The demographics and other characteristics of the community also play a role, including income levels, social capital, level of education, and institutional and political contexts, such as relative power and influence in policy-making. In the Great Lakes Basin, local governments taking the most action to manage water resources were in suburban areas around urban centers and in the Finger Lakes region (Gross, 2003). They tended to have larger populations that were more educated and economically healthier. Thus certain populations will be more prepared than others for climate impacts, and strategies to build adaptive capacity need to be locally tailored.

4.5.2 Equity and Flooding

Adaptation to flooding is another key challenge for local and regional governments. There are a variety of adaptations available, ranging from developing or expanding levees and other flood-control structures, to moving homeowners and public infrastructure out of high-risk, flood-prone areas. (See Case Study A for a more detailed discussion.) Each of these options is associated with varying levels of risk exposure and expense burdens, which have strong equity implications. For example, land-use controls that would modify property rights or values, such as remapping a floodplain, enacting new zoning and building codes, relocating infrastructure, or limiting developments, need to be weighed carefully against the burden on property owners and whether compensations are being distributed fairly, relative to market value. Among local governments in the Great Lakes Basin, 28 percent control floodplain development on a case-by-case basis

(Gross, 2003). Case-by-case decision-making that is not informed by established plans, policies, or regulations may increase the probability that the process is co-opted by those with greater power or elites or biased against ill-informed owners or low-income residents. In addition, cost-sharing responsibilities and cascading effects at the local and regional levels need to be considered prior to making these modifications. Ensuring an open and fair process is essential.

4.5.3 Equity and Water Supply

Communities that have limited water storage will be more vulnerable to periods of reduced water availability as the frequency and duration of summer droughts increases. Specifically, some of the 1.9 million people who rely on domestic well water and several hundred thousand others connected to small public water systems may experience periods of scarce or no water. Many of these people, including farmers, are located in more rural areas and may be economically more vulnerable. Conversely, rapidly developing, higher-income exurban communities (those that are located outside the city and suburbs) may also experience water scarcity as increasing demands overwhelm local supplies. Management options to address water supply limitations include conservation programs, water-pricing schedules, or infrastructure development. Policy choices should carefully consider the costs associated with each option relative to economic capacity of the specific community.

4.5.4 Equity and Water Quality

Water quality is already a serious concern for most regions of New York State. Problems associated with pollutant and nutrient loading, toxic and waste runoff, and disease-causing pathogens affect water quality in the state's rivers and lakes. Water-quality vulnerabilities include a number of equity-related issues. For example, nitrogen loading from New York City wastewater treatment plants has impacts on fishing and ecosystem management in places such as Jamaica Bay, Queens. In addition to geographic differences (e.g., upstream versus downstream), poor water quality can be associated with socioeconomic and racial status (Calderon, 1993). Lower socioeconomic groups are more likely to access contaminated waterways for swimming and fishing (Evans and Kantrowitz, 2002). Lower-income or non-English-speaking populations may be particularly vulnerable to increasing levels of pathogens in water or contaminants in groundwater wells, both from lack of insurance and lack of awareness about government programs and warnings. Recreational fishing, particularly among Latinos, is common from the piers and shores of Brooklyn (Corburn, 2002). The Water Working Group of the New York State Environmental Justice Interagency Taskforce unanimously rated combined sewage overflow (CSO) improvement as an urgent priority (NYS Department of Environmental Conservation, 2008). The confluence of these water quality vulnerabilities within specific water bodies in New York State, including Newtown Creek, the Gowanus Canal, and the Bronx River, is, in the current climate, a noteworthy environmental justice concern.

The specific impacts that climate change will have on water quality across the state are less certain. The degree to which increasing rainfall amounts will translate to increased pollutant loads remains unknown. Existing research documenting temperature controls on pathogen survival is also ambiguous. More research on the impacts of climate change on water quality is seriously needed. However, any climate change policies regarding water quality management will need to take into account education, literacy, ethnicity, and income characteristics of the relevant communities, as these factors will drive the success of the programs. One example of a successful water quality management strategy is the adoption of "Green Infrastructure" approaches designed to divert or slow storm water. Such approaches, which fall under a broad spectrum of low-impact development strategies and practices (LIDs), have gained increasing attention and support of environmental justice communities in New York State. Environmental justice communities are supportive of green infrastructure measures because of the ancillary benefits they can provide, including beautification, open space, air quality improvements, and shade, as well as the mitigation of CSOs.

4.6 Conclusions

This section highlights the key points from this ClimAID chapter, summarizing them under key existing and future climate risks, vulnerabilities and opportunities, adaptation options, and knowledge gaps.

4.6.1 Key Existing and Future Climate Risks

Although there are several water-quality issues directly linked to higher average air temperatures, in general, hydrologic processes are dependent on multiple interacting climate factors. In addition to temperature, possible future changes in timing and quantity of snow, rainfall, and evaporation will all have impacts on the state's water resources.

- Rising air temperatures intensify the water cycle by driving increased evaporation and precipitation. The resulting altered patterns of precipitation include more rain falling in heavy events, often with longer dry periods in between. Such changes can have a variety of effects on water resources.
- Heavy downpours have increased over the past 50 years, and this trend is projected to continue, causing an increase in localized flash flooding in urban areas and hilly regions.
- Flooding has the potential to increase pollutants in the water supply and inundate wastewater treatment plants and other vulnerable development within floodplains.
- Less-frequent summer rainfall is expected to result in additional, and possibly longer, summer dry periods, potentially impacting the ability of water supply systems to meet demands.
- Reduced summer flows on large rivers and lowered groundwater tables could lead to conflicts among competing water users.
- Increasing water temperatures in rivers and streams will affect aquatic heath and reduce the capacity of streams to assimilate effluent from wastewater treatment plants.

Water resources in New York State are already subject to numerous human-induced stresses, and these pressures are likely to increase over the next several decades. For instance, water supplies are more likely to be stressed by increasing demands and insufficient coordination of supplies rather than by a dramatic downward shift in the availability of water. Water quality is more likely to be harmed by aging wastewater treatment plants, continued combined sewer overflow events, and excess polluting nutrient loading in agricultural regions. Therefore, nearly all the suggested adaptation strategies are intended to address these non-climate-change factors in tandem with the challenges posed by climate change.

4.6.2 Main Findings on Vulnerabilities and Opportunities

In **Table 4.5**, vulnerabilities have been divided into categories that parallel the major sections in this chapter: flooding, drinking water supply, commercial and agricultural water availability, and water quality. No single vulnerability takes precedence, since no vulnerability can be identified at this time as having a disproportionate societal or economic impact on the state. Additionally, some items listed are only potential vulnerabilities that require additional time and information before a more definitive determination of their importance can be made.

The New York City water supply stands as a special case when considering potential climate change impacts. Neither current hydrologic trends nor climate model projections suggest that a dramatic decline in water availability is likely (particularly since the system has a large storage capacity that provides resilience to increased intra-annual variability in stream flows to its reservoirs). However, since this single system serves such a large number of people and since it is also strongly impacted by several factors external to climate (population growth, interstate agreements on discharges to the Delaware River, aging infrastructure), even slight decreases in water availability could couple with other factors to constrain available water supplies to a large amount of the state's population.

Notably, the New York City Department of Environmental Protection has already been highly proactive in assessing its system reliability and is in the process of conducting additional in-depth studies of impacts on water quantity and quality at a level of detail far beyond the scope of this chapter. The Department of Environmental Protection will presumably maintain this proactive stance in dealing with the uncertainties of climate change. Some vulnerabilities and adaptation strategies in **Table 4.5** loosely encompass the New York City supply (i.e., increased frequency of deficits in systems with moderate-to-large storage volumes).

4.6.3 Adaptation Options

In this section, adaptation strategy options for the State are summarized and potential recommendations put forward. To make it easier for stakeholders and

Section	Vulnerability	Adaptation
Flooding	1. Uncertain changes in flooding in large basins	Consideration of moving development from flood-prone areas when infrastructure reaches end of life span
	2. Increased flooding in smaller, urbanized watersheds	Implementation of infrastructure that replicates natural hydrologic processes
	3. Uncertain potential for increased flooding of wastewater treatment plants	Design modification of new WWTPs
Drinking water supply	1. Likely increased frequency of deficits in homeowner wells, small community well systems, and run-of-the-river systems	i. Enhanced monitoring of groundwater levels; ii. Stockpiling of equipment for emergency withdrawals; iii. Water conservation
	2. Possible increased frequency of deficits in systems with moderate-to-large storage volumes	i. Enhanced monitoring of reservoir and aquifer levels; ii. Use of rule curves in reservoir or aquifer operation; iii. New basin commissions; iv. Water conservation
Commercial & agricultural water availability	1. Increased demand from additional agricultural irrigation	i. Establish minimum streamflow requirements; ii. Statewide inventory of water withdrawals
	2. Competition for water among human consumption, commercial uses, and ecological needs	i. New basin commissions; ii. Establish minimum streamflow requirements; iii. Statewide inventory of water withdrawals
Water quality	1. Decreased stream low-flows and higher water temperatures decrease assimilative capacity of waterbodies receiving waste	i. Modify waste discharge permits (given further study of likely changes); ii. Further study of low-flow characteristics of streams
	2. Uncertain changes in pathogen levels	Long-term monitoring and data analysis
	3. Uncertain changes in nutrient loading with no land use change	Long-term monitoring and data analysis
	4. Possible changes in CSO frequency (particularly in systems with high thresholds for CSO initiation)	Monitoring of possible changes in CSO frequency and implementation of scalable CSO mitigation plans
	5. Increased sediment and nutrient loads due to expanded agricultural production in water-rich region	Better targeted water and soil conservation measures

Table 4.5 Summary of vulnerabilities and adaptation options

decision-makers to evaluate, the strategies are grouped according to robustness.

Resources and Current Status of Implementation

There is considerable natural variability in hydrologic systems even without climate change. Water resource managers have long dealt with this innate climate variability, as well as changes in other factors such as land use and population. Nearly all suggested adaptation options are an extension or expansion of existing strategies for managing this variability. While most adaptation options are not yet being formally implemented, many are related to ongoing water-resource-related projects at the federal, State, and local levels.

Several adaptation options are extensions of existing State and interstate institutions and policies. For instance, a Hudson River Commission could be developed modeling the format and successful strategies of the existing Delaware and Susquehanna River Basin commissions and expanding on the powers of the Hudson River/Black River Regulating District to control releases from the upper Hudson Basin. Minimum streamflow requirements already exist in the Susquehanna and Delaware River basins, and guidelines could similarly be established throughout the rest of the state if the Generic Environmental Impact Statement for shale gas extraction is accepted (although State laws would need to be changed for them to apply to all water users). Water use reporting by industrial and commercial users (although not permitting) is required as of February 1, 2010, under a new State law. The National Weather Service already operates an effective flood-warning system, and the U.S. Geological Survey already measures numerous water bodies throughout the state.

Other possible adaptation options follow from existing operating protocols at the municipal level. In terms of water supply, most water utilities already make some attempt to encourage water conservation, although many appear to have only an *ad hoc* approach to reducing demands in time of drought. In terms of urban flooding and water-quality issues, many municipalities have long had laws restricting heightened peak runoff following new development, and nearly all have had to comply recently with federal stormwater management regulations.

Potential adaptation options

There are many adaptation pathways that can reduce the potential detrimental consequences of climate change for water resources. A challenge to building resilience, however, is the lack of certainty in the degree, pace, and even direction of water-related climate changes anticipated for New York State. Water resource managers will need to make decisions based on climate projections that reflect this uncertainty, which is very different from current approaches that rely on the historic record and the assumption of a stationary climate. One approach to making such decisions under uncertain conditions is to apply the concept of robustness. In the context of decision-making science, the term robustness is defined as a strategy that is effective (in terms of cost, societal impact, and risk reduction) under a range of possible future outcomes (Lempert et al., 2006). Recent research has identified several robust decision-making strategies (Hallegatte, 2009), and we supplement them with several categories of our own; although the adaptations are placed in distinct groups for organizational purposes, many have relevance in more than one category:

Strategic expenditures on adaptation options with co-benefits that result in a net public benefit with or without climate change.

1) Continue to encourage the development of cost-effective stormwater management infrastructure for use in urban and suburban landscapes that enhances natural hydrologic processes (infiltration, recharge, evaporation), instead of rapidly conveying stormwater to receiving water bodies.
2) The New York State Department of Health currently maintains a supply of equipment (mobile pumps, water tanks, filters, etc.) that can be used by municipalities to assist in temporarily supplementing critically low water supplies. The New York State Department of Health should assess and augment the adequacy of its inventory of emergency equipment if needed.
3) Encourage water conservation strategies that guarantee water sufficiency without increasing supplies through building reservoirs or other new infrastructure. Ultimately, increased water availability may provide economic benefits to New York State by increasing the viability and sustainability of water export or virtual trade through agricultural and industrial products.
4) Establish minimum flow requirements using biology-based criteria to determine and permit the maximum amount of water that can be withdrawn from a water body during different times of the year. Such minimum-flow criteria will also be important to make water allocation decisions for water bodies.

Taking advantage of low-cost margins of safety in new construction to avoid more expensive retrofits and modifications in the future.

1) Devise wastewater treatment plant upgrades and combined sewer overflow mitigation strategies (for communities that do not have one in place) to address possible changes in flood risk, sea level rise, and increases in large rainfall events. Modest water infrastructure design changes at the planning stage will avoid more costly modifications to constructed infrastructure later.
2) Consider moving development out of floodplains as buildings, infrastructure, and flood-protection structures age and it becomes time to rebuild.

Soft strategies (in contrast to hard infrastructure solutions) that seek to build new institutional or organizational frameworks.

1) Basin-level commissions could be established for major rivers in the state without them. The Hudson and Mohawk Rivers would likely be high-priority areas, as other major rivers (the Genesee River, Black River, Alleghany River, Oswego River/Finger Lakes region) have fewer users, already fall into the Great Lakes Basin, and will be subject to some oversight due to the Great Lakes Compact. Such basin commissions could provide oversight of supplies as well as water quality and fisheries issues.
2) Many adaptations to climate variability are implemented and coordinated at a local scale. The presence of a lead town or other entity can play a key role in mobilizing efforts in the surrounding region by providing leadership in education, best management practices, and fundraising. State-level recognition as well as funding support to leading local entities could enhance adaptation activities at

the local scale, which will build adaptation capacity that will help address future climate change.

3) Water demand data could be more widely reported even during non-drought periods through mass communication, but also through innovative technologies, such as smart water metering. Such efforts might help break the entrenched mentality that water supply systems should supply 100 percent of demand at all times (Rayner et al., 2005) and shift public behavior to recognize variations in demand and appreciate years when water is plentiful and conserve it in years when it is scarce. This also falls under the more general adaptation option of increasing water conservation.
4) Public water suppliers could establish formal rule curves for water supply reservoirs and aquifers. This involves no new infrastructure but entails establishing a new framework for system operation (likely to include the enactment of well-defined conservation measures at certain drought levels) understood by water managers, other municipal decision-makers, and residential and commercial water users.

Extensive monitoring efforts that expand the collection of environmental data, which are critical for making informed management decisions.

1) Starting in 2010, New York State Environmental Conservation Law Article 15 Title 33 will require all water withdrawals that exceed 0.1 million gallons per day to be reported to the New York Department of Environmental Conservation. Consider developing a publicly accessible online system for tracking water allocations to all users (water supply, industrial, thermoelectric, agricultural) across the entire state. A more open system would enhance accountability and potentially allow for more input from additional stakeholders groups (e.g., recreational, habitat).
2) Consider developing an automated gauging network or, at a minimum, a formal reporting network (e.g., routine manual measurements submitted to a central online clearinghouse) of water levels in public water supply reservoirs and aquifers. This would assist the New York State Department of Environmental Conservation and Department of Health in making drought declarations. More importantly, it would encourage water suppliers to more systematically track water storage and would complement the suggestion for suppliers to develop rule curves to regulate reservoir and aquifer operations.
3) Expand the extent and types of monitoring by the U.S. Geological Survey or other entities to provide additional data for decision-making. This could include additional measurements of groundwater levels, low streamflows, temperatures, and dissolved oxygen.
4) Additionally, there may be the possibility of leveraging widespread Internet connectivity and inexpensive data storage in order to enlarge informal data collection networks. The GLOBE data project (www.globe.gov) set a precedent for students collecting assorted weather data. With nominal funding, a water-quality program could be developed among community colleges, colleges, universities, public-interest groups, and watershed organizations to collect and analyze water data for specific water bodies, feeding it into a central clearinghouse. Such efforts would help engage the public, which is essential for building resilience.

4.6.4 Knowledge Gaps

There are several areas that require additional fundamental research to make educated policy and management decisions. These research areas are discussed below.

There are several fundamental hydrologic processes that need more in-depth assessment. In particular, groundwater recharge, stream low-flows, evaporation, and flooding need to be better understood in light of a changing climate. Such studies need to be process based, instead of simply drawing conclusions from historic data. Additionally, they need to specifically look at processes within the region and avoid making generalizations from other areas, a typical limitation of many existing studies. Such region-specific studies would benefit from additional data, much of which could come from a refinement of existing monitoring networks. The existing rain-gauge network could be expanded to ensure there is a satisfactory density of rain gauges in each basin with a stream gauge, providing a better understanding of the hydrologic response in gauged basins. Potential evapotranspiration as well as soil moisture could also be measured at several sites across the state to better understand how evaporation is affected by changing climate factors. Finally, snow depth measurements could start to include snow-water equivalents (which report snow as a depth of liquid water to account for snow compaction and differences in density) in order to provide a more objective measure of how snowfall and snowpacks are changing over time.

To protect water quality, research that identifies additional critical pollutant-contributing areas and processes is needed. Adaptation measures for water-quality protection in a changing climate should be targeted to the critical areas and processes rather than intervening with multiple management options across entire watersheds without regard to the primary pollution source (Garbrecht et al., 2007). At small scales, a critical field-based assessment of the effectiveness of best-management practices is needed. At larger scales, improved monitoring of primary polluting nutrients, turbidity (cloudiness of water caused by suspended sediment), and pathogen indicators on major rivers (Chemung, upper Susquehanna, and Delaware) would enable a clearer picture to emerge of the relationships among climate factors, land use, and water quality in New York State.

Many pollutant discharge permits for wastewater treatment plants are based on streamflow and temperature data from decades ago. An assessment is needed to estimate future streamflow and water temperature scenarios and to model what impact these changes will have on the quality of water bodies receiving treatment plant effluent.

There is often a desire for actionable future climate information for making decisions on infrastructure needs or policy changes. Results could be provided by downscaling global climate model projections and using these as inputs to a hydrologic or ecology model, but the reality is that this model estimate would be far from certain. Therefore, there is a need for a fundamental shift in the way engineers, planners, and policymakers make decisions. Instead of devising a strategy optimized for one outcome, the strategy should instead perform effectively (in terms of cost, societal impact, and risk reduction) across many outcomes (robustness). There needs to be both basic research as well as educational outreach to decision-makers to expand the concept of robust decision-making.

More severe climate changes in other parts of the United States (relative to the Northeast) could shift population growth and water-intensive economic activities to New York. A statewide water plan could provide guidance to commercial and industrial entities as well as homebuilders and homebuyers. The plan could detail which communities in the state have the most excess water supplies, even with the additional stress of climate change. Additionally, a state water plan could initiate thinking into potential economic opportunities, and the private sector would presumably have an incentive to further investigate these possibilities.

Case Study A. Susquehanna River Flooding, June 2006

Flooding is already a major problem across New York State and it may be exacerbated by climate change. Currently flood damage costs an average of $50 million a year in the state (Downton et al., 2005). The majority of the flood events consistently occur in the ten Southern Tier counties. The June 2006 Susquehanna River flood provides insights into the pros and cons associated with different strategies that can be used to reduce future flooding risks and impacts. Record precipitation during June 2006 culminated in significant flooding throughout the Susquehanna and Delaware River basins in New York and in northeastern Pennsylvania. Twelve counties in New York and thirty in Pennsylvania were declared disaster areas. This ClimAID analysis and summary focuses on Broome County, New York, which incurred the largest portion of damages throughout the entire flooded area.

Climatological and Hydrologic Drivers

The flood resulted from significant, intense precipitation falling on already-saturated soils throughout the 4,000-square-mile Upper Susquehanna basin. A total of 4.29 inches of rain fell in the 25 days before the most intense rain began, with varying amounts of rainfall almost daily, which prevented the soils from drying out. A stalled low-pressure system began contributing intense rainfall on June 25, and from June 25 to 28 total rainfall ranged from 3 to 11 inches throughout the basin (Suro et al., 2009).

Runoff from the steep hillslopes led to a record rise in river water levels. On the Susquehanna River at Conklin, river levels were less than 5 feet in elevation on June 26 (**Figure 4.8**). Flood stage was reached at 2:15 p.m. on June 27. Nine hours later, it reached the

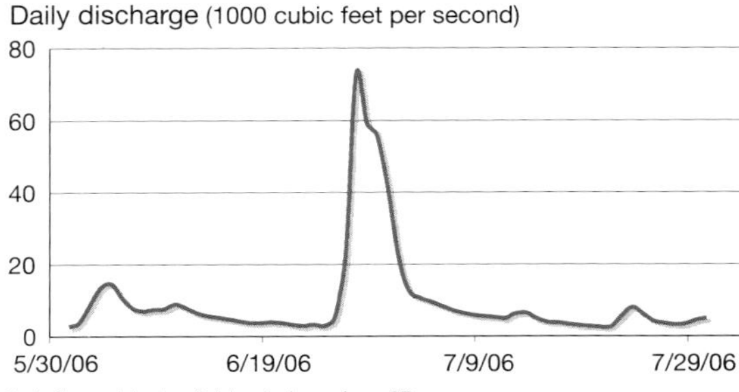

Note the rapid rate of rising limb on June 27.

Figure 4.8 Water level of the Susquehanna River at Conklin, New York, during the June 2006 flood

previous high-water record of 20.83 feet. Rainfall stopped early in the morning Wednesday, June 28, but the river level continued to rise as a result of the water already in the basin, reaching its peak of 25 feet by 11:30 a.m. This flood is the largest recorded on the Susquehanna River at Conklin since gauging began in 1912. However, Broome County had experienced only slightly smaller floods in 2004, associated with Hurricane Ivan, and again in April 2005 when a combination of extreme rainfall and snowmelt very quickly flooded parts of the upper Susquehanna River.

Social and Economic Impacts

The city of Binghamton and a number of smaller rural towns in Broome County were flooded during the June 2006 event. This area included the majority of the FEMA-designated 100-year floodplain and portions of the 500-year floodplain (**Figure 4.10**).

Source: E. Aswald, used with permission.

Figure 4.9 Aerial photograph of Endicott Sewage Treatment Plant during June 2006 flood

There are thousands of properties at risk from flooding in Broome County (**Table 4.6**). Though it is difficult to determine exactly which properties were damaged by the flood and to what degree, an estimate was made by overlaying the flood extent on parcel level data from Broome County's 2004 property tax register. Because of data restrictions, the City of Binghamton is excluded from this dataset and from all of the associated analysis. Also, since there is no way to judge how well individual properties fared during the 2006 flood, aggregate market values should be taken as a maximum estimate of risk. Actual flood damage was a fraction of these estimates. Approximately 3,350 properties were in the flood zone, distributed largely among commercial (10 percent) and residential (58 percent) uses (**Table 4.6**). Two sewage treatment plants, a public works facility, a hospital, and several hundred miles of roads were also in the flood zone (**Figure 4.9**). Approximately 8 percent of the aggregate value of property in the county was at risk, amounting to nearly $563 million.

Despite the rural nature of the county as whole, less than 1 percent of the flooded parcels was agricultural. The Susquehanna River has been a historic beacon for growth, with significant and disproportionate development occurring along its banks. A large amount of commercial property value was within the flood inundation zone, accounting for about 19 percent of the county's total (**Table 4.6**). This helps explain why a number of critical commercial classes were flooded in greater number than would be expected given the size of the county as a whole. For example, nearly 30 percent of the neighborhood shopping area in the county was within the flood zone, as was more than a quarter of the county's warehouse and storage facilities.

	Number of Parcels		Aggregate Market Value	
	Not flooded	Flooded	Not flooded	Flooded
Agricultural	675	12	$39,883,239	$933,359
Commercial	2,870	319	$876,156,504	$210,199,932
Community services	763	55	$1,405,660,897	$61,987,704
Industrial	197	44	$277,401,671	$34,041,765
Public services	353	69	$75,659,552	$110,578,336
Recreational	158	30	$52,536,817	$11,270,494
Residential	47,134	1,954	$3,929,664,223	$123,055,714
Vacant	13,392	804	$169,609,828	$9,748,270
Wild/forest	250	21	$16,173,424	$1,033,142
No data	488	40	$0	$0
Total	66,280	3,348	$6,842,746,155	$562,848,716

Table 4.6 Land use and value of properties in Broome County, broken out by flooded and non-flooded parcels in June 2006

Of all the towns that experienced flooding, Conklin was hit the hardest, with 30 percent of its properties flooded, followed by 13 percent in Kirkwood and 10 percent in Port Dickinson. Of the remaining eight towns, flooding ranged from 6.9 percent in Johnson City to 0.1 percent in Binghamton. The difference in flooding across localities is a reminder of the range of exposures that local governments face during a regional flood event and the importance of cost-sharing mechanisms in the aftermath.

Flood Insurance and FEMA Designation

The extent of the flood draws attention to the limitations of current flood insurance and the uncertainty of FEMA's modeled 100-year floodplain versus actual flooding. The FEMA boundaries are important, not just because they indicate areas where insurance is federally mandated, but also because these boundaries often become the definitive communication

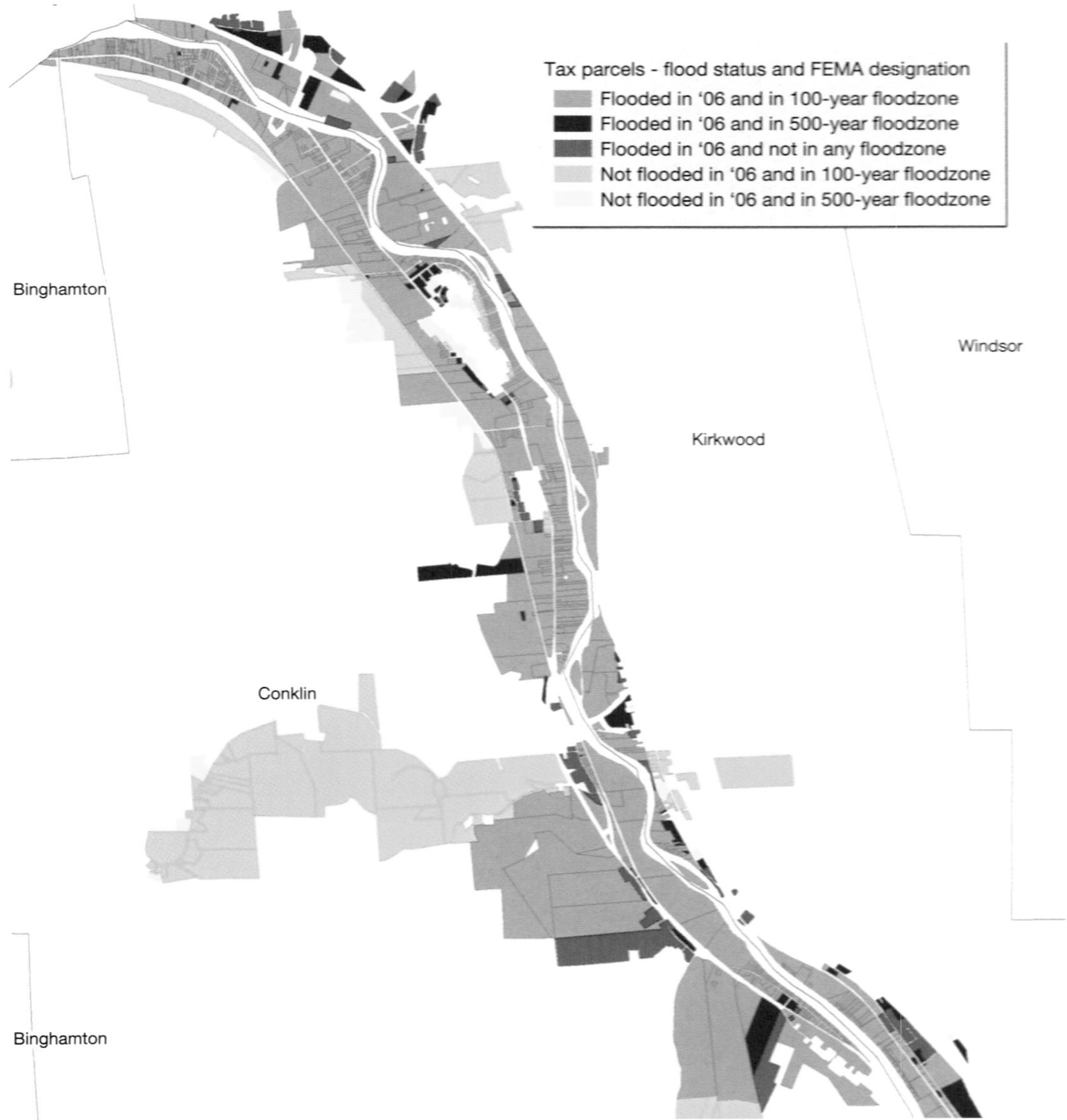

Figure 4.10 Distribution of flood risks in a select area in Broome County: Properties flooded in 2006 relative to FEMA designation

of perceived risk and thereby serve to define the range and limits of a homeowner's or community's response to potential flooding. In total, 1,020 properties were located within the 2006 flood extent but not included within FEMA's Special Flood Hazard Area (the 100-year floodplain) (**Table 4.7**). Of these, 723 were residential and comprised about $46,316,088 worth of property that was exempt from the federally mandated insurance requirements. Large numbers of homeowners outside of FEMA's Special Flood Hazard Area who were devastated by the flood did not have flood insurance (**Figure 4.10**).

Approximately 6,200 people were living within the extent of the flood. Across the entire flood extent, more than 30 percent of housing units were renter-occupied, amounting to more than 920 in total (**Table 4.8**). Conklin and Dickinson had especially high rates of renters in the area inundated, with blocks located on the western side of the river containing 50 to 100 percent renters. Displacement of renters had widespread impacts, including difficulty finding suitable affordable accommodations near places of work or their children's schools. Availability of rental units also decreased as displaced homeowners and renters competed for viable alternative housing.

Patterns of seniors at risk, defined as individuals 65 years or older, varied among communities. Seniors are considered particularly vulnerable due to higher rates of impaired mobility, difficulties with communication, and potential lack of awareness of warning and evacuation systems. More than 30 percent of households in the flood area had at least one member who was 65 years or older, but these households were not concentrated in any one area. The dispersed nature of this vulnerable population may complicate evacuation and response.

There was not a distinct relationship between the flooded area and where nonwhite residents lived. The highest densities of the nonwhite populations were located in Binghamton, but very few were located in the flooded area. In Conklin, there were several blocks composed of nonwhite communities, and more than 25 percent of that population resided within the flooded areas.

In order to gain some measure of systematic inequities between flooded tracts and concentrations of vulnerable populations, we compared block groups within the extent of the flood to those block groups in the rest of the county outside of the flood. Apart from slightly higher rates of renters in the floodplain (perhaps a reflection of the more urban housing context along some parts of the river), the profile of the two populations did not differ significantly in terms of demographics.

	Number of Parcels	Aggregate Market Value
Total		
Flooded and within 100-year flood zone	2,328	$373,050,159
Flooded but outside 100-year flood zone	1,020	$189,798,557
Not flooded but within 100-year flood zone	4,651	$499,763,169
Residential		
Flooded and within 100-year flood zone	1,231	$76,739,626
Flooded but outside 100-year flood zone	723	$46,316,088
Not flooded but within 100-year flood zone	2,717	$184,005,238

Table 4.7 Distribution of flooded and unflooded properties within FEMA's flood zones

	Population
Total population within flood extent	6234
Latino	87
African American or black	101
Non-white	317
Households	2650
Households with one or more members over 65 years of age	845
Housing units, renter-occupied	921

Table 4.8 Select demographic profile of population within the 100-year flood[4]

Adaptation Response Options

Various options are available to increase resilience and reduce risks from the flood-related impacts of climate change, each with its own benefits and disadvantages. A few options are presented here. A complete assessment should evaluate damages due to structural losses or costs of new construction, costs due to loss of work-related productivity, and offsets provided by insurance. However, there are impacts, such as from losses in human lives or subsequent short-term and long-term illness, as well as loss of ecosystem services, to which it is much more difficult to assign dollar values.

Option A: Maintain Status Quo

Despite the very rapid onset of the flood and the thousands of properties that were inundated, there were only four deaths (Suro et al., 2009), which is sometimes viewed as the real measure of a disaster. The success was due to the excellent warning-and-response system coordinated by NOAA's National Weather Service that is linked to local communities. This system depends on the availability of real-time data on streamflow for the Susquehanna River and several tributaries measured at USGS gauges. Notably, some of these gauge sites were recently at risk of being eliminated due to budget constraints, but currently most have been continued with other funding sources. The response included pre-flood community-wide warnings and evacuations, water pumping and sand bag efforts, and emergency evacuations and medical services during the flooding. Such flood-warning systems are not cheap. For example, it costs about $17,000 per year to operate and maintain a single USGS streamflow station in New York (Ward Freeman, personal communication, 2009). Additional costs for manpower, communications, computer resources, vehicles, and other emergency-response equipment total in the hundreds of thousands of dollars or more (Ward Freeman, personal communication, 2009).

If the status quo approach is taken, similar levels of damage are likely to occur when the next flood of similar magnitude occurs on the Susquehanna River. Properties that are within the 2006 flood area constitute an estimated 8 percent of the value of all property in the county. And while the rates of socially at-risk populations are not extremely high, the raw numbers suggest that thousands of people with potential vulnerabilities were located within the flood zone. A no-action scenario is likely to lead to similar or worsening impacts, especially if floods become more frequent or severe. Uncertainty about flood forecasting, due to climate change or inadequate emergency systems elsewhere in the state, will also raise the likelihood that people and their property will be caught off guard.

Option B: Increase Levees, Dams, and Other Barriers to Reduce the Flooding Risk

Extensive levees and dams were built in the 1950s. This system has been highly successful at preventing flooding for several decades along the Susquehanna River. However, in some locations the current system is no longer adequate to deal with potential higher-magnitude floods. Additionally, development within the floodplains behind these barriers has intensified, perhaps due to an artificial perception of safety, making communities more vulnerable and damages greater when floods do occur. Developing or upgrading existing levees to support floodplain development is extremely expensive. This is being demonstrated in parts of Broome County, where recent updates to FEMA flood mapping will soon require upgrades to existing levees. If levees are not updated, thousands of residents will have to purchase flood insurance policies for the first time. Costs for the planning stage and subsequent renovation of the levees by the U.S. Army Corps of Engineers are estimated to be in the tens of millions of dollars (William Nechamen, personal communication, 2009).

Both options A and B are strongly influenced by the availability of flood insurance to property owners as it offsets a significant portion of flood-damage expenses. Current FEMA regulations require some level of flood "proofing" (e.g., constructing elevated buildings to reduce damages within insured communities). However, the global insurance industry has recently recognized the potential risk increase given the impacts of climate change (Geneva Group, 2009). Future insurance options will have significant ramifications for homeowners' willingness to accept risks, housing values, and future land uses within high flood-risk areas.

Option C: Phased Withdrawal from the Highest Flood-risk Areas Over Time

Moving out of the highest flood-risk areas has been successfully accomplished by using homeowner buyouts following floods in multiple places nationwide, including Conklin in 2007. Payouts have been set at pre-flood fair market value or at pre-flood value minus the estimated costs of damage repairs. The local government takes over the land, which is deeded for recreation or open space. At a minimum, towns may consider moving infrastructure, such as wastewater treatment plants, out of floodplains. The withdrawal strategy reduces all subsequent flood risk, both to human lives and buildings. Monetary costs can be comparable to or less than costs to expand levees. It has the added benefit of expanding natural flood-control processes by increasing floodplain storage and not "bouncing" floodwaters downstream. It also improves water quality and aquatic ecosystem health. However, one of the strongest deterrents to greater adoption of a withdrawal policy has been the strong sense of place or "roots" that people feel for their locations (homes and communities); another is a desire to live near water.

Option D: Improve Watershed Management

Improving watershed management should be considered in conjunction with any of the options listed above and can play a significant role in reducing the amount of runoff that contributes to flooding. Best-management practices, including improving soil infiltration capacity, expanding vegetated surfaces, decreasing impervious cover, and uncoupling roadside ditch systems, have been documented to reduce downstream flooding.

In all of these options, the critical question is, How will decisions about new infrastructure, buyouts, or land management be made? The largest infrastructure projects are possible only with regional and national collaboration, while local and individual property protections can be cobbled together by local governments or individual property owners. Whichever path is pursued will determine who bears responsibility for the costs of these measures, including impacts on livelihoods, and who controls the safety nets in the event of system failure. Whichever option is chosen, the process must be inclusive of those living in affected neighborhoods and sensitive to underlying socioeconomic conditions and indirect economic impacts

Case Study B. Orange County Water Supply Planning

A portion of future population growth in New York State will occur in higher-income exurban regions. Unlike traditional suburbs, these areas are farther from established major city centers and in recent decades have grown mainly through low-density residential development. Much of the growth in Rockland, Orange, and Putnam counties, located 20 to 40 miles outside of New York City, has been in exurban areas. In terms of water resources, these areas frequently have higher per capita water usage due to greater landscaping demands. These areas may face major challenges in developing new larger-scale sources of water, particularly surface water reservoirs, in part due to regulatory constraints. In some parts of the state, exurban areas have emerging regional governmental entities that are just becoming established enough to plan and finance centralized water supplies. In contrast, Orange County has had a legally established and active water authority (Orange County Water Authority, or OCWA) for more than two decades. Orange County is in many ways representative of other areas of exurban growth in the state that face water-resource supply issues, many of which could be complicated or exacerbated by climate change. However, it is also one of the few locations to have proactively undertaken a recent study of water resources that has included the impacts of future climate change in addition to socioeconomic, geographic, and political constraints. Much of the following information in this ClimAID analysis is drawn from the proposed Orange County Water Master Plan Amendment of July 2009 (OCWA, 2010).

Water Supply Planning in the County

In 2007, Orange County's population was 377,000. It was partially concentrated within several small cities (Middletown, Newburgh, and Port Jervis) and in smaller villages, but also was widely dispersed throughout the county. By 2018, the population is projected to grow to 436,000 (OCWA, 2010). With this increase in population, water demand is expected to rise from 29.9 million gallons per day in 2008 to 34.1 million gallons per day in 2018 (OCWA, 2010). While there are no estimates of longer-term growth, given the projected rise in the U.S. population it would seem reasonable to assume that the county would continue at its current growth rate, adding about 45,000 additional people per decade and thereby increasing the demand for water by about 5 million gallons per day each decade. These projections and others in the OCWA Water Master Plan are based on an assumption that per capita demand will remain approximately constant over the next 10 years (OCWA, 2010).

As of 2007, approximately 101,000 people in the county (spread over 80 percent of the land area) were served by individual homeowner wells (OCWA, 2009). The remaining people were supplied from groundwater and surface-water sources by 63 municipally operated water districts and 89 community water suppliers. These groundwater and surface-water supplies are almost exclusively owned and operated by the individual entities, with limited sharing of water-supply infrastructure. Thus, there are more than 150 independent and decentralized water suppliers in Orange County alone.

The OCWA is the county's primary agency responsible for planning and development of drinking water supply resources and infrastructure. Other county and State agencies—notably the Department of Health and Department of Environmental Conservation—have key roles and authority over certain aspects of drinking water supplies as well. The OCWA was formed in 1987 to implement a county-wide wholesale water system that would be composed of new reservoirs, treatment plants, and a looped transmission pipeline throughout the central part of the county ("the water loop"). The water loop project was the culmination of several decades of planning for a centralized supply. As early as 1959, an engineering firm created plans for damming creeks and rivers to create reservoirs. Based on a 1977 water supply study, the county purchased three possible reservoir sites: Black Meadow, Dwaar Kill, and Indigot. Then, a 1987 study laid the foundation for the OCWA and loop project. In addition to the water loop centralized distribution system, this study called for construction of 1) the Dwaar Kill Reservoir with flood skimming from the Shawangunk Kill in order to provide an increased capacity of 18 million gallons per day, and 2) a siphon to the Catskill Aqueduct that could provide 21 million gallons per day. However, the water loop project was never implemented, largely due to the unwillingness of local municipalities to commit to long-term contracts to purchase water from the OCWA. These proposed contracts were the basis of the OCWA's plan to finance the project, which was estimated to cost $142 million in 1990 dollars ($236 million in 2010 dollars). Despite not having its own facilities, the OCWA has certain legal and financial powers not available to municipalities. The OCWA is still establishing its place in coordinating the water supply in the county, with the recent report an example of the evolving process of trying to provide centralized planning.

Instead of a large centralized project, smaller-scale projects have been developed by local municipalities to provide additional supplies. These smaller projects have largely been based on groundwater sources, primarily because of their much lower capital costs and their ability to be phased in as demand increases. Orange County is currently evaluating the potential for additional groundwater supplies from two basins originally planned for reservoir development (Dwaar Kill and Indigot). It is estimated that each of these two basins could supply 600 to 800 gallons per minute, enough water for 2,000 homes. Additionally, interconnections have been suggested among several municipal systems to help better distribute supplies. The OCWA's water conservation programs, especially a leak detection program, have contributed to controlling demand.

County Vulnerability to Water Supply Shortages

The current drinking water supply capacity is 50 million gallons per day with a demand of 30 million gallons per day. In 2018, the projected supply capacity is 53 million gallons per day with a demand of 32 million gallons (OCWA, 2010; Table 2). Supply estimates are based on the so-called safe-yield projection, the amount of water available during the drought of record. In Orange County and other parts of the region, the drought of record is based on the 1960s drought. Thus, according to this supply estimate, in 2018 there will be a county-wide surplus of 20 million gallons per day.

However, since the water supply sources are localized, certain municipalities are projected to have deficits in supply. Some of these deficits would be eliminated if planned or existing back-up systems come online. But based on projections, the Village of Goshen, the City of Middletown, and the Village of Kiryas Joel would have a deficit in water supply by 2018 (OCWA, 2010; Table 2).

While there is a significant range of possible futures in current modeling projections, climate change projections that are within the most-likely range suggest that average annual surface-water supplies will stay near their current levels (Frei et al., 2009). But these projections also indicate reduced groundwater and soil moisture, and there is certainly the possibility of decreases in surface-water supplies as well. With certain systems in the county already approaching a threshold at which demand may exceed supply in the next decade, additional population growth, combined with even slight changes due to climate, could lead to deficits in these already-stressed systems. During the drought of 2001–2002, five municipal water systems had to activate emergency supplies (including using water from New York City's Catskill Aqueduct). And even if average annual supplies remain relatively constant, increased variability in summer precipitation and evaporation could temporarily stress small water systems (which largely serve small residential developments) that have little storage. During the drought of 2001–2002, four water districts (Walton Lakes Estates, Arcadia Hills, Hambletonian Park, and Pheasant Hill) serving residential developments (with a total population of only several hundred each) needed to truck in water from neighboring communities and empty it into district wells in order to replenish the supply.

Case Study Conclusions

The examination of drinking water planning in Orange County provides a glimpse at some of the bureaucratic, political, and financial challenges that are intertwined with any assessment of the impacts and possible adaptations to climate change in an exurban region.

Orange County as a whole probably has sufficient drinking water supplies to meet demand in the near future, but certain areas within the county do not. While the development of major centralized water sources now (such as new reservoirs) could reduce the risk of water shortages from a changing climate in the future, the economic costs and regulatory challenges involved suggest that this is not likely. The water loop plan of the late 1980s or other large-scale centralized supply and distribution plans may not be reinvigorated given the high overall cost and lack of demonstrated benefits to many of the numerous communities that would have to agree to participate.

Future strategies to ensure there is sufficient water in all communities would benefit from addressing population growth and potential climate change impacts. They will most likely be similar to past and current efforts of the OCWA. The proposed Water Master Plan recommends planning limited interconnections between nearby municipalities to address localized shortfalls, the ongoing use of water from New York City's aqueducts, and studying new small-scale sources, such as groundwater wells. The plan also recommends increased conservation programs, though it does not factor potential water-efficiency gains into projected water demand. This is a conservative approach that seems aimed at ensuring adequate supplies in case conservation programs have a limited impact. The county's ongoing ownership of several potential reservoir sites may become more important in the future if water availability becomes more limited and there is a heightened political willingness to fund reservoir development.

References

Alley, W.M. 1981. "Estimation of Impervious Area Wash-off Parameters." *Water Resources Research* 17: 1161-1166.

Barnett, T.P. and D.W. Pierce. 2009. "Sustainable water deliveries from the Colorado River in a changing climate." *Proceedings of the National Academy of Sciences of the United States of America* 106(18): 7334-7338.

Berg, P., J.O. Haerter, P. Thejll, C. Piani, S. Hagemann, and J.H. Christensen. 2009. "Seasonal characteristics of the relationship between daily precipitation intensity and surface temperature." *Journal of Geophysical Research* 114:D18102.

Brookes, J.D., J. Antenucci, M. Hipsey, M.D. Burch, N.J. Ashbolt, and C. Ferguson. 2004. "Fate and Transport of Pathogens in Lakes and Rivers." *Environment International* 30:741-759.

Brutsaert, W. 2006. "Indications of increasing land surface evaporation during the second half of the 20th century." *Geophysical Research Letters* 33:L20403.

Burns, D.A., J. Klaus, and M.R. McHale. 2007. "Recent climate trends and implications for water resources in the Catskill Mountain Region, New York, USA." *Journal of Hydrology* 336:155-170.

Busciolano, R., 2005. "Statistical Analysis of Long-Term Hydrologic Records for Selection of Drought-Monitoring Sites on Long Island, New York." *U.S. Geological Survey Scientific Investigations Report 2005* 5152, 14 p.

Buxton, H.T. and D.A. Smolensky. 1998. "Simulation of the Effects of Development of the Ground-Water Flow System of Long Island, New York." *U.S. Geological Survey Water-Resources Investigations Reports* 98-4069.

Calderon R., C. Johnson, G. Craun, A. Dufour, R. Karlin, et al. 1993. "Health risks from contaminated water: Do class and race matter?" *Toxicology and Industrial Health* 9(5):879–900.

Choi, O. and A. Fisher. 2003. "The impacts of socioeconomic development and climate change on severe weather catastrophe losses: Mid-Atlantic Region and the US." *Climatic Change* 58:149-170.

Chowdhury, S., P. Champagne, and P.J. McLellan. 2009. "Models for predicting disinfection byproduct formation in drinking waters: A chronological review." *Science of the Total Environment* 407:4189-4206.

Coon, W. 2005. "Hydrologic Evidence of Climate Change in Monroe County, NYS." *U.S. Geological Survey Open-File Report* 2008:1199.

Corburn, J. 2002. "Combining community-based research and local knowledge to confront asthma and subsistence-fishing hazards in Greenpoint/Willamsburg, Brooklyn, NYS." *Environmental Health Perspective* 110(suppl. 2):241–248.

Crain, L.J. 1966. "Ground-water Resources of the Jamestown Area." *U.S. Geological Survey, Bulletin* 58B.

DeGaetano, A. 1999. "A temporal comparison of drought impacts and responses in the New York City Metropolitan area." *Climatic Change* 42: 539-560.

DeGaetano, A. 2009. "Time-dependent changes in extreme precipitation return-period amounts in the continental U.S." *Journal of Applied Meteorology and Climatology* 48:2086-2099.

De Vries, M. and L.A. Weiss. 2001. "Salt-front movement in the Hudson River Estuary, New York – Simulations by one-dimensional flow and solute-transport models." *U.S. Geological Survey Water-Resources Investigations Reports* 00-4024.

Dorfman, M. and K.S. Rosselot. 2009. "A Guide to Water Quality at Vacation Beaches, 19th Edition." *Natural Resources Defense Council.*

Downton, M.W., J.Z.B. Miller, and R.A. Pielke, Jr. 2005. "A reanalysis of the U.S. National Weather Service flood loss database." *Natural Hazards Review* 6: 13–22.

Easton, Z.M., M.T. Walter, and T.S. Steenhuis. 2008. "Combined monitoring and modeling indicate the most effective agricultural best management practices." *Journal of Environmental Quality* 37:1798-1809.

Ehlke, M.H. and L.A. Reed. 1999. "Comparison of Methods for Computing Streamflow Statistics for Pennsylvania Streams." *U.S. Geological Survey Water-Resources Investigations Reports* 99-4068.

Eissler, B.B. 1979. "Low-flow frequency analysis of stream in New York." *New York State Department of Environmental Conservation Bulletin* 74.

Eng, K. and P.C.D. Milly. 2007. "Relating low-flow characteristics to the base flow recession time constant at partial record stream gauges." *Water Resources Research* 43:W01201.

Evans, G.W. and E. Kantrowitz. 2002. "Socioeconomic status and health: The potential role of environmental risk exposure." *Annual Review of Public Health* 23:303-331.

Feeley, T.J., T.J. Skone, G.J. Stiegel, A. McNemar, M. Nemeth, B. Schimmoller, J.T. Murphy, and L. Manfredo. "Water: A critical resource in the thermoelectric power industry." *Energy* 33:1-11.

Freeman, W. 2009. New York Water Science Center, U.S. Geological Survey, personal communication.

Frei, A., R.L. Armstrong, M.P. Clark, and M.C. Serreze. 2002. "Catskill Mountain water resources: Vulnerability, hydroclimatology, and climate-change sensitivity." *Annals of the Association of American Geographers* 92:203-224.

Frei, A., S. Gruber, C. Molnar, J. Zurovchak, and S.Y. Lee. 2009. "Progress Report: Potential Impacts of Climate Change on Sustainable Water Use in the Hudson River Valley, May 2009." Report from a grant to the City University of NYS Institute for Sustainable Cities from the NYS Water Resources Institute, FY2008. http://www.geo.hunter.cuny.edu/~afrei/freigrubermay09OCprogreport.pdf

French, T.D., and E.L. Petticrew. 2007. "Chlorophyll a seasonality in four shallow eutrophic lakes (northern Britsh Columbia, Canada) and the role of internal phosphorus loading and temperature." *Hydrobiologia* 575:285-299.

Futter, M.N., and H.A. de Wit. 2008. "Testing seasonal and long-term controls of streamwater DOC using empirical and process-based models." *Science of the Total Environment* 407:698-707.

Garbrecht, J.D. J.L. Steiner, and C.A. Cox. 2007. "The times they are changing: soil and water conservation in the 21st century." *Hydrological Processes* 21:2677-2679.

Groisman, P.A., and R.W. Knight. 2007. "Prolonged dry episodes over the conterminous United States: New tendencies emerging during the last 40 years." *Journal of Climate* 21:1850–1862.

Gross, D. L. 2003. "Assessing Local Government Capacity for Ecologically Sound Management of Flowing Water in the Landscapes of the Great Lakes Basin." Unpublished M.S. thesis, Cornell University Department of Natural Resources.

Hallegatte, S. 2009. "Strategies to adapt to an uncertain climate." *Global Environmental Change* 19:240-247.

Han, H.J., J.D. Allan, and D. Scavia. 2009. "Influence of climate and human activities on the relationship between water nitrogen input and river export." *Environmental Science and Technology* 43:1916-1922.

Hayhoe, K., C.P. Wake, T.G. Huntington, L. Luo, M.D. Schwartz, J. Sheffield, E. Wood, B. Anderson, J. Bradbury, A. DeGaetano, T.J. Troy, and D. Wolfe. 2007. "Past and Future Changes in Climate and Hydrological Indicators in the US Northeast." *Climate Dynamics* 28:381-407.

Hennessy, K.J., J.M. Gregory, and J.F.B. Mitchell. 1997. "Changes in Daily Precipitation Under Enhanced Greenhouse Conditions." *Climate Dynamics* 13:667-680.

Hodgkins, G.A., R.W. Dudley, T.G. Huntington. 2003. "Changes in the timing of high river flows in New England over the 20th century." *Journal of Hydrology* 278:244-252.

Howarth, R.W., D.P. Swaney, T.J. Butler, R. Marinaro, N. Jaworski, and C. Goodale. 2006. "The influence of climate on average nitrogen export from large watersheds in the Northeastern US." *Biogeochemistry* 79:163-186.

Huber, W.C. 1993. "Contaminant Transport in Surface Water." In *Handbook of Hydrology*, edited by D.R. Maidment, Chapter 14. New York, NY: McGraw-Hill.

Intergovernmental Panel on Climate Change (IPCC). 2007. "Climate Change 2007: The Physical Science Basis." Contribution of Working Group I to the Fourth Assessment Report of the Intergovernmental Panel on Climate Change. S. Solomon, D. Qin, M. Manning, Z. Chen, M. Marquis, K.B. Averyt, M. Tignor, and H.L. Miller, Eds. Cambridge University Press, Cambridge, United Kingdom and New York, NY, USA.

International Lake Ontario – St. Lawrence River (ILOSLR) Study Board. 2006. *Options for Managing Lake Ontario and St. Lawrence River Water Levels and Flows – Final Report.*

Jyrkama, J.I., and J.F. Sykes. 2007. "The impact of climate change on spatially varying groundwater recharge in the Gran River watershed (Ontario)." *Journal of Hydrology* 338:237-250.

Kenney, D.S., C. Goemans, R. Klein, J. Lowrey, and K. Reidy. 2008. "Residential water demand management: Lessons from Aurora, Colorado." *Journal of the American Water Resources Association* 44:192-208. *Proceedings of the National Academy of Sciences* 106:7334-7338.

Koster, R.D. H. Wang, S.D. Schubert, M.J. Suarez, and S. Mahanama. 2009. "Drought-induced warming in the continental United States under different SST regimes." *Journal of Climate* 22:5385–5400.

Leathers, D.J., M.M. Malin, D.B. Kluver, G.R. Henderson, and T.A. Bogart. 2008. "Hydroclimatic variability across the Susquehanna River Basin, USA, since the 17th century." *International Journal of Climatology* 28:1615-1626.

Lempert, R.J. and M.T. Collins. 2007. "Managing the risk of uncertain thresholds responses: comparison of robust, optimum, and precautionary approaches." *Risk Analysis* 27:1009-1026.

Lenderink, G. and E. Van Meijgarrd. 2008. "Increase in hourly precipitation extremes beyond expectations from temperature changes." *Nature Geosciences* 1:511-514.

Louis, V.R., E. Russek-Cohen, N. Choopun, I.N.G. Rivera, B. Gangle, S.C. Jiang, A. Rubin, J.A. PAtz, A. Huq, and R.R. Colwell. 2003. "Predictability of Vibrio Cholerae in Chesapeake Bay." *Applied and Environmental Biology* 69:2772-2785.

Lumia, D.S. and K.S. Linsley. 2005. "NYS Water Use Program and Data 2000." *U.S. Geological Survey Open File Report* 1352.

Major, D.C., and O'Grady, M.C. 2010. "Adaptation Assessment Guidebook," in New York City Panel on Climate Change, *Climate Change Adaptation in New York City: Building a Risk Management Response*, edited by Cynthia Rosenzweig and William Solecki. New York: *Annals of The New York Academy of Sciences* 1185.

McCabe, G.J. and D.M. Wolock. 2002. "A step increase in streamflow in the conterminous United States." *Geophysical Research Letters* 29:2185.

McKenney, M.S. and N.J. Rosenberg. 1993. "Sensitivity of some potential evapotranspiration estimation methods to climate change." *Agricultural and Forest Meteorology* 64:81-110.

Mauget, S.A. 2006. "Intra- to multi-decadal terrestrial precipitation regimes at the end of the 20th century." *Climatic Change* 78:317-340.

Miller, T. S. 2004. "Hydrogeology and Simulation of Ground-Water Flow in a Glacial Aquifer System at Cortland County, NYS." *U.S. Geological Survey Fact Sheet* 054-03.

Misut, P.E., C.E. Schubert, R.G. Bova, and S.R. Colabufo. 2004. "Simulated effects of pumping and drought on ground-water levels and the freshwater-saltwater interface on the North Fork of Long Island, NY." *U.S. Geological Survey Water-Resources Investigations Reports* 03-4184.

Milly, P.C.D., K.A. Dunne, and A.V. Vecchia. 2005. "Global pattern of trends in streamflow and water availability in a changing climate." *Nature* 438:347-350.

Montalto, F., C. Behr, K. Alfredo, M. Wolf. M. Arye, and M. Walsh. 2007. "Rapid assessment of the cost-effectiveness of low impact development for CSO control." *Landscape and Urban Planning* 82:117-131.

Najjar, R., L. Patterson, and S. Graham. 2009. "Climate Simulations of Major Estuarine Watersheds in the Mid-Atlantic Region of the US." *Climatic Change* 95:139-168.

Nechamen, W. 2009. Floodplain Management Section, New York State Department of Environmental Conservation, personal communication.

Neff, R., H. Chang, C.G. Knight, R.G. Najjar, B. Yarnal, and H.A. Walker. 2000. "Impact of climate variation and change on Mid-Atlantic Region hydrology and water resources." *Climate Research* 14:207-218.

New, M., A. Lopez, S. Dessai, and R. Wilby. 2007. "Challenges in using probabilistic climate change information for impact assessments: an example from the water sector." *Philosophical Transactions of the Royal Society* A 365:2117-2131.

New York City Department of Environmental Protection (NYC DEP). 2008. *Climate Change Task Force -Assessment and Action Plan, Report 1*. May.

New York City Department of Environmental Protection (NYC DEP) and New York City Department of City Planning. 2006. *Population projections by age/sex and borough*. December.

New York City Municipal Water Finance Authority (NYCMWFA). 2009. "Comprehensive Annual Report for the Fiscal Year ended June 30, 2009." New York, New York.

New York State Department of Environmental Conservation (NYS DEC). 2008. "Water Working Group Policy Recommendations to the NYS Environmental Justice Interagency Taskforce." September 17. Accessed May 30, 2009. http://www.dec.ny.gov/docs/permits_ej_operations_pdf/EJ_08.04.08_Water_working_group_notes_FINAL_1.pdf.

New York State Department of Environmental Conservation (NYS DEC). 2008. "The Final New York State 2008 Section 303(d) List of Impaired Waters Requiring a TMDL/Other Strategy." May 26. http://www.dec.ny.gov/docs/water_pdf/303dlist08.pdf.

New York State Department of Environmental Conservation (NYS DEC). 2008a. "Wastewater infrastructure needs of New York State." http://www.dec.ny.gov/docs/water_pdf/infrastructurerpt.pdf.

OCWA. 2010. "Water Master Plan: Proposed County Comprehensive Plan Amendment July 2009." prepared for Orange County Water Authority, Goshen, NYS by Henningson, Durham & Richardson Architecture and Engineering, PC. http://waterauthority.orangecountygov.com/

Pielke, R.A., and M.W. Downton. 2000. "Precipitation and Damaging Floods: Trends in the United States, 1932-97." *Journal of Climate* 13:3625-3637.

Rajagopalan, B. K. Nowak, J. Prairie, M. Hoerling, B. Harding, J. Barsugli, A. Ray, and B. Udall. 2009. "Water supply risk on the Colorado River: Can management mitigate?" *Water Resources Research* 45:W08201.

Randall, A.D. 1977. "The Clinto Street- Ballpark Aquifer in Binghamton and Johnson City, New York." *U.S. Geological Survey Bulletin 73*.

Rayner, S., Lach, D., and Ingram, H. 2005. "Weather Forecasts are for Wimps: Why Water Resource Managers Do Not Use Climate Forecasts." *Climatic Change* 69:197-227.

Risser, D.W., W.J. Gburek, and G.J. Folmar. 2005. "Comparison of Methods for Estimating Ground-Water Recharge and Base Flow at a Small Watershed Underlain by Fractured Bedrock in the Eastern United States." *U.S. Geological Survey Scientific Investigations Report* 5038.

Sartor, J.D., G.B. Boyd, and F.J. Agardy. 1974. "Water pollution aspects of street surface contaminants." *Journal of the Water Pollution Control Federation* 46:458-467.

Schaefer, S.C. and M. Alber. 2007. "Temperature controls a latitudinal gradient in the proportion of watershed nitrogen exported to coastal systems." *Biogeochemistry* 85:333-346.

Shaw, S.B., J.R. Stedinger, and M.T. Walter. 2010. "Evaluating urban pollutant build-up/wash-off models using a Madison, Wisconsin catchment." *Journal of Environmental Engineering* 136:194-203.

Suro, T.P., G.D. Firda, and C.O. Szabo. 2009. "Flood of June 26–29, 2006, Mohawk, Delaware, and Susquehanna River Basins" *U.S. Geological Survey Open-File Report* 1063.

Trenberth, K.E., A. Dai, R.M. Rasmussen, and D.B. Parsons. 2003. "The Changing Character of Precipitation." *Bulletin of the American Meteorological Society* 1205-1217.

Tryhorn, L. and A. DeGaetano. 2010. "A comparison of techniques for downscaling extreme precipitation over the Northeastern United States." Submitted to International Journal of Climatology.

U.S. Environmental Protection Agency (US EPA). 2008. A Screening Assessment of the Potential Impacts of Climate Change on Combined Sewer Overflow (CSO) Mitigation in the Great Lakes and New England Regions (Final Report). U.S. Environmental Protection Agency, Washington, D.C., EPA/600/R-07/033F.

U.S. Global Change Research Program (USGRCP). 2009. *Global Climate Change Impacts in the United States*. Edited by Thomas R. Karl, Jerry M. Melillo, and Thomas C. Peterson. Cambridge University Press.

Whitman, R.L. and M.B. Nevers. 2008. "Summer *E. Coli* Patterns and Responses Along 23 Chicago Beaches." *Environmental Science and Technology* 42:9217-9224.

Winslow, J.D., H.G. Stewart, R.H. Johnson, and L.J. Crain. 1965. "Ground-water Resources of Eastern Schenectady County, New York with Emphasis on Infiltration from the Mohawk River." *U.S. Geological Survey Bulletin* 57.

Zhao, C., Z. Yu, L. Li and G. Bebout. 2010. "Major shifts in multidecadal moisture variability in the Mid-Atlantic region during the last 240 years." *Geophysical Research Letters* 37:L09702.

Appendix A. Stakeholder Interactions

The major issues of vulnerability and associated potential adaptations were identified in conjunction with seven stakeholder workshops involving about 200 stakeholders held during the ClimAID project. These workshops included presentations and discussions with lake association leaders at the annual conference of the Federation of Lake Associations, with engineers and planners at the annual conference of the American Public Works Association, with forestry and wildlife professionals in a regional ForestConnect webinar, and with multiple workshops for Cornell Cooperative Extension educators and rural landowners. Given the breadth of professions and stakeholders that are involved with water resources across New York State, these efforts need to be viewed as the beginning of an ongoing and expanding engagement with all water stakeholders statewide as New York prepares to cope with the challenge of climate change.

List of Workshops

1) **Date:** 4 May 2009
Venue: New York State Lake Associations Annual Conference, Hamilton, NY
Lead: Rebecca Schneider, Cornell University, Dept. Natural Resources
Participants: 16 directors of lake associations across NY

2) **Date:** 27 March 2009
Venue: American Public Works Association – Ann. Conference, Canandaigua, NY
Lead: Rebecca Schneider, Cornell University, Dept. Natural Resources
Participants: 28 municipal engineers, town planners, watershed council program managers, private engineering consultants

3) **Date:** 10 March 2009
Venue: Cornell Cooperative Extension Advisory Council Workshop for Natural Resources and Environment
Lead: Rebecca Schneider, Cornell University, Dept. Natural Resources
Participants: 20 CCE directors, members – American Wildlife Conservation Foundation, member – Federation of NYS Solid Waste Assoc., biologist – Ontario Dune Coalition, assoc. director of state wetland managers and state floodplain managers; president – NY Forest Owners Assoc.

4) **Date:** 18 March 2009
Venue: ForestConnect Webinar
Lead: Kristi Sullivan, Cornell University, Dept. Natural Resources
Participants: 149 including 46 percent landowners, 29 percent foresters, 20 percent educators and 12 percent specialists responsible for ~11,000,000 acres of land in 21 states

5) **Date:** 12 November 2009
Venue: Rural Landowner Workshop on Climate Change, Arnot Forest, Newfield, NY
Lead: Rebecca Schneider, Kristi Sullivan, Cornell University, Dept. Natural Resources.
Participants: 12 private landowners

6) **Dates:** 11 November and 8 December 2009
Venues: Ecosystems Climate Change Workshop, Ithaca, NY Ag and Food Systems CALS In-service, Ithaca, NY
Lead: David Wolfe, Cornell University, Dept. of Horticulture

Stakeholders

Adirondack Mountain Club
Akwesasne Task Force on the Environment
Alley Pond Environmental Center
American Wildlife Conservation Foundation, Inc.
Association of State Wetland Managers
Au Sable River Association
Basha Kill Area Association Hudson Basin River Watch
Battenkill Conservancy
Beacon Sloop Club
Black Creek Watershed Coalition
Boquet River Association (BRASS)
Bronx River Working Group
Buffalo Niagara Riverkeeper
Building Watershed Bridges in the Mid-Hudson Valley
Butterfield Lake Association
Cornell Cooperative Extension
Canandaigua Lake Improvement Association
Canandaigua Lake Watershed Task Force
Catskill Center for Conservation and Development
Cayuga Lake Watershed Network
Cedar Eden Environmental

Central New York Watershed Consortium
Chautauqua Lake Conservancy
Chautauqua Watershed Conservancy, Inc.
Coalition to Save Hempstead Harbor
Coalition to Save the Yaphank Lakes
Columbia County Lakes Coalition
Community Water Watch Program
Cornell Cooperative Extension - all counties
Croton Watershed Clean Water Coalition
Dutchess County Soil and Water Conservation District
Esopus Creek Conservancy, Inc.
Executive Director of Catskill Watershed Corporation
Fed. of NY Solid Waste Associations
Finger Lakes Land Trust, Inc.
Friends of Jerome Park Reservoir
Friends of the Bay
Genesee Land Trust
Groundwork Yonkers/Saw Mill River Coalition
Honeyoe Valley Association
Horseshoe Pond/Deer River Flow Association
Hudson River Foundation
Hudson River Environmental Society
Hudson River Sloop Clearwater
Hudson River Watershed Alliance
Jamaica Bay Watershed Alliance
Java Lake Colony, Inc.
Keep Putnam Beautiful
Lake Colby Association
Lake Erie Alliance
Lake George Association
Lake George Land Conservancy
Land Trust of the Saratoga Region
Metropolitan Waterfront Alliance
Mirror Lake Watershed Association
Mohawk River Research Center
Mohegan Lake Improvement District
Monroe County Stormwater Coalition
Natural Resources Defense Council
Nature Conservancy Great Swamp Program
Nature Conservancy Neversink River Program
New York Agricultural Land Trust
New York Forest Owners Association
New York Rural Water Association
New York Rivers United
New York State Federation of Lake Associations, Inc.
New York State Lakes
North River Community Environmental Review
Oatka Creek Watershed Committee
Onesquethaw/Coeymans Watershed Council
Onondonaga Creek Revitalization Committee
Ontario Dune Coalition
Peconic Bay
Peconic Baykeeper
Peconic Estuary Program
Plymouth Reservoir Association
Protect the Plattekill Creek & Watershed
Quassaick Creek Coalition
Riverkeeper, Inc.
Saranac Lake River Corridor Commission
Saranac Waterkeeper/Upper Saranac Lake Foundation
Save Our Seashore
Sawkill Watershed Alliance Scenic Hudson, Inc.
Saw Mill River Coalition
Seneca County
Skaneateles Lake Watershed Agricultural Program
Snyder Lake Association
South Bronx River Watershed Alliance
Sparkill Watershed Conservancy
St. Regis Mohawk Tribe
The River Project
The Urban Divers Estuary Conservancy
Upper Delaware Council, Inc.
Upper Saranac Lake Association
Upper Susquehanna Coalition
Wallkill River Task Force
Westchester Land Trust
Western New York Land Conservancy

Appendix B. New York State Flood Analysis

To better understand flood processes in New York State, linkages between stream discharge and precipitation and snowmelt were examined for three moderately sized watersheds in three different regions of the state: Ten Mile River in the lower Hudson Valley, Fall Creek in the Finger Lakes Region, and the Poultney River in the Lake Champlain Valley. The three water bodies were selected because they have at least 50 years of stream gauge and precipitation records and do not have any major impoundments or diversions.

For each water body, the following indicators were examined: the annual maximum daily average flow, the daily average flow associated with the annual maximum two-day precipitation event, and the annual maximum daily flow associated with the maximum three-day snowmelt event. The intent was to investigate the relationship between maximum stream discharges and two of the most important causes of flooding, large precipitation events and melt of a sizable snowpack (**Table 4.9**).

The three watersheds generally reveal the same patterns in flooding. Most notably, less than 40 percent of annual maximum daily discharges correspond to either two-day maximum rainfalls or maximum snowmelts. To explain this outcome in terms of snowmelt, the largest snowpacks are usually only on the order of 20 inches in depth (it would be preferable to know the actual water content of the snowpack, but this is not routinely measured), and they typically melt over at least several days (the largest one-day melt in any of the watersheds was 13 inches but the median was only 6 inches). An inch of snow typically contains about 0.1 inch of water, so melting of a large snowpack is only equivalent to a moderate rainfall event and not sufficient to result in very large stream discharges.

To explain this outcome in terms of precipitation maximums, most two-day maximum rainfall events occurred between May and October. During this time of year, moisture-laden air from the south reaches New York State and causes two-day rainfall amounts that can exceed 5 inches. However, counteracting these larger rainfall amounts is an increase in available soil-water storage capacity due to the drier soil conditions and lowered water tables that are common in the same timeframe. Using data from the Fall Creek watershed in central New York State, **Figure 4.11** shows the correlation between the two-day storm precipitation amount and the resulting average daily discharge in the stream. Although there is a general upward trend (larger stream discharges occur with larger precipitation amounts), rainfall is clearly not the only factor related to peak discharge. For instance, there are three days with discharges around 6,000 cubic feet per second, but these correspond to medium-sized storm rainfall amounts ranging anywhere from 3 to 5 inches. Though the rainfall amounts of the largest storms will likely increase, impacts on peak flows are uncertain due to hydrologic buffering from possible increases in soil dryness.

	% Annual Maximum Discharge Events Occurring When:		
Watershed	2-day annual max rainfall	3-day annual max snowmelt	May to October
Ten Mile River	15	10	14
Fall Creek	20	20	5
Poultney River	17	9	10

Table 4.9 Causative conditions of annual maximum discharges on three watersheds representative of conditions in New York State

Rather than snowmelt or large precipitation events, approximately 60 percent (**Table 4.9**) of the annual maximum discharges in these New York watersheds typically result from a combination of limited soil-moisture storage capacity (i.e., wet soils) and moderate rainfall events (1-to-3-inch two-day events). In all three watersheds only 15 percent of annual maximum daily discharges occur between May and October—at most—because the soils are relatively dry (**Table 4.9**, column 3). Ultimately, the degree of change in flooding will likely be dependent on the timing of projected increases in spring rainfall when soils tend to be saturated. If it entails moderate amounts of rainfall on more days, streamflows will be higher more often but will not necessarily reach new maxima. If, however, the number of spring rainfall events remains the same but their maximum potential size increases, flood magnitudes could increase. Such an increase could be partially offset by a lengthened growing season that narrows the window in which a large storm event on wet soils could occur. In brief, given the number of interacting factors, it remains uncertain whether the magnitude of annual maximum flows will increase with climate change. If it does, it would seem probable that wetter spring conditions would likely increase the number of moderate floods (10- to 25-year return periods).

The largest floods of record (50- to 100-year return periods) in the three watersheds can be attributed to distinct hydrometeorological conditions. In the

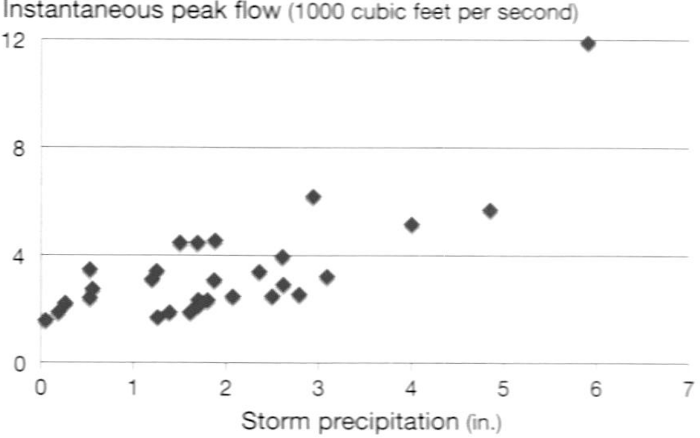

Figure 4.11 Storm rainfall amount and resulting instantaneous peak discharge on Fall Creek, Tompkins County, for the annual peak discharges from 1974 to 2007

Poultney River watershed—the watershed expected to have the most snow—the largest flood (7,010 cubic feet per second) was caused by a combination of snowmelt and rain on presumably still frozen soils (snow remains at the end of the flood event). Surprisingly, this is one of only five annual maximum discharges with rain on snow in this watershed. In Fall Creek, the largest discharge (7,060 cubic feet per second) was caused by a 5.9-inch rainfall event during a month with unusually wet antecedent conditions. In the Ten Mile River watershed, the two largest maximum discharges were associated with a hurricane (10,700 cubic feet per second) and several consecutive days of rainfall (9,930 cubic feet per second). In all these cases, the large flow events were caused by relatively rare conditions. The historical record is too short to observe a trend in the occurrence of these conditions, and projecting future trends is currently beyond the skill of coarse-scale climate models. Additional research is needed to determine the change in probability of these large floods.

[1] The salt front is not a sharp division between freshwater and ocean water; rather it is the point at which chloride concentration exceeds 100 mg/L. This concentration has been selected to minimize the negative impacts of high salt intake on human health, but it is far lower than a level where it would be entirely unfit to drink for temporary periods (ocean water is 200 times saltier). In extenuating circumstances, withdrawing water even if the salt front had moved upriver beyond the Poughkeepsie intake could still be done with minimal short-term consequences.

[2] The vulnerability of reservoirs to drought was assessed by dividing storage volume by daily demand, and a similar analysis can be accomplished for aquifers by comparing subsurface storage to daily demand. Demand data were determined from the annual water quality reports mandated by the Safe Drinking Water Act. Days of supply with no recharge may be exaggerated because water will not be perfectly redistributed to well fields and the area around a well field may experience localized depletion. Since only the surface area of an aquifer is typically given, stored water is determined by assuming 30 percent of aquifer volume consists of recoverable water and by taking an average aquifer thickness (typically ~40 feet).

[3] This assessment of wells includes the following: A 16-foot-deep sand and gravel well in Madison County (M-178 Valley Mills) had its water level drop to 11 feet below the surface in the summer of 1999, the lowest point reached in its 27-year history. A 31-foot well in sand and gravel in Oneida County (Oe-151 Woodgate) dropped to 30 feet below ground surface in late 2002, the lowest point in its 18-year history; if subject to pumping, it is presumed that this well would probably have run dry. A 79-foot sand and gravel well (Re-703 East Greenbush) dropped to 42 feet below ground surface in 1986, the lowest point in its 18-year history. A 126-foot bedrock well (364TRNN) (Du-321 Hyde Park) dropped to 72.5 feet below land surface in late 1981, the lowest point in its 24-year history.

[4] Table 4.8 was constructed by comparing 2000 census block data to areas of inundation to determine the vulnerability of certain populations to flooding, specifically renters, seniors, and non-white populations. Demographic data were weighted by the area of the block that was inundated in 2006. The results provide some measure of the human profile of flooding and highlight the presence of a few populations with potential vulnerabilities.

Chapter 5

Coastal Zones

Authors: Frank Buonaiuto,[1,2] Lesley Patrick,[1,3] Vivien Gornitz,[4] Ellen Hartig,[5] Robin Leichenko,[6] and Peter Vancura[6]

[1] Sector Lead
[2] City University of New York, Hunter College
[3] City University of New York, CUNY Institute for Sustainable Cities
[4] Columbia University Earth Institute, Center for Climate Systems Research
[5] New York City Department of Parks and Recreation
[6] Rutgers University, Department of Geography

Contents

Introduction 122
5.1 Sector Description 122
5.1.1 Economic Value 123
5.1.2 Non-Climate Stressors 124
5.2 Climate Hazards 125
5.2.1 Sea Level Rise 125
5.2.2 Increasing Coastal Water Temperatures 125
5.2.3 Change in Regional Precipitation 126
5.2.4 Other Climate Factors 126
5.3 Vulnerabilities and Opportunities 130
5.3.1 Coastal Erosion 130
5.3.2 The Hudson River Estuary 132
5.3.3 Freshwater Resources 134
5.3.4 Coastal Ecosystems 134
5.3.5 Fish and Shellfish Populations 137
5.4 Adaptation Strategies 139
5.4.1 Adaptations for Key Vulnerabilities 140
5.4.2 Co-Benefits, Unintended Consequences, and Opportunities 141
5.5 Equity and Environmental Justice Considerations 142
5.5.1 Vulnerability 142
5.5.2 Equity Issues in Adaptation 145
5.6 Conclusions 146
5.6.1 Main Findings on Vulnerabilities and Opportunities 146
5.6.2 Adaptation Options 148
5.6.3 Knowledge Gaps 148
Case Study A. 1-in-100-year Flood and Environmental Justice 149
Case Study B. Modeling Climate Change Impacts in the Hudson River Estuary 155
Case Study C. Salt Marsh Change at New York City Parks and Implications of Accelerated Sea Level Rise 156
References 157
Appendix A. Stakeholder Interactions 161

Introduction

The anticipated global sea level rise due to climate warming will greatly amplify risks to the coastal population of New York State, leading to permanent inundation of low-lying areas (including wetlands), more frequent flooding by storm surges, and potential for increased beach erosion. Saltwater could reach farther up estuaries, such as the Hudson River, potentially contaminating urban water supplies, while increased water depth could alter the propagation of both the tide and storm surges up the Hudson River to the Federal Dam in Troy. These hazards will continue to be exacerbated by development in the coastal zone.

The Intergovernmental Panel on Climate Change (IPCC) concluded in 2007 that global sea level will likely rise between 7 and 23 inches by the end of the century (2090–2099), relative to the base period (1980–1999), not counting unexpected rapid changes in ice flow from the Greenland and Antarctic ice sheets. However, these projections may be too low, as they do not consider the uncertainty associated with ice sheet melting processes or cover the full likely temperature range given in the Fourth Assessment Report (up to 6.4°C) (Rahmstorf et al., 2007; Rohling et al., 2008; Pfeffer et al., 2008; Horton et al., 2008). Regional sea level rise projections used in the assessment explicitly include a "rapid ice-melt scenario" based on acceleration of recent rates of ice melt in the Greenland and West Antarctic ice sheets and paleoclimate studies (See Chapter 1, "Climate Risks," for a complete description of this method). Most of the observed current climate-related rise in global sea level over the past century can be attributed to expansion of the oceans as they warm; however, it is anticipated that the melting of land-based ice may become the dominant contributor to global sea level rise in the future.

Historically, the rise in regional (or relative) sea level —a measurement of sea level height that includes local effects such as the vertical movement of land, local ocean temperatures, atmospheric pressure, and tides— has varied through time and accelerated during the 20th century as global temperatures have increased (Gornitz et al., 2002; Gehrels et al., 2005; Donnelly et al., 2004; Holgate and Woodworth, 2004; IPCC, 2007). Regional sea level was rising at rates of 0.34 to 0.43 inch per decade over the past thousand years; however, current rates are ranging between 0.86 and 1.5 inches per decade with a 20th century average rate of 1.2 inches per decade (see Chapter 1, "Climate Risks").

5.1 Sector Description

The U.S. Coastal Zone Management Act of 1972, as amended in 1996, defines the coastal zone as the land inward of the shoreline needed to control or manage uses that are likely to directly and significantly impact coastal waters or are likely to be "affected by or vulnerable to sea level rise." New York State considers coastal waters to extend three miles into the open ocean, and up to the state lines of Connecticut and New Jersey along the shore.

In the ClimAID assessment, we consider the coastal zone to include the shoreline of New York State, including coastal wetland areas and inland areas adjacent to the shoreline that are likely to be affected by sea level rise and coastal storms. We also consider impacts and adaptation strategies for Great Lakes coastlines were not included in this assessment even though these regions are clearly part of the coastal zone, as they could not be properly analyzed given scheduling and budgeting constraints. Additional resources should be made available to conduct a comprehensive assessment of climate-change-related impacts and adaptation strategies specifically targeted at the Great Lakes regions (an investigation that would require multi-state collaboration). In particular, this assessment effort focuses on identifying 1) climate change risks affecting the coastal zone, arising from sea level rise, storm surges, increased water temperatures, and changes in precipitation; 2) critical vulnerabilities (populations, ecosystems, and regional coastal communities); and 3) potential adaptation strategies for coastal communities.

The New York State coastline is composed of a combination of glacial bluffs, pocket beaches, and extensive barrier island/bay systems. Long Island is particularly vulnerable to the effects of shoreline erosion since it is largely formed of sand and gravel deposits left by the retreating glaciers, after the end of the last Ice Age around 20,000 years ago. The South Shore of Long Island is a sandy environment consisting largely of barrier islands, spits, and back-barrier salt marshes that are very erodible and subject to inundation.

Coastal ecosystems include nearshore subtidal areas, the low marsh intertidal zone (**Figure 5.1**, top), high marsh (**Figure 5.1**, bottom), beaches, dunes, stream channels, rocky platforms (**Figure 5.2**), seagrass meadows, algal beds, and tidal flats (Nordstrom and

Roman, 1996). Even in a densely populated urban environment such as New York City these coastal ecosystems provide numerous functions and values. These include wildlife habitat, storm surge protection, wave attenuation, pollution absorption, and aesthetic appeal. More than 300 bird species spend part of their life cycle in New York's coastal shores, feeding, resting, or nesting. Every May and June, thousands of horseshoe crabs come to spawn on the sandy beaches of Long Island, New York City, and Westchester County. Many bird species depend on the horseshoe crab eggs or other invertebrates of the tidal zone to replenish their fatty reserves and continue on migration routes along the Atlantic flyway.

New York State's coastal marshes are limited to the north and south shores of Long Island (Suffolk and Nassau Counties), New York City (Queens, Brooklyn, Staten Island, and the Bronx), Westchester County, and up the Hudson River. In the tidally influenced portion of the Hudson River Estuary (up to the Troy Dam), the dominant ecological communities are freshwater and brackish tidal marshes, freshwater tidal swamps, tidal creeks, mud and sand flats, and freshwater subtidal aquatic beds (Edinger, 2002). However, these are limited to north of the Tappen Zee Bridge as there is little or no break in shoreline armoring (bulkheads and riprap) from Manhattan to the bridge.

Location: Idlewild Park, Queens, NY. Photo credit: Mike Feller

Location: Pelham Bay, Bronx, NY. Photo credit: Ellen Kracauer Hartig

Figure 5.1 Salt marshes in the urban environment

5.1.1 Economic Value

The coastal zone is not a category in the North American Industrial Classification System (NAICS) (U.S. Bureau of Economic Analysis, n.d.), since values produced by economic activity in the coastal zone are distributed among a wide variety of industry, government, commercial, and private activities. One way to consider value is the estimated insured value of properties in coastal counties in the state in 2004. This was nearly $2 trillion: $1,901.6 billion, or 61 percent of the total insured value in the state of $3,123.6 billion (AIR, 2005; see Annex II, "Economics," of full ClimAID Report).

Insured losses from previous storms can give a general idea of the current costs of climate-related impacts to the sector. This information is available for hurricanes, winter storms, and thunderstorms. The losses from winter storms and hurricanes are principally located in the coastal zone, whereas losses from thunderstorms occur throughout the state. The largest insured loss since 1990 (in 1992) was approximately $1 billion (in present dollars) from winter storms; from 1990 to 2010 there are nine other years with losses of more than $0.4

Photo credit: Mike Feller

Figure 5.2 Rocky shoreline at Pelham Bay Park, Bronx

billion (ISO, 2010). Additionally, Pielke et al. (2008, p. 35) adjusted the losses from the 1938 hurricane to account for inflation, changes in population density (and thus exposures), and asset value, and estimated that the 1938 storm, if it occurred today, would cause $39.2 billion (2005 dollars) in economic damages (see Annex III, "Economics," of full ClimAID Report).

This information gives a picture of the order of magnitude of coastal zone storms losses without further adaptation measures. As sea level rises, the probability of any given amount of flooding rises so that a storm in the future may cause a larger amount of flooding than the same storm today. See Annex III, "Economics," of full ClimAID Report for further details.

5.1.2 Non-Climate Stressors

The ClimAID assessment is focused on climate-related stresses influencing the coastal zone of New York State. However, as in any complex system, there are multiple forces interacting to produce the observed behavior. Often it is difficult to attribute the response of the system to any particular forcing function; therefore, it is necessary to briefly mention some of the non-climate stressors impacting the various coastal components described in this sector. Many of these stresses are associated with human consumption of natural resources and land-use practices. For example, coastal development, construction of drainage alterations, and impervious surfaces have led to a reduction in groundwater recharge and degraded coastal water quality (Bavaro, 2005). The interconnection among precipitation, land use, and local fish populations has also been documented, suggesting increased urbanization may lead to a reduction in stream biodiversity and migratory fish run sizes (Limburg and Schmidt, 1990). A number of human-induced factors (including sewage discharges and contaminated stormwater runoff from developed and agricultural areas) cause pollution and pathogen outbreaks that can lead to closures of shellfish harvesting areas. The relationship between agricultural lands, storm-water

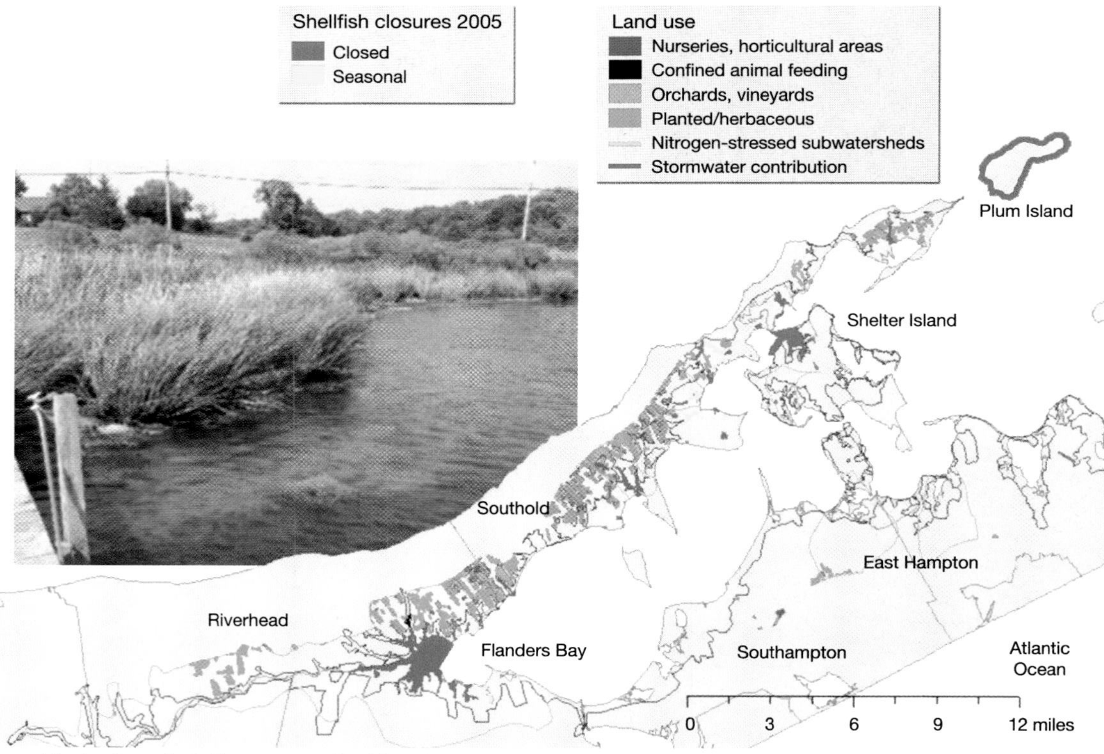

Photo of Flanders Bay wetlands. Figure courtesy of Frank Buonaiuto

Figure 5.3 Peconic River Estuary shellfish closures and land use practices

Tappan Zee (along the boundary between Rockland and Westchester counties), is influenced by rising sea levels and storm surge up to the dam at Troy. The total length of exposed, eroding river bank along the Hudson appears to be small, only a few miles in aggregate. Much of the shoreline is rock or it has been stabilized by the construction of the railroad lines. The Hudson River Estuary's shoreline has been dramatically altered over the last 150 years to support industry and other development, contain channel dredge spoils, and to withstand erosive forces of ice, wind, and waves. About half of the natural shoreline has been engineered with revetments, bulkheads, or cribbing, or reinforced with riprap. Many shorelines contain remnant engineered structures from previous human activities. The remaining "natural" shorelines (which have been affected by human activities such as disposal of dredge spoil, invasive species, and contaminants) include a mix of wooded, grassy, and unvegetated communities on mud, sand, cobbles, and bedrock.

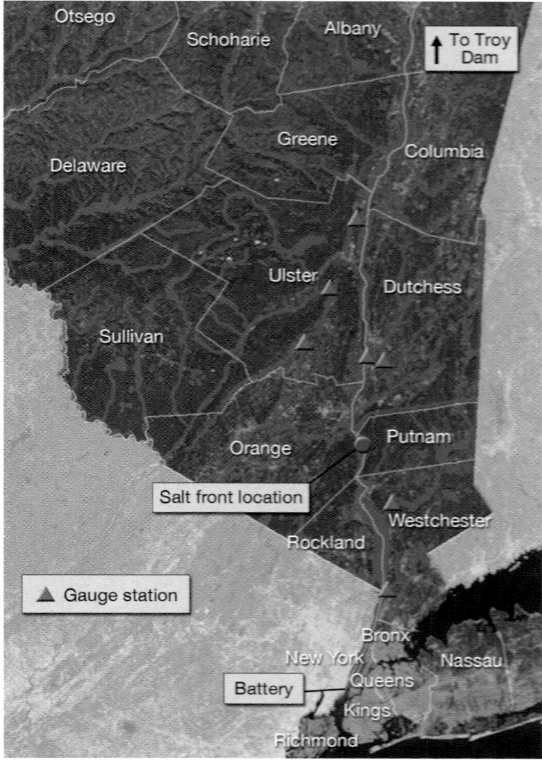

Note: Approximate distance to salt front is 53 river miles from the Battery at New York City (USGS)

Figure 5.6 Hudson River Estuary with location of salt front on 10/30/2009

The average tidal range along the Hudson River is about 4 feet, peaking at 5 feet at either end of the estuary. The transition from freshwater to saltwater occurs in the lower half of the river, and the position of the salt front (interface between saltwater and freshwater) depends in part on the deposition of sediment on the river bed and the flow of freshwater down the Hudson River (discharge). As the climate changes and sea levels rise, the position of the saltwater and propagation of tide and storm surge throughout the estuary will be altered (see Case Study B. Modeling Climate Change Impacts in the Hudson River Estuary).

Climate change could affect the location of the salt front in three ways: 1) reduction in precipitation can reduce stream flow, allowing the salt front to move upstream, 2) increase in temperature can increase evaporation, reducing freshwater runoff, which in turn would cause the salt front to migrate upstream, and 3) rising sea level may push the mean position of the salt front upstream (Rosenzweig and Solecki, 2001). The rates of northward salt front migration could be higher especially for the rapid ice-melt scenario. However, even in the face of recent, post-glacial sea level rise, there is evidence that saltwater has retreated out of the estuary slightly over the last 6,000 years (Weiss, 1974).

Vertical land movements, tributary inputs, and channel characteristics influence local rates of sea level rise and the propagation of tides and storm surge in the estuary. In addition, changes in channel characteristics associated with increased water levels would alter shoaling (wave transformation) processes, which might lead to changes in tide and surge amplitudes.

Hurricanes and nor'easters are generally accompanied by strong winds, surge, and heavy rain. Not only is the influence of surge propagation throughout New York Harbor and the Hudson River Estuary critical, but the impact of increased freshwater discharge to the coastal regions during storm events is also critical. The timing of freshwater input from the New York and northern New Jersey rivers and direct runoff from the surrounding urban landscape will influence overall water levels around New York City as well as supply nutrients and other land-deposited pollutants to coastal waters. Depending on the time of the year, the excess nutrients and pollutants could lead to further degradation of water quality.

Major river systems discharging into New York Harbor include the Hackensack, Passaic, and Raritan Rivers in New Jersey, as well as the Hudson River. Combined, these river systems constitute a drainage area of approximately 16,640 square miles. This complicated drainage system can influence harbor water levels for several days following a rainfall event, a process that was well documented for Hurricane Floyd (Bowman et al., 2004).

5.3.3 Freshwater Resources

In addition to influencing the position of the salt front and storm surge propagation within the Hudson River, sea level rise and changes in precipitation will impact Long Island's water table. The water table will gradually rise at approximately the same rate as sea level. As a result, depending on local conditions, the geographic extent of ponds and wetlands and the carrying capacity of streams may change. This will depend partly on the amount and timing of precipitation in the region. Saltwater entering the aquifers (salt water intrusion and salinization) is a slow process. It is likely that much less than 1 percent of the freshwater reserves would be affected in 100 years. Quantifying the impact that changes in sea level and precipitation patterns will have on Long Island's water table is difficult given the much larger effects of anthropogenic forces such as flood control, groundwater withdrawal, and sewering.

5.3.4 Coastal Ecosystems

Coastal ecosystems are at risk from rising sea levels. Already many tidal marshes are receding in horizontal extent and appear to be collapsing internally as if they are drowning in place (Hartig et al., 2002). Indications include a "Swiss cheese" appearance as they become increasingly ponded (see the Mississippi Delta for an example). While the exact cause of wetland loss is not known, future sea level rise will exacerbate the losses. Current losses are being blamed on multiple stressors, including channelization and armoring of the shoreline (causing sediment starvation), boat waves, excess nutrient loadings (e.g., nitrogen from treated sewage effluent), changes in tidal range (Swanson and Wilson, 2008), excessive bird grazing, overabundance of mussels and sea lettuce, as well as sea level rise.

For coastal ecosystems north of the Tappan Zee Bridge, substantial marsh loss has not been recently documented, although inventories do show loss of native subtidal aquatic beds. These aquatic beds, which are strongly light-limited in the turbid Hudson, are also likely to be sensitive to rising sea levels.

New challenges in protecting remaining coastal ecosystems come from accelerated sea level rise. Where slopes are gradual and land can accommodate the change (even at the expense of forested habitat), under an accelerated sea level rise regime vegetated tidal habitats will shift inland. However, where squeezed between rising sea levels and either human infrastructure or steep slopes, these systems will diminish in size or disappear. The effect is that a previously diverse habitat lying between the deeper waters and uplands (that included the beaches, coastal shoals, mudflats, or marshes) becomes converted to a more simplified, deeper water habitat. A recently released report on coastal sensitivity in the Mid-Atlantic region included the following findings (US EPA/CCSP, 2009):

- Rising water levels are already an important factor in submerging low-lying lands, eroding beaches, converting wetlands to open water, and exacerbating coastal flooding. All of these effects will be increased if the rate of sea level rise accelerates in the future.
- Most coastal wetlands in the Mid-Atlantic would be lost if sea level rises 3 feet in this century. Even a 20-inch rise would threaten most wetlands.

In the New York region, tidal marshes developed over the last 5,000 years after the last glaciers melted and the rate of sea level rise slowed down. Coastal wetlands usually maintain a delicate balance among rates of sea level rise, upward accretion, wave erosion, and sediment deposition, any changes in which could affect the stability of the marsh (Burger and Shisler, 1983; Orson et al., 1985; Allen and Pye, 1992; Varekamp et al., 1992; Nydick et al., 1995; Nuttle, 1997). A salt marsh lies very close to mean sea level and experiences frequent inundation by the tides, which provide nutrients and suspended sediments for accretion. If the marsh grows too high, tidal inundation decreases, with a corresponding decrease in nutrient and sediment supply, thus slowing down accretion and upward growth. Given a sufficient inorganic sediment and nutrient supply, as well as accumulated organic material, accretion rates for some marshes along the U.S. Gulf of Mexico and Atlantic Coasts can match or exceed present-day local sea level rise (Dean et al., 1987; Titus et al., 1988; Nuttle, 1997). However, where relative rates of sea level rise are too rapid and exceed rates of mineral sedimentation and/or organic accretion, the marsh may begin to drown in place, a process observed in many East Coast wetlands (Downs et al., 1994; Wray et al., 1995; Leatherman and Nicholls, 1995; Kearney et al., 2002; DeLaune et al., 1994; Anisfeld and Linn, 2002; Warren and Niering, 1993). Shifts in marsh vegetation distributions are also sensitive indicators of sea level rise and accretion rates (Bertness, 1991; Donnelly and Bertness, 2001).

In New York, while some marshes are thriving, recent studies indicate dramatic losses of other salt marshes over the last several decades (**Table 5.5**). At Jamaica Bay (Gateway National Recreation Area), island salt marsh area declined by 20 percent between the mid-1920s and mid-1970s; since then this trend has accelerated and close to 30 percent has subsequently been lost (Hartig et al., 2002; Rosenzweig and Solecki, 2001; Gornitz et al., 2002; NYSDEC, 2003). Only 7 out of 13 salt marsh islands in Shinnecock Bay (southeastern Long Island) that were present in 1974 remained by 1994 (Fallon and Mushacke, 1996). The apparent submergence of these islands was partially compensated by inland migration of salt marshes, an indicator that sea level rise is a contributing factor.

Location	Acres	Acres and Year of Observation		Change Since 1974	
	(1974)	(1995 to 2000)		(% Loss)	(% Loss / Year)
North Shore and Long Island Sound					
Alley Pond Park/Flushing Bay	18	17	1999	3	0.1
Manhasset Bay	25	9	1994	60	3.0
Marshlands Conservancy	35	24	2000	31	1.2
Pelham Bay Park					
Hutchinson River near Coop City	51	28	1999	45	1.8
Orchard Beach/City Island	77	51	1999	33	1.3
Stony Brook Harbor Area	299	190	1999	36	1.5
Udalls Cove Park	20	13	1999	38	1.5
South Shore					
Jamaica Bay	1969	1223	1999	38	1.5
Oyster Bay Area	1300	1016	1998	22	0.9
Shinnecock Bay–Islands only	30	17	1995	40	1.9

Note: Coastal marsh acreage in New York State observed in 1974 and the late 20th century, and percentage change in area. Source: (Hartig et al., 2002, 2004; Fallon and Mushacke, 1996; and NYSDEC, 2004)

Table 5.5 Reductions in extent of vegetated salt marshes, NY, between 1974 and 2000

More recently, Mushacke (NYSDEC, 2004) has documented additional marsh loss on the north and south shores of Long Island. Multiple factors, including dredging, bulkheading, and excessive nutrient enrichment, may be dominant at these sites. The New York State Department of State (NYSDOS), under the Coastal Zone Management Act, is also recording coastal wetland changes on Long Island's south shore. In the areas examined, there were losses as well as gains, although gains were not enough to compensate for losses (Jeffrey Zappieri, personal communication, 2003).

Marsh loss in Long Island and New York City has been documented through GIS analysis of historic aerial photographs (**Table 5.5**). By comparing 1974 images with those from between 1994 and 2000, a percent loss per year was derived. Marsh losses over the period were mainly between 1 and 2 percent per year. These losses were unexpected prior to the analysis, and the exact causes have yet to be determined. While sea level rise is among several stressors that may be acting together on vulnerable marshes, it may become the dominant factor in future decades as it outpaces sedimentation and vertical accretion.

Table 5.6 indicates the rate of sea level rise according to local tide gauges, the oldest of which was installed in 1856 and is located at the Battery in New York City. According to this tide gauge, sea level rise has been approximately 0.109 inch per year, or almost 1 foot per century over this period. In order for marshes to be sustainable, they need to at least keep pace with sea level rise. As indicated in **Table 5.5**, many marshes are already not keeping pace and are receding. The wetland loss rate, for example, at Jamaica Bay, Udalls Cove Preserve, and Stony Brook Harbor is 1.5 percent per year.

Station	Sea Level Rise (inches/yr)	Sea Level Rise (ft/century)	Record Length (years)
Bridgeport, CT	0.101	0.84	1964–2006
New London, CT	0.089	0.74	1938–2006
Montauk, NY	0.109	0.90	1947–2006
New York City, NY	0.109	0.90	1856–2006
Port Jefferson, NY	0.096	0.80	1957–1992
Willets Point, NY	0.093	0.78	1931–2006
Atlantic City, NJ	0.157	1.31	1911–2006
Cape May, NJ	0.160	1.33	1965–2006
Sandy Hook, NJ	0.154	1.28	1932–2006

Source: http://co-ops.nos.noaa.gov/sltrends/sltrends.shtml

Table 5.6 Relative sea level rise for the Atlantic coastal areas of New York, New Jersey and Connecticut according to local tide gauges

Table 5.7 gives known accretion rates for different marshes together with the rate of sea level rise according to the nearest tide gauge station. This provides a first-order guide in assessing the ability of these particular marshes to keep pace with increasing water levels. It should be noted that this analysis does not include subsurface compaction, which may be occurring.

Additional measurements are being taken at some marshes using sediment elevation tables (SETs) together with marker horizons. At Fire Island National Seashore SETs were placed together with feldspar markers to measure both shallow subsidence and accretion at the surface (Roman et al., 2007). As measured over a five-year period at three different marsh locations, there was a net loss of elevation. At two of the locations, surface accretion was greater than the rate of sea level rise; nevertheless, the rate of accretion was not enough to compensate for the overall land subsidence plus sea level rise. Likewise, in Jamaica Bay, initial data analysis indicates that while accretion is occurring beyond the rate of sea level rise, low marsh-dominated areas are subsiding at accelerated rates (Elders Point Marsh), while high marsh areas are experiencing more minimal loss rates (JoCo Marsh) (Jim Lynch, personal communication, 2009). As described in Case Study C: Salt Marsh Change at New York City Parks and Implications of Accelerated Sea Level Rise, Elders Point Marsh is being supplied with sediment supplements to raise the marsh elevation artificially.

The impacts of GCM-based and rapid ice-melt sea level rise scenarios on tidal wetlands for the 2020s, 2050s, and 2080s (**Table 5.1**) are evaluated for Long Island, New York City and Lower Hudson Valley, and separately for the Mid-Hudson Valley and Capital Region (**Table 5.8**). A sensitivity study was performed in order to evaluate the ability of marshes to keep pace with sea level rise alone (not accounting for subsurface

State	Marsh Zone	Accretion Rate (in/yr)	SLR (in/yr)	Source
Alley Pond (Queens, NY)	high	0.14	0.09	Cochran et al. (1998)
Caumsett Park (Nassau, NY)	high	0.16	0.09	Cochran et al. (1998)
Goose Creek (Bronx, NY)	high	0.09	0.09	Cochran et al. (1998)
Hunter Island (Bronx, NY)	high	0.04	0.09	Cochran et al. (1998)
Jamaica Bay (Queens, NY)		0.11-0.17	0.11	Kolker (2005)
Jamaica Bay (Queens, NY)	high	0.2	0.11	Zeppie (1977)
Jamaica Bay (Queens, NY)	low	0.31	0.11	Zeppie (1977)
Stony Brook, Youngs Island (Suffolk, NY)	high to low	0.09-0.11	0.09	Cademartori (2000)
Stony Brook, Youngs Island (Suffolk, NY)		0.14-0.19	0.09	Cochran et al. (1998)

Note: Lead 210 was used as method for determining accretion rates from soil cores. As SET/marker horizon data become available marsh accretion rates will be accompanied by rate of subsurface subsidence to determine change in net marsh elevation.

Table 5.7 Surface accretion rates measured in the salt marshes of the New York city region compared with the mean rate of sea level rise

Decade	2020s			2050s			2080s		
Accretion Rate (in)	L	M	H	L	M	H	L	M	H
Lower Hudson Valley, New York City & Long Island									
Central Range									
lower range	-0.50	**1.00**	**2.50**	-2.50	**2.00**	**6.50**	-4.50	**3.00**	**10.50**
upper range	-3.50	-2.00	-0.50	-7.50	-3.00	**1.50**	-15.50	-8.00	-0.50
Rapid Ice-Melt									
lower range	-3.50	-2.00	-0.5	-14.50	-10.00	-5.50	-33.50	-26.00	-18.50
upper range	-8.50	-7.00	-5.50	-24.50	-20.00	-15.50	-47.50	-40.00	-32.50
Mid-Hudson Valley & Capital Region									
Central Range									
lower range	**0.50**	**2.00**	**3.50**	-0.50	**4.00**	**8.50**	-0.50	**7.00**	**14.50**
upper range	-2.50	-1.00	**0.50**	-4.50	0.00	**4.50**	-10.50	-3.00	**4.50**
Rapid Ice-Melt									
lower range	-2.50	-1.00	**0.50**	-12.50	-8.00	-3.50	-29.50	-22.00	-14.50
upper range	-7.50	-6.00	-4.50	-21.50	-17.00	-12.50	-42.50	-35.00	-27.50

Note: L = Low (0.1 inch/yr), M = Medium (0.2 inch/yr), and H= High (0.3 inch/yr) accretion rates. Negative numbers indicate drowned marshes; positive numbers in bold indicate marsh survival. This simple model accounts only for sea level rise and accretion; subsurface compaction, subsidence, and other potential causes of marsh loss are neglected. Accretion calculated from 2010 to 2020s (15 years); 2010 to 2050s (45 years); 2010 to 2080s (75 years). Numbers represent the difference in assumed accretion rates and rates of sea level rise from Table 5.1.

Table 5.8 Chances for marsh survival given projected sea level rise (inches) and low, medium, and high rates of accretion for the 2020s, 2050s, and 2080s

compaction). The range of sea level rise projections from **Table 5.1** is used in combination with observed low (0.1 inch per year), medium (0.2 inch per year), and high (0.3 inch per year) accretion rates to offer a simple model of marsh survival in the coming decades. In order to construct the projections, each accretion rate was multiplied by the time span from the year 2010 to the mid-point of each decade (i.e., 2025, 2055, and 2085), and the accumulated accreted elevation is compared with the projected sea level rise. If sea level is greater than the potential accreted marsh, the marsh is considered drowned (negative numbers). If the potential accreted marsh has an elevation at least as great as the projected sea level, the marsh survives (positive numbers).

For Long Island, New York City, and Lower Hudson Valley, results indicate that only given medium or high accretion rates would marshes survive under the central range of projected sea level rise by the 2020s. None would survive the rapid ice-melt scenario. According to the analysis, the marshes have a slightly better chance in the Mid-Hudson Valley and Capital Region, with more marshes surviving in the GCM-based scenarios. In addition, marshes could survive a rapid ice-melt scenario only until the 2020s if they had a high-enough marsh accretion rate.

The marshes of New York State provide a dense urban population close access to a natural ecosystem rich in wildlife and recreation opportunities, function as buffering protection against coastal storm damage, furnish important habitat for migratory birds along the Atlantic Flyway, and act as productive nurseries for local fisheries (Wells, 1998). Losing marsh acreage to submergence at a rate of 1 percent and more per year compromises their function and value. Further research is needed to improve our understanding of the underlying causes of marsh loss in New York State in order to be able to conserve these marshes for future generations.

5.3.5 Fish and Shellfish Populations

New York has an extensive marine coastline, composed of Long Island Sound, all of Long Island's south shore, and portions of New York City. The lower Hudson River also is saline, grading toward freshwater northward. New York's marine waters lie in the northern portion of the Virginian Zoogeographic Province (Cape Cod to Cape Hatteras). This province is situated between the colder Acadian and the warmer Carolinian provinces. Fish diversity in New York's marine waters is high. Briggs and Waldman (2002) list 326 recorded marine and estuarine species.

The Marine Environment of New York

The high biodiversity in New York State marine waters is due in part to the great variety of habitats (e.g., estuaries, coastal bays, tidal straits, ocean beaches, continental shelf) and to the pronounced seasonal temperature changes that occur (Briggs and Waldman, 2002). The inner New York Bight has a range of about 25°C between summer and winter surface temperatures (from 1°C to 26°C) and bottom temperatures vary from a maximum of about 21°C in summer to less than 1°C in winter. Although many temperature-tolerant fish occur in New York waters year-round, these large seasonal temperature changes favor migratory rather than sedentary fish fauna (Grosslein and Azarovitz, 1982).

A relevant ecological feature of the lower Hudson River and Long Island's south shore bays is that they receive warm waters as eddies that pinch off from Gulf Stream meanders. These warm-core rings, approximately 100 kilometers in diameter, also carry early life stages of "tropical" fish that mature in New York waters through summer and early autumn. Although species composition varies, this is an annual phenomenon, and regularly includes groupers (*Epinephelus*), snappers (*Lutjanus*), butterflyfishes (*Chaetodon*), and jacks (*Carangidae*).

New York also lies at the juncture of the commercially viable ranges of two important crustaceans: American lobster and blue claw crab. Lobster is a cold water species that is harvested in numbers as far south as Long Island Sound and the New York Bight. Blue claw crab is a southern species that is abundant as far north as Long Island Sound.

To date, little attention has been focused on the biological effects of sea temperature changes in New York waters. Circumstantial evidence indicates that some faunal shifts already have occurred, most notably the extirpation late in the 20th century of a boreal species, rainbow smelt, in the Hudson River (Waldman, 2006) and in tributary streams to Long Island Sound. Another cold water species, Atlantic tomcod, also is showing declines in the Hudson River (Waldman, 2006) and is rare or absent in other former New York habitats. In contrast, a euryhaline (tolerates fresh to salt waters) species, gizzard shad, once found north only to Sandy Hook, has since the 1970s colonized the Hudson River and has become established as far north as the Merrimack River, Massachusetts.

Predicted Effects of Climate Change on Temperatures of New York Waters

Water temperatures in the Hudson River already have shown substantial warming. Although their data from Poughkeepsie do not include the past 20 years, Ashizawa and Cole (1994) found a statistically significant trend of 0.22°F per decade between 1920 and 1990, a change they believed was consistent with global increases.

For the New York Bight, our forecasted sea surface temperature changes for the 2050s in comparison with a 1980s baseline derived from the CCSM and GFDL models under two emissions scenarios all show substantial increases for its near-shore waters. These increases are on the order of 1.8 to 3.2°F, depending on the model and emissions combination. Visual inspection of sea surface isotherms from the mid-1900s at 1°F resolution (Fuglister, 1947) indicates that differences of these magnitudes between Long Island and warmer waters to its south correspond geographically with points between the southern tip of the Delmarva Peninsula and Delaware Bay, varying by month. Thus, the present-day fish community of the Delaware coast provides a glimpse of what the fish community of New York may resemble in the 2050s.

Likely Responses of Fish and Shellfish to Temperature Changes in New York Waters

There is considerable overlap in marine fish communities between the Delmarva region and New York, but there also are differences. Warm-water fish frequently seen in this southern region (Hildebrand and Schroeder, 1928) that are only rarely observed in New York include tarpon, cobia, and cownose ray. A higher-order difference is the greater prominence of members of the drum family (Sciaenidae) rarely seen in New York, including croaker, spotted seatrout, and red drum. The Delmarva region also does not support inshore winter fisheries for gadoids seen in New York Bight waters in cold months, such as Atlantic cod, pollock, silver hake, and squirrel hake.

Among important macrocrustaceans, blue claw crabs flourish in the warmer waters of the mid-Atlantic and should not decline because of higher temperatures. However, lobsters are at the southern edge of their inshore range in New York and have already shown declines that may be linked to warming waters (Howell et al., 2005).

Other fish whose northern ranges have extended to New York in the past include black drum and sheepshead. Both were recreationally and commercially harvested in New York Harbor and New York Bight waters in the 1800s but have been exceedingly scarce since. The reason may be habitat loss: Both are closely associated with oyster reefs, which declined sharply at the same time (Waldman et al., 2006). Both warming waters and increasing numbers of oysters (naturally occurring and through restoration projects) may result in increased abundances of these fish in New York.

A difficulty in discerning climate-driven changes in marine fish distributions is that the signal from the climatic effects may be highly confounded by other factors. Even under nearly constant environmental conditions, fish distributions are not static. Population theory and observations indicate that fish populations occupy the most optimal habitats under low abundances but also disperse into less optimal habitats at high abundances (MacCall, 1990). This means that mainly mid-Atlantic species that are only rarely or periodically seen in numbers in New York waters may occur there largely as a function of density dependence (relative population size within an area) and not because of favorable temperatures. Primarily southerly fishes that have appeared in New York during high population abundances include spot and Spanish mackerel (Waldman et al., 2006). Bluefish and weakfish are two other economically important fishes that are numerous in New York waters only during periods of high coast-wide abundances.

Another source of complexity is changes in fish and crustacean communities that occur because of ecological regime shifts, of which climate change may be a major driver. At nearly the same latitude as Long Island Sound in the waters of Rhode Island, Oviatt (2004) showed that modest increases in water temperatures caused large ecological shifts, in which macrocrustaceans (e.g., crabs, lobster) and southern pelagic fish (e.g., bay anchovy, butterfish) were favored at the expense of boreal demersal fishes (e.g., winter flounder, red hake).

A challenge in assessing changes in New York's marine fish community in the future will be to parse the effects of climate change from the normal seasonal and density-dependent vagaries of fish population dynamics. Annual long-term monitoring of fish and macrocrustaceans is critical to detecting climate-associated faunal changes in New York's marine waters. Both tracking fish community assemblages over time and observing the annual abundances of certain key species that are on the edges of their northern or southern distributions are important.

Impact of Increased CO_2 Concentrations and Ocean Acidification in New York Waters

The ocean is becoming more acidic as increasing atmospheric carbon dioxide is absorbed at the sea surface. Models and measurements suggest that surface pH has decreased by 0.1 pH unit since 1750 (Bindoff et al. 2007). It has been estimated that approximately half of the increased carbon dioxide emissions due to burning of fossil fuels since the Industrial Revolution has been absorbed in the ocean's surface waters (Sabine et al. 2004). However, continued acidification will reduce the ability of the ocean to take up atmospheric CO_2 and have potential negative impacts on finfish, shellfish, and plankton populations.

Much of the early research has been focused on calcifiers, which are believed to be most vulnerable during early developmental and reproductive stages of their life cycles. Kurihara (2008) notes that ocean acidification has negative impacts on the fertilization, cleavage, larva, settlement, and reproductive stages of several marine calcifiers, including echinoderm, bivalve, coral, and crustacean species. In addition, this research suggests that future changes in ocean acidity will potentially impact the population size and dynamics as well as the community structure of these species, influencing the overall health of marine ecosystems.

For New York coastal water, the relatively minor increases in ocean acidity brought about by high levels of carbon dioxide are likely to have significant detrimental effects on the growth, development, and survival of hard clams, bay scallops, and Eastern oysters (Talmage and Gobler, 2009). Recent research has shown that the larval stages of these shellfish species are extremely sensitive to enhanced levels of carbon dioxide in seawater; under carbon dioxide concentrations estimated to occur later this century, clam and scallop larvae showed a more than 50 percent decline in survival (Talmage and Gobler, 2009). These larvae were also smaller and took longer to develop into the juvenile stage. Oysters also grew more slowly at this level of carbon dioxide, but their survival was only diminished at carbon dioxide levels expected next century. The more time these organisms spend in the water column, the greater their risk of being eaten by a predator. A small change in the timing of the larval development could have a large effect on the number of larvae that survive to the juvenile stage and could dramatically alter the composition of the entire population (Talmage and Gobler, 2009).

Although it appears that fish are able to maintain their oxygen consumption under elevated carbon dioxide levels, the impacts of prolonged CO_2 exposure on reproduction, early development, growth, and behavior of marine fish are important areas that need urgent investigation (Ishimatsu et al., 2008). Changes in ocean chemistry might also affect marine food webs and biogeochemical cycles but are less certain because of their complexity (Haugan et al., 2006). Important global biogeochemical cycles (e.g., of carbon, nutrients, and sulfur) and ecosystem processes (changes in community structure and biodiversity) other than calcification may be vulnerable to future changes in carbonate chemistry and to declining pH.

5.4 Adaptation Strategies

As beaches retreat, wetlands disappear, and storm damage becomes more severe, coastal development and infrastructure will face increasing threats, regional tourism and fishing industries could suffer, and the insurance industry will increasingly be called upon to buffer economic losses (Frumhoff et al., 2007). Communities and industries must be able to adapt to these changes over the long term, in a manner that is economically, socially, and environmentally sustainable.

5.4.1 Adaptations for Key Vulnerabilities

It is difficult to determine an effective course of action, since natural processes within these dynamic systems operate on different time scales and are poorly understood. Implementation of adaptation strategies is further complicated by the division of power and jurisdiction in the coastal zone between various levels of government and different agencies. Regional-scale adaptation strategies presented in the ClimAID report assume that the legal and institutional changes necessary for implementation can be achieved. However, there will likely be competing and/or conflicting adaptation strategies depending on the objective. This will require a public process to achieve resolution. This section introduces some basic adaptation strategies and frameworks for evaluating the most effective methods to reduce vulnerability.

Coastal Storms, Coastal Floods, and Coastal Erosion

For coastal flooding and storm damage reduction, regional adaptation strategies will depend on economic, social, and environmental factors such as the desired level of protection, level of development, presence of critical infrastructure and natural resources, and consequences to the environment and neighboring communities. An example framework for evaluating possible adaptation strategies from the perspective of storm damage reduction to infrastructure is provided in **Table 5.9**. For example, beach nourishment (addition of sand from offshore or inland areas) is often used to protect coastal communities from flooding, and the level of protection depends on the design criteria, which are often constrained by financial resources and stakeholder/sponsor requirements. Sand can be placed on beaches relatively quickly (less than one year) and the projects are usually designed to last around five years. The actual life of the project depends strongly on the rate of erosion, which is associated with storm activity. As the rate of sea level rise increases, the rates of erosion increase and sand placement projects will become more expensive. Approximately 1 million cubic yards of sand are placed on New York beaches each year (Lynn Bocamazzo, personal communication, 2009). It has been estimated that the additional sand volume needed to compensate for sea level rise could range between 2.3 percent and 11.5 percent of the total current placement for the 2020s and 18 percent to 26 percent by the 2050s (Gornitz and Couch, in Rosenzweig and Solecki, 2001). A substantial volume of suitable sand (approximately 10 billion cubic yards) is present on the continental shelf and could be mined for this purpose (Bliss et al., 2009).

Depending on the level of development, communities may choose to implement a slow retreat or phased withdrawal from the coast. This could entail the use of hard (e.g., seawalls, storm surge barriers, rip rap) and soft (e.g., beach nourishment and beach drainage, beach vegetation) engineering solutions as well as the adoption of more policy-based strategies. For example, coastal communities may periodically place sand on beaches or use seawalls and groins to protect critical infrastructure. Coastal development and storm damage could be reduced by re-evaluating the delineation of coastal erosion hazard areas, improving building codes to promote more storm-resistant structures, increasing

Adaptation Strategy	Level of Development	Level of Protection	Time Imp/Life*	Potential Consequences	SLR**	
					GCM	RIM
Beach Nourishment	All	Up to 1-in-100 yr flood	< 1-yr/ 3–7 yrs	Steepen profile, reduce overwash sediments to bay, habitat disruption	X	X
Moderate Engineering Solutions#	Urban to suburban	Up to 1-in-100 yr flood	2–5 yrs/ 20–50 yrs	Reduced littoral sediments creates downdrift erosion	X	
Macro Engineering Solutions##	Urban	> 1-in-100 yr flood level	10–15 yrs/ 75–150 yrs	Alter regional hydrodynamics, habitat disruption		X
Slow Retreat	Suburban to rural	NA	NA	Depends on strategies used	X	
Rapid Retreat	Suburban to rural	NA	NA	Loss of equity, decreased property values		X
Do Nothing	All	NA	NA	Catastrophic loss of property and natural resources	X	X

* Time necessary for implementation of adaptation measure and life expectancy of project. Estimates do not include the political/legal/scientific processes necessary for design and implementation.
** Sea level rise scenarios, GCM for central range and RIM for rapid ice-melt.
\# Moderate engineering structures such as seawalls, revetments, groins, and bulkheads.
\## Macro engineering structures such as storm surge barriers and dikes.

Table 5.9 Example adaptation strategy framework for flood-damage reduction

building setbacks, and implementing rolling easements in regions of new development. In addition, depending on the financial resources and land availability, communities may institute buyout or land swap programs to encourage migration out of flood-prone regions. These strategies could be coupled with re-establishment of natural shoreline habitats to promote tourism and ecosystem services.

For urbanized areas, phased withdrawal may not be possible and communities may choose to use micro- (e.g., bulkheads, groins, seawalls)and macro-engineering (e.g., storm surge barriers, system of levees and dikes) solutions to prolong the use of coastal properties and infrastructure. For example, lower Manhattan is home to such critical transportation infrastructure as FDR Drive, West Street, the West Side Highway, the Port Authority Trans-Hudson (PATH) tunnels linking Manhattan and New Jersey, and the Brooklyn-Battery auto tunnel entrance.

The 2010 report of New York City Panel on Climate Change (Rosenzweig and Solecki, 2010) recommends that sea level rise projections should be incorporated into regulatory maps of coastal areas, including FEMA Flood Insurance Rate Maps (FIRMs) and their A- and V- Zones, the SLOSH model, and the delineation of the Coastal Zone Boundary and Coastal Erosion Hazard Areas.

Currently concrete bulkheads and seawalls protect much of this region; however, higher projected ocean levels may mean that these structures will need additional protection. Rather than armor the coastline of New York City and the surrounding boroughs, it may be more cost effective to construct storm surge barriers and dikes. Initial hydrodynamic studies have explored the feasibility of such a project; however, the economic, social, and environmental impacts have not been assessed (Bowman et al., 2004). Regardless of the adaptation strategy a community chooses to institute, a strong public outreach component should be undertaken for successful implementation.

Coastal Ecosystems

The framework presented in **Table 5.9** could be expanded to include various sustainable technologies or criteria for evaluation, or developed for other coastal components. For example, some possible adaptation strategies for saltwater wetlands are illustrated in **Table 5.10**. These strategies are applicable for a wide range of rates of sea level rise and therefore may not include sea level rise as a critical evaluation criterion.

5.4.2 Co-Benefits, Unintended Consequences, and Opportunities

Coastal systems are dynamic, and a loss of beach or habitat in one area may result in the gain or opportunity for expansion of habitats in another area. For example, increased beach erosion may lead to dune overwash and, in more severe cases, barrier island breaching and migration. Breaching and the creation of new inlets could adversely impact existing marshes but could also provide areas for the potential creation of new marshes under the right conditions. However, not enough is known about the sedimentary systems or budgets within the state's bays to make predictions about the relative importance of these features in terms of supplying inorganic sediments necessary for marsh maintenance and accretion.

Adaptation Strategy	Level of Develop-ment	Level of Protection	Time Imp/Life*	Potential Consequences
Wetland restoration	All	Appropriate waterfront sites where tidal inundation can be restored	2–5 yrs/ 10–100 yrs	Convert former wetlands along the waterfront that had been used for other land uses
Wetland creation	All	In newly established flood zones as SLR continues	2–10 yrs/ 10–100 yrs	Where new flooding is occurring, convert upland sites along the waterfront to wetlands
Sediment augmentation in submerging marshes	All	Up to mean higher high water (MHHW) line	2–5 yrs/ 10–100 yrs	Sediment must be diverted/taken from elsewhere
Wetland regulations: Use maximum allowable buffer area to allow for inland migration of marshes	All	150–300 feet beyond the wetland boundary	5–10 yrs 10–40 yrs	Development community will challenge tighter controls on wetland adjacent area permits

* Time necessary for implementation of adaptation measure and life expectancy of project. Estimates do not include the political/legal/scientific processes necessary for design and implementation.

Table 5.10 Example strategy framework for wetland adaptation

Warmer waters will also be more hospitable to fish species that normally range near New York, but rarely reach it under current conditions. For example, blue claw crabs are becoming more abundant in New York waters, and this particular fishery should be enhanced as the climate changes.

5.5 Equity and Environmental Justice Considerations

As climate change progresses, New York State's coastal ecosystem will undergo physical, chemical, and biological transformations. These transformations will have uneven impacts on coastal residents and coastal communities.

5.5.1 Vulnerability

The coastal vulnerability index (CVI) provide a comprehensive summary of vulnerability arising from geologic and hydrodynamic processes; however, they do not incorporate socioeconomic interactions (Thieler and Hammar-Klose, 1999). A combination of both physical and social factors can provide some measure of coastal resilience of a population or region and begin to identify potential inequities.

Flooding and natural hazards can disproportionately impact certain socioeconomic groups, such as people of color and low-income communities (Wu et al., 2002; Fothergill et al., 1999). Often this is an expression of physical vulnerability, such as pre-Katrina New Orleans, where low-lying areas at risk of inundation were home largely to African Americans. Frequently, physical vulnerability to a hazard is compounded by intrinsic individual vulnerabilities—related to age and physical immobility, for example—as well as a host of contextual vulnerabilities that can surface in every phase from prevention to relief, recovery, and reconstruction (Morrow, 1999). Contextual vulnerabilities are frequently an expression of underlying socioeconomic inequities and barriers: Low-income communities are less likely to have access to a full range of preventative strategies, such as resources to fortify property, prepare emergency provisions, and acquire insurance (Morrow, 1999; Yarnal, 2007).

Discriminatory practices and policies—from insurance redlining to constrained transportation options—may create systematic barriers to communities of color (Wright and Bullard, 2007). Other groups, such as renters, may lack the proper incentives to make precautionary investments, while people who speak English as a second language may be particularly vulnerable to miscommunications about preventative strategies and risks (Fothergill et al., 1999). Even when the risks have been made clear, housing discrimination or lack of affordable options may prevent certain groups from accessing the full range of relocation options (Wright and Bullard, 2007).

Some subsets of women may also have particular vulnerabilities. In general, women earn lower average incomes than men, tend more than men to be single parents, and are more likely to perform the labor of childcare, housework, and caring for elderly family members (Root et al., 2000). During extreme events and post-disaster recovery, these burdens and responsibilities may manifest as a disproportionate amount of hardship, lost income, increased labor, and emotional stress related to family care (Bolin et al., 1998; Morrow, 1999).

Relief, recovery, and reconstruction efforts are often associated with unequal access to emergency and recovery loan assistance and inadequate resources for compensation of health and property losses. At a community or city level, planning for sea level rise and increased coastal flooding at this stage requires an inherently strong equity framework: Both real and perceived inequities have plagued rebuilding following past hurricanes and coastal storms, when victims have often found themselves confronted with pre-planned packages of redevelopment doled out in top-down fashion to a handful of influential corporations (Wright and Bullard, 2007). Furthermore, federal disaster funds often focus first on issues of critical regional connectivity (restoring major arteries and highways), which increases the likelihood that local jurisdictions with little capacity will have to take responsibility for the finer-grained service restoration, the scale at which inequities often play out.

Table 5.11 shows an estimated breakdown of the population living in the 100-year floodplain in New York City and Long Island. Population estimates were generated using data from the 2000 census aggregated at the block group level and weighted by the area of each block group located within FEMA's 100-year floodplain boundaries. These estimates likely underestimate both the current total population in the floodplain and the affected subpopulations.

While it is difficult to discern precisely how a 1-in-100-year storm event would impact the regional economy of Long Island and New York City, **Table 5.11** gives some indication of the stakes. It also offers a snapshot of fundamental regional differences in household economies and demographics. Estimated aggregate value of all owner-occupied housing located within the 100-year floodplain in ClimAID Region 4 (New York City and Long Island) topped $27.5 billion in the 2000 census (not adjusted for inflation). In contrast to the distribution of vulnerable renters, which is skewed heavily toward the urban centers of New York City, approximately half the regional value of owner-occupied housing is located in Nassau County. This graduated pattern of suburban homeownership and urban renting parallels differences in population density across the coastal region of Region 4. Of the more than 500,000 people estimated to reside in the 100-year floodplain, a majority lives in coastal New York City with density decreasing gradually across Nassau and Suffolk (**Figure 5.7**).

Many coastal communities on the south shore of Long Island are fairly affluent. Indeed, **Table 5.12** indicates that residents within FEMA's 100-year floodplain tend

	New York City	Nassau County	Suffolk County	Total
Population				
Total population	286,374	159,644	70,523	516,541
Over 65	41,305	24,188	9,882	75,375
Below poverty line	61,260	8,895	4,550	74,703
African American	72,559	7,932	2,013	82,504
Latino	64,447	13,652	4,745	82,844
Foreign born	77,036	21,542	6,114	104,691
Housing				
Occupied housing units	110,194	58,206	27,103	195,503
Renter occupied housing units	77,003	13,930	5,428	96,360
Aggregate value of owner-occupied housing (Millions)	$8,255	$13,342	$6,171	$27,768

Source: US Census 2000; authors' calculations as described above

Table 5.11 Profile of the population residing in the 100-year floodplain (ClimAID Region 4—New York City and Long Island)

	In Floodplain	Out of Floodplain
Median Income	$56,132	$48,551
Median housing value, owner occupied housing	$235,297	$229,149
Female head of households as percentage of total	20.2	26.5
% in poverty	12.8	17.1
% less than high school	18.7	25.5
% over 65	14.2	12.2
% African American	12.6	23.0
% Hispanic	15.4	22.1
% renter	41.5	54.1
% vacant housing	9.5	5.2
% foreign born	18.9	30.2

Source: U.S. Census 2000; weighting by authors' calculations

Table 5.12 Area-weighted characteristics of population in census block groups in and out of the 100-year floodplain for New York City and Suffolk and Nassau Counties

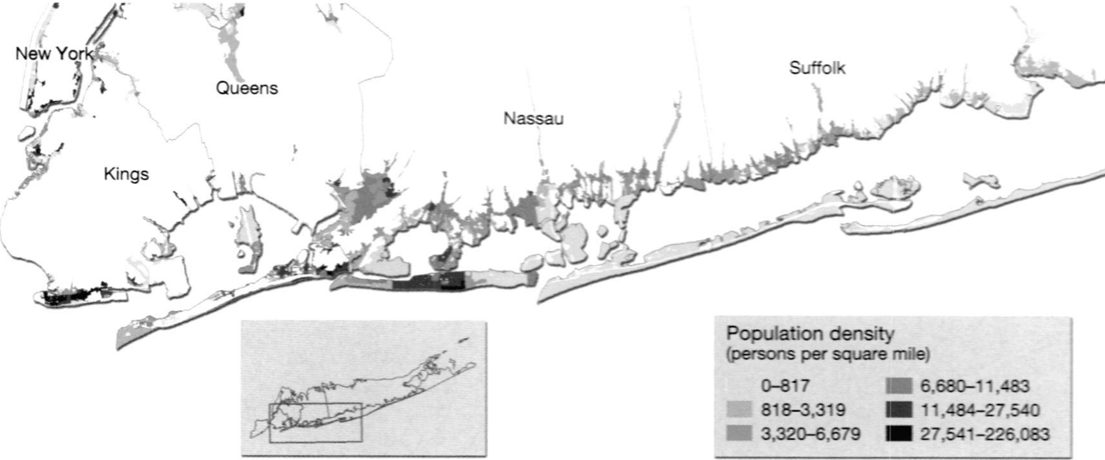

Figure 5.7 Population density within the FEMA 100-year floodplain boundaries

to have higher incomes, live in more expensive homes, and represent a lower minority population than those outside the floodplain. Examining the distribution of certain higher-risk subsets within this population can help locate potential environmental injustice effects. Low-income households, for example, are confronted with constrained resource options for both long-term adaptation and immediate coping (Wu et al., 2002). In the coastal floodplain of ClimAID Region 4, nearly 75,000 people live under the poverty line. More than 80 percent of this population resides in New York City. In New York City in particular, wealthier and poorer

Queens
Nassau
Suffolk
Kings
Population below the poverty line
1 Dot = approximate location of 50 residents living below the poverty line
FEMA 100-year flood zone
Source: US Census 2000

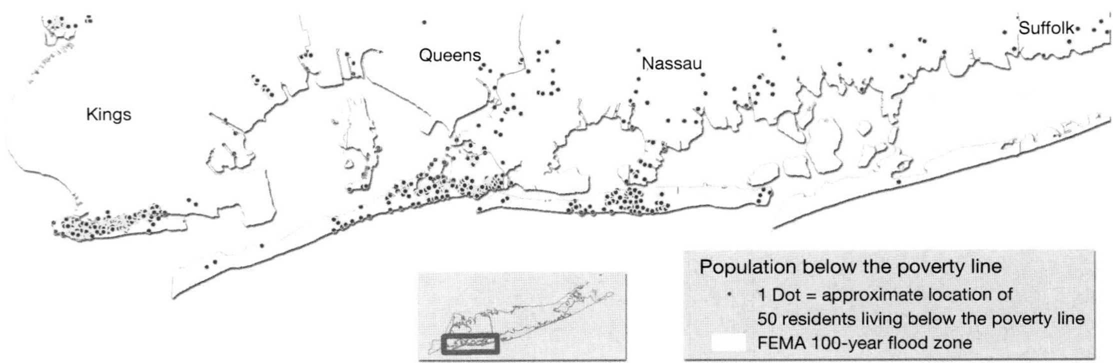

Figure 5.8 Population in FEMA's 100-year floodplain living below the poverty line

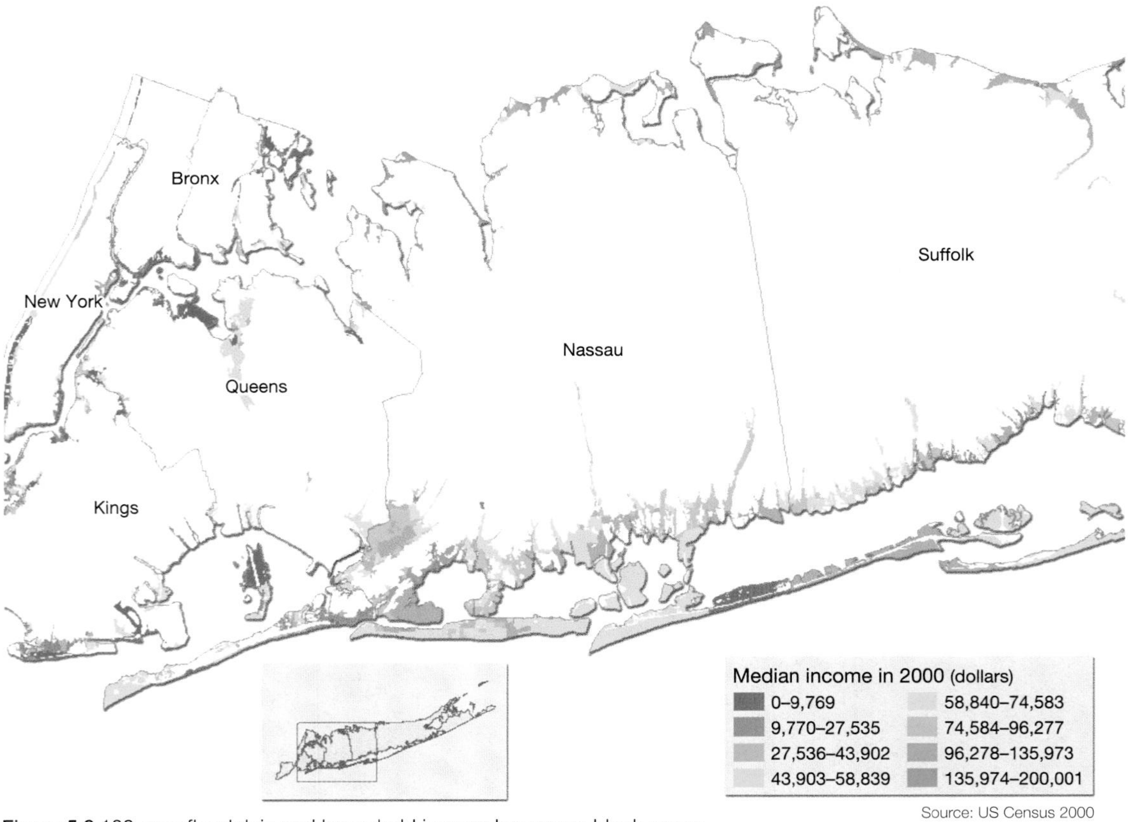

Bronx
Suffolk
New York
Nassau
Queens
Kings
Median income in 2000 (dollars)
0–9,769
9,770–27,535
27,536–43,902
43,903–58,839
58,840–74,583
74,584–96,277
96,278–135,973
135,974–200,001
Source: US Census 2000

Figure 5.9 100-year floodplain and household income by census block group

neighborhoods often co-exist in close proximity near the shore (e.g., Coney Island, Brighton Beach, the Rockaways) and are thus potentially equally exposed to the physical consequences of flooding from a major storm or hurricane (**Figure 5.8**). Equity issues may arise in the form of structural damage associated with variations in construction, ease of timely evacuation and availability of transportation, or the ability to recover after a storm.

Previous research also has suggested that racial and ethnic minorities are more vulnerable when exposed to similar events than non-minority populations (see for example, Fothergill et al., 1999). Hurricane Katrina provided a vivid reminder of this uneven burden in 2005 (Yarnal, 2007). In coastal New York City and Long Island, just over 82,000 African Americans and nearly the same number of Latinos live in the 100-year floodplain. Examination of **Table 5.11** suggests that African Americans and Latinos are significantly overrepresented in New York City's flood zone relative to the distribution of the total population in coastal New York, which likely reflects a legacy of suburban settlement patterns on Long Island (i.e., fewer minorities the further away from New York City). This population distribution, in combination with the disproportionately high concentration of poverty and the greater proportion of renters, suggests that New York City would face fundamentally different equity challenges than Nassau and Suffolk counties. In contrast, proportionally higher rates of home ownership and greater income may signal a measure of resilience across more wealthy regions of Long Island (**Figure 5.9**).

5.5.2 Equity Issues in Adaptation

Several alternative adaptation strategies—managed retreat, beach nourishment, and engineering solutions—have varying economic impacts. Earlier studies have shown that beach nourishment preserves the recreational values of coastal beaches, while engineering solutions may be needed to maintain fixed structures. Landry et al. (2003) and Kriesel et al. (2005) estimate the relative value (willingness to pay) for alternative adaptation strategies. Their basic finding is that the relative value of the three basic adaptation strategies is a function of the value of coastal property to be protected. Preemptive planning for flood security should evaluate the specific distributional burdens and benefits of each adaptation strategy. For illustrative purposes a few adaptation strategies are discussed in the following section, along with a review of critical equity issues.

Infrastructure

Building climate-secure hard infrastructure offers an amenity that may create new patterns of winners and losers. Which communities will be protected and in what ways? Who bears the cost of building hard infrastructure, such as seawalls or levees? Where are they placed and whom do they protect? What areas of a city or town are treated as critical while others are deemed non-priorities? These equity issues extend into strategies that include "softer" design. Choosing which wetlands to restore, beaches to fortify with additional sand, or structures and lands to elevate are not simple issues of exposure to risk. They involve making difficult decisions about distributing benefits and costs among communities and prioritizing some areas potentially at the expense of others.

Managed Relocation

Managed relocation from floodplains is another adaptation strategy that is accompanied by a portfolio of equity concerns related to the specific measures employed in the policies, from the relocation incentives to the environmental restoration of reclaimed lands. For example, if retreat will be a rapid buy out of highest risk areas, how does one choose these areas and the specific properties within them? What mechanisms are in place to hedge against the risk of redlining and inequitable selection of properties for priority buy-out?

Upland areas could be transformed by migration and localized population pressures. These communities may experience gentrification, increased cost of services from in-migration, and burdens of displacement from lowland areas. The viability and cohesion of low-income communities tend to be vulnerable under these conditions. Retreat from the southern coast of Long Island, for example, where housing issues are already a critical concern, may displace households, increase housing demand, and push up property values, a process that may indirectly burden the low-income population.

Managed retreat may materialize less as proactive planning and more as reactive incrementalism or planned obsolescence, such as service cutbacks, squeezing areas into shrinkage, or "choking" growth. In effect, such strategies outsource adaptation planning to individuals, meaning that those with the widest range of job and residence options and the ability to forecast policy changes would be the most quick to migrate. Lower-income populations could find themselves at an adaptive disadvantage, because they lack either the capital to invest in new housing or the socioeconomic flexibility allowing them to transfer jobs and livelihoods locations. Relocation can be difficult for any business, but minority-owned businesses may be especially vulnerable. They tend to be smaller, less well capitalized, less able to get loans, and subject to discrimination.

Adaptation, Insurance, and Equity

Existing risk-spreading mechanisms are in flux. The insurance industry is undergoing changes in the way it approaches insuring high-risk regions, and in some areas, such as parts of Long Island, companies have withdrawn their products, limiting insurance options for property owners (Insurance Journal, 2007). What options will be available in their place, and for what level of risk government will bear responsibility are long-range distributional issues. Coastal erosion and increasing inundation of areas with infrastructure will take their toll in repair, recovery, and replacement costs. Similarly, the need for increased emergency management has a price tag that has to be distributed. Whether monies are distributed as community-level preparedness programs versus federally budgeted disaster aid, for example, has different implications in terms of the distribution of responsibility, the allocation of labor, and differential resource burdens.

5.6 Conclusions

This section details the basic findings on vulnerabilities and opportunities associated with climate change for the coastal zone of New York State, summarizes potential adaptation measures, and identifies some critical knowledge gaps that hinder more effective planning efforts, as identified in the ClimAID assessment.

5.6.1 Main Findings on Vulnerabilities and Opportunities

Coastal Storms

- Because of the highly developed nature of the coast in New York State, a considerable portion of population, private property, and infrastructure will be potentially at risk of enhanced inundation and flooding due to sea level rise associated with climate change.
- While permanently lost land is projected to involve a relatively narrow coastal strip by the 2080s, the higher storm surges associated with higher sea levels could periodically engulf a much greater area. Also, wave action will erode and reshape the shoreline, affecting the location and extent of storm surge inundation.

Coastal Floods and Recurrence Intervals

- Moderate flooding events may become more frequent. Sea level rise projections of 5, 12, and 23 inches at the Battery for the 2020s, 2050s, and 2080s would result in 4, 16, and 136 moderate flooding events annually, respectively. Under a rapid ice-melt scenario, New York State could experience between 200 and 275 moderate flooding events each year by the 2080s.
- As sea levels rise, coastal flooding associated with storms will very likely increase in intensity, frequency, and duration. By the end of the 21st century, sea level rise alone would increase the frequency of coastal flood levels that currently occur on average once per decade to once every one to three years for New York City and once every one to two years for Westhampton Beach, with water levels associated with these events increasing by 1 to 2 feet. The more severe current 1-in-100-year event may occur on average approximately four times as often by the end of the century.
- The greater likelihood of coastal flooding (as well as heavier rainfall) would result in an increase in street, basement, and sewer flooding; an increase in flood risk to low-elevation transportation, energy, and communications infrastructure; more frequent delays on low-lying highways and public transportation; increased structural damage and saltwater exposure to infrastructure, commercial, and residential property; increased inflow of seawater to storm sewers and wastewater treatment

plants and reduced ability of gravity discharge of sewer-effluent overflows; encroachment of saltwater into freshwater sources and ecosystems; and increased beach erosion and sand placement needs.

- Sea level rise and coastal inundation could affect the heights of tide gates designed to prevent the inflow of seawater and backing up of outfall sewers.

Coastal Erosion

- Waves, currents, and tides constantly reshape the shoreline, and as sea level rise accelerates, these forces have the potential to dramatically alter New York State's coast.
- Sea level rise may increase the frequency of wave attack at the base of glacial bluffs, accelerating recession and erosion and increasing the rate of supply of sediment and other material to the near-shore system. This change in sediment supply could have different impacts that are difficult to predict given the limited understanding of sediment budgets and transport pathways.
- Shoreline change rates measured over the last 100 years show that generally the shoreline is receding at relatively low average rates of 1 to 2 feet per year, but that some areas are stable and others are accreting.
- Accelerated sea level rise will tend to exacerbate barrier island erosion. At low-to-moderate increases, the effects of sea level rise will still be of lesser magnitude than storm events and disruptions in longshore sediment transport. At higher rates of projected sea level rise, the migration of barrier islands landward should accelerate, but this migration may not be initiated in some sections of the barrier island for hundreds of years. At the most extreme rates of increased rate of rise, the barrier islands may not be able to maintain themselves if sea level rise outpaces the ability of the system to supply sediment naturally

The Hudson River Estuary

- Climate change could affect the location of the salt front of the Hudson River Estuary in three ways: 1) reduction in precipitation could reduce stream flow, allowing the salt front to move upstream, while an increase in precipitation would have the opposite effect; 2) increase in temperature could increase evaporation, reducing freshwater runoff, which in turn would cause the salt front to migrate upstream; and 3) rising sea level may push the mean position of the salt front upstream.
- Sea level rise and storm surge will continue to affect the entire Hudson Estuary up to the dam at Troy.

Freshwater Resources

- Sea level rise and changes in precipitation will cause Long Island's water table to rise at approximately the same rate as sea level. As a result, the geographic extent of ponds and wetlands could change along with the carrying capacity of streams.
- Saltwater entry into aquifers would be slow, with less than 1 percent of the freshwater reserves affected over 100 years.

Coastal Ecosystems

- Coastal ecosystems are at risk from rising sea levels, which may impose additional stress and exacerbate wetland losses in some sensitive regions.
- At Jamaica Bay Gateway National Recreation Area, New York, island salt marsh area declined by 20 percent between the mid-1920s and mid-1970s. Since then this trend has accelerated, and close to 30 percent has subsequently been lost. The wetland loss rate at Jamaica Bay, Udalls Cove Preserve, and Stony Brook Harbor is 1.5 percent per year. Only 7 out of 13 salt marsh islands in Shinnecock Bay (southeastern Long Island) that were present in 1974 remained by 1994. Since the 1970s, rate of march losses on Long Island and New York City have been between 1 and 2 percent per year. According to the tide gauge at The Battery in New York City, sea level has been rising at a rate of approximately 1 foot per century. While sea level rise is among several stressors that may be acting together on vulnerable marshes, it may become the dominant factor in future decades as it outpaces sedimentation and vertical accretion.
- For Long Island, New York City, and Lower Hudson Valley, marshes would only survive under the central range of projected sea level rise by the 2020s, given medium or high accretion rates. None would survive the rapid ice-melt scenario. Marshes have a slightly better chance in the Mid-Hudson Valley and Capital Region, with more marshes surviving in the central-range set of scenarios and

some potentially surviving a rapid ice-melt scenario until the 2020s.

Fish and Shellfish Populations

- Water temperatures in the Hudson River already have shown substantial warming on the order of 0.22°F per decade between 1920 and 1990, which is projected to be consistent with global increases.
- For the New York Bight, projected sea surface temperature changes for the 2050s in comparison with a 1980s baseline show substantial increases on the order of 1.8 to 3.2°F, depending on the climate model and greenhouse gas emissions combination.
- Blue claw crabs flourish in the warmer waters of the mid-Atlantic and should not decline because of higher temperatures. However, lobsters are at the southern edge of their inshore range in New York and have already shown declines that may be linked to warming waters.
- Warming waters and increasing numbers of oysters (naturally occurring and through restoration projects) may result in increased abundances of black drum and sheepshead in New York.
- Warmer waters around New York will be more hospitable to fish species that normally range near New York but rarely reach it under current conditions.
- For New York coastal waters the relatively minor increases in ocean acidity brought about by high levels of carbon dioxide may have significant detrimental effects on the growth, development, and survival of hard clams, bay scallops, and Eastern oysters.

5.6.2 Adaptation Options

This section briefly introduces some basic adaptation options that could increase coastal community and ecosystem resilience to climate-induced hazards.

- Incorporate climate change and sea level rise information into State and local adaptation strategies and planning related to coastal land use, waterfront development, open space and natural habitat preservation, and emergency response and evacuation.
- Identify coastal area responses to sea level rise impacts at multiple timescales, such as more frequent and extensive storm flooding, areas of permanent inundation, land loss due to erosion, various wetland responses, barrier island migration and breaching, and the migration of the salt front in estuarine environments. Evaluate the level of risk to human and natural systems, infrastructure, and population in these areas to prioritize and guide risk reduction and adaptation responses.
- Compile a detailed inventory of shoreline assets located in at-risk areas, their elevations, and the design lifetime for all sectors of coastal communities throughout New York State.
- Acquire currently vacant shorefront property in high-risk areas to serve as buffer zones against coastal flooding and sea level rise. Re-zone these for low-density use, recreation, and/or potential wetlands migration.
- Encourage responsible shoreline development in view of increasing sea level rise and coastal storm risks by providing guidance, incentive programs, and financial assistance to localities and sectors most at risk.
- Develop tools such as flood maps to effectively communicate sea level rise risks and community vulnerability to decision makers and the public.
- Establish a network of stakeholders and volunteers to assist in monitoring for sea level rise impacts and to coordinate outreach and education efforts.
- Coordinate regional efforts to update and re-evaluate periodically the range of risks associated with sea level rise and coastal storms, and modify existing environmental regulations and permitting accordingly.

In addition, the New York State Sea Level Rise Task Force (SLRTF) offers a comprehensive set of policy recommendations to reduce vulnerability from sea level rise and coastal hazards. Its recommendations include legal and regulatory changes as well as strategies for developing funding mechanisms for research, monitoring, and adaptation, and an evaluation of the public health risks associated with sea level rise and coastal hazards (New York State Sea Level Rise Task Force, 2010).

5.6.3 Knowledge Gaps

Climate change assessment efforts are often limited by the level of understanding of natural processes (barrier island evolution, ecosystem functions, and interactions), the lack of spatial and temporal monitoring data, the availability and quality of existing data, and modeling capabilities. As new research methods emerge and scientific understanding of natural systems evolves, climate adaptation strategies and existing regulations (building codes, setbacks) should be reconsidered. Recommendations to improve assessment tools and understanding of natural processes operating in coastal regions specific to New York State are listed below.

- The responses of barrier islands and tidal wetlands to accelerated rates of sea level rise, such as the rapid ice-melt scenario, are currently unknown. Monitoring barrier island and tidal wetland evolution and determining the influence of regional geologic controls on their spatial variability would improve process-level understanding of these systems.
- Regional sediment management strategies require an understanding of transport processes along the coast as well as across the continental shelf. Presently the quantity of sand and processes by which it moves from the inner shelf to the littoral zones are unknown. This will influence the selection of "borrow" sites for sand that may be placed on beaches.
- Quantifying and monitoring land use and coastal water quality will help determine the most suitable land-use practices and adaptation strategies to improve coastal environments and increase resiliency.
- Assessment of ecosystem services for natural and engineered shorelines will aid in identifying potential adaptation strategies for more urbanized sections of the coastline.
- Establishing a monitoring program for submarine groundwater discharge throughout Long Island with particular focus on low-lying areas will allow tracking of the influence of submarine groundwater discharge on submerged aquatic vegetation.
- Systematic and standardized protocols (every two to five years) for all New York State coastal regions are needed. Mapping could include bathymetry, topography to the 500-year floodplain, and the extent of existing wetlands.
- Development of a comprehensive, easily accessed GIS-based data repository will facilitate interagency collaboration and future assessment efforts. The repository should include all monitored and modeled data, such as an inventory of hardened shorelines, land use, critical infrastructure, sea level rise rates, distribution of habitats and species, historic shorelines, and storm water level recurrence intervals.
- Hydrodynamic modeling capability for the Hudson River is required to investigate the effect of climate change on the position of the salt front and on nutrient loads associated with extreme precipitation events.
- Research on climate-related impacts and adaptation strategies for Great Lakes coastlines is critical, since these regions were not included in this assessment.

Case Study A. 1-in-100-year Flood and Environmental Justice

New York coastal communities are vulnerable to both tropical and extra-tropical storms. As the climate changes there is a potential for more-intense storm systems to impact New York State, and coupled with an accelerating rate, of sea level rise the likelihood of experiencing what is currently considered a 1-in-100-year event is increasing. The highly developed nature of the coast, the large population, and considerable private property and infrastructure at risk require society to develop holistic adaptive management strategies that promote community resilience. The implementation of various strategies will depend strongly on population and critical infrastructure density, as well as societal priorities. This particular ClimAID case study is focused on flood adaptation strategies for the urban and suburban regions of Long Beach (**Figure 5.10**) and communities along the mainland coastline of Great South Bay (**Figure 5.11**). In particular, a severe coastal storm consistent with the 1-in-100-year event (the theoretical storm that produces the 100-year floodplain) is considered for this analysis. The purpose is to illustrate where New York State and coastal communities may need to transition from phased withdrawal or managed relocation to fortification strategies, while highlighting community vulnerabilities associated with socioeconomic conditions. This case study suggests that managed relocation might be the appropriate strategy for agricultural or low-density residential land; engineering strategies might be required for urbanized lands; and an intermediate strategy (beach nourishment, for example), for moderate-density residential areas.

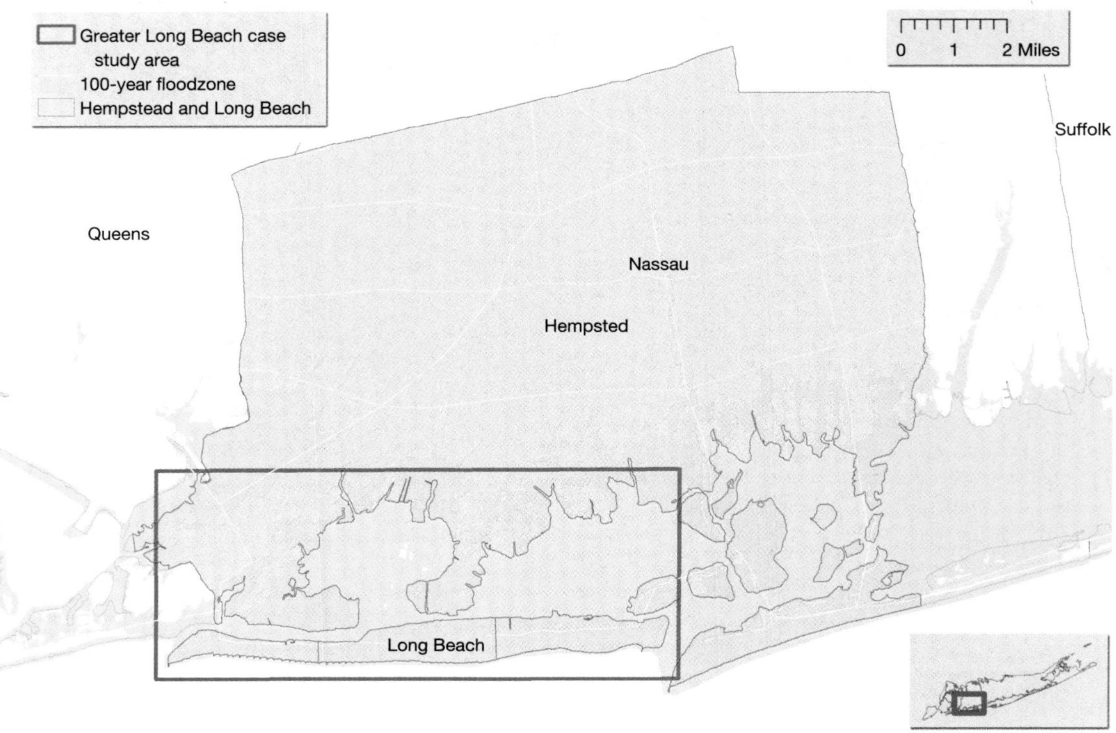

Figure 5.10 Long Beach and surrounding Bay communities

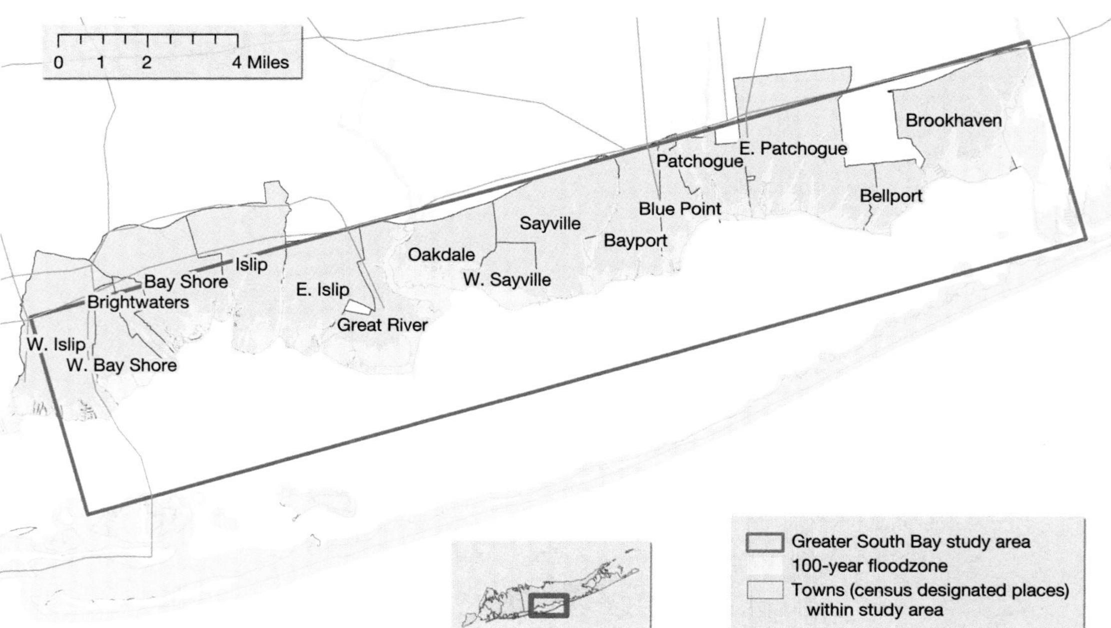

Figure 5.11 Mainland coast of Great South Bay

Social vulnerabilities are generally expressed at a more local or household level. Land use and coastal decision making is also done at a local scale. Still, one of the unique challenges of climate change is that it frequently is regional in exposure, so climate change adaptation strategies require a wider regional planning focus. Being attuned to who is excluded by the telescoping of scale and regionalization of focus will help make the planning process more inclusive, valid, and responsive.

Analysis of Vulnerability to Storm Events

A number of variables generally associated with vulnerability were chosen from the 2000 census at the census block group level.[1] A comparison was made of mean values in the 100-year floodplain and means outside the floodplain. For Long Beach, the 100-year floodplain was compared to block groups within the Town of Hempstead, as defined in Census 2000 as a Minor Civil Division (**Figure 5.10**). For Great South Bay, the present-day 100-year floodplain was compared to block groups falling within the Census Designated Places of the case study area: Bayport, Bay Shore, Bellport, Blue Point, Brightwaters, Brookhaven, East Islip, East Patchogue, Great River, Islip, Oakdale, Patchogue, Sayville, West Bay Shore, West Islip, and West Sayville (**Figure 5.11**).

In general, for both case study areas, differences between the populations in the floodplain and those outside were relatively small (**Tables 5.13** and **5.14**). However, median household incomes and the values of homes were slightly higher within the floodplain in both case studies, which likely reflects the amenity value of living by the coast. Key indicators of vulnerability or inequity in the distribution of burdens and benefits, such as race, poverty, and educational attainment, showed slight differences, but are not concentrated in flood-prone populations.

At a finer scale within the case study regions, there nevertheless is a wide range in the incidence of potential vulnerability from neighborhood to neighborhood. For example, **Figure 5.12** suggests that much of the disabled population in and around Long Beach is clustered in a few distinct locations. Patterns such as these may present opportunities for targeted emergency planning. Other social indicators, such as percent poverty, percent non-white, and the number of female-headed households, tend to cluster spatially and occur concurrently, which may indicate concentrated populations that are likely to be more sensitive to the impact of flood events (**Figure 5.13**, for example).

Social disparities are evident within Great South Bay as well. The highest rates of poverty and greatest proportion of renters, minorities, and foreign-born residents tend to center in Bay Shore and Patchogue (**Figures 5.14** and **5.15**). In general, the less densely populated areas between these centers are wealthier, better educated, and enjoy higher rates of home ownership.

	In Floodplain	Out of Floodplain
Median income	$75,653	$72,014
Median housing value, owner occupied housing	$287,262	$224,955
Female head of households as percentage of total	14.7	16.5
% in poverty	6.7	5.2
% less than high school	11.1	14.1
% over 65	14.7	14.2
% African American	5.4	15.1
% Hispanic	9.6	10.6
% renter	24.7	16.7
% vacant housing	5.7	1.8
% foreign born	14.1	17.4

Source: U.S. Census 2000; authors' calculations

Table 5.13 Characteristics of population in census block groups in and out of 100-year floodplain for the Long Beach case study area

	In Floodplain	Out of Floodplain
Median income	$73,323	$65,834
Median housing value, owner occupied housing	$238,480	$176,489
Female head of households as percentage of total	13.3	15.1
% in poverty	6.2	4.8
% less than high school	11.4	12.4
% over 65	14.1	12.0
% African American	2.5	4.7
% Hispanic	7.5	8.7
% renter	22	21.1
% vacant housing	5.4	2.8
% foreign born	8.4	7.4

Source: U.S. Census 2000; authors' calculations

Table 5.14 Characteristics of population in census block groups in and out of 100-year floodplain for the Great South Bay case study area

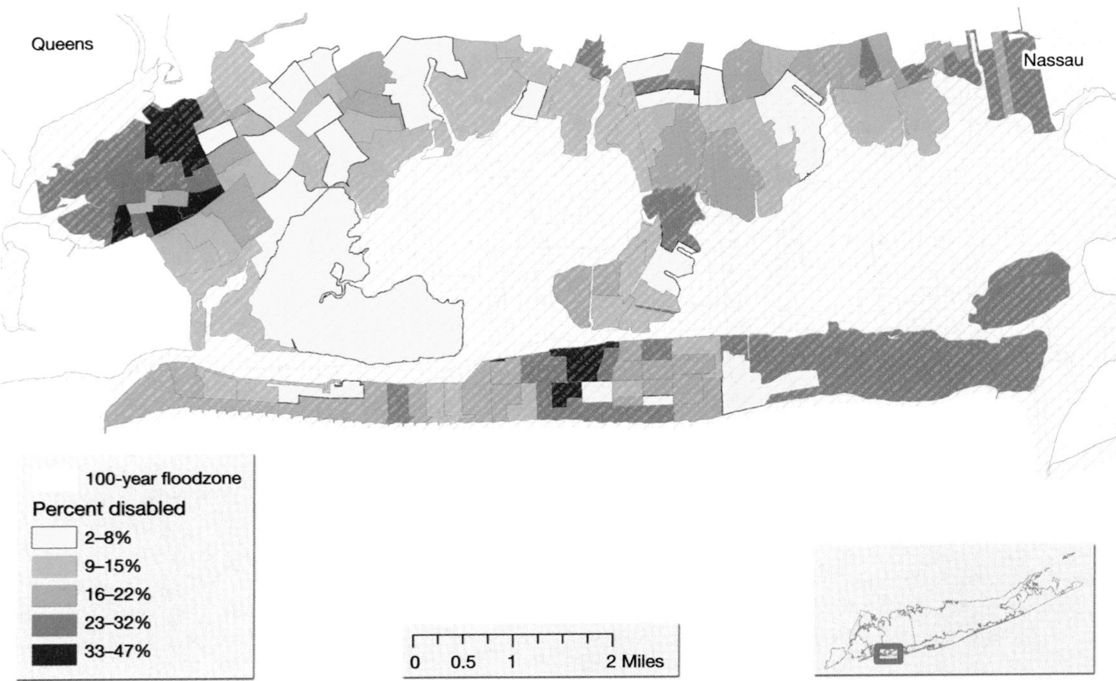

Figure 5.12 Concentration of disabled population in Long Beach region

Figure 5.13 Female-headed households in Long Beach region

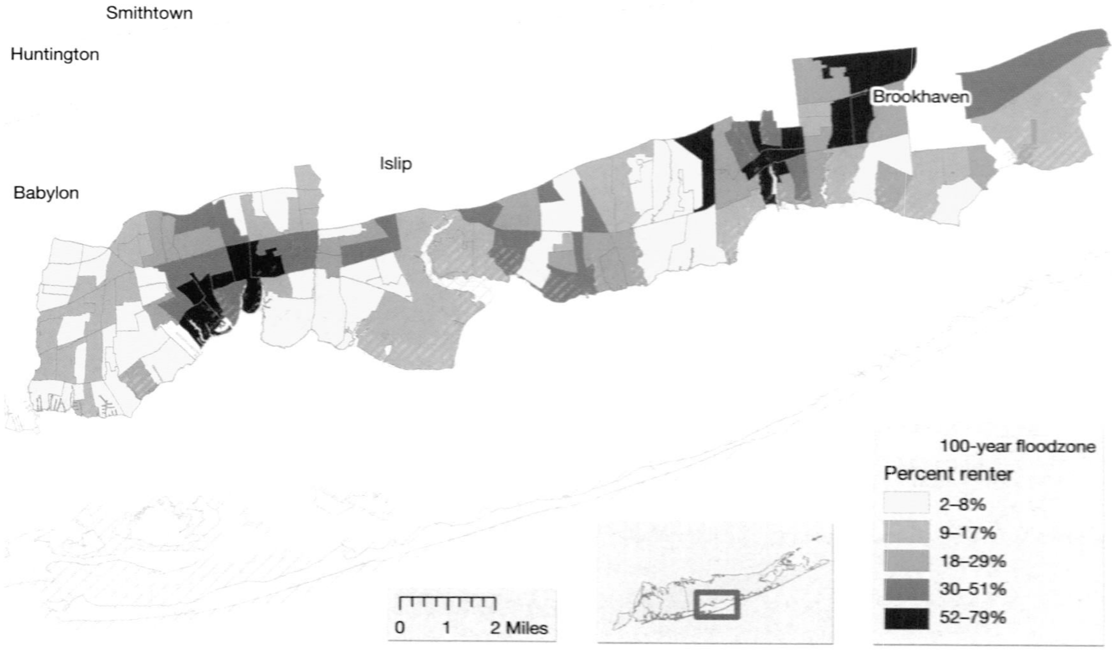

Figure 5.14 Local variation in density of renters in Great South Bay

Figure 5.15 Concentrated poverty in Great South Bay

Adaptation, Economic Impacts, and Distributional Inequities

Identifying and understanding how economic impacts associated with severe coastal storms will change temporally under different sea level rise scenarios is critical for developing effective adaptation and sustainable management strategies. For the Long Beach and Great South Bay study regions, two sea level rise scenarios were considered, the GCM-based central

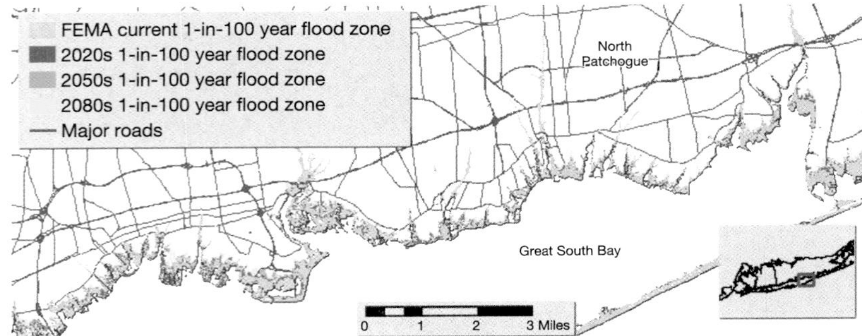

Figure 5.16 1-in-100-year flood zone for Great South Bay based on the GCM-based central range sea level rise scenario

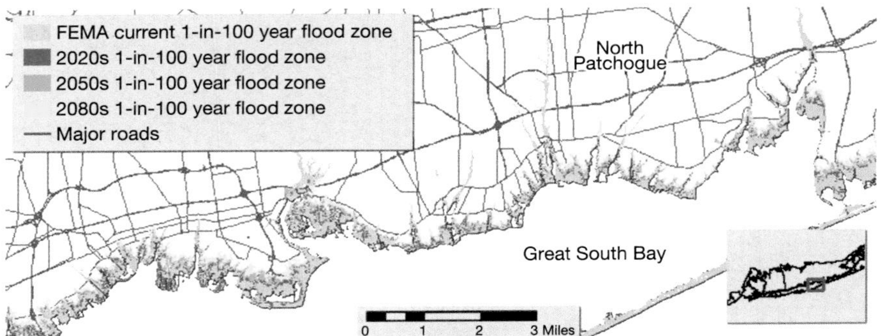

Figure 5.17 1-in-100-year flood zone for Great South Bay based on the rapid ice melt sea level rise scenario

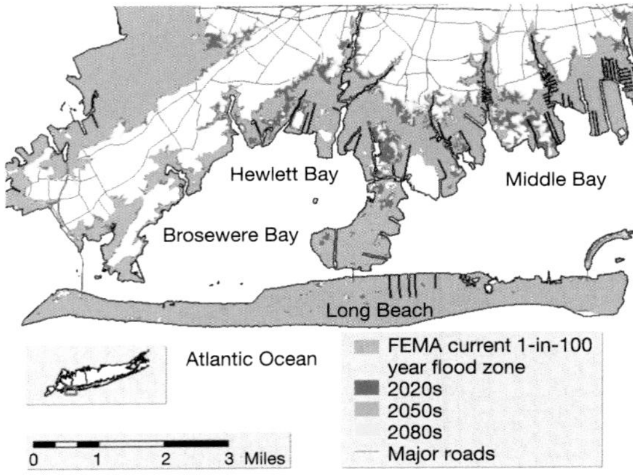

Figure 5.18 1-in-100-year flood zone for Long Beach based on the GCM-based central range sea level rise scenario

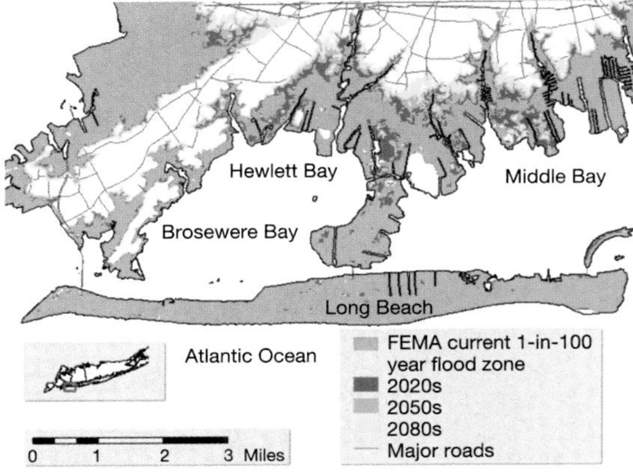

Figure 5.19 1-in-100-year flood zone for Long Beach based on the rapid-ice-melt sea level rise scenario

range forecast and a rapid ice-melt scenario. The GCM-based and rapid ice-melt scenarios are consistent with approximate 2-foot and 4-foot sea level rise by the 2080s, respectively. Corresponding 100-year floodplains for Long Beach and Great South Bay for each of the scenarios are shown in **Figures 5.16** through **5.19**.

Over the 2000–2080 forecast period, sea level rise is expected to place a growing population and increasing property at risk from flood and storm damage. These base-case analyses take as their starting point the 2000 U.S. Census estimates for population and property values within the study areas (U.S. Census Bureau Population Division, 2005). **Table 5.15** lists the base-case forecasts for the Long Beach and Great South Bay coastal regions.

Case Study B. Modeling Climate Change Impacts in the Hudson River Estuary

The Hudson River extends unimpeded from The Battery in New York Harbor north to the Federal Dam at Troy just above Albany. The river coastline is highly populated. As the river has become cleaner over the last several decades, development pressure along its shores has increased. The goal of this case study is to describe the relative impact of sea level rise, storm surge, and large precipitation events on estuary water levels using a publicly available hydrodynamic computer model to determine which of the impacts of climate change are likely to be of greater significance to the planners, regulators, and communities along the estuary shoreline.

Risk of Sea Level Rise: Long Beach Case Study Area			
	2020s	2050s	2080s
Population at risk			
GCM-based forecast	94,526	101,188	107,934
Rapid-ice-melt forecast	95,859	105,836	114,515
Property at risk (millions)			
GCM-based forecast	$6,266	$6,485	$6,739
Rapid-ice-melt forecast	$6,376	$6,814	$7,163
Risk of Sea Level Rise: Great South Bay Case Study Area			
	Year 2020	Year 2050	Year 2080
Population at risk			
GCM-based forecast	17,387	20,512	24,606
Rapid-ice-melt forecast	18,822	25,222	33,560
Property at risk (millions)			
GCM-based forecast	$1,159	$1,348	$1,585
Rapid-ice-melt forecast	$1,262	$1,669	$2,162

Table 5.15 Population and property at risk for GCM-based and rapid ice-melt scenarios

The National Oceanic and Atmospheric Administration (NOAA) currently makes short-term forecasts of water levels using predictions of tides and watershed inflows. Cornell researchers used a variant of the NOAA National Weather Service model that employs the U.S. Army Corps of Engineers software HEC-RAS (Hydrologic Engineering Center-River Analysis System) to predict water level rise due to conditions outside those normally addressed by NOAA. This ClimAID assessment included three scenarios: 1) a scenario with 2 or 4 feet of sea level rise on top of tidal fluctuations, 2) high freshwater inflow scenarios, and 3) a storm surge scenario. The study also considered the relative value of improved topographic data (bathymetry, land elevation, hydraulic channel characteristics, and tributary flows) for understanding the impacts of climate change.

Sea Level Rise and Impact on Tidal Range

Because of the low topographic gradient along the river, not counting the effect of land subsidence south of Kingston, a change in sea level at New York Harbor results in nearly the same change in water level at Albany. For example, a 3-foot increase in water level at The Battery would coincide with a roughly 3-foot increase in water levels at Albany. Thus, any change in mean water level due to sea level rise would be imposed upon the regular tidal fluctuation of 4 to 5 feet in Albany. Additionally, an increase in sea level of 2 or 4 feet will result in deeper water in the Hudson River estuary, allowing the estuary to better transmit tidal energy from The Battery to Albany. Model simulations suggested that this would effectively increase the tidal range at Albany by as much as 0.3 feet.

Large Rainfall Events

The high freshwater inflow scenario used the 2008 annual peak flow at the Troy Dam (an approximately 5-year return period flow) and found that only water levels in the uppermost part of the tidal river above Castleton-on-Hudson changed appreciably. This conclusion is corroborated by discharge data directly measured at different points on the river. A measured 2008 peak flow of 104,000 cubic feet per second at the Troy Dam (a measure of the majority of watershed inflows to the river) is only one-third the normal peak tidal flow in the Hudson near Poughkeepsie (300,000 cubic feet per second peak flow). The annual mean flow at Troy of 14,000 cubic feet per second is 20 times less than the 2008 peak flow. The tidal flow in the river is much larger than the largest freshwater inflows to the river. This suggests that changes in precipitation and resulting freshwater inflows to the river associated with climate change will have a more localized influence on the water levels.

Storm Surge

A trial storm surge scenario assumed a 10-foot increase in water level on top of normal tidal fluctuations at The Battery over a 36-hour period. Storm surge is often slow relative to the dynamics of the estuary and the model indicated that the surge would travel up the river to the Troy Dam with relatively little diminishment or increase in magnitude. Thus, a storm surge can be thought of as a temporary sea level change, with all areas of the tidal Hudson affected nearly equivalently. However, Albany and the Battery are likely to have a slightly greater tidal maximum than areas at mid-river due to channel characteristics. The results indicated that these conclusions were insensitive to modest changes in the bathymetry of shallow regions of the river or expansion of wetlands. However, finer-scale modeling along with detailed elevation and topographic data is still a critical need in order to determine which areas of which communities will be most vulnerable to the impacts of climate change on the estuary.

Case Study C. Salt Marsh Change at New York City Parks and Implications of Accelerated Sea Level Rise

For the ClimAID assessment case study, historical aerial photographs were used to help evaluate marsh sustainability at two New York City Department of Parks and Recreation (NYCDPR) salt marshes: Udalls Cove Park Preserve (Queens) and Pelham Bay Park (Bronx). Prior evidence of New York State-wide marsh losses during the 25-year time span 1974 to 1999 was documented by NYSDEC (2004), including at these parks (**Table 5.5**). The current research used aerial photography obtained from 1951, 1974, 1999 (panchromatic), and 2005 (infrared) to quantify progressive marsh loss over the last half century. On-the-ground observations, sampling, and monitoring were used to gain an understanding of the observed

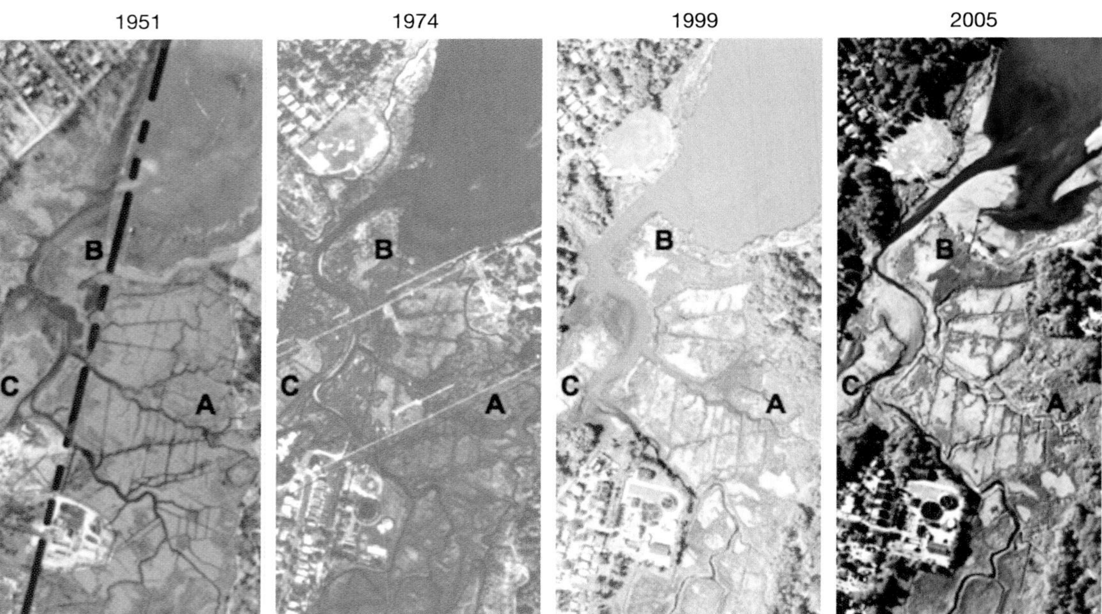

Photo credit 1951: Aero Service Corp. Photogrammetric Engineers, Phila. Pa; Photo credit 1974, 1999, 2005: NYSDEC

Figure 5.20 Marsh loss comparisons at Udalls Cove Park Preserve, Queens

rates of loss. The results were compared with marsh loss elsewhere along the eastern United States.

For long-term monitoring, NYCDPR installed sediment elevation tables (SETs) in clusters of three at Udalls Cove Park Preserve and Pelham Bay Park in cooperation with NYSDEC and U.S. Geological Survey (USGS). These platforms have been used internationally and are effective at separating the components of surface accretion and shallow subsidence in the marshes (together with feldspar markers placed on the surface in 0.25-meter squares near the SETs). Over the next several years and decades, NYCDPR will be comparing the results from the selected parks with accretion rate data at Jamaica Bay and Fire Island, New York (Roman et al., 2007), Mashomack Preserve (Shelter Island, New York), Hackensack Meadowlands (New Jersey), and Narragansett Bay (Rhode Island).

While **Table 5.8** offers a sensitivity study on marsh survival using low, medium, and high rates of accretion in the face of projected sea level rise, on-the ground determinations of accretion and subsidence rates from SETs will offer data on how to manage specific marsh sites. The combination of aerial photo analysis and SET data from sampling stations can aid park managers, scientists, and public advocates in managing, and thereby perhaps minimizing, salt marsh loss in the coming decades.

Udalls Cove Park, Queens

At Udalls Cove, initial analysis indicates significant land loss, including breaking up of previously contiguous marshland (see **Figure 5.20**, point A), eroding embankments (see point B), and widening of channels (see point C) (**Figure 5.20**). The amount of loss already under way was compared with projections of future loss over the next century (**Table 5.8**).

Jamaica Bay Wildlife Refuge, Queens

Since 1998 there has been much speculation as to the cause of salt marsh deterioration and submergence at Jamaica Bay National Wildlife Refuge, part of Gateway National Recreation Area in New York City and New Jersey (Hartig et al., 2002; NYSDEC, 2006). While the exact cause is unknown, the marsh loss at Jamaica Bay has been attributed to multiple stressors, including nutrient inputs from WPCPs (water pollution control plants for sewage treatment), deepening of navigation channels, shoreline armoring, increased tidal range (Swanson and Wilson, 2008), sea level rise, and more. Whatever the cause (or causes acting synergistically), the loss was extreme and action was taken to stem the loss.

In a pilot project at Big Egg Marsh conducted in part by local activists the Jamaica Bay Ecowatchers, the National Park Service, and many volunteers, a degraded marsh was restored by spraying sediment at a thickness of up to 3 feet and replanting with Spartina plugs. More recently, using sand from maintenance dredging, the U.S. Army Corps of Engineers conducted large-scale restoration at Elder's Point East for $13 million. At both sites the artificially elevated *Spartina alterniflora* stands are thriving. A priority list has been generated through a Jamaica Bay Task Force for follow-up locations; the next restoration with sediment supplements is planned for Elder's Point West, to be followed by Yellow Bar Hassock.

References

Allen J.R.L. and K. Pye. 1992. "Coastal saltmarshes: their nature and importance." In *Saltmarshes: Morphodynamics, Conservation and Engineering Significance*, edited by J.R.L. Allen and K. Pye, 1-18. Cambridge, UK: Cambridge University Press.

Aldrich, S., M. Dunkle, and J. Newcomb. 2009. "Executive Summary - Rising Waters: Helping Hudson River Communities Adapt to Climate Change Scenario Planning 2010 – 2030." The Nature Conservancy Eastern New York Chapter. Accessed January 2010.http://www.nature.org/wherewework/northamerica/states/newyork/files/rw_070509_exec.pdf.

Anisfeld, S. and J. Linn. 2002. "Wetland Loss in the Quinnipiac River Estuary: Baseline Assessment." Final Report to The Community Foundation for Greater New Haven, CT.

Ashizawa, D. and J. Cole. 1994. "Long-term temperature trends of the Hudson River: a study of the historical data." *Estuaries* 17(1B):166-171.

Bamber, J.L., R.E.M. Riva, B.L.A. Vermeersen, and A.M. LeBrocq. 2009. "Reassessment of the Potential Sea Level Rise from a Collapse of the West Antarctic Ice Sheet." *Science* 324 (5929): 901 doi:10.1126/science.1169335.

Bavaro, L. 2005. "Paving Paradise." Suffolk County Department of Health Services, Peconic Estuary Program. Vol 2. Iss. 4.

Bertness, M. D. 1991. "Zonation of Spartina patens and Spartina alterniflora in a New England salt marsh." *Ecology* 72:138-148.

Bindoff, N. L., et al. 2007. Observations: Oceanic Climate Change and Sea Level. Climate Change 2007: The Physical Science Basis. Contribution of Working Group I to the Fourth Assessment Report of the Intergovernmental Panel on Climate Change., S. Solomon, et al., Eds., Cambridge University Press, 386–432.

Bird, E., 2008. *Coastal Geomorphology—an Introduction. 2nd ed.* Chichester, UK: Wiley & Sons, Ltd.

Blake, E.S., E.N. Rappaport, and C.W. Landsea. 2007. "The Deadliest, Costliest, and most Intense United States Tropical Cyclones from 1851 to 2006 (and other frequently requested hurricane facts)." NOAA Technical Memorandum NWS TPC-5. http://www.nhc.noaa.gov/Deadliest_Costliest.shtml

Bliss, J.D., S.J. Williams, and M.A. Arsenault. 2009. "Mineral resource assessment of marine sand resources in cape- and ridge-associated marine sand deposits in three tracts, New York and New Jersey, United States Atlantic continental shelf." In *Contributions to Industrial-Minerals Research*, Ch. N. *U.S. Geological Survey Bulletin* 2209-N.

Bocamazzo, L. 2009. US Army Corps of Engineers, personal communication.

Bolin R., M. Jackson, and A. Crist. 1998. "Gender inequality, vulnerability and disaster: issues in theory and research." In *The gendered terrain of disaster*, edited by E. Enarson and B.H. Morrow, 27-44. Westport, CT: Praeger.

Bowman, M.J., B. Colle, R. Flood, D. Hill, R.E. Wilson, F. Buonaiuto, P. Cheng, and Y. Zheng. 2004. *Hydrologic Feasibility of Storm Surge Barriers to Protect the Metropolitan New York–New Jersey Region:* Final Report to HydroQual, Inc., Marine Sciences Research Center, State University of New York, Stony Brook, NY.

Briggs, P. T. and J. R. Waldman. 2002. "Annotated list of fishes reported from the marine waters of New York." *Northeastern Naturalist* 9(1):47-80.

Bullard, Robert D., ed. 2007. *Growing Smarter: Achieving Livable Communities, Environmental Justice, and Regional Equity*. Cambridge, MA:MIT Press.

Buonaiuto, F.S. and H.J. Bokuniewicz. 2005. "Coastal Bluff Recession and Impacts on Littoral Transport: Special Reference to Montauk, NY." *Shore and Beach* 73(4): 24 – 29.

Burger, J. and J. Shisler. 1983. "Succession and productivity on perturbed and natural Spartina salt marsh areas in New Jersey." *Estuaries* 6:50-56.

Cademartori, E.A. 2000. "An assessment of salt marsh vegetation changes in southern Stony Brook Harbor: implications for future management." M.A. Thesis. Marine Environmental Science, SUNY, Stony Brook, NY, USA.

Cahoon, D.R. and J. Lynch. 2009. "Surface Elevation Table." *Patuxent Wildlife Research Center* Accessed November 1, 2009. http://www.pwrc.usgs.gov/set/.

Coch, N.K. 1993. "Hurricane hazards along the northeastern Atlantic Coast of the United States." *Journal of Coastal Research, Special Issue* 12:115-147.

Cochran, J.K., D.J. Hirshberg, J. Wang, and C. Dere. 1998. "Atmospheric deposition to metals to coastal water (Long Island Sound, New York, USA.) Evidence from salt marsh deposits." *Estuarine, Coastal and Shelf Science* 46:503-522.

Colle, B.A., F. Buonaiuto, M.J. Bowman, R.E. Wilson, R. Flood, R. Hunter, A. Mintz, D. Hill. 2008. "New York City's vulnerability to coastal flooding." *Bulletin of the American Meteorological Society* 89:829–841.

Colle, B. 2009. Stony Brook University, personal communication.

Colle, B.A., K. Rojowsky, F.S Buonaiuto. 2010. "New York City Storm Surges: Climatology and Analysis of the Wind and Cyclone Evolution." *Journal of Applied Meteorology and Climatology* 49(1):85-101.

Colucci, S.J. 1976. "Winter cyclone frequencies over the eastern United States and adjacent western Atlantic, 1964–1973." *Bulletin of the American Meteorological Society* 57:548–553.

David Yang, USACE, August 2010, personal communication.

Davis, R.E., R. Dolan, and G. Demme. 1993. "Synoptic climatology of Atlantic coast northeasters." *International Journal of Climatology* 13:171–189.

Dean, R.G., A.R. Dalrylmple, R.W. Fairbridge, S.P. Leatherman, D. Nummedal, M.P. O'Brien, O.H. Pilkey, W. Sturges III, and R.L. Wiegel. 1987. "Responding to Changes in Sea Level: Engineering Implications." Washington D.C.:National Academy Press.

DeGaetano, A.T. 2008. "Predictability of seasonal East Coast winter storm surge impacts with application to New York's Long Island." *Meteorological Applications* 15(2): 231–242.

DeGaetano, A.T., M.E. Hirsch, and S.J. Colucci. 2002. "Statistical prediction of seasonal East Coast winter storm frequency." *Journal of Climate* 15:1101–1117.

DeLaune, R.D., J.A. Nyman, and W.H. Patrick, Jr. 1994. "Peat collapse, ponding and wetland loss in a rapidly submerging coastal marsh." *Journal of Coastal Research* 10:1021-1030.

Doland, R. and R.E. Davis. 1992. "An intensity scale for Atlantic coast northeast storms." *Journal of Coastal Research* 8:352–364.

Donnelly, J.P. and M.D. Bertness. 2001. "Rapid shoreward encroachment of salt marsh cordgrass in response to accelerated sea-level rise." *Proceedings of the National Academy of Sciences* 98:14218-14223.

Donnelly, J.P., Cleary, P., Newby, P., and Ettinger, R. 2004. "Coupling instrumental and geological records of sea-level change: Evidence from southern New England of an increase in the rate of sea level rise in the late 19th century." *Geophysical Research Letters* 31: L05203. doi:10.1029/2003GL018933.

Downs, L.D., R.J. Nicholls, S.P. Leatherman, and J. Hautzenroder. 1994. "Historic evolution of a marsh island: Bloodsworth Island, Maryland." *Journal of Coastal Research* 10:1031-1044.

Edinger, G.J., D.J. Evans, S. Gebauer, T.G. Howard, D.M. Hunt, and A.M. Olivero, eds. 2002. *Ecological Communities of New York State. Second Edition.* A revised and expanded edition of Carol Reschke's Ecological Communities of New York State. (Draft for review). New York Natural Heritage Program, New York State Department of Environmental Conservation, Albany, NY.

Fallon D. and F. Mushacke. 1996 (unpublished). "Tidal Wetlands Trends in Shinnecock Bay, New York 1974 to 1995." Division of Fish, Wildlife and Marine Resources, New York State Department of Environmental Conservation, East Setauket, NY, USA.

Fothergill, A., and L.A. Peek. 2004. "Poverty and disasters in the United States: a review of recent sociological findings." *Natural Hazard* 32:89–110.

Fothergill, A., E.G.M. Maestas, and J.D. Darlington. 1999. "Race, ethnicity and disasters in the United States: a review of the literature." *Disasters* 23:156–73.

Frumhoff, P.C., J.J. McCarthy, J.M. Melillo, S.C. Moser, and D.J. Wuebbles. 2007. "Coastal Impacts." In *Confronting Climate Change in the U.S. Northeast: Science, Impacts, and Solutions.* Report of the Northeast Climate Impacts Assessment: 15-32.

Fuglister, F.C. 1947. "Average monthly sea surface temperatures of the Western North Atlantic Ocean." *Papers in Physical Oceanography and Meteorology.* Massachusetts Institute of Technology and Woods Hole Oceanographic Institution 10(2):25.

Gehrels, R.W., J.R. Kirby, A. Prokoph, R.M. Newnham, E.P. Achertberg, H. Evans, S. Black, and D.B. Scott. 2005. "Onset of recent rapid sea level rise in the western Atlantic Ocean." *Quaternary Science Reviews* 24:2083-2100.

Gornitz, V. and S. Couch. 2001. *Chapter 3*. In *Climate Change and a Global City: The Potential Consequences of Climate Variability and Change – Metro East Coast*, Edited by Rosenzweig, C. and W.D. Solecki. Report for the U.S. Global Change Research Program, National Assessment of the Potential Consequences of Climate Variability and Change for the United States. New York: Columbia Earth Institute.

Gornitz, V., S. Couch, and E. K. Hartig. 2002. "Impacts of sea level rise in the New York City metropolitan area." *Global and Planetary Change* 32:61-88.

Gornitz, V.M., R.C. Daniels, T.W. White, and K.R. Birdwell. 1994. "The development of a coastal assessment database: vulnerability to sea level rise in the U.S. Southeast." *Journal of Coastal Research, Special Issue* 12:327-338.

Grosslein, M.D., and T.R. Azarovitz. 1982. "Fish distribution." Marine Ecosystems Analysis Program New York Bight Atlas Monograph 15.

Hartig, E.K. 2001. *Chapter 5*. In *Climate Change and a Global City: The Potential Consequences of Climate Variability and Change – Metro East Coast*, Edited by Rosenzweig, C. and W.D. Solecki. Report for the U.S. Global Change Research Program, National Assessment of the Potential Consequences of Climate Variability and Change for the United States. New York: Columbia Earth Institute.

Hartig E.K., V. Gornitz, A. Kolker, F. Mushacke, and D. Fallon. 2002. "Anthropogenic and climate-change impacts on salt marsh morphology in Jamaica Bay, New York City." *Wetlands* 22:71-89.

Hartig, E.K. and V. Gornitz. 2005. "Salt marsh change, 1926-2003 at Marshlands Conservancy, New York." Long Island Sound Research Conference Proceedings.

Haugan, P.M., C. Turley, H.O. Pörtner. 2006. "Effects on the marine environment of ocean acidification resulting from elevated levels of CO2 in the atmosphere." OSPAR intersessional correspondence group. DN-utredning. www.dirnat.no.

Hayden, B.P. 1981. "Secular variation in Atlantic Coast extratropical cyclones." *Monthly Weather Review* 109:159–167.

Hildebrand, S.F., and W.C. Schroeder. 1928. "Fishes of Chesapeake Bay." *Bulletin of the Bureau of Fisheries* 43(1):1-366.

Hirsch, M.E., A.T. DeGaetano, and S.J. Colucci. 2001. "An East Coast winter storm climatology." *Journal of Climate* 14: 882–899.

Holgate, S.J. and P.L. Woodworth. 2004. "Evidence for Enhanced Coastal Sea Level Rise During the 1990s." *Geophysical Research Letters* 31:L07305. doi:10.1029/2004GL019626.

Holland, D.M., et al. 2008. "Acceleration of Jakobshavn Isbrae triggered by warm subsurface ocean water." *Nature Geoscience* 1:659-664.

Horton, R., C. Herweijer, C. Rosenzweig, J.P. Liu, V. Gornitz, and A.C. Ruane. 2008. "Sea level rise projections for current generation CGCMs based on the semi-empirical method." *Geophysical Research Letters*, 35, L02715, doi:10.1029/2007GL032486. http://pubs.giss.nasa.gov/cgi-bin/abstract.cgi?id=ho05300q

Horton, R., V. Gornitz, M. Bowman, and R. Blake. 2010. "Climate observations and projections." *Annals of the New York Academy of Sciences*, 1196, 41-62, doi:10.1111/j.1749-6632.2009.05314.x. http://pubs.giss.nasa.gov/cgi-bin/abstract.cgi?id=ho08400z

Howell, P., J. Benway, C. Giannini, K. McKown, R. Burgess, and J. Hayden. 2005. "Long-term population trends in American lobster (Homarus americanus) and their relation to temperature in Long Island Sound." *Journal of Shellfish Research* s24:849-857.

Hu, A., G.A. Meehl, W. Han, and J. Yin. 2009. "Transient response of the MOC and climate to potential melting of the Greenland ice sheet in the 21st century." *Geophysical Research Letters* 36:L10707.

Insurance Journal. 2007. "Agents Fear N.Y. Coast Insurance Market Worse; Pols Weigh Options". *Insurance Journal*. Accessed May 2009. http://www.insurancejournal.com/news/east/2007/10/11/84189.htm.

Ishimatsu A., M. Hayashi, T. Kikkawa. 2008. "Fishes in high-CO_2, acidified oceans." *Marine Ecology Progress Series* 373:295-302.

ISO (Insurance Services Office) Property Claims Department, n.d. http://www.isopropertyresources.com/Products/Property-Claims-Service/Property-Claim-Services-PCS.htm.

Kana, T.W. 1995. "A mesoscale sediment budget for long island, New York." *Marine Geology* 126:87-110.

Kearney, M.S., A.S. Rogers, J.R.G. Townsend, E. Rizzo, D. Stutzer, J.C. Stevenson, and K. Sundborg. 2002. "Landsat imagery shows decline of coastal marshes in Chesapeake and Delaware Bays." *Eos, Transactions, American Geophysical Union* 83(16):177-178.

Klein, R.J.T. and R.J. Nicholls. 1998. "Coastal Zones." In *Handbook on Methods for Climate Change Impact Assessment and Adaptation Strategies, Version 2.0*, edited by J.F. Feenstra, I. Burton, J.B. Smith, and R.S.J. Tol. Nairobi and Amsterdam: United Nations Environment Programme and Institute for Environmental Studies, Vrije Universiteit: 7.1-7.35.

Kocin, P.J, and L.W. Uccelini. 1990. "Snowstorms along the Northeastern Coast of the United States: 1955–1985." *Meteorological Monograph No. 44*, American Meteorological Society.

Kolker, A.S. 2005. "The Impacts of Climate Variability and Anthropogenic Activities on Salt Marsh Accretion and Loss on Long Island." Dissertation. Marine Sciences Research Center, Stony Brook University, Stony Brook, NY.

Kriesel, W., C.E. Landry, and A. Keeler. 2005. "Coastal erosion management from a community economics perspective: The feasibility and efficiency of user fees." *Journal of Agricultural and Applied Economics* 37(2):451-461.

Kurihara H. 2008. "Effects of CO_2-driven ocean acidification on the early developmental stages of invertebrates." *Marine Ecology Progress Series* 373:275–284.

Landry, C.E., A.G. Keeler, and W. Kriesel. 2003. "An economic evaluation of beach erosion management alternatives." *Marine Resource Economics* 18:105-127.

Leatherman, S.P. and J.R. Allen, eds. 1985. *Geomorphic Analysis of South Shore of Long Island Barriers*. U.S. Army Corps of Engineers.

Leatherman, S.P. and R.J. Nicholls. 1995. "Historic and future land loss for upland and marsh islands in the Chesapeake Bay, Maryland, USA." *Journal of Coastal Research* 11:1195-1203.

Limburg, K.E. and R.E. Schmidt. 1990. "Patterns of fish spawning in Hudson River tributaries: Response to an urban gradient?" *Ecology* 71(4):1238-1245. Report to Hudson River Foundation. Hudsonia Limited. http://www.hudsonriver.org/ls/reports/Schmidt_005_87R_peer1.pdf

Ludlum, David M. 1963. "Early American Hurricanes: 1492-1870." *American Meteorological Society*.

Lynch, J. 2009. National Park Service, personal communication.

MacCall, A.D. 1990. *Dynamic geography of marine fish populations*. Seattle: University of Washington Press.

Metropolitan Transportation Authority (MTA). 2007. *August 8, 2007 Storm Report*.

Morrow, B. H. 1999. "Identifying and Mapping Community Vulnerability." *Disasters* 23(1):11–18.

Naparstek, A. 2005. "A history of hurricanes in New York—including the day in 1893 that Hog Island disappeared for good." *New York Magazine*, Sept. 4.

New York City Department of Environmental Protection (NYC DEP). 2008. *Jamaica Bay Watershed Protection Plan*. http://www.nyc.gov/html/dep/html/dep_projects/jamaica_bay.shtml

New York City Department of Environmental Protection (NYC DEP). 2008. *Report 1: Assessment and Action Plan - A Report Based on the Ongoing Work of the DEP Climate Change Task Force*. New York City Department of Environmental Protection Climate Change Program. With contributions by Columbia University Center for Climate Systems Research and Hydroqual Environmental Engineers & Scientists, P.C., New York, NY.

New York City Panel on Climate Change (NPCC). 2010. *Climate Change Adaptation in New York City: Building a Risk Management Response*. C. Rosenzweig and W. Solecki, Eds. Prepared for use by the New York City Climate Change Adaptation Task Force. *Annals of the New York Academy of Sciences* 1196. New York, NY.

New York State Department of Environmental Conservation (NYS DEC). 2006. *Welcome to Marine Habitat Protection*. Website no longer available. Portion retrieved May 12, 2008, from: http://www.dec.ny.gov/lands/5489.html.

New York State Department of Environmental Conservation (NYS DEC). 2003. *Tidal Wetlands Losses in Nassau and Suffolk Counties*. Prepared by F. Mushacke. Accessed on August 7, 2003, portion retrieved May 12, 2008, website no longer available. www.dec.state.ny.us/website/dfwmr/marine/wetlands/index.html.

New York State Department of Environmental Conservation (NYS DEC). 2004. *1974 – 1999 Vegetated Tidal Wetland Trends of the New York Metropolitan Area and the Lower Hudson River*. Final report prepared by Mushacke, F. and Picard, E.

New York State Sea Level Rise Task Force, 2010. New York State Sea Level Rise Task Force Report to the Legislature. http://www.dec.ny.gov/docs/administration_pdf/slrtffinalrep.pdf

Nick, F.M., A. Vieli, I.M. Howat, and I. Joughlin. 2009. "Large-scale changes in Greenland outlet glacier dynamics triggered at the terminus." *Nature Geoscience* 2:110-114.

NOAA. 2004. *Historic Sea Level Trends*. Accessed on September 30, 2004. http://co-ops.nos.noaa.gov/sltrends/sltrends.shtml

Nordstrom, K.F. and C.T. Roman. 1996. "Environments, processes and interactions of estuarine shores." In *Estuarine Shores; Evolution, Environments and Human Alterations* 1-12. New York: John Wiley & Sons.

Nuttle, W.K. 1997. "Conserving coastal wetlands despite sea-level rise." *EOS, Transactions, American Geophysical Union* 78: 260-261.

Nydick, K.R., A.B. Bidwell, E. Thomas, and J.C. Varekamp. 1995. "A sea level rise curve from Guilford, Connecticut, USA." *Marine Geology* 124:137-159.

Onishi, N. 1997. "The little island that couldn't." *New York Times*, March 18.

Orson, R.A., W. Panageotou, and S.P. Leatherman. 1985. "Response of tidal salt marshes of the U.S. Atlantic and Gulf Coasts to rising sea levels." *Journal of Coastal Research* 1:29-38.

Oviatt, C. 2004. "The changing ecology of temperate coastal waters during a warming trend." *Estuaries* 27:895-904.

Pfeffer, W.T., et al. 2008. "Kinematic constraints on glaciers contributions to 21st century sea level rise." *Science* 321:1340-1343.

Rahmstorf, S., A. Cazenave, J.A. Church, J.E. Hansen, R.F. Keeling, D.E. Parker, and R.C.J. Somerville. 2007. "Recent climate observations compared to projections." *Science* 316:709.

Reitan, C.H. 1974. "Frequencies of cyclones and cyclogenesis for North America, 1951–1970." *Monthly Weather Review* 102:861–868.

Reschke, C. 1990. "Ecological Communities of New York State." *New York Natural Heritage Program, NYSDEC*. Latham, NY.

Rignot, E., et al. 2008. "Recent Antarctic ice mass loss from radar interferometry and regional climate modeling." *Nature Geoscience* I:106-110.

Rohling, E.J., K. Grant, C. Hemleben, M. Siddall, B.A.A. Hoogakker, M. Bolshaw, and M. Kucera. 2008. "High rates of sea level rise during the last interglacial period." *Nature Geoscience* 1:38-42.

Roman, C.T., J.W. King, D.R. Cahoon, J.C. Lynch, and P.G. Appleby. 2007. "Evaluation of marsh development processes at Fire Island National Seashore (New York): recent and historic perspectives." *Technical Report NPS/NER/NRTR* 2007/089. National Park Service, Boston, MA.

Root, A., L. Schintler and K.J. Button. 2000. "Women, travel and the idea of 'sustainable transport'." *Transport Reviews* 20(3):369-383.

Rosenzweig, C. and W. D. Solecki, eds. 2001. *Climate Change and a Global City: The Potential Consequences of Climate Variability and Change*. Metro East Coast Report for the U.S. Global Change Research Program, National Assessment of the Potential Consequences of Climate Variability and Change for the United States. Columbia Earth Institute, New York. 224 pp.

Sabine C.L., R.A. Feely, N. Gruber, R.M. Key, K. Lee, J.L. Bullister, R. Wanninkhof, C.S. Wong, D.W.R. Wallace, B. Tilbrook, F.J. Millero, T.H. Peng, A. Kozyr, T. Ono, and A.F. Rios. 2004. The oceanic sink for anthropogenic CO2. Science 305, 367–371.

Salmun, H., A. Molod, F.S. Buonaiuto, K. Wisniewska, K.C. Clarke. 2009. "East Coast Cool-weather Storms in the New York Metropolitan Region." *Journal of Applied Meteorology and Climatology* 48(11):2320-2330.

Steig, E.J., D.P. Schneider, S.D. Rutherford, M.E. Mann, J.C. Comiso, and D.T. Shindell. 2009. "Warming of the Antarctic ice-sheet surface since the 1957 International Geophysical Year." *Nature* 457:459-463.

Swanson R. L. and R E. Wilson. 2008. "Increased Tidal Ranges Coinciding with Jamaica Bay Development Contribute to Marsh Flooding." *Journal of Coastal Research* 24(6):1565-1569.

Sze, J., and J.K. London. 2008. "Environmental justice at the crossroads." *Sociology Compass* 2(4):1331-1354.

Talmage, S.C. and C.J. Gobler. 2009. "The effects of elevated carbon dioxide concentrations on the metamorphosis, size, and survival of larval hard clams (Mercenaria mercenaria), bay scallops (Argopecten irradians), and Eastern oysters (Crassostrea virginica)." Limnology and Oceanography 54(6):2072-2080.

Tanski, J. November 2009. New York Sea Grant Cornell Extension, personal communication.

Thieler, E.R., and E.S. Hammar-Klose. 1999. *National Assessment of Coastal Vulnerability to Future Sea-Level Rise: Preliminary Results for the U.S. Atlantic Coast*. U.S. Geological Survey, Open-File Report 99–593, 1 sheet.

Titus, J.G. 1988. "Greenhouse Effect, Sea Level Rise and Coastal Wetlands." U.S. Environmental Protection Agency, Office of Policy, Planning and Evaluation. Washington D.C. July. EPA 230-05-86-013.

U.S. Census Bureau Population Division. 2005. "Interim State Population Projections, 2005, Table 1: Interim Projections: Ranking of Census 2000 and Projected 2030 State Population and Change: 2000 to 2030."

U.S. Environmental Protection Agency (EPA)/CCSP. 2009. *Coastal Sensitivity to Sea-Level Rise: A Focus on the Mid-Atlantic Region.* A report by the U.S. Climate Change Science Program and the Subcommittee on Global Change Research. J.G. Titus, K.E. Anderson, D.R. Cahoon, D.B. Gesch, S.K. Gill, B.T. Gutierrez, E.R. Thieler, and S.J. Williams. Accessed on May 5, 2010. http://www.epa.gov/climatechange/effects/coastal/sap4-1.html.

Varekamp, J.C., E. Thomas, and O. Van de Plassche. 1992. "Relative sea level rise and climate change over the last 1500 years." *Terra Nova 4: Global Change, Special Issue*: 293-304.

Waldman, J. 2006. "The diadromous fish fauna of the Hudson River: life histories, conservation concerns, and research avenues." In *The Hudson River Estuary*, edited by J.S. Levinton and J.R. Waldmans, 171-188. New York: Cambridge University Press.

Waldman, J.R., T.R. Lake, and R.E. Schmidt. 2006. "Biodiversity and zoogeography of the fishes of the Hudson River: watershed and estuary." *American Fisheries Society Symposium* 51:129-150.

Warren, R.S. and W.A. Niering. 1993. "Vegetation change on a northeast tidal marsh: interaction of sea level rise and marsh accretion." *Ecology* 74:96-103.

Weiss, D. 1974. "Late Pleistocene Stratigraphy and Paleoecology of the Lower Hudson River Estuary." *GSA Bulletin*, v 85(10), 1561-1570.

Wells, J. V. 1998. *Important Bird Areas of New York State*. New York State Office of National Audubon Society. Albany, NY.

Wray, R.D., S.P. Leatherman, and R.J. Nicholls. 1995. "Historic and future land loss for upland and marsh islands in the Chesapeake Bay, Maryland, USA." *Journal of Coastal Research* 11:1195-1203.

Wright, B. and R. Bullard. 2007. "Rebuilding a 'new' New Orleans," in *Growing Smarter: Achieving Livable Communities, Environmental Justice, and Regional Equity*. Cambridge, MA: MIT Press.

Wu, S.Y., B. Yarnal, and A. Fisher. 2002. "Vulnerability of coastal communities to sea-level rise: A case study of Cape May County, New Jersey, USA." *Climate Research* 22(4): 255-270.

Yang, D. August 2010. US Army Corps of Engineers, personal communication.

Yang, D. May 2011. US Army Corps of Engineers, personal communication.

Yarnal, B. 2007. "Vulnerability and all that jazz: Addressing vulnerability in New Orleans after Hurricane Katrina." *Technology in Society* 29:249-255.

Yin, J., M.E. Schlesinger, and R.J. Stouffer. 2009. "Model projections of rapid sea level rise on the Northeast coast of the United States." *Nature Geoscience* 15:1-5. (see also Hu et al., 2009, above).

Zappieri, J. 2003. NYS Department of State, personal communication.

Zeppie, C.R. 1977. "Vertical profiles and sedimentation rates of Cd, Cr, Cu, Ni, and Pb in Jamaica Bay, New York." M.S. Thesis. Marine Environmental Sciences Program, State University of New York, Stony Brook, NY, USA.

Zielinski, G.A. 2002. "A Classification Scheme for Winter Storms in the Eastern and Central United States with an emphasis on Nor'easters." *Bulletin of the American Meteorological Society* 83:37–51.

Zishka, K.M. and P.J. Smith. 1980. "The climatology of cyclones and anticyclones over North America and surrounding ocean environs for January and July, 1950–77." *Monthly Weather Review* 108:387–401.

Appendix A. Stakeholder Interactions

Stakeholder interaction is a key component of the ClimAID assessment design, integrating scientific knowledge with local experience and allowing the prioritization of vulnerabilities and provision of tangible adaptation strategies that decision-makers can use. The Coastal Zones Sector interacted with relevant stakeholders through meetings and phone conferences, and through a more regularly engaged focus group.

Meetings

The Coastal Zones Sector held its first stakeholder meeting on January 9, 2009, at the City University of New York Graduate Center in New York City. The following agencies, stakeholders, and academic institutions were represented: National Park Service, Stony Brook University, New York State Department of State, Suffolk and Nassau Counties, the New York District Army Corps of Engineers, New York City Department of Parks and Recreation, Department of Environmental Conservation Hudson River Estuary Program and New York State Climate Change Office, The Nature Conservancy, New York State Emergency Management Office, and the New York Sea Grant Extension. The meeting included a presentation by the New York State Department of State on its emerging post-storm redevelopment plan.

We held our second stakeholder meeting as a webinar on February 10, 2010. The following agencies, stakeholders, and academic institutions were represented: CUNY Institute for Sustainable Cities, City University of New York, New York State Department of State, NASA Goddard Institute for Space Studies, New York City Department of Parks and Recreation, Department of Environmental Conservation Hudson River Estuary Program and New York State Climate Change Office, Suffolk County Department of Environment and Energy, The Nature Conservancy, New York State Emergency Management Office, and New York Sea Grant Extension.

Focus Group and Related Assessment Efforts

From the initial stakeholder meeting a focus group was constructed from members of The Nature Conservancy (point of contact [POC] Sarah Newkirk), New York Sea Grant (POC Jay Tanski), New York City Department of Parks (POC Ellen Hartig), New York State Department of Environmental Conservation (NYS DEC) Region 2 (POCs Kristin Marcell, Betsy Blair), and Stony Brook University (POC Henry Bokuniewicz). The members of the focus group served as regional experts, contributed text to the Coastal Zone chapter, and provided feedback on the overall sector assessment progress and methodology. Several of the focus group members also act as liaisons between ClimAID and related regional assessment efforts in order to coordinate work products and recommendations to policy makers.

New York City Climate Change Adaptation Task Force and the New York City Panel on Climate Change

In August 2008, Mayor Michael Bloomberg launched the Climate Change Adaptation Task Force and the New York City Panel on Climate Change (NPCC) as part of his PlaNYC 2030, to develop adaptation strategies to protect the city's infrastructure from climate change impacts (Rosenzweig and Solecki, 2010). Experts on the NPCC from academic institutions and from legal, engineering, and insurance industries advised the Adaptation Task Force in developing comprehensive and inclusive strategies to protect the city's infrastructure against the effects of climate change. Of the many products developed from the NPCC work, the sea level rise information and mapping strategies were most critical to the development of the ClimAID Coastal Zones chapter, forming the foundation of case study flood projections and illustrations.

New York State Sea Level Rise Task Force

The New York State Sea Level Rise Task Force, established by the state legislature and chaired by the NYSDEC Commissioner, was charged with providing New York State with the best available science as to sea level rise and its anticipated impacts. Its tasks were to develop inventories of at-risk assets, describe the impacts of sea level rise and prepare guidance for the development of risk and adaptation strategies. The final report and website includes recommendations for protective standards and adaptive measures to be used by state and local governments as they move forward with planning for sea level rise and climate change. The Task Force adopted projections developed by the NPCC for sea level rise and coastal inundation (Rosenzweig and Solecki, 2010). These projections, which were also adopted for the ClimAID assessment, were refined for the Hudson River (see section 5.2.1 and Chapter 1, "Climate Risks") and included a rapid ice-melt scenario.

Rising Waters

The Rising Waters project was a multi-stakeholder scenario planning project to prepare for climate change in the Hudson Valley (Aldrich et al., 2009). The Nature Conservancy was the lead on the effort along with five major partners: the Cary Institute for Ecosystem Studies, the NYS DEC Hudson River Estuary Program, NYS DEC/NOAA Hudson River National Estuarine Research Reserve, the Cornell University Water Resources Institute, and Sustainable Hudson Valley. The process, based on a scenario planning process developed by Royal Dutch Shell aimed to develop realistic plausible scenarios or stories of the future based on the best available information today on the drivers of environmental, social, economic, and technological change and how they relate to one another (Aldrich et al., 2009). The scenarios are designed to serve as a tool to evaluate adaptation strategies that will work best across the range of possible futures. Four future scenarios were developed for the Hudson Valley for the year 2030. Two primary variables were explored in the scenarios. The first was whether the Hudson Valley opted to adapt to climate change in a way that tends to work with nature (using greener, non-structural solutions) or more engineered structural solutions. The second variable is the level of effort (large or small) in preparing for climate change. Climate information for the scenarios was based upon the best available scientific projections at the time and was the same for all four scenarios. A list of adaptation strategies was developed and evaluated based on criteria set by stakeholders and performance of the strategy in each scenario.

[1] Census tracts are small, relatively permanent statistical subdivisions of a county delineated by local participants as part of the U.S. Census Bureau's Participant Statistical Areas Program. A census block group is a cluster of census blocks having the same first digit of their four-digit identifying numbers within a census tract.

Chapter 6

Ecosystems

Authors: David W. Wolfe,[1,2] Jonathan Comstock,[2] Holly Menninger,[2] David Weinstein,[2] Kristi Sullivan,[2] Clifford Kraft,[2] Brian Chabot,[2] Paul Curtis,[2] Robin Leichenko,[3] and Peter Vancura[3]

[1] Sector Lead
[2] Cornell University
[3] Rutgers University, Department of Geography

Contents

Introduction 164
6.1 Sector Description 165
6.1.1 Terrestrial Ecosystems (forests, shrublands, and grasslands) 165
6.1.2 Aquatic Ecosystems 166
6.1.3 Fish and Wildlife 167
6.1.4 Non-climate Stressors 169
6.1.5 Economic Value and Ecosystem Services 172
6.2 Climate Hazards 174
6.2.1 Temperature 174
6.2.2 Precipitation 174
6.2.3 Other Climate Factors 175
6.3 Vulnerabilities and Opportunities 176
6.3.1 Criteria for Determining Vulnerability of Species, Communities, and Ecosystems 176
6.3.2 Forest, Grassland, and Alpine Communities 176
6.3.3 Aquatic Ecosystems and Wetlands 179
6.3.4 Fish and Wildlife 180
6.3.5 Pests, Pathogens, and Invasive Species of Concern 184
6.3.6 Effects on Natural Resource Use and Human Communities 187
6.4 Adaptation Strategies 188
6.4.1 Forest, Grassland, and Alpine Communities 189
6.4.2 Aquatic Ecosystems and Wetlands 189
6.4.3 Fish and Wildlife 191
6.4.4 Invasive Species 192
6.4.5 Larger-scale Adaptations 193
6.5 Equity and Environmental Justice Considerations 194
6.5.1 Forests, Parkland, and Urban Ecosystems 194
6.5.2 Winter Recreation, Resource Dependency, and Equity 195
6.5.3 Maple Syrup Industry: Vulnerability and Inequity 195
6.6 Conclusions 196
6.6.1 Main Findings on Vulnerabilities and Opportunities 196
6.6.2 Adaptation Options 198
6.6.3 Knowledge Gaps 199
Case Study A. Hemlock: Cascading Effects of Climate Change on Wildlife and Habitat 199
Case Study B. Creative Approaches to Monitoring and Adaptive Management: New York's Invasive Species Program as a Model 201
Case Study C. Maple Syrup Industry: Adaptation to Climate Change Impacts 202
Case Study D. Brook Trout: Reduction in Habitat Due to Warming Summers 203
References 209
Appendix A. Stakeholder Interactions 214
Appendix B. Relevant Ongoing Adaptation-planning Efforts 215

Introduction

Valuable ecosystem services provided by New York's landscapes include harvested products (food, timber, biomass, maple syrup), clean water and flood control, soil conservation and carbon sequestration, biodiversity support and genetic resources, recreation, and preservation of wild places and heritage sites. Ecosystems recharge groundwater supplies and reduce soil erosion by creating catchments that enhance rainwater infiltration into soils as opposed to allowing rapid runoff of storm water into streams. The healthy vegetation of landscapes helps to stabilize and conserve soils, and also sequesters carbon above ground in the standing biomass of trees and perennial plants and below ground in the form of roots and soil organic matter. The diverse flora and fauna supported by New York landscapes play a role in maintaining Earth's biological heritage, and the complex interactions among species benefit society in many ways, such as natural control of insect pests and disease. Genetic diversity will be essential for the natural adaptation of our ecosystems to environmental stresses such as high temperatures and drought that will be exacerbated by climate change. In addition, genetic diversity has potential economic value for new pharmaceuticals, or for organisms or compounds with biotechnology applications.

Figure 6.1 depicts a conceptual framework for how these services are related to ecosystem function, species composition, and habitat integrity. As this framework indicates, the impacts of climate change cannot be viewed in isolation, as other stressors are also affecting ecosystems and will affect vulnerability to climate change. While society and policy-makers are likely to focus on ecosystem services, adaptation interventions by natural resource managers often will be implemented at the level of species, communities, and habitats. As climate changes and the habitable zones of wild species continue to shift northward and/or upward in elevation throughout the century, natural resource managers will face new challenges in maintaining ecosystem services and difficult decisions regarding change in species composition.

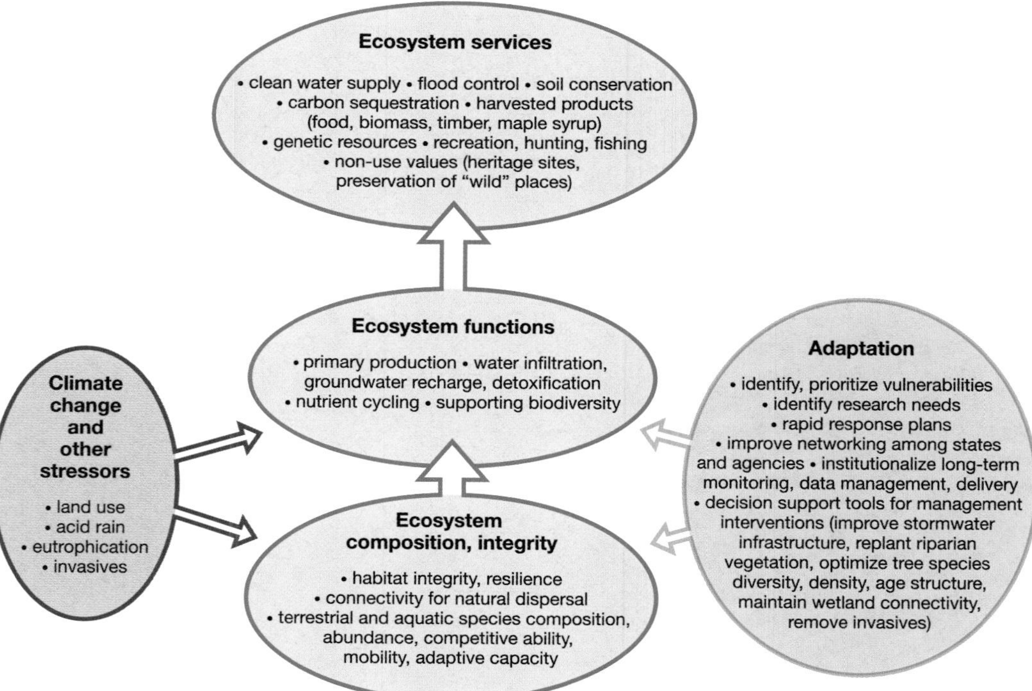

Figure 6.1 Ecosystem services in relation to climate change and adaptation

6.1 Sector Description

New York State covers an area of 54,556 square miles comprised of 47,214 square miles of land and 7,342 square miles of inland waters, including extensive lake and river systems throughout the state as well as substantial portions of Lake Erie and Lake Ontario. Variation in topography and proximity to bodies of water cause large climatic variations and distinct ecological zones (**Figure 6.2**) that support the complex web of biological diversity and provide important ecosystem services.

Ecosystems, as defined in this ClimAID report, encompass the plants, fish, wildlife, and resources of all natural and managed landscapes (e.g., forests, grasslands, aquatic systems) in New York State except those land areas designated as agricultural, coastal, or urban. This sector includes timber and maple syrup industries and tourism and recreation businesses conducted within natural and managed ecosystems. It also encompasses interior wetlands, waterways, and lakes as well as their associated freshwater fisheries and recreational fishing. Water resources per se are covered in Chapter 4, "Water Resources." Marine fisheries are covered in Chapter 5, "Coastal Zones," as are coastal wetlands and marine shoreline ecosystems.

6.1.1 Terrestrial Ecosystems (forests, shrublands, and grasslands)

Sixty-one percent of New York's land area (18.5 million acres) is covered by forest canopy. This reflects considerable forest regrowth since the late 1800s when forest cover was at a low point (about 25 percent of

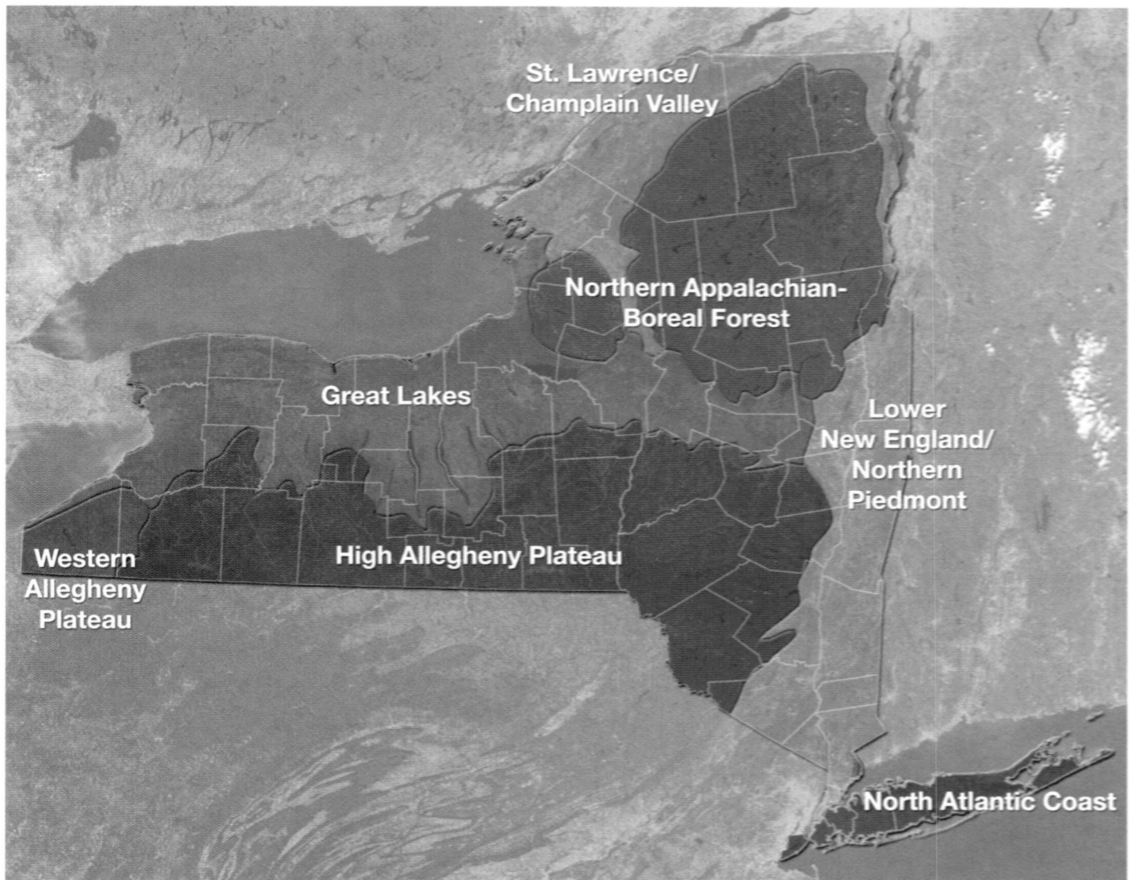

Figure 6.2 New York State Department of Environmental Conservation ecoregions

total land area) due to agricultural expansion during European settlement. Those tree species categorized as northern hardwoods by the U.S. Forest Service form the most common type of forest in New York, occupying 7.4 million acres or 40 percent of total forested area, but many other tree species are important (**Figure 6.3**). The state also is home to many shrub and woodland acres, representing various stages of forest succession on abandoned farmland and recently harvested forestlands.

Among the tree species inhabiting these forests, some have particularly important functional roles. Spruce and fir trees are key components of the unique and cherished high-elevation forests of the Adirondacks, although they occupy just 1 percent of the state's forested land. White pine and hemlock are important evergreen species found throughout the state. Hemlock trees often provide shade to stream banks (which is important for coldwater fish species) and are essential habitat to many species. While hemlock stands have largely recovered from heavy logging during the previous centuries (when they were used in the tanning industry), more recently they are under threat by infestations in some areas by the hemlock wooly adelgid insect pest (Paradis et al., 2008; and see Case Study A: Hemlock).

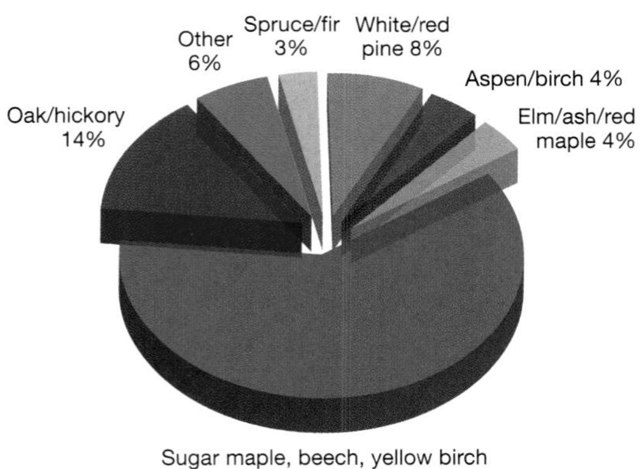

Note: Oak/hickory forest is defined as containing a mixture of red oak, black oak, scarlet oak, white oak, chestnut oak, pignut hickory, bitternut hickory, shagbark hickory, flowering dogwood, blueberry, mountain laurel, and hawthorn. The other categories are defined as containing high proportions of the two or three species named in the type title. The "Other" category includes oak/pine, exotic softwood, loblolly/shortleaf, pinyon/juniper, and oak/gum/cypress trees. Source: Data for figure were taken from the USDA Forest Inventory and Analysis 2005 webpage http://fia.fs.fed.us

Figure 6.3 New York State forest types

New York's terrestrial ecosystems also include meadows, grasslands, and wetlands. The wetlands in particular are home to many vulnerable species. The mountainous high elevations of the Adirondack State Park and the Catskills are the only regions of the state with a cool climate suitable for alpine boreal communities and alpine bogs, containing many specialist species that are limited to habitats within 5°F of current temperatures (Jenkins, 2010). The Adirondacks are home to unique alpine tundra communities with additional specialist species found nowhere else within New York State.

6.1.2 Aquatic Ecosystems

New York's rich assemblage of water resources provides a wide array of habitat types and supports a high diversity of plant and animal species. There are 70,000 miles of streams and rivers and 4,000 lakes and ponds spread over New York's 17 major watersheds (DEC website, www.dec.ny.gov/61.html) and seven ecoregions (**Figure 6.2**). There are more than 2.4 million acres of wetlands widely distributed throughout the state, with 1.2 million acres legally protected and administered by the Department of Environmental Conservation and 0.8 million administered by the Adirondack Park Agency.

Wetlands are distinguished from stream and lake habitats by the presence of emergent vegetation (e.g., cattails, sedges, shrubs, and trees). Wetlands are most extensively developed in the more level topography of the western Lake Plains and in the Adirondacks, which together account for 74 percent of all New York's wetlands. Seventy-five percent of New York's wetlands have a forested cover, but this figure does not reflect the full diversity of the different wetland types, which have distinctive flora and fauna (Edinger et al., 2002) and differing levels of vulnerability to climate change. Wetlands are distinguished by the degree to which they are fed directly by precipitation, runoff, and/or groundwater seeps and by their hydroperiod, the length of time each year that the soils are submerged. Wetlands with short or intermediate hydroperiods, such as forest vernal pools (shallow seasonal pools in woodland depressions where wood frogs and some salamanders breed) and intermittent headwater streams, lack fish and are extremely important for the reproductive success of some amphibians. Small, isolated wetlands are home to a disproportionate number of rare and endangered species.

6.1.3 Fish and Wildlife

New York's diverse ecosystems are habitat for abundant wildlife, including 165 freshwater fish species, 32 amphibians, 39 reptiles, 450 birds, 70 species of mammals, and a variety of insects and other invertebrates. The Comprehensive Wildlife Conservation Strategy is a collaborative effort led by New York State Department of Environmental Conservation's Division of Fish, Wildlife, and Marine Resources. The Comprehensive Wildlife Conservation Strategy lists 537 "Species of Greatest Conservation Need," which includes federally endangered or threatened vertebrate and invertebrate species occurring in New York State, as well as state-listed species of special concern (**Table 6.1**) (for species added by Department of Environmental Conservation staff based on status, distribution, and vulnerability, visit www.dec.ny.gov/animals/9406.html).

In all, 70 mammal species inhabit the state (NYSDEC, 2007). Two mammals—the New England cottontail (*Sylvilagus transitionalis*) and the small-footed bat (*Myotis leibii*)—are state species of concern. In addition, the Indiana bat (*Myotis sodalis*) is federally endangered.

The breeding bird atlas (McGowan and Corwin, 2008) lists 251 species that breed in the state and 125 additional species that spend the winter or visit occasionally. Several forest and grassland bird species are area-sensitive and depend upon large, unfragmented areas of habitat to breed and successfully raise young (Herkert, 1994). Important migratory and stopover habitats occur for waterfowl, raptors, and songbirds. The Shawangunk Ridge is a well-known raptor migration route. Waterfowl and other birds migrate along the shores of Lakes Ontario and Erie. Similarly, the Montezuma National Wildlife Refuge is significant regionally as a major staging, feeding, and resting area for an estimated 1 million migratory birds.

Information on amphibians and reptiles is found in the New York State "Herp Atlas" (Gibbs et al., 2007). Diverse habitats support 32 amphibian and 33 native reptile species (excluding sea turtles). The amphibians include 18 salamander species and 14 frogs and toads

Table 6.1 Endangered (E), threatened (T) and special concern (SC) fish and wildlife species in New York State (continued on next page)

Common Name	Scientific Name	Federal Status	State Status	Primary Habitat
Amphibians				
Hellbender	*Cryptobranchus alleganiensis*		SC	Streams and rivers
Marbled salamander	*Ambystoma opacum*		SC	Forest habitat, seasonal pools
Jefferson salamander	*Ambystoma jeffersonianum*		SC	Forest habitat, seasonal pools
Blue-spotted salamander	*Ambystoma laterale*		SC	Forest habitat, seasonal pools
Eastern tiger salamander	*Ambystoma tigrinum*		E	Pine barrens, seasonal or permanent pools
Long-tailed salamander	*Eurycea longicauda*		SC	Forest, shale banks, streams, springs
Eastern spadefoot toad	*Scaphiopus holbrookii*		SC	Sandy soils, seasonal pools
Northern cricket frog	*Acris crepitans*		E	Shallow ponds, slow-moving water
Southern leopard frog	*Lithobates sphenocephala*		SC	Freshwater ponds
Reptiles				
Eastern mud turtle	*Kinosternon subrubrum*		E	Fresh or brackish water with vegetation
Spotted turtle	*Clemmys guttata*		SC	Bogs, swamps, marshy meadow
Bog turtle	*Clemmys muhlenbergii*	T	E	Open, wet meadow, shallow water
Wood turtle	*Clemmys insculpta*		SC	Forest, riparian areas
Eastern box turtle	*Terrapene carolina*		SC	Fields and forest
Blandings turtle	*Emydoidea blandingii*		T	Shrub swamps, open field
Eastern spiny softshell	*Apalone spinifera*		SC	Rivers, lakes
Northern fence lizard	*Sceloporus undulatus*		T	Rocky areas surrounded by forest
Queen snake	*Regina septemvittata*		E	Streams with rocky bottoms
Eastern hog-nosed snake	*Heterodon platirhinos*		SC	Barrens, woodlands
Eastern worm snake	*Carphophis amoenus*		SC	Barrens, woodlands
Eastern massasauga	*Sistrurus catenatus*		E	Bog, swamps, barrens
Timber rattlesnake	*Crotalus horridus*		T	Deciduous forest, rocky ledges

Common Name	Scientific Name	Federal Status	State Status	Primary Habitat
Mammals				
New England cottontail	*Sylvilagus transitionalis*		SC	Shrubland, early successional forest
Small-footed bat	*Myotis leibii*		SC	Caves, rock crevices, forest
Indiana bat	*Myotis sodalis*	E		Caves, forest
Allegheny woodrat*	*Neotoma magister*		E	Rocky outcrops, oak forest
Birds				
Spruce grouse	*Falcipennis canadensis*		E	High elevation spruce/fir forest
Common loon	*Gavia immer*		SC	Lakes
Pie-billed grebe	*Podilymbus podiceps*		T	Ponds, marshes, estuarine wetlands
Manx shearwater	*Puffinus puffinus*	T		Pelagic, small islands
American bittern	*Botaurus lentiginosus*		SC	Marsh
Least bittern	*Ixobrychus exilis*		T	Marsh
Osprey	*Pandion haliaetus*		SC	Lakes, rivers, marshes
Bald eagle	*Haliaeetus leucocephalus*	T	T	Lakes, rivers
Northern harrier	*Circus cyaneus*		T	Grasslands
Sharp-shinned hawk	*Accipiter striatus*		SC	Forest
Cooper's hawk	*Accipiter cooperii*		SC	Forest
Northern goshawk	*Accipiter gentilis*		SC	Extensive mature forest
Red-shouldered hawk	*Buteo lineatus*		SC	Forest near water
Golden eagle	*Aquila chrysaetos*		E	Grassland
Peregrine falcon	*Falco peregrinus*		E	Cliffs, buildings
Black rail	*Laterallus jamaicensis*		E	Coastal marshes
King rail	*Railus elegans*		T	Coastal and freshwater marshes
Piping plover	*Charadrius melodus*	E	E	Beaches
Upland sandpiper	*Bartramia longicauda*		T	Grasslands
Roseate tern	*Sterna dougallii*	E	E	Beaches, salt marsh islands
Common tern	*Sterna hirundo*		T	Beaches, grassy uplands
Least tern	*Sterna antillarum*		T	Beaches, river sandbars
Black tern	*Chlidonias niger*		E	Wetlands, lakes, river edges
Black skimmer	*Rynchops niger*		SC	Coastal
Short-eared owl	*Asio flammeus*		E	Grasslands
Common nighthawk	*Chordeiles minor*		SC	Rooftops, open habitats
Whip-poor-will	*Caprimulgus vociferus*		SC	Open forest
Red-headed woodpecker	*Melanerpes erythrocephalus*		SC	Open forest, forest edge, beaver meadows with dead standing trees
Loggerhead shrike	*Lanius ludovicianus*		E	Hedgerows, hayfields, pasture
Horned lark	*Eremophila alpestris*		SC	Grassland
Sedge wren	*Cistothorus platensis*		T	Damp meadows and marshes
Bicknell's thrush	*Catharus bicknelli*		SC	High elevation spruce/fir forest
Golden-winged warbler	*Vermivora chrysoptera*		SC	Early successional forest
Cerulean warbler	*Dendroica cerulea*		SC	Large deciduous forests, tall trees
Yellow-breasted chat	*Icteria virens*		SC	Shrubland
Vesper sparrow	*Pooecetes gramineus*		SC	Grasslands
Grasshopper sparrow	*Ammodramus savannarum*		SC	Grasslands
Henslow's sparrow	*Ammodramus henslowii*		T	Grasslands
Seaside sparrow	*Ammodramus maritimus*		SC	Marsh

*The Allegheny woodrat, classified as Endangered, has not been found in New York State since the mid-1980s and is already considered to be extirpated at this point. http://www.dec.ny.gov/animals/6975.html
Source: www.dec.ny.gov/animals/7494.html

(NYSDEC, 2007). Six salamanders and three of the frogs and toads are endangered or of special concern (**Table 6.1**). The reptiles include four lizards, 17 snakes, 11 species of freshwater or land turtles, and one turtle that inhabits saltwater or brackish water. Seven of the 12 turtles, one of the 4 lizards, and five of the 17 snakes are endangered, threatened, or of special concern at the state and/or federal levels. Amphibians and reptiles exhibit the greatest species richness values (number of species per given area) in the Hudson River Valley, which is globally significant for its high diversity of turtles (www.dnr.cornell.edu/ gap/land/land.html).

New York is currently home to approximately 165 freshwater fish species, dominated by north temperate species living in watersheds draining to the Great Lakes and the St. Lawrence River. The coldwater fish range throughout lakes and rivers in the northern United States and Canada. New York also has many freshwater fish species representative of southern fauna found in watersheds that extend southward to the Gulf of Mexico and Middle Atlantic.

6.1.4 Non-climate Stressors

Several factors currently negatively impact natural ecosystems in New York State with various levels of severity. Some of these may be exacerbated by climate change, or may reduce the adaptive capacity of ecosystems or certain species to respond to climate change.

Invasive Species

As a major port of entry, New York State, with its vast natural and agricultural resources, is particularly vulnerable to damage from many invasive species (**Table 6.2**, **Figure 6.4**). Increases in global commerce and human travel have led to increasing rates of species invasion (Mack et al., 2000; Liebhold et al., 2006) that show no sign of slowing down in the years to come (Levine and D'Antonio, 2003; Liebhold et al., 2006; McCullough et al., 2006; Tatem, 2009) and pose serious threats to the integrity of the state's lands and waters. Most recently, the devastating emerald ash borer (*Agrilus planipennis* Fairmaire), an invasive forest pest from Asia, was detected in Cattaraugus County in western New York in June 2009 (NYSDEC, 2009), and the invasive aphid-like insect pest, hemlock wooly adelgid (*Adelges tsugae*), has been observed in some hemlock stands of the state.

Invasive species have altered and continue to alter the ecological structure and function of New York's ecosystems. Invasive understory shrubs and plants, like Amur honeysuckle (*Lonicera maackii* (Rupr.) Herder) and pale swallow-wort (*Vincetoxicum rossicum* (Kleopow) Barbar.), commonly crowd out or smother native vegetation, impeding forest regeneration (Gorchov and Trisel, 2003) and reducing understory plant diversity (DiTommaso et al., 2005). Invasive pests and pathogens, including gypsy moth (*Lymantria dispar*) and beech bark disease, can intensely impact the productivity, nutrient cycling, and food-web structure of the forests (Lovett et al., 2006).

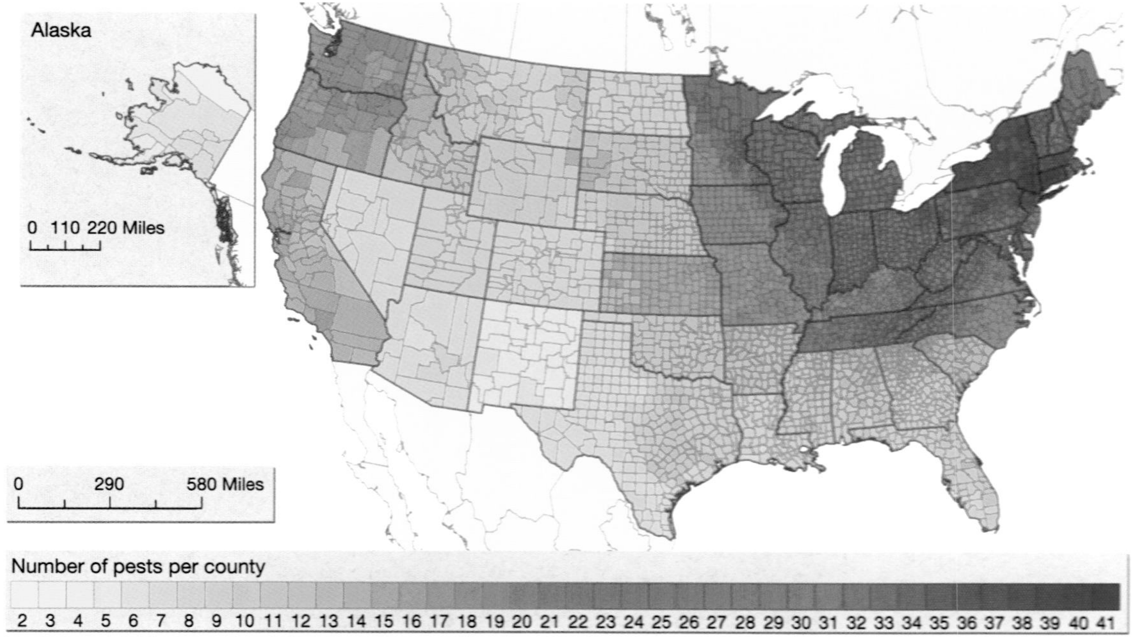

Figure 6.4 Number of invasive forest pest and pathogen species established per county throughout the United States

Table 6.2 Invasive species of management concern across New York ecosystems and the predicted direct impacts of climate change on those species based on current scientific information

Species	Habit	Origin (date introduced or detected)	Habitat	Impact	Management Options	Possible Direct Impacts of Climate Change on Species
Plants						
Common reed *Phragmites australis* (Haplotype M)	Perennial clonal grass	Europe (late 1800s)	Freshwater & brackish tidal wetlands	↓ native biodiversity & habitat; alters nutrient cycling & hydrology	M, C, B*	↑ CO_2 may stimulate growth (Farnsworth & Meyerson 2003; Meyerson et al., 2009); sea level rise may aid restoration of *Phragmites*-invaded coastal wetlands (Hellmann et al., 2008)
Eurasian watermilfoil *Myriophyllum spicatum* L.	Submerged aquatic perennial herb	Europe (~1900)	Freshwater ponds, lakes, & pools	Displaces native vegetation; negative impacts on macroinvertebrate & fish communities; impedes recreation	M, C, B	Higher water temperatures may ↑ growing season & require control actions to be implemented earlier & longer (Rahel & Olden, 2008)
Giant hogweed *Heracleum mantegazzianum* Sommier & Levier	Biennial or perennial herb	Caucasus Mtns, between Black & Caspian Seas (1917)	Wet areas (e.g., stream & river banks, along RRs & roads)	Displaces native vegetation; toxic sap causes severe photodermatitis and burns	M, C	Requires low winter temperatures for seeds to germinate in the spring (Pyšek et al., 1998)
Japanese knotweed *Polygonum cuspidatum* Siebold & Zucc.	Perennial herbaceous shrub	Japan (late 1800s)	Riparian areas, ditches & disturbed areas	Spreads rapidly, forming dense thickets that crowd and shade out native vegetation; adversely affects species diversity and wildlife habitat	M, C, B*	Milder winters may result in increased seedling survival (Forman and Kesseli, 2003)
Mile-a-minute *Persicaria perfoliata* (L.) H. Gross	Annual herbaceous vine	India, East Asia, Japan to Philippines (1930s)	Open & disturbed areas	Crowds out native species	M, C, B, G	No information available
Swallow-wort Black: *Cynanchum louiseae* Pale: *Cynanchum rossicum*	Perennial herbaceous vine	Europe (mid-1800s)	Upland areas, including old fields & woodland ground layers	Crowds out native vegetation & adversely affects native wildlife, including grassland birds and monarch butterflies	M, C, B*	No information available
Water chestnut *Trapa natans* L.	Annual aquatic herb	Western Europe, Africa to Asia (late 1870s)	Ponds, shallow lakes & river margins	Displaces native vegetation; impedes recreation; reduces dissolved oxygen	M, C, B*	No information available
Invertebrates						
Asian long-horned beetle *Anoplophora glabripennis*	Generalist wood-boring beetle	China & Korea (1996)	Urban & natural forests	Attacks and kills hardwood trees including: maples (*Acer* spp.), horsechestnut (*Aesculus hippocastanum*), willows (*Salix* spp.), American elm (*Ulmus americana*) birches (*Betula* spp.) and poplars (*Populus* spp.)	Tree removal, C (limited), B*	No information available
Emerald ash borer *Agrilus planipennis* Fairmaire	Specialist metallic wood-boring beetle	Eastern Russia & Asia, including Japan & Taiwan (2002)	Urban & natural forests	Attacks and kills all North American ash (*Fraxinus* spp.) trees	Tree removal, C (limited), B*	No information available
Hemlock woolly adelgid *Adelges tsugae*	Aphid-like insect	Southern Japan (1951, eastern United States)	Deep-shade riparian forests	Attacks and kills eastern hemlock trees (*Tsuga canadensis*)	C, B	↑ temperatures may release hemlock woolly adelgid from overwintering constraints and promote range expansion (Paradis et al., 2008)
Zebra mussel *Dreissena polymorpha* Pallas Quagga mussel *Dreissena rostriformis bugensis*	Bi-valve mollusks	Black, Caspian & Aral Seas; Ural drainage in Eurasia (1988)	Zebra: hard substrates along lakeshores & river bottoms; Quagga: deeper waters & softer substrates	Excess removal of plankton & detritus from water column, resulting in changes to food web, lake productivity & water clarity; displacement of native mussel communities; colonization & obstruction of water pipelines and canals; ship hull fouling	M, C, B*	Increased water temperatures may ↑ growing season & require control actions to be implemented earlier & longer (Rahel & Olden, 2008)

Below ground, invasive earthworms from Europe and Asia alter soil structure and nutrient retention, with cascading impacts on the soil food web and native plant communities that can be detrimental (Bohlen et al., 2004). Invasive plants, such as purple loosestrife (*Lythrum salicaria*) and common reed (*Phragmites australis*), have

Species	Habit	Origin (date introduced or detected)	Habitat	Impact	Management Options	Possible Direct Impacts of Climate Change on Species
Vertebrates						
Feral swine *Sus scrofa*	Mammals	Eurasia (1500s)	Rural & natural areas	Damage to croplands & sensitive natural areas including riverbanks & springs; degrades wildlife habitat; competition with & predation of native species; can transmit diseases to domestic swine, including pseudorabies & swine brucellosis	Trapping & shooting	No information available
Northern snakehead fish *Channa argus*	Air-breathing freshwater fish	China, Russia & Korea (2002)	Shallow ponds, swamps & slow streams	Voracious predator; competes with native species for food & habitat	C	No information available
Pathogens						
Beech bark disease	A complex syndrome involving attack by beech scale (*Cryptococcus fagisuga* Lind.) followed by invasion of *Nectria* fungi	Europe (1890)	Deciduous forests	Decline and death of American beech, *Fagus grandifolia* (Ehrh.)	Tree removal, C (at local scales)	Fungal cankering may be worse after mild winters, favoring survival & spread of the scale insect and infection (Harvell et al., 2002)
Viral hemorrhagic septicemia	Aquatic rhabdovirus	Eastern & Western Pacific coasts; Atlantic Coast of North America	Freshwater & saltwater	Infects at least 50 freshwater & saltwater fish species, including commercially & recreationally important brook trout, Chinook salmon, lake trout, rainbow trout, walleye, smallmouth bass, northern pike, yellow perch & muskellunge	None but regulations to prevent spread (e.g., prohibiting transport of live fish, restricting use of baitfish	VHS virus is less active in warmer water (higher than 59°F) (Meyers & Winton, 1995)

Note: While this is not an exhaustive list, it provides a selection of species of concern that are the focus of management efforts statewide. Abbreviations for management options are: M, mechanical; C, chemical; B, biological control; B*, biological control in development; G, grazers. See http://nyis.info for additional information.

replaced diverse wetland plant communities with monocultures, leading to cascading consequences on wetland food webs and biogeochemical cycles. Over the last century, invasive aquatic plants like Eurasian water milfoil (*Myriophyllum spicatum* L.) and water chestnut (*Trapa natans* L.) have spread extensively throughout New York's lake and river systems (Boylen et al., 2006), displacing native vegetation (Boylen et al., 1999), negatively impacting fish and invertebrate communities (Keast, 1984), and impeding recreational activities like swimming, boating, and fishing (U.S. Congress, 1993). The filter-feeding zebra and quagga mussel species (*Dreissena polymorpha* and *Dreissena rostriformis bugensis*), introduced to the Great Lakes from the Pontic-Caspian region via ballast water, have transformed the food webs in Lakes Erie and Ontario from largely pelagic systems (where fish and other organisms thrive throughout the water column) to benthic systems (where fish and other organisms are all concentrated near the lake bottom).

The economic impacts of invasive species are equally as profound as the ecological impacts, with a cost to the United States by one estimate of $120 billion per year in damage and control expenditures (Pimentel et al., 2005). The economic impact of a single species, the emerald ash borer (*Agrilus planipennis* Fairmaire), which is now established in 13 states including New York, is projected to amount to $10.7 billion from urban tree mortality alone over the next 10 years (Kovacs et al., 2009). Specifically in New York State, invasive species pose serious economic threats to agriculture, forestry, maple sugar production, and recreation.

Increasing Deer Populations

High deer populations in many areas of New York State cause concern for resource managers, farmers, and homeowners. In addition to damage caused to residential landscape plants and agricultural crops, selective feeding of white-tailed deer alters plant community structure and can negatively affect the health and diversity of forests and other natural areas. Through their direct effects on plants, deer have cascading effects on many other wildlife species.

Many of the preferred forage species of deer, such as sugar maple and oaks, are valued for timber or as food-producing trees for wildlife. Deer also feed on wildflowers like trillium and lady slipper, but they tend to avoid ferns, invasive species like garlic mustard and barberry, and native tree species such as American beech and striped maple. Selective feeding of deer has led to dominance of ferns and grasses (Horsley and Marquis, 1983), along with invasive species and monocultures of beech in some New York forests (Stromayer and Warren, 1997). Over-browsing by deer leads to loss of forest understory vegetation that is an important habitat and food source for many songbirds and other forest wildlife.

Land Use Change, Land Ownership, and Habitat Fragmentation

Management of New York's "natural" ecosystems ranges from minimal to intensive depending on land use and ownership. While public lands are important habitats for abundant birds, wildlife, and fish, private land owners and nonprofit organizations control the vast majority of non-agricultural land. For example, 90.2 percent of the 15.8 million acres available for timber production is privately owned (NEFA, 2007). Less than 10 percent of terrestrial vertebrates in New York State are on public lands. This has important implications for developing adaptive management strategies for coping with climate change or other environmental changes. In addition, land in New York supporting natural plant and animal communities is becoming increasingly urbanized and suburbanized, altering its ability to support these communities and the water and other resources supplied by these lands to neighboring habitats.

Urbanization and other forms of human land-use change threaten some habitats and lead to fragmentation—the breaking up of large, connected terrestrial or aquatic habitats. Habitat fragmentation constrains plant and animal dispersal patterns across habitats, alters plant and wildlife community composition, and increases vulnerability to pathogens, insect pests, and invasive species. It can also reduce nesting habitat for forest interior birds and area-sensitive grassland bird species, and increase rates of predation and parasitism on nesting songbirds.

Acid Rain, Nitrogen Deposition, and Ozone

Acid rain is produced when nitrogen and sulfur compounds, emitted primarily from power plants and automobiles, react with water in the atmosphere and are deposited as acidic precipitation and dry deposition. The Adirondacks, Catskills, Hudson Highlands, Rensselaer Plateau, and parts of Long Island are particularly sensitive to acid deposition because they lack the capacity in the soil to neutralize the acid (Adams et al., 2000). Acidic compounds damage leaf tissue, leach vital nutrients from the soil (Rustad et al., 1996; Fernandez et al., 2003), and mobilize toxic aluminum that damages roots and impairs decomposition in forests (USEPA, 2010). Acid rain also negatively affects some fish and other aquatic species and can increase the sensitivity of both aquatic and terrestrial species to other stresses, such as high temperatures. Extended periods of nitrogen deposition can lead to saturation and consequent leaching of nitrogen from soils with negative effects on water quality (Stoddard, 1994). While environmental regulations have reduced emissions of contributing air pollutants in recent years and enabled substantial recovery of many forest and aquatic systems (NYSERDA, 2009), acid rain remains an important stressor in some parts of the state.

Excess quantities of nitrogen deposition also can disrupt ecosystems by fertilizing the growth of a few plant species to the detriment of others (Howarth et al., 2006; Aber et al., 2003). The most common examples of this are stream and lake eutrophication, where algal and other populations grow rapidly to the detriment of many others. Ozone is also a product of high nitrogen emissions reacting in the atmosphere. High levels of ozone impede the growth of key plant species, disrupting the normal competitive relationships among species (Krupa, 2001).

6.1.5 Economic Value and Ecosystem Services

Linking ecosystem goods and services to ecosystem structure and function and identifying the best approach for placing a value on those goods and services is a major challenge of this century. **Figure 6.1** describes a conceptual framework for placing ecosystem services, values, and functions into context with adaptation to climate change and multiple stressors.

Valuation Challenges

The economic value of some ecosystems goods and services is relatively straightforward, such as recreational value and value of commodities including timber and maple syrup (see more details, below). However, many services fall under the category of ecological functions, which have indirect value, such as carbon sequestration, water storage and water quality maintenance, flood control, soil erosion prevention, nutrient cycling and storage, species habitat and biodiversity, and dispersal/migration corridors for birds and other wildlife. These functions clearly have value, but quantifying them is much more complex. Even more difficult to quantify are the existence or non-use values associated with concepts such as preservation of cultural heritage, resources for future generations, charismatic species, and wild places.

The National Research Council recently commissioned a review of ecosystem value by experts in the field (NRC, 2005). It lays out the challenges of valuation in the context of uncertainty. It also describes various approaches such as nonmarket valuation, revealed- and stated-preference methods, and the use of production functions. The review also discusses how the results of valuation analysis can be linked to policy. More recently, new modeling tools are being developed that use ecological production functions and valuation methods to examine the impact of projected changes in land use and land cover on ecosystem services, conservation, and the market value of commodities produced by the landscape (Daily et al., 2009; Nelson et al., 2009). A recent study conducted in New Jersey used several approaches and concluded ecosystem services within the state had a value of $11.6–19.4 billion per year (Costanza et al., 2006).

Recreation and Tourism

Hunting, fishing, and wildlife viewing have a significant impact on the economies of New York State. More than 4.6 million state residents and nonresidents fish, hunt, or watch wildlife in the state (USFWS, 2006), spending $3.5 billion annually on items such as equipment, trip-related expenditures, licenses, contributions, land ownership, and leasing and other items. The 2007 New York State Freshwater Angler Survey (www.dec.ny.gov/outdoor/56020.html) indicated that there were more than 7 million visitor-days fishing for warmwater game fish (predominantly smallmouth and largemouth bass, walleye, and yellow perch) and nearly 6 million days spent in pursuit of coldwater game fish (predominantly brook, brown, and rainbow trout). About 20 percent of the freshwater angling effort was directed toward Great Lakes fisheries, with the remainder directed toward inland fisheries.

Winter recreation is another major component of the economic value of the state's natural ecosystems. New York has more ski areas than any other state in the nation. Lake Placid in the Adirondacks is known internationally as a former winter Olympics site. Combined, the state's ski areas host an average of 4 million visitors each year, contributing $1 billion to the state's economy and employing 10,000 people (Scott et al., 2008). New York is also part of a six-state network of snowmobile trails that totals 40,500 miles and contributes $3 billion each year to the Northeast regional economy.

The local economies of the Adirondacks, Catskills, Finger Lakes, coastal, and other recreation areas are dominated by tourism and recreation. The Northeast State Foresters Association, using U.S. Forest Service statistics for 2005, found that forest-based recreation and tourism provided employment for ~57,000 people and generated a payroll of $300 million in the region (NEFA, 2007).

Timber and Forest-based Manufacturing

In 2005, the estimated value of timber harvested in the state exceeded $300 million (NEFA, 2007). The manufactured conversion of these raw timber components into wood products such as commercial-grade lumber, paper, and finished wood products adds considerably to the value of this industry to the state. The total forest-based manufacturing value of shipments in 2005 was $6.9 billion (NEFA, 2007). Each 1,000 acres of forestland in New York supports three forest-based manufacturing, forestry, and logging jobs. This industry is particularly important to the regional economies of areas like the Adirondacks, where wood- and paper-product companies employ about 10,000 local residents (Jenkins, 2010).

Maple Syrup Industry

Sugar and red maple are New York's most abundant forest tree species and, historically, the state's climate has been conducive to profitable maple syrup production. It is estimated that less than 1 percent of New York's maple trees are currently used for maple syrup production (compared to about 2 percent in Vermont) (personal communication, Michael Farrell, Director, Uihlein Forest). In 2007, New York produced 224,000 gallons of syrup (making New York second in the United States, after Vermont) at a value of $7.5 million (New York State Agriculture Statistics Service, www.nass.usda.gov/ny).

6.2 Climate Hazards

Several climate change factors that are particularly relevant to New York's ecosystems are highlighted and briefly introduced below. These factors are discussed in more detail in section 6.3 and in the case study analyses. See Chapter 1, "Climate Risks," for further information about climate change factors.

6.2.1 Temperature

Increased temperatures will have numerous effects on both plants and animals. Some effects are very direct, like the physiological tolerances of different organisms to specific temperature ranges. Some are indirect, such as increased water requirements at higher temperatures or changes in habitats due to less snow and ice cover.

Warmer Summer Temperatures and Longer Growing Seasons

Warmer summer temperatures and longer growing seasons will affect plant and animal species non-uniformly, and thus will affect species composition and interactions. Primary productivity of some ecosystems could potentially increase if other environmental factors do not limit plant growth. Changes in ecosystem processes are expected, such as the timing and magnitude of the depletion of soil water and nutrients by vegetation. Some insect pests and insect disease vectors will benefit in multiple ways, such as more generations per season and increased over-winter survival, and weaker resistance of stressed host plants (Rodenhouse et al., 2009).

In aquatic systems, warmer waters and a longer summer season could increase vegetative productivity, but also increase the risk of algal blooms and other forms of eutrophication, leading to low dissolved oxygen (Poff et al., 2002) and negative effects on fish and other aquatic species. Many aquatic organisms mature more quickly but reach smaller adult sizes at higher temperatures. Rising temperatures are likely to be particularly harmful to coldwater fish, including brook and lake trout, while favoring warmwater species, such as large-mouth bass.

Increased Frequency of Summer Heat Stress

Increased frequency of summer heat stress will negatively affect many plant and animal species, constraining their habitable range and influencing species interactions. Temperature increases will drive many changes in species composition and ecosystem structure, most notably leading to eventual complete loss or severe degradation of high-elevation spruce and fir, and alpine bog and tundra habitats.

Warmer Winters

Warmer winters will have substantial effects on species composition, as the reproductive success and habitable ranges of many plant, animal, and insect species currently south of New York are now constrained by winter temperatures. Warmer winters will also increase the winter survival and spring populations of some insect, weed, and disease pests that today only marginally overwinter in the New York region. If climate change leads to more variable winter temperatures, perennial plants may be negatively affected. Variable winter temperatures may make them more vulnerable to mid-winter freeze damage (due to de-hardening) or spring frost (due to premature leaf out and bud break). Variable winters could also have negative effects on hibernating animal species, including some threatened and endangered species.

6.2.2 Precipitation

Changes in precipitation can include changes in total annual precipitation, its seasonal distribution, how much of it comes as rain versus snow, and the intensity of individual storms.

Reduced Snow Cover

Reduced snow cover will have numerous cascading effects on species and habitats. Winter survival of many small mammals (e.g., voles) that depend on snow for insulation and protective habitat will be at risk. This could protect some trees and other vegetation from winter damage by these mammals, but it will have negative implications for predators that depend on them as a winter food source (e.g., fox). In contrast, reduced snow cover will favor herbivores such as deer by exposing more winter vegetation for browsing, to the detriment of those plant species preferred by the herbivores. Less snow-cover insulation in winter will affect soil temperatures, with complex effects on soil microbial activity, nutrient retention (Rich, 2008; Groffman et al., 2001), and winter survival of some insects, weed seeds, and pathogens (see section 6.3.4 on pests).

Changes in Rainfall, Evapotranspiration, and Hydrology

Changes in rainfall, evapotranspiration, and hydrology are described in detail in Chapter 1 ("Climate Risks") and Chapter 4 ("Water Resources") and in Case Study C: Drought in Chapter 7 ("Agriculture"). Increased frequency of high rainfall events and associated short-term flooding is currently an issue and is projected to continue. This leads to increased runoff from agricultural and urban landscapes into waterways, which can lead to pollution or eutrophication effects, erosion and damage to riparian zones, flood damage to plants, and disturbance to aquatic ecosystems. Summer water deficits are projected to become more common by mid- to late-century, and the impacts on ecosystems could include reduced primary productivity (vegetation growth), and reduced food and water availability for terrestrial animals. Summer water deficits could lead to a reduction of total wetland area, reduced hydroperiods of shallow wetlands, conversion of some headwater streams from constant to seasonal flow, reduced summer flow rates in larger rivers and streams, and a drop in the level of many lakes. Late winter and spring will continue to be the seasonal period of peak groundwater recharge and stream flow rates, but the total snowpack accumulation will be lower, so stream and river flows directly associated with spring thaw are likely to decrease. If spring rainfall increases, however, this could compensate for low snowpack. Thus, it is uncertain whether spring flood events will be more or less common than they are today.

6.2.3 Other Climate Factors

The lack of robust projections for some climate factors makes assessment of some vulnerabilities and planning adaptive management for them difficult. (See Chapter 1, "Climate Risks," for further discussion.) Factors of particular concern are discussed here.

Climate Variability and Frequency of Extreme Events

Most climate scenarios assume no change in climate variability per se, but there is not a high degree of certainty that this will be the case. Changes in winter temperature variability could have profound effects on hibernating animals and on the risk of cold damage to plants. The frequency of crossing environmental thresholds (e.g., freezing temperatures) and storms and extreme events can cause a cascade of effects leading to disruption of entire communities and ecosystem function (Fagre et al., 2009), particularly if they occur in clusters. We currently are not able to determine whether such events are part of a long-term climate change trend, and climate models cannot yet project these trends reliably.

Changes in Cloud Cover

Current climate models cannot reliably project changes in cloud cover, yet such changes can have profound effects on the surface radiation balance (the net balance of solar radiation and exchanges of thermal radiation between the Earth's surface and the sky), which influences vegetation water use and total photosynthetic production.

Higher Atmospheric Carbon Dioxide Levels

Higher atmospheric carbon dioxide levels can potentially increase growth of many plants, particularly those with the C_3 photosynthetic pathway growing under optimum conditions. The magnitude of the carbon dioxide effect varies widely among species and, even without climate change, could alter species composition in some ecosystems by favoring some species over others. Many fast-growing species, including many invasive plants and aggressive weed species, tend to show greater growth stimulation than slow-growing species and can gain a competitive advantage at high carbon dioxide concentrations (Ziska, 2003). An analysis by Mohan et al. (2007) suggested that in the understory of temperate forest ecosystems some late successional, shade-tolerant species benefit more than shade-intolerant species. In general, when plant growth is constrained by nutrients, high or low temperature stress, or environmental factors, the absolute magnitude of the carbon dioxide benefit is reduced or not apparent (Wolfe, 1995).

6.3 Vulnerabilities and Opportunities

The initial impacts of climate change on species are already apparent, with documented accounts of changes in phenology (i.e., seasonal timing of events like bud-break or flowering) and species range shifts across the Northern Hemisphere (Backlund et al., 2008; Parmesan and Yohe, 2003; Parmesan, 2007). Within the northeastern United States, researchers have documented earlier bloom dates of woody perennials (Wolfe et al., 2005; Primack et al., 2004), earlier spring arrival of migratory birds (Butler, 2003), and other biological and ecological responses discussed in more detail below. Species and ecosystems are responding directly to climate drivers and indirectly to secondary effects, such as changes in timing and abundance of food supply, changes in habitat, and increased pest, disease, and invasive species pressure. Ultimately, biodiversity, net primary productivity, vegetation water use, and biogeochemical cycles could be affected by climate change. To date, however, there is not unequivocal evidence of climate change impacts on ecosystem services such as carbon sequestration or water storage and quality in New York State. The certainty in projecting climate change impacts diminishes as projections are scaled up from individual species and ecosystem structure to ecosystem function and services.

6.3.1 Criteria for Determining Vulnerability of Species, Communities, and Ecosystems

Criteria for determining vulnerability of species, communities, and ecosystems to climate change have been discussed by a number of studies (e.g., Bernardo et al., 2007; Foden et al., 2008; Pörtner and Farrell, 2008; Kellerman et al., 2009). The vulnerability criteria encompassed in the ClimAID analysis include:

- location currently near the southern border of habitable range;
- low tolerance for environmental change or stress;
- specialized habitat requirements;
- specialized food requirements;
- specialized interactions with other species that will be disrupted by climate change;
- poor competitor with species infringing on range;
- susceptibility to new pests or disease infringing on range;
- poor dispersal ability;
- limited genetic diversity; and
- low population levels or current status as an endangered species or species of concern.

Species and Communities Identified as Highly Vulnerable

Species and communities identified as highly vulnerable to climate change projected for New York, as defined by the metrics above, include:

- boreal and spruce- and fir-dominated forests;
- high-elevation alpine tundra communities of the Adirondacks;
- brook trout, Atlantic salmon, and other coldwater fish;
- snow-dependent species such as the snowshoe hare;
- moose;
- some bird species, such as Bicknell's thrush, Baltimore oriole, and rose-breasted grosbeak; and
- amphibians and other wetland species.

Species Likely to Benefit

Species likely to benefit include habitat and food generalists that are currently constrained by cold temperatures, as well as some invasive species. Examples include:

- white-tailed deer;
- warmwater fish species such as bass;
- some bird species such as northern cardinal, robin, and song sparrow;
- invasive insect pests such as the hemlock wooly adelgid; and
- invasive plant species such as kudzu.

See below for a more detailed discussion of each of these.

6.3.2 Forest, Grassland, and Alpine Communities

The distribution of most vegetation types is strongly influenced by the interactions of climatic variation with elevation, latitude, lake effects, topography, etc. Climate

when they leave the pool, hamper dispersal from the pool, and strand young that have not yet metamorphosed (Rodenhouse et al., 2009). Decreasing soil moisture may also limit surface activity of terrestrial and stream salamanders, reduce feeding opportunities, and increase competition for refugia.

On the other hand, increased winter and early spring temperatures may lead to increased foraging opportunity for salamanders and other amphibians early in the year, provided that their prey respond similarly and are available earlier. Also, earlier springs associated with climate change could lead to earlier breeding and larger amphibians with competitive advantages. In Ithaca, an analysis of historical records documented that four frog species are initiating spring mating calls an average of 10 to 13 days earlier now than they did in the early 1900s (Gibbs and Breisch, 2001). Earlier breeding will not compensate for possible negative effects associated with drier conditions and loss of aquatic habitat, but it further complicates attempts to project the magnitude of climate change effects on growth and reproductive success of amphibians (Rodenhouse et al., 2009).

Reptiles

The physiology of reptiles is temperature sensitive and could be influenced profoundly by climate change. For example, painted turtles (*Chrysemys picta*) grow larger during warmer years and reach sexual maturity more quickly (Frazer et al., 1993). Therefore, increasing temperatures may result in a higher rate of reproduction. However, for some species (like the painted turtle), the sex ratio of hatchlings is determined by the average July temperature in the nest. A change of as little as 3–4°F could skew the sex ratio in favor of female hatchlings (Janzen, 1994), with very few or possibly no males being produced. In addition, a decrease in the amount of snow cover (which serves as insulation) could lower overwinter survival of turtle hatchlings (Breitenback et al., 1984).

Many species of turtle in the state are already of special concern. Their limited dispersal abilities, combined with relatively small, isolated populations of animals, make them more prone to local extirpations than larger, more widespread populations of animals. Landscape changes that alter or fragment habitats will limit the potential for these animals to move across the landscape in response to environmental stresses such as climate change.

Fisheries

Temperature plays a primary role in governing most life processes in fish (Brett, 1971). The potential for climate change impacts on freshwater fisheries in New York has generally focused on coldwater fish species, which require year-round access to water temperatures below 68°F. The most prominent New York coldwater fisheries target both native (e.g., brook trout, lake trout, Atlantic salmon) and non-native (e.g., brown trout, rainbow trout, and Chinook salmon) trout and salmon. Fish populations in rivers and shallow lakes will experience relatively significant reductions in coldwater refuges with continued warming and, thus, will be particularly vulnerable to climate change. Coldwater fish in many New York streams and shallow lakes currently require coldwater refuges provided by shaded stream banks, upwelling groundwater, and lakes with sufficient depth to stratify (maintain a stable zone of cold water) during summer. Any reduced availability of these refuges during warm summer conditions will reduce the future distribution and abundance of coldwater fish in New York (see Case Study D: Brook Trout, this chapter). Although New York coldwater fish communities in cooler, high-elevation regions of the Adirondacks and Catskills have already suffered population declines due to acid rain, these well-documented impacts—and subsequent 30-year efforts to reduce those losses of coldwater fish—provide a useful foundation to address the negative impacts of climate change on freshwater fisheries

In contrast to the climate warming effects on rivers and shallow lakes, sufficient bottom coldwater regions are likely to be maintained in deep, large lakes (such as the Great Lakes, Finger Lakes, Lake Champlain, and the larger Adirondack lakes) and be able to support breeding populations of coldwater species even after decades of projected warming climate trends. However, other aspects of these large lake habitats will be affected by other stressors, such as eutrophication and changes in water chemistry.

Two native coldwater fish species appear to be particularly susceptible to climate-risk factors: brook trout and round whitefish. Brook trout (*Salvelinus fontinalis*) are popular for recreational fishing and have been designated as New York's state fish. Brook trout have disappeared from many New York waters in response to non-native fish introductions, acid rain, habitat destruction, and hydrological disruption. The thermal preferences and effects of temperature on brook trout and closely related species are well known (Baldwin, 1956; Hokanson et al., 1973; Reis and Perry, 1995; Schofield et al., 1993; Selong et al., 2001). Brook trout populations are particularly vulnerable because this species requires cool water temperatures and relies on upwelling groundwater for reproduction and thermal refuge during hot summers (Curry and Noakes, 1995; Borwick et al., 2006). Brook trout populations have already been greatly reduced in their native range, and changes in thermal regimes are one of the greatest threats to their continued persistence (Hudy et al., 2005). Several studies have provided information regarding the potential impact of temperature increases on stream (Meisner, 1990a; Meisner, 1990b; Wehrly et al., 2007) and lake populations of brook trout (Robinson, 2008). Water temperature changes associated with a warming climate and human modifications to watersheds have been shown to reduce brook trout growth (Reis and Perry, 1995; King et al., 1999), available thermal habitat (Meisner, 1990a), and range (Meisner, 1990b).

Round whitefish (*Prosopium cylindraceum*) is another key coldwater species that could suffer from projected climate changes, though the single largest reason for the disappearance of New York round whitefish populations to date is the presence of non-native species, such as smallmouth bass and yellow perch. If changing climate conditions favor bass and perch by providing more abundant warmwater habitat, round whitefish would be affected indirectly, even without the loss of suitable coldwater refuges.

New York's threatened and endangered species include both southern species that were never widely distributed in the state (e.g., bluebreast darter and mud sunfish) and northern species (e.g., round whitefish and deepwater sculpin) that have been disturbed by habitat changes and introductions of non-native species. The coldwater threatened and endangered species are likely to be susceptible to projected warming trends in climate conditions. In contrast, a few southern species that were historically rare in New York, but remain abundant in suitable southern habitats (e.g., longear sunfish), could increase in distribution in New York if warmer conditions prevail.

6.3.5 Pests, Pathogens, and Invasive Species of Concern

It is likely that New York wildlife and land managers will experience new challenges with insect and disease management as longer growing seasons increase the number of insect generations per year, warmer winters lead to larger spring populations of marginally over-wintering species, and earlier springs lead to earlier arrival of migratory insects. New invasive species (discussed below) will also be an issue as habitat ranges of some pests shift northward. Numerous studies throughout the northern hemisphere have already documented changes in spring arrival and/or geographic range of many insect and animal species due to climate change (Parmesan and Yohe, 2004). Also, those plants and wildlife negatively affected by changes in climate will become more vulnerable to insect pests and disease, which could increase both individual mortality and in some cases promote widespread outbreaks. This is of particular concern with regard to forest stands made up of potentially long-lived individuals, because climate changes are likely to be much faster than adaptive changes in species composition through natural dispersal, competition, and gradual replacement (Dukes et al., 2009).

Climate factors such as warmer temperatures, increased frequency of heavy rainfall events, and wet soils in spring will tend to favor some leaf and root pathogens (Coakley et al., 1999). However, increases in short- to medium-term drought during some summer seasons would tend to decrease the duration of leaf wetness and wet soils and reduce some forms of pathogen attack on leaves and roots, respectively.

While there is not a high level of certainty regarding projections of humidity and precipitation events, it is possible to make some generalizations for New York: 1) higher winter temperatures are likely to result in larger populations of pathogens surviving the winter that can initially infect plants, 2) increased temperatures are likely to result in the northward expansion of the range of some diseases because of earlier appearance and more generations of pathogens per season, and 3) more frequent and more intense rainfall events will tend to favor some types of pathogens over others and also cause wash-off from leaves of fungicide or other pesticides.

Climate change may have serious implications for diseases affecting wildlife and people. Vector species, such as mosquitoes, ticks, midges, and other biting insects, respond dramatically to small changes in climate, which in turn alters the occurrence of diseases they carry. For example, Lyme disease, erlichiosis, and other tick-borne diseases are spreading as temperatures increase, allowing ticks to move northward and increase in abundance. Epizootic hemorrhagic disease, a viral disease affecting white-tailed deer, spread to New York State in 2007. Epizootic hemorrhagic disease is transmitted by the bites of infected midges, commonly referred to as gnats. During periods of drought, animals congregate around limited water sources where midges occur in greatest numbers, allowing for the rapid spread of the virus (Sleeman et al., 2009). Outbreaks end with the onset of the first hard frost in fall. The combination of drought and delayed first frost allows for the spread of this disease.

Snow Cover Effects on Overwintering

Minimum winter air temperature has often been used to assess the potential overwinter survival of insect pests, weed seeds, and disease pathogens. However, this does not account for possible climate change effects on snow cover, which has an insulating effect on soil temperatures. **Figure 6.6** shows simulations predicting the annual minimum soil temperature at ground level underneath snow for three locations in New York. Currently, these locations vary in the number of their average annual snow cover days (DeGaetano et al., 2001). At Riverhead, the southernmost and least snowy location, temperatures at the soil surface show a projected increase that is similar to that of the air temperature. At the snowier Binghamton location, the increase in soil surface temperature is muted relative to the air temperature, with air temperature increasing more quickly than soil temperature (0.04°F per year for the soil surface versus 0.07°F per year increase in air temperature). The difference in the soil-temperature relationship between Riverhead and Binghamton is presumably a result of the greater impact of the reduction in winter snow cover at the snowier Binghamton relative to less-snowy Riverhead. At Plattsburgh, the northernmost (and snowiest) location, air temperature increases are similar to the other locations, but the ground surface temperature decreases through time at a rate of -0.05°F per year. Thus, winter soil temperatures at Plattsburgh are projected to actually become colder than they are today because the air temperature warming trend is not enough to compensate for the loss of snow cover depth and duration and, thus, the insulating effect of snow cover.

The results of **Figure 6.6** illustrate the complexities of projecting climate change impacts on survival of insects and pathogens overwintering in the soil. For regions of New York that currently have low snow cover, using projections of winter air temperature to project winter survival of insects and pathogens in soil may be reliable. In these locations, overwintering insect populations may increase. However, for historic high-snow regions in which snow cover is projected to decline during the coming decades, winter soil temperatures could remain the same or actually become colder than they are today despite a trend for warming winter air temperatures because of the loss of the snow-cover insulation effect. In locations where soil temperatures decrease, overwintering insect populations may decrease.

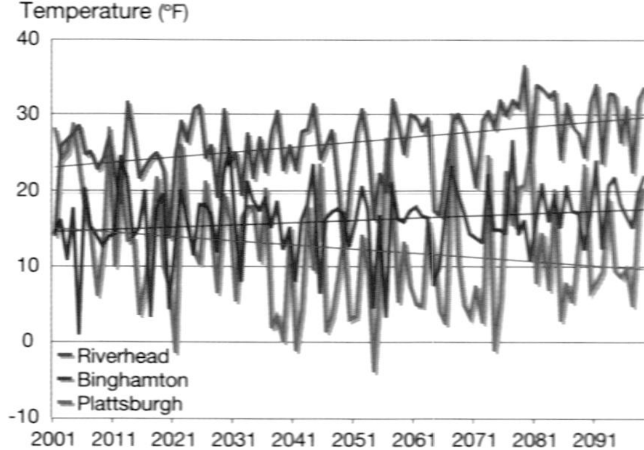

Note: These sites differ in current winter snow cover, which affects the response of future soil temperatures to rising air temperatures. Riverhead is the southernmost and least snowy location; Plattsburgh is the northernmost and snowiest location; Binghamton is between the two, both in terms of its location and amount of snow. As snow depth decreases in Plattsburgh, ground-level temperatures are projected to decrease as air temperature increases, because the ground will lose some of the warming effect of the insulating snow cover. National Center for Atmospheric Research, USA (PCM) model simulations for the A1F1 emission scenario. These projections are broadly consistent with those of other climate models used in ClimAID.

Figure 6.6 Minimum annual temperature (°F) at ground level under ambient snow cover for grids near Riverhead (blue), Binghamton (red), and Plattsburgh (green)

Invasive Species

Invasive species are defined as those species that are not currently native to New York's ecosystems and cause harm to the economy, environment, or human health (U.S. Executive Order 13112, 1999; Laws of New York, 2008, Chapter 26). For the analyses reported here, we are primarily concerned with "transformer" invasive species (sensu Richardson et al., 2000)—those species not native to North America that have the capacity to profoundly change the structure and function of ecosystems, as the chestnut blight did in the early 1900s (Gravatt, 1949; Anagnostakis, 1987) and as the emerald ash borer threatens to do now. Furthermore, we suggest that spending significant management effort on native species migrating within the continent in response to climate change may not be a good use of limited resources. Climate change is already resulting, and will continue to result, in the northward range expansion of some native southern species, and efforts to halt the movement of these species would be counterproductive. Strategically directing attention and prevention/ management actions toward those species known to be aggressively invasive elsewhere, and that will increase ecosystem vulnerability to climate change (Crooks, 2002), would be more sensible.

There is some recent evidence regarding the impacts of climate change on invasive species. Predictions that the hemlock wooly adelgid (*Adelges tsugae*), an invasive insect whose range is largely constrained by overwintering temperatures, would spread more rapidly throughout the Northeast with a warming climate (Paradis et al., 2008) have already come to pass in New York's Finger Lakes Region (USDA Forest Service, 2008). Recent work examining the flowering time of native and non-native species over 150 years in Concord, Massachusetts, indicates that non-native plants—particularly invasive species—have adapted better to long-term temperature increases than native plants. Over the last 100 years, invasive plants, on average, are flowering 11 days earlier than native plants. This may confer greater advantage to the invasive species (Willis et al., 2010).

Native communities stressed by climate change and other elements of global change (e.g., land-use change, habitat fragmentation, and nitrogen deposition) may become even more vulnerable to species invasions (Dukes and Mooney, 1999; Shea and Chesson, 2002). Further, invasive species may stand poised to exploit the changing climate via new transport pathways, overcoming previous environmental constraints, expanding ranges, and increasing competitive abilities. Although the specific outcomes of invasive species/climate-change interactions may be difficult to predict, it is certain that the combinations of species composing New York's ecosystems will look and interact differently than they do presently (Williams and Jackson, 2007).

By changing patterns of tourism and commerce, climate change may alter mechanisms of transport and introduction of invasive species (Hellmann et al., 2008). For example, expected air traffic increases and climatic convergence between China and parts of northern Europe and North America may result in increased invasion risk (Tatem, 2009). Loss of Arctic sea ice could open new shipping channels, shorten transport time, and connect new geographic regions via the Northwest Passage (Hellmann et al., 2008). Particularly relevant for New York State, climate change could allow for longer shipping seasons in the Great Lakes and, thus, more opportunities for detrimental species introductions, such as the monkey goby, an invasive fish species (Kolar and Lodge, 2002).

Before a new invasive species can become established and spread, it must first overcome a number of environmental and ecological constraints. Projected warmer winters and hotter summers facing the Northeast in the coming century ("Climate Risks," Chapter 1) will allow invasive species previously unable to persist to overcome temperature constraints. For example, kudzu (*Pueraria montana*), a prevalent invasive plant species in the southeastern United States, may expand its range northward (Wolfe et al., 2008). Additionally, increasing temperatures, precipitation, and humidity may benefit invasive forest pathogens (Dukes et al., 2009). Elevated carbon dioxide concentration, temperature, and precipitation may all contribute to increasing the competitive ability and dominance of some invasive plants over native species (Dukes and Mooney, 1999; Song et al., 2009). Furthermore, climate stress, the loss of species poorly adapted to future climate changes, and altered biotic interactions between species may open new niches and increase a native ecosystem's vulnerability to invasion. Increased incidence of extreme weather events, such as floods and drought, may also create additional windows of opportunity for the establishment and spread of invasive species, many of which are well-adapted to disturbed environments (Hobbs and Mooney, 2005).

The impacts of species currently invading New York's ecosystems will be exacerbated by climate change (Table 6.2). Some invaders, particularly herbivorous insects that have a physiology sensitive to temperature, may increase in abundance and impact within their range as a result of faster development times and longer growing seasons for plant hosts (Dukes et al., 2009). Climate change may also affect the phenology and efficacy of natural enemies to invasive species (e.g., parasites) and introduced biological control agents, with the potential to indirectly benefit invasive species (Burnett, 1949; Dukes et al., 2009). Increased water temperatures could result in earlier and longer growing seasons for aquatic invaders like Eurasian watermilfoil (*Myriophyllum spicatum*) and zebra mussels (*Dreissena polymorpha*), which in turn would require more frequent (and costly) implementation of control actions (Rahel and Olden, 2008). Additionally, there is some evidence to suggest that climate change may lead to increased per capita impact of invasive species. For example, one study observed that Japanese beetles (*Popillia japonica*) increased their feeding on soybean plants grown under elevated carbon dioxide concentrations, because plant leaves had increased sugar levels that served as a feeding stimulant (Hamilton et al., 2005). Undoubtedly, the impacts of invasive species under climate change will interact, perhaps synergistically, with other elements of global change in unpredictable ways. Thus, it is important to keep in mind that there remains very high uncertainty in the ability to predict what, how, and where new species will invade and existing invaders will spread.

6.3.6 Effects on Natural Resource Use and Human Communities

Climate change will also make products and activities based on natural resources, such as timber, maple syrup, and winter recreation, more vulnerable.

Timber Industry

Those managing forests for timber harvest will be faced with new challenges as climate change favors the competitive ability of some tree species over others and as range shifts occur in potential insect, disease, and invasive plant pests. Foraging and selective feeding by increasing deer populations will remain a problem and could become exacerbated by climate change. Some hardwoods currently grown in the region will not be suited to the new climate emerging this century (**Table 6.3**) (Iverson et al., 2008). However, there will be considerable variability among hardwoods. A modeling effort discussed previously (Section 6.3.1) suggests that longer growing seasons and increasing atmospheric carbon dioxide concentrations could increase productivity and growth rates of some hardwoods (Ollinger et al., 2008), while spruce and fir would not benefit because of their sensitivity to projected high temperatures.

Maple Syrup Industry

Although one study projected that the distribution of sugar maple will largely shift out of New York and into Canada during this century (Iverson et al., 2008), trees managed for sugar production are protected from competition, much as are agricultural crops, and are likely to remain part of the New York landscape. The majority of sugar maple in unmanaged forests could have a different future.

Maple sap flow requires days with alternating freezing and thawing. Currently the period with the greatest likelihood of such days is mid-March to early April, but this period is gradually shifting to occur earlier in the year. One study of sap production in four northeastern states, including New York, shows that as average winter temperature increases, sap production decreases (Rock and Spencer, 2001). Another study, which examined 40 years of weather records, found significant increases in potential sap-flow days for three Quebec stations and non-significant trends in the same direction for other Canadian stations, two sites in Vermont, and one site at Watertown, New York (MacIver et al., 2006). A study that compiled sap production records for the past 30 years in four northeastern states, including New York, found a trend for fewer sap-flow days, because the end of sap-flow has advanced by more days (come earlier in the year) than the onset of sap-flow (Perkins, personal communication). This study projects adverse impacts from climate warming on sap production should these trends continue. A study that drew on similar evidence concluded that climate warming is already contributing to a northward shift in maple sugar production (Frumhoff et al., 2007). A more recent analysis considering all these factors suggests that impacts of climate change on sap production in New York will vary greatly by region (Skinner et al., 2010), but that the industry should remain strong in many parts of New

York State through the end of this century and beyond. For a more detailed analysis, see Case Study C: Maple Syrup Industry, this chapter.

Winter Recreation and Lake-Effect Snows

The ski industry in New York will be vulnerable to climate change and the reductions in snow cover projected for the region. Increasing the use of artificial snowmaking is an adaptation that already is being used by the industry. However, a recent analysis concluded that the number of ski resorts that could continue to maintain a reasonable profit margin using this strategy will diminish to only those located at the highest elevations by end of century as snowfalls and snow cover duration continue to decline (Scott et al., 2008). Snowmaking may provide a sufficient number of years of buffer for some resorts to diversify and survive, such as by developing alternative winter activities and expanding summer recreation offerings. Even with adaptation, certain communities and individual operations that rely on ski tourism are likely to suffer. Those communities with economies linked with snowmobiling recreation will be particularly vulnerable because of their inability to compensate by making snow.

Although reductions in snow cover have already been occurring and are expected to continue for much of the state, the analysis is more complex for those regions subject to lake-effect snows. In the near term, warming lake temperatures and decreased ice cover will increase air humidity above the lakes, with the potential to cause increases in lake-effect snow during cold events that trigger snowfall, and this is consistent with observed increases in lake-effect snow in the recent years (Burnett et al., 2003). **Figure 6.7** presents a new analysis that illustrates how increasing lake temperatures above those recorded during a recent historical Lake Ontario snow event (November 9, 2008) would increase water-equivalent precipitation from the event by as much as 0.35 inches. The modeling study projected that this phenomenon of increased lake-effect snow with climate change would continue in the short term; by the end of this century, however, lake-effect snows are expected to decline by 50 to 90 percent, becoming lake-effect rain events as winter air temperatures become too warm to trigger snow (Kunkel et al., 2002).

6.4 Adaptation Strategies

New York ecosystems have adapted to climate change in the past, but the pace of change projected for this century is faster by several orders of magnitude than that of the most recent ice age transition and other historical events in the paleobiological record. There is a lack of reliable information and consensus regarding the future resilience and capacity of ecosystems to maintain

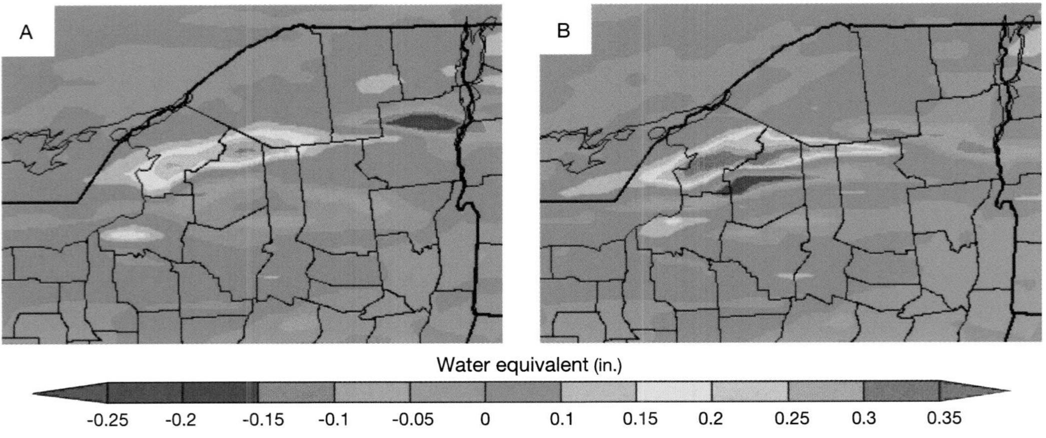

Note: Areas of increased and decreased lake effect snow are color coded showing inches of water equivalent. A) Weather conditions of wind and temperature gradients identical to an historic event recorded Nov. 9, 2008, but with lake and air temperatures uniformly increased by 1.8°F. B) The same conditions as A, but lake temperatures (not air) increased an additional 1.8°F (3.6°F total). Areas of red color show increases in lake-effect snow. These increase with further warming of water temperatures (B). (Weather Research and Forecasting model)

Figure 6.7 Simulations of the effects of climate change on lake-effect snow, in inches of water equivalent[1]

function through the replacement of lost species with new species that serve similar functions or by redundancy of function among species currently present.

The capacity of resource managers to facilitate ecosystem adaptation to rapid climate change is uncertain. A concern is that, to date, prior to the confounding effects of climate change, we have had only limited success with management interventions attempting to control species declines or invasions or undesirable damage by individual species. Many potential management interventions for coping with climate change exist, but most of these have not been tested on a wide scale and some are controversial even among experts in the field. The adaptation strategies proposed below are generally supported in the science literature and among the experts consulted for this study, but some may be considered too expensive or not cost-effective by policy-makers, unless better and more persuasive methods for documenting the value of ecosystems can be developed.

A few fundamentals for building the adaptive capacity of communities and ecosystems have emerged in this analysis:

- Maintain healthy communities and ecosystems more tolerant or better able to adapt to climate change by minimizing other biotic (e.g., insect infestations) and abiotic (e.g., acid rain, nitrogen deposition, drought) stressors.
- Manage primarily for ecosystem function and biodiversity rather than attempting to maintain indefinitely the current mix and relative abundance of species present today.
- Facilitate natural adaptation to climate change by improving connectivity among habitats to allow species dispersal, migration, and range shifts.

Below, we first describe adaptation options for specific habitats (Sections 6.4.1 to 6.4.4), followed by adaptations that would be implemented at the institutional or agency level (Section 6.4.5).

6.4.1 Forest, Grassland, and Alpine Communities

A recent review suggests that, in the context of climate change in the Northeast, it will be preferable to focus on future desired ecosystem function rather than aiming for specific species mixes (Evans and Perschel, 2009). Management strategies might, therefore, emphasize maintaining a diverse suite of species with some redundancy in function to hedge against loss of individual species. Diversity in species and tree age distribution will also help buffer against losses due to biotic or abiotic disturbance. Thinning and planting of trees can be designed to reduce the dominance and dependence of ecosystem function on tree species that are most vulnerable. However, the majority of older, intact forests should be maintained and allowed to evolve in their own way because of their ability to resist invasive species. Goals might include retaining selected legacy trees with heritage value, or habitats that can provide a seed source or refuge for plant and animal communities that are underrepresented in the landscape and are under stress due to climate change.

A key to adaptation is maintaining healthy tree stands, and from this standpoint many "best management practices" already suggested will be beneficial. This includes emphasis on low-impact harvest techniques, such as minimizing soil compaction (e.g., harvest when soils are relatively dry or frozen), and directional felling and careful removal of harvested trees. Biological or chemical control may be warranted in some cases for rapid-response containment of pests, disease, or invasive species, particularly for protection of unique habitats or species with irreplaceable function. Intervention solutions in alpine systems, however, will likely be problematic because of the multitude of sensitive and unique species.

6.4.2 Aquatic Ecosystems and Wetlands

Adaptation options exist for aquatic ecosystems and wetlands and include restoring and expanding riparian buffer zones, improving habitat connectivity, restoring legal protection, limiting water withdrawls, limiting invasives, and minimizing eutrophication.

Restoration and Expansion of Riparian Buffer Zones

Riparian (streamside) zones provide natural corridors for dispersal and migration of terrestrial and aquatic species and thus are vital to species shifting range in response to climate change. Other co-benefits of riparian zones include providing a unique and valuable terrestrial habitat, moderating flood and erosion damage, contributing to the energy and food web of adjacent aquatic communities, and shading streams and pools and thus providing cool-water refuges for coldwater fish in summer. The goals should be both to protect currently intact riparian zones and to restore those that have been degraded wherever practical. Options to accomplish this could include support of local governments with model ordinances, education and outreach, and support of voluntary conservation easement efforts. The New York State Open Space Conservation plan recommends a 100–300 foot (or more) zone around all streams that is free from physical development or high-impact activities, such as forestry, farming, or animal husbandry (www.dec.ny.gov/lands/47990.html).

Improve Habitat Connectivity by Removing Dams, Replacing Culverts

Dams and culverts (pipe-like constructions passing under roads) fragment habitats and limit dispersal potential for both animals and plants, which may make it difficult for them to shift their ranges in response to climate change. Programs at the federal and State level to develop inventories of abandoned and derelict obstructing dams that are barriers to fish and wildlife could be further developed. It would be beneficial to remove dams that are no longer necessary. Most culverts were not designed with consideration of their effects on aquatic and terrestrial species. Many are too long, some do not carry water year-round, and some are set at an elevation the wildlife and fish cannot access (L. Zicari, U.S. Fish and Wildlife Service, personal communication). For high-priority regions or species affected by climate change, redesign and replacement of these culverts to minimize barriers to aquatic and terrestrial species will be an important approach to building ecosystem adaptation capacity.

Restore Legal Protection to Isolated Wetlands

The New York State Department of Environmental Conservation protects wetlands larger than 12.4 acres, but this does not cover many small isolated wetlands, particularly fens and vernal pools that support a disproportionate amount of biological diversity relative to their total acreage (Comer et al., 2006). As a result of their scattered distribution across the landscape, these smaller wetlands also provide connectivity for the dispersal of many wetland species. These isolated wetlands need protection, as called for in the New York State Open Space Conservation Plan of 2009.

Limit Water Withdrawals that Affect Wetlands

Many wetland systems may be negatively affected by increased agricultural water use as summer soil water deficits intensify with climate change (see "Agriculture," Chapter 7, Case Study C: Drought). Land-use change and groundwater depletion by rural populations may also adversely affect many wetlands. A current high priority is to develop an inventory of wetlands and their landscape position in relation to hydrology and current and projected land and water use.

Limit Transport of Aquatic Invasive Species

Given that climate change is likely to increase the number of invasive species that will be able to survive and spread throughout New York's waters, limiting the transport of invasive species via infested boats and angling gear and from bait and aquarium releases will be increasingly important. A number of boat launch steward programs are currently in place in the Adirondacks (Adirondack Park Invasive Plant Program, personal communication); similar programs should be considered statewide. Regulatory approaches, such as enforceable aquatic invasive species transport laws, may be warranted.

Minimize Eutrophication

The impact of climate change on pollutant and nutrient loads to New York waterways is uncertain (see "Water Resources," Chapter 4). While increase in pollutant-laden runoff is possible in winter and early spring, algal growth response will be constrained by low temperatures during this time of the year. Of more concern might be diminished low-flows in late summer that lower dilution potential, increase summer water temperatures, and reduce dissolved oxygen. The State Pollutant Discharge Elimination System permitting processes and guidelines for combined sewer overflow releases may need to be revised in recognition of lower dilution potential in summer and fall. Excluding cattle from riparian zones can sometimes be more effective than more costly manure-management options (Easton et al., 2008).

iMap Invasives is an online, GIS-based, invasive species mapping tool (http://imapinvasives.org). This website now provides real-time information on the locations of numerous invasive species in New York State and allows individuals to report new locations of invasive pests. Private landowners, volunteers, and State and federal agencies all can play a role in monitoring for the hemlock woolly adelgid.

Adaptations for dealing with hemlock woolly adelgid include monitoring the spread of hemlock woolly adelgid and its impacts on forests and dependent wildlife species, education on control options as they emerge, and managing to reduce other stressors currently affecting hemlock forests, including overabundant deer populations and invasive plant species, both of which threaten forest regrowth following hemlock mortality.

Case Study B. Creative Approaches to Monitoring and Adaptive Management: New York's Invasive Species Program as a Model

The comprehensive adaptive management approach New York State has employed toward invasive species may serve as a useful model for adaptation to a wider range of emerging climate change challenges. The State's invasive species program provides a framework for coordination among local, State, and regional efforts; a broad educational outreach program; and research, information management, and regulatory policy recommendations.

In 2003, Governor George Pataki signed legislation convening the Invasive Species Task Force (ISTF, Laws of New York, 2003; Chapter 324). The Task Force was composed of representatives from diverse stakeholder groups, including key State agencies, environmental advocacy and non-profit organizations, academia, and trade and industry groups. In November 2005, the Invasive Species Task Force released a final report that outlined the invasive species problem, identified existing efforts and, most significantly, provided 12 strategic recommendations for action (ISTF, 2005). These recommendations have been codified into New York State law (Laws of New York, 2008; Chapter 26) and have significant funding from the state's Environmental Protection Fund.

To coordinate all invasive species efforts at the State level, a permanent leadership structure, which was modeled after the federal approach to invasive species, was established. It consists of an agency executive-level council and an advisory committee of non-government stakeholders. The council, advisory committee, and day-to-day statewide coordination are supported by the Office of Invasive Species Coordination at the Department of Environmental Conservation.

Building on existing grassroots partnerships that formed to address local invasive species concerns, the Invasive Species Task Force recommended the formation of eight Partnerships for Regional Invasive Species Management (PRISMs) (**Figure 6.9**). These partnerships coordinate local invasive species management functions, including engaging partners, recruiting and training citizen volunteers, delivering education and outreach, establishing early-detection monitoring networks, and implementing direct eradication and control efforts—all within the context of the local landscape. The Adirondack PRISM, also known as the Adirondack Park Invasive Plant Program, has served as a successful model for the other PRISMs, delivering educational programs and coordinating volunteer monitoring programs for terrestrial and aquatic invasive species since 1998 (http://www.adkinvasives.com). Due to the State fiscal crisis, most PRISMs have not yet received intended State funds, but do benefit from voluntary

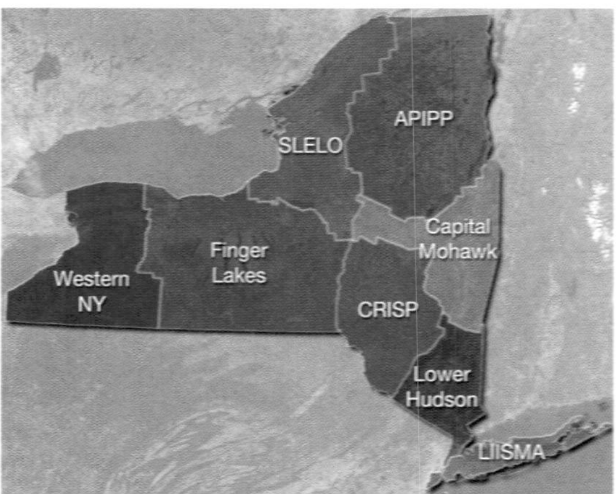

Note: Abbreviations are as follows: APIPP—Adirondack Park Invasive Plant Program; CRISP—Catskills Regional Invasive Species Partnership; LIISMA—Long Island Invasive Species Management Area; SLELO—St. Lawrence-Eastern Lake Ontario. Source: Brad Stratton, The Nature Conservancy.

Figure 6.9 The eight Partnerships for Regional Invasive Species Management (PRISMs)[3]

coordination and the in-kind support of partners. A strong communication network has also developed within and among PRISM partners to share educational resources, promote outreach events, and rapidly disseminate information about new invasions.

Other key Invasive Species Task Force recommendations now implemented as part of the State invasive species program include the following:

- The New York Invasive Species Research Institute, located at Cornell University. This group serves the scientific research community, natural resource and land managers, and State offices and State-sponsored organizations by promoting information-sharing and developing recommendations and implementation protocols for research, funding, and management of invasive species (http://nyisri.org).
- Use of iMapInvasives, an online, GIS-based, all-taxa invasive species mapping tool, coordinated by the New York Natural Heritage Program. The tool aggregates species records and locations from new observations and previously existing databases to provide a real-time, fully functional tool to serve the needs of volunteers and professionals working to manage invasive species (http://imapinvasives.org).
- The New York Invasive Species Information Clearinghouse, which is coordinated by the New York Sea Grant and Cornell Cooperative Extension. The Clearinghouse website is a comprehensive, online information portal (http://nyis.info) that provides stakeholders with links to scientific research, State and federal invasive species management programs and policy information, outreach education, and grassroots invasive species action in and around New York.

Case Study C. Maple Syrup Industry: Adaptation to Climate Change Impacts

Production of maple sugar products is based on sap flow from maple trees caused by positive internal sap pressures. These pressures are mostly from a physical process caused by freezing and thawing of a tree's woody tissues (Tyree, 1983). One analysis used historical data and climate models for individual states to project maple distribution and sugar production (Rock and Spencer, 2001). The study predicted an end to both the presence of sugar maple and to the maple industry in the northeastern United States by the end of this century. Another analysis, which used historical data from four northeastern states, concluded that, over the past 30 years, trees are being tapped for sap increasingly earlier and that sap flow is also ending earlier (Perkins, personal communication). The sap flow season is becoming shorter; the movement of the end of the season to earlier in the year is outpacing its earlier onset. A more recent study coupled a simple model for sap flow with downscaled global climate model results to project the number of sap flow days during the spring period and annually for about 10,000 locations across the northeastern United States (Skinner et al., 2010). This fine-scale analysis revealed that different parts of New York are likely to experience different impacts of climate warming on sugar production (**Figure 6.10**). Areas in New York at lower elevations and in southern counties have fewer days with freezing temperatures. In these areas, climate warming will force a continuing decrease in freezing temperatures with a resulting loss of sap production. In contrast, cooler parts of the state, at higher elevations and in northern New York, currently have fewer thawing days. The model predicts that, with warming, the number of days with sap flow will initially increase in these areas through the end of this century,

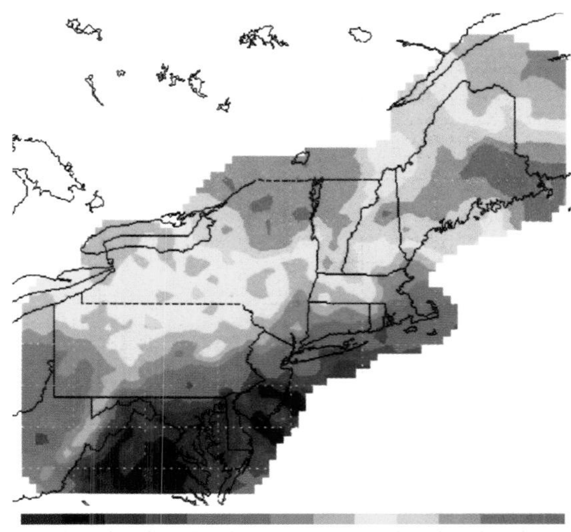

Note: The average change shown here is based on climate projections from the HadCM3 climate model (one of the 16 used in ClimAID), using the B1 emissions scenario. Northern areas in New York show an increase in sap flow days and southern areas a small decrease. Source: Based on data from Skinner et al., 2010

Figure 6.10 Average change in the total number of days (see color-coded scale at bottom) of modeled sap flow per season comparing the 1969–1999 historical climate data with projections for 2069–2099 period

followed by a decrease of days with sap flow with further warming after the end of this century. This analysis also shows that the sap flow season is moving earlier in the year such that by the end of the century tapping will begin in January rather than March. Eventually, it will merge with temperature conditions in November and December that are favorable for sap production.

Contrary to the prediction that the maple industry in New York will disappear by the end of the century (Rock and Spencer, 2001), this ClimAID analysis suggests that with adaptation to climate change the industry can remain viable for at least the next 100 years. There are several approaches to adaptation:

1) *Maintain attention on tree health through good forest management.* Competition from other tree species and pest impacts can be substantially reduced by existing management options. Research projects are under way to examine the optimal tree spacing for maximal growth and sugar production. Effective methods to control competing woody vegetation are also being studied.
2) *Begin tapping trees earlier in the year.* It is both essential and possible to move the sap production period to earlier in the season as the climate warms. Maple producers already pay considerable attention to weather forecasts to determine when to begin tapping. One analysis mentioned above (Skinner et al., 2010) predicts that the loss of production could amount to 14 days, if tapping begins at traditional times; normal seasons are 24 to 30 days long. If tapping begins earlier, there could be no net loss in number of sap flow days in warmer areas and there could be a net gain of sap flow days in cooler areas.
3) *Increase the sap yield from trees.* Recent research regarding why tap holes "dry up" has led to the introduction of a new type of spout. The main cause for loss of production from a tap hole relates to microorganisms plugging the xylem elements, which are the water-conducting elements of the tree. This is accelerated by increases in temperature and, thus, could be affected by a warming climate. The new spout has a check valve that prevents backflow of sap from the tubing into the tree, thus reducing the rate of microbial plugging. Initial results show a substantial production increase that could offset declining production from climate warming.
4) *Bring more maple trees into production.* One study, which uses U.S. Forest Service Forest Inventory Analysis data, estimates that in New York there are about 138 million sugar and 151 million red maples that are the correct size for tapping (Farrell, 2009). About 0.5 percent of these are currently used in sugar production. Vermont taps about 2 percent of its potential trees; Quebec taps about 30 percent of its trees. Thus, the potential to compensate for loss of production by bringing more trees into production and better utilizing red maples is enormous. Increasing the number of trees tapped seems to be occurring in response to economic incentives, as the price of syrup has increased dramatically in recent years.
5) *Increase use of red and silver maples for sugar production.* Whereas producers are currently tapping roughly 80 percent of the sugar maples on their own property, they are only using 20 percent of the available red maples (Farrell and Stedman, 2009). One of the main objections to using red and silver maples has been the lower sugar concentrations in the sap. However, with increased use of reverse osmosis to remove 80 to 90 percent of the water before boiling, this concern is not as great as it once was. Red maple (*Acer rubrum*) has a broader environmental tolerance than does sugar maple and is becoming the dominant tree species throughout the Northeast. It will be affected less by climate warming and tends to grow faster than sugar maple on a variety of sites. Thus, even if sugar maple disappears from New York's forests, syrup production could continue with better use of red maples.

Case Study D. Brook Trout: Reduction in Habitat Due to Warming Summers

The historical abundance of brook trout, New York's state fish, is likely to be severely reduced by climate warming, since it is currently located near the southern extent of its habitable range.

To examine the effects of regional warming on brook trout populations, three classes of water bodies in the Adirondack region were considered by ClimAID: 1) unstratified lakes, which have extensive water mixing during the summer and minimal temperature gradients with depth, 2) stratified lakes, which have deep zones that remain cold and unmixed with surface waters throughout mid-summer, and 3) streams and rivers.

Details of the analysis, including economic and social equity issues are provided below.

Unstratified Lakes

Primary findings are that brook trout in unstratified lakes, which represent about 41 percent of brook trout lakes in the Adirondacks (Scofield et al., 1993), will be most vulnerable to continued warming associated with climate change because of the lack of cold water refugia. Brook trout in streams and rivers will also be vulnerable, but may be less vulnerable than those in unstratified lakes. Least vulnerable will be those brook trout in stratified lakes where large, deep coldwater refugia are maintained (e.g., Great Lakes, Finger Lakes). However, the deep coldwater refugia in large stratified lakes can become oxygen depleted, and this stress may be exacerbated in many lakes by the lengthening summer season as a result of global warming.

A brook trout seasonal heat-stress index (Robinson et al., 2008 and 2010) was developed based on Rock Lake, an unstratified lake in the Adirondacks. The index uses a water degree-day metric that sums daily average lake-bottom temperatures above 68°F (e.g., a daily water temperature of 67°F would contribute 0 to the total, 68.5°F would contribute 0.5, and 71°F would contribute 3 degree-days). Annual reproductive success correlates with cumulative water degree-days over the summer ($r^2 = 0.85$). Reproductive success drops to zero at a water degree-day value of 365, i.e., years in which the average temperature at the lake bottom is much higher than 68°F for prolonged periods. Full mortality of the oldest age classes of brook trout was also observed in years with this heat index level. **Figure 6.11** illustrates that, for this class of unstratified lakes, the average air temperature observed from June 1 to September 30 accurately predicts lake temperature water degree-days. This is important because it indicates that climate model projections of air temperature can be reliably

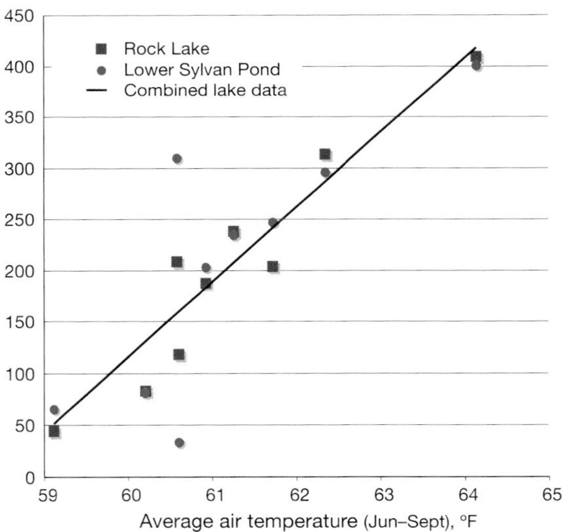

Note: Water degree days are a measure of predicted temperature stress on brook trout that takes into account both the amount of warming on single days and the total amount of time spent at the high temperatures (see text for more details). The y-axis is calculated from daily water temperature data throughout the summer using temperatures at maximum lake depths—6 meters (about 20 feet) for Rock Lake and 4 meters (about 13 feet) for Lower Sylvan Pond. Air temperature data are taken from the nearby Indian Lake weather station and daily values have been averaged into seasonal values on the x-axis. The regression line shown is fit to all data from both lakes and shows that the seasonal stress index can be accurately predicted from the average summer temperature of the air. Brook trout are predicted to be free of high temperature stress (degree days = 0) when average summer air temperature is below 58.4°F, and increases by 73 degree days for every one degree rise in the average summer temperature.

Figure 6.11 Cumulative water degree days related to seasonal air temperature for two Adirondack lakes: Rock Lake and Lower Sylvan Pond

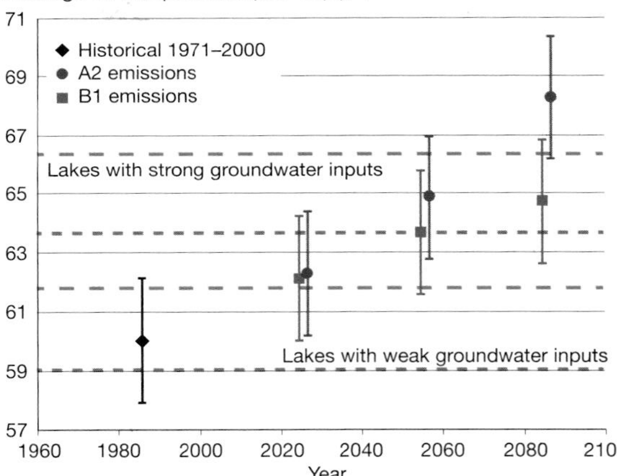

Note: Temperature projections for the lower-emissions B1 scenario and the higher-emissions A2 scenario by year, for the Adirondacks region. Projections are based on the B1 and A2 greenhouse gas emissions scenarios as indicated in the legend and utilizing five global climate models (GFDL, GISS, MIROC, CCSM and UKMO), a subset considered broadly representative of the full suite of 16 GCMs used by ClimAID. The green and brown horizontal lines represent the upper and lower boundary of the air temperature range where injury to brook trout will occur for unstratified lakes with strong groundwater inputs (green) and weak groundwater inputs (brown). Air temperatures exceeding the upper boundary of either range would lead to complete mortality for that lake class.

Figure 6.12 Climate projections for air temperatures under two emissions scenarios and potential damage to brook trout populations

used as indicators of trends in water temperature for unstratified lakes in the region.

Figure 6.12 illustrates the summer air temperature changes predicted for the Adirondack region for two different greenhouse gas emission scenarios and the effect this is likely to have on brook trout reproduction and survival. The lower threshold is the temperature where negative effects on brook trout reproduction would first be detected and the upper threshold is the temperature at which there would be complete elimination of reproduction and lethal effects on adult fish. The lower and upper thresholds for two unstratified lakes with differing levels of cold groundwater inputs are compared (groundwater inputs will have an overall cooling effect). The magnitude of groundwater inputs is controlled by soil depth in the surrounding basins, as determined by the thickness of till from past glaciations (Newton and Driscoll, 1990). Rock Lake is an example of an unstratified lake formed in thin glacial till, which results in weak groundwater inputs. Such lakes represent 56 percent of all unstratified Adirondack brook trout lakes. The vulnerability of these lakes to climate change is indicated by brown threshold lines in **Figure 6.12**. Thermal regimes in most years during the historical record from 1971 to 2000 were warm enough to adversely affect reproduction, but even the hottest years would not have caused full adult mortality. In contrast, by the 2020s, the hottest years will produce full mortality. While one single such year in isolation will not extirpate brook trout from a lake (because first-year fish can find thermal refuges in small shoreline groundwater seeps), two or three such years in succession would effectively eliminate all age cohorts. After the 2050s, even the average year will result in lethal temperatures, and brook trout will most likely not be viable in these lakes.

Temperatures monitored in lakes formed in thick glacial till and having high groundwater inputs (e.g., Panther Lake) indicated that cold groundwater was able to reduce average lake temperatures by 3.0°F relative to lakes in areas with thin till. This class represents only 20 percent of all unstratified lakes. The vulnerability of resident brook trout in lakes with high groundwater inputs to climate change is indicated by the green threshold lines in **Figure 6.12**. Under a high emissions scenario, none of these lakes would retain viable brook trout habitat, but under a low emissions scenario lethal temperatures occur in only the most extreme years, which would allow some brook trout populations to persist.

Stratified Lakes

Deep lakes and lakes with more color from algae and dissolved organic compounds develop a thermocline, which separates warm surface water from cold deeper water (i.e., become stratified). Weakly and strongly stratified lakes represent 59 percent of Adirondack brook trout lakes. Stressful warm temperatures are unlikely to occur below the thermocline in these lakes. However, these lakes are prone to oxygen depletion in deep waters that lack contact with the lake surface (Schofield et al., 1993). Oxygen levels often drop throughout the summer, and this stress may be exacerbated in many lakes by the lengthening summer season under global warming. Such dynamics require further study to determine how many lakes may develop serious oxygen deficiencies in the zones favorable to coldwater fish.

Rivers and Streams

Finally, rivers and streams make up a large fraction of the Adirondack waters fished for trout, though many of these bodies are stocked with hatchery-reared brown trout. One study, which examined brook trout that were released into a fifth-order river with radio transmitters and temperature sensors, showed that brook trout maintained body temperatures that averaged 4°F cooler than the temperature of the bulk river water; this difference increased to more than 7°F during periods when bulk river water was more than 68°F (Baird and Krueger, 2003). The brook trout were able to maintain lower body temperatures than that of the bulk river water by using cool refuges where tributary streams fed the larger river or pool bottoms were fed by groundwater seeps. Studies such as this emphasize the ability of brook trout to use thermal refugia when available. These studies also indicate, however, that bulk river temperatures are similar to the unstratified lakes discussed above and already are crossing thermal stress boundaries in mid-summer, with possible effects on brook trout reproduction success and adult mortality. Another study showed a similar pattern by which stocked brown trout also used thermal refuges during mid-summer in the Hudson River upstream from North Creek (Boisvert, 2008). Both surface-flow waters (e.g., where tributary streams feed the larger river) and groundwater seeps will increase in temperature with regional warming, and many rivers are likely to become too hot for brook or brown trout. More thermal monitoring is needed to define the prognosis for Adirondack rivers through the coming century.

Groundwater seeps are crucial to the thermal properties of the thick-till lakes discussed above and for the presence of thermal refugia in rivers and streams. Leaving aside the direct effects of climate change on air temperatures, groundwater supply is likely to become less reliable in the Adirondacks as global warming progresses. While the Adirondacks is likely to remain the wettest region of the state, it may nonetheless experience greater and more frequent levels of soil drying in the coming century. As a result, it may have a decrease in the abundant groundwater resource that supports thermal refugia.

Adaptation Options

Possible adaptations to ameliorate rising temperature effects on brook trout include maintaining or increasing vegetation that provides shade along stream, river, and lake shorelines, and minimizing disturbances that would impede water flows and groundwater inputs. More elaborate interventions for high-priority regions could include piping cold water from springs or lakes located at higher elevations to shoreline locations of thermally stressed lakes, and manipulations that might darken the "color" of the water in order to darken the propensity to form stable thermal stratification. Adding lime to some Adirondack lakes has already been practiced to partially compensate for pollutant acidity and promote primary production; primary production and a healthy level of natural algae also tends to darken water color and, thus, also shades the depths and promotes thermal stratification. This practice has not been approved in the context of thermal modification and could only be implemented if justified by further evaluation and after lake policy review.

Economics, Equity, and Environmental Justice Issues

Trout fishing is prominent in most of the state's major fishing areas, and trout is the second most popular group of species for recreational fishing in the state after black bass (Connelly and Brown, 2009a). To highlight the economic and equity issues associated with possible reduction of brook trout with climate change, a geographic region in the Adirondacks where brook trout are a key species for recreational fishing is analyzed (**Figure 6.13**). As described in Chapter 3 ("Equity and Economics"), the economy of the Adirondacks region depends heavily upon natural resource-related activities and tourism. Among the counties in the case study region, Herkimer, Lewis, and St. Lawrence are especially dependent on natural resources and agriculture as a share of total county employment (see Chapter 3, **Figure 3.4**). It also is important to note that all counties in the Adirondacks region have relatively high poverty rates and lower median income levels than the state overall (see Chapter 3, **Figures 3.1** and **3.2**), suggesting that these regions may face significant challenges adapting to all types of climate-change-related stresses.

Concerning fishing-related economic activities, **Figure 6.14** illustrates total fishing-related expenditures across all counties in New York in 2007 for all fish species. The map reveals that nearly all counties in the state benefit from fishing-related revenue, but counties in the case study region generally tend to have higher fishing-related expenditures than other counties. As illustrated in **Figure 6.15**, which estimates expenditures related specifically to trout fishing (based on estimates of percentage of angler days devoted to trout), trout represent an important component of fishing-related expenditures in the case study region. While the data used to construct **Figure 6.15** combine brook, brown,

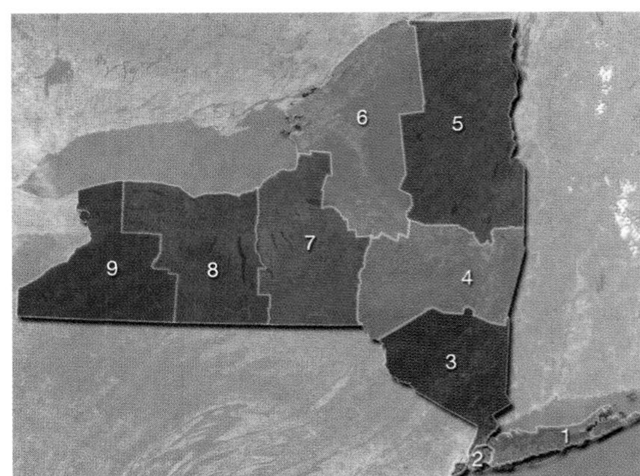

Figure 6.13 New York State Department of Environmental Conservation fishery management regions used for regional classification; the Adirondacks are located within regions 5 and 6

and rainbow trout, brook trout represent the most popular species of trout for anglers within the Adirondacks region. Moreover, anglers who are fishing specifically for brook trout are often willing to travel significant distances to lakes where this species is plentiful. Although other species are likely to replace brook trout under warmer temperatures, such species (e.g., bass) may not have the same type of appeal for out-of-town anglers—and particularly out-of-state anglers—who are willing to travel to the region for brook trout, but who would be able to fish for warm water species, such as bass, in areas closer to home.

Total expenditures in New York by anglers fishing in the Adirondacks case study region was estimated at $112 million in 2007 (**Table 6.5**) (personal communication, Nancy Connelly, based on 2007 New York Statewide Angler Survey). To determine how much of this was associated with the trout lakes identified above as being most vulnerable to loss of brook trout, this analysis assumes the following: 1) the fraction of the total expenditure related to trout fishing is proportional to the days spent fishing for trout (32.2 percent), and 2) trout fishing is equally divided in the Adirondacks between rivers and lakes. Together, there was an estimated $17.8 million in economic activity in 2007 associated with fishing for trout in Adirondack lakes. Forty percent of these lakes have been identified above as unstratified lakes, which are likely to lose their brook trout populations by the 2050s. The loss of brook trout in these lakes is associated with a total economic activity loss of $7.2 million annually, of which $4.8 million is spent at or near the fishing locations. Brook trout in summer-stratified lakes and in the river and stream systems are also threatened by rising temperatures as previously discussed, but specific predictions for these trout are not yet available.

The counties within the case study region may be especially vulnerable to loss of tourism revenue, as each has a significant presence of anglers from other regions in the state as well as from other states. Nearly half of the total angler days spent in the region are accounted for by anglers who live outside the region (**Table 6.5**). In terms of fishing-related expenditures within the region, which were estimated at approximately $74.5 million in 2007, local expenditures by anglers from other regions in the state and out-of-state regions represented more than 85 percent of this total (**Table 6.5**). The loss of revenue that is associated with anglers from other regions and states would represent a significant economic blow to the area's tourism-related industries, such as hotels, gas stations, and restaurants.

While loss of brook trout would hurt the region's fishing economy overall, such losses may have a

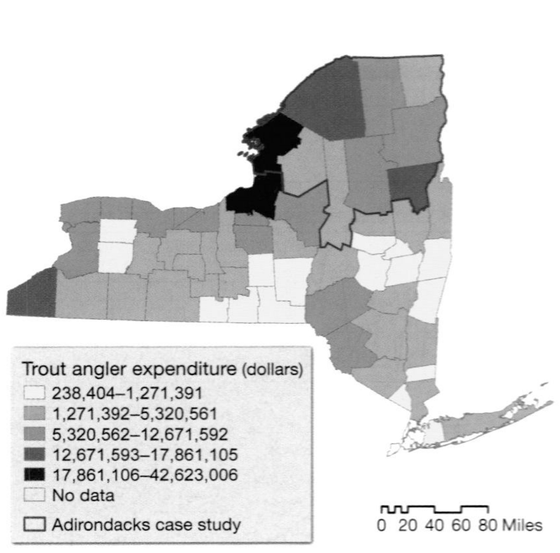

Source: Connelly and Brown (2009b) Statewide Angler Survey, NYSDEC

Figure 6.14 Total angler expenditure by county

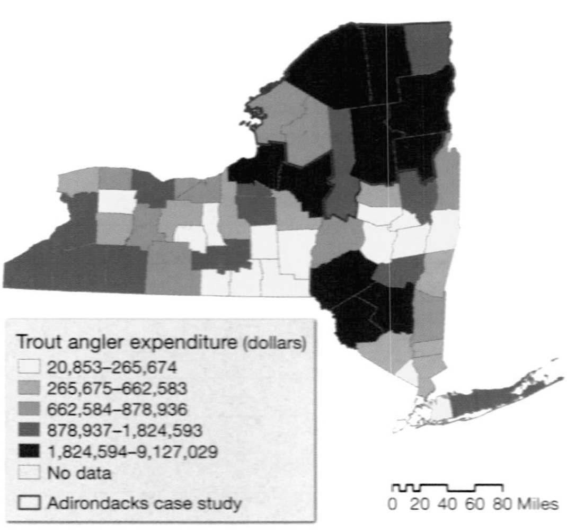

Note: Total fishing expenditures in the survey were translated into an estimate of expenditures for trout fishing by assuming that the percent of days spent fishing for different kinds of fish was equal to the percent of expenditure attributable to each kind of fish. Source: Connelly and Brown (2009b), Statewide Angler Survey, NYSDEC (authors' calculations)

Figure 6.15 Angler expenditure by county for trout fishing

Residence Areas of Anglers	Angler days		At-location Expenditures		En-route Expenditures	
	Number	Confidence limit ±	1,000s of $	Confidence limit ±	1,000s of $	Confidence limit ±
Total	2,912,938	200,203	$74,564	$6,613	$45,464	$4,761
Live in selected Adirondack region	1,241,905	150,836	$10,602	$2,601	$6,761	$1,431
Regions 5, 6—outside selected region	380,184	58,798	$6,878	$1,743	$5,030	$1,129
Regions 1, 2	59,995	9,826	$4,199	$1,551	$1,325	$292
Regions 3, 4	421,745	60,643	$14,698	$3,231	$9,762	$2,602
Regions 7, 8, 9	509,327	80,505	$17,763	$2,930	$11,902	$2,511
Out-of-state	299,794	36,001	$19,455	$3,134	$9,012	$2,256

Note: The selected Adirondack region was defined as the Department of Environmental Conservation regions 5 and 6, not including Washington, Saratoga, Fulton, and Oneida counties and not including fishing effort originating in the region on Lake Ontario, Lake Champlain, and the St. Lawrence River.

Table 6.5 Estimated number of angler days with at-location and en-route expenditures for fishing in the selected Adirondack region in 2007

disproportionate effect on small, fishing-dependent communities. Those areas that are dominated by unstratified lakes (which are likely to lose all of their trout) may also be particularly hard hit. Within fishing communities of the region, smaller tourism operators (e.g., fishing guides) may be most affected. They are likely to have limited ability to withstand any reduction in angler visits and may have limited capital to shift to other types of recreational businesses. Small, independently owned restaurants and hotels may be similarly vulnerable to reductions in angler expenditures by those living outside the region.

References

Aber, J., C. Goodale, S.V. Ollinger, M. Smith, A.H. Magill, M.E. Martin, R. Hallet, and J.L. Stoddard. 2003. "Is nitrogen deposition altering the nitrogen status of northeastern forests?" *BioScience* 53: 375–389.

Adams, M.B., J.A. Burger, A.B. Jenkins, et al. 2000. "Impact of harvesting and atmospheric pollution on nutrient depletion of eastern U.S. hardwood forests." *Forest Ecology and Management* 138(1–3): 301–319.

Anagnostakis, S. L. 1987. "Chestnut Blight: The Classical Problem of an Introduced Pathogen." *Mycologia* 79:23–37.

Arctic Climate Impact Assessment. 2004. "Impacts of a Warming Arctic: Arctic Climate Impact Assessment." Cambridge University Press, Cambridge, U.K.

Backlund, P., A. Janetos, and D. Schimel. 2008. "The Effects of Climate Change on Agriculture, Land Resources, Water Resources, and Biodiversity in the United States." In *Synthesis and Assessment Product 4.3*, U.S. Climate Change Science Program. Washington, D.C. www.climatescience.gov.

Baird, O.E. and C.C. Krueger. 2003. "Behavioral Thermoregulation of Brook and Rainbow Trout: Comparison of Summer Habitat Use in an Adirondack River, New York." *Transactions of the American Fisheries Society* 132:1194–1206.

Baldwin, N. S. 1956. "Food consumption and growth of brook trout at different temperatures." *Transactions of the American Fisheries Society*. 86:323–328.

Bernardo, J., R. Ossola, J. Spotila, and K.A. Crandall. 2007. "Validation of interspecific physiological variation as a tool for assessing global warming-induced endangerment." *Biology Letters* 3:695–698. DOI: 10.1098/rsbl.2007.0259.

Birdsey, R.A., J. Hayes, L.A. Joyce, and A.M. Solomon. 2009. "Forest Service Global Change Research Strategy 2009–2019." *USDA Forest Service/FS-917a*, 20 pp.

Bohlen, P.J., S. Scheu, C.M. Hale, M.A. McLean, S. Migge, P.M. Groffman, and D. Parkinson. 2004. "Non-native invasive earthworms as agents of change in northern temperate forests." *Frontiers in Ecology and the Environment* 2:427–435.

Boisvert, B.A. 2008. "The effect of pulsed discharge events on thermal refugia use by brown trout in thermally marginal streams." Master's thesis. Cornell University, Ithaca, New York.

Borwick, J., J. Buttle, and M.S. Ridgway. 2006. "A topographic index approach for identifying groundwater habitat of young-of-year brook trout (Salvelinus fontinalis) in the land-lake ecotone." *Canadian Journal of Fisheries and Aquatic Sciences*. 63:239–253.

Boylen, C.W., L.W. Eichler, and J.D. Madsen. 1999. "Loss of native aquatic plant species in a community dominated by Eurasian watermilfoil." *Hydrobiologia* 415:207–211.

Boylen, C.W., L.W. Eichler, J.S. Bartkowski, and S.M. Shaver. 2006. "Use of Geographic Information Systems to monitor and predict non-native aquatic plant dispersal through north-eastern North America." *Hydrobiologia* 570:243–248.

Breitenback, G.L., J.D. Congdon, and R.C. Sels. 1984. "Winter temperaturesof Chrysemys picta nests in Michigan: effects on hatchling survival." *Herpetologica* 40:76–81.

Brett, J.R. 1971. "Energetic responses of salmon to temperature a study of some thermal relations in the physiology and fresh water ecology of sockeye salmon *Oncorhynchus nerka*." *American Zoologist* 11:99113.

Brook, R.W., R.K. Ross, K.F. Abraham, D.I. Fronczak, and J.C. Davies. 2009. "Evidence for Black Duck Winter Distribution Change." *Journal of Wildlife Management* 73:98-103.

Brooks, R.T. 2009. "Potential impacts of global climate change on the hydrology and ecology of ephemeral freshwater systems of the forests of the northeastern United States." *Climatic Change* 95:469–483.

Burnett, A.W., M.E. Kirby, H.T. Mullins and W.P. Patterson. 2003. "Increasing Great Lake–effect snowfall during the twentieth century: A regional response to global warming?" *Journal of Climate* 16:3535–3542.

Burnett, T. 1949. "The effect of temperature on an insect host-parasite population." *Ecology* 30:113–134.

Butler, C.J. 2003. "The disproportionate effect of global warming on the arrival dates of short-distance migratory birds in North America." *Ibis* 145: 484–495.

Campbell, J.L., L.E. Rustad, E.W. Boyer, S.F. Christopher, C.T. Driskoll, I.J. Fernandez, P.M. Groffman, D. Houle, J. Kiekbusch, A.H. Magill, M.J. Mitchell, and S.V. Ollinger. 2009. "Consequences of climate change for biogeochemical cycling in forests of northeastern North America." *Canadian Journal of Forest Research* 39(2):264–284.

Chatrchyan, A., S. Broussard Allred, and R. Schneider. 2010. Resilience in the Face of Climate Change: Empowering Natural Resource Managers and Professionals through Extension Education. 7th Natural Resource Extension Professionals Conference, Fairbanks, AK, June 27–30, 2010. Available online at: http://www.slideshare.net/Shorna_Allred/resilience-in-the-face-of-climate-change-empowering-natural-resource-managers-and-professionals-through-extension-education

Clark, J.S., M. Lewis, J.S. McLachlan, and J. HilleRisLambers. 2003. "Estimating population spread: what can we forecast and how well?" *Ecology* 84:1979–1988.

Coakley, S.M., H. Scherm, and S. Chakraborty. 1999. "Climate change and plant disease management." *Annual Review of Phytopathology* 37:399–426.

Comer, P., K. Goodin, A. Toamino, G. Hammerson, G. Kittel, S. Menard, C. Nordman, M. Pyne, M. Reid, L. Sneddon, and K. Snow. 2006. "Biodiversity Values of Geographically Isolated Wetlands in the United States." *NatureServe* Arlington, VA.

Connelly, N. and T. Brown. 2009a. "New York Statewide Angler Survey 2007. Report 1: Angler Effort and Expenditures." New York State Department of Environmental Conservation, Bureau of Fisheries.

Connelly, N. and T. Brown. 2009b. "New York Statewide Angler Survey 2007. Report 1: Estimated Angler Effort and Expenditures in New York State Counties." New York State Department of Environmental Conservation, Bureau of Fisheries.

Costanza, R., M. Wilson, A. Troy, A. Voinov, S. Liu, and J. D'Agostino. 2006. "The Value of New Jersey's Ecosystem Services and Natural Capital." New Jersey Department of Environmental Protection. Trenton, NJ.

Curry, R.A., and D.L.G. Noakes. 1995. "Groundwater and the selection of spawning sites by brook trout, *Salvelinus fontinalis*." *Canadian Journal of Fisheries and Aquatic Sciences* 52:1733–1740.

Crooks, J.A. 2002. "Characterizing ecosystem-level consequences of biological invasions: the role of ecosystem engineers." *Oikos* 97:153–166.

Daily, G.C., S. Polasky, J. Goldstein, P.M. Kareiva, H.A. Mooney, L. Pejchar, T.H. Ricketts, J. Salzman, and R. Shallenberger. 2009. "Ecosystems services in decision making: time to deliver." *Frontiers in Ecology and the Environment* 7(1):21–28.

DeGaetano, A.T., M.D. Cameron and D.S. Wilks. 2001. "Physical simulation of maximum seasonal soil freezing depths in the United States using routine weather observations." *Journal of Applied Meteorology* 40:546–555.

DiTommaso, A., F.M. Lawlor, and S.J. Darbyshire. 2005. "The biology of invasive alien plants in Canada. 2. Cynanchum rossicum (Kleopow) Borhidi [= Vincetoxicum rossicum (Kleopow) Barbar.] and Cynanchum louiseae (L.) Kartesz & Gandhi [= Vincetoxicum nigrum (L.) Moench]." *Canadian Journal of Plant Science* 85:243–263.

Dukes, J.S. and H.A. Mooney. 1999. "Does global change increase the success of biological invaders?" *Trends in Ecology & Evolution* 14:135–139.

Dukes, J.S., J. Pontius, D. Orwig, J.R. Garnas, V.L. Rodgers, N. Brazee, B. Cooke, K.A. Theoharides, E.E. Stange, R. Harrington, J. Ehrenfeld, J. Gurevitch, M. Lerdau, K. Stinson, R. Wick, and M. Ayres. 2009. "Responses of insect pests, pathogens, and invasive plant species to climate change in the forests of northeastern North America: What can we predict?" *Canadian Journal of Forestry Research* 39:231–248.

Dunn, P.O., and D.W. Winkler. 1999. "Climate change effect on breeding date in tree swallows." *Proceedings of the Royal Society of London* 266:2487–2490.

Easton, S.M., M.T. Walter, and T.S. Steenhuis. 2008. "Combined monitoring and modeling indicate the most effective agricultural best management practices." *Journal of Environmental Quality* 37:1798–1809.

Edinger, G.J., D.J. Evans, S. Gebauer, T.G. Howard, D.M. Hunt, and A.M. Olivero, eds. 2002. "Ecological Communities of New York State. Second Edition: A revised and expanded edition of Carol Reschke's Ecological Communities of New York State." (Draft for review). New York Natural Heritage Program, New York State Department of Environmental Conservation, Albany, NY.

Eschtruth, A.K. and J.J. Battles. 2009. "Assessing the relative importance of disturbance, herbivory, diversity, and propagule pressure in exotic plant invasion." *Ecological Monographs* 79(2):265–280.

Evans, A.M. and R. Perschel. 2009. "A review of forestry mitigation and adaptation strategies in the Northeast U.S." *Climatic Change* 96:167–183.

Fagre, D.B., D.W. Charles, C.D. Allen, F.S. Chapin, P.M. Groffman, G.R. Gunenspergen, A.K. Knapp, A.D. McGuire, P.J. Mulholland, D. Peters, D.D. Roby, and G. Sugihara. 2009. "Thresholds of Climate Change in Ecosystems." In *US Climate Change Science Program, Synthesis and Assessment Product 4.2*. Washington, D.C.

Farnsworth, E.J. and L.A. Meyerson. 2003. "Comparative ecophysiology of four wetland plant species along a continuum of invasiveness." *Wetlands* 23:750–762.

Farrell, M. 2009. "Assessing the growth potential and future outlook for the US maple syrup industry." In *Agroforestry Comes of Age: Putting Science into Practice,* edited by M.A. Gold and M.M. Hall, 99–106. Proceedings of the 11th North American Agroforestry Conference, Columbia, Missouri, May 31–June 3.

Farrell, M.L. and R.C. Stedman. 2009. "Survey of New York State Landowners." *Report to the Steering Committee of the Lewis County Maple Syrup Bottling Facility*. 20 p.

Fernandez, I.J., L.E. Rustad, S.A. Norton, J.S. Kahl, and B.J. Cosby. 2003. "Experimental acidification causes soil base-cation depletion at the Bear Brook watershed in Maine." *Soil Science Society of America Journal* 67:1909–1919.

Frick, W.S., D.S. Reynolds and T.H. Kunz. 2010. "Influence of climate and reproductive timing on demography of little brown myotis, *Myotis lucifugus*." *Journal of Animal Ecology* 79:128–136.

Foden, W., G. Mace, J.C. Vié, A. Angulo, S. Butchart, L. DeVantier, H. Dublin, A. Gutsche, S. Stuart and E. Turak. 2008. "Species susceptibility to climate change impacts." In *The 2008 Review of The IUCN Red List of Threatened Species*, edited by J.C. Vié, C. Hilton-Taylor and S.N. Stuart. IUCN Gland, Switzerland.

Forman, J. and R.V. Kesseli. 2003. "Sexual reproduction in the invasive species *Fallopia japonica* (Polygonaceae). Am. J. Bot. 90:586–592.

Frazer, N.B., J.L. Greene and J.W. Gibbons. 1993. "Temporal variation in growth rate and age at maturity of male painted turtles, *Chrysemys picta*." *American Midland Naturalist* 130:314–324.

Frumhoff P.C., J.J. McCarthy, J.M. Melillo, S.C. Moser, D.J. Wuebbles. 2007. *Confronting Climate Change in the U.S. Northeast: Science, Impacts and Solutions. Synthesis report of the Northeast Climate Impacts Assessment (NECIA)*. Union of Concerned Scientists, Cambridge, MA.

Gandy, M. 2002. *Concrete and Clay: Reworking Nature in New York City*. Cambridge: Massachusetts Institute of Technology Press.

Gibbs, J.P., A.R. Breisch, P.K. Ducey, G. Johnson, J.L. Behler, and R.C. Bothner. 2007. *The amphibians and reptiles of New York State: identification, natural history and conservation*. Oxford: Oxford University Press.

Gibbs, J.P., and A.R. Breisch. 2001. "Climate warming and calling phenology of frogs near Ithaca, New York, 1900–1999." *Conservation Biology* 15:1175–1178.

Gorchov, D.L. and D.E. Trisel. 2003. "Competitive effects of the invasive shrub, *Lonicera maackii* (Rupr.) Herder (Caprifoliaceae), on the growth and survival of native tree seedlings." *Plant Ecology* 166:13–24.

Gordon, D.R., D.A. Onderdonk, A.M. Fox and R.K. Stocker. 2008. "Consistent accuracy of the Australian weed risk assessment system across varied geographies." *Diversity and Distributions* 14:234–242.

Gravatt, G.F. 1949. "Chestnut blight in Asia and North American." *Unasylva* 3:3–7.

Grime, J.P., J.D. Fridley, A.P. Askew, K. Thompson, J.G. Hodgson and C.R. Bennett. 2008. "Long-term resistance to simulated climate change in an infertile grassland." *Proceedings of the National Academy of Sciences* 105(29):10028–10032.

Groffman, P.M., C.T. Driscoll, T.J. Fahey, J.P. Hardy, R.D. Fitzhugh and G.L. Tierney. 2001. "Effects of mild winter freezing on soil nitrogen and carbon dyamics in a northern hardwood forest." *Biogeochemistry* 56:191–213.

Gu, L., P.J. Hanson, W.M. Post, D.R. Kaiser, B. Yang, R. Nemani, S.G. Pallardy, T. Meyers. 2008. "The 2007 Eastern U.S. spring freeze: Increased cold damage in a warming world?" *BioScience* 58(3):253–262.

Hamilton, J.G., O. Dermody, M. Aldea, A.R. Zangerl, A. Rogers, M.R. Berenbaum and E.H. Delucia. 2005. "Anthropogenic Changes in Tropospheric Composition Increase Susceptibility of Soybean to Insect Herbivory." *Environmental Entomology* 479–485.

Hatre, J. and R. Shaw. 1995. "Shifting Dominance within a montane vegetation community: results of a climate-warming experiment." *Science* 267:876–880

Harvell, C.D., C.E. Mitchell, J.R. Ward, S. Altizer, A.P. Dobson, R.S. Ostfeld and M.D. Samuel. 2002. "Climate warming and disease risks for terrestrial and marine biota." *Science* 296:2158–2162.

Hayhoe, K., C. Wake, T. Huntington, L. Luo, M. Schwartz, J. Sheffield, E. Wood, B. Anderson, J. Bradbury, A. DeGaetano, T. Troy and D. Wolfe. 2007. "Past and future changes in climate and hydrological indicators in the U.S. Northeast." *Climate Dynamics* 28:381–407.

Hellmann, J.J., J.E. Byers, B.G. Bierwagen and J.S. Dukes. 2008. "Five potential consequences of climate change for invasive species." *Conservation Biology* 22:534–543.

Herkert, J.R. 1994. "The Effects of Habitat Fragmentation on Midwestern Grassland Bird Communities." *Ecological Applications* 4(3):461–471.

Heynen, N., H.A. Perkins and P. Roy. 2006. "The political ecology of uneven urban green space – The impact of political economy on race and ethnicity." In "Producing Environmental Inequality in Milwaukee." *Urban Affairs Review* 42:3–25.

Hobbs, R.J. and H.A. Mooney. 2005. "Invasive species in a changing world: The interactions between global change and invasives." In *Invasive alien species: A new synthesis,* edited by H.A. Mooney, R.N. Mack, J.A. McNeely, L.E. Neville, P.J. Schei and J.K. Waage, 310–331. Washington, D.C., Island Press.

Hokanson, K.E.F., J.H. McCormick, B.R. Jones and J.H. Tucker. 1973. "Thermal requirements for maturation, spawning, and embryo survival of the brook trout, *Salvelinus fontinalis*." *Journal of the Fisheries Research Board of Canada* 30:975–84.

Horsley, S.B. and D.A. Marquis. 1983. "Interference of weeds and deer with Allegheny hardwood reproduction." *Canadian Journal of Forest Research* 13:61–69.

Howarth, R.W., D.P. Swaney, E.W. Boyer, R. Marino, N. Jaworski and C. Goodale. 2006. "The influence of climate on average nitrogen export from large watersheds in the Northeastern U.S." *Biogeochemistry* 79(1–2):163–186.

Hudy, M., T.M. Thieling, N. Gillespie and E.P. Smith. 2005. "Distribution, status, and threats to brook trout within the eastern United States." Report submitted to the Eastern Brook Trout Joint Venture, International Association of Fish and Wildlife Agencies, Washington, D.C.

Hubacek, K., J.D. Erickson and F. Duchin. 2002. "Input–output modeling of protected landscapes: the Adirondack Park." *The Review of Regional Studies* 32(2):207–222.

Huggett, R.J. Jr., E.A. Murphy and T.P. Holmes. 2008. "Forest disturbance impacts on residential property values." In *The Economics of Forest Disturbances: wildfires, storms, and invasive species,* edited by T. Holmes, J. Prestemon, K. Abt, 209–228. Dordrecht, The Netherlands: Springer.

Inkley, D.B., M.G. Anderson, A.R. Blaustein, V.R. Burkett, B. Felzer, B. Griffith, J. Price and T.L. Root. 2004. "Global climate change and wildlife in North America." *Wildlife Society Technical Review* 04–2. The Wildlife Society, Bethesda, Maryland.

Inouye, D.W. and A.D. McGuire. 1991. "Effects of snowpack on timing and abundance of flowering in *Delphinium nelsonii* (Ranunculaceae): implications for climate change." *American Journal of Botany* 78(7): 997–1001.

ISTF. 2005. Final Report of the New York State Invasive Species Task Force. New York State Department of Environmental Conservation.

Iverson, L., A. Prasad and S. Matthews. 2008. "Potential changes in suitable habitat for 134 tree species in the northeastern United States." *Mitigation and Adaptation Strategies of Global Change* 13: 487–516.

Iverson, L.R. and A.M. Prasad. 2002. "Potential tree species shifts with five climate change scenarios in the eastern United States." *Forestry Ecology and Management* 155:205–222.

Janzen, F.J. 1994. "Climate change and temperature-dependent sex determination in reptiles." *Proceedings of the National Academy of Sciences of the United States of America* 91:7487–7490.

Jenkins, J. 2010. *Climate Change in the Adirondacks: The Path to Sustainability*. Ithaca, NY: Cornell University Press.

Keast A. 1984. "The introduced aquatic macrophyte, *Myriophyllum spicatum*, as habitat for fish and their invertebrate prey." *Canadian Journal of Zoology-Revue Canadienne De Zoologie* 62:1289–1303.

Kellermann, V., B.v. Heerwaarden, C.M. Sgro, and A.A. Hofmann. 2009. "Fundamental evolutionary limits in ecological traits drive Drosophila species distributions." *Science* 325:1244–1246.

King, J.R., B.J. Shuter and A.P. Zimmerman. 1999. "Empirical links between thermal habitat, fish growth, and climate change." *Transactions of the American Fisheries Society* 128:656–665.

Kolar, C.S. and D.M. Lodge. 2002. "Ecological predictions and risk assessment for alien fishes in North America." *Science* 298:1233–1236.

Kovacs, K.F., R.G. Haight, D.G. McCullough, R.J. Mercader, N.W. Siegert and A.M. Liebhold. 2009. "Cost of potential emerald ash borer damage in U.S. communities, 2009–2019." *Ecological Economics* (In Press).

Krupa, S., M.T. McGrath, C. Andersen, F.L. Booker, K.O. Burkey, A. Chappelka, B. Chevone, E. Pell and B. Zillinskas. 2001. "Ambient ozone and plant health." *Plant Disease* 85:4–17.

Kunkel, K.E., N.E. Westcott and D.A.R. Kristovich. 2002. "Assessment of potential effects of climate change on heavy Lake-effect snowstorms Near Lake Erie." *Journal of Great Lakes Research* 28:521–536.

Levine, J.M. and C.M. D'Antonio. 2003. "Forecasting biological invasions with increasing international trade." *Conservation Biology* 17:322–326.

Liebhold, A.M., T.T. Work, D.G. McCullough and J.F. Cavey. 2006. "Airline baggage as a pathway for alien insect species invading the United States." *American Entomologist* 52:48–54.

Lovett, G.M., C.D. Canham, M.A. Arthur, K.C. Weathers and R.D. Fitzhugh. 2006. "Forest ecosystem responses to exotic pests and pathogens in eastern North America." *Bioscience* 56:395–405.

Mack, R.N., D. Simberloff, W.M. Lonsdale, H. Evans, M. Clout and F.A. Bazzaz. 2000. "Biotic invasions: causes, epidemiology, global consequences, and control." *Ecological Applications* 10:689–710.

MacIver D.C., M. Karsh, N. Comer, J. Klaassen, H. Auld and A. Fenech. 2006. "Atmospheric influences on the sugar maple industry of North America." Environment Canada, Adaptation and Impacts Research Division. Occasional Paper 7.

McCullough, D.G., T.T. Work, J.F. Cavey, A.M. Liebhold and D. Marshall. 2006. "Interceptions of nonindigenous plant pests at US ports of entry and border crossings over a 17-year period." *Biological Invasions* 8:611–630.

McGowan, K.J. and K. Corwin, eds. 2008. *The Second Atlas of Breeding Birds in New York State*. Ithaca, NY: Cornell University Press.

McMahon, S.M., G.G. Parkera and D.R. Millera. 2010. "Evidence for a recent increase in forest growth." *Proceedings of the National Academy of Sciences of the United States of America Early Edition*. www.pnas.org/cgi/doi/10.1073/pnas.0912376107.

Mehlman, D.W. 1997. "Change in avian abundance across the geographic range in response to environmental change." *Ecological Applications* 7(2):614–624.

Meisner, J.D. 1990a. "Potential loss of thermal habitat for brook trout, due to climatic warming, in two southern Ontario streams." *Transactions of the American Fisheries Society*. 119:282–291.

Meisner, J.D. 1990b. "Effect of climate warming on the southern margins of the native range of brook trout, *Salvelinus fontinalis*." *Canadian Journal of Fisheries and Aquatic Sciences*. 47:1065–1070.

Meyers, T.R. and J. R. Winton. 1995. "Viral hemorrhagic septicemia virus in North America." *Annual Review of Fish Diseases* 5:3–24.

Meyerson, L.A., K. Saltonstall and R.M. Chambers. 2009. "*Phragmites australis* in eastern North America: a historical and ecological perspective." In *Salt marshes under global siege*, edited by B.R. Silliman, E. Grosholz and M.D. Bertness, 57–82. Berkeley, CA: University of California Press.

Michaels, S., R. Mason and W. Solecki. 1999. "Motivations for ecostewardship partnerships: examples from the Adirondack Park." *Land Use Policy* 16:1–9.

Mohan, J.E., R.M. Cox and L.R. Iverson. 2009. "Composition and carbon dynamics of forests in northeastern North America in a future warmer world." *Canadian Journal of Forest Research* 39(2):213–230.

Mohan, J.E., J.S. Clark and W.H. Schlesinger. 2007. "Long-term CO_2 enrichment of a forest ecosystem: implications for forest regeneration and succession." *Ecological Applications* 17:1998–1212.

Murray, D.L., E.W. Cox, W.B. Ballard, H.A. Whitlaw, M.S. Lenarz, T.B. Custer and T.K. Fuller. 2006. "Pathogens, Nutritional Deficiency, and Climate Influences on a Declining Moose Population." *Wildlife Monographs* 166:1–30.

NEFA (North East Foresters Association). 2007. "The Economic Importance of Wood Flows from New York's Forests." www.nefa.org.

Nelson, E., G. Mendoza, J. Regetz, S. Polasky, H. Tallis, D.R. Cameron, K. Chan, G.C. Daily, J. Goldstein, P.M. Karelva, E. Lonsdorf, R. Naidoo, T.H. Ricketts and M.R. Shaw. 2009. "Modeling multiple ecosystems services, biodiversity conservation, commodity production, and tradeoffs at landscape scales." *Frontiers in Ecology and the Environment* 7(1):4–11.

New England Regional Assessment Group. 2001. "Preparing for a Changing Climate, The Potential Consequences of Climate Variability and Change, New England Regional Assessment Overview." *U.S. Global Change Research Program*, University of New Hampshire, Durham, NH

NYSDEC (New York State Department of Environmental Conservation). 2009. Press Release: Emerald Ash Borer Found in New York State. http://www.dec.ny.gov/press/55725.html.

NYSDEC (New York State Department of Environmental Conservation). 2007. *Checklist of the amphibians, reptiles, birds and mammals of New York, including their protective status.* Eighth revision.

New York State GAP Analysis Program. www.dnr.cornell.edu/gap/land/land.html.

Newton, R.M. and C.T. Driscoll. 1990. "Classification of ALSC lakes." In *Adirondacks Lake Survey: An Interpretive Analysis of Fish Communities and water chemistry 1984–87*, 2-70 to 2-91. Adirondack Lakes Survey Corporation, Ray Brook, NY.

NRC (National Research Council). 2005. *Valuing Ecosystem Services: Toward Better Environmental Decision-Making.* Washington D.C.: National Academies Press.

NYSERDA (New York State Energy Research and Development Authority). 2009. "Response of Adirondack Ecosystems to Atmospheric Pollutants and Climate Change at the Huntington Forest and Arbutus watershed: Research Findings and Implications for Public Policy." Report 09–08, NYSERDA 4917. http://www.nyserda.org/publications/09-08response_of_adirondack_ecosystems.pdf.

Ollinger, S.V., C.L. Goodale, K. Hayhoe and J.P. Jenkins. 2008. "Potential effects of climate change and rising CO_2 on ecosystem processes in northeastern U.S. forests." *Mitigation and Adaptation Strategies of Global Change* 13:467–486.

Paradis, A., J. Elkinton, K. Hayhoe and J. Buonaccorsi. 2008. "Role of winter temperature and climate change on the survival and future range expansion of the hemlock woolly adelgid (*Adelges tsugae*) in eastern North America." *Mitigation and Adaptation Strategies for Global Change* 13:541–554.

Parmesan, C. 2007. "Influences of species, latitudes and methodologies on estimates of phenological response to global warming." *Global Change Biology* 13:1860–1872.

Parmesan, C. and G. Yohe. 2003. "A globally coherent fingerprint of climate change impacts across natural systems." *Nature* 421:37–41.

Petit, R.J., F.S. Hu and C.W. Dick. 2008. "Forests of the past: a window to future changes." *Science* 320(5882):1450–1452.

Pimentel, D., R. Zuniga and D. Morrison. 2005. "Update on the environmental and economic costs associated with alien-invasive species in the United States." *Ecological Economics* 52:273–288.

Poff, N.L., M.M. Brinson, J.W. Day Jr. 2002. *Aquatic ecosystems & Global climate change: Potential Impacts on Inland Freshwater and Coastal Wetland Ecosystems in the United States.* Prepared for the Pew Center on Global Climate Change.

Pörtner, H.O. and A.P. Farrell. 2008. "Physiology and climate change." *Science* 322:690–692.

Primack, D., C. Imbres, R.B. Primack, A.J. Miller-Rushing and P. Del Tredici. 2004. "Herbarium species demonstrate earlier flowering times in response to warming in Boston." *American Journal of Botany* 91(8):1260–1264.

Pyŝek, P., M. Kopecky, V. Jarosik and P. Kotkova. 1998. "The role of human density and climate in the spread of *Heracleum mantegazzianum* in the Central European landscape." *Diversity and Distributions* 4:9–16.

Rahel F.J. and J.D. Olden. 2008. "Assessing the effects of climate change on aquatic invasive species." *Conservation Biology* 22:521–533.

Reis, R.D. and S.A. Perry. 1995. "Potential effects of global climate warming on brook trout growth and prey consumption in central Appalachian streams, USA." *Climate Research* 5:197–206.

Renecker, L.A. and R.J. Hudson. 1986. "Seasonal energy expenditures and thermoregulatory responses of moose." *Canadian Journal of Zoology* 64:322–327.

Rich, J. 2008. "Winter nitrogen cycling in agroecosystems as affected by snow cover and cover crops." M.S. Thesis, Cornell University. Ithaca, NY.

Richardson, D.M., P. Pysek, M. Rejmanek, M.G. Barbour, F.D. Panetta and C.J. West. 2000. "Naturalization and invasion of alien plants: concepts and definitions." *Diversity and Distributions* 6:93–107.

Robinson, J.M. 2008. "Effects of summer thermal conditions on brook trout (*Salvelinus fontinslis*) in an unstratified Adirondack Lake." Master's thesis. Cornell University, Ithaca, New York.

Robinson, J.M., D.C. Josephson, B.C. Weidel and C.E. Kraft. 2010. "Influence of variable interannual summer water temperatures on brook trout growth, consumption, reproduction and mortality in an unstratified Adirondack lake." *Transactions of the American Fisheries Society*. (In press.)

Rock, B and S. Spencer. 2001. "Preparing for a Changing Climate, The Potential Consequences of Climate Variability and Change, New England Regional Overview." *U.S. Global Change Research Program*. University of New Hampshire, Durham, NH.

Rodenhouse, N.L., L.M. Christenson, D. Parry and L.E. Green. 2009. "Climate change effects on native fauna of northeastern forests". *Canadian Journal of Forest Research* 39:249–263.

Rodenhouse, N.L., S.N. Matthews, K.P. McFarland, J.D. Lambert, L.R. Iverson, A. Prasad, T.S. Sillett and R.T. Holmes. 2008. "Potential effects of climate change on birds of the Northeast." *Mitigation and Adaptation Strategies for Global Change* 13:517–540.

Rustad, L.E., I.J. Fernandez, M.B. David, M.J. Mitchell, K.J. Nadelhoffer and R.B. Fuller. 1996. "Experimental soil acidification and recovery at the Bear Brook watershed in Maine." *Soil Science Society of America Journal* 60:1933–1943.

Schaberg, P.G. and D.H. DeHayes. 2000. "Physiology and environmental cause of freezing injury in red spruce." In *Responses of northern U.S. forests to environmental change*, edited by R.A. Mickler, R.A. Birdsey, and J. Hom. Springer-Velag. *Ecological Studies* 139, New York.

Schofield, C.L., D. Josephson, C. Keleher and S.P. Gloss. 1993. "Thermal stratification of dilute lakes—an evaluation of regulatory processes and biological effects before and after base addition effects: effects on brook trout habitat and growth." U.S. Fish and Wildlife Service Biological Report NEC-93/9.

Schwaner-Albright, O. 2009. "As Maple Syrup Prices Rise, New York Leaders See Opportunity." *New York Times*, March 10. Accessed May 23, 2009. http://www.nytimes.com/2009/03/11/dining/11maple.html?_r=1&scp=1&sq=maple%20syrup&st=cse.

Scott, D., J. Dawson and B. Jones. 2008. "Climate change vulnerability of the US northeast winter recreation tourism sector." *Mitigation and Adaptation Strategies for Global Change* 13(5–6):577–596.

Selong, J.H., T.E. McMahon, A.V. Zale and F.T. Barrows. 2001. "Effect of temperature on growth and survival of bull trout, with application of an improved method for determining thermal tolerance in fishes." *Transactions of the American Fisheries Society* 130:1026–1037.

Shea, K and P. Chesson. 2002. "Community ecology theory as a framework for biological invasions." *Trends in Ecology & Evolution* 17:170–176.

Skinner, C.B., A.T. DeGaetano and B.F. Chabot. 2010. "Implications of twenty-first century climate change on Northeastern United States maple syrup production: impacts and adaptations." *Climatic Change* 100:685–702.

Sleeman, J.M., J.E. Howell, W.M. Knox and P.J. Stenger. 2009. "Incidence of hemorrhagic disease in white-tailed deer is associated with winter and summer climatic conditions." Ecohealth 6: 11–15.

Song, L.Y., J.R. Wu, C.H. Li, F.R. Li, S.L. Peng and B.M. Chen. 2009. "Different responses of invasive and native species to elevated CO_2 concentration. Acta Oecologica-International." *Journal of Ecology* 35:128–135.

Stager, J.C. and M. Thill. 2010. *Climate Change in the Champlain Basin*. Report commissioned by Vermont and Adirondack chapters of The Nature Conservancy http://www.nature.org/wherewework/northamerica/states/vermont/science/art31636.html

Stoddard, J.L. 1994. "Long-term changes in watershed retention of nitrogen: Its causes and aquatic consequences. Environmental Chemistry of Lakes and Reservoirs pp 223–284.

Stromayer, K.A.K. and R.J. Warren. 1997. "Are overabundant deer herds in the eastern United States creating alternative stable states in forest plant communities?" *Wildlife Society Bulletin* 25:227–233.

Tatem, A.J. 2009. "The worldwide airline network and the dispersal of exotic species: 2007–2010." *Ecography* 32:94–102.

Thomas, R.Q., C.D. Canham, K.C. Weathers, C.L. Goodale. 2010. "Increased tree carbon storage in response to nitrogen deposition in the US." *Nature Geoscience* 3:13–17.

Tingley, M., D. Orwig, R. Field, and G. Motzkin. 2002. "Avian response to removal of a forest dominant: consequences of hemlock woolly adelgid infestations." *Journal of Biogeography* 29:1505–1516.

Tyree, M.T. 1983. "Maple sap uptake, exudation, and pressure changes correlated with freezing exotherms and thawing endotherms." *Plant Physiology* 73(2)277–285.

United States Congress, Office of Technology Assessment. 1993. *Harmful Non-Indigenous Species in the United States*, OTA-F-565. U.S. Government Printing Office, Washington, DC.

United States Department of the Interior, Fish and Wildlife Service, and United States Department of Commerce, and United States Census Bureau. 2006. *National Survey of Fishing, Hunting, and Wildlife-Associated Recreation*.

United States Environmental Protection Agency (USEPA). 2010. "Acid rain." www.epa.gov/acidrain/effects/index.html

United States Fish and Wildlife Service, Northeast Region. www.fws.gov/northeast/white_nose.html

United States Forest Service, 2011. Climate Change Atlas. http://www.nrs.fs.fed.us/atlas/

Waite, T.A. and D. Strickland. 2006. "Climate change and the demographic demise of a hoarding bird living on the edge." *Proceedings of the Royal Society B* 273:2809–2813.

Walker, M. D., W.A. Gould and F.S. Chapin, III. 2001. "Scenarios of Biodiversity Changes in Artic and Alpine Tundra." In *Scenarios of Global Biodiversity*, edited by F.S. Chapin, III, O. Svala and A. Janetos. New York: Springer.

Walther, G.R. 2000. "Climatic forcing on the dispersal of exotic species." *Phytocoenologia* 30(3–4):409–430.

Wehrly, K.E., L.W. Wang and M. Mitro. 2007. "Field-based estimates of thermal tolerance limits for trout: incorporating exposure time and temperature fluctuation." *Transactions of the American Fisheries Society* 136:365–374.

Williams, J.W. and S.T. Jackson. 2007. "Novel climates, no-analog communities, and ecological surprises." *Frontiers in Ecology and the Environment* 5:475–482.

Willis, C.G., B.R. Ruhfel, R.B. Primack, A.J. Miller-Rushing, J.B. Losos and C.C. Davis. 2010. "Favorable climate change response explains non-native species' Success in Thoreau's woods." *PLoS ONE* 5:e8878.

Wittenberg, R. and M.J.W. Cock, editors. 2001. "Invasive Alien Species: A Toolkit of Best Prevention and Management Practices." CAB International xvii – 228. Wallingford, Oxon, UK.

Wolfe, D.W., M.D. Schwartz, A.N. Lakso, Y. Otsuki, R.M. Pool and N.J. Shaulis. 2005. "Climate change and shifts in spring phenology of three horticultural woody perennials in Northeastern USA." *International Journal of Biometeorology* 49:303–309.

Wolfe, D.W. 1995. "Physiological and growth responses to atmospheric carbon dioxide concentration." In *Handbook of Plant and Crop Physiology*, edited by M. Pessarakli, chapt. 10. Marcel Dekker, Inc., New York.

Wolfe, D.W., L. Ziska, C. Petzoldt, A. Seaman, L. Chase and K. Hayhoe. 2008. "Projected change in climate thresholds in the Northeastern U.S.: Implications for crops, pests, livestock, and farmers." *Mitigation and Adaptation Strategies for Global Change* 13:555–575.

Yamasaki, M., R.M. DeGraaf and J.W. Lanier. 2000. "Habitat associations in Eastern hemlock birds, smaller mammals, and forest carnivores." *Proceedings, Symposium on Sustainable Management of Hemlock Forests in Eastern North America* 135–141.

Ziska, L.H., J.R. Teasdale and J.A. Bunce. 1999. "Future atmospheric carbon dioxide may increase tolerance to glyphosate." *Weed Science* 47:608–615.

Ziska, L.H. 2003. "Evaluation of the growth response of six invasive species to past, present and future carbon dioxide concentration." *Journal of Experimental Botany* 54: 395–404.

Appendix A. Stakeholder Interactions

The Ecosystems team gathered information and enlisted participation from key stakeholders in this sector through existing relationships and collaboration with the New York State Department of Environmental Conservation; other State and federal governmental organizations (e.g., U.S. Geological Survey, U.S. Fish and Wildlife Service); Cornell Cooperative Extension (natural resources specialists); non-governmental organizations (e.g., The Nature Conservancy, National Wildlife Federation, Audubon NY, Wildlife Conservation Society, Adirondack Mountain Club); business associations (e.g., New York Forest Landowners Association, Empire State Forest Products Association, Olympic Regional Development Authority); land, fish, and wildlife managers; and maple growers.

Meetings and Events

On December 8, 2008, a meeting was held with over 50 stakeholders, including representatives of State and federal government organizations, leaders of non-government organizations, leaders of recreational-user organizations, representatives from affected industries, and academics. After a series of presentations, there was a two-hour breakout session with small groups. Each group provided its input regarding high-priority vulnerabilities and potential opportunities; feasible adaptation strategies; and needs for additional information, decision tools, and/or resources to help stakeholders cope with climate change and protect the state's natural resources. These data were summarized and sorted into groups of statements with thematic similarity, and contributed to the development of the chapter.

On August 6, 2009, the Ecosystems and Water Resources sectors and representatives of the ClimAID team at Columbia University met with stakeholders at the New York State Department of Environmental Conservation headquarters in Albany for an all-day workshop. This meeting was used to update stakeholders on ClimAID activities and progress and, especially, to collect input on needs and current relevant activities and planning by Department of Environmental Conservation and related stakeholder groups.

On November 6, 2009, an expert panel was assembled to meet with the Ecosystems sector team in Albany to review initial findings and provide suggestions regarding the project. The meeting included introductory presentations, followed by discussions focused on climate factors and key vulnerabilities, adaptation strategies, prioritization, and broad issues and recommendations. The 25 people in attendance included scientists from non-governmental organizations, State and government agencies, and research institutes within the state.

Web-based Survey Tool and Analyses

The results from early-phase stakeholder input were used to create a Web-based survey that cast a wider net among stakeholders and gathered expert opinion regarding the current state of knowledge regarding climate change; evidence of climate change impacts; high-priority vulnerabilities; high-priority climate change factors; importance and feasibility of various adaptation strategies; current efforts to adapt to climate change; research, monitoring, and communication gaps; and needed decision tools (Chatrchyan et al., 2010).

The survey was reviewed by several experts and stakeholders before dissemination in November 2009. The survey was sent to research scientists; land and water resource managers, educators, and others from State and federal government agencies; elected officials; private industry and landowners; non-government organizations; and universities and other research institutes. One section of the survey allowed participants to choose among several areas of specialization: water resources; forests, grassland, wetland, and riparian zones; fish and wildlife; and invasive species.

After survey responses were collected, the analysis characterized how issues were conceptualized by stakeholders and identified issues of priority/importance, using an approach similar to that described by Cabrera et al. (2008). Results were integrated into this report.

Renewable Fuels Roadmap, intended to guide State policy on renewable fuels, was recently issued by the New York State Energy Research and Development Authority (NYSERDA, 2010).

Water Issues

New York has historically been characterized as a humid region with significant summer rainfall that, in most years, provides for acceptable productivity of rain-fed grain and forage crops. In the context of a changing climate, however, the state lacks an inventory of drought-vulnerable locations, clearly defined agricultural water rights, and regional infrastructure for water delivery to farmland in dry years.

An analysis of historical data for New York reveals that even with today's climate in an average year, summer rainfall does not completely meet seasonal crop water requirements; supplemental irrigation is required for maximum productivity (Wilks and Wolfe, 1998), particularly on sandy or compacted soils with low water-holding capacity. Only a small percentage of farm acreage is irrigated in the state, most of this occurring on the relatively high-value vegetable and fruit acreage that accounts for about 6.5 percent of total cultivated land area. However, even farmers producing high-value fruit and vegetable crops often lack sufficient irrigation capacity to meet water needs of their entire acreage during extended periods of summer drought. Such drought events are projected to increase in frequency (Hayhoe et al., 2007; and see Chapter 1, "Climate Risks," and Case Study C. Drought). The substantial rain-fed grain crop, corn silage, and hay acreage of the state (often providing low-cost feedstock for dairy and other livestock) would be particularly vulnerable to potential increases in summer drought frequency because the value of such crops is not likely to warrant investment in irrigation equipment.

Too much as well as too little rainfall is currently a recurrent problem for farmers in New York. The recent historical trend for increased frequency of high rainfall events (see Chapter 1, "Climate Risks") has adversely affected some vegetable growers in recent years by direct reductions in yields and also by delaying spring planting or other farm operations. Additionally, use of heavy farm equipment on wet soils is detrimental to soil structure and quality and further limits crop yield.

7.2 Climate Hazards

Below are aspects of climate change projected for New York that will be particularly relevant to the agriculture sector (see Chapter 1, "Climate Risks"). Several high-priority vulnerabilities and opportunities associated with these factors are discussed in more detail in section 7.3.

7.2.1 Temperature

Warmer summer temperatures and longer growing seasons may increase yields and expand market opportunities for some crops. Some insect pests, insect disease vectors, and disease-causing pathogens may also benefit in multiple ways, such as having more generations per season and, for leaf-feeding insects, an increase in food quantity or quality.

Increased frequency of summer heat stress will be damaging to the yield and quality of many crops and will adversely affect health and productivity of dairy cows and other livestock.

Warmer winters will affect the suitability of various perennial fruit crops and ornamentals for New York. The habitable range of some invasive plants, weeds, and insect and disease pests will have the potential to expand into New York, and warmer winters will increase survival and spring populations of some insects and other pests that currently marginally overwinter in the state.

7.2.2 Precipitation

Projections of future precipitation patterns are inherently less certain than projections of future temperature. ClimAID analyses for New York suggest total annual precipitation may increase somewhat, primarily in the winter months, but the magnitude of this change is quite uncertain. Of greater certainty are expected changes in qualitative aspects such as the fraction of precipitation coming as snow and the intensity of individual rainfall events (see Chapter 1, "Climate Risks").

Less snow cover insulation in winter will affect soil temperatures and depth of freezing, with complex effects on root biology, soil microbial activity, and nutrient retention (Rich, 2008), as well as winter survival of some insects, weed seeds, and pathogens.

Snow cover also will affect spring thaw dynamics, levels of spring flooding, regional hydrology, and water availability.

Increased frequency of late-summer droughts will adversely affect productivity and quality and will increase the need for irrigation (see Case Study D. Drought). Rain-fed crops, for which irrigation is not economically feasible, would be particularly vulnerable. Despite new challenges with water deficits, New York is not threatened with the severity of drought projected for many other agricultural regions in the United States and internationally.

Increased frequency of heavy rainfall events is already being observed with adverse consequences, such as direct crop flood damage, non-point source losses of nutrients and sediment via runoff and flood events, and costly delays in field access.

7.2.3 Sea Level Rise

Sea level rise will have few direct effects on agriculture in most parts of the state. Issues such as increased potential for saltwater intrusion into groundwater or coastal flooding in agricultural areas in Long Island and the Hudson Valley are discussed in Chapter 5, "Coastal Zones."

7.2.4 Other Climate Factors

There are some climate factors, such as increased frequency and clustering of extreme events, that could potentially have severe negative impacts on the agriculture industry, but our current level of certainty about these factors is low (see Chapter 1, "Climate Risks"). Although not a climate factor, the continued increase in atmospheric carbon dioxide has direct effects on plants separate from its influence on climate, as described briefly below with other factors of particular concern.

Most climate models project little change in climate variability per se, but there is not a high degree of certainty that this will be the case and, in fact, there is observational evidence of increased winter variability in recent years. More variable winter temperatures can adversely affect perennial plants and winter crops by making them more vulnerable to mid-winter freeze damage (due to de-hardening) or spring frost (due to premature leaf out and bud break). There is a need for new climate research and monitoring to determine whether such events are part of a long-term climate change trend, and there is a need for new extreme-event early warning systems for farmers.

Current climate models cannot project changes in cloud cover reliably, yet cloud cover changes can have profound effects on crop productivity and quality and on crop water demands.

While great strides have been made by climate modelers and computing power in improving the spatial resolution of climate projections, it will be important for farmers to have even higher resolution to encompass microclimate effects.

Higher atmospheric carbon dioxide levels can potentially increase growth and yield of many crops under optimal conditions. However, research has shown that many aggressive weed species benefit more than cash crops, and that weeds also become more resistant to herbicides at higher carbon dioxide concentrations (see Section 7.3.2).

7.3 Vulnerabilities and Opportunities

Warmer temperatures, a longer growing season, and increased atmospheric carbon dioxide could create opportunities for farmers with enough capital to take risks on expanding production of crops adapted to warmer temperatures (e.g., European red wine grapes, peaches, tomatoes, watermelon), assuming a market for new crops can be developed. However, many of the high-value crops that currently dominate the state's agriculture economy (e.g., apples, cabbage, potatoes), as well as the dairy industry, benefit from the state's historically relatively cool climate. Some crops may have yield or quality losses associated with increased frequency of late-summer drought, increased summer high temperatures, increased risk of freeze injury as a result of more variable winters, and increased pressure from weeds, insects, and disease. Dairy milk production per cow will decline in the region as temperatures and the frequency of summer heat stress increase, unless farmers adapt by increasing the cooling capacity of animal facilities. Below are some high-priority vulnerabilities for New York.

Expanded Irrigation Capacity and Other Major Capital Investments

Climate change could require significant capital investment to ensure survival of agricultural businesses or to take advantage of new opportunities. Examples include new irrigation or drainage systems, new planting or harvesting equipment for new varieties, new crop storage facilities, new equipment to allow more timely management, and improved cooling facilities for livestock. The challenge will be strategic investment in relation to the timing and magnitude of climate change.

7.4.3 Adaptation Beyond the Farm: Institutions, Agencies, and Policy

Climate change impacts on crops and livestock will have human health and societal impacts beyond the individual farmer. For this reason, adaptations that involve societal investment or private industry responses are also likely to be necessary.

Technological/Applied Research Developments

Technological/applied research developments might involve seed company development of new varieties and university development of decision-support tools and of cooling and irrigation technologies.

Information Delivery/Extension Systems

Examples of an information delivery/extension system might include delivery of real-time local weather data for integration into farm-management decision-support tools and better integrated pest management (IPM) monitoring of potential invasives. Improved delivery of state-of-the-art weather forecasts will be needed to prepare growers for extreme weather events and can be used for various farm management decision tools. A state- and grower-funded, weather-based pest-prediction network (NEWA) is active in parts of the state providing near real-time pest forecasts (http://newa.cornell.edu). However, many more than the current 50 stations will be needed for adequate coverage of all agricultural areas of the state. Current IPM programs will need to be strengthened and better linked at the regional level.

Locally Available Design and Planning Assistance

Assistance could be made available for farmers or for farm regions to help in designing new heat-resistant barns and on-farm drainage systems.

Disaster Risk Management and Insurance

Current crop insurance programs are not adequate for accurate and uniform assessment of economic losses associated with weather-related disasters. This is particularly true for high-value fruit and vegetable crops, where insurance personnel are not adequately trained on the diverse range of crops grown in the state.

Financial Assistance

Examples of financial assistance include low-cost loans and State and federal cost-share programs for adaptation investments. Many aspects of adaptation are potentially expensive even when solutions are clearly available, such as capital investments for new water management systems or livestock facility renovations to improve cooling capacity.

Major Capital Investments

Major capital investments could be required at a regional or State level and might include new dams or reservoirs and new large-scale flood-control and drainage systems.

Policy and Regulatory Decisions

These could be designed to facilitate adaptation by farmers, to alter regulations, to create financial incentives for adaptation investment, and/or to stimulate local renewable energy production. For example, Section 18 of the Federal Insecticide, Fungicide, and Rodenticide Act (FIFRA) currently allows the U.S. Environmental Protection Agency to approve emergency use of an unregistered pesticide in cases where new pests create several specific types of crises. Section 24C discusses Special Local Need applications, which are a second method to address crisis pesticide situations under the Act. Both of these processes, which can be initiated by land-grant universities, faculty, or industry groups, are likely to be used to address new agricultural pest arrivals under climate change conditions. A 24C pesticide application is reviewed on a state-by-state basis and requires an environmental risk assessment by a State agency (e.g., the New York State Department of Environmental Conservation), thus adding burden to State regulatory agencies in addition to adding pesticide load to the New York State environment.

Research on New Crops and Pests

Building adaptive capacity for the agriculture sector will require investment in new information, crops, and adaptation strategies (See Knowledge Gaps, Section 7.6.3, in Conclusions).

7.4.4 Co-benefits, Unintended Consequences, and Opportunities

Adaptations made to address specific climate change vulnerabilities may have additional effects beyond their primary intentions. In some cases these may raise new problems, while in others it is possible to design actions with multiple simultaneous benefits and to provide opportunities to New York State farmers.

Co-benefits

Climate change may provide an incentive for farmers and consumers to take advantage of some adaptation strategies that benefit both the farmer and the environment. Some of these may eventually be applicable to carbon-offset payments in emerging carbon-trading markets. New York State farmers could consider any or all of the following actions:

- Conserve energy and reduce greenhouse gas emissions (increase profit margin and minimize contribution to climate change).
- Increase soil organic matter (this not only improves soil health and productivity, but because organic matter is mostly carbon derived from carbon dioxide via plant photosynthesis, it reduces the amount of this greenhouse gas in the atmosphere).
- Improve nitrogen use efficiency (synthetic nitrogen fertilizers are energy intensive to produce, transport, and apply, and soil emissions of the greenhouse gas nitrous oxide increase with nitrogen fertilizer use).
- Improve manure management (reduces nitrous oxide, methane, and carbon dioxide emissions; also can be used as renewable energy in manure digesters).

Unintended Consequences

Described here are potential unintended consequences of adaptation strategies, which could potentially have cascading negative effects on rural economies.

Increased Water Use and Chemical Loads to the Environment
Increases in water and chemical inputs will not only increase costs for the farmer, but may also have society-wide impacts in cases where the water supply is limited, by increasing the reactive nitrogen and pesticide loads to the environment or by increasing the risks to food safety and increasing human exposure to pesticides.

Increased Energy Use
Higher energy use (and its attendant greenhouse gas emissions) may be associated with some adaptation strategies. Examples include increased running of cooling fans in livestock facilities, more energy to pump irrigation water as more farmers expand irrigation capacity (and in some cases pump from deeper wells), and increased energy use associated with greater use of products that are energy intensive to manufacture, such as some fertilizers and pesticides.

Changes in Land Use
Such shifts could result from changes in cropping systems and other farm adaptations. Harvesting of wooded areas for biofuel crops is possible, or increased diversion of corn acreage for biofuel markets. Such effects can be averted with appropriate strategic planning, and efforts towards this end have been initiated in the *Renewable Fuels Roadmap* (NYSERDA, 2010). Land clearing for expansion of food or forage crop acreage may occur, particularly if other production regions of the country are harder hit by climate change than New York due to water shortages or other factors.

Cascading Negative Effects on Rural Economies
These may be likely where farmers lack capital for adaptation (see Equity and Environmental Justice Considerations, Section 7.5).

Opportunities

Opportunities for NYS farmers could include the following:

- Possible extension of agricultural production on idle and under-used agricultural lands due to shifts in comparative advantage vis-à-vis other regions (see Chapter 4, "Water Resources").
- Enter the expanding market for renewable energy using marginal land (e.g., wind energy, solar, biomass fuels, energy through anaerobic digestion of livestock manures and food processing wastes).
- Increase consumer support—from households to large institutional food services—of local "foodshed" networks, which can reduce greenhouse gas emissions from transportation of agricultural goods.

7.5 Equity and Environmental Justice Considerations

In New York State, there is a range of equity and environmental justice issues at the intersection of climate change and agriculture. Particular agricultural sectors, regions, and crops will be more at risk from exposure to climate change and burdened by the effort and costs associated with adaptation measures. Meeting the costs of adaptation to climate change will put additional stresses on the fragile and economically important dairy industry in the state. Regional vulnerabilities include farmers on Long Island facing a disproportionate risk of crop damage from sea level rise, saltwater intrusion, and coastal flooding. Finally, certain crops have disproportionate vulnerabilities, such as perennials for which the cost and economic risk of changing crops as an adaptation strategy is sometimes much higher than for annual crops.

Of these regions and groups, those most vulnerable to climate change include small family farms with little capital to invest in on-farm adaptation strategies, such as new infrastructure, stress-tolerant plant varieties, new crop species, or increased chemical and water inputs. Small family farms[2] also are less able to take advantage of cost-related scale economies associated with such measures. Small farmers, particularly those in the dairy sector, already face severe competitive pressures due to rising production costs and flat or declining commodity prices. Indeed, as noted earlier, current trends suggest that the total number of dairy farms will decline from approximately 7,900 in 2000 to 1,800 in 2020, with most of this decline resulting from closure or consolidation of smaller farms[3] in New York State (LaDue, Gloy, and Cuykendall, 2003; USDA, 2007). Climate change is likely to exacerbate cost pressures on small farmers, particularly if adaptation requires significant capital investments, thus accelerating trends toward consolidation within the industry. Survival for many smaller farms will hinge, in part, on making good decisions regarding not only the type of adaptation measures to take but also in the timing of the measures. The most vulnerable farmers will be those without access to training about the full range of strategies or those who lack adequate information to assess risk and uncertainty.

In addition to supply-side dimensions, climate change also may impact agricultural demand. These effects can be associated with both long-term regional disinvestment such as out of high-risk areas (floodplains), or one-time extreme events in areas with high demand for New York State produce (like a hurricane in the New York metropolitan region). These conditions may disrupt supply chains, close retail centers, or otherwise cut consumer access to markets, with especially detrimental effects on low-income or mobility-constrained residents. Low-income farmers with insufficient information and training or without access to credit or infrastructure are particularly at risk when conditions demand immediate flexibility, such as requiring quickly lining up alternative supply lines and retail locations.

Under such conditions, rural, resource-dependent communities may feel pressure to supplement incomes or diversify their business beyond agriculture, but may lack the training or capital necessary to engage such strategies. Decreasing yields and the high costs of adaptation may translate into significant downstream job losses and cascading economic effects across rural communities. Low-wage, temporary, seasonal, and/or migrant workers are particularly exposed to these shifts.

Examining equity in adaptation involves evaluating existing vulnerabilities, but it also requires evaluating the unintended outcomes, externalities (secondary consequences), and emergent processes of specific adaptation strategies. Successful adaptation by individual farmers or regions may create downstream inequities. As some farmers successfully adapt, other farmers may experience relative increases in inequality related to rural income and agricultural productivity. Certain industries (such as the grape and wine industries) also may consolidate in such ways that it becomes difficult for smaller businesses to enter the market. Increasing chemical inputs, such as fertilizers and pesticides, may create or exacerbate inequitable distributions of human health burdens, or negatively affect waterways, disproportionately impacting low-income or natural resource-dependent communities involved in hunting- and fishing-related revenue. Furthermore, degrading land and community health could drive down property values, exacerbating geographic inequities. Finally, increasing natural resource use, whether it is water for irrigation or energy for cooling, is likely to raise utility prices. These increases are felt the most by low-income families who proportionally spend more on these basic goods than middle- and upper-income families.

Addressing and avoiding spillover effects in the implementation of adaptation measures requires engaging local communities and agricultural managers in each stage of the planning process. This includes mechanisms for expressing and addressing property disputes and conflicting claims to resources, collaborative regional planning across sectors and communities, and training or retraining to provide information regarding strategies and best practices. In particular, adaptation strategies focused at regional or state scales have the capacity to marginalize local actors who are unable to capitalize on social or economic networks or access policymaking procedures.

Equity issues should be considered along every part and process of the agriculture food-supply chain. For low-income communities throughout the state, the connection between climate change and issues of food justice is an area of growing concern. Food justice issues, including lack of access to grocery stores in lower-income urban and rural communities, and inability of lower-income individuals to afford healthy, fresh foods, may be exacerbated by climate change. For example, climate stress on agriculture could affect the quality, accessibility, and affordability of local produce. This has implications for food security among low-income groups, those communities with fragile connections to markets offering nutritional options, or those otherwise burdened by pre-existing poor nutrition. Increased incidence of extreme heat or prolonged droughts may also affect the cost structures and productivity of community gardens and other local food production systems that serve lower-income urban areas.

7.6 Conclusions

Those aspects of climate change already occurring in New York or anticipated within this century that have known effects on crops, livestock, weeds, insects, and disease pests have been the primary focus of this ClimAID analysis. **Table 7.2** summarizes selected climate factors as linked to vulnerabilities or opportunities for the agriculture sector and adaptation strategies. A qualitative level of certainty is assigned to each of these components. The relative timing of when climate change factors and impacts are anticipated to become pronounced is also indicated, as this will be critical in setting priorities for adaptation. The table illustrates an approach and a possible useful tool for setting priorities and for climate action planning, but is not meant to be comprehensive. It can and should be modified for specific purposes and as new information and expertise become available.

Below, key findings regarding vulnerabilities and opportunities, adaptation options, and knowledge gaps are highlighted and discussed in more detail.

7.6.1 Main Findings on Vulnerabilities and Opportunities

The climate risks, crop or livestock responses, and relative certainties indicated in **Table 7.2** have been integrated to develop this list of main vulnerabilities and opportunities.

- *Summer heat stress.* Warmer summers will bring an increase in the frequency of days that exceed high temperature thresholds negatively affecting crop yields, crop quality and livestock productivity. The ClimAID analysis for the dairy industry indicates significant milk production declines by mid- to late century; the high milk-producing cows being used today are particularly vulnerable.
- *Increased weed and pest pressure* associated with longer growing seasons (allowing more insect generations per season and more weed seed production) and

warmer winters (allowing more over-wintering of pests) will be an increasingly important challenge. New York farmers are already experiencing earlier arrival and increased populations of some insect pests, such as corn earworm.

- *Risk of frost and freeze damage* continue, and these risks are exacerbated for perennial crops in years with variable winter temperatures. For example, midwinter-freeze damage cost Finger Lakes wine grape growers millions of dollars in losses in the winters of 2003 and 2004. This was likely due to de-hardening of the vines during an unusually warm December, increasing susceptibility to cold damage just prior to a subsequent hard freeze. Another avenue for cold damage, even in a relatively warm winter, is when there is an extended warm period in late winter or early spring causing premature leaf out or bloom, followed by a frost event. This latter phenomenon may explain, in part, the lower apple yields in summers following warm winters. There is a low level of certainty regarding whether variability per se associated with recently observed freeze damage is a component of overall climate change in New York State (**Table 7.2**). This, however, will be a concern for tree fruit crops and other perennial species, at least in the short term (the next few decades).
- *Increased risk of summer drought* (defined here as crop water requirements exceeding water available from rainfall plus stored soil water) is projected for New York by mid- to late century. Compared to some agricultural regions, such as the western U.S., however, New York State is likely to remain relatively water rich. As indicated in **Table 7.2**, projections for future rainfall and drought severity are not as certain as those for temperature.
- *Increased frequency of heavy rainfall events and flood damage*. In addition to direct crop damage, wet springs delay planting and subsequently delay harvest dates. For some fresh market vegetable growers, much of their profit is based on early season production so this can have substantial negative economic effects. Use of heavy equipment on wet soils leads to soil compaction, which subsequently reduces soil water-holding capacity, water infiltration rates, root growth, and yields.
- *New crop options*. While climate change will add to the physical and economic challenges of farming in New York, there are likely to be new opportunities as well as vulnerabilities, such as developing new markets for new crop options that will come with longer growing seasons and warmer temperatures. The expansion in New York of the non-native and cold-sensitive European (*Vitis vinifera*) white wine industry over the past 40 years has benefited from the reduced frequency of severe cold winter temperatures over this time period. European red grape varieties such as Merlot could benefit with additional warming, as could other crops such as peaches, watermelon, and tomato. Some New York field corn growers are already experimenting with slightly longer growing-season varieties that produce higher yields.

Table 7.2 Climate factors, vulnerabilities and opportunities, and adaptation strategies for agriculture in New York State

Climate Factor	Climate Certainty	Associated Vulnerabilities/Opportunities	Certainty*	Timing	Adaptation Strategies	Adaptation Capacity
Increasing carbon dioxide	High	Variable plant response affecting growth, competitiveness, yield. Under optimum conditions, yield increases are possible. Some C_3 weeds will benefit more than crops and be more resistant to herbicides.	High, but large variation in effects depending on other environmental constraints to plant growth	Now	Minimize water, nutrient constraints to crop growth to take full advantage of any beneficial effects. Develop varieties that take advantage of the effect of increases of carbon dioxide concentrations. Increased weed control and new approaches to minimize chemical inputs.	Moderate
Warmer summers; longer growing seasons	High	*Crops and weeds* Opportunities to obtain higher yields with current crops and grow higher-yielding varieties and new crops. Eventual double-cropping opportunities. Weeds will grow faster and will have to be controlled for longer periods. Increased seasonal water and nutrient requirements.	Moderate to high	Now, with some effects occurring later this century	Cautiously explore new varieties, new crops; develop markets for new crops. Increased weed control and new approaches to minimize chemical inputs. Increased water and fertilizer applications.	High
		Insects More generations per season; shifts in species range.	Moderate to high	Now	Better regionally coordinated monitoring through integrated pest management. Increased pest control. Proactively develop new approaches to minimize chemical inputs.	Moderate

Climate Factor	Climate Certainty	Associated Vulnerabilities/Opportunities	Certainty*	Timing	Adaptation Strategies	Adaptation Capacity
Increased frequency of summer heat stress	High	*Livestock (dairy)* Reduced milk production; reduced calving rates.	High	Serious by mid-century	Increase cooling capacity of existing dairy barns. Increase use of fans and sprinklers. Change feed rations. Provide plenty of water. Design new barns based on projected future heat loads.	Moderate to High
		Crops Could negatively affect yield or quality of many cool-season crops that currently dominate the agricultural economy, such as apple, potato, cabbage, and other cole crops.	High	Serious by mid-century	New heat-tolerant varieties when available. Change plant dates to avoid stress periods. Explore alternative crops.	Moderate to high
Warmer winters	High	*Crops* Could increase productivity or quality of some woody perennials (e.g., European wine grapes), while by mid to late century negatively affecting those adapted to current climate (e.g., Concord grape, some apple varieties). More winter cover crop options. Depending on variability of winter temperatures, can lead to increased freeze or frost damage of woody perennials	High	Now, with some effects occurring later in century	Explore new cash crops and varieties; explore new cover crop options.	High
					Better freeze and frost warning systems for farmers; new winter pruning strategies.	Moderate
		Insect and weed pests Increased spring populations of marginally overwintering insects. Northward range expansion of invasive weeds.	High	Now	Better regionally coordinated monitoring through integrated pest management. Increased pest control. Proactively develop new approaches to minimize chemical inputs.	Moderate
Increased frequency high rainfall, flooding	High	Delays in spring planting and harvest, negatively affecting market prices. Increased soil compaction, which increases vulnerability to future flooding and drought. Increased crop root disease, anoxia and reduced yields. Wash-off of applied chemicals.	High	Now	Increase soil organic matter for better drainage. Shift production to more highly drained soils. Install tile drains. Shift to flood-tolerant crops. Change plant dates to avoid wet periods. Increased disease control and new approaches to minimize chemical inputs.	Low to moderate; some options are expensive
Increased summer drought	Moderate	Reduced yields and crop losses, particularly for rain-fed agriculture. Inadequate irrigation capacity for some high-value crop growers.	Moderate to high	Mid to late century	Increase irrigation capacity. Shift to drought-tolerant varieties. New infrastructure for regional water supply.	Moderate, assuming capital available and economics warrant investment
Changes in hydrology, groundwater	Moderate	Dry streams or wells in drought years. Increased pumping costs from wells.	Moderate	Mid to late century	Deeper wells, new pumps.	Moderate
Frequency of extreme events	Low	Major crop and profit loss due to hail, extreme temperatures, flooding, or drought. Particularly devastating if extreme events occur in clusters.	Moderate to high	Unknown	New climate science research to determine current trends and develop early-warning systems for farmers.	Moderate
Increased seasonal variability	Low	Crop damage due to sudden changes, such as increased freeze damage of woody plants as a result of winter variability and loss of winter hardiness or premature leaf-out and frost damage.	Moderate	Now, but not clear if part of climate change	New climate science to determine relation to climate change and better predict variations.	Low
Changes in cloud cover and radiation	Low	Important factor affecting plant growth, yields and crop water use. Cloudy periods during critical development stages reduces yields.	High	Unknown	New climate science research to determine current trends and better model these factors.	Low to moderate

* Climate certainty in this table is qualitatively consistent with more quantitative assessments in Chapter 1, "Climate Risks," and formulated from expert opinion from chapter authors and stakeholder groups.

7.6.2 Adaptation Options

Adaptation options are available for many of the vulnerabilities summarized above and listed in **Table 7.2**. A challenge for farm managers, however, will be uncertainties regarding the optimum timing of adaptation investment, and the optimum magnitude of adaptation investment relative to the risks. Also, adaptations will not be cost- or risk free, and inequities in availability of capital or information for strategic adaptation may become an issue to resolve at the policy level (see also Knowledge Gaps, 7.6.3, below).

- *Improved cooling capacity of livestock facilities.* Increasing the summer use of fans and sprinklers for cooling will be an early adaptation strategy for the dairy industry. New barns should not be designed based on the 20th century climate, but rather for the increased heat loads anticipated in the 21st century.

- *Increased pest control and new approaches to minimize chemical inputs*. While we can look to more southern regions for control strategies for weeds and pests moving northward, these may not always be directly transferable or desirable for our region, particularly if they involve substantial increases in chemical loads to the environment. New policies and regulatory frameworks may become necessary, involving good communication among farmers, IPM specialists, and State agencies.
- *Supplemental irrigation* will be a first-step adaptation strategy in New York, and investment in expanded irrigation capacity will likely become essential for those growing high-value crops by mid- to late-century. This assumes that summer droughts do not become so severe as to dry up major surface and groundwater supplies. Since New York does not currently have a significant regional irrigation water supply infrastructure, state-wide investments in such may need to be considered by mid- to late-century.
- *Drainage for wet conditions*. Adaptations for wet conditions include maintaining high soil organic matter and minimizing compaction for good soil drainage. In some cases this will not be sufficient and installation of tile drainage systems will be warranted, a costly adaptation strategy. Shifting crop production to highly drained soils is an effective adaptation, but would then require irrigation for the expected drought periods.

7.6.3 Knowledge Gaps

With timely and appropriate proactive investment in research, as well as support for monitoring and information delivery systems, and policies to facilitate adaptation, the agriculture sector of the New York economy will have the necessary tools for strategic adaptation to meet the challenges and take advantage of any opportunities associated with climate change. Some relevant needs include the following:

- *Non-chemical control strategies for looming weed and pest threats* are needed, as well as enhanced regional IPM coordination, and monitoring and rapid-response plans for targeted control of new weeds or pests before they become widespread.
- *New economic decision tools for farmers* are needed that will allow exploration of the costs, risks, benefits, and strategic timing of various adaptation strategies (e.g., the timing of investment in new irrigation equipment) in relation to various climate change scenarios and potential impacts on crops and livestock.
- *Sophisticated real-time weather-based systems for monitoring and forecasting stress periods and extreme events* are needed. Current guidelines for many agricultural practices are based on outdated observations and the assumption of a stationary climate.
- *Crops with increased tolerance to climate stresses* projected for our region, with emphasis on horticultural or other crops important to the New York economy but not currently being addressed by commercial seed companies, will be needed, and can be developed using conventional breeding, molecular-assisted breeding, or genetic engineering.
- *New decision tools for policy-makers* are needed that integrate economic, environmental, and social equity impacts of agricultural adaptation to climate change.
- *Regional climate science and modeling research* is needed to help farmers discern between adverse weather events that are part of normal variability and those that are indicative of a long-term climate shift warranting adaptation investment. There are some climate factors, such as increased climate variability and increased frequency and clustering of extreme events, that could potentially have severe negative impacts on the agriculture industry, but our current level of certainty about these climate factors is low.

Case Study A. Frost Damage to Grapes

Warmer winters bring opportunities with the potential to introduce higher-value but less cold-hardy fruit varieties and may in the long term be beneficial to European wine grapes (*V. vinifera*) that are not native to the region. However, particularly in the near term, challenges associated with cold injury to crops may be problematic, as explored in this ClimAID analysis. In recent years these events have cost the New York agriculture industry millions of dollars (Levin, 2005). Warmer temperatures at the beginning of winter reduce cold hardiness and can raise the probability of mid-winter damage. In late winter or early spring (after the winter-chilling requirement has been met), a prolonged warm period may lead to premature bud break and increased spring frost vulnerability. Decisions related to variety selection thus require information on recent trends in winter-chill accumulation and projections of these values into the future. Assessing changes in spring-frost vulnerability is also necessary; typically, the lower the winter-chill requirement, the higher the risk of early bud break. Projecting such changes is important for New York State agriculture to meet its full economic potential in the context of a changing climate.

The date of the last spring freeze is a potential hazard for plants that have broken bud dormancy and begun active growth. **Figure 7.2** shows historical and projected values for last freeze dates at Fredonia. Fredonia's climate is currently moderated by its proximity to Lake Erie, making it a favorable location for tree fruit production and concord grapes. Since 1971, the date of the last occurrence of 28°F (the last spring freeze) at Fredonia has shifted from approximately April 25 (day 115) to April 15 (day 105). Overnight temperatures less than 28°F are now less likely to occur during April. This trend toward earlier last-freeze dates is expected to continue into the future. For example, based on downscaled minimum temperatures from the Hadley Centre Coupled Model, version 3 (HadCM3), under the low-emissions B1 scenario, the steady shift in the date of the last freeze reaches April 5 (day 95) by the end of this century. Under the high-emissions A2 scenario, this date moves into March, with the last freeze expected to occur on day 85 (March 26). This is nearly a month earlier than the 1971 date.

The projected trend for an earlier date of last frost does not necessarily reduce risk of spring frost damage if grapes are responding to an earlier spring with earlier leaf out and bloom (Wolfe et al., 2005). In fact, frost risk could possibly increase with climate warming because leaf and flower emergence are driven by a cumulative factor—the accumulation of daily average temperatures above 50°F (degree-days)—but it just takes a single frost event, occurring within the bounds of natural spring temperature variability, to cause severe damage.

Figure 7.3 shows the recent historical and projected growing degree-day accumulation in the interval preceding the last spring freeze for a region in western New York. It compares this to a threshold line of 133 degree-days, the average growing degree-days required for bud break of Concord grapes. Historical data from the 1971–2007 period indicate that, on average, only 50 growing degree-days accumulate prior to the last spring freeze, and the 133 degree-day threshold leading to bud break before the last frost was observed in only two growing seasons. There is some indication that the average value of pre-frost growing degree-days begins to increase in the post-2060 period, and there is a

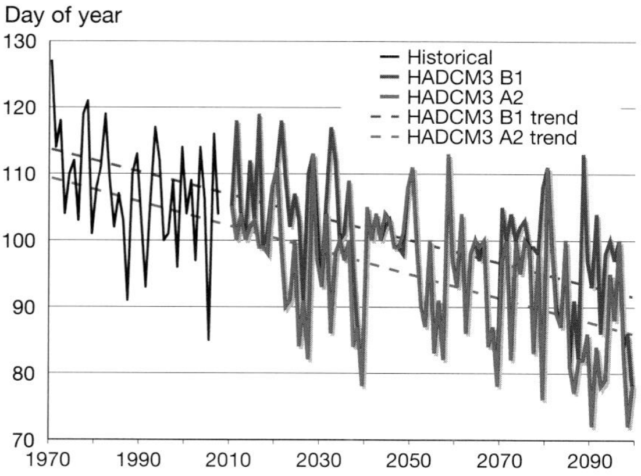

Note: Black datapoints are the observed patterns from 1970 to 2007. Red trends are simulated based on the lower-emissions B1 scenario projections; green trends are simulated based on the higher-emisisons A2 scenario projections. Results are broadly consistent with other GCMs used in ClimAID.

Figure 7.2 Changing date of the day of last frost; vertical axis indicates the number of days after January 1 (Julian Day)

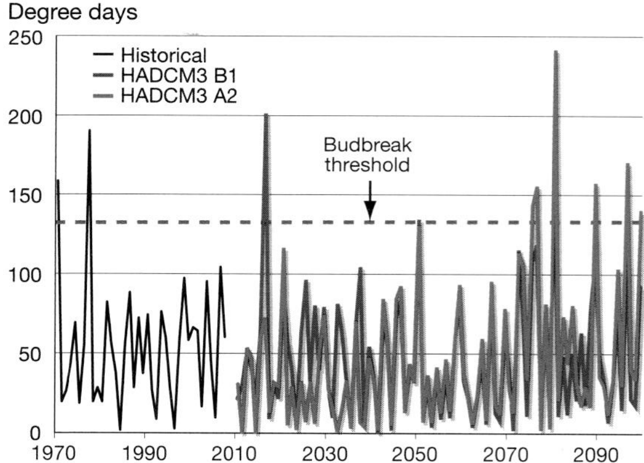

Note: The dashed blue horizontal line represents a threshold cumulative degree-day threshold that would lead to bud break prior to the last spring frost for Concord grapes, based on a 28-year phenology dataset in the Fredonia region. Those years exceeding the threshold are years with high risk of frost damage. Degree days for budbreak. Results are broadly consistent with other GCMs used in ClimAID. Source: Alan Lakso, personal communication, October 2009

Figure 7.3 Degree-day accumulations above 50°F occurring prior to the last frost dates shown in Figure 7.2

notable increase in year-to-year variability. For the higher-emissions scenario (A2), this results in a significant increase in the frequency of years near the end of the century with risk of frost damage—sufficient degree-day accumulation prior to the last frost to cause bud break (i.e., the 133 degree-day accumulation threshold line is crossed).

The projections in **Figure 7.3** reflect the interaction between climate change effects on earliness of bud and fruit development and the date of last spring frost, within the context of spring temperature variability. Results suggest that spring frost risk will not only persist, but could even increase by late century. Numerous strategies for avoiding damage from spring frost events are well tested and reviewed (Poling, 2008). Section 7.4.2 provides on-farm crop adaptation strategies and more details regarding freeze- and frost-protection strategies for perennial fruit crops.

Case Study B. Potato Late Blight

The potato late blight disease is a severe disease caused by the pathogen *Phytophthora infestans* (Fry, 2008). This is the same disease that caused the Irish potato famine starting in the 1800s. The disease is most severe in moderately cool, wet weather. Extended periods (typically more than 10 hours) of leaf wetness with moderate temperatures (54–72°F) are particularly favorable to the pathogen and lead to severe disease. This disease is a problem all over the world where potatoes are grown. There are about 20,000 acres of potatoes in New York. Based on the estimate that chemical costs are $250 to $500 per acre per year (Haverkort et al., 2008), New York growers spend $5–10 million annually on fungicides to protect their crops from this disease.

Climate change could influence the severity of potato late blight disease in a variety of ways. Elevated temperatures could have the indirect effect of reducing the duration of wet periods, thus lessening disease severity. Less frequent rainy periods might also reduce the number and duration of wet periods, also lessening disease severity. Alternatively, heavier rainfall events would remove protective fungicide from the foliage and thus increase the disease severity. Also, disease might begin earlier and/or be more prolonged with climate change.

This ClimAID case study uses an extensively tested mechanistic simulation model (Andrade-Piedra et al., 2005) of potato late blight to estimate the impacts of New York climate change on fungicide use for control of this disease. The model uses weather data to predict pathogen development and is currently used to provide disease severity forecasts for farmers. The model also contains a sub-model of fungicide dynamics (Bruhn and Fry, 1982a; Bruhn and Fry, 1982b), so that the amount of fungicide necessary to suppress disease in any given environment can be assessed. We compare the fungicide load for protecting potato plants under current weather conditions with the fungicide load required for a similar level of control under weather conditions projected during the coming century, under the business-as-usual (A2) and lower (B1) emissions scenarios.

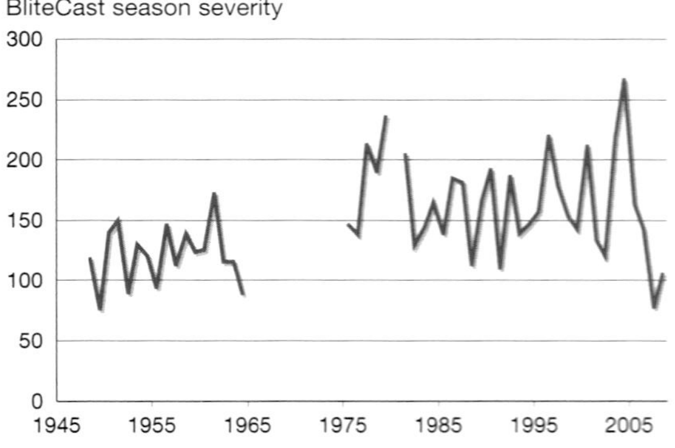

Note: Higher values are associated with wet and humid conditions and indicate greater danger of disease development.

Figure 7.4 BliteCast severity values based on historical weather records for Rochester

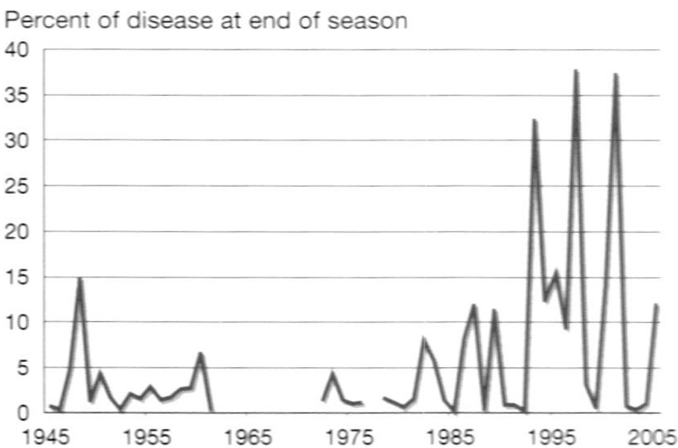

Figure 7.5 Likely prevalence of potato late blight disease forecast by BliteCast at the end of the season, considering both weather conditions and expected crop-management responses

First, weather data for Rochester from 1947 to 2008 were used to investigate the impact of historical weather on severity index values for potato late blight (i.e., potential for disease) predicted by the disease severity model, BliteCast (**Figure 7.4**). In general, during the latter part of this period (1977–2008) weather conditions led to higher disease severity. This period also showed greater year-to-year variability compared to 1947–1966.

The percent of potatoes with disease at the end of the growing season and the predicted amount of fungicide necessary to suppress the disease were also examined for the historical period. These predictions were obtained using the complex simulation model of the potato late blight disease (Andrade-Piedra et al., 2005). This model identifies the impact of weather on disease development and also identifies the impact of fungicide on disease development (Bruhn and Fry, 1982a). In agreement with the BliteCast severity index values (**Figure 7.4**), the percent of potatoes with disease was generally more severe with greater variance in the later (1977–2008) period compared to the earlier (1947–1966) period (**Figure 7.5**). Additionally, the amount of fungicide necessary to achieve adequate suppression of disease in the later period was greater than in the earlier period (**Figure 7.6**).

Using the same statistical models and approach as for the historical analysis describe above, projections of future disease severity and fungicide application needed for control were computed for the period 2040–2065. The models consist of three climatological input parameters: hourly temperature, hourly relative humidity, and daily precipitation to predict potato blight severity. Except for humidity, these variables were available from the standard suite of ClimAID climatological parameters discussed in Chapter 1, "Climate Risks." For humidity, the ClimAID Climate Team employed a statistical downscaling technique similar to that used for precipitation applied to global climate model grid-scale projections of specific humidity from five models (GISS, GFDL, UKMO, CCSM, and MIROC). Observed values of temperature and relative humidity at Rochester were converted to specific humidity and the delta change (1970–1995 versus 2040–2065) method applied to the specific humidity projections from the global climate models. The delta change in specific humidity was then applied to the three-hourly observations. The corresponding downscaled three-hourly temperatures were also obtained and used to calculate relative humidity projections. A cubic spline was fit to the three-hourly data to obtain the hourly resolution required by the potato late blight model.

Averaged across the five models, the projected BliteCast seasonal severity index for the A2 and B1 emissions scenarios for 2040–2065 (data not shown) was similar to the observed values in the 1995–2008 period of **Figure 7.6**. Despite an increase in temperature in both scenarios (favoring disease), relative humidity actually decreases slightly in the projections and, as a result, disease severity shows little change. Nonetheless, projected fungicide application rates required for adequate control (based on the models by Andrade-Piedra et al., 2007 and Bruhn and Fry, 1982a) significantly increased in most years in the higher-emissions A2 scenario (**Figure 7.7**). On average, the

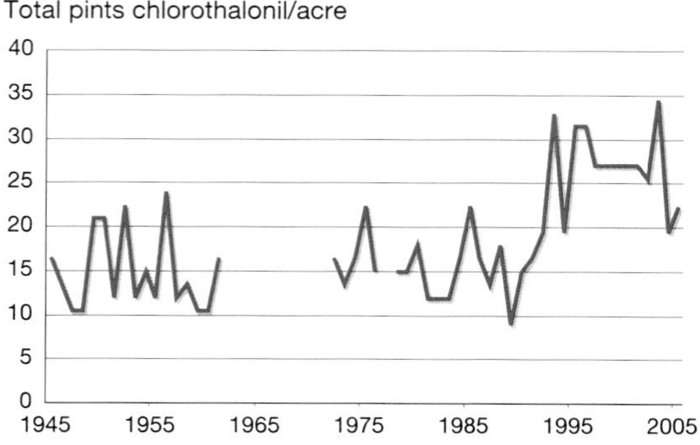

Figure 7.6 Total fungicide application recommended by BliteCast for each season of the historical record

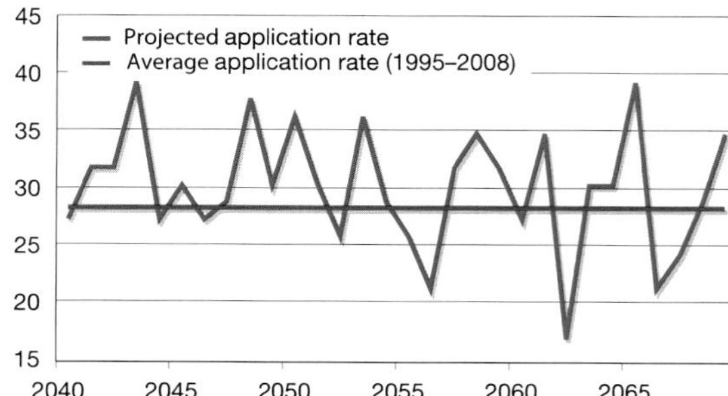

Figure 7.7 Projected total seasonal fungicide (chlorothalonil) application rate required for control of late blight for years 2040–2065 in comparison to the average application rate required for control during the 1995–2008 period

application rates for the 2040–2065 period under the A2 scenario increased to 34 pints per season, i.e., higher than the average of 28 pints for the latter half of the historical period (1995–2008, **Figure 7.6**). For the lower-emissions B1 scenario, the simulations suggest that application rates will remain similar—less than 30 pints—to the rates observed during the 1995–2008 historical period.

This analysis projects a significant increase in fungicide application required for control of late blight in 2040–2065 under the higher-emissions A2 scenario compared to today. There are several possible explanations regarding why the simulation projects an increased need for fungicide application, despite little change in the projected BliteCast severity index. Warmer temperatures in the A2 scenario may speed up pathogen development and, perhaps, cause disease outbreak to occur earlier, thus expanding the duration of required fungicide application without necessarily affecting severity values. Fungicide effectiveness is particularly sensitive to the occurrence and amount of precipitation and resulting wash-off of residual fungicide from the plant surface. This is not captured by the BliteCast seasonal severity index, but it is captured by the fungicide application models.

Case Study C. Drought

New York currently benefits from a moderately humid climate with a relatively uniform distribution of precipitation throughout the year. However, a considerable amount of winter precipitation is lost as runoff from saturated soils, and summer precipitation is not, on average, adequate to meet all potential evapotranspiration (PET) of a fully developed crop canopy or other dense vegetation (**Figure 7.8**). Depending on soil-water storage capacity, timing of rainfall, and crop growth stage, supplemental irrigation is currently warranted in many years to fully meet crop water requirements for maximum yield (Wilks and Wolfe, 1998). This case study examines the effect of climate change on future summer water deficits in seven climatic regions (**Table 7.3**) chosen because they had more than 100 years of weather records for evaluating drought frequency. **Figure 7.9** provides a graphical representation of three of these regions: Indian Lake (Adirondacks, northern New York with a relatively wet climate), Elmira (southern New York), and Rochester (western New York, an area with major production of high-value fruit and vegetable crops as well as dairy).

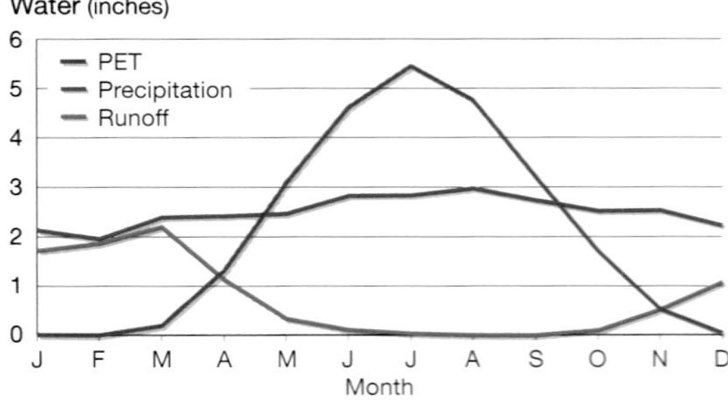

Figure 7.8 Historical (1901–2006) average monthly soil-water balance parameters at Rochester[6]

For this analysis, the ClimAID Climate Team provided a tailored product, in which climate projections from five global climate models (GFDL, GISS, MIROC, CCSM, and UKMO) were used for calculation of the Palmer Drought Severity Index (Palmer, 1965). These results are used to estimate seasonal water deficits. The water deficit index values (in inches of water) in **Table 7.3** and **Figure 7.9** were calculated from PET (June to September) minus precipitation (Pcp), Runoff, and available soil water (ASW) (the amount of total soil water stored that plants can extract without negative effects on growth):

$$Deficit_{jun\text{-}sep} = PET_{jun\text{-}sep} - (Pcp_{jun\text{-}sep} - Runoff_{jun\text{-}sep} + ASW)$$

It is important to note that PET provides an estimate of water demand by mature plants at or near full canopy ground cover (i.e., maximum light interception and transpiration potential). This analysis assumes that actual evapotranspiration is equal to potential water demand for the entire June through September period. This is most applicable to perennial plants, grasslands, and ground covers, but tends to overestimate water deficits for early (June) or late (September) parts of the growing season for annual row crops, when actual crop water demand is less than PET because plants have reduced transpirational surface (leaf) area. Future planned analyses, discussed in more detail at the end of this section, will address this issue for row crops.

Maximum soil-water storage of 6 inches was assumed in the original Palmer Drought Severity Index calculations, and, in New York State, soils often begin the growing season in June near this level. Maximum available soil water was assumed to be half of total stored water, or 3 inches. This was based on prior work from many regions that has documented that, for many crops, depletion of soil water below 50 percent of maximum is a threshold at which plants become stressed and irrigation is recommended to maintain growth and productivity.

Warm season water deficits vary across the state, primarily due to variations in summer precipitation and summer temperatures used to calculate PET (**Table 7.3**). Current June through September cumulative precipitation averages 14.4 inches across the state for the seven weather stations used in this analysis, with a high of 16.1 inches in Port Jervis and a low of 11.4 inches in Rochester. In general, summer precipitation decreases from east to west across the state and is particularly low along the shoreline of Lake Ontario in the western half of the state. The cumulative deficit from June through September currently averages 2.1 inches for the seven stations representing the state,

Region	Station	Historical	B1; 2020s	A2; 2020s	B1; 2050s	A2; 2050s	B1; 2080s	A2; 2080s
		June to September Average Temperature °F						
1	Rochester	66.9	69.2	69.2	70.7	71.9	71.8	75.5
2	Port Jervis	68.5	70.5	70.7	72	73.1	72.9	76.2
3	Elmira	67.3	69.5	69.6	71	72.2	72.1	75.7
4	NYC	72.6	74.6	74.7	76.1	77.2	77.1	80.3
5	Albany	67.4	69.4	69.6	70.9	72.1	72	75.3
6	Watertown	65.6	67.8	67.9	69.3	70.5	70.4	74.1
7	Indian Lake	60.2	62.3	62.5	63.9	65.1	64.9	68.5
All	Average	66.9	69.1	69.2	70.5	71.7	71.6	75.1
		June to September Total Precipitation, inches						
1	Rochester	11.4	11.4	11.6	11.5	11.6	11.5	10.8
2	Port Jervis	16.2	17.1	16.8	16.9	17.3	17.2	16.9
3	Elmira	14.1	14.6	14.6	14.9	14.6	14.8	14.2
4	NYC	16	16.7	16.4	16.8	16.9	16.6	16.9
5	Albany	14	14.7	14.3	14.7	14.9	14.9	14.6
6	Watertown	13.9	14.5	14.3	14.4	14.5	14.5	13.8
7	Indian Lake	15	15.8	15.5	15.6	15.7	16	15.5
All	Average	14.4	15	14.8	14.9	15.1	15.1	14.7
		June to September Cumulative Water Deficits, inches						
1	Rochester	4.7	6.2	6	7.1	7.8	7.7	11.1
2	Port Jervis	1.2	1.7	2	2.7	3	2.9	5.3
3	Elmira	2.2	3.2	3.2	3.9	4.8	4.4	7.5
4	NYC	2.9	3.7	4	4.8	5.5	5.6	7.9
5	Albany	2.4	3.1	3.4	3.9	4.5	4.5	7
6	Watertown	1.8	2.5	2.7	3.5	4	4	6.9
7	Indian Lake	-0.6	-0.2	0	0.6	1	0.8	2.9
All	Average	2.1	2.9	3.1	3.8	4.4	4.3	7
		Absolute Increase in Precipitation Deficit Relative to Historical Average, inches						
1	Rochester		1.54	1.35	2.39	3.14	3.03	6.46
2	Port Jervis		0.44	0.77	1.43	1.75	1.68	4.06
3	Elmira		0.91	0.99	1.62	2.52	2.21	5.24
4	NYC		0.8	1.11	1.9	2.56	2.7	4.95
5	Albany		0.68	1.03	1.5	2.15	2.07	4.61
6	Watertown		0.74	0.91	1.68	2.25	2.2	5.16
7	Indian Lake		0.35	0.59	1.16	1.54	1.42	3.53
All	Average		0.78	0.97	1.67	2.27	2.19	4.86

Figures for representative stations from all seven climate regions of New York, assuming maximal plant water demand (i.e., potential evapotranspiration), calculated assuming full canopy cover for the entire period; see text for more discussion). Historical temperature and precipitation values represent averages of weather station data over the period 1901–2006 using 5 GCMs (GFDL, GISS, MIROC, CCSM, and UKMO) of the 16 used in ClimAID.

Table 7.3 Current and projected summer (June to September) water deficits and related temperature and precipitation

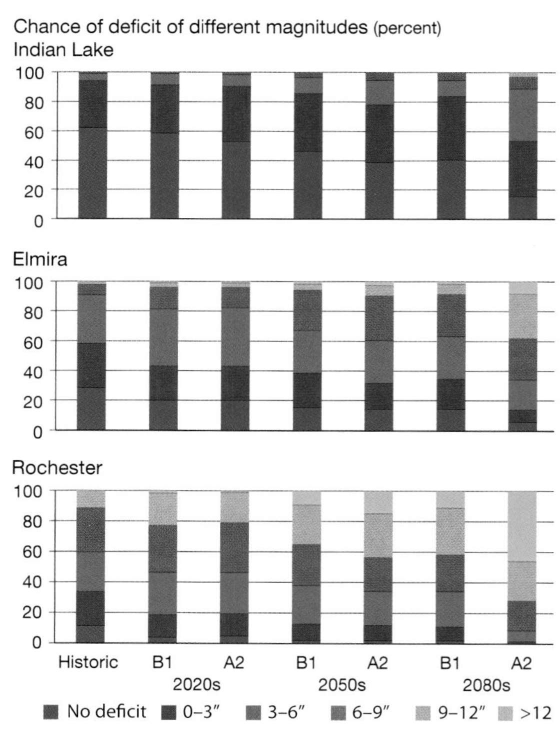

Note: Each multi-colored bar adds up to 100 percent probability of what a given year will be like, and the chance of a water deficit of a particular size is shown by the lengths of different colored segments within each bar. The distribution of colors within a single bar illustrates the underlying variability of weather patterns from year to year during a particular current or future time period. Different bars moving from left to right show how climate change will progressively alter these distributions, increasing the likelihood of years with larger summer water deficits.

Figure 7.9 Magnitude (inches) of total summer water deficit (June through September) under current and projected future conditions[5]

- holding area (area where cows wait to enter the milking parlor)
- milking area
- close-up dry cows (cows within three to four weeks of calving)
- calving area
- fresh cows (cows that have recently calved)
- high-producing cows
- low-producing cows

Cows in the holding area typically are very close together, touching each other. Even though they may be in this area for only a short time, heat stress and the thermal heat index (THI) can be very high. The preferred option for this area is to provide both fans and sprinklers. One report indicated that cows cooled in the holding pen produced 1.7 to 4 pounds more milk per day than cows not cooled in the holding area. A 1993 trial in Arizona indicated that cows cooled in the holding area produced 1.9 pounds more milk per cow per day than cows that were not cooled (Armstrong, 2000).

Priorities can be set for where the fans should be placed in barns. If funds are limited, the first choice would be to place fans over the feed bunk area. In addition, fans could be placed over the cow resting area (stalls). Ideally, fans would be placed over both areas. Tunnel ventilation is sometimes a good option (Gooch, 2008), but several other styles of ventilation system may be more appropriate depending on barn structure and site configurations.

Herd Size and Economies of Scale in Ventilation Systems

Many ventilation systems are inherently more cost-effective when deployed for larger animal housing situations. An interactive program available from Cornell University's Prodairy website can be used to calculate the costs and pounds of milk needed to break even with a tunnel ventilation system. This program calculates initial investment, operating costs, loan payments, days of fan operation, and the pounds of milk needed to break even. Using a five-year loan period and a milk price of $15 per 100 pounds of milk, model runs for both a small, tie-stall barn (50 or 100 cows) and a free-stall barn (300 or 600 cows), and assuming that the fans would operate 50, 100, or 150 days per year, provide the following results for initial investment (not including loan interest):

- 50 cow tie-stall barn = $262 per cow
- 100 cow tie-stall barn = $132 per cow
- 300 cow free-stall barn = $144 per cow
- 600 cow free-stall barn = $72 per cow

The degree to which milk production must be increased through avoidance of heat stress effects in order for the cooling systems to pay for themselves over a five-year payback period is shown in **Table 7.6** for each combination of barn style and herd size and considering three different scenarios of how many days per year reached stressful temperatures. Larger numbers for milk production in the same column imply that higher, more stressful outside temperatures would have to be experienced before installing a cooling system for a given barn style and herd size represented a cost-effective investment. In both styles of barns, there is a distinct economy of scale, with larger herds reaching cost-effectiveness at smaller minimum savings in milk production.

To summarize, to adequately ameliorate the effects of high temperatures, both adequate ventilation at high airspeeds directly over the cows and appropriately deployed sprinkler systems will be needed in the future. Many dairy barns already have both fans and sprinklers, but a significant number do not. The greatest cost in this configuration lies in the fans, but sprinklers without fan systems are not effective. While these cooling systems represent added investments, the literature shows that they have a high likelihood of paying for themselves over time through increased milk production. With projected climate change, adequate cooling systems will be increasingly important for the future of New York's dairy industry.

	Milk Production Savings (lbs/day) Needed to Pay Back Investment in Cooling Fans		
	50 heat stress days/yr	100 heat stress days/yr	150 heat stress days/yr
50 cow tie-stall barn	10.94	7.12	5.75
100 cow tie-stall barn	5.47	3.63	2.87
300 cow free-stall barn	5.9	3.9	3.1
600 cow free-stall barn	2.97	1.97	1.56

The values in the body of the table are levels of milk production increase (lbs per cow per day) that must be realized when cooling fans are employed in order to make the fan investment cost-effective, as described in the text. Three scenarios are considered showing that if fewer days of heat stress are experienced per year, the impact per day must be high to make installation cost effective. Tie-stall and free-stall barns are compared, each at two relevant herd sizes.

Table 7.6 Magnitude of potential heat stress losses required for fan and cooling systems to pay for their own installation and operation over a five-year pay-back period

Equity and Environmental Justice Issues

Vulnerability and capacity to adapt to climate change may vary substantially across different dairy regions in New York State (**Figure 7.10**) due to differences in climate change exposure, regional cost structures, farm sizes, and overall productivity. Should climate change have a highly detrimental effect on dairy farming in the state overall, those regions with higher concentrations of dairy farms are likely to experience a more substantial economic disruption. On the other hand, farmers in regions with higher concentrations of farms may also have some advantages associated with external economies of scale that facilitate adaptation to climate change, such as ability to learn from other farmers in the area regarding best adaptation practices or pooling of resources for different types of services that are needed to foster adaptation.

Regional comparison of the location of dairy operations in New York in 2007 (**Figure 7.11**) reveals that dairy farms are particularly abundant in the western parts of the Northern New York and Central Valleys Dairy Regions and in the Western and Central Plateau Region. Measured in terms of annual sales of milk and dairy products (**Figure 7.12**), the regional pattern is slightly different. The counties with the highest concentrations of dairy sales are located in the western portion of the Northern New York Region (also a region with the highest number of operations) and in the Western and Central Plain Region. According to the U.S. Census of Agriculture, the three New York counties with the highest sales in milk and other dairy products in 2007 are Wyoming County ($179 million) and Cayuga County ($140 million), both located in the Western and Central Plain Region, followed by St. Lawrence County ($113 million in sales) in the Northern New York Region.

In addition to differences in numbers of farms and total sales, the major dairy regions within the state also exhibit different characteristics in terms of size and profitability. Detailed data that permit comparisons

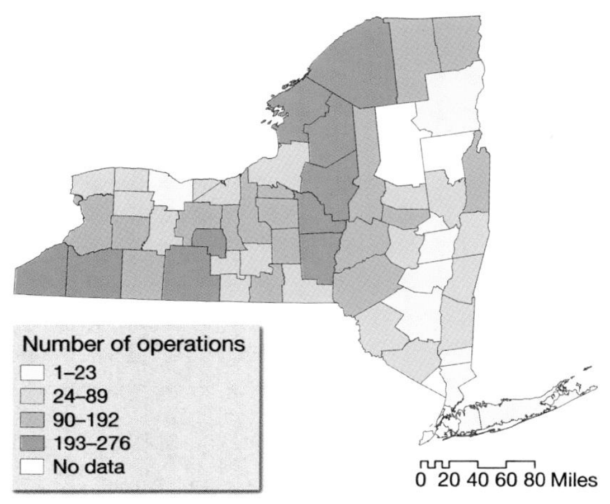

Source: U.S. Agricultural Census, 2007

Figure 7.11 Locations of dairy operations in New York State

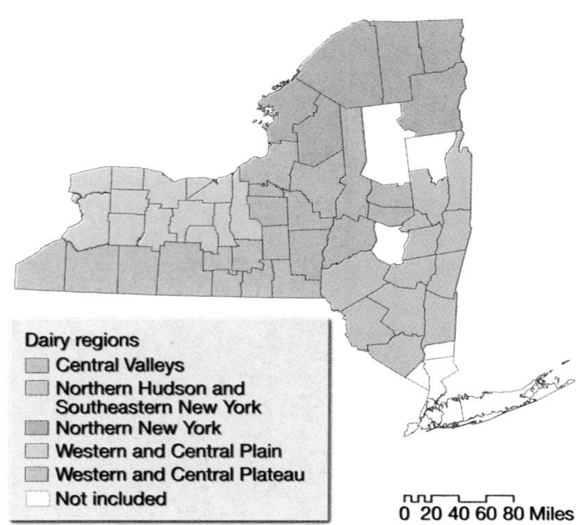

Figure 7.10 Dairy regions in New York State

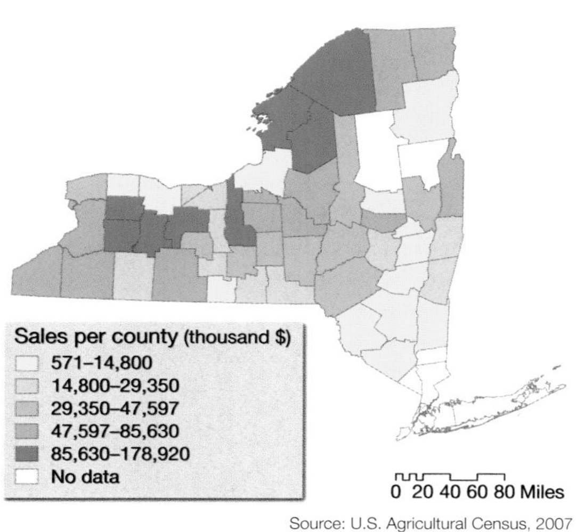

Source: U.S. Agricultural Census, 2007

Figure 7.12 Dairy sales by county for New York State

among regions in New York are available from the Cornell Cooperative Extension's Dairy Farm Business Summary and Analysis Project. Because participation in the survey is voluntary, these data do not represent a statistically robust sample. In particular, the data may contain some overrepresentation of farms with better organization and record keeping, as these farms are more likely to participate in the survey on a continuing basis. Nonetheless, the data provide useful insights into some of the major differences in the characteristics of dairy farms among the regions of the state (**Table 7.7**).

Examination of **Table 7.7** suggests that larger farms are concentrated in the Western and Central Plain region, where the average farm size within the Cornell sample has 673 cows and more than 1,200 tillable acres. The smallest average farm size is in the Western and Central Plateau region, where the average farm within the sample has 168 cows on 417 tillable acres. Costs of milk production range from $17.04 per hundredweight of milk in Northern New York to $18.63 per hundredweight in the Northern Hudson and Southeastern New York Region. Some factors that account for these regional differences include land and labor costs, which are likely to be higher in the areas closer to metropolitan New York. The Western and Central Plateau Region (which also has the smallest farms) is another region with relatively high costs of $18.03 per hundredweight of milk sold. These differences in milk production costs across regions may influence capacity to adapt to climate change, particularly in cases where adaptation requires additional expenditures for energy and pest control due to higher summer temperatures.

Differences in farm and herd size are also potentially significant factors in determining vulnerability and capacity to adapt to climate change. Comparison of small versus large farms throughout the state reveals significant differences in costs, milk sales per cow, capital efficiency, income, and profitability (**Table 7.8**). All of these differences may affect the overall capacity of smaller farms to adapt to climate change, particularly if such adaptation requires significant new outlays of capital for purchase and installation of ventilation systems in dairy barns, as well as additional costs associated with energy for operating this equipment.

While it is difficult to know precisely how different-sized dairy farms will be affected by climate change, the

	West & Central Plateau	West & Central Plain	Northern New York	Central Valley	North Hudson & Southeastern NY
Average number of cows per establishment	168	673	372	289	210
Tillable acres	417	1241	848	707	495
Total cost of producing milk ($ per hundredweight)	$18.03	$17.16	$17.04	$17.33	$18.63
Average price received ($ per hundredweight)	$20.62	$20.09	$20.06	$20.77	$20.95

Source: Cornell University, Dairy Business Summary and Analysis Project (Knoblauch et al., 2008)

Table 7.7 Characteristics of dairy regions in New York State

	Small Farms (39 farms)			Large Farms (83 farms)		
Farm Size	2007	2008	% chg	2007	2008	% chg
Average number of cows	52	52	0.0	773	797	3.1
Total tillable acres	40	41	2.5	1,482	1,595	7.6
Milk sold (lbs)	975,626	976,710	0.1	18,500,129	19,671,976	6.3
Costs						
Grain and concentrate purchases as a percent of milk sales	24%	31%	29.2	24%	30%	25.0
Total operating expenses per hundredweight sold	$16.29	$17.51	7.5	$16.32	$17.87	9.5
Capital Efficiency						
Farm capital per cow	$11,880	$12,576	5.9	$7,981	$8,772	9.9
Machinery and equipment per cow	$2,152	$2,382	10.7	$1,309	$1,467	12.1
Income and Profitability						
Gross milk sales per cow	$3,817	$3,678	-4.4	$4,870	$4,753	-2.4
Net farm income (w/o appreciation)	$54,680	$28,117	-48.6	$939,605	$483,799	-48.5
Income per operation per manager	$20,267	-$5,257	-126	$388,494	$128,755	-67
Farm net worth	$498,120	$502,664	0.9	$4,421,159	$4,658,105	5.4

Source: Cornell University, Dairy Business Summary and Analysis Project, Knoblauch et al., 2009 and Karszes et al., 2009

Table 7.8 Small versus large dairy farms in 2007 and 2008

effects of other types of shocks can help to illustrate which types of farmers might be more or less vulnerable to climate change. Comparison of how farms in the Cornell study performed in 2007 (a relatively profitable year) versus 2008 (a more challenging year due to spikes in input prices including feed, energy, and fertilizer) provides a glimpse into how shocks affect farms of different sizes. Similar types of input price shocks may also occur under climate change, because more frequent extreme weather conditions could lead to higher feed and energy prices. Policies intended to reduce emissions may also contribute to higher energy prices, though such effects are likely to be more gradual as taxes or other mechanisms to mitigate climate change are put into place.

Data from the Cornell dairy survey reveals that both small and large farms experienced significant challenges in coping with these conditions in 2008, but small farms appear to have fared worse (**Table 7.8**). Small farms experienced no increase in milk sold and had a 4.4 percent decline in gross milk sales per cow from 2007 to 2008. By contrast, large farms increased sales by 6.3 percent and experienced a 2.4 percent decline in gross milk sales per cow. Small farms also experienced relatively larger increases in purchased input costs (29.2 percent for small farms compared to 25 percent for large farms), but smaller increases in total production costs (7.5 percent for small farms compared to 9.5 percent for large farms). While net farm income for both small and large farms declined by approximately 48 percent between 2007 and 2008, income per farm operator or manager declined much more precipitously for small farms. Overall, the typical small farm operator experienced an income loss of 126 percent, while large farms operators experienced losses of 67 percent.

Collectively, these ClimAID results suggest that small farms may be less able to withstand shocks related to climate change without some type of adaptation assistance.

References

Allen, R.G., L.S. Pereira, D. Raes and M. Smith. 1998. "Crop Evapotranspiration: Guidelines for computing crop water requirements." *Food and Agriculture Organization (FAO) Irrigation and Drainage Paper No 56*. Rome:FAO.

Andrade-Piedra, J., R.J. Hijmans, G.A. Forbes, W.E. Fry and R.J. Nelson. 2005. "Simulation of potato late blight in the Andes: I, Modification and parameterization of the late blight model." *Phytopathology* 95:1191-1199.

Anisko, T., O.M. Lindstrom and G. Hoogenboom. 1994. "Development of a cold hardiness model for deciduous woody plants." *Physiologia Plantarum* 91:375-382.

Armstrong, D.V. 2000. "Methods to reduce heat stress in cows." Proceedings of the Heart of America Dairy Management Conference. Kansas State University, Manhattan, Kansas, June 21-22.

Baumgard, L.H. and R.P. Rhoads. 2009. "The effects of heat stress on nutritional and management decisions." Proceedings of the Western Dairy Management Conference, Reno, Nevada, March 11-13.

Bennet, E.M. and P. Balvanera. 2007. "The future of production systems in a globalized world." *Frontiers in Ecology and Environment* 5(4):191-198.

Berman, A. 2005. "Estimates of heat stress relief needs for Holstein dairy cows." *Journal of Animal Science* 83:1377-1384.

Blaney, D. 2002. "The changing landscape of U.S. milk production." *United States Department of Agriculture Statistical Bulletin 978*. www.ers.usda.gov

Brouk, M., D. Armstrong, J. Smith, M. VanBaale, D. Bray and J. Harner. 2005. "Evaluating and selecting cooling systems for different climates." Proceedings of the Western Dairy Management Conf., Reno, Nevada, March 9-11.

Bruhn, J.A. and W.E. Fry. 1982a. "A mathematical model of the spatial and temporal dynamics of chlorothalonil residues in a potato canopy." *Phytopathology* 72:1306-1312.

Bruhn, J.A. and W.E. Fry. 1982b. "A statistical model of fungicide deposition in a potato canopy." *Phytopathology* 72:1301-1305.

Chase, L.E. January 2007. Cornell University, personal communication.

Coakley, S.M., H. Scherm and S. Chakraborty. 1999. "Climate change and plant disease management." *Annual Review of Phytopathology* 37:399-426.

Coviella, C. and J. Trumble. 1999. "Effects of elevated atmospheric carbon dioxide on insect-plant interactions." *Conservation Biology* 13:700-712.

Dermody, O., B.F. O'Neill, A.R. Zamgerl, M.R. Berenbaum and E.H. DeLucia. 2008. "Effects of elevated CO_2 and O_3 on leaf damage and insect abundance in a soybean agroecosystem." *Arthropod-Plant Interactions* 2:125-135.

Dhuyvetter, K.C., T.L. Kastens, M.J. Brouk, J.F. Smith and J.P. Harner III. 2000. "Economics of Cooling Cows." Proceedings of the Heart of America Dairy Management Conference. Kansas State University, Manhattan, Kansas, June 21-22.

Fox, D.G. and T.P. Tylutki. 1998. "Accounting for the effects of environment on the nutrient requirements of dairy cattle." *Journal of Dairy Science* 81:3085-3095.

Fry, W.E. 2008. "*Phytophthora infestans*, the crop (and *R* gene) destroyer." *Molecular Plant Pathology* 9:385-402.

Garrett, K.A., S.P. Dendy, E.E. Frank, M.N. Rouse, and S.E. Travers. 2006. "Climate Change Effects on Plant Disease: Genomes to Ecosystems." *Annual Review of Phytopathology* 44:489-509.

Glenn, D.M. 2009. "Particle film mechanisms of action that reduce the effect of environmental stress in 'Empire' apple." *Journal of the American Society for Horticultural Science* 134:314-321.

Goho, A. 2004. "Gardeners anticipate climate change." *American Gardener* 83(4):36-41.

Gooch, C.A., M.B. Timmons and J. Karszes. 2000. "Economics of tunnel ventilation for freestall barns." Paper No. 004101, Proceedings of the American Society of Agricultural Engineers' annual international meeting, Milwaukee, Wisconsin, July 9-12.

Gooch, C. 2008. "Dairy freestall barn design - a northeast perspective." Ninth Annual Fall Dairy Conference, Cornell Pro-Dairy Program.

Greene, D.W. 2002. "Chemicals, timing, and environmental factors involved in thinner efficacy on apple." *HortScience* 37:477-481.

Gu, S., S. Dong, J. Li and S. Howard. 2001. "Acclimation and deacclimation of primary bud cold hardiness in 'Norton, 'Vignoles' and St. Vincent grapevines." *Journal of Horticultural Science & Biotechnology* 76:655-660.

Gu, L., P.J. Hanson, W.M. Post, D.R. Kaiser, B. Yang, R. Nemani, S.G. Pallardy and T. Meyers. 2008. "The 2007 Eastern U.S. spring freeze: Increased cold damage in a warming world?" *BioScience* 58(3):253-262.

Hamilton, J.G., O. Dermody, M. Aldea, A.R. Zangerl, A. Rogers, M.R. Berenbaum and E. Delucia. 2005. "Anthropogenic Changes in Tropospheric Composition Increase Susceptibility of Soybean to Insect Herbivory." *Environmental Entomology* 34(2):479-485.

Hanninen, H. 1991. "Does climatic warming increase the risk of frost damage in northern trees?" *Plant, Cell and Environment* 14:449-454.

Hatfield, J.L. 1990. "Methods of estimating evapotranspiration." In *Irrigation of Agricultural Crops. Agronomy Society of America*, edited by B.A. Stewart and D.R. Nielsen. Madison, WI.

Haverkort, A.J., P.M. Boonekamp, R. Hutten, E. Jacobsen, L.A.P. Lotz, G.J.T. Kessel, R.G.F. Visser and E.A.G. Van der Vossen. 2008. "Societal costs of late blight in potato and prospects of durable resistance through cisgene modification." *Potato Research* 51:47-57.

Hayhoe, K., C. Wake, T. Huntington, L. Luo, M. Schwartz, J. Sheffield, E. Wood, B. Anderson, J. Bradbury, A. DeGaetano, T. Troy and D. Wolfe. 2007. "Past and future changes in climate and hydrological indicators in the U.S. Northeast." *Climate Dynamics* 28:381-407.

Howell, G.S. 2000. "Grapevine cold hardiness: mechanisms of cold acclimation, mid-winter hardiness maintenance, and spring deacclimation." Proceedings of the American Society of Enology and Viticulture Annual Meeting, Seattle, Washington, June 19-23.

Hunter, M.D. 2001. "Effects of elevated atmospheric carbon dioxide on insect-plant interactions." *Agricultural and Forest Entomology* 3:153-159.

Jackson, J.E. and P.J.C Hamer. 1980. "The causes of year-to-year variation in the average yield of Cox's orange pippin apple in England." *Journal of Horticultural Science* 55:149-156.

Jackson, J.E., P.J.C. Hamer and M.F. Wickenden. 1983. "Effects of early spring temperatures on the set of fruits of cox's orange pippin apple and year-to-year variation in its yields." *Acta Horticulturae* 139:75-82.

Jones, G.V., M.A. White, O.R. Cooper and K. Storchmann. 2005. "Climate change and global wine quality." *Climatic Change* 73(3):319-343.

Karszes, J., W. Knoblauch and L. Putnam. 2009. "New York Large Herd Farms, 300 Cows or Larger, 2008." *Dairy Farm Business Survey*, Department of Applied Economics and Management, College of Agricultural and Life Sciences, Cornell University.

Kaukoranta, T. 1996. "Impact of global warming on potato late blight: risk, yield loss and control." *Agriculture and Food Science in Finland* 5(3):311-327.

Klinedinst, P.L., D.A. Wilhite, G.L. Hahn and K.G. Hubbard. 1993. "The potential effects of climate change on summer season dairy cattle milk production and reproduction." *Climatic Change* 23:21-36.

Knoblauch, W., L. Putnam, M. Kiraly and J. Karszes. 2009. "New York Small Herd Farms, 80 Cows or Fewer, 2008." *Dairy Farm Business Survey*. Department of Applied Economics and Management, College of Agricultural and Life Sciences, Cornell University.

Knoblauch, W., L. Putnam, J. Karszes, D. Murray and R. Moag. 2008. "Business Summary, New York State. Dairy Farm Business Summary and Analysis." Department of Applied Economics and Management, College of Agricultural and Life Sciences, Cornell University.

Kondo, S. and Y. Takahashi. 1987. "Effects of high temperature in the nighttime and shading in the daytime on the early drop of apple fruit 'Starking Delicious'." *Journal of the Japanese Society for Horticultural Science* 56:142-150.

LaDue, E., C. Gloy and B. Cuykendall. 2003. "Future Structure of the Dairy Industry: Historical Trends, Projections, and Issues." Cornell Program on Agriculture and Small Business Finance, College of Agricultural and Life Sciences, Cornell University.

Lakso A.N. 1987. "The importance of climate and microclimate to yield and quality in horticultural crops." In *Agrometeorology*, edited by F. Prodi, F. Rossi, G. Cristoferi, 287-298. Cesena, Italy: Commune di Cesena.

Lakso, A.N. 1994. "Apple." In *Environmental physiology of fruit crops; Vol 1, Temperate crops*, edited by B. Schaffer and P.C. Andersen, 3-42. Boca Raton: CRC Press.

Lakso A.N., M.D. White and D.S. Tustin. 2001. "Simulation modeling of the effects of short and long-term climatic variations on carbon balance of apple trees." *Acta Horticulturae* 557:473-480.

Lakso, A.N. October 2009. Cornell University, personal communication.

Levin M.D. 2005. "Finger Lakes freezes devastate vineyards." *Wines and Vines* July.

McDonald, A., S. Riha, A. Ditommaso and A. DeGaetano. 2009. "Climate change and the geography of weed damage: analysis of US maize systems suggests the potential for significant range transformations." *Agriculture Ecosystems and Environment* 130:131-140.

MKF Research. 2005. *Economic Impact of New York Grapes, Grape Juice and Wine*. MKF Research LLC, St. Helena, CA. www.mkf.com.

Montaigne, F. 2004. "The heat is on: eco-signs." *National Geographic* 206(3):34-55.

New York State Department of Agriculture and Markets (NYS DAM). 2010. http://www.agmkt.state.ny.us/

New York State Energy Research and Development Authority (NYSERDA). 2010. *Renewable fuels roadmap and sustainable biomass feedstock supply for New York*. Report 10-05. http://www.nyserda.org/publications/renewablefuelsroadmap/default.asp

Patterson, D.T., J.K. Westbrook, R.J.C. Joyce, P.D. Lingren and J. Rogasik. 1999. "Weeds, insects and diseases." *Climatic Change* 43:711-727.

Palmer, W.C. 1965. "Meteorological drought." Research Paper No. 45, U.S. Department of Commerce Weather Bureau, Washington, D.C.

Parmesan, C. 2006. Ecological and evolutionary responses to recent climate change. Annual Review Ecological Systems 37:637-69.

Penman, H.L. 1948. "Natural evaporation from open water, bare soil, and grass." *Proceedings Royal Society of London* 193:120-146

Poling, E.B. 2008. "Spring cold injury to winegrapes and protection strategies and methods." *HortScience* 43(6):1652-1662.

Reiners, S. and C. Petzoldt, eds. 2009. *Integrated Crop and Pest Management Guidelines for Commercial Vegetable Production*. Cornell Cooperative Extension publication #124VG http://www.nysaes.cornell.edu/recommends/

Rich, J. 2008. "Winter nitrogen cycling in agroecosystems as affected by snow cover and cover crops." M.S. Thesis. Cornell University. Ithaca, NY.

Rochette, P., G. Belanger, Y. Castonguay, A. Bootsma and D. Mongrain. 2004. "Climate change and winter damage to fruit trees in eastern Canada." *Canadian Journal of Plant Science* 84:1113-1125.

Salinari, F., S. Giosue, V. Rossi, F.N. Tubiello, C. Rosenzweig and M.L. Gulino. 2007. "Downy mildew outbreaks on grapevine under climate change: elaboration and application of an empirical-statistical model." *EPPO Bulletin* 37:317-26.

Sasek, T.W. and B.R. Strain. 1990. "Implications of atmospheric CO_2 enrichment and climatic change for the geographical distribution of two introduced vines in the USA." *Climatic Change* 16:31-51.

Seaman, A.J. January 2007. Cornell University, personal communication.

Seeley, S.D. 1996. "Modelling climatic regulation of bud dormancy." In *Plant Dormancy*, edited by G. Lang, 361-376. CAB International.

Stacey, D.A. 2003. "Climate and biological control in organic crops." *International Journal of Pest Management* 49:205-214.

St. Pierre, N.R., B. Cobanov and G. Schnitkey. 2003. "Economic losses from heat stress by U.S. livestock industries." *Journal of Dairy Science* 86:E52-E77.

Staples, C.R. 2007. "Nutrient and feeding strategies to enable cows to cope with heat stress conditions." Proceedings of the Southwest Nutrition and Management Conference, Tempe, Arizona, February 22-23.

Thornthwaite, C.W. 1948. "An approach toward the rationale classification of climate." *Geography Review* 38:55-94.

Trumble, J.T. and C.D. Butler. 2009. "Climate change will exacerbate California's insect pest problems." *California Agriculture* 63:73-78.

Turner, L.W., R.C. Warner and J.P. Chastain. 1997. "Micro-sprinkler and Fan Cooling for Dairy Cows: Practical Design Considerations." University of Kentucky; Department of Agriculture. Cooperative Extension Service. AEN-75. http://www.ca.uky.edu/agc/pubs/aen/aen75/aen75.pdf

United States Department of Agriculture (USDA) Census of Agriculture 2007. U.S. Census, United States Department of Agriculture, National Agriculture Statistical Service. www.nass.usda.gov/ny.

Walther, G.R. 2002. "Ecological responses to recent climate change." *Nature* 416:389-395.

Ward, N.L. and G.J. Masters. 2007. "Linking climate change and species invasion: and illustration using insect herbivores." *Global Change Biology* 13:1605-15.

White, M.A., N.S. Diffenbaugh, G.V. Jones, J.S. Pal and F. Giorgi. 2006. "Extreme heat reduces and shifts United States premium wine production in the 21st century." *Proceedings National Academy of Sciences* 103(30):11217-11222.

Wilks, D.S., D.W. Wolfe. 1998. "Optimal use and economic value of weather forecasts for lettuce irrigation in a humid climate." *Agricultural and Forest Meteorology* 89:115-129.

Wolfe, D.W., M.D. Schwartz, A.N. Lakso, Y. Otsuki, R.M. Pool and N.J. Shaulis. 2005. "Climate change and shifts in spring phenology of three horticultural woody perennials in northeastern USA." *International Journal of Biometeorology* 49:303-309.

Wolfe, D.W., L. Ziska, C. Petzoldt, A. Seaman, L. Chase and K. Hayhoe. 2008. "Projected change in climate thresholds in the Northeastern U.S.: Implications for crops, pests, livestock, and farmers." *Mitigation and Adaptation Strategies for Global Change* 13:555-575.

Wolfenson, D., I. Flamenbaum and A. Berman. 1988. "Dry period heat stress relief effects on prepartum progesterone, calf birth weight and milk production." *Journal of Dairy Science* 71:809-818.

Yamamura, K. and K. Kiritani. 1998. "A simple method to estimate the potential increase in the number of generations under global warming in temperate zones." *Applied Entomology and Zoology* 33:289-298.

Zimbelman, R.B., R.P. Rhoads, M.L. Rhoads, G.C. Duff, L.H. Baumgard and R.J. Collier. 2009. "A re-evaluation of the impact of temperature humidity index (THI) and black globe humidity index (BGHI) on milk production in high producing dairy cows." Proceedings of the Southwest Nutrition and Management Conference, Tempe. Arizona, February 26-27.

Ziska, L.H. 2003. "Evaluation of the growth response of six invasive species to past, present and future carbon dioxide concentrations." *Journal of Experimental Botany* 54:395-404.

Ziska, L.H., and K. George. 2004. "Rising carbon dioxide and invasive, noxious plants: potential threats and consequences." World Resource Review 16: (4) 427-447.

Ziska, L.H. and G.B. Runion. 2006. "Future weed, pest and disease problems for plants." In *Agroecosystems in a Changing Climate*, edited by P. Newton, A. Carran, G. Edwards, P. Niklaus. New York:CRC Press.

Ziska, L.H., J.R. Teasdale and J.A. Bunce. 1999. "Future atmospheric carbon dioxide may increase tolerance to glyphosate." *Weed Science* 47:608-615.

Appendix A. Stakeholder Interactions

The ClimAID Agriculture team gathered information and enlisted participation from key stakeholders in the agriculture sector through existing relationships and collaboration with the New York State Department of Agriculture and Markets, Cornell Cooperative Extension and Integrated Pest Management specialists, crop consultants, farmer commodity groups (e.g., Sweet Corn Growers and Finger Lakes Grape Growers Associations), and individual farmer collaborators. Below are some specific aspects of stakeholder involvement associated with the project.

Meetings and Events

Two important half-day meetings were held with stakeholders early on to gather expert opinions, with formal presentations followed by an opportunity to provide feedback:

1) Conference with approximately 35 Cornell Cooperative Extension staff from across the state, held in Ithaca on November 11, 2008. The expertise of these specialists ranged from fruit and vegetable crops to dairy and grain crops. The conference included presentations and breakout group input on high-priority vulnerabilities and potential opportunities, feasible adaptation strategies, and needs for additional information, decision tools, and/or resources to help farmers cope with climate change.
2) Briefing at New York State Agriculture and Markets headquarters (Albany, November 12, 2008) attended by the Agriculture Commissioner and approximately 15 other key department leaders.

Other presentations of preliminary results have included the 2009 November "In-Service" training for Cornell Cooperative Extension staff in Ithaca, and the NYSERDA Agriculture Innovations Conference in December, 2009, held in Albany.

Focus Group and Technical Working Groups of the Climate Action Plan

A focus group of several stakeholders has been used for frequent feedback as this project proceeds. In addition, results have been shared with individuals at New York State Agriculture and Markets.

[1] Abbreviations are for counties in geographic areas: N = Northern (Jefferson, Lewis, St. Lawrence); NE = Northeastern (Clinton, Essex, Franklin, Hamilton, Warren); W = Western (Erie, Genesee, Livingston, Monroe, Niagara, Ontario, Orleans, Seneca, Wayne, Wyoming, Yates); C = Central (Cayuga, Chenango, Cortland, Herkimer, Madison, Oneida, Onondaga, Oswego, Otsego); E = Eastern (Albany, Fulton, Montgomery, Rensselaer, Saratoga, Schenectady, Schoharie, Washington); SW = Southwestern (Allegany, Cattaraugus, Chautauqua, Steuben); S = Southern (Broome, Chemung, Schuyler, Tioga, Tompkins); SE = Southeastern (Columbia, Delaware, Dutchess, Greene, Orange, Putnam, Rockland, Sullivan, Ulster, Westchester); LI = Long Island (Nassau, New York City, Queens, Richmond, Suffolk).

[2] Small farms in New York State are defined as those with total acreage of less than 100 acres and/or annual sales of less than $50,000. Approximately 51 percent of farms in New York State are less than 100 acres in size, and approximately 75 percent of New York farms have revenue of less than $50,000 (USDA, 2007).

[3] Within the dairy sector, small farms are defined as having 80 or fewer cows and no milking parlor (Knoblauch et al., 2009).

[4] Precipitation is in liquid-water equivalents for rain or snow; PET is the potential evapotranspiration (evaporative water loss from soil and plants); runoff is the fraction of precipitation that exceeds soil holding-capacity and passes either into deep groundwater or into streams. PET calculations in this figure assume full leaf-area development throughout the growing season and are not specific to any particular crop's growth and development.

[5] Predictions are shown for Indian Lake in the Adirondacks, Elmira, and Rochester, and include historic values (based on the period 1901–2006) and climate change projections for two different carbon dioxide emissions scenarios, B1 (low emissions) and A2 (high emissions). Calculations were derived from the same dataset as Table 7.3, which assumes maximal plant water demand (i.e., potential evapotranspiration, calculated assuming full canopy cover for the entire period; see text for more discussion). Projected changes in monthly temperature and precipitation used to calculate deficit probabilities were derived from ClimAID data generated from (GFDL, GISS, MIROC, CCSM, and UKMO) GCMs as appropriate for each of the timeslices and emissions scenarios.

Chapter 8

Energy

Authors: Stephen A. Hammer,[1,5] Lily Parshall,[2] Robin Leichenko,[3] Peter Vancura,[3] and Marta Panero[4]

[1] Sector Lead
[2] Columbia University
[3] Rutgers University, Department of Geography
[4] New York University
[5] Massachusetts Institute of Technology, Department of Urban Studies and Planning (formerly at Columbia University)

Contents

Introduction ... 256
8.1 Sector Description ... 256
8.1.1 Brief Profile of the New York State Energy System ... 256
8.1.2 Economic Value ... 258
8.1.3 Non-climate Stressors ... 259
8.2 Climate Hazards ... 259
8.2.1 Temperature ... 261
8.2.2 Precipitation ... 261
8.2.3 Sea Level Rise ... 261
8.2.4 Other Climate Factors ... 261
8.3 Vulnerabilities and Opportunities ... 262
8.3.1 Energy Supply ... 262
8.3.2 Energy Demand ... 268
8.4 Adaptation Strategies ... 277
8.4.1 Key Adaptation Strategies ... 278
8.4.2 Larger-scale Adaptations ... 280
8.4.3 Co-benefits, Unintended Consequences, and Opportunities ... 281
8.5 Equity and Environmental Justice Considerations ... 281
8.6 Conclusions ... 282
8.6.1 Main Findings on Vulnerabilities and Opportunities ... 282
8.6.2 Adaptation Options ... 282
8.6.3 Knowledge Gaps ... 283
Case Study A. Impact of Climate Change on New York State Hydropower ... 283
Cast Study B. Climate-change-induced Heat Wave in New York City ... 286
References ... 291
Appendix A. Stakeholder Interactions ... 295
Appendix B. Relationship between NYISO Load Zones and ClimAID Regions ... 296

Introduction

This ClimAID chapter considers how global climate change may improve or exacerbate existing weather-related stresses on the energy sector and reviews possible short- and long-term adaptation strategies.[1] The chapter broadly groups specific vulnerabilities and opportunities into supply-side issues and demand-side issues, with a particular emphasis on the power sector. Transport-related energy considerations are covered in Chapter 9, "Transportation," of this report.

Research for this chapter was conducted both as a literature review and through direct stakeholder engagement with a range of energy companies operating in different parts of the state (see Appendix A).

8.1 Sector Description

Reliable energy systems are critical to commerce and quality of life. New York State's electricity and gas supply and distribution systems are highly reliable, but weather-related stressors can damage equipment, disrupt fuel supply chains, reduce power plant output levels, or increase demand beyond the energy system's operational capacity.

8.1.1 Brief Profile of the New York State Energy System

Energy is derived from a wide variety of fuel sources and technologies in New York State. Roughly 49 percent of the state's electricity is generated in-state using fossil fuels; nuclear power (30 percent) and renewables[2] (21 percent) account for the balance (NYISO, 2009) (**Figure 8.1**). The generation mix varies widely in different parts of the state. For example, approximately 50 percent of the fossil-fired power plant capacity is located in New York City and Long Island, while most hydropower capacity is located in the northern and western part of the state (USEPA, 2009). **Table 8.2** presents generation capacity by fuel type in each ClimAID region.

New York State is divided into 11 electricity load zones, which are managed by the New York Independent System Operator (NYISO) (**Figure 8.2**). These zones are drawn by primarily administrative boundaries and do not reflect unique geographic or operating characteristics. They do differ significantly from the seven ClimAID regions highlighted in this report (see Appendix B, and Chapter 1, "Climate Risks").

New York City is by far the largest load zone in the state, responsible for approximately one-third of total annual electricity demand statewide (**Table 8.3**). Between 2002 and 2008, the period for which data are readily available, load growth (in total gigawatt-hours) has occurred in nine of the 11 zones around New York State. Growth has not been consistent, as load has declined in one or more years in most zones, but on average the total annual electricity load has increased by 4.3 percent each year statewide (NYISO, 2009a).

Thermal energy needs are satisfied in a variety of ways. New York State is home to more than a dozen district energy systems, which centrally generate steam, hot

Energy Unit	Description
Kilowatt (kW)	A measure of electrical power equal to 1,000 watts
Megawatt (MW)	A measure of electrical power equal to 1,000 kW (or 1 million watts)
Gigawatt (GW)	A measure of electrical power equal to 1,000 MW (1 billion watts)
Kilowatt or Megawatt peak (kWp or MWp)	Peak power plant generation capacity
Kilowatt hours (kWh), Megawatt hours (MWh), or Gigawatt hours (GWh)	A time-related measure of electrical energy. Running a 3,000 MWp power plant at 100% capacity for 1 hour would produce 3,000 MWh of energy.

Table 8.1 Definitions of key energy terms used in this chapter

	Number of Power Plants (by fuel type) and Peak Generation Capacity (MWp)			
ClimAID Region	Fossil fuel	Nuclear power	Renewables	Total
Region 1	19	1	14	34
	2,761 MWp	517 MWp	2,628 MWp	5,905 MWp
Region 2	11		13	24
	3,548 MWp		1,106 MWp	4,654 MWp
Region 3	10		4	14
	775 MWp		11 MWp	786 MWp
Region 4	49		7	56
	12,996 MWp		137 MWp	MWp
Region 5	14	2	50	66
	1,350 MWp	2,339 MWp	594 MWp	4,283 MWp
Region 6	13	2	42	57
	3,968 MWp	2,784 MWp	304 MWp	7,056 MWp
Region 7	6		52	58
	566 MWp		1,263 MWp	1,829 MWp
New York State	122	5	182	309
	25,964 MWp	5,640 MWp	6,043 MWp	37,647 MWp

Table 8.2 New York State power plant data by ClimAID Region

thermoelectric power plants and those affecting different renewable power sources. Thermoelectric power plants generate electricity by converting heat into power. Conversion processes vary based on fuel sources at the power plant (e.g., nuclear, gas, oil, coal). Renewable power technologies harness naturally occurring resource flows (e.g., solar power, flowing water, wind) to generate electricity. Some forms of biomass—often considered a renewable resource—may be combusted in thermoelectric power plants, converted to liquid fuels, or used to generate heat for buildings or industry.

Impacts on Thermoelectric Power Generation and Power Distribution

Thermoelectric power plants are vulnerable to increases in flooding, droughts, water temperature, air temperature, and other extreme weather events. Plants located along coastal areas may be affected by rising sea levels and storm surges.

Flooding

Vulnerability is largely a function of the elevation of power plants and their proximity to the path that any storm-related tidal surge would follow during extreme weather events. To get a sense of the scale of vulnerability, this analysis overlaid New York City power plant locations obtained from the U.S. Environmental Protection Agency's (USEPA) eGrid database on a U.S. Geological Survey digital elevation model and identified power plants within 5 meters (about 16.4 feet) of current sea level. **Figure 8.4** shows that a majority of the city's largest power plants are at an elevation below 5 meters, which means they currently could potentially be affected by Category 3 or higher hurricane-induced storm surges.

A different flooding risk involves the elevation of the cooling water intake and outflow pipes at thermoelectric power plants. To the extent that these pipes become clogged by debris during flooding or storm surges, power plants may be forced to shut down (Aspen Environmental Group and M Cubed, 2005; Union of Concerned Scientists, 2007). One power plant operator with a facility fronting on a large lake in the northwestern part of the state noted that high winds can stir up debris that clogs their intakes located at the shoreline. The operator contrasted this situation with another of its plants, which has intakes extending much farther into an adjacent lake. The latter facility is far less vulnerable to this type of debris problem.

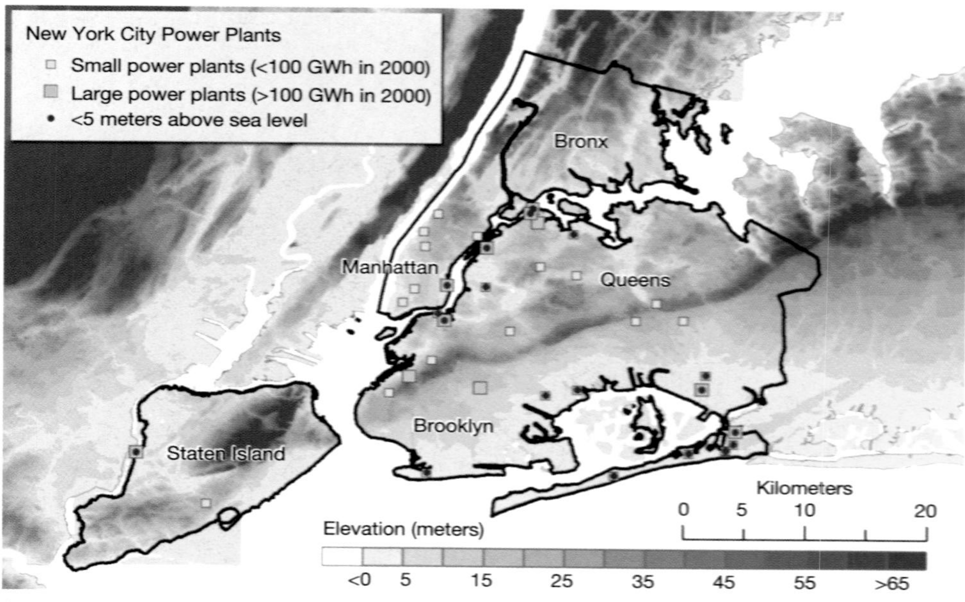

Sources: Data: Power plant data for 2000 were extracted from CARMA 2008; New York City digital elevation model is from the USGS 1999, which has a vertical error of +/- 4 feet. Map credit: Lily Parshall 2009.

Figure 8.4 Location and elevation of power plants in New York City

Table 8.8 presents the intake pipe depth at a number of power plants around New York State (NETL, 2009). Whether these intake depths may prove vulnerable to debris problems from flooding events is unclear, although deeper intakes are presumably less vulnerable than shallower intakes.

Drought

A recent U.S. Department of Energy study seeking to highlight potential drought-related water intake problems across the United States provided data on 12 large power plants around New York. The report did not pass judgment on whether current intake pipe depth levels were inadequate, because this is largely a location-specific issue (NETL, 2009).[4] New York State facilities tend to have shallower intake depths when compared to other plants around the U.S.; whether this will create problems at these facilities in the future is unclear. To date, there do not appear to be any instances where drought has created problems of this nature.

Water Temperature

The DEC thermal discharge rules may create challenges during extended heat waves, when the receiving waters may already be close to the upper temperature limit defined in the facility's operating permit. This situation may force the power station to reduce production to decrease the heat content of the water leaving the condenser (ICF, 1995). During Europe's deadly 2003 summer heat wave, several nuclear power plants in Spain and Germany closed or cut output to avoid raising the temperature of rivers cooling the reactors. The French government allowed nuclear power plants to discharge cooling waters at above-normal temperatures as an emergency measure to avoid blackouts (Jowit and Espinoza, 2006).

The five nuclear power plants in New York State are located either on the Great Lakes or the Hudson River. In both cases, these facilities draw cooling waters from deep-water sources less vulnerable to dramatic temperature rises. The situation at other thermoelectric power plants around state, several of which draw from shallower water sources, is less clear.

A different, and under-researched, topic relates to how climate change may affect biota levels in New York waterways currently used for power plant cooling. To the extent that biota levels increase, changes may be required in the screening processes currently employed at these facilities to ensure that the water flow into the facility is not inhibited in any way.

Air Temperature

Changes in ambient air temperature and air density levels resulting from climate change may affect power plant output levels. One potential temperature-related impact occurs at combined-cycle gas turbine facilities (Hewer, 2006). These units are designed to fire at a specific temperature, and when ambient air temperatures rise, air density declines, which reduces the amount of oxygen available to achieve peak output (ICF, 1995). Similar problems exist at steam turbine facilities.

Three studies discount the importance of these impacts, arguing that capacity and/or output reductions will be less than 1 percent under most climate scenarios (Stern,

Facility Name	Primary Fuel	Water Source	Intake Depth Below Surface	Intake Depth for this Type of Facility[1] (feet)		Intake Depth for this Type of Water Source[1] (feet)	
			(feet)	Mean	Median	Mean	Median
AES Cayuga	Coal	Cayuga Lake	44	16.1	12	21.6	17
AES Greenridge	Coal	Seneca Lake	11	16.1	12	21.6	17
AES Somerset	Coal	Lake Ontario	16	16.1	12	21.6	17
Dunkirk Generating Station	Coal	Lake Erie	21	16.1	12	21.6	17
Danskammer Generating Station	Coal	Hudson River	5	16.1	12	13.2	10
Roseton Generating Station	Gas	Hudson River	29	14.4	12	13.2	10
Fitzpatrick	Nuclear	Lake Ontario	12	16.8	13.5	21.6	17
CR Huntley Generating Station	Coal	Niagara River	10	16.1	12	13.2	10
Oswego Harbor Power	Oil	Lake Ontario	20	16.1	12	21.6	17
PSEG Albany Generating Station	Gas	Hudson River	24	14.4	12	13.2	10
Ginna	Nuclear	Lake Ontario	15	16.8	13.5	21.6	17
Rochester 7	Coal	Lake Ontario	36	16.1	12	21.6	17

[1] Based on nationwide data.
Source: NETL/DOE 2009, pp 20, A-10

Table 8.8 Cooling water intake depth at selected New York State power plants compared to national data

1998; Bull et al., 2007; and Linder et al., 1987). Several New York State utilities and power plant operators interviewed for this report also noted that the impacts of changing temperature levels are likely to be negligible, because the equipment is already designed to handle wide temperature swings between the winter and summer months. It may well be that changes in extreme temperatures are more relevant, since during such conditions the equipment's design parameters are more likely to be breached (David Neal, personal communication, October 30, 2009; Victoria Simon, personal communication, October 15, 2009). Moreover, depending on the rate at which climate change progresses, many vulnerable facilities will reach the end of their useful lives and be replaced with better-adapted ones before these long-term power generation impacts are felt (ICF, 1995; Victoria Simon, personal communication, October 15, 2009).

The New York State Reliability Council[5] reports that there is some decline in power output levels at higher temperatures, but the Council also characterizes the impact as rather small. As part of their technical assessment of the reliability of the state's power system, the New York State Reliability Council quotes the NYISO research finding that for each degree above 92°F, combustion power plants around the state collectively lose approximately 80 megawatts in production output (New York State Reliability Council, 2004). This decrease is built into their estimates of how much power will be available around the state under certain operating conditions. Given that the state has more than 37,000 megawatts of generation capacity overall and roughly 26,000 megawatts of fossil-fired combustion facilities (USEPA, 2009), this decrease is relatively minor (**Table 8.2**).

Rising ambient air temperatures may also affect the electricity transmission and distribution system. Because transmission and distribution lines and electrical transformers are rated to handle certain amounts of voltage for a given period of time, climatic conditions can lead to equipment failure by driving energy demand beyond the rated capacity. For instance, an extended heat wave in the summer of 2006 led to the failure of thousands of transformers in southern and northern California. Sustained high nighttime temperatures meant that the transformers could not cool down sufficiently before voltage levels increased again the next morning. Insulation materials within the transformers burned and circuit breakers tripped, knocking out the devices and causing more than a million customers around the state to lose power (Miller et al., 2008; Vine, 2008).

Power lines both above and below ground may also suffer mechanical failure as a result of higher ambient air temperatures. Power lines naturally heat up when conducting electricity; ordinarily, relief is provided by the cooler ambient air. Lines below ground rely on moisture in the soil to provide this cooling function. In both cases, as temperatures increase, the cooling capacity of the surrounding air or soil decreases, potentially causing above-ground lines to fail altogether or sag to levels where the public is placed at risk (Hewer, 2006; Mansanet-Bataller et al., 2008). The extent to which this is a problem in New York State is unclear. The New York Power Authority (NYPA) reports it regularly conducts sophisticated aerial surveys to assess hazards presented by sagging transmission lines (Victoria Simon, personal communication, October 15, 2009), but no data were available on how distribution line conductivity may change as a result of climate change.

The most newsworthy blackouts in New York City in recent years have tended to occur when heat waves extend over several days (Revkin, 1999; Waldman, 2001; Chan and Perez-Pena, 2006; Newman, 2006). In the past, two different State agency analyses have expressed concern about the age of local distribution network equipment and how this compounds system vulnerabilities on hot days when peak load levels increase dramatically (NYS Attorney General, 2000; NYS Department of Public Service, 2007). Little information has been published on this topic, however, so the extent of the problem is unknown.

One area where additional research may be beneficial is the link between the average temperature of extreme heat events and the duration of the heat event. For example, one distribution utility provided anecdotal information suggesting that the frequency of distribution system service interruptions appeared to be higher for multi-day heat events above 95°F than for multi-day heat events above 90°F. The company did not have evidence about the statistical significance of this finding, although an analysis of such tipping points (beyond which the likelihood of distribution system service interruptions significantly increases) might prove helpful in terms of system design, equipment ratings, or the development of operating procedures during extreme heat events (see Chapter 1, "Climate Risks").

Ice and Snow Storms

New York State's energy system has long been vulnerable to impacts of ice and snow storms. The great blizzard of 1888 led to the decision to bury most electric wires around New York City (*New York Times*, 1888). Ice storms typically affect a wide geographic area, making repair work a sizable task (John Allen, personal communication, September 29, 2009; James Marean, personal communication, September 29, 2009). With more than 15,000 miles of electric transmission lines and 200,000 miles of distribution lines across the state (New York State Public Service Commission, 2008), ice storms are particularly problematic.

In 1998, a massive multi-day ice storm resulted in more than $1 billion in damage across the northeastern United States and eastern Canada. In New York State alone, dozens of high-voltage transmission towers, 12,500 distribution poles, 3,000 pole-top transformers, and more than 500 miles of wire conductor required replacement, affecting 100,000 customers from Watertown to Plattsburgh. Most of the repairs were completed within two months, although some areas were not completely repaired for four months (EPRI, 1998). Subsequent research found that much of the equipment was not rated for a storm of that magnitude. Another major ice storm in December 2008 resulted in the loss of power to 240,000 customers in the state's capital region (Gavin and Carleo-Evangelist, 2008).

Impacts on Natural Gas Distribution Infrastructure

Ninety-five percent of the state's natural gas supply is imported via grid pipeline from other states and Canada. Underground storage facilities (primarily depleted gas wells) in western New York and Pennsylvania are important features of the state's natural gas system, ensuring that adequate supplies are available during the peak-demand winter months. They also provide some level of insurance against natural disasters that may disrupt the production or delivery of natural gas to the state at other times of the year (State Energy Planning Board, 2009b), although the extent of this benefit is unclear. For example, an extensive amount of underwater pipeline damage occurred in the Gulf Coast region during hurricanes Katrina and Rita in 2005. Buried onshore pipelines were also damaged (Cruz and Krausmann, 2008). As a result of these supply chain disruptions, natural gas prices spiked to unprecedented levels in New York in 2005 (State Energy Planning Board, 2009b).

The impacts of climate change on in-state gas distribution infrastructure are unclear. Gas distribution pipes are buried for safety reasons. Although this does not make them immune to flooding risks associated with extreme weather events (Associated Press, 1986; *New York Times*, 1994), there is little published evidence that this has been a significant problem in New York in recent decades. Gas pipelines are also vulnerable to frost heaves (Williams and Wallis, 1995), although the extent to which climate change may alter current frost heave risks is unclear. Both of these subjects may require additional research, although the research need not examine all regions of the state. Currently, large swaths of ClimAID regions 2, 5, and 7 (Catskill Mountains and West Hudson River Valley, East Hudson and Mohawk River Valleys, and Adirondack Mountains) lack gas distribution service because of the low population levels in these areas.

Impacts on Renewable Power Generation

Climate change may also affect renewable power output around the state by affecting the timing or level of the natural resource responsible for power generation.

Hydropower

New York's 338 conventional hydropower facilities collectively generate more hydropower than any other state east of the Rocky Mountains. With a peak generation capacity of 5,756 megawatts, they currently satisfy 15 percent of the state's total annual electricity requirements (State Energy Planning Board, 2009). Three facilities operated by the New York Power Authority are responsible for 80 percent of the state's total hydropower capacity. Two are fed by the Great Lakes watershed, while the third is a pumped storage facility located in the Catskills. The potential exists to deploy another 2,500 megawatts of hydropower around the state by 2022, but "environmental, siting, financial, and regulatory barriers suggest that relatively little new development is likely to occur" (State Energy Planning Board, 2009).

In a changing climate, power supply availability must be considered. In projecting power supply availability from different sources, the New York Independent System Operator assumes that non-New York Power

Authority hydropower generators around the state—which represent approximately 1,000 megawatts of installed capacity—experience power generation output declines of approximately 45 percent in July and August due to reduced water availability during the summer months (NERC, 2008; NYISO, 2004). This 45 percent de-rate factor (the output decline) assumes the state is not experiencing drought conditions; under such conditions, the de-rate figure might be even higher. (For comparison purposes, when the northeastern United States suffered from drought in 2001, actual output from these same non-New York Power Authority facilities declined by 65 percent during summer months compared to their peak-rated capacity (NYISO, 2004).)

Case Study A examines how climate change may affect hydropower output levels at two large New York Power Authority-owned facilities near Niagara Falls and on the St. Lawrence River in Massena, New York, noting the correlation between precipitation levels and the level of power produced by these facilities. To the extent precipitation levels are expected to increase across the state by 2080 (see Chapter 1, "Climate Risks"), hydropower production levels may actually increase over time, although there are likely to be seasonal differences.

As Chapter 1 notes, however, New York State is also expected to experience more frequent late-summer drought conditions over the coming decades, which could lead to sizable reductions in hydropower output levels. This would have significant cost repercussions around the state, as lost capacity would likely be replaced by more expensive forms of power generation (Morris et al., 1996). Moreover, because the impacts of climate change are likely to be felt at hydropower in surrounding states (and Canada) as well, New York may not be able to rely on the same level of electricity imports it has previously, exacerbating already tight local power supply markets and raising prices even higher.

Solar Power

Although there is relatively little solar photovoltaic technology currently deployed around New York (approximately 14.6 MW), estimates are that the state enjoys significant solar resources, exceeding that of any other renewable energy source in the state (State Energy Planning Board, 2009). Whether climate change will enhance or hinder local solar resources is unknown. One study modeled solar radiation in the United States through 2040, projecting that increased cloud cover attributable to rising carbon dioxide levels could reduce solar radiation levels by as much as 20 percent, particularly in the western United States (Pan et al., 2004). No clear trends were projected for the Northeast. Another study focused on Nordic (Scandinavian) cities estimates that a 2-percent decrease in solar radiation could reduce solar cell output by 6 percent (Fidje and Martinsen, 2006). A solar expert at SUNY Albany, Dr. Richard Perez, reviewed this literature but discounted these impacts, noting that because of differences in latitude between New York, Nordic areas, and other parts of the United States, New York State should "expect, in the worst case, a 1 to 2 percent decrease in [solar] PV yield, and the best case, no change at all" (Dr. Richard Perez, personal communication, September 9, 2009).

More research is necessary to examine the potential impacts of climate change on solar power, as decreases in solar photovoltaic system output in New York State would increase the per-kilowatt cost of solar power, reducing the cost competitiveness of photovoltaic systems compared to other forms of electricity. Research should also examine the extent to which such losses may be offset by advances in solar panel efficiency that will likely occur over time.

Wind Power

New York State's proximity to the Atlantic Ocean and Great Lakes places it close to excellent conditions to support wind power development. According to the American Wind Energy Association (2009), New York ranks 15th nationally in terms of its overall wind power potential, although support from New York State's Renewable Portfolio Standard and favorable federal tax rules have helped the state achieve a seventh-place ranking with regard to its current wind power deployment. Already, there are 791 large wind turbines installed around the state, with a peak generation capacity of 1,264 megawatts (American Wind Energy Association, 2009), and forecasts are that this could increase to more than 8,500 MW by 2015 (State Energy Planning Board, 2009). The majority of this capacity will come in the form of large wind turbine installations, as opposed to small rooftop turbines that are more scale-appropriate for urban areas.

Because wind turbine power output is a function of the cube of the wind speed, small changes in wind speed can translate into large changes in output. For example, a recent study notes how a 10-percent change in wind speed can lead to a 30-percent change in energy output (Pryor and Barthelmie, 2010). The consequences of changing wind patterns can thus be sizable in terms of the state's ability to rely on large quantities of wind power.

The same study also notes a dearth of research examining extreme wind speeds and gusts and their relationship to wind turbine design protocols (Pryor and Barthelmie, 2010). Given that current industry design criteria generally call for turbines to withstand 1-in-50-years wind speed events lasting no more than 10 minutes, more research may be necessary to assess whether these standards should be upgraded or whether they can be relaxed in the coming decades.

Biomass

Forestry and agricultural products currently make a very minor contribution to the state's overall electricity picture, combusted in biomass-only facilities near Utica and Chateaugay or co-combusted with coal at a power plant near Niagara Falls. These facilities have a peak generation capacity of 65 megawatts, and collectively generate approximately 440,000 megawatt hours of power per year (State Energy Planning Board, 2009a).

Biomass is also used as a primary fuel for heating purposes in some New York homes and businesses. In 2007, approximately 94 trillion British thermal units (TBtus) of heat were generated statewide from wood resources (State Energy Planning Board, 2009a), the vast majority of which was consumed in homes. Some of this wood is shredded and then reprocessed into uniform-sized pellets, which are designed for use in more efficient boilers and wood stoves.

Recently, interest has grown in the conversion of biomass into liquid fuels, some of which is blended with fuel oil for use in residential or commercial heating systems. As of 2008, there was one biodiesel manufacturing facility in the state, although there are more than a dozen fuel oil companies around the state that blend biodiesel with their fuel oil to sell to customers (State Energy Planning Board, 2009a). Most of their biodiesel is purchased from refiners located outside of the state.

The effects of climate change on New York's biomass-based energy systems is unclear. As the "Ecosystems" chapter (Chapter 6) notes, some species of trees may do better than others, a function of the level of temperature change, vulnerability to vectors, and level of drought conditions. Homes and businesses that rely on downed trees as the source of their wood may or may not find changes in the level of available supply; the impacts of climate change on wood resources is likely to be very local in nature. The effects on power plants or pellet manufacturers, which rely on managed forests or waste wood from manufacturing operations, are similarly unclear and may ultimately depend on the characterizations of different wood species, as the impacts are expected to vary (see Chapter 6, "Ecosystems"). These facilities tend to source their material many months or even years in advance from a range of suppliers, which may help offset any adverse impacts attributable to climate change, although this is an area where more research would be beneficial.

The impacts of climate change on biodiesel production or blending operations in New York State is similarly unclear, as production facilities tend to source material from an international feedstock market. The extent to which the supply chain will be affected is uncertain, and as the market for biodiesel fuel grows across the state, this may also be an area where further research would be beneficial.

8.3.2 Energy Demand

Climate change may affect energy demand for space heating and cooling. Electricity demand is most sensitive to changes in summer climate, whereas heating fuel demand is most sensitive to changes in the winter climate. Impacts may have multiple dimensions, including changes in total demand, seasonal variability, and peak demand (Amato et al., 2005; Wilbanks, 2007; Scott and Huang, 2007). Overall, in the northern United States, net energy demand is likely to decrease as a result of warmer winters. This effect is expected to outweigh air conditioning-related increases in summertime energy demand in the southern United States, leading to a net national reduction in total energy demand (Scott and Huang, 2007).[6]

To understand how climate change may affect energy demand in New York State, the ClimAID climate team first provided current trends and expected changes in heating degree days and cooling degree days, metrics that affect demand for space heating and cooling, respectively (see Chapter 1, "Climate Risks," for an analysis of current trends and expected changes in average temperature).[7] Next, the sensitivity of energy demand to changes in temperature was analyzed. This analysis was carried out only for electricity demand, as data are not available to analyze heating fuels. Finally, to project changes in electricity demand the analysis combined the information on projected changes in climate with information on the sensitivity of demand to those changes. The focus is on the 2020s because non-climate drivers dominate energy planning in the medium and long terms. The 2020s is defined as the 2011–2039 time period (consistent with Chapter 1, "Climate Risks"), so projections for the 2020s are an average of the projections for each of the 30 years in that time period.

Linear regression was used to estimate historical changes in temperature. Linear forecasts for future climate are then based on the assumption that the observed trend in climate over the historical period will continue into the future. This assumption may be reasonable over short time periods (up to 10 years). An advantage of linear forecasts is the ability to make projections for a specific geographic location using hourly, daily, and/or monthly data from a local weather station. The benefit of temporal specificity is offset by inclusion of only a limited number of variables. Therefore, these forecasts are compared with global climate model (GCM) projections that account for the dynamic relationships among many different climate variables.

New York State Heating and Cooling Seasons

In New York State, the heating season is longer than the cooling season, and 50 to 55 percent of heating degree days occurs during the winter peak months of December, January, and February, with the other 50 percent occurring during the fall and spring "shoulder" seasons. There is substantial variation among different regions of the state. For example, Binghamton, Utica, and Watertown have more than 7,000 heating degree days per year, whereas New York City has fewer than 5,000 (**Table 8.9**). On the other hand, New York City has more than 1,000 cooling degree days, whereas Binghamton, Albany, and Buffalo have 400 to 600 cooling degree days.[8] The direction and magnitude of changes in energy demand depend on changes in heating degree days, cooling degree days, and other climate-related changes as well as the sensitivity of demand to climate factors. In some cases, sensitivities may be nonlinear, for example if higher temperatures in the summertime lead to a significant increase in air-conditioning saturation rates (the number of households with some form of air conditioning).

Projected Changes in Heating Degree Days

In all regions of the state, heating degree days have significantly declined over the past few decades (**Figure 8.5**). Annual heating degree days are expected to decline by between 5 and 8 percent in the 2020s compared to the current (1970–2007) average; expected changes are relatively consistent across all regions of the state. Global climate model projections for the number of heating degree days in the 2020s are broadly consistent with the linear forecasts. Agreement between these two methods should help to address the skepticism with which global climate models have historically been viewed by energy sector stakeholders. The two methods would not necessarily be expected to agree over the medium to long term.

Warmer winters may reduce electricity demand for heating, although just 10% of New York State's heating demand is met with electricity. Declining heating degree days may also put downward pressure on demand for utility gas and fuel oil, the two primary sources of space heat in the state (**Figure 8.2**), although climate is just one of many drivers of demand for these resources. Additional research is needed to better understand how climate change may affect the breakdown of demand for natural gas for building heat versus power generation.

Weather Station	ClimAID Region	NYISO Zone	Heating Degree Days (per year)	Cooling Degree Days (per year)
Buffalo	Region 1	Zone A	6,654	557
Rochester	Region 1	Zone B	6,663	585
Elmira	Region 3	Zone C	6,904	479
Binghamton	Region 3	Zone C	7,211	409
Utica	Region 5	Zone E	7,229	483
Watertown	Region 6	Zone E	7,457	521
Albany	Region 5	Zone F	6,813	567
NYC (Central Park)	Region 4	Zone J	4,740	1,158

Note: Maps showing the relationship between NYISO zones and ClimAID regions are shown in Appendix B. Note that these seven stations were selected for the analysis presented in this table as well as for Figures 5 and 6 because global climate model projections for heating degree days and cooling degree days were available for each of these stations. Source: Historical climate data obtained from NOAA

Table 8.9 Average annual heating degree days and cooling degree days, 1970 to 2007

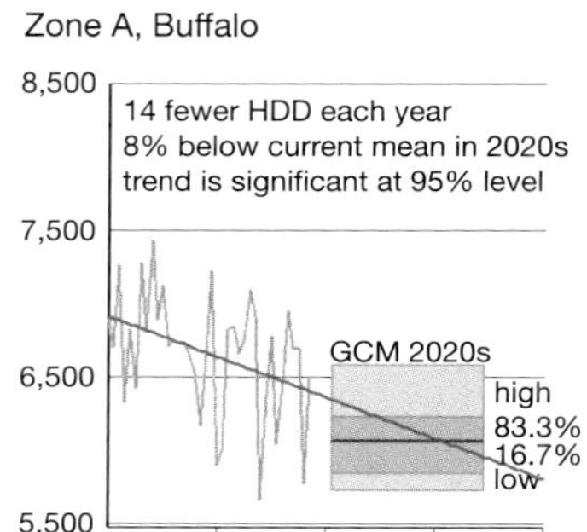

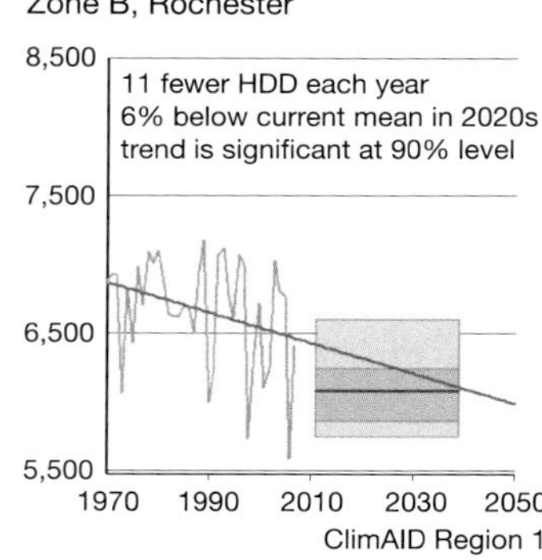

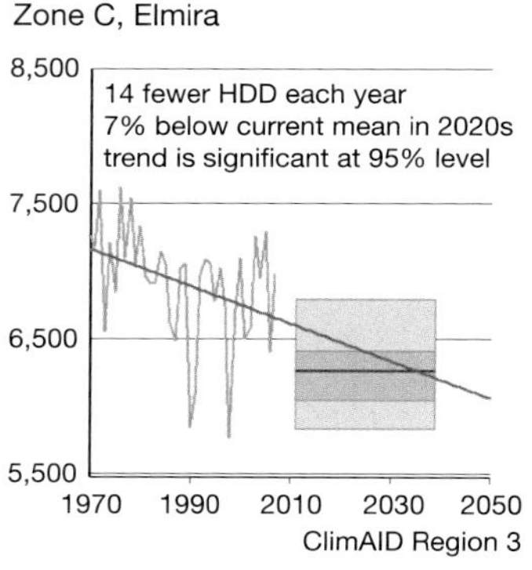

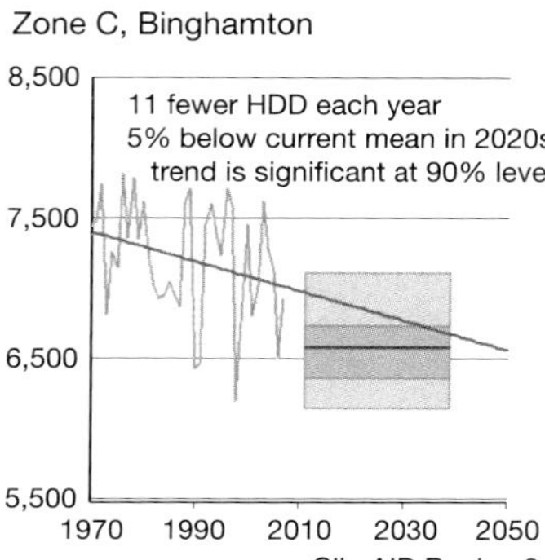

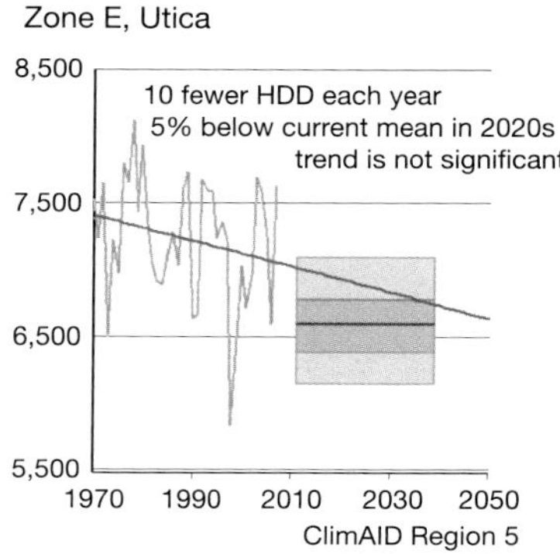

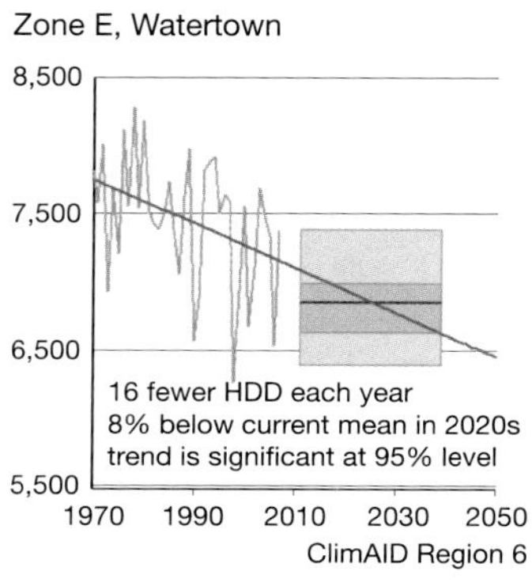

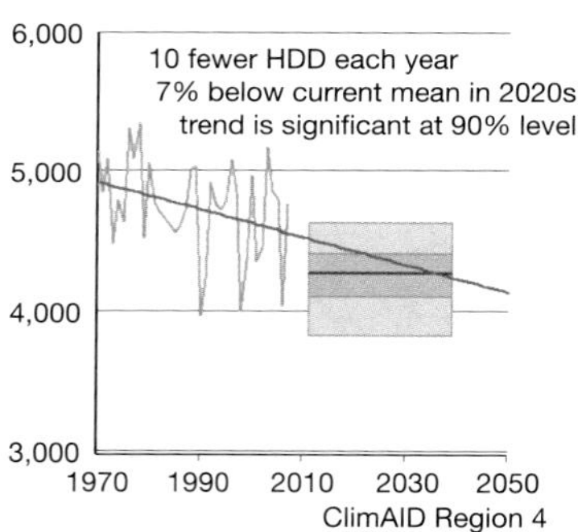

Note: The jagged line on each graph represents historical annual HDD, based on actual weather station data. The straight line indicates the trend in historical HDD, extrapolated into the future. The number of HDD in the 2020s was estimated by extrapolating the linear trend out to 2039 and then averaging the values for 2011–2039. The grey boxes show the range of GCM projections for the 2020s, with the black line in the center of each grey box indicating the mean GCM projection. Note that the trend is considered significant if the p-value is <0.1. Not all NYISO zones and ClimAID regions are represented; only those stations for which GCM projections for HDD were available are included. Data source: Historical climate data from NOAA; GCM projections from ClimAID.

Figure 8.5 Projected changes in heating degree days in selected regions of the state. The projections are for the 16 GCMs and 3 emissions scenarios used in ClimAID. Shown are the minimum, central range, and maximum values across the GCMs and emissions scenarios.

Projected Changes in Cooling Degree Days

New York State has a relatively short cooling season. In New York City, 79 percent of annual cooling degree days occur during the summer months; in many cities in northern and western New York, the figure is closer to 85 percent. Although the global climate models used in ClimAID project increases in cooling degree days on the order of 24–47 percent, depending on the region, these projections generally exceed forecasts based on linear extrapolation of current trends (**Figure 8.6**). Also, in most regions, historical trends in cooling degree days are not statistically significant, reducing confidence in the linear extrapolations.[9] Of the weather stations for which data were obtained, only Elmira has a statistically significant upward trend over

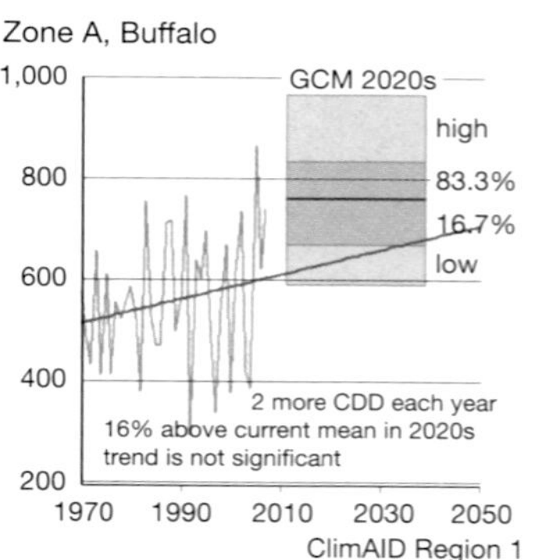

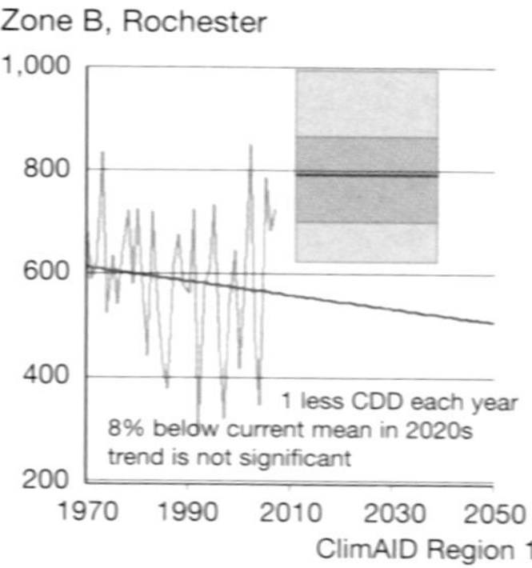

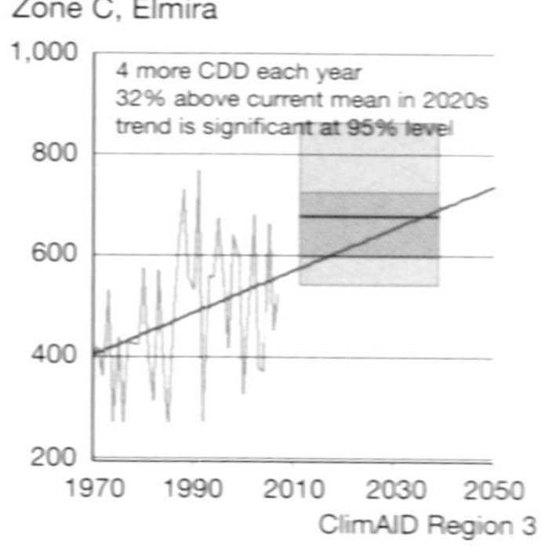

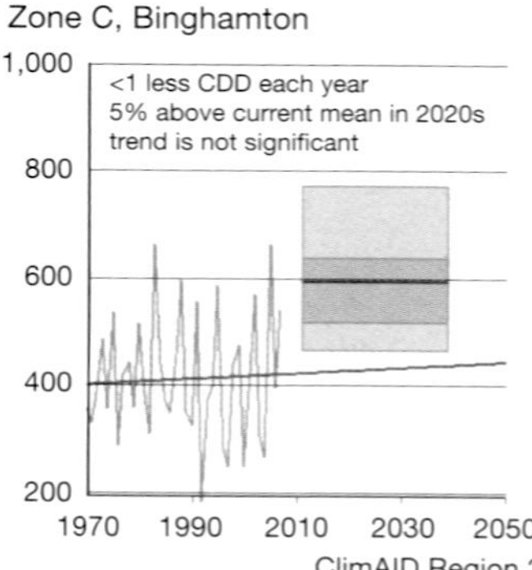

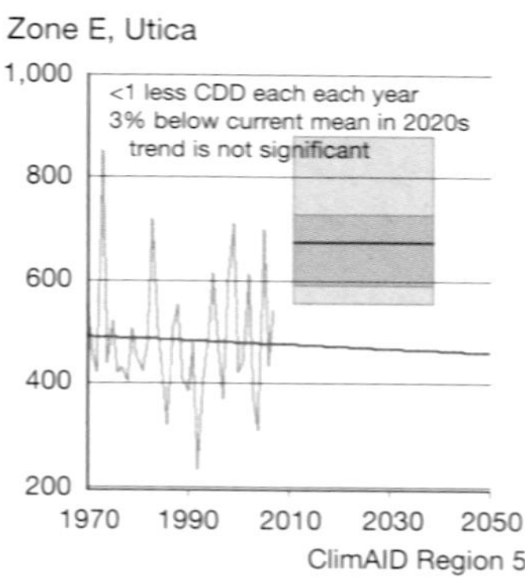

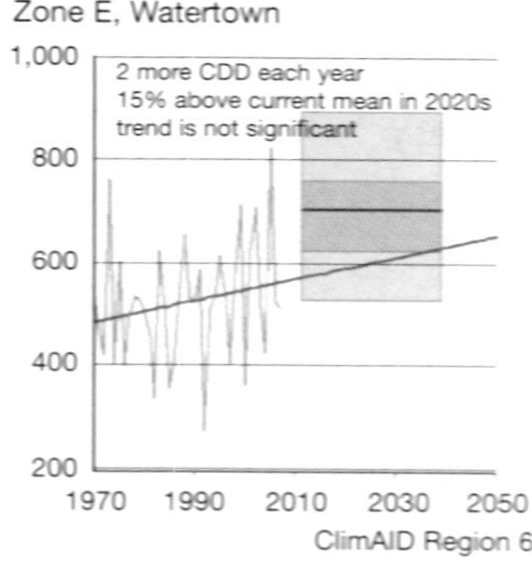

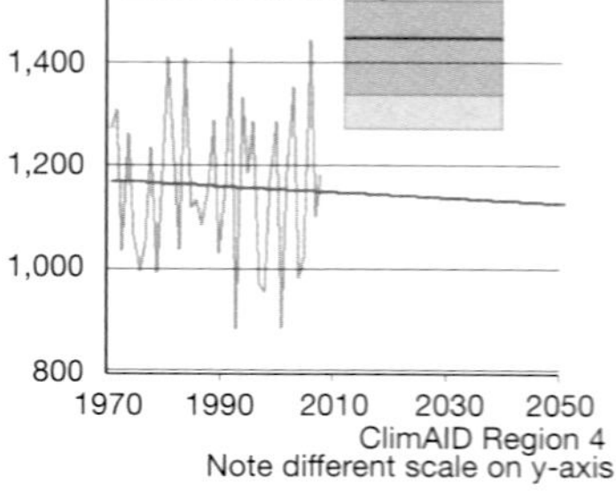

Note: The jagged line on each graph represents historical annual CDD, based on actual weather station data. The straight line indicates the trend in historical CDD, extrapolated into the future. The number of CDD in the 2020s was estimated by extrapolating the linear trend out to 2039 and then averaging the values for 2011–2039. The grey boxes show the range of GCM projections for the 2020s, with the black line in the center of each grey box indicating the mean GCM projection. Note that the trend is considered significant if the p-value is <0.1. Not all NYISO zones and ClimAID regions are represented; only those stations for which GCM projections for CDD were available are included. Data source: Historical climate data from NOAA; GCM projections from ClimAID.

Figure 8.6 Projected changes in cooling degree days in selected regions of the state. The projections are for the 16 GCMs and 3 emissions scenarios used in ClimAID. Shown are the minimum, central range, and maximum values across the GCMs and emissions scenarios.

the 1970–2007 period. Note that patterns of urban development can affect local temperature trends through heat island formation, an effect that was not accounted for in the data analysis.[10]

Electricity Demand

A key question for the power sector is whether climate change will require a significant shift in energy planning or will remain a small demand driver relative to population and economic growth, efficiency projects, and other factors. The need for new generation and/or transmission capacity depends on the geographic location and timing of the increases. All else being equal, warmer nighttime temperatures in the summer and/or a longer cooling season would not necessarily require new generation capacity.[11] However, if summertime peak demand increases at a faster rate than overall demand, the likelihood of brownouts or blackouts increases (Miller et al., 2008).[12] Also, although increases in average daily demand might not require new capacity, they can still affect energy prices, since more expensive generation sources may need to be online more frequently.

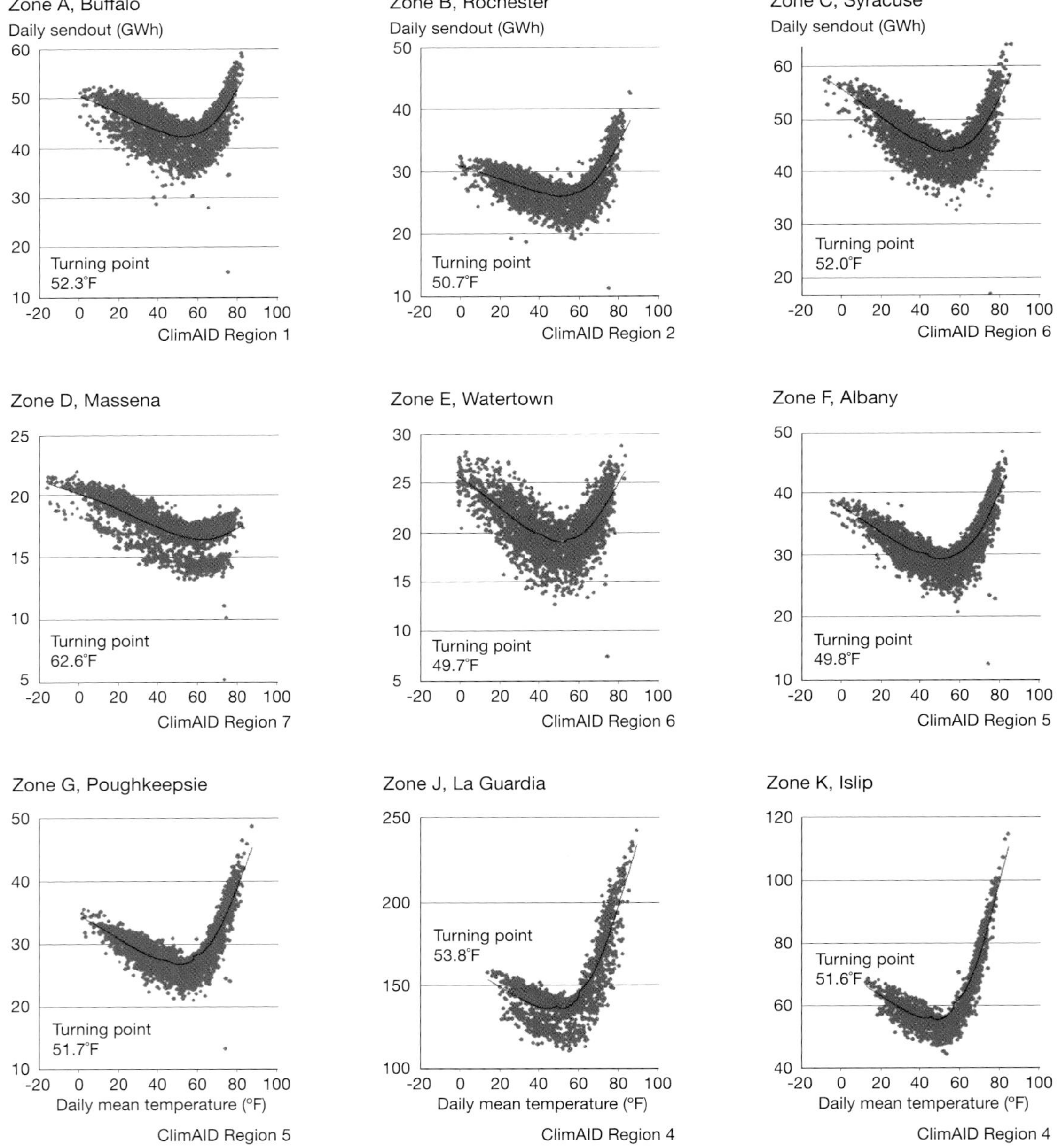

Figure 8.7 Daily average temperature versus daily electricity demand (GWh) for each NYISO Zone

such as a carbon tax (Overbye et al., 2007) or improved public education programs (Vine, 2008), as well as those more narrowly focused on reducing air conditioning demand growth.

Table 8.15[18] presents a wide range of adaptation strategies included in the literature, broken out by whether they focus on energy supply or demand and by which stakeholders are in a position to implement these strategies. Most articles and reports detailing these ideas offer little insight into such matters. Several studies do note barriers to the implementation of adaptation strategies, such as cost, the number of actors involved in specific decisions (Vine, 2008), and market structure (Audin, 1996), but these studies largely ignore governance concerns.

Source	Adaptation Strategy	Agency/Organization with Primary Responsibility for Implementation						
		Other NYS Agencies	NYISO	NYS PSC	NYSERDA	Consumers	Power Plant Owners	Distribution Utilities
Energy Supply								
(Mansanet-Bataller et al., 2008)	Protect power plants from flooding with dykes/berms.			X			X	
	Bury or re-rate cable to reduce failures.			X				X
(Stern, 1998)	Establish new coastal power plant siting rules to minimize flood risk.	X		X				
	Change water management rules to protect hydropower supply availability.	X						
(Sanstad, 2006)	Install solar PV technology to reduce effects of peak demand.				X	X		X
(Aspen Environmental Group and M Cubed, 2005)	Use increased winter stream flow to refill hydropower dam reservoirs.	X						
	Develop non-hydropower generation resources to reduce need for hydropower generation during winter.	X		X	X		X	
(Hill and Goldberg, 2001)	Construct additional transmission line capacity to bring more power to New York City to address peak demand periods.		X					X
	Upgrade existing local transmission and distribution network to handle increased load.		X					X
(Overbye et al., 2007)	Retrofit/reinforce existing energy infrastructure with more robust control systems that can better respond to extreme weather and load patterns.		X				X	X
	Automate restoration procedures to bring energy systems back on line faster after weather-related service interruptions.		X				X	X
Energy Demand								
(Miller et al., 2008)	Design new buildings and retrofits with improved flow-through ventilation to reduce air conditioning use.	X			X	X		
	Use fans for cooling to decrease air conditioning use.					X		X
(Commonwealth of Australia, 2007)	Increase use of insulation in new buildings and retrofit existing buildings with more insulation and efficient cooling systems.					X		X
	Reduce lighting and equipment loads.					X		
(Vine, 2008)	Improve information availability on climate change impacts to decision makers and the public.	X	X	X	X			X
	Use multi-stage evaporative coolers to reduce energy consumption in new buildings.					X		X
	Establish stricter window-glazing requirements in new buildings.	X						
	Plant trees for shading and use reflective roof surfaces on new and existing buildings.				X	X		X
	Establish price-response programs to achieve behavioral response on energy use.		X					X
(Stern, 1998)	Reduce or eliminate energy subsidies so prices reflect true cost.	X	X					
	Establish new air-conditioning efficiency standards.	X						
(Morris and Garrell, 1996)	Improve and rigidly enforce energy-efficient building codes.	X						
(Audin, 1996)	Install power management devices on office equipment.				X	X		X
	Upgrade building interior lighting efficiency.				X	X		X
	Improve domestic hot water generation and use.				X	X		X
	Improve HVAC controls.				X	X		X
	Upgrade elevator motors and controls.				X	X		X
	Design HVAC improvements (e.g. variable flow, thermostats on individual radiators).				X	X		X
	Install more efficient HVAC equipment.				X	X		X
	Improve steam distribution.			X	X		X	X
(Hill and Goldberg, 2001)	Weatherize low-income households.	X			X	X		X

Table 8.15 Selected climate-change adaptation strategies for the energy sector

8.4.2 Larger-scale Adaptations

The ClimAID team's interactions with stakeholders including a range of energy utilities and power generation firms made clear that there is wide divergence in the level of attention paid to climate change issues by New York's energy sector. Climate change mitigation has been on most of these firms' radar screens for some time, because of the requirements of the Regional Greenhouse Gas Initiative, audit filings such as the Carbon Disclosure Project, or their need to interconnect with new renewable-energy installations proposed in their service territories (see www.rggi.org and www.cdproject.net). In contrast, many of the energy companies characterized climate change adaptation as a relatively new area of focus. Climate change does not appear to be identified as the source of any current operating challenges or changes in operating conditions. Few have engaged in comprehensive assessments of their potential climate-change-related operating vulnerabilities. There were some exceptions, principally among companies with operations in New York City, as many of the firms were involved in the climate change adaptation initiative spearheaded by the city's Office of Long Term Planning and Sustainability (NPCC, 2010). Those companies were more likely to have convened internal working groups, hired or appointed a climate change coordinator, developed new policies and procedures, or actually begun to make operational changes or procurement decisions with adaptation considerations in mind. New York State might similarly benefit from multiple regional climate working groups or a comprehensive statewide initiative aimed at ensuring key utilities and large-scale power generation facilities are taking steps to reduce their climate-change-related vulnerabilities. A regional approach might allow for better targeting of localized issues or challenges. However, because many energy companies operate in multiple regions of the state, a statewide approach might be logistically easier for the climate teams at each company by avoiding unnecessary repetition.

Additionally, stakeholders expressed interest in an authoritative climate-risk database that could be used by a regional working group. Regardless of the organizational structure chosen for statewide climate change adaptation planning for the energy sector, such a database is central to this planning work. Stakeholders agreed that such data would be most helpful if it were updated on a regular schedule, and if it were officially sanctioned by State officials as the basis on which operating plans and investment strategies are to be made. This would eliminate the potential for disagreements by officials at different regulatory agencies over the quality of data, methodology, etc.

Finally, New York might also benefit from a formal review process that examines whether the state's currently regulatory and market policies for electric, gas, and steam utilities will continue to be appropriate in the wake of future climate change. Several issues arose in the course of this chapter's research that suggest the need for thoughtful consideration of this question.

First, because of expected long-term reductions in heating degree days around the state, there may be a disproportionate economic impact on natural gas customers in some regions, as the full cost of maintaining the system may ultimately be shouldered by a smaller rate base. Understanding the extent of this problem and how it might be addressed would likely prove important both to local ratepayers and the utilities involved.

Second, State regulators and distribution utilities may increasingly find themselves in situations where, because of uncertainty over the exact severity or timeliness of climate risks in different parts of the state, it is unclear whether capital investments proposed by utilities to enhance the climate resilience of their distribution system will be eligible for rate reimbursement. State regulators must balance the need for a safe and reliable system with the imperative of keeping prices at reasonable levels. Guidelines clarifying this matter might prove helpful for utility capital investment and maintenance planning purposes.

Similarly, the current NYISO wholesale market dispatch system satisfies statewide electricity demand based on a formula that essentially prioritizes the lowest cost sources of power. In the future, the reliability of a provider may prove equally important, particularly during extreme weather events. Power generators may be more willing to make capital investments that enhance their climate resilience if they knew there was a way to account for these expenditures in the dispatch system.[19]

In all of these cases, the issues link directly to the fundamental nature of the market and regulatory system in New York. A comprehensive review may find that no significant structural changes are necessary, but it may also uncover specific issues that can be addressed more satisfactorily under an amended market or regulatory regime.

The final area where the state may benefit from some type of policy review or activity is demand-side management. This chapter highlights the impacts changing temperatures may have on the state electricity system by the 2020s, some of which may be disproportionately felt in certain ClimAID regions. It was beyond the scope of this analysis to assess the efficacy of NYSERDA's current demand-reduction initiatives or funding programs, but it may prove informative to assess whether climate change should be more explicitly factored into the agency's program model. For example, given that climate-change-related temperature increases are likely to have the greatest impact on electricity demand in ClimAID Region 4 (New York City and Long Island), NYSERDA might consider prioritizing demand-side funding in that region because of the sizable system-wide benefits that would be achieved. Conversely, because air conditioning saturation rates are likely to grow at a faster rate in certain sections of northern, central, and western New York State, NYSERDA may decide to dedicate funds aimed at addressing this growth rate.

8.4.3 Co-benefits, Unintended Consequences, and Opportunities

Prioritization of efficiency and demand-side management to reduce the impacts of climate change on the energy sector will reduce greenhouse gas emissions, yielding mitigation co-benefits. Shading buildings and windows, use of highly reflective roof paints and surfaces, and green roofs will also create adaptation and mitigation synergies. These actions will keep building occupants and residents cooler while reducing the use of air conditioners, thereby lowering fossil fuel emissions from power plants. However, adoption of such programs needs to be distributed across the state and its citizens in order to avoid unintended consequences to vulnerable groups. The existing equipment replacement cycle provides opportunities to increase system resiliency, while climate change may provide New York State with opportunities in regard to biomass, hydropower, and other renewable energy sources.

8.5 Equity and Environmental Justice Considerations

Although large-scale blackouts are relatively rare, these events typically occur during the summer months, when electricity demand is highest. The effects of climate change on the frequency of large-scale blackout events is uncertain, yet examination of such events nonetheless highlights important equity concerns. For example, although not solely heat related, an analysis of the 2003 blackout that affected much of the Northeast revealed that even in a case where a very large region is affected, the impacts are felt unevenly across sectors and households (Anderson et al., 2007). Using a modified input-output analysis to model the effects of the 2003 blackout, Anderson et al. found that apart from the utilities themselves, retail trade suffered the greatest aggregate financial loss.

Larger businesses with backup energy sources are more likely to withstand the shock associated with a large-scale outage or a major blackout. Of those businesses that suffered losses in 2003, perhaps 10 to 15 percent had supplementary insurance to cover the damage; the smallest businesses were less likely to hold such insurance, meaning they had to absorb the losses and hope for government loans (Treaster, 2003). Another important consideration is workforce impact. In the Anderson et al. analysis, loss of labor was estimated to account for two-thirds of the total financial losses in the blackout. The people most likely to bear these losses are those living farther from their jobs or more dependent on inoperable forms of transportation, which tends to be people of color and low-income individuals (Bullard, 2007). Those who can afford to take a few days or weeks off and absorb lost wages are most likely to be resilient (Chen, 2007).

In addition to business closures, an important cross-cutting element with the health sector involves increased health risks and the vulnerability of health services (see Chapter 11, "Public Health"). The Northeast blackout significantly increased EMS calls and ambulance responses, as well as high rates of failure on respiratory devices (Prezant et al., 2005). Anderson et al. estimated that the health services sector had the second highest workforce losses in the blackout due to business closures. Decreased availability of health workers at times of increased service needs raises further questions about the capacity of the health sector to care for the infirm, elderly, and disabled in the event of a blackout. Especially critical is care of heat-related health stress, since power outages are most likely to occur during extreme heat events. Heat-related health vulnerabilities are detailed in Chapter 11, "Public Health".

8.6 Conclusions

ClimAID's main findings on vulnerabilities and opportunities, adaptation strategies, and knowledge gaps are described below.

8.6.1 Main Findings on Vulnerabilities and Opportunities

- Impacts of climate change on energy demand are likely to be more significant than impacts on supply. Climate change will adversely affect system operations, increase the difficulty of ensuring adequate supply during peak demand periods, and exacerbate problematic conditions, such as the urban heat island effect.
- More frequent heat waves will cause an increase in the use of air conditioning, increasing peak demand loads and stressing power supplies.
- Increased air and water temperatures may affect the efficiency of power plants, with impacts varying across the state.
- Energy infrastructure in coastal areas of southern New York State is vulnerable to flooding as a result of sea level rise and severe storms.
- Hydropower, located primarily in northern and western New York State, is vulnerable to drought and changes in precipitation patterns.
- The availability and reliability of solar power systems are vulnerable to changes in cloud cover, although this may be offset by advances in technology; wind power systems are similarly vulnerable to changes in wind speed and direction. However, changes in cloud cover and wind speed and direction are uncertain.
- Transformers and distribution lines for both electric and gas supply are vulnerable to extreme weather events, temperature, and flooding.
- Decreases in heating demand will primarily affect natural gas markets, while increases in cooling demand will affect electricity markets; such changes will vary regionally.
- The indirect financial impacts of climate change may be greater than the direct impacts of climate change. These indirect impacts include those to investors and insurance companies as infrastructure becomes more vulnerable and those borne by consumers due to changing energy prices and the need to use more energy.

8.6.2 Adaptation Options

- Equipment replacement cycles present opportunities to improve system resiliency.
- Transformers and wiring may require derating to ensure they continue to function as expected at higher temperatures.
- Berms and levees can protect infrastructure from flooding. It may also help to raise the elevation of sensitive energy technology in flood-prone locations.
- Saltwater-resistant transformers may help protect against electric system damage from sea level rise and saltwater intrusion.
- Tree-trimming programs are of critical importance to protect power lines from wind, ice, and snow damage.
- Reservoir release policies may need to be adjusted to ensure sufficient late-summer hydropower capacity.
- Demand-side management and energy efficiency initiatives may provide "no regrets" benefits to the state energy system in the near term, regardless of how climate change ultimately manifests itself across the state. Monitoring of impacts on the energy system is needed in the long term.
- Solar gain in buildings can be reduced by shading buildings and windows, using highly reflective roof paints and surfaces, and installing green roofs.
- Regional or statewide working groups may help increase the level of attention paid to climate change issues by power generators and utilities around the state.
- Power generators and utilities may benefit from the creation of an authoritative climate risk database to ensure that State regulators and other agencies rely on the same information in their rulemaking.
- New York may benefit from a formal review of how well climate change considerations are factored into the State's regulatory and market programs for electric, gas, and steam utilities.

8.6.3 Knowledge Gaps

Throughout the chapter, areas where additional research is needed have been noted. These include:

Energy Supply

- Potential vulnerabilities associated with cooling waters at thermoelectric power stations around the state. These include vulnerabilities associated with water temperature increases during heat waves; blockages to cooling water intakes during other extreme weather events; and impacts on biodiversity in waterways used for cooling water purposes that might necessitate changes in the cooling system design. Such a review would help policymakers considering whether intake or discharge rule changes are in order.
- The existence of temperature tipping points, beyond which the likelihood of distribution system service interruptions significantly increases. Given anticipated changes in the number and duration of heat waves around the state, this information could prove helpful in identifying deficiencies in current equipment rating or system design practices.
- Potential impacts of climate change on wind patterns and speeds in selected areas of the state currently used or proposed for wind farm development. Given anticipated growth in wind system deployment around the state, this information would be helpful for energy planning purposes.
- Potential impacts of climate change on biomass-based heat production around the state (either at a large central station or co-firing facilities) and on a more localized basis in regions of the state that depend heavily on biomass combustion for heat production in residential and commercial facilities.
- Potential impacts of climate change on ice storm frequency in different parts of the state over the coming decades. This information would be useful in assessing whether design rule changes are required for electricity transmission and distribution towers and poles.
- Potential impacts of climate change on hydropower availability in different parts of the state. This information could also be helpful in informing policymakers about the potential need for rule changes regarding water releases from hydropower facilities at different times of the year or day. This information might also prove important in assessing the need to pursue rule changes governing releases on the Niagara River, given the priority currently placed on the allocation for Niagara Falls during tourist season.

Energy Demand

- Potential impacts on the demand for natural gas and other heating fuels around the state, given anticipated decreases in heating degree-days over the coming decades. Such information would prove helpful in determining the economic impact on individual customers and local gas distribution utilities in different regions of the state.
- Ways to better incorporate climate change into demand forecasts for each load zone, and to enhance models' incorporation of the impacts of extreme events on electricity demand. Such information would be helpful to State energy planners, because this will clarify how much additional generation capacity must be developed over the coming decades or whether it can be addressed by other means, such as demand-reduction initiatives.

Case Study A. Impact of Climate Change on New York State Hydropower

There are nearly 370 large and small hydropower developments in New York, and their collective output gives the state more hydropower than any other state east of the Rockies (EIA, 2009).

Two projects are responsible for the lion's share of the state's hydropower production; both are operated by the New York Power Authority (NYPA), the largest state-owned power operation in the United States. The Niagara Power Project, located on the Niagara River between Lake Erie and Lake Ontario, is the hydropower leader in the state, generating more than 13,000 gigawatt-hours (GWh) of electricity in 2007. The second project, the St. Lawrence-FDR Project[20], generated another 6,600 GWh that same year (NYPA, 2007).

Both projects are fed by water from the massive Great Lakes Basin, a 300,000-square-mile watershed that extends 2,000 miles from end to end (Croley, 2003). Because four of the five Great Lakes are bisected by the U.S.-Canada border, the governance of the lakes (and thus operations at these two large hydropower systems) is bound up in a web of international treaties and bi-lateral and multi-lateral agreements designed to satisfy the competing interests of two countries, eight states, and one Canadian province.

The Boundary Waters Treaty of 1909 established the International Joint Commission (IJC), an important adjudicator in Great Lakes hydropower issues. Under the Boundary Waters Treaty, the IJC acts on applications for hydropower dams and other projects in waters along the U.S.-Canadian border, seeking to balance the impacts of the projects on different stakeholders. The IJC has jurisdiction over Great Lake water management issues, with day-to-day responsibilities for water flow levels and other important operating decisions delegated to different IJC-created Boards.

Some of the most important jurisdictional decisions arise from a 1950 treaty between Canada and the United States that establishes baseline guarantees on how much water must flow over Niagara Falls during daytime hours in the tourist season. The IJC's International Niagara Board of Control oversees implementation of the 1950 treaty. Key decisions about the Robert Moses Power Dam are handled by the International St. Lawrence River Board of Control.

Both Boards have "Orders of Approval" that guide water-release planning at their respective facilities. The goals are relatively straightforward: to balance river or lake height at different locations to generate hydropower, satisfy municipal water system needs, accommodate commercial navigation, and protect private property and wildlife from flooding and erosion (International Lake Ontario/St. Lawrence River Study Board, 2006). In practice, however, this means regular fine-tuning of water release levels at different hydro system assets. On a weekly basis, orders are sent out to NYPA and other hydro dam owners/operators to open or close water intake and release gates to meet water height and release targets. Factors influencing these decisions include local climate circumstances, including wind, rain, snow, ice, drought, etc.

At Niagara, guidance comes from a 1993 Board of Control Directive focused on maintaining a mean surface elevation of 171.16 meters (562.75 feet) in the Chippawa/Grass Island Pool upstream of Niagara Falls, balancing this target against treaty obligations for water release over the falls a few miles downstream that vary between day and night and tourist/non-tourist seasons (FERC, 2006). On the St. Lawrence River, Plan 1958-D calls for reduced flow rates during ice formation in early winter to allow more stable ice covers to form on Lake St. Lawrence, reducing the potential for ice jams that would lead to upstream flooding problems on the St. Lawrence River and Lake Ontario[21] (FERC, 2003).

Effects of Potential Changes in Great Lakes on Hydropower

Understanding how climate change may affect the Great Lakes is a topic of increasing interest to stakeholders around the region. Because of the interconnected nature of the lakes—water from Lake Superior eventually finds its way to the Atlantic Ocean via the other lakes and the Niagara and St. Lawrence Rivers—climate studies must necessarily examine the entire Great Lakes Basin.

The earliest studies dating back to the 1980s and 1990s all note the likelihood that temperatures in the Great Lakes Basin will gradually warm and that precipitation and water levels will change. (For example, see Croley, 1983; Cohen, 1986; Quinn, 1988; USEPA, 1989; Mortsch and Quinn, 1996; Chao, 1999). For example, the EPA analysis applied three different general circulation models (GCMs) to assess future impacts on the basin. Under all three climate models, the EPA projected that precipitation levels would stay relatively constant, but that snowmelt and runoff would decline and lake evaporation levels would increase, resulting in a net decrease in overall lake levels.

Lofgren et al. (2002) found more variable results. Under one model (CGCM1)[22], lake levels were expected to drop by an average of 0.72 meters by 2030 and 1.38 meters by 2090 compared to a 1989 baseline. Another model (HADCM2)[23], in contrast, forecast sizeable precipitation increases, which ultimately lead to lake level increases of 0.01 meters by 2030 and 0.35 meters by 2090. Croley's (2003) simulation using four different climate models found high levels of absolute variability, although the trends clearly fall in the same downward direction under the majority of the scenarios (see **Table 8.16**).

More recent work carried out for the International Joint Commission has begun to look at both annual impacts (in terms of lake level changes and outflow rates) and more discrete seasonal impacts. For instance, in the case of Lakes Erie, Ontario, and Superior, Fay and Fan (2003) note that mean annual lake outflow may decline by 5 to 24 percent on Lake Ontario and 5 to 26 percent on Lake Erie, depending on which climate model is applied. Mean lake level changes also decline, by 0.10 to 0.85 meters on Lake Erie and by 0.04 to 0.54 meters on Lake Ontario (see **Table 8.17**).

Given the depth of the Great Lakes, such changes appear quite modest in terms of absolute elevation, but there are implications for the New York State energy sector.

A 2006 IJC report examining alternatives to the 1958-D Order of Approval estimated that the economic impact of climate change on hydropower production at NYPA's St. Lawrence/FDR project could vary from -$28.5 million to $5.86 million, depending on which GCM is employed (personal communication. Victoria Simon, New York Power Authority, February 19, 2010). The "not-as-warm-and-wet" scenario was the only one of the four models to produce a positive economic impact. Data are not presented in that study to explain what this translates into in terms of increases or decreases in overall power production. However, NYPA has developed two alternative estimates, calculating that a 1-meter decrease in the elevation of Lake Ontario would result in a loss of roughly 280,000 megawatt-hours (MWh) of power production at the St. Lawrence/FDR project. NYPA also estimates that a 5–24 percent reduction in water flow from Lake Ontario would result in production losses of approximately 340,000 to 1,650,000 MWh/year (Victoria Simon, personal communication, June 9, 2010).

	Base Case	Warm & Dry	Not-as Warm & Dry	Warm &Wet	Not-as Warm & Wet
		HadCM3A1FI	CGCM2A21	HadCM3B22	CGCM2B23
Superior	841	-180	54	-161	-80
Michigan	818	-273	-232	-232	-59
Huron	572	-173	-135	-168	-21
Erie	843	-350	-330	-266	45
Ontario	1926	-272	-223	-254	21

Note: CGCM2 is a global climate model from the Canadian Center for Climate Modeling and Analysis. Results from ensemble simulations related to the SRES A2 greenhouse gas scenario (A21—warm and dry) and the SRES B2 greenhouse gas scenario (B23—not as warm but dry) are shown. HADCM3 is a global climate model from the United Kingdom Meteorological Office's Hadley Centre. Results from the SRES A1FI greenhouse gas scenario (A1FI—warm and wet) and from the SRES B2 greenhouse gas scenario (B22—not as warm but wet) are shown. Source: Crowley, 2003, p. 62

Table 8.16 Projected changes in Great Lakes net basin supply (mm) for four climate change scenarios, through the 2050s

There is evidence that during times of drought, power output at the Niagara Project has been curtailed because of the pre-eminence of the obligation to ensure adequate flow over Niagara Falls. According to the New York Power Authority, in the 1960s, when the Great Lakes basin endured one of the most severe droughts of the century, generation levels at the Niagara Power Project dropped "dramatically while [Niagara] Falls retained its full flow" (Victoria Simon, personal communication, December 10, 2009). **Figure 8.10** compares the annual power output levels at the Niagara Power Project[24] with the Niagara River's mean monthly discharge level near Buffalo. Although the annual power output data make exact month-to-month comparisons difficult, there are discernable changes in power production levels that correlate closely (r=0.89) to periods when the river's discharge rates increase or decrease. To the extent climate change increases the incidence of drought in the Great Lakes Basin, hydropower production levels across the state will likely decline.

NYPA's hydropower is sold through contracts to business customers participating in NYPA economic development programs, municipal and rural electric

	Base Case	Warm & Dry	Not-as Warm & Dry	Warm &Wet	Not-as Warm & Wet
		HadCM3A1FI	CGCM2A21	HadCM3B22	CGCM2B23
Lake outflow (annual mean, in cubic meters/second)					
Lake Erie	6576	4867 (-26%)	5410 (-18%)	5153 (-22%)	6263 (-5%)
Lake Ontario	7770	5890 (-24%)	6460 (-17%)	6170 (-21%)	7420 (-5%)
Change of lake level from base case (m)					
Lake Erie					
Winter		-0.79	-0.55	-0.69	-0.15
Spring		-0.79	-0.53	-0.62	-0.10
Summer		-0.83	-0.54	-0.64	-0.13
Autumn		-0.85	-0.57	-0.73	-0.21
Annual		-0.81	-0.55	-0.67	-0.15
Lake Ontario					
Winter		-0.45	-0.27	-0.32	-0.07
Spring		-0.54	-0.30	-0.29	-0.04
Summer		-0.49	-0.23	-0.30	-0.08
Autumn		-0.40	-0.19	-0.36	-0.12
Annual		-0.47	-0.25	-0.32	-0.08

See note on models for Table 8.16. Source: Fay and Fan 2003 in Mortsch, Croley and Fay, 2006

Table 8.17 Lake outflows and change of lake levels from base case (m) for various climate scenarios

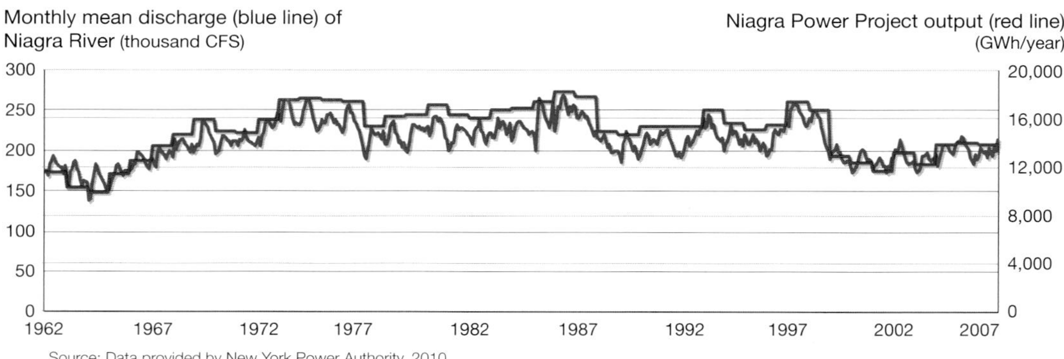

Figure 8.10 Comparison of power output levels of Niagara Power Project and monthly mean discharge rate of Niagara River (1962–2006)

cooperatives, investor-owned utilities, and other contractual arrangements. Any substantial reduction in water levels in the Great Lakes could potentially have an impact on these customers and others throughout the state. For "firm"[25] hydropower service customers, low water levels mean NYPA satisfies production shortfalls with higher-priced electricity purchased on the NYISO wholesale markets. For "interruptible"[25] service customers, low water levels mean that 100 percent of their interruptible power needs will be met through wholesale market purchases. The economic impact of a significant drought may also extend to non-NYPA customers, as greater demand for non-hydropower sources will tend to drive up prices across New York and in adjacent wholesale markets.

Cast Study B. Climate-change-induced Heat Wave in New York City

Coping with summer heat waves is a key challenge for the energy sector in New York State. Under climate change, heat waves affecting New York are likely to become more frequent and to increase in duration (see Chapter 1, "Climate Risks"). Within New York City, where urban heat island effects are already prominent during warm periods of the summer, worsening heat waves under climate change pose a challenge for the city's energy sector (Rosenzweig et al., 2006). With these worsening heat waves, it is likely that blackouts may occur somewhat more frequently (although to an extent reduced by the regular, ongoing investment of the electricity industry). This cross-cutting example considers the social equity and economic implications of energy outages associated with summer heat waves in New York City, although the effects will likely be similar in urban regions around the state. This ClimAID case study is specifically designed to illustrate equity and economic issues that have arisen in the past during heat-wave-related outage events, in order to highlight those that may potentially arise under climate change. (The public health effects of heat waves in New York State are addressed in Chapter 11, "Public Health.")

Sustained high temperatures contribute to increased energy usage during heat wave events, primarily for cooling of indoor space and industrial equipment. When high temperatures persist overnight during these extended heat waves, the likelihood of outages increases. The design of the local grid system will affect whether the outages will be geographically isolated or more widespread. However, heat waves can also be associated with multiple outages across the city under conditions of prolonged heat stress.

Equity and Environmental Justice Issues

In considering potential equity and environmental justice issues associated with heat-wave and outage events in New York City, we consider three types of impacts: 1) effects of sustained high temperatures, 2) effects of outages, and 3) effects of adaptation measures.

Heat waves place a physical and financial burden on nearly all segments of the population in New York City. Concerning the spatial distribution of heat wave impacts, heat waves under climate change are likely to intensify existing urban heat island patterns, meaning that areas that are already warmer due to heat island effects will become relatively hotter during a heat wave (Rosenzweig et al., 2005). While heat island effects occur in many parts of the city, a NYSERDA study of heat effects in New York City noted that heat island effects are prominent in many lower-income neighborhoods, such as Fordham in the Bronx and Crown Heights in Brooklyn (Rosenzweig et al., 2006). Such areas tend to have fewer street trees than other neighborhoods, leading to hotter conditions at the sidewalk level. Researchers in other cities have also noted similar correlations between locations of poor neighborhoods and more severe urban heat island effects due to higher settlement density, lack of open space, and sparse vegetation (e.g., Harlan et al., 2006).

Differential prevalence of indoor air conditioning may also exacerbate the effects of extreme heat. As noted earlier, 84 percent of housing units in New York City had some form of air conditioning in 2003. However, these rates are not uniform across the city. Results of the New York City Community Health Survey indicate that higher poverty areas, particularly in northern Manhattan, the South Bronx, and areas of Brooklyn, have lower rates of home air conditioning than many other parts of the city (see **Figure 8.11**).

Heat waves mean higher energy costs for all consumers, but these costs are not borne equally by all residents.

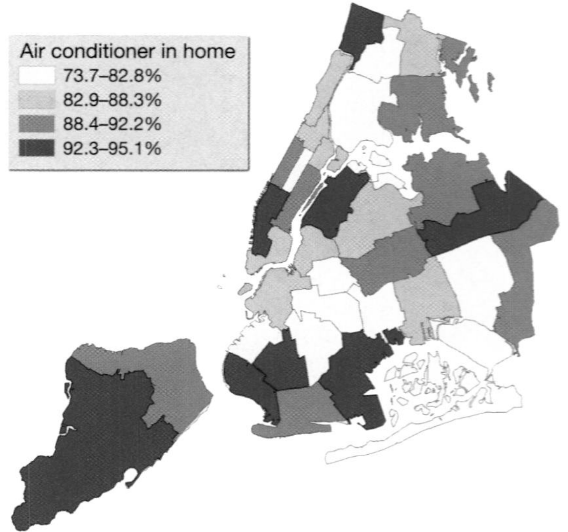

Note: Percentages are age adjusted.
Source: NYC Community Health Survey 2007, Bureau of Epidemiology Services, NYC DOHMH

Figure 8.11 Home air conditioner use in New York City, 2007

These costs represent a larger share of household income for lower-income customers. As a result, lower-income households with air conditioners may be more reluctant to use them in times of extreme heat. During the Chicago heat wave of 1995, reluctance by low-income households to use air conditioning due to concerns about energy costs was a significant factor contributing to mortality (Klineberg, 2003). Furthermore, while heat wave events lead to increased energy usage throughout New York City, locations in the city with greater heat island effects (i.e., the hottest locations) have been found to experience greater increases in energy demand (Gaffin et al., 2008). These spatial differences may exacerbate energy cost burdens on those lower-income areas that are subject to heat island effects.

Higher energy usage due to sustained high temperatures may also contribute to increased air pollution during heat wave events. Under heat wave events, less efficient and more highly polluting sources of power may be used to meet peak demand. High levels of ozone due to the combination of high temperatures and air pollution are particularly harmful for the elderly and ill, as discussed in Chapter 11, "Public Health".

Historically, heat waves in New York City have been associated with sustained power outages in some neighborhoods. For example, the Washington Heights/Inwood blackout of July 1999 was a summertime, heat-related outage that affected more than 200,000 residents living north of 155th Street in Manhattan (Office of the Attorney General, New York State, 2000). Within the affected region, which is dominated by high-poverty areas, among those hardest hit by the outage were elderly residents of high-rise apartments, where elevator service failed and fans and air conditioners for cooling were inoperable (Office of the Attorney General, New York State, 2000).

Concerning adaptation of the energy sector to heat waves, some current options are expansion of smart grid initiatives, demand management, load reduction efforts, and on-site generation. All of these measures have the potential to raise social equity issues. For example, different households will have different capacity to invest in the equipment needed for on-site generation. Such differences in capacity to adapt represent an important type of equity issue that needs to be taken into account as adaptation strategies are put into place.

Economic Analysis of Heat Wave Impacts

Electric power generation, transmission, and distribution systems play an important role in supporting the economy of the United States. Hence, power outages and other disruptions are likely to negatively affect economic activity, mainly by restricting infrastructure and other services on which the economy relies. Power failures, which may take place when electricity demand exceeds supply such as during a heat wave, have both direct and indirect impacts on the economy, national security, and the environment.

Economic losses from electric service interruptions are not trivial, as illustrated by different studies. A 2001 report that extrapolated from surveyed businesses the losses due to poor power quality, outages, and other disruptions (referred to collectively as "reliability events") estimated costs to U.S. consumers to range from $119 billion to $188 billion per year (EPRI, 2004). The Pacific Gas & Electric Company (PG&E) used direct costs of reliability events (based on a combination of direct cost measures and willingness-to-pay indicators) to assess that such power disruptions cost its customers approximately $79 billion per year (USEPA, 2010). A 2004 Berkeley National Laboratory comprehensive study of end-users focusing on power outages alone[26] estimated annual losses to the national economy of approximately $80 billion (LaCommare and Eto, 2004). The figures provided by these studies coincide with estimates by the U.S. Department of Energy, ranging from $25 billion to $180 billion per year (USDOE, 2009).

Given the number of major power outages, including those in the Northeast in 1965, 1977, and 2003, different methodologies have been developed to estimate their associated economic costs. While much of the earlier research has focused on calculating physical damage and cost of replacement of major infrastructure systems, fewer studies have been conducted to assess the overall economic impacts.

Estimates of the economic impact of the 25-hour blackout that affected most of New York City on July 13 and 14 of 1977 are sketchy, with damage costs assessed at $60 million. More information is available on the costs of the cascading blackout that started on August 14, 2003, and affected 55 million people. Initial reports projected that economic losses would range from $4 billion to $6 billion. Others estimated that this major power outage translated into a $10 billion loss for the national economy, and an ICF Consulting report put the price tag between $7 billion and $10 billion (Knowledge@Wharton, 2003; USEPA, 2010; The Public Record, 2008; ICF, 2003; Anderson and Geckil, 2003; ELCON, 2004). Moreover, this blackout contributed to at least eleven fatalities, including six in New York City (Knowledge@Wharton, 2003).

Certain sectors of the economy were particularly affected during the 2003 blackout, with the airline industry losing an estimated $10–$20 million, mostly because of grounded flights. In New York City, where over 14 million people were affected, it has been estimated that approximately 22,000 restaurants collectively lost $75–$100 million in foregone business and wasted food. In addition, the City of New York reported losses of $40 million in lost tax revenue and $10 million in overtime payments to city workers (Knowledge@Wharton, 2003). Adding to the losses was the cost of using "defensive measures" such as backup generators as well as servicing them, given that half of New York City's 58 hospitals experienced some kind of failure during the blackout (USEPA, 2009).[27]

While cascading blackouts have significant impacts, the majority of power outages are localized blackouts and brownouts, and the cumulative impact to the national economy may be quite large.[28]

Localized service outages in New York City include the July 3–9, 1999, blackout that affected 170,000 Con Edison customers, including 70,000 in Washington Heights, as well as the nine-day blackout that started on July 16, 2006, in Long Island City (in Queens) and affected 174,000 residents (New York State Public Service Commission, 2000; Chan, 2007). Most reports of economic losses focus on customer claims, which for the 1999 blackout amounted to $100 each to compensate residents for spoilage of food and medicine and $2,000 each to business customers. These fees were raised to $350 and $7,000, respectively, in 2006. Total claims paid by Con Edison in 2006 amounted to $17 million; an additional $100 million was estimated to be spent by the utility on recovery costs to repair and replace damaged equipment (Cuomo, 2007).

However, economic compensation paid by utilities to affected customers represents only a portion of total economic losses to society, and does not even take into account the value of unsold (or unserved) electricity to communities and businesses. Several approaches have been developed to attempt to estimate the overall economic cost of blackouts. In general, most methods focus on calculating the value added that customers place on power reliability, which can be quantified by the consumers' willingness to pay, taking into account their income, or in the case of businesses, their revenues net of economic losses due to power failures. Nevertheless, the value-added approaches do not account for all the societal benefits that result from reliability improvements, as they fail to estimate the associated improvements in public safety and health or environmental benefits. These societal benefits must be incorporated separately.

The value added of electricity reliability is often presented as customer damage functions that may take into account a number of variables. Such values may be estimated by 1) calculating the direct costs of power outages based on customers' experience, 2) conducting surveys to estimate the consumer's willingness-to-pay (WTP) or willingness-to-accept compensation (WTA) to avoid such outages, and 3) estimating by indirect analytic methods.

The first approach attempts to estimate the value that electricity services represent to each customer, based on losses experienced to particular facilities operations. What is referred to as the customer's value of service (VOS) can be measured in terms of the direct costs of an outage, which may include damaged plant equipment and/or replacement costs, spoiled products, additional maintenance costs, production losses/lost revenue, costs of idle labor, and potential liabilities.

The WTP/WTA approach provides another measure of the "cost of reliability" of electrical services considered in terms of how consumers value such services, or more precisely the value assigned to the lack of survey interruptions. Various studies provide survey-based estimates of the WTP for different groups of electric power customers. While economic losses to commercial and industrial facilities from power interruptions may be monetized in a straightforward manner (e.g., on the basis of lost profits), assessing the direct costs to residential customers may be more complicated, in part because surveyed customers do not always describe economic losses in monetary terms but rather as disruptions or hassles. Rather than assigning values to such inconveniences (which go

Sector	Annual kWh
Medium and large C&I	7,140,501
Small C&I	19,214
Residential	13,351

Table 8.18 Average kWh usage per year by customer class

Interruption Cost	Interruption Duration				
	Momentary	30 min.	1 hour	4 hours	8 hours
Medium and Large C&I					
Morning	$8,133	$11,035	$14,488	$43,954	$70,190
Afternoon	$11,756	$15,709	$20,360	$59,188	$93,890
Evening	$9,276	$12,844	$17,162	$55,278	$89,145
Small C&I					
Morning	$346	$492	$673	$2,389	$4,348
Afternoon	$439	$610	$818	$2,696	$4,768
Evening	$199	$299	$431	$1,881	$3,734
Residential					
Morning	$3.7	$4.4	$5.2	$9.9	$13.6
Afternoon	$2.7	$3.3	$3.9	$7.8	$10.7
Evening	$2.4	$3.0	$3.7	$8.4	$11.9

Note: C&I = Commercial and Industrial. Source: Lawrence Berkeley National Laboratory (2009), Estimated Value of Service Reliability for Electric Utility Customers in the United States; prepared by Michael J. Sullivan, Ph.D., Matthew Mercurio, Ph.D., Josh Schellenberg, M.A, Freeman, Sullivan & Co.; June, 2009. Accessed online on 1/12/10 from: http://certs.lbl.gov/pdf/lbnl-2132e.pdf

Table 8.19 Estimated average electric customer interruption costs per event in US 2008$ by customer type, duration, and time of day

Interruption Cost	Interruption Duration				
	Momentary	30 min.	1 hour	4 hours	8 hours
Medium and Large C&I					
Agriculture	$4,382	$6,044	$8,049	$25,628	$41,250
Mining	$9,874	$12,883	$16,368	$44,708	$70,281
Construction	$27,048	$36,097	$46,733	$135,383	$214,644
Manufacturing	$22,106	$29,098	$37,238	$104,019	$164,033
Telecommunications & Utilities	$11,243	$15,249	$20,015	$60,663	$96,857
Trade & Retail	$7,625	$10,113	$13,025	$37,112	$58,694
Fin., Ins., & Real Estate	$17,451	$23,573	$30,834	$92,375	$147,219
Services	$8,283	$11,254	$14,793	$45,057	$71,997
Public Administration	$9,360	$12,670	$16,601	$50,022	$79,793
Small C&I					
Agriculture	$293	$434	$615	$2,521	$4,868
Mining	$935	$1,285	$1,707	$5,424	$9,465
Construction	$1,052	$1,436	$1,895	$5,881	$10,177
Manufacturing	$609	$836	$1,110	$3,515	$6,127
Telecommunications & Utilities	$583	$810	$1,085	$3,560	$6,286
Trade & Retail	$420	$575	$760	$2,383	$4,138
Fin., Ins., & Real Estate	$597	$831	$1,115	$3,685	$6,525
Services	$333	$465	$625	$2,080	$3,691
Public Administration	$230	$332	$461	$1,724	$3,205

Note: C&I = Commercial and Industrial. Source: Lawrence Berkeley National Laboratory (2009), Estimated Value of Service Reliability for Electric Utility Customers in the United States; op. cit.

Table 8.20 Estimated average electric customer interruption costs per event in US 2008$ by duration and business type (summer weekday afternoon)

beyond the cost of food and medicine spoilage), economists often rely on WTP or WTA surveys in order to assess loses to residential customers (Lawton et al., 2003). Such surveys describe different scenarios and ask residential customers how much they would be willing to pay for power reliability or the amount of money they would require to accept service interruptions.

A 2009 report (Sullivan et al., 2009) that conducted a metadata analysis using 28 different customer-value-of-service reliability surveys that had been carried out by 10 major U.S. electric utilities between 1989 and 2005 provides average estimates of the value of service reliability for electricity customers in the United States (except in the Northeast). The information collected is classified by customer types surveyed, including both medium and large commercial and industrial (C&I) non-residential consumers with sales greater than 50,000 kilowatt-hours (kWh) per year, with an average of 373 employees; small commercial and industrial non-residential customers with sales ≤50,000 kWh per year; and residential customers. The metadata analysis provides an average kWh usage per customer type, as summarized in **Table 8.18**.

Interruption Cost	Interruption Duration				
	Momentary	30 min.	1 hour	4 hours	8 hours
Medium and Large C&I					
Summer Weekday	$11,756	$15,709	$20,360	$59,188	$93,890
Summer Weekend	$8,363	$11,318	$14,828	$44,656	$71,228
Winter Weekday	$9,306	$12,963	$17,411	$57,097	$92,361
Winter Weekend	$6,347	$8,977	$12,220	$42,025	$68,543
Small C&I					
Summer Weekday	$439	$610	$818	$2,696	$4,768
Summer Weekend	$265	$378	$519	$1,866	$3,414
Winter Weekday	$592	$846	$1,164	$4,223	$7,753
Winter Weekend	$343	$504	$711	$2,846	$5,443
Residential					
Summer Weekday	$2.7	$3.3	$3.9	$7.8	$10.7
Summer Weekend	$3.2	$3.9	$4.6	$9.1	$12.6
Winter Weekday	$1.7	$2.1	$2.6	$6.0	$8.5
Winter Weekend	$2.0	$2.5	$3.1	$7.1	$10.0

Note: C&I = Commercial and Industrial. Source: Lawrence Berkeley National Laboratory (2009), Estimated Value of Service Reliability for Electric Utility Customers in the United States; op. cit.

Table 8.21 Estimated average electric customer interruption costs per event in US 2008$ by customer type, duration, season, and day type

Summary results for the cost of power interruptions are given in **Tables 8.19–8.21**, including estimates of the costs of power interruptions per event by customer class, business type, size of the facility and time of the event, and geographical location. Information is also available on the expected cost of unserved energy, which is a metric widely used for expressing interruption costs, as shown on **Table 8.22**, which provides another example of the value of service (VOS) direct cost estimation approach.

The information summarized in the tables shows that large commercial and industrial customers experience losses averaging $20,000 and $8,166 for a 1-hour power interruption during a winter afternoon and summer afternoon, respectively. As the power outage increases in duration, so do costs, sharply during the winter and significantly in the summer.

Heat Wave and Power Outage Adaptation Measures

According to a 2009 report by the American Society of Civil Engineers, electricity demand since 1990 has grown approximately 25 percent but construction of transmission facilities has declined by roughly 30 percent (American Society of Civil Engineers, 2009). In 2003, other reports estimated that investment in high-voltage transmission lines had decreased by 45

Facility Outage Impacts			Annual Outages		Annual Cost	
Power Quality Disruptions	Outage Duration per Occurrence	Facility Disruption per Occurrence	Occurrences per Year	Total Annual Facility Disruption	Outage Cost per Hour*	Total Annual Costs
Momentary Interruptions	5.3 Seconds	0.5 Hours	2.5	1.3 Hours	$45,000.00	$56,250.00
Long-Duration Interruptions	60 Minutes	5.0 Hours	0.5	2.5 Hours	$45,000.00	$112,500.00
Total			3	3.8 Hours		$168,750.00
Unserved kWh per hour (based on 1,500 kW average demand)			1,500 kWh			
Customer's Estimated Value of Service, $/unserved kWh			$30 /unserved KWh			
Normalized Annual Outage Costs, $/kW-year			$113 $/kW-year			

Note: Outage costs per hour estimated based on facility data and include production losses, increased labor, product spoilage, and other costs.
Source: USEPA – Combined Heat and Power Partnership; Calculating Reliability Benefits, http://www.epa.gov/CHP/basic/benefits.html

Table 8.22 Value of service direct cost estimation

percent over the previous 25-year period (ELCON, 2004). Moreover, the Energy Department expects that electricity use and production will increase by 20 percent over the next decade but the nation's high-voltage electric network will only increase by 6 percent in the same time period. After the major blackout of 2003, there have been calls for investments ranging from $50 billion to $100 billion to reduce severe transmission bottlenecks and increase capacity (Knowledge@Wharton, 2003).

While long-term planning and investments are necessary, significant improvements are needed over the next few years to ensure that operators can have access to the necessary information to properly manage power flows and transmission systems. Investments to upgrade the grid can provide network operators with clearer metrics of the potential risks in order to avoid major power outages (Apt et al., 2004). The costs of installing sensors nationwide are much smaller than those for a single blackout event. A recent report made the case for installing sensors every 10 miles over the existing 157,000 miles of transmission lines in the United States and found that, at a cost of $25,000 per sensor, total costs would amount to $100 million if all sensors were replaced every five years. Such investment would increase the average residential electricity bill by 0.004 cents per kilowatt hour. The total would be roughly one-tenth the estimated annual cost of blackouts (Apt et al., 2004).

Other adaptation measures to prevent power outages include reducing demand and distributed generation. Load-shedding strategies may be used during heat waves in advance of peak-demand episodes and include broad calls for consumers to reduce demand as well as voluntary and mandatory load reduction programs, for which customers receive a number of incentives. Customers participating in voluntary options such as the "Distribution Load Relief" program must reduce at least 50kW or 100kW (for individuals or aggregators respectively) to receive compensation of at least $0.50 per kWh after each event. Other mandatory programs are similarly structured with additional incentives such as reservation (capacity) fees and bonus payments (Con Edison). Other, long-term strategies to increase overall network capacity include demand-side management, which decreases the need for investments in additional power generation.[29]

References

Allen, J. September 29, 2009. Rochester Gas & Electric, personal communication.

Amato, A.D., M. Ruth, P. Kirshen and J. Horwitz. 2005. "Regional Energy Demand Responses to Climate Change: Methodology and Application to the Commonwealth of Massachusetts." *Climatic Change* 71:175-201.

American Society of Civil Engineers. 2009. Report Card for America's Infrastructure - Energy Page. Available online at: http//www.infrastructurereportcard.org/fact-sheet/energy

American Wind Energy Association. 2009. "U.S. Wind Energy Projects - New York." June 27, 2009.

Anderson, C.W., J.R. Santos and Y.Y. Haimes. 2007. "A Risk-Based Input-Output Methodology for Measuring the Effects of the August 2003 Northeast Blackout." *Economic Systems Research* 19(2):183-204.

Anderson, P.L., and I.K. Geckil. 2003. Northeast Blackout Likely to Reduce US Earning by $6.4 Billion. AEG Working Paper 2003-2. August 19, 2003

Apt, J. , L.B. Lave, S. Lukdar, M. Granger Morgan, and M. Ilic. 2004. Electrical Blackouts: A Systemic Problem." Issues in Science and Technology

Ascher, K. 2005. *The Works: Anatomy of a City*. New York: Penguin.

Aspen Environmental Group and M Cubed. 2005. "Potential Changes in Hydropower Production from Global Climate Change in California and the Western United States." Prepared in support of the 2005 Integrated Energy Policy Report Proceeding (Docket #04-IEPR-01G). California Energy Commission.

Associated Press. 1986. "6 western states affected by flooding." *New York Times*, February 25.

Audin, L. 1996. "Growing" Energy Efficient Physical Plants in the Greenhouse Era." In *The Baked Apple: Metropolitan New York in the Greenhouse*, edited by D. Hill. New York: *Annals of the New York Academy of Sciences*.

Baxter, L.W. and K. Calandri. 1992. "Global warming and electricity demand: a study of California." *Energy Policy* 233-244.

Bevelhymer, C. 2003. "Steam." *Gotham Gazette*. Citizens Union Foundation.

Breslow, P. B. and D.J. Sailor. 2002. "Vulnerability of wind power resources to climate change in the Continental United States." *Renewable Energy* 27:585-598.

Bull, S.R., D.E. Bilelo, J. Ekmann, M.J. Sale and D.K. Schmalzer. 2007. "Effects of climate change on energy production and distribution in the United States in Effects of Climate Change on Energy Production and Use in the United States." A Report by the U.S. Climate Change Science Program and the Subcommttee on Global Change Research: Washington D.C.

Bullard, R.D., ed. 2007. "Growing Smarter: Achieving Livable Communities, Environmental Justice, and Regional Equity." Cambridge, MA: MIT Press.

Changnon, S.A. and J.M. Changnon. 2002. "Major ice storms in the United States 1949-2000." *Environmental Hazards* 4:105-111.

Chao, P. 1999. "Great Lakes Water Resources: Climate Change Impact Analysis with Transient GCM Scenarios." *Journal of the American Water Resources Association* 35(6):1499-1507.

Chan, S. 2007. "Con Ed Seeks to raise Electricity Rates." New York Times, May, 4, 2007.

Chen, D. 2007. "Linking Transportation Equity and Environmental Justice with Smart Growth." in *Growing Smarter: Achieving Livable Communities, Environmental Justice, and Regional Equity*, edited by R.D. Bullard. Cambridge, MA: MIT Press.

City of New York. 2007. *PlaNYC: A Greener, Greater New York*. Mayor's Office of Long Term Planning and Sustainability.

Cohen, S.J. 1986. "Impacts of CO_2-induced climatic change on water resources in the Great Lakes basin." *Climatic Change* 8:135-153.

Commonwealth Of Australia. 2007. *Climate Change Adaptation Actions for Local Government*. Australian Greenhouse Office, Department of the Environment and Water Resources.

Company of New York, Inc. Case 07-E-0523. Prepared Testimony of Kin Eng, Utility Analyst 3, Office of Electric, Gas, and Water. September.

Con Edison. Demand Response Program Comparison Guide – Business Customers http://www.coned.com/energyefficiency/DR_comparison_chart.pdf.

Croley, T.E. 2003. "Great Lakes Climate Change Hydrologic Impact Assessment; IJC Lake Ontario-St. Lawrence River Regulation Study." NOAA Great Lakes Environmental Research Laboratory. September.

Cruz, A.M. and E. Krausmann. "Damage to offshore oil and gas facilities following hurricanes Katrina and Rita: an overview." *Journal of Loss Prevention in the Process Industries* 21:620-626.

Cuomo, A. 2007. RE: Case 06-E-0894 – Proceeding on Motion of the Commission to Investigate the Electric Power Outages in Consolidated Edison Company of New York, Inc.'s Long Island City Electric Network; State of New York, Office of the General Attorney. March 2, 2007.

Dao, J. 1992. "The Storm's Havoc: Failure of two outdated generators cited in disruption of subway's safety signals." *New York Times*, December 13.

de Nooij, M., R. Lieshoutand and C. Koopmans. 2009. "Optimal blackouts: Empirical results on reducing the social cost of electricity outages through efficient regional rationing." *Energy Economics* 31(3):342-347.

Electric Power Research Institute (EPRI). 1998. "Ice Storm '98: Characteristics and Effects." In *EPRI Report*, edited by M. Ostendorp.

Electric Power Research Institute (EPRI). 2004. Power Delivery System of the Future: A Preliminary Estimate of Costs and Benefits. Final Report 1011001.

Electricity Consumers Resource Council (ELCON). 2004. "The Economic Impacts of the August 2003 Blackout." Available online at: http://www.elcon.org/Documents/EconomicImpactsOfAugust2003Blackout.pdf

Fan, Y. and D. Fay. 2003. "Lake Erie – Climate Change Scenarios" and "Lake Ontario – Climate Change Scenarios" cited in Mortsch, L., T. Croley, and D. Fay. 2006. "Impact of Climate Change on Hydro-electric Generation in the Great Lakes." Presentation to C-CIARN-Water Hydro-Power and Climate Change Workshop, Winnipeg, Manitoba, March 2-3, 2006.

Federal Energy Regulatory Commission. 2003. Draft Environmental Impact Statement – Niagara Project (FERC/EEIS-0155D) St. Lawrence-FDR Project – New York (FERC Project No. 2000-036). June.

Federal Energy Regulatory Commission. 2008. "Existing and Proposed North American LNG Terminals." Washington DC: FERC Office of Energy Projects.

Federal Energy Regulatory Commission. 2006. Final Environmental Impact Statement – Niagara Project (FERC/FEIS-0198F) Niagara Project – New York (FERC Project No. 2216-066). December.

Fidje, A. and T. Martinsen. 2006. "Effects of climate change on the utilization of solar cells in the Nordic region." European Conference on Impacts of Climate Change on Renewable Energy Resources, Reykjavik, Iceland.

Franco, G. and A.H. Sanstad. 2006. "Climate change and electricity demand in California." California Climate Change Center.

Franco, G. and A.H. Sanstad. 2008. "Climate Change and Electricity Demand in California." *Climatic Change* 87:S139-S151.

Gaffin, S.R., C. Rosenzweig, R. Khanbilvardi, L. Parshall, S. Mahani, H. Glickman, R. Goldberg, R. Blake, R.B. Slosberg and D. Hillel. 2008. "Variations in New York City's urban heat island strength over time and space." *Theoretical and Applied Climatology* 94:1-11.

Gaffin, S.R., R. Khanbilvardi, and C. Rosenzweig. 2009. Development of a green roof environmental monitoring and meteorological network in New York City. *Sensors*, 9, 2647-2660, doi:10.3390/s90402647.

Gavin, R. and J. Carleo-Evangelist. 2008. "Icy Disaster." *Albany Times-Union*, December 13.

Hallegate, S. 2008. "Adaptation to Climate Change: Do Not Count on Climate Scientists to Do Your Work." Reg-Markets Center.

Harlan, S.L., A.J. Brazel, L. Prashad, W.L. Stefanov and L. Larsen. 2006. "Neighborhood microclimates and vulnerability to heat stress." *Social Science and Medicine* 63:2847–2863.

Hewer, F. 2006. Climate Change and Energy Management. *UK Met Office*.

Hill, D. and R. Goldberg. 2001. "Chapter 8: Energy Demand." In Climate Change and Global City: The Potential Consequences of Climate Variability and Change — Metro East Coast, edited by C. Rosenzweig and W.D. Solecki. Report for the US Global Change Research Program, National Assessment of the Potential Consequences of Climate Variability and Change for the United States. New York:Columbia Earth Institute.

ICF. 1995. "Potential Effects of Climate Change on Electric Utilities." TR105005, Research Project 2141-11. Prepared for Central Research Institute of Electric Power Industry (CRIEPI) and Electric Power Research Institute (EPRI).

ICF Consulting. 2003. The Economic Cost of the Blackout: An issue paper on the Northeastern Blackout, August 14, 2003.

International Lake Ontario-St. Lawrence River Study Board. 2006. "Options for Managing Lake Ontario and St. Lawrence River Water Levels and Flows." Final Report to by the International Lake Ontario – St. Lawrence River Study Board to the International Joint Commission. March.

IPCC. 2007. "Climate Change 2007: The Physical Science Basis, Summary for Policymakers." Contribution to Working Group 1 to the Fourth Assessment Report of the Intergovernmental Panel on Climate Change, Cambridge, UK, Cambridge University Press.

Jowit, J. and J. Espinoza. 2006. "Heatwave shuts down nuclear power plants." *The Observer*. London.

Kirschen, D.S., 2007. "Do Investments Prevent Blackouts?" In Power Engineering Society General Meeting, Institute of Electrical and Electronics Engineers, 1-5. DOI: 10.1109/PES.2007.385653.

Klineberg, E. 2003. "Heat wave: A social autopsy of disaster in Chicago." Chicago: University of Chicago Press.

Knowledge@Wharton. 2003. "Lights Out: Lessons from the Blackout." August 27, 2003. Available online at: http://knowledge.wharton.upenn.edu/article.cfm?articleid=838

La Commarre, K.H., and J.H. Eto. 2004. Understanding the Cost of Power Interruptions to U.S. Electricity Consumers. Lawrence Berkeley National Laboratory.

Lawton, L., M. Sullivan, K. Van Liere, A. Katz, J.H. Eto. 2003. Framework and Review of Customer Outage Costs : Integration and Analysis of Electric Utility Outage Cost Surveys. November 2003, Lawrence Berkeley National Laboratory.

Linder, K.P., M.J. Gibbs and M.R. Inglis. 1987. "Potential Impacts of Climate Change on Electric Utilities." New York State Energy Research and Development Authority, Edison Electric Institute, Electric Power Research Institute, US Environmental Protection Agency.

Lofgren, B.M., F.H. Quinn, A.H. Clites, R.A. Assel, A.J. Everhardt and C.L. Luukkonen. 2002. "Evaluation of potential impacts on Great Lakes water resources based on climate scenarios of two GCMs." *Journal of Great Lakes Resources* 28(4)537-554.

Maniaci, A. August 2009. New York Independent System Operator (NYISO), personal communication.

Mansanet-Bataller, M., M. Herve-Mignucci and A. Leseur. 2008. "Energy Infrastructures in France: Climate Change Vulnerabilities and Adaptation Possibilities." Mission Climate Working Paper. Paris:Caisse des Depots.

Marean, J. September 29, 2009. NYSEG, personal communication.

Miller, N., K. Hayhoe, J. Jin and M. Auffhammer. 2008. "Climate, Extreme Heat, and Electricity Demand in California." *Journal of Applied Meteorology and Climatology* 47:1834-1844.

Morris, S.C. and M.H. Garrell. 1996. "Report of the Scenario Planning Group for Accelerated Climate Change: Apple Crisp." In *The Baked Apple? Metropolitan New York in the Greenhouse*, edited by D. Hill. *Annals of the New York Academy of Sciences*.

Morris, S.C., G. Goldstein, G., A. Singhi and D. Hill. 1996. "Energy Demand and Supply in Metropolitan New York with Global Climate Change." In *The Baked Apple? Metropolitan New York in the Greenhouse*, edited by D. Hill. *Annals of the New York Academy of Sciences*.

Mortsch, L.D. and F.H. Quinn. 1996. "Climate Change Scenarios for the Great Lakes Basin Ecosystem Studies." Limnology and Oceanography 41(5):903-911.

National Energy Technology Laboratory (NETL). 2009. "Impact of Drought on U.S. Steam Electric Power Plant Cooling Water Intakes and Related Water Resource Management Issues." National Energy Technology Laboratory, US Department of Energy.

Neal, D. October 30, 2009. NRG Energy, personal communication.

New York Independent System Operator (NYISO). 2002. "New Electricity Demand Forecast Warns of Continued Risk of Energy Crisis if Power Plant Development Lags."

New York Independent System Operator (NYISO). 2004. "Adjusting for the Current Overstatement of Resource Availability in Resource Adequacy Studies."

New York Independent System Operator (NYISO). 2009. "2009 Load and Capacity Data: Gold Book."

New York Independent System Operator (NYISO). 2009a. "Load Data, Market and Operations." http://www.nyiso.com/public/markets_operations/market_data/ load_data/index.jsp.

New York Power Authority (NYPA). 2007. "NYPA Annual Report: Planning for the Future."

New York State Department of Environmental Conservation (NYS DEC). 2010a. Part 704: Criteria Governing Thermal Discharges. Environmental Conservation Law §§15-0313, 17-301.

New York State Department of Environmental Conservation (NYS DEC). 2010b. "2008 Annual Gas and Oil Production Data." Viewed on 2/15/10 at http://www.dec.ny.gov/energy/36159.html.

New York State Department Of Public Service. 2007. In the Matter of Consolidated Edison New York City Department of Environmental Protection (NYC DEC). 2008. "Assessment and Action Plan: A Report Based on the Ongoing Work of the DEP Climate Change Task Force." New York City Department of Environmental Protection Climate Change Program.

New York State Energy and Research Development Authority (NYSERDA). 2010. "About NYSERDA." Accessed February 23. http://www.nyserda.org/About/default.asp.

New York State Energy and Research Development Authority (NYSERDA). 2010b. "Patterns and Trends. New York State Energy Profiles: 1994-2008." January 2010.

New York State Office of the Attorney General. 2000. "Con Edison's July 1999 Electric Service Outages: A Report to the People of the State of New York." Accessed December 2009. *http://www.oag.state.ny.us/media_center/2000/mar/mar09a_00.html.*

New York State Public Service Commission. 2008. "Utility Watchdog Agencies Stress Need to Trim Trees — Trees and Power Lines Don't Mix." Press release. New York State Public Service Commission and New York State Consumer Protection Board.

New York State Public Service Commission. 2009. http://www.askpsc.com

New York State Public Service Commission 2000. CASE 99-E-0930 - Proceeding on Motion of the Commission to Investigate the July 6, 1999, Power Outage of Con Edison's Washington Heights Network. March 15, 2000.

New York State Reliability Council. 2004. "New York Control Area Installed Capacity Requirements for the Period May 2005 through April 2006 — Executive Committee Resolution and Technical Study Report."

New York Times. 1994. "Texans striving to contain pipeline spills" 22 October.

North American Electric Reliability Corporation (NERC). 2008. "2008 Summer Reliability Assessment."

Overbye, T., J. Cardell, I. Dobson, W. Jewell, M. Kezunovic, P.K. Sen and D. Tylavsky. 2007. "The Electric Power Industry and Climate Change: Power Systems Research Possibilities." Power Systems Engineering Research Center.

Pan, Z., M. Segal, R.W. Arritt and E.S. Takle. 2004. "On the potential change in solar radiation over the U.S. due to increases of atmospheric greenhouse gases." *Renewable Energy* 29:1923-1928.

Perez, R. September 9, 2009. SUNY Albany, personal communication.

Potomac Economics, Ltd. 2009. "2008 State of the Market Report: New York ISO." Prepared for the New York State Independent System Operator.

Prezant, D.J., J. Clair, S. Belyaev, et al. 2005. "Effects of the August 2003 blackout on the New York City healthcare delivery system: A lesson for disaster preparedness." Critical Care Medicine 33(1):S96-S101.

Pryor, S.C. and R.J. Barthelmie. 2010. "Climate change impacts on wind energy: a review." *Renewable and Sustainable Energy Reviews* 14:430-437.

The Public Record. 2008. "5 Years After Blackout, Power Grid Still in 'Dire Straits'." August 7, 2008. Available online at: http://pubrecord.org/nation/394/5-years-after-blackout-power-grid-still-in-dire-straits/

Quinn, F.H. 1988. "Likely effects of climate changes on water levels in the Great Lakes." In Proceedings of the First North American Conference on Preparing for Climate Change, 481-487. Climate Change Institute, Washington D.C.

Ramage, J. 2004. "Chapter 5: Hydroelectricity" in *Renewable Energy – Power for a Sustainable Future (2nd Edition)*, edited by G. Boyle. Oxford Press.

Risk Management Solutions. 2008. "RMS Special Report: The 1998 Ice Storm — 10 year retrospective."

Rosenthal, D.H. and H.K. Gruenspecht. 1995. "Effects of global warming on energy use for space heating and cooling in the United States." *Energy Journal* 16.

Rosenzweig, C. and W.D. Solecki, eds. 2001. "Climate Change and a Global City: The Potential Consequences of Climate Variability and Change—Metro East Coast." Report for the US Global Change Research Program, National Assessment of the Potential Consequences of Climate Variability and Change for the United States. New York: Columbia Earth Institute.

Rozenzweig, C. and W.D. Solecki. 2001. "Climate change and a global city: Learning from New York." *Environment* 43:8-18.

Rosenzweig, C., W.D. Solecki, L. Parshall, M. Chopping, G. Pope and R. Goldberg. 2005. "Characterizing the urban heat island in current and future climates in New Jersey." *Environmental Hazards* 6:51–62.

Rosenzweig, C., W.D. Solecki and R.B. Slosberg. 2006. "Mitigating New York City's Heat Island with Urban Forestry, Living Roofs, and Light Surfaces." New York City Regional Heat Island Initiative Final Report 06-06. New York State Energy Research and Development Authority.

Sailor, D.J. and A.A. Pavlova. 2003. "Air conditioning market saturation and long-term response of residential cooling energy demand to climate change." *Energy* 28:941-951.

Sailor, D.J., M. Smith and M. Hart. 2008. "Climate change implications for wind power resources in the Northwest United States." *Renewable Energy* 33:2393-2406.

Schoeberl, K. October 28, 2009. Central Hudson Gas & Electric, personal communication.

Scott, M. J. and Y.J. Huang. 2007. "Effects of climate change on energy use in the United States in Effects of Climate Change on Energy Production and Use in the United States." A Report by the US Climate Change Science Program and the Subcommittee on Global Change Research.

Segal, M., Z. Pan, R.W. Arritt and E.S. Takle. 2001. "On the potential change in wind power over the US due to increases in atmospheric greenhouse gases." *Renewable Energy* 24:235-243.

Smith, J.B. and D. Tirpak, eds. 1989. "The Potential Effects of Global Climate Change on the United States: Report to Congress." U.S. Environmental Protection Agency (EPA), Office of Policy, Planning and Evaluation, Office of Research and Development. EPA-230-05-89-050. Washington D.C.

Simon, V. October 15, 2009. New York Power Authority, personal communication.

Simon, V. December 10, 2009. New York Power Authority, personal communication.

Simon, V. June 9, 2010. New York Power Authority, personal communication.

State Energy Planning Board. 2009a. "Renewable Energy Assessment — New York State Energy Plan 2009." New York State Energy Planning Board.

State Energy Planning Board. 2009b. "Natural Gas Assessment — New York State Energy Plan 2009." New York State Energy Planning Board.

State Energy Planning Board. 2009c. "2009 State Energy Plan, Volume 1. New York State Energy Planning Board. December 2009.

Stern, F. 1998. "Chapter 11: Energy." In *Handbook on Methods for Climate Change Impact Assessment and Adaptation Strategies*, edited by J.F. Feenstra, I. Burton, J. Smith and R.S.J. Tol. United Nations Environment Program/Institute for Environmental Studies.

Sullivan, M.J., M. Mercurio, and J. Schellenberg. 2009. Estimated Value of Service Reliability for Electric Utility Customers in the United States. June 2009. Lawrence Berkeley National Laboratory.

Treaster, J.B. 2003. "The Blackout: Business Losses; Insurers Say Most Policies Do Not Cover Power Failure." *New York Times*, August 16. Accessed May 2009. http://www.nytimes.com/2003/08/16/business/blackout-business-losses-insurers-say-most-policies-not-cover-power-failure.html?scp=3&sq=blackout%20regrigeration&st=cse

Union of Concerned Scientists. 2007. "Got Water?"

United States Census Bureau. 2004. "American Housing Survey for the New York-Nassau-Suffolk-Orange Metropolitan Area: 2003." U.S. Department of Commerce, Economics and Statistics Administration, U.S. Census Bureau and U.S. Department of Housing and Urban Development, Office of Policy Development and Research.

United States Census Bureau. 2010. "CenStats: 2008 County Business Patterns (NAICS) — New York." Accessed on August 14. http://censtats.census.gov/cgi-bin/cbpnaic/cbpsect.pl.

United States Department of Energy (USDOE). 2009. National Electric Transmission Congestion Study

United States Energy Information Administration (EIA). 2007. "Annual Energy Review 2007." Table 2.3 Manufacturing Energy Consumption for Heat, Power, and Electricity Generation by End Use, 2002. U.S. Department of Energy. DOE/EIA-0384(2007).

United States Energy Information Administration (EIA). 2009a. "Annual Energy Outlook 2009 Early Release." Appendix A: Table A4 Residential Sector Key Indicators and Consumption. U.S. Department of Energy. DOE/EIA-0383(2009).

United States Energy Information Administration (EIA). 2009b. "Annual Energy Outlook 2009 Early Release." Appendix A: Table A5 Commercial Sector Key Indicators and Consumption. U.S. Department of Energy. DOE/EIA-0383(2009).

United States Energy Information Administration (EIA). "State Energy Profile – New York." Accessed 27 August. http://tonto.eia.doe.gov/state/state_energy_profiles.cfm?sid=NY.

United States Environmental Protection Agency (US EPA). 1989. "The Potential Effects of Global Climate Change on the United States. Report to Congress," edited by J.B. Smith and D.A. Tirpak. EPA-230-05-89-050, EPA Office of Policy, Planning and Evaluation, Washington D.C.

United States Environmental Protection Agency (US EPA). 2009. "eGrid." April ed.

United States Environmental Protection Agency (USEPA). 2010. Calculating Reliability Benefits. Accessed online at 1/5/2010 at: http://www.epg.gov/chp/basic/benefits.html

United States Geological Survey (USGS). 1999. "New York City Area Digital Elevation Model, 1/3 Arc Second." EROS Data Center.

Vine, E. 2008. "Adaptation of California's Electricity Sector to Climate Change." In *Preparing California for a Changing Climate*. Public Policy Institute of California.

Wilbanks, T.J. 2007. "Introduction in Effects of Climate Change on Energy Production and Use in the United States." A Report by the U.S. Climate Change Science Program and the subcommittee on Global Change Research.

Williams, P.J. and M. Wallis. 1995. "Permafrost and Climate Change: Geotechnical Implications." *Philosophical Transactions: Physical Sciences and Engineering* 352(1699):347-358.

Appendix A. Stakeholder Interactions

The ClimAID Energy team interacted with relevant stakeholders around the state through meetings and one-on-one interviews. Drafts were shared with selected stakeholders to obtain their feedback on different topics and to ensure the accuracy of specific information contained in the report.

The first stakeholder meeting was held at NYSERDA's office in Albany in March 2009. Stakeholders invited to the meeting included a range of power plant operators, officials from New York-based energy and environmental organizations, distribution utilities, and New York State officials, including the New York Independent System Operator. Of the 38 invitations sent out, 18 individuals from 15 organizations were represented. A list of participating stakeholders is included at the end of this appendix.

The first meeting introduced the ClimAID project and solicited feedback on the first draft of the energy sector analysis that was completed in early 2009. A draft stakeholder survey was also shared to obtain feedback on its length and content. Based on feedback provided by the stakeholders, the survey was shortened considerably and tailored to reflect the unique perspective of each sector participant (e.g., utility, power plant operator, etc.).[30]

Energy demand forecasting was also discussed at the meeting, with the stakeholders providing important information regarding their concerns about the ClimAID team's efforts to forecast climate-change-related demand impacts beyond a 20–30 year timeframe, arguing that longer-term forecasts were subject to other factors (e.g., technology changes, population changes, climate change mitigation policies) that made it difficult to forecast demand with a high level of certainty. As a result, a decision was made by the ClimAID Energy team to concentrate on demand impacts, taking into account only those climate change impacts projected for the 2020s, and to convene a separate demand modeling working group.

Following the initial meeting, individual meetings and phone calls were conducted with six different stakeholders representing distribution utility and/or power generation operations in different parts of the state. These conversations were in-depth, lasting between 45 minutes and two hours. In some cases, a single company representative was interviewed, while in other cases there were six company participants, each with a different area of specialization.

In most cases, follow-up questions were submitted to these companies to clarify information raised in the original meeting or to solicit additional information. These interviews were helpful both in validating many of the conclusions drawn by the literature review, and in identifying nuanced differences or more recent information specifically relevant to New York State.

The demand modeling working group met in June 2009 to solicit input from stakeholders on priorities with respect to understanding how climate change may affect energy demand in New York State. After this meeting, a follow-up call was held to discuss methodological issues and further refine the objectives. During this call and subsequent communications, the group determined that additional statistical analysis of historical climate data should be prioritized over producing demand forecasts for the state. There are some efforts to incorporate climate change into demand forecasts, so the group saw an opportunity for the ClimAID team to provide data and analysis to support these efforts. The results of the demand modeling research are included in the Energy Demand section of the chapter (section 8.3.2).

Stakeholder Meeting Participants, March 2009

- AES
- Alliance for Clean Energy New York
- Environmental Energy Alliance of New York
- Cogentrix
- Con Edison
- Dynegy
- FirstLight Power/Suez GDF
- Long Island Power Authority
- National Grid
- NRG Energy
- New York Independent System Operator
- New York Power Authority
- TransCanada/Ravenswood
- US PowerGen

Demand Modeling Meeting Participants, June 2009

- Con Edison
- New York State Department of Public Service
- National Grid
- New York City Office of Long-Term Planning and Sustainability
- New York Independent System Operator
- New York Power Authority
- New York State Department of Environmental Conservation
- New York State Energy Research and Development Authority

Appendix B. Relationship between NYISO Load Zones and ClimAID Regions

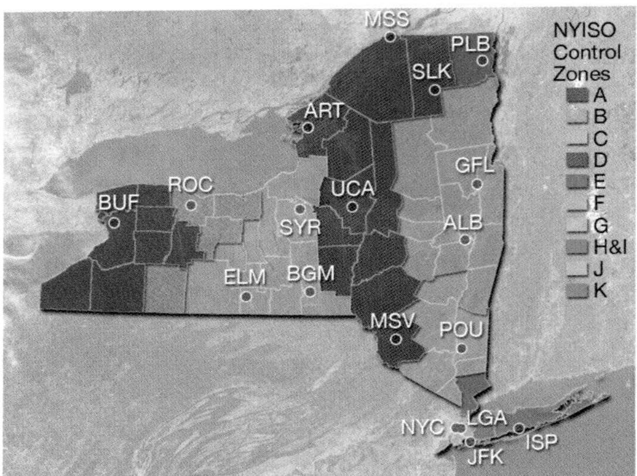

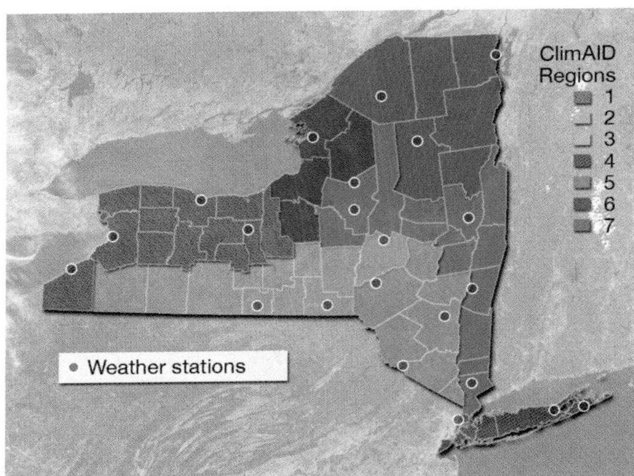

ClimAID Regions: 1. Western New York Great Lakes Plain; 2. Catskill Mountains and West Hudson River Valley; 3. Southern Tier; 4. New York City and Long Island; 5.East Hudson and Mohawk River Valleys; 6. Tug Hill Plateau; 7. Adirondack Mountains. Source: NYISO (2009a), basemap NASA

Figure 8.12 Locations of weather stations used in ClimAID climate analysis related to NYISO load zones

---

1 Interactions with out-of-state infrastructure may be discussed, but are not a direct focus of the ClimAID report.

2 The New York Independent System Operator (NYISO) Gold Book characterizes conventional hydropower plants as a renewable resources, although it acknowledges this does not match the definition used in other New York State policies, including the Renewable Portfolio Standard.

3 There has not been any follow-up analysis examining the accuracy of these projections.

4 In general, the DOE report suggested there is a heightened vulnerability at power plants with shallow intake depths, because of the risk that water levels may be inadequate, exposing the intake pipe or resulting in limits in how much water the power plant may siphon off. Drought conditions may also result in higher water temperature levels at depths close to the intake, creating problems at facilities requiring specific intake water temperatures.

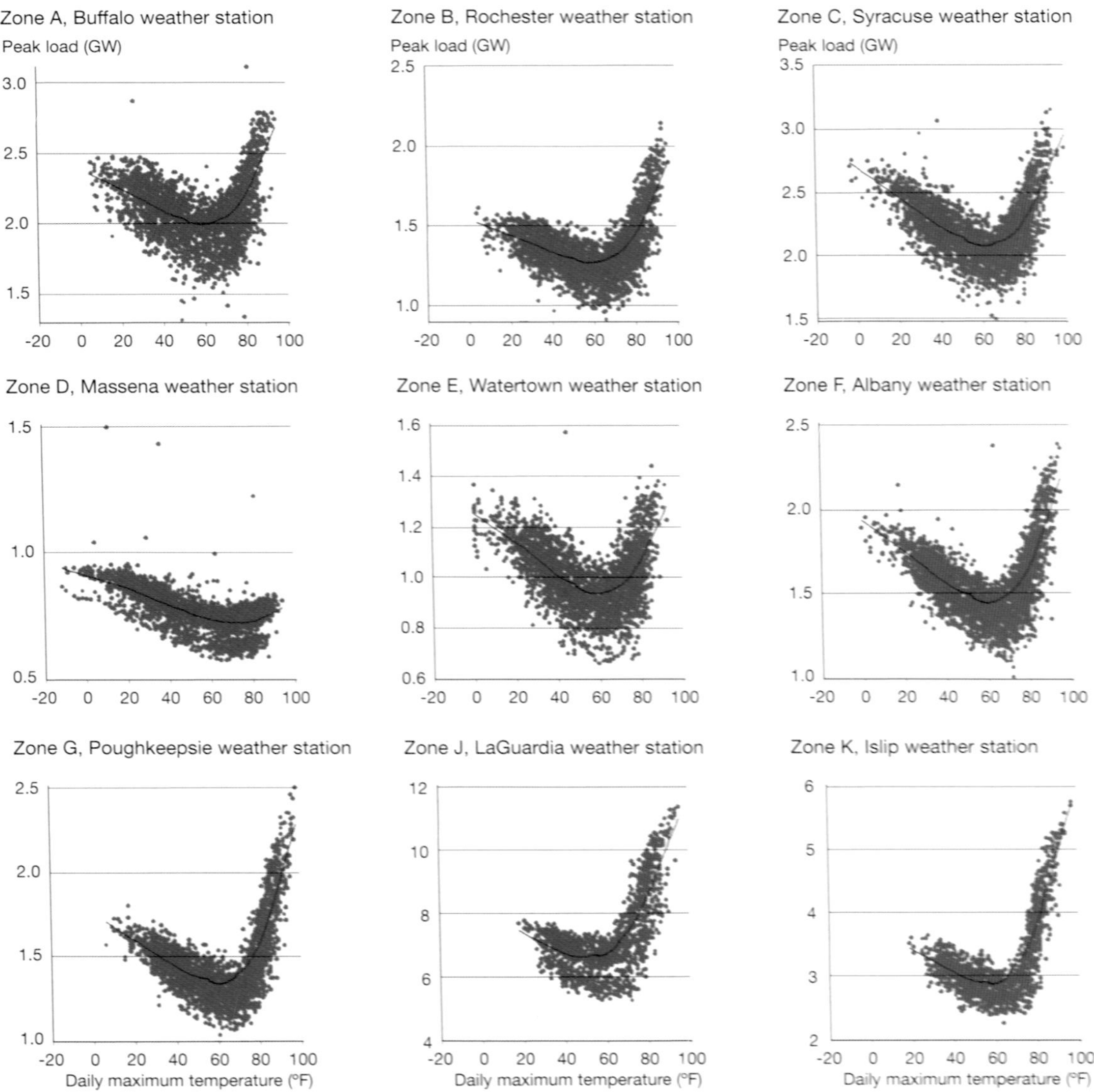

Time period is 2002-2008, except for Zones J and K, for which the time period is February 2005-2008.

Figure 8.13 Maximum daily temperature (°F) versus daily peak electricity demand (mw) for each NYISO Control Zone

[5] The Reliability Council is the official entity authorized by the Federal Energy Regulatory Commission to analyze supply and demand levels in New York State on a periodic basis, identifying conditions that may affect future system reliability and issuing rules that the New York Independent System Operator and other entities must abide by when making supply and power distribution decisions.

[6] Assessment of the overall net energy demand impact is clouded by the wide range of scenarios and assumptions used in different studies, as well as different approaches to energy accounting. For example, some studies assess impacts on delivered (on-site) energy consumption whereas other studies assess impacts on primary energy demand, after accounting for generation, transmission, and distribution losses.

[7] HDD=65- T_{mean} if T_{mean}<65°F. CDD= T_{mean}-65 if T_{mean}>65°F, where T_{mean} is the mean daily temperature. Total annual HDD/CDD is the sum of daily HDD/CDD.

[8] Cooling degree days (CDD) are calculated as the mean daily temperature minus 65 deg F. For example, if the mean temperature is 75 deg F, then there are 10 CDD. Total annual CDD are the sum of daily CDD. Similarly, heating degree days (HDD) are calculated as 65 deg F minus mean daily temperature, and total annual HDD is the sum of daily HDD.

[9] Chapter 1, "Climate Risks," provides some additional analysis of historical climate data that is in general agreement with our findings. Historical temperature trends for different weather stations around the state are shown for several different periods: 1900–1999, 1970–2000,
and 1970–2008 (see Table 1.2 in the climate chapter). In general, significant upward trends in mean annual temperature are driven by significant increases in winter temperature, although significant increases in summer temperature are observed for some stations and time periods. The 1970–2008 period is comparable to the CDD trends covering 1970–2007 shown in Figure 6; over this period, with the exception of Elmira, trends in summertime temperature are not significant (see Table 2c in the climate chapter).

[10] Gaffin et al. (2008) estimate that one-third of the observed increase in mean annual temperature in New York City of 2.7°F is attributable to a strengthening urban heat island, with two-thirds of the increase attributable to global climate change. Urban development can lead to a higher concentration of heat-trapping built surfaces and a lower concentration of vegetation, which can increase local temperature. Heat island mitigation strategies include urban forestry, planting of street trees, and incorporating more reflective surfaces into the urban environment.

[11] This is true because nighttime demand levels will remain lower than afternoon demand levels and because "shoulder season" peak demand will still be lower than the summertime peak. Shoulder season refers to the months between peak demand and low demand (late spring and early fall in New York).

[12] This is partly a function of where a city or region derives its power. Because most cities can and do draw on power generated outside of the city limits, it is common for areas with surplus capacity to sell power to areas experiencing a shortfall. (For example, Morris et al., 1996 noted that Con Edison's summertime peak demand was 40 percent higher than its winter peak demand, freeing up winter-time generating capacity in New York City.) To the extent warming temperatures drive up peak summer demand in traditional winter-peaking areas (and vice-versa), there may be less power available to share, creating the need for additional generation capacity across the system.

[13] Baxter and Calandri (1992) and Franco and Sanstad (2008) analyzed impacts on electricity sales in California. ICF (1995) analyzed the service territories of six utilities in different parts of the U.S. and Japan.

[14] To carry out this study, NYSERDA partnered with the U.S. Environmental Protection Agency, the Electric Power Research Institute, and the Edison Electric Institute. Climate change impacts on both "upstate" and "downstate" electric systems were examined. There has not been any follow-up analysis examining the accuracy of these projections.

[15] The MEC report's study region was comprised of 31 counties in the New York Metropolitan area, which extends into Connecticut and New Jersey, so results are not directly comparable to estimates for New York City or New York State. Also, Hill and Goldberg (2001) focused on projecting impacts on peak demand, rather than annual demand.

[16] Note that a small portion of the rise in electricity demand in the winter, relative to shoulder seasons, may be related to additional lighting demand on shorter, winter days.

[17] Turning points were computed by running locally weighted (Lowess) regressions of demand on temperature and saving the Lowess smoothed estimate for each temperature observation. The temperature value corresponding to the minimum of the Lowess smoothed variable was defined as the turning point.

[18] NYSERDA has long been active in funding research and deployment of many of the strategies listed in Table 12. Since its inception, NYSERDA has provided support for renewable energy technology deployment and market development efforts, including solar PV techology. For example, by the end of 2006, NYSERDA had provided financial support for nearly three-fourths of all of the solar PV systems installed outside of the Long Island Power Authority (LIPA) service territory, although the number is likely even higher now. (A separate funding program sponsored by LIPA targets PV deployment on Long Island.) Demand-side management efforts are another long-time focus of NYSERDA, and most of the demand-side strategies listed in Table 12 have recently been or are currently eligible for funding from various NYSERDA programs.

[19] Under the current system, suppliers are penalized if they fail to deliver supply they had formally committed to the NYISO system, meaning the system suggested here might prove redundant. Such penalties do little to protect against climate-related system failures, however, and may encourage firms to underbid their capacity to deliver power during extreme events, artificially increasing prices beyond levels otherwise justified.

[20] The St. Lawrence–FDR Project includes the Moses-Saunders power dam (a single structure featuring 32 turbines divided equally between the New York Power Authority and Ontario Power Generation), the Long Sault Dam, and the Iroquois Dam.

[21] Since 2000, the IJC has been examining alternatives to Plan 1958-D, and one plan known as Plan 2007 is currently awaiting final approval; its prospects are unclear.

[22] Canadian Center for Climate Modeling and Analysis (model version CGCM1)

[23] United Kingdom Meteorological Office's Hadley Centre (model HadCM2)

[24] The Niagara Power Project includes generation output from both the Robert Moses Niagara Power Plant and the adjacent Lewiston Pump Generation Plant.

[25] "Firm" power customers can expect power to be available at all times, except possibly in emergencies. "Interruptible" power customers may pay a lower rate, but the utilities have the right to curtail their power for periods of time if necessary (e.g., due to high demand and/or reduced power availability).

[26] Excluded from this calculation are estimated losses due to power-quality events.

[27] As reported in the *New York Times*, August 16, 2003.

[28] As described above, nationwide costs may reach up to $180 billion annually, much more than the cost of the 2003 major blackout.

[29] This may include investments in distributed generation, which has been defined as the electricity production that is on-site or close to the load center and is interconnected to the utility distribution system (http://www.energy.ca.gov/papers/2004-08-30_rawson.pdf).

[30] Because the survey was primarily aimed at soliciting New York State-specific information to supplement the original literature review that formed the basis for much of this chapter, the Energy Team decided to narrow the stakeholder survey to power plant operators and distribution utilities in different regions of the state.

Chapter 9

Transportation

Authors: Klaus Jacob,[1,2] George Deodatis,[2] John Atlas,[2] Morgan Whitcomb,[2] Madeleine Lopeman,[2] Olga Markogiannaki,[2] Zackary Kennett,[2] Aurelie Morla,[2] Robin Leichenko,[3] and Peter Vancura[3]

[1] Sector Lead
[2] Columbia University
[3] Rutgers University, Department of Geography

Contents

Introduction....300
9.1 Sector Description....300
9.1.1 Economic Value....300
9.1.2 Statewide Overview....300
9.1.3 Metropolitan Transportation Authority....301
9.1.4 Port Authority of New York and New Jersey....304
9.1.5 Other Transportation Operators Serving the New York Metropolitan Area....306
9.1.6 Freight Railway Services in New York State....306
9.2 Climate Hazards....306
9.2.1 Temperature and Heat Waves....307
9.2.2 Precipitation....308
9.2.3 Sea Level Rise and Storm-Surge Hazards in Coastal Regions, Tidal Estuaries, and Rivers....308
9.2.4 Other Climate Factors....309
9.3 Vulnerabilities and Opportunities....310
9.3.1 Ground Transportation....310
9.3.2 Aviation....312
9.3.3 Marine Transportation, Hudson River, and Great Lakes/St. Lawrence River Seaway Shipping....312
9.4 Adaptation Strategies....313
9.4.1 Key Adaptation Strategies....314
9.4.2 Large-Scale Adaptations....315
9.5 Equity and Environmental Justice Considerations....320
9.5.1 Social Vulnerability and Equity....320
9.5.2 Adaptation and Equity....321
9.6 Conclusions....321
9.6.1 Main Findings on Vulnerabilities and Opportunities....321
9.6.2 Adaptation Options....322
9.6.3 Knowledge Gaps....322
Case Study A. Future Coastal Storm Impacts on Transportation in the New York Metropolitan Region....322
References....354
Appendix A. Stakeholder Interactions....356
Appendix B. Method of Computation of Area-Weighted Average Flood Elevations for Nine Distinct Waterways in New York City....357
Appendix C. Method to Compute Economic Losses....359

Introduction

The transportation sector, as defined in the context of the ClimAID report, consists of the built assets, operations, services, and institutions that serve public and private needs for moving goods and people within, to, and from the State of New York. The transportation sector and the energy and communications sectors are highly interdependent (see Chapter 8, "Energy," and Chapter 10, "Telecommunications").

Transportation occurs by different *modes*: land, air, and water. On land, it can be divided into road, rail, and pipeline systems. Transported goods are people and freight (the latter includes raw materials, supplies, finished products, and waste). In urban areas, mass transit systems serve commuting populations traveling to and from daily work, school, shopping, etc. In suburban and rural areas, largely private vehicular transportation on roads and highways dominates, but this also reaches the central business districts of cities. Long-distance and interstate traffic on roads is complemented by railway, water, and air transport.

The purposes of this chapter are 1) to provide a comprehensive overview of the vulnerabilities of the state's transportation system to changing climate, and 2) to present the adaptation options that can turn the challenges posed by the changing climate into opportunities to revitalize and modernize the state's transportation systems while at the same time improving their climate resilience. This chapter is structured based on climate hazards and risks. This means that regions with the highest concentration of transportation assets located in the most vulnerable places, and hence representing the largest risks for potential losses from climate change, will be scrutinized in much greater detail than those regions with fewer assets at risk and with lesser climate change impacts on the state's economy.

9.1 Sector Description

Transportation is a lifeline fundamental to modern developed societies. Provided in this section is an overview of the transportation sector in New York State. This section includes a description of the many transportation systems in the state and discusses the agencies that are responsible for managing them.

9.1.1 Economic Value

Nationally, transportation contributes on the order of 10 percent to the economy. Translated to New York State's annual gross state product (in excess of $1 trillion), this would correspond to a contribution of about $100 billion per year to the state's economy.[1] Without an effective transportation infrastructure, the economy of a state cannot function and grow.

9.1.2 Statewide Overview

Transportation in New York State is a complex system in which the public and private sectors interface by different transportation modes, including roads, rails, aviation, and shipping.[2] The New York State Department of Transportation (NYSDOT) is the state's transportation lead agency and has the following functions:[3]

- Developing and coordinating comprehensive transportation policy for the State; assisting in and coordinating the development and operation of transportation facilities and services for highways, railroads, mass transit systems, ports, waterways, and aviation facilities; and formulating and keeping current a long-range, comprehensive statewide master plan for the balanced development of public and private commuter and general transportation facilities.
- Administering a public safety program for railroads and motor carriers engaged in intrastate commerce;

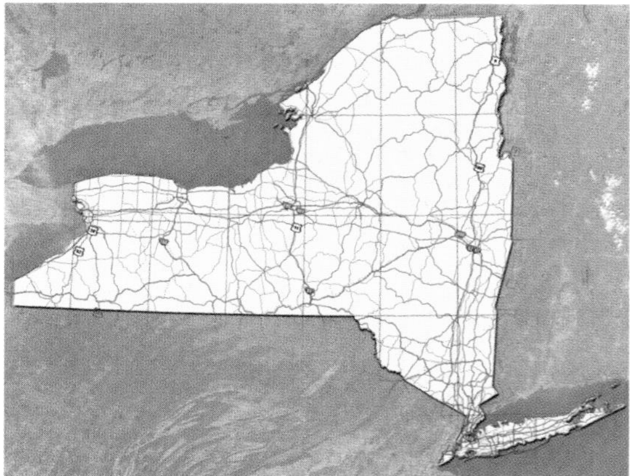

Source: National Atlas, modified

Figure 9.1 Interstate and major state highways in New York State

The Port Authority has four operating divisions: 1) Aviation; 2) Tunnels, Bridges, and Terminals; 3) Rail Transit; and 4) Port Commerce.

Major ground transportation facilities include the following Hudson River and other water crossings (**Figure 9.3**):

- PATH commuter rail (ridership about a quarter-million people per day)
- George Washington Bridge (GWB)
- Lincoln and Holland Tunnels
- Bayonne Bridge, Goethals Bridge, and Outerbridge-Crossing

Total eastbound vehicle volume on these tunnels and bridges in 2008 was about 124 million per year, with GWB alone accounting for 53 million vehicles per year.

The Port Authority owns three regional bus terminals:

- George Washington Bridge Bus Station
- Mid-town Manhattan Port Authority Bus Terminal
- Journal Square Transportation Center Bus Terminal in Jersey City

These are used by private and public bus operators. Total combined passenger volume (in 2008) was nearly 72 million passengers per year. The total interstate (NY/NJ) ground transportation network produced gross operational revenues (largely tolls and fares) of about $1.1 billion, of which the George Washington Bridge (GWB) contributed about 40 percent.

The Port Authority operates three major international / national airports (JFK, Newark, and LaGuardia), and two smaller airports (Teterboro and Stewart). Combined total passenger volume at these airports fluctuates between 100 and 110 million passengers per year. Of these, JFK (47 million passengers in 2008) and Newark airport (35.4 million passengers in 2008) are important

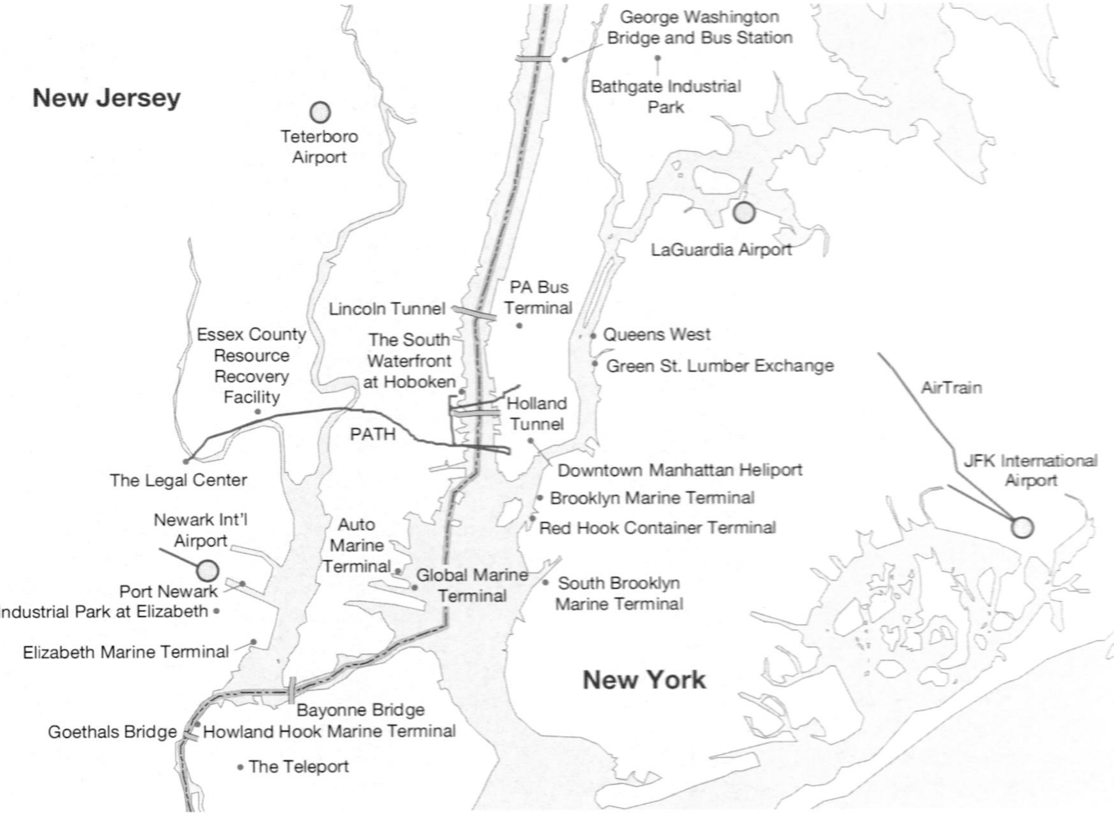

Source: The Port Authority of New York and New Jersey, 2001; Bureau of Transportation Statistics, Transportation Atlas of the United States

Figure 9.3 Facilities operated by the Port Authority of New York and New Jersey (PANYNJ)

gateways for international flights to and from the U.S. The combined air cargo for 2008 was 2.4 million tons.

Combined airport gross operating revenues in 2008 were about $2 billion (with JFK accounting for $0.951 billion, Newark about $0.718 billion, and LaGuardia about $0.307 billion).

The Port Authority operates major marine port facilities and container terminals in the NY/NJ harbor. In 2008 the port facilities handled 5.27 million TEU (20-foot Trailer Container Equivalent Units), or 33.6 million metric tons, with a value of about $190 billion (about $51 billion in exports and $139 billion in imports). The ports' gross operating revenues in 2008 were about $0.21 billion.

The Port Authority owns the World Trade Center (WTC) site in downtown Manhattan, and owns and operates many other facilities (**Figure 9.3**).

The Port Authority had a $6.7 billion budget for 2009, which provided for $3.3 billion in capital projects; this was set at $3.1 billion for 2010.

9.1.5 Other Transportation Operators Serving the New York Metropolitan Area

NJ TRANSIT brings commuters by rail from New Jersey into Penn Station on Manhattan's midtown West Side via tunnels under the Hudson that are also used by Amtrak for its Washington, D.C.–New York–Boston rail passenger service. NJ TRANSIT, with funding participation by the Port Authority, is in the process of increasing trans-Hudson transportation capacity by constructing a new rail tunnel under the river between New Jersey and Manhattan. This Access to the Region's Core (ARC) project also includes a new underground station that will have a pedestrian connection to Penn Station, New York, where there will be no interconnection at track level.

The ARC project will more than double commuter rail capacity between New Jersey and New York. The availability of more and improved train service is expected to remove 22,000 cars from the region's highways. Additionally, NJ TRANSIT and the private bus carriers it supports transport 127,000 people every weekday for a total of 254,000 passenger trips each weekday into and out of New York City. The ARC project was put temporarily on halt in 2010; alternatives to increase trans-Hudson commuter rail capacity at reduced capital spending are being explored.

The City of New York operates the Staten Island Ferry and all toll-free bridges between four of the five boroughs of New York City, including the Brooklyn, Manhattan, Williamsburg, and Queensboro bridges, and several smaller bridges crossing the Harlem River between Manhattan and the Bronx.

New York Waterway and other private ferry and water taxi services provide growing passenger service between points in and to the central business districts of New York City and on routes connecting them to communities along the lower Hudson River, Long Island Sound, Great South Bay within New York State, and to nearby Connecticut and New Jersey shore points.

9.1.6 Freight Railway Services in New York State

Freight services by railroads are in resurgence (see NYSDOT, 2009).[7] According to Railroads of New York (RONY), a trade association of New York State freight railroads, and data collected by NYSDOT, approximately 45 railroads operate in the state, although only four are Class-1 freight railroads (CSX, CN, CP, NS), in addition to the four commuter/intercity railroads (Amtrak, LIRR, Metro-North, NJ TRANSIT).

According to the American Association of Railroads (AAR),[8] in 2005, total miles of track operated in New York were about 3,600 miles, of which 65 percent is Class-1[9] railroad mileage. Amtrak owns about 150 miles of track in New York. In comparison, Metro-North and LIRR operate nearly 800 and 600 miles of track, respectively. According to the AAR, in 2005 carload tons originating in New York totaled almost 10.5 million, transporting major products including chemicals, waste and scrap, and nonmetallic minerals. Tons terminated in New York totaled over 25.3 million, including coal, chemicals, and food products.

Actual rail carloads originating and terminating within the state totaled 196,000 and 375,000, respectively. A map of all rail lines currently operating in the state is depicted in **Figure 9.4**. Major freight rail facilities and yards are located in Buffalo, Rochester, Albany, Binghamton, and New York City. Smaller yards and facilities are distributed throughout the rest of the state.

Source: New York State Department of Transportation, Office of Integrated Modal Services, Freight Bureau

Figure 9.4 Operating rail lines in New York State in 2008

9.2 Climate Hazards

The impacts of climate change (see Chapter 1, "Climate Risks") have significant consequences for the transportation sector. Sea level rise, the intensity and frequency of some extreme weather events, mean precipitation, flooding, and coastal erosion are all projected to increase, putting transportation infrastructure and operations at risk. (For an assessment by transportation mode, see Section 9.3.)

9.2.1 Temperature and Heat Waves

Increases in both the annual average temperature (see Chapter 1, "Climate Risks," Section 1.3) and the number of days per year with extreme high temperatures will affect transportation systems in several ways. Materials such as asphalt pavements; other road, bridge, and runway surfaces; and railroad tracks, electrified third rail, and catenary wires will need new performance specifications to cope with higher extremes and more frequent high temperatures. Air conditioning requirements for rolling stock and stations and ventilation requirements for tunnels will increase. Some runways of airports may need to be lengthened, since hotter air provides less lift and hence requires higher speeds for safe takeoff and landing.

A good example of the impact of heat waves on transportation systems is given by the European heat wave of 2003:[10]

- Britain's transport system suffered during the heat wave, particularly the railways. Widespread speed restrictions were imposed because of rail buckling, which becomes a problem when rail temperatures reach 36°C (97°F). Official figures show 137 cases of rail buckling in 2003/4, compared with 36 the year before and 42 the year after. However, the authors caution that confounding factors such as maintenance cannot be discounted.

- The resulting delays are estimated to have cost passengers £2.2 million ($3.6 million) in lost time, while the National Network Rail had to pay £6.5 million ($10.7 million) to the train companies in compensation. The researchers also found that disruptive fires at the side of the tracks jumped 42 percent in 2003 compared to the following year, which also might be due to the hot weather.
- Britain's road network bore the brunt of the searing heat. Sections of the M25 highway melted, and the total costs of repairs across the country are estimated at £40 million ($66 million), of which the government contributed £23 million ($38 million). The rest of the burden fell on local authorities.
- Temperatures on the London Underground passed 41°C (106°F) and passenger numbers dropped 1–1.5 percent during the hottest two weeks, reducing revenue by £500,000 ($0.8 million).

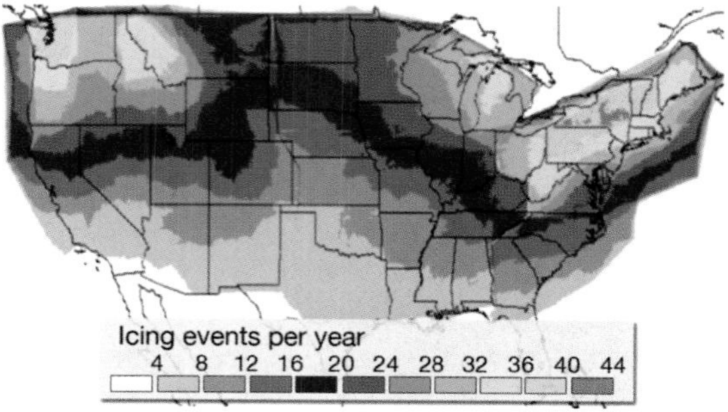

Note: Icing hazards are particularly severe in New York State, with the highest icing hazard in the central and western regions of the state. Note that warmer colors in this map indicate a greater icing hazard. Source: NOAA 2004

Figure 9.5 Estimated rendering of the likelihood of icing events across the United States

9.2.2 Precipitation

The central and northern regions of New York (with elevations that exceed 5,000 feet) currently are prone to more frequent and severe ice and snowstorms than near-coastal regions of the state. Air- and land-based transportation systems and operations are susceptible to freezing rain (icing) and snow. In fact, New York State is the most vulnerable to icing of all of the lower 48 states (**Figure 9.5**) (NOAA, 2004).

Icing can affect transportation systems in many different ways. It is a direct, serious hazard for aviation and for vehicular traffic on the ground. Indirectly, icing can also affect transportation by loss of electric power and/or, to a lesser degree, communication systems.

Freezing rain, black-ice conditions, and severe snow pose hazards to highway transportation and increase accident rates under current climate conditions. Climate change is likely to bring changes to these hazards. For instance, increasing winter temperatures are likely to shorten the duration of ice cover of the Great Lakes and, therefore, potentially allow more moisture to be drawn from the ice-free lakes, which would then fall as snow in western New York during the cold season (see Chapter 1, "Climate Risks").

While the severity of such extreme snowfalls is likely to increase, the number of days per year with snow on the ground is likely to decrease. In the estuary and coastal regions, nor'easter storms, which in the past caused blizzards, may more often turn into severe rainstorms rather than severe snowstorms. On the benefit side, it is more likely than not that the need for snow removal and salting of highways will gradually decrease for low-elevation, southern, and coastal areas of the state. The need for snow removal and salting under future climate conditions may change little in northern New York, though it may increase in western New York in the next couple of decades in areas that are subject to episodes of extreme winter lake effects (see Chapter 1, "Climate Risks").

9.2.3 Sea Level Rise and Storm-Surge Hazards in Coastal Regions, Tidal Estuaries, and Rivers

All transportation systems—roads, tunnels, railways, subways, airports, and seaports—are at risk from coastal storms and related coastal storm-surge flooding hazards. In New York, a number of these systems are located along the water at low elevations, and some subways, railroads, and highways are located in tunnels below sea level.

Storm-surge hazards along New York's shores (and the tidal Hudson River from New York Harbor to the Federal Dam at Troy) arise from tropical cyclones—

hurricanes, tropical storms, tropical depressions—during the summer and fall, and from nor'easter storms during winter and early spring. Coastal storm surges have caused damage in the past, and based on their historic frequency and severity of occurrence, these hazards have been quantified for the historic record.[11]

Climate change, especially its effect on sea level rise, will significantly raise coastal storm-surge hazard levels, as described in Chapter 1, "Climate Risks." Many near-shore transportation systems are at risk already (e.g., to coastal storm surges that reach the 100-year base flood elevations in coastal zones as currently mapped by FEMA). Sea level rise will increase the probability of flooding dramatically. Projections show[12] that the storm elevations now reached by the 100-year flood (i.e., a 1-percent annual probability of occurrence) will be reached before the end of the century by a flood with an approximately 3 to 10 percent annual probability of occurrence—about a three- to ten-fold increase. These changes will require the flood maps in near-shore areas to be updated to reflect new flood elevations that account for sea level rise. The flood-risk zones will need to be extended farther inland accordingly. These updates will place many transportation facilities that are currently safely located above and/or outside designated flood zones and related flood elevations within the newly assessed coastal flood zones. Additional details are discussed in Case Study A.

Sea level rise will eventually inundate low-lying areas permanently if no mitigation or adaptation measures are taken, and it may also accelerate saltwater intrusion in some areas. For most transportation facilities, the increased coastal storm surge hazard, however, will dominate over these permanent inundation hazards for most of this century.

9.2.4 Other Climate Factors

Additional climate hazards that impact the transportation sector are extreme storms events and droughts. These hazards and how they are projected to change in the future are described here.

Increased Storm Intensities

While it is unclear whether the total number of storms (hurricanes, nor'easters, thunderstorms, tornados, wind storms) will significantly change, it is more likely than not that the most extreme hurricanes and nor'easters will become more frequent. (see Chapter 1, "Climate Risks," and Chapter 5, "Coastal Zones"). The increase in intensity will affect air transportation: More storms (of any kind) may increase the number of delayed or cancelled flights, cause the temporary shutdown of airports, and/or result in flight detours to alternate airports. High winds may result in more frequent temporary closures or restricted use of larger bridges.

Intense storms redistribute existing sediments in the periodically dredged New York Harbor and Hudson River shipping lanes and bring increased sediment loads into them. These processes may increase the frequency of needed dredging operations. On the other hand, sea level rise tends to increase the available water depth. However, sediment transport in the Hudson and New York/New Jersey harbor is not sufficiently understood, and the understanding of sediment transport for these waterways under future climate conditions is even less well understood. Thus, it is not known whether sediment clogging or sea level rise will dominate over time or over which spatial distribution.

Urban Flash Flooding and Inland River Flooding

ClimAID projections show that the number of days per year with extreme precipitation (e.g., more than 2 inches per day) is likely to increase.[13] Projections for annual average precipitation rates (inches per year), however, show no clear trends in New York State for some time. An increase in extreme precipitation events will increase the hazards for urban and river flooding, with associated risks for transportation in cities and in rural areas along many rivers. This will necessitate increases in street stormwater drainage and processing peak capacity and/or result in environmentally undesirable combined sewer overflow events in those communities (including New York City) where street runoff is channeled into the public sewage system. The scouring potential for bridge foundations in some rivers is also likely to increase.

Droughts and Great-Lakes Climate Effects

Droughts can affect New York State's transportation systems in several ways. Extended droughts may lower the water levels of the Great Lakes and canals, and reduce the shipping capacity to the Atlantic coast via the St. Lawrence River Seaway (Millerd, 2007). For the Great Lakes, climate change is expected to result in lower water levels, higher surface water temperatures, and shorter duration of ice cover—all of which will affect shipping.

To maintain sufficient water depth along shipping routes (i.e., keel clearances), vessels may need to reduce the total weight of cargo carried on each voyage to mitigate the effects of reduced water levels. On average, shipping between Lake Ontario and Montreal (passing through the Welland Canal that connects Lake Ontario and Lake Erie) amounts to about 2,700 transits, carrying about 31 million tons per year. Transporting a given weight of a commodity with reduced under-keel clearance will require additional trips, thus increasing total shipping costs. Lake Erie's water level has been projected to decrease by 1.97 feet by 2030 and 2.62 feet by 2050, using the Canadian Centre for Climate Modelling and Analysis (CCCMA) climate model, and by 4.59 feet assuming stabilization of atmospheric carbon dioxide concentration after it doubles (Millerd, 2007). The water level of Lake Ontario has been projected to drop by 1.15 feet by 2030 and by 1.64 feet by 2050, and by 4.27 feet under the same stabilization conditions (these lake level changes are relative to the International Great Lakes Datum of 1985.) The decrease in load capacity from these reduced water levels in the navigable channels may require an increase in the number of trips needed to ship the same tonnage, resulting in increased shipping costs. For example, the cost to ship grain under these lower-water conditions is projected to increase by 5 to 10 percent per ton of grain.

On the other hand, a warming climate may increase the shipping season since the duration of ice cover in the winter will be shortened. Ice breaking is currently shared between two Canadian and one U.S. Coast Guard ice breaker. Due to warmer temperatures, the time at which winter ice is cleared at the beginning of the shipping season may occur earlier, but no quantitative estimates are currently available. The closure of shipping in the winter has been used in past decades for lock maintenance of the Welland Canal. If year-round shipping becomes possible, then consistent twinning (doubling up the number of locks in each direction) may be needed to allow maintenance without impeding shipping.

Droughts can also affect land transportation by leading to fires along railroad tracks and interstate and state highways. They can cause temporary closures, traffic delays, and slowdowns, and can increase highway traffic accidents because of reduced visibility (apart from undesirable pulmonary health effects; see Chapter 11, "Public Health").

Extended droughts may affect the availability of water for washing buses and mass transit rolling stock fleets—a water-intensive operation. These activities may be curtailed during extended droughts that lead to water shortages. Measures to mitigate this consequence may include recycling gray water.

9.3 Vulnerabilities and Opportunities

Earlier reports have addressed the vulnerabilities of transportation systems to climate change on national,[14] regional,[15] and some New York City[16] scales. The national and regional reports provide an excellent background to major vulnerabilities, but need to be modified for statewide climate projections and transportation systems across New York State. Lessons learned from extreme weather events at other locations across the United States (e.g., Hurricane Katrina and other major storms along the Gulf Coast)[17] and Canada (e.g., the ice storm of 1998)[18] also provide useful information for New York, if modified to meet the needs of the state. This section of the ClimAID analysis addresses climate change vulnerabilities of transportation systems by mode of transportation. In Section 9.4 the risks from climate change are described from the perspective of the type of climate hazards. For each transportation mode, it is important to distinguish between the vulnerabilities of operations and those of physical assets. Information on generic vulnerabilities to climate change is largely based on the Transportation Research Board's report on the potential impacts of climate change on the transportation sector in the United States (TRB, 2008a).

9.3.1 Ground Transportation

One specific area of the transportation sector that is vulnerable to climate is ground transportation. This section discusses the vulnerabilities of ground transportation systems, which include roads, highways, and railways.

Roads and Highways (including bridges, tunnels, drainage, and signal systems)

The physical assets and structures of the transportation system are vulnerable to climate change amplified precipitation and flooding and related erosion of road embankments near inland rivers and streams. Gradually increasing severe coastal storm surge flooding (because of anticipated sea level rise) along coasts and estuaries, including the tidal portions of the Hudson River, also put transportation structures at risk. Heavy rains can also cause mud and landslide hazards. High temperatures require heat-resilient asphalt mixtures for road and highway pavements.

There are also a number of other structure-related vulnerabilities. Drainage systems may have insufficient capacity to cope with the heavier precipitation events. Bridge foundations in some streams will likely experience increased scour potential. Clearances of some bridges across waterways subject to sea level rise may be diminished below the limits set by the U.S. Coast Guard or other jurisdictions. Bridge access ramps, tunnel entrances and ventilation shafts, and highway beds may need to be raised in coastal zones to prevent frequent coastal storm-surge flooding, amplified over time by sea level rise. The same hazards may make ineffective the collision fenders protecting bridge foundations in navigable rivers from impacts of ships or barges during high-water events; the fenders may have to be vertically extended to accommodate sea level rise (e.g., for the Tappan Zee Bridge[19] main span, relying on the buoyancy of caissons vulnerable to impact by out-of-control ships or barges). Road surface materials and bridge decks will need to be resilient to virtually certain higher and more frequent peak temperatures. Roadbeds and surfaces may experience winter temperatures nearer the freeze and thaw cycle, rather than steady below-freezing conditions (TRB, 2008a).

For highway operations and construction activities, more extreme weather events will increase traffic interruptions, may increase the number of extreme-weather-related traffic accidents, and may slow down or interrupt summer construction activities at temperatures above 105°F, largely because of worker heat exhaustion. Heat-resistant pavements will need to be used where they were not needed before as the number of days per year with average temperatures above certain thresholds increases substantially (Chapter 1, "Climate Risks"). Power outages during summer heat waves may affect signals, and hence slow traffic, especially in urban areas.

Freezing rains at higher elevations are more likely than not to become more frequent, and so may snow hazards, mostly in western New York. Both snow and freezing rains, however, may diminish in the southern portions of the state and along the coast, thereby reducing snow removal and salting costs. Closures of roads due to wildfires and related diminished visibility from smoke during extreme and extended droughts are likely to increase in frequency and geographic extent. High winds are likely to require more frequent temporary closures of major bridges, may cause more damage to traffic signs, and may call for increased fallen-tree and debris removal from roads and highways.

Coastal evacuation routes may have to be prepared to accommodate reverse traffic flow to speed up evacuations out of coastal flood zones by using all traffic lanes to direct flow from coastal to safe inland or higher locations. Road tunnels and sub-grade underpasses in coastal areas and other flood-prone zones relying on pumped drainage will very likely need increased pump capacity and back-up power, especially if they serve as designated evacuation routes and/or need to stay open for first-responder emergency services.

Railways (subways and commuter, passenger, and freight railroads)

Rail systems in coastal zones and tidal estuaries are subject to storm surges, whether at grade or partially elevated, or running in tunnels below grade and/or below sea level when crossing bodies of coastal or estuary waters. These vulnerabilities will become ever more amplified by sea level rise. To reduce or remove these vulnerabilities in the coastal and estuary zones will require large long-term investments and, in some instances, either vertical or even horizontal relocation. For the latter option, this may require new rights-of-way and related land-use decisions for communities served by rail services.

Vulnerabilities to flooding, washouts and erosion, mud- and landslides in steep terrain of some railroads running along inland rivers and streams, and insufficient or marginal drainage capacity of culverts and catch basins will need attention. Increased river flooding is not always due to more extreme climate events, but can be caused by changed land use, i.e., developments that increase rapid runoff and reduce infiltration of rain into natural ground cover and soils.

Extreme heat events also increase the vulnerability of railroads. Extreme heat can cause rail buckling. Routes along wooded areas may see increased wildfire hazards during extended droughts and heat waves. Power and related signal and/or communication failures during heat waves, floods, or windstorms can contribute to interruptions in rail and commuter services, with related economic effects.

In the New York City metropolitan area, coastal emergency evacuation plans partly rely on mass transit to provide evacuation capacity in the hours before severe coastal storms make landfall.

9.3.2 Aviation

Another area of the transportation sector that is vulnerable to climate is aviation. Vulnerabilities to aviation structures and facilities and operations are discussed in this section.

Structures and Facilities

Airports and related technical aviation facilities located in coastal areas at low elevations (e.g., La Guardia, Newark, JFK) and serving the greater New York City metropolitan region are all to some degree vulnerable to coastal storm-surge flooding amplified over time by sea level rise. Existing flood-protection levees (e.g., for LaGuardia) may have to be raised or new ones installed, to the extent that raised levee elevations are compatible with the clearance height required for takeoffs and landings. Over time, some runways and other airport facilities located at low elevations above sea level, such as fuel-storage farms, terminals, sewage treatment plants, and maintenance sheds, may have to be raised or protected in place to keep up with sea level rise and increased coastal storm-surge hazards. Drainage of runways is generally designed such that it is likely to keep up with increased intense precipitation events.

More frequent weather-related power failures might require improved back-up capacity at airports. Runway materials will need to resist higher and more frequent peak temperatures. Indoor airport facilities may need additional air conditioning capacity to deal with more extreme hot days. To determine effective adaptation strategies, each facility will need to conduct its own evaluation to assess its respective vulnerabilities (see TRB, 2008a).

Operations

Aviation operations will more likely than not have to cope with more severe weather conditions (high winds, thunderstorms, extreme precipitation, high temperatures) that generally lead to flight delays, cancellations, or detours to unscheduled landing destinations. These outcomes have economic implications for airlines, airports, and travelers alike. Loaded planes waiting excessive times for takeoff under extreme heat conditions can cause passenger discomfort and health emergencies. Extreme high air temperatures reduce the lift capacity of planes during takeoff and landing (TRB, 2008a), thus requiring, in some locations, longer runways, lower passenger or freight loads, or lower fuel loads that reduce distance range and reserve safety margins. In-flight icing conditions or deicing needs before takeoff could become more acute for airports and flight routes, especially in western and central New York.

9.3.3 Marine Transportation, Hudson River, and Great Lakes/St. Lawrence River Seaway Shipping

In coastal and estuary ports, including along the tidal portions of the Hudson River, vulnerabilities to coastal storm surges, amplified over time by sea level rise, will need to be assessed and addressed. Sea level rise, tides, and coastal storm surges propagate up the Hudson River estuary to Albany and the Federal Dam in Troy. The magnitude of the inland effects of sea level rise on the estuary is the same as for the coast; the inland effects of storm surge and tides decrease very little in force and amplitude.[20] This virtually certain increase in hazard related to sea level rise may affect pier heights, base elevation of loading cranes, power supply substations, access roads and rail tracks, open air storage (for containers or automobiles), and warehouse facilities located at low elevations along all shores subject to tides. In particular, the frequency of the 1-in-10-year coastal flood may triple over the next century, depending on sea level rise (see Chapter 1, "Climate Risks", and Chapter 5, "Coastal Zones").

On the other hand, for Great Lake ports (and related St. Lawrence River Seaway shipping lanes), increased lake evaporation under severe and prolonged drought conditions and extended heat waves are likely to lower the lake levels to such a degree that it may impede shipping capacity to the Atlantic Ocean and, via out-of-state routes, to the upper Great Lake states and Canada.[21] During extended droughts, the canal and lock systems in central, western, and northern New York, which currently serve largely recreational purposes (Erie-Mohawk and St. Lawrence-Lake Champlain-Hudson systems), may also not be able to accommodate as much traffic in the future as a result of periodic water scarcity needed to operate the locks. On the benefit side, the expected climate warming is likely to prolong the ice-free shipping season on the Great Lakes and St. Lawrence Seaway and make the navigable portions of the Hudson River less prone to the ice floes or shore-to-shore freezes that occurred more commonly in past centuries and on occasion interrupted the transport of fuel and other supplies to Albany and mid-Hudson terminals.

9.4 Adaptation Strategies

Adaptation to climate change involves a complex multi-dimensional array of options (See Chapter 2, "Vulnerability and Adaptation"). Typically adaptation is specific to a particular mode of transportation and to the specific climate hazards that pose the threats. Options may differ across the geographical, land use, and climatic zones within the state. They can differ in scale and granularity, from statewide to regional to local and site-specific solutions. Short-, medium-, and long-term solutions must be balanced against each other. Adaptation should be risk-based and consider benefits versus costs. In this context, the questions of who pays the costs and who gets the benefits raises social and environmental (and intergenerational) justice issues with fiscal, economic, and ecological consequences (see Chapter 3, "Equity and Economics"). How and where current investments in infrastructure are planned, engineered, and constructed affects their future vulnerabilities. If existing infrastructure is not upgraded and adapted to the new demands posed by climate change (just as infrastructure needs periodic upgrades to demographic and economic demands) it will put the neglected regions, their economies, and, in the worst cases, lives in jeopardy.

Transportation adaptation strategies are intertwined with land-use issues. The question of whether land use leads to transportation demands or transportation capacities lead to land use must be approached holistically (TRB, 2008a). Land use has implications for both climate change mitigation (i.e., limiting greenhouse gas emissions) and for climate change adaptation (e.g., of transportation corridors along flood-prone coasts and inland rivers). There exists a vast literature that has detailed the relationship between land use and natural disaster risk management (see e.g., Mileti, 1999; Godschalk et al., 1999). Climate change add an additional dimension to managing natural hazard risks in the context of land use over the long term.

The connection of climate change adaptation to land use is clearest in the coastal zones that are at an increasing risk from sea level rise and related coastal storm surge inundations. This connection is discussed in the coastal storm surge case study. The issue of "home rule"[22] is embedded in the culture and legal foundations of the nation, states, counties, cities, and villages and puts local communities in the critical position of primary decision-maker. As a tool to guide states and, in some cases, local communities toward an environmentally sustainable path, the federal government can attach conditions to transportation financing. The actual authority for designs and planning generally lies, however, with the state or local community. Hence, states and local communities are key partners for sustainability. Federal guidance via the financing option is limited, and the project-by-project approval process, including how environmental impact statements are prepared, reviewed, and approved, is not yet well suited to sustainable adaptation to climate change and sea level rise.

In this context, transportation agencies having active roles in the state's coastal zones (including New York State Department of Transportation, MTA, the Port Authority, and many others in the public and private transportation sector) will need to balance their adaptation efforts to cope with sea level rise on a project-by-project basis with a more regional approach. Such balancing will include difficult decisions for communities and, consequently, transportation agencies. Such decisions include determining whether engineered defensive levees, pumping stations, and estuary-wide protective storm surge barriers are sustainable adaptation solutions, or if such defensive structures are only temporary solutions. Such structures could be combined with long-term exit strategies involving carefully staged and equitable retreats from, and relocation of assets in, communities that are at risk. FEMA's National Flood Insurance Program includes an option to buy out properties; a potential strategy would be to extend such buyout programs beyond the National Flood Insurance Program to include critical infrastructure systems exposed to repetitive risks. This may require new federal initiatives, but states could help to bring about such changes.

Transportation agencies will be at the center of a systematic river and coastal flood-risk assessment, land-use planning, and ultimately a consensus-forming decision process. Without such an overarching process—and with challenges to the *status quo* on home rule and other land-use practices—it will be difficult to shape a sustainable future for communities and for the transportation systems that serve them, and to build resilience to river flooding and sea level rise. At-risk communities may be given some assurance that government will assist in creating a safer future, if the communities recognize and act upon managing responsibly their exposure to the risks from sea level rise and increased coastal, estuary, and river flooding.

A likely outcome will be that well-organized, large transportation organizations (such as the New York State Department of Transportation, MTA, the Port Authority, several New York City agencies, and others, including some county and community governments) will initially plan for, seek financing, and implement interim adaptation measures at their existing facilities, often as part of their regular maintenance plans, capital budgets, and operations. Private operators and owners of properties will do the same, sometimes motivated by the availability (or lack) and pricing of insurance in high-risk coastal zones. This insurance effect is already starting to become operative in local development projects by the private sector. Over time, it will gradually affect future demographic and transportation patterns and related demands for infrastructure. The public sector may be supportive of these self-regulating market forces. Transportation planning agencies should collaborate with these positive developments, even if they occur only on a project-by-project basis.

With time, as climate stresses increase, more central, coordinated, regionally planned yet grassroots-supported, integrated planning will be needed to more cost-effectively and safely address coastal adaptation measures. Local decisions may eventually be replaced by a comprehensive approach that aims at flexible, adaptive solutions with sustainable outcomes.

There are precedents for such overarching efforts, some successful. The Netherlands' Delta Waterworks and the London/Thames Estuary Project (TE2100) are well-planned, flexible, and foresighted projects. Both protect land already at or below sea level.

Currently, New York State has virtually no land with built-at-grade structures at elevations below sea level. However, a large and often critical portion of its transportation (and some other) infrastructure—largely in the New York City metropolitan area—is already well below sea level and, therefore, increasingly at risk. Agencies such as MTA and others are in the initial stages of an evolving process to include climate adaptation principles in their planning, design, capital construction, and financing procedures. A similar planning process needs to include the transport infrastructure along the Hudson River below the Federal Dam in Troy and the Great Lakes and St. Lawrence River shipping routes. Any initial administrative and exploratory steps that have been undertaken (e.g., MTA, 2009a) require additional attention in the pertinent institutions. They also require endorsements from their governing boards and by society at large. This will require corporate leadership. Even then, however it may take time before climate change adaptation is firmly embedded into the normal functioning and decision making of the transportation institutions. The seeds are sown,[23] but in order to take root, sustained leadership, financing, political and public support, and implementation is required.

The rising technical awareness of changing climate conditions in the transportation community will need to be echoed by public and its representative political institutions. Their strong support for a broadly based sharing, for the common good, of the costs for safeguarding the public transportation infrastructure—especially in the coastal and estuarine risk zones—will be important to ensure effectiveness of agencies' adaptation efforts. The State of New York is in a position to provide leadership. The formation of the State's Sea Level Rise Task Force (SLRTF)[24] and the Climate Action Plan are good first steps. Near-term implementation of adaptation measures is the next step.

9.4.1 Key Adaptation Strategies

The technical and procedural tasks for climate change adaptation at hand will include the following steps:

- A full inventory of the hazards as a function of time related to climate change (e.g., NPCC, 2010; and Horton and Rosenzweig, 2010; also see Chapter 1 of this report, "Climate Risks").
- A full inventory of the transportation infrastructure at risk to these climate change hazards (and benefits where applicable) and a systematic assessment of transportation system vulnerabilities to these hazards.
- A well-planned effort of technical and fiscal evaluation of adaptation options and their local, regional, social, and environmental implications. An important part of developing these multiple options is to allow flexible implementation along multiple, time-staggered decision paths (e.g., see NPCC, 2010).
- This approach requires, in turn, institutionalization of a scientifically based monitoring and decision-support system and process that can inform decision-makers of when the climate risks reach trigger (or tipping) points where decisions cannot be any longer delayed without potentially dire consequences (NPCC, 2010).
- The above steps need to be reassessed regularly until it is clear that full consideration of short- and long-term effects of climate change are effectively embedded in infrastructure planning and decision making at all levels of government and by the operating transportation agencies.

9.4.2 Large-Scale Adaptations

The following sections provide medium- to long-term technical options for adaptation to various types of climate change, i.e., those that go beyond temporary emergency measures (such as sandbagging or pumping by mobile units). They are organized by type of climate hazard.

Adaptation for Coastal Hazards

- Constructing local flood proofing by building local levees, sea walls, floodgates, and pumping facilities. For truly low-lying areas such measures may be only temporarily effective (in some instances only for several decades). Site-specific studies for different time horizons will be needed.
- Raising structures or rights of way. For instance, commuter rail tracks could be put on elevated structures as part of a regional rejuvenation to a new generation of commuter and intercity rail systems, like those already implemented in Japan, Taiwan, and parts of Europe or as currently under widespread construction in China. Privately owned freight rail systems (e.g., along the west shore of the Hudson River) may need to consider equivalent options.
- Sealing of ventilation grates of belowground facilities (e.g., NYCT subway system) only in those locations that are in potential and future storm surge inundation zones. These sealed tunnel sections will need a newly engineered, forced ventilation system not open to the normal street grade, with consideration of fire safety.
- Designing innovative gates at subway and road/rail tunnel entrances, unless other options to extend the entrances to higher elevations exist, are practical, and can be implemented.
- Designing road and rail embankments as super-levees that could provide a double function: flood protection and transportation corridors.
- Conducting a feasibility study for a system of storm-surge barriers to assess their potential position and ability to provide protection for New York State's waterfront and transportation systems.
- Retreating and relocating critical systems out of and/or above flood zones.
- Raising bridge landings along shorelines to ensure there is sufficient clearance for the transportation systems (highways, roads, rail systems) they cross over, given the need to potentially raise these systems as a result of sea level rise and related storm surge inundation hazards. Site-specific studies are needed to develop solutions for this seemingly intractable problem. Preventive solutions (sufficient clearances) need to be planned for any new bridge structures that cross bodies of water controlled by tides and rising sea levels (e.g., the currently planned new Tappan Zee Crossing).
- Vertically extending collision fenders to higher elevations on bridge foundations in tidal waters.

Adaptation for Heat Hazards

- Confirming that currently used heat-resistant road surfacing and rail track materials are capable of withstanding additional, more extreme heat conditions.

- Upgrading air conditioning of rolling stock (trains, subways, buses) to meet the demand on extreme hot days.
- Inspecting bridge expansion joints, since they tend to lock with age, imposing extra stresses under extreme heat. This condition needs attention during bridge inspections. Ensuring adequate bridge clearances, as very large bridges tend to sag during extreme heat. Sea levels will rise, and modern ships often stack containers to heights that use as much of the clearance available, so it must be ensured that the available clearances continue to conform to U.S. Coast Guard limits for bridges across tide-controlled waterways. New height limitations may have to be imposed.
- Modifying airport and airplane functions. The aviation industry may encounter more frequent extreme weather events, with respective travel delays for airlines and their customers, and related economic impact. Airport runway lengths, extreme high air temperatures (hot air provides less lift), and required takeoff speeds of airplanes must be in balance to provide sufficient safety margins for takeoff. New generations of planes with more powerful engines are able to overcome this issue, but older planes may have to face load limitations or be phased out.
- Preparing for power and communication failures. Transportation agencies may need to be prepared for more frequent power failures (and related potential communication failures (see Chapter 10, "Telecommunications"). This applies especially during extended summer heat waves, when peak power demands exceed what electric utilities can supply due to increased need for air conditioning, unless the utilities' adaptation plans cover these needs or plans are in place to reduce public demand during such times (see Chapter 8, "Energy").

Adaptation for Precipitation Hazards

- Increasing the carrying capacities of culverts, retention basins, and other drainage systems in accordance with future precipitation normals (i.e., new averages and extremes, to be issued on a regional basis by NOAA's National Climatic Data Center). It may also require changes in American Association of State Highway and Transportation Officials (AASHTO) drainage guidelines and other applicable engineering standards.
- Raising road and rail embankments and/or strengthening their slopes to be resilient to flow dynamics and bank erosion in river flood zones prone to high flow velocities.
- Relocating rights of way out of new and future flood zones.
- Monitoring and remediating scour action at bridge foundations in rivers as flood and related flow conditions become more severe and frequent as a consequence of more extreme precipitation events (often further amplified by inappropriate upstream land use and development).
- Working with local agencies to reduce runoff from nearby properties and other rights of way onto transportation systems. This may involve creation of permeable surfaces and retention basins, restoring marshlands, increasing sewer or pumping capacities, and regrading slopes to direct runoff away from critical transportation infrastructure.

Adaptation for Winter Storms (Snow and Ice)

Overall snowfall and days per year with snow cover, especially in the more southern portions of New York, are expected to decrease gradually as snow will be more frequently replaced by rain. On the other hand, individual snowstorms and ice storms (with freezing rain) may become more intense, especially in higher elevations, in more northern regions, and those in areas prone to the lake effect, which may be amplified by a shorter duration of ice cover on the Great Lakes. These geographically diverse trends across the state may require potential reallocation of operational resources for snow clearing and sanding/salting. One alternative includes increasing the amount of intelligent signage that warns drivers about high-hazard road conditions. On average, across the state, a net reduction in snow hazard is more likely than not, but no clear trend is yet forecast for future freezing rain and icing conditions (see Chapter 1, "Climate Risks").

Other Adaptation Options

Other climate-related risks that require adaptation measures may originate from more frequent extreme winds (characterized as hard to quantify, see Chapter 1, "Climate Risks"). Transportation agencies may want to keep track of whether the design wind speeds need to be adjusted with time on a regional basis. A practical adaptation measure to cope with higher wind speeds is operational. Anemometers measuring wind conditions may be installed on bridges of a certain length and height above ground or water, and wind velocity limits may need to be set, above which bridge traffic will be allowed only at reduced speeds or, for higher wind speeds, will need to be suspended entirely to avoid excessive accident rates. Such limitations are already in place on some bridges in New York State and are, for instance, included as constraints in the New York City Office of Emergency Management hurricane evacuation plan.[25] MTA Bridges and Tunnels, and operators of other large bridges in the region, also have protocols in place for traffic restrictions during high winds.

Adaptation Options Related to Federal/State/Agency Policies and Cooperation

Intrastate Cooperation

At this time the major transportation agencies, authorities, owners, or operators in New York State do not yet have publicly accessible, internally approved master plans for how to adapt to those aspects of climate change that are currently known. The New York City Climate Change Adaptation Task Force (City of New York, 2011) in conjunction with the NPCC (2010) assessments come close to producing a roadmap and a technical/scientific foundation on which such a master plan can be based. This ClimAID project contributes to fulfilling a similar goal statewide. The MTA's *Greening Mass Transit & Metro Regions* report (MTA, 2009) provides the recommendations for such a plan. Actionable and internally approved plans can become an integral part of a long-term capital-spending budget to which the respective entity is committed.

For the private sector, and for the first time, the Securities and Exchange Commission (SEC) issued in January 2010 a statement[26] that "public companies should warn investors of any serious risks that global warming might pose to their businesses." An equivalent rule (or even law) may be developed by the State to go one step further and request that each transportation operator doing business in New York State produce every few years an updated actionable plan on how it intends to manage the emerging risks from climate change over short to medium time horizons. A less detailed but mandatory long-term outlook for up to a century should also be included.

The federal and state governments could use such agency plans as a precondition for financing climate change adaptation assistance. There are many precedents for such conditional financial assistance, ranging from the multi billion dollar federal sponsorship to reform state and local education systems, to DHS/FEMA's disaster mitigation assistance grants given to states (and in earlier times, directly to communities), conditional on their having developed a FEMA-approved disaster mitigation plan.

Given this situation, the State could consider establishing a ruling that each Transportation Agency operating in the State of New York should develop by a certain deadline (say, 2015) a climate change adaptation master plan with an institutional management, operational, engineering, and capital spending project component, for short (years), medium (a few decades), and long-term (50–100 years) time horizons laid out in various degrees of detail, respectively. The basis of the report should be a science-based hazard assessment pertinent to the transport agency's assets and operations, an engineering-based vulnerability (fragility) and risk assessment, and a ranking of options to manage these risks, with estimates of costs for adaptation measures and of potential costs (risks) for incurring gradual and/or potential catastrophic losses if no action is taken. Such plans should be updated on a regular basis, perhaps on the order of, say, 5 years, or commensurate with agency-specific planning cycles.

In many of the major urban or metropolitan centers across New York State, multiple agencies are responsible for operating various modes of transport systems, whether they are public or private entities. Since transportation is a networked system, delays, failures, or (at worst) catastrophic failures in one system can affect the other systems, and in such cases the customer may not be able to get from point A to point B within a reasonable time at reasonable cost.

Especially in the case of floods in connected underground structures, system vulnerability is often determined by hydraulic connectivity between tunnels, stations, and other structures. Any effort by one agency to adapt to a certain climate change performance standard can be made ineffective by others adhering to a lower standard. The weakest link in the system may critically control the system's overall performance, even if it is a very diverse and redundant system.

There are examples of how transportation agencies have worked together to coordinate joint planning, set performance standards, or solve other coordination issues for the benefit of the public at large. The EZ-Pass is one such example of an interagency practical, successful solution.

Another example of interagency coordination occurred in the post-9/11 cleanup and recovery phase. Due to the urgency of rebuilding several high-priority projects in Lower Manhattan, the Federal Transit Administration (FTA) set up an FTA Lower Manhattan Recovery Office[27] that worked with project sponsors on innovative, streamlined project delivery processes in the areas of development, oversight, and environmental management. This approach was developed early, with a consensus among federal and local partners. In the arena of environmental oversight it led to a memorandum of understanding between EPA and other federal agencies defining roles and response times. It also developed agreement among project sponsors to a common Environmental Analysis Framework and Environmental Performance Commitments as well as to coordinated cumulative effects analysis.

An organization potentially suited to take on this regional coordination for climate change adaptation standard and performance goals in the state's coastal region could be the New York Metropolitan Transportation Council, an association of governments, transportation providers, and environmental agencies that is the Metropolitan Planning Organization for New York City, Long Island, and the lower Hudson Valley.

In regions of the state with dense and diverse transport systems operated by multiple agencies and owners, an alliance of operators should be formed to coordinate climate change adaptation measures to ensure a coherent systematic approach with mutually agreed-upon performance goals and standards. In addition, the alliance may coordinate policy, oversight, and other issues with federal and state agencies to streamline a regional approach and to put the region in a better competitive position when applying for federal technical and financial support for climate change adaptation.

A particular task for coordination could be delegated to an "adaptation moles" technical working group. This group should be charged to ensure that the underground connectivity between multi-agency below-ground rail-based transportation systems in the NY/NJ metro region (including NYCT subway, LIRR, MNR, the Port Authority, New Jersey Transit, Amtrak, and others) will become flood-resilient as a whole. The working group would also engage with experts from vehicular tunnel operators (Port Authority and MTA Bridges and Tunnels), and state, county, and city agencies including DOTs, and power and communications utilities to ensure that a flood protection and general adaptation plan, with special emphasis on sea level rise, is comprehensive, system-wide, and performs in accordance with an agreed-upon performance standard for the benefit of all agencies and the public at large.

National Cooperation

Of course, New York State is not isolated, which is particularly relevant in the transportation sector. Not all regions of the nation will be affected equally by climate change. Those regions that are population centers and vital drivers of the national and global economy have generally the highest concentration of transportation infrastructure. If these major nodes of the transport systems fail and become unreliable, redundancy and diversity of the transport links between such centers cannot maintain the system capacity. These centers also serve to maintain a large state and federal tax base that needs to be stable. Their gradual or catastrophic failure could bring disproportionately large losses to state and national economies. Therefore these centers deserve special scrutiny and attention to sustain the economic viability of the state and nation at large, especially in the context of global economic competition. Without a climate-change-resilient transportation infrastructure, these economic centers cannot fulfill their role as reliable engines for the state and national economies, and hence warrant state and national support. Assessment of priorities is most effective when ranking is risk-based. Consequently, New York State may want to work closely with the federal government to pursue the following adaptation options:

- Set priorities for policies for providing sound knowledge and data, and direct financial support, to strengthen the nodes of transport infrastructures to make them climate-change-resilient.
- Consider a comprehensive program of research and technological development for advancing innovative and cost-effective climate-resilient urban and inter-urban transportation infrastructure.
- Devise incentives for states, regions, and cities with vital nodes and concentrations of transport

infrastructure to partner and exchange best practices in climate change adaptation, and to help set the national agenda for sustainable, energy-efficient transport systems.

Ground transport systems (roads and rails) of coastal population centers and estuaries (controlled by tides and brackish waters), are often placed underground in tunnels very close to or below sea level. Such systems, especially when built many decades ago without anticipating rising seas, are vulnerable to the combination of accelerating sea level rise and coastal storm surges. It is vital to make these low-lying transportation systems flood proof and to avoid systemic damage from saltwater intrusion before it is too late. To relocate such systems would require exorbitant resources. This poses new technological challenges and requires adequate resources to find innovative engineering solutions to protect these underground systems from the rising and encroaching seas. Consequently, it would be helpful for the federal government, in cooperation with states, to sponsor a technology assistance program to develop and install engineered protective measures targeting underground and near-shore transport systems that are under threat from sea level rise and saltwater damage.

Other transport facilities near New York State's coastline, and along the nation's coasts, including harbor facilities and their interfacing ground transportation links such as road, rail, storage, and freight transfer facilities, and many industries such as refineries and chemical plants that rely on marine shipping access, are also at risk from coastal storm-surge flooding amplified by accelerated rising seas. Inundation would not only damage these ports, ground transport, and industrial facilities, but also pose potentially severe environmental risks from spreading debris and toxic substances to nearby coastal population centers. Consequently, it would be helpful for the federal government to provide assistance to regions like New York State, with major port facilities and related industries that serve the nation's import/export demands. Such assistance should be aimed to develop and implement cooperative solutions among port operators, connecting transport systems, proximal industries, and nearby population centers and communities, with a goal of safeguarding them from coastal storm-surge flood hazards that will increase with rising sea levels. Federal assistance would greatly foster the development and installation of technical solutions that can reduce related environmental and health risks from potentially toxic materials and debris being carried by flood waters into communities, natural and developed land, ground-water, beaches, and/or fisheries.

FEMA flood insurance rate maps (FIRM), whether near rivers or coasts, have become an important guiding tool for local zoning, planning, land-use, construction permits, environmental impact statements, etc. These now-widespread uses are far beyond the original intent of FIRM maps for guiding the National Flood Insurance Program (NFIP) aimed mostly at residential housing in flood-prone areas. FIRM maps are based on *past* data and information, in terms of land use and climate. Therefore they are not suited for planning *future* sustainable development of communities and the transport systems that need to serve these communities under new and *changing* climate conditions.

Consequently, the federal government could establish a technical assistance program to help states and communities and their transportation agencies develop sound science-based flood zoning tools that allow forward-looking adaptation to climate change, including the associated engineering guidelines. Such guidance tools would be more appropriate than the FIRM maps for coastal zones and other flood-prone areas to cost-effectively plan and design new, or to modify existing, transportation and other critical infrastructure, and to support future community development that is sustainable for periods of time not shorter than the expected lifetime of the respective infrastructure.

Other needs exist, as well, that could best be met with coordination between New York and federal organizations. For example, accurate, high-resolution LIDAR (light detection and ranging) surveys need to be flown to facilitate the development of digital elevation models (DEM) of sufficiently high vertical and horizontal resolution to perform forward-looking flood risk assessments and regional planning of sustainable developments.

A similar need exists for forward-looking climate normals (in contrast to traditional climate normals, which are produced by NOAA based on past climate data). Future temperature normals are needed to guide the design of transportation cooling and ventilation systems that can meet increased demand, for heat resistant pavements on roads and airport runways, and for designing airport runways with sufficient length to ensure safe takeoff during extremely hot days. Future precipitation normals are needed to design drainage systems that can handle future extreme rainfalls. New York State should undertake formal steps to work with the respective federal agencies to produce these products in a timely fashion, with clear presentation of uncertainties and regular updates as new climate projections are produced.

Regional Cooperation

Regional transportation agencies own and operate assets that are often fully or partly self-insured. Insurance against climate-related disaster losses works best when the risk is spread geographically, by diversity in asset ownership, and by exposure to diverse, independent, and uncorrelated hazards and risks. The risks to regional transportation agencies from climate change are instead highly concentrated geographically and exposed to process-related climate hazards. Therefore, the principles for effective self-insurance are violated since all assets can be hit by the same event. Furthermore, one event may entail a number of correlated perils (e.g., wind, lightning, flood, debris impact, power outages, and saltwater damage), which may strike at the same time caused by the same event (e.g., the same hurricane).

Consequently, regional transportation authorities may want to spread their risks from climate-related weather events by entering insurance pools of transportation owners spread over diverse geographical regions across the nation. This may be achieved by a blend of mutual and self-insurance (with or without participation of federal or state governments and/or the private insurance and reinsurance sectors), or by floating catastrophe bonds on capital/equity markets. The federal and/or state governments may provide the regulatory framework for such sharing of the risks to public transportation lifelines across the entire nation, and set standards by which the insured and insurers shall abide. Another option is for federal and/or state governments (i.e., the taxpayers) to become the ultimate bearers of climate-change-induced risks for regional public transportation systems. In either case, a federal or joint federal/state program for assessing the climate change risk exposure of regional transportation agencies and of insurance options vis-à-vis climate change risks appears to be a desirable and much needed risk management measure.

9.5 Equity and Environmental Justice Considerations

Transportation planning is a longstanding priority of environmental justice advocates. In transportation analyses, core equity concerns often include unequal access to different types of transportation, the spatial mismatch of jobs and residences, the disproportionate health burden of automobile pollution, and a commitment to affordable public transportation (Bullard, 2007; Sze and London, 2008; Chen, 2007). Constructing adaptive, climate-secure transportation provides opportunities to build social equity into the infrastructure, but with less care it may exacerbate some of these existing inequities as well as create emergent burdens.

9.5.1 Social Vulnerability and Equity

Social, economic, and geographic marginality add to the challenges of transportation planning. For the United States as a whole, the poorest 20% of households spend more than 13 percent of their income on transport (U.S. Bureau of Labor Statistics, 2010). In urban centers, and increasingly the inner suburbs, lower-income people of color are disproportionately dependent on public transportation to get to their jobs (Pucher et al., 2003). African Americans and Latinos, in particular, are less likely than whites to own a car (Sanchez et al., 2004). Across most cities in the country, including New York City, there is a correlation between carless populations and poverty and minority status (Milligan, 2007). While reliance on public transport has positive implications for environmental quality and mitigation of climate change, reliance on public transport also creates vulnerabilities in times of natural disasters and climate-stress events. In one extreme example, Hurricane Katrina exposed the severity of this transport disadvantage: Upper-income populations left New Orleans by car, while disabled, low-income, and African American populations were stranded (Litman, 2005).

Some of the largest urban centers in the United States, including New York City, have detailed evacuation plans incorporating varied levels of social disadvantage. The strengths and limitations of New York City's plans are discussed in the case study later in the chapter. In contrast, one analysis discovered that central cities elsewhere in New York State were even less prepared to deal with transport disadvantage (Hess and Gotham,

of the base flood elevations (BFE) for the 100-year flood probability (**Figure 9.7**, red areas). Importantly, these base flood elevations are not scenario-specific. Thus, variations from the portrayed FEMA flood map estimates can and will occur. The reach of the added flood zones due to sea level rise and the added impact of sea level rise on the transportation infrastructure are also uncertain.

The FEMA maps are used as a starting point for a number of reasons, despite the fact that some locations have been flooded more than once during the last 100 years at flood levels higher than depicted by the maps, and other locations have not been flooded during the last 100 years to the degree that the maps predict. A number of the deviations can be explained by changes that have occurred since the FEMA maps were created. For example, many of the floods that are more severe than expected by FEMA's baseline flood elevation standards have occurred along inland rivers, where upstream development and changes in land use have increased runoff since the FEMA maps were produced. Along some coastal locations, beach erosion and, in a few cases, ill-conceived coastal management practices, have increased coastal flood hazards since FEMA completed its flood mapping. Adding to these hazards, local sea level along New York State shorelines has been rising at a rate of almost 1 foot during the last century. On the other hand, new sea walls and other protective structures may have reduced flood hazards in some locations. Lastly, the 100-year flood is a statistical

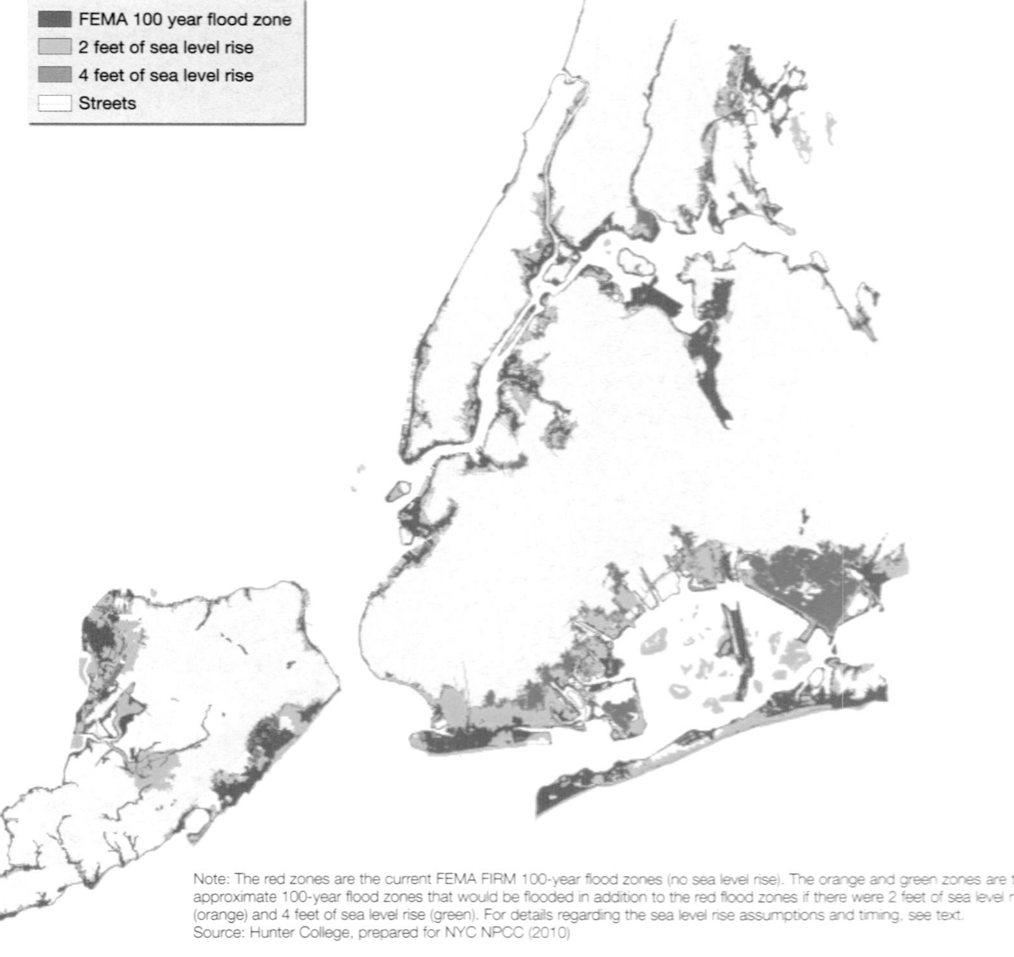

Note: The red zones are the current FEMA FIRM 100-year flood zones (no sea level rise). The orange and green zones are the approximate 100-year flood zones that would be flooded in addition to the red flood zones if there were 2 feet of sea level rise (orange) and 4 feet of sea level rise (green). For details regarding the sea level rise assumptions and timing, see text.
Source: Hunter College, prepared for NYC NPCC (2010)

Figure 9.7 100-year flood zones in New York City (i.e., with a probability of being flooded of 1 percent per year) for current and two different ClimAID sea level rise scenarios

estimate that describes the *average* occurrence of a randomly distributed sample of flood occurrences; each location has only a 63-percent chance of experiencing a 100-year flood within the 100-year timeframe.[30]

Methods: Averaged 100-year Flood Elevations and Sea Level Rise Added

This case study uses sea level rise estimates of 2 and 4 feet, which are added to FEMA's 100-year base flood elevations (for details, see Appendix B). For convenience, the base flood elevations were rounded, within the boundaries of New York City only, to the nearest full foot. The method involves averaging the flood heights into an average flood elevation[31] for each of the waterways surrounding New York City, as depicted in **Figure 9.8** and listed in **Table 9.13** in Appendix B.

To determine the risk that flooding poses to transportation infrastructure, the elevation of the structures relative to the elevation of the floodwaters according to FEMA's 100-year flood maps were analyzed. The new flood zones that account for the anticipated 2- and 4-foot sea level rise were then used to assess the vulnerabilities of transport structures and systems.

Note that the original base flood elevations from FEMA's Flood Insurance Rate Maps are generally (at least for New York) referenced to the National Geodetic Vertical Datum of 1929 (NGVD, 1929). The investigators chose, however, for their newly computed, averaged sea level rise-dependent flood-zone elevations to reference to the more recent, and now more commonly used, North American Vertical Datum (NAVD, 1988). Note that in contrast to FEMA maps in New York, FEMA maps for New Jersey use the NAVD 1988 datum. A constant difference of 1.1 feet between the two datums was applied throughout the New York City area such that the numerical elevations above the two vertical datums relate to each other by Equation 1:

Equation 1. Elevation (ft) above NAVD'88 = Elevation (ft) above NGVD'29 - 1.1 ft

The area-weighted average base flood elevations (in the NAVD 1988 reference frame) were, for the New York City waterways, rounded to the nearest integer foot for assessing the flood and sea level rise impact on transport in the region. The averaged flood elevations, Z_i, were then compared to the lowest critical elevations (LCE) of the transportation systems.

In the regions outside New York City, including Long Island (Nassau and Suffolk Counties), Westchester County and the Lower Hudson Valley, and Connecticut, more generalized approaches were used, for a number of reasons. First, no high-resolution digital elevation model with a 1-foot vertical resolution was uniformly available for these regions outside of New York City (Suffolk County is an exception). Additionally, for these areas, the lowest critical elevations are not known for many of the transportation systems and related structures as well as they are known within New York City. The New York City estimates were largely obtained from the Hurricane Transportation Study (USACE, 1995), and the metropolitan east coast (MEC) climate change infrastructure study (Jacob et al., 2000, 2001, and 2007).[32]

This lack of basic information points to the need for accurate, accessible digital elevation models in all the storm-surge-prone coastal zones of New York State. These models need a vertical resolution of substantially less than 1 foot.

Case Study Results for General Inundation Patterns

A 2-foot rise in sea level would have significant impacts in many parts of New York City, and especially along the Brooklyn and Queens shorelines, around Jamaica Bay, and on the Rockaway Peninsula. As shown in **Figure 9.7,** the increase in additionally flooded area from a 2-foot rise to a 4-foot rise in sea level is less significant than the increase in flooded areas from current sea level to the first 2-foot sea level rise. This is a result of the topography of the area and has to do with the presence of glacial landforms. In the subject regions, the terrain slopes between the 100-year base flood elevations (at current sea level) and the next 2 feet of higher elevations tend to be minimal, while terrain slopes tend to become steeper at elevations above base flood elevations of 2 feet. This is typical for former flat glacial-outwash regions. They are interspersed with remnants of glacial end moraines that stand above the plains and are now coastal flats or marshes, after more than 400 feet of sea level rise during the last 18,000 years of glacial retreat.

The areas indicated as additionally flooded zones under the 2- and 4-foot sea level rise scenarios in **Figure 9.7** will be flooded only if protective measures such as levees and/or sea walls are not kept in good repair where available or newly constructed. Such measures could diminish the additional flooding, but issues of sustainability (discussed below) will need to be considered.

A sea level rise of 2 or 4 feet will cause more streets to be flooded during a coastal storm surge (**Figure 9.9**).[33] The increase in total length of streets flooded during the first 2 feet of sea level rise over the current sea level is almost twice as much as the increase in the total length of streets flooded during the second 2 feet of sea level rise (from 2 feet to 4 feet of sea level rise).

Flooding of Transportation Infrastructure and Expected Impacts

Flooding of city streets affects the flow of vehicular and pedestrian traffic, parking patterns, and many of the

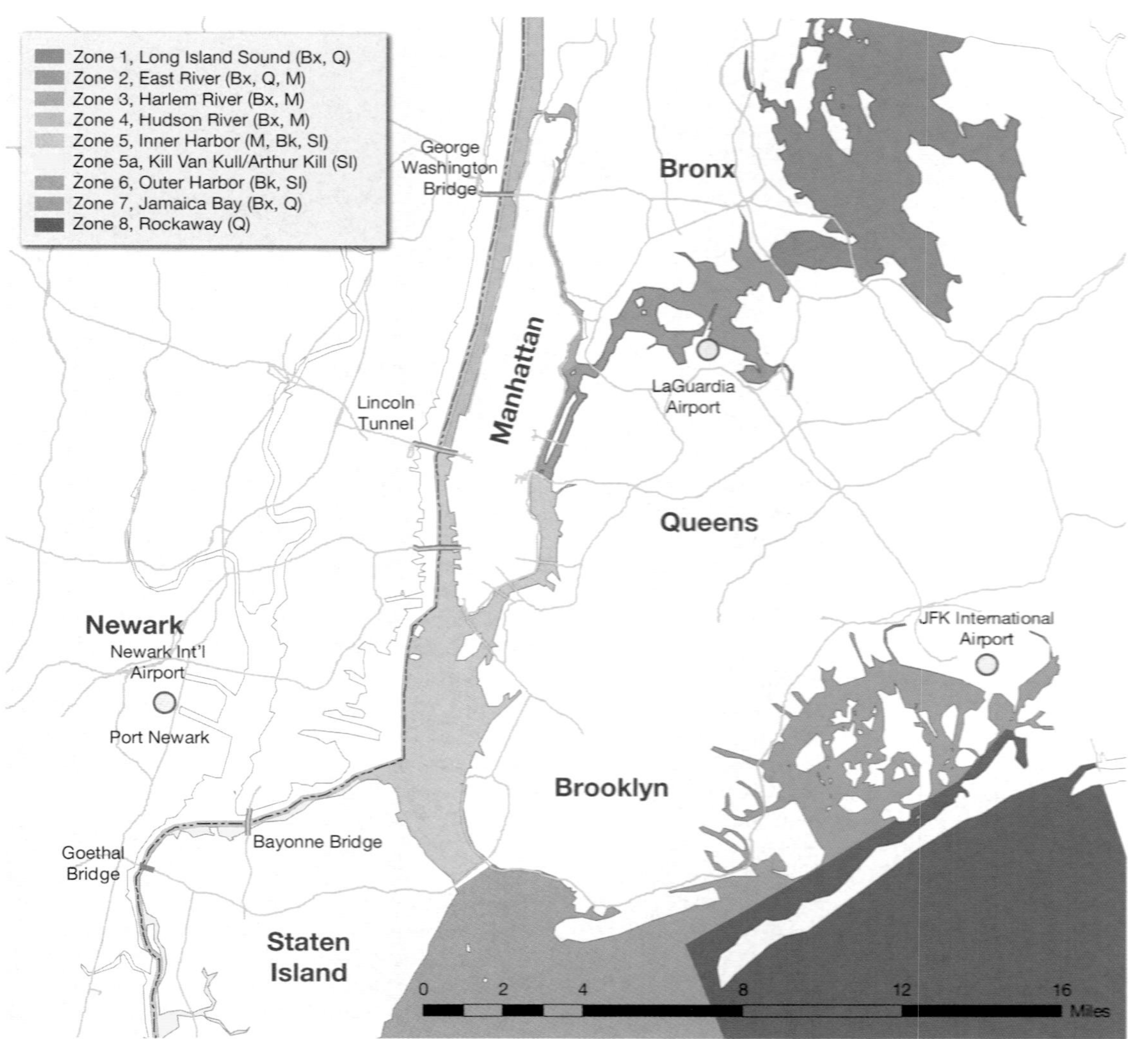

Note: Legend gives color code and ID number for the nine zones, and lists the boroughs (in parenthesis) that share a shoreline with the zone. The obtained numerical AW BFEs (Z values) are listed in Table 9.13 (Appendix B). Legend abbreviations are: Bx = Bronx, Q = Queens, M = Manhattan, Bk = Brooklyn, SI = Staten Island. Sources: Base map, The Port Authority of New York and New Jersey, 2001; Overlays, by ClimAID team

Figure 9.8 Delineation of waterway zones for which area-weighted base flood elevations (AW BFE) are calculated

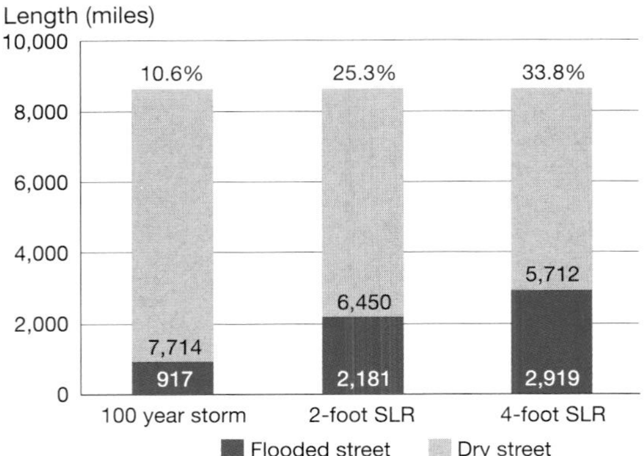

Note: The length of flooded streets that fall into the three flood zones increases from 917 to 2,181 to 2,919 miles, or from 10.6, to 25.3 and 33.8 %, respectively, of the total NYC street length, which measures about 8,600 miles.

Figure 9.9 Total length (miles) of NYC streets that fall inside (blue) and outside (gray) the respective flood zones as a function of sea level rise, for current sea level and with 2-ft and 4-ft sea level rise, respectively

MTA-NYCT bus routes. It also can slow access by first responders and emergency vehicles.

There are currently no centralized GIS- or CAD-based (computer-aided design) or other digital databases that show the as-built elevations of engineered transportation systems, including subways, railroads, state highways and/or major bridge access ramps, roads, toll plazas, tunnels, and airports and seaports in the New York City metropolitan region. It is generally very difficult to compile the needed information from hardcopy blueprint plans that must be retrieved from archives one by one. ClimAID transportation stakeholders, and in particular the Port Authority, MTA, NYSDOT, and NJ TRANSIT, provided the needed information, in addition to data already available from the U.S. Army Corps of Engineers (USACE, 1995), including tunnel elevations, volumes, pumping capacity, etc. As climate adaptation efforts advance, the assembly of such a database on a statewide basis is urgently needed, but should at least be developed on an agency-by-agency basis. This points to data and information needs we address later.

Focus of Impact Analysis and Assumptions

The only structures considered in this case study are those that are near or below sea level and are potentially vulnerable to coastal storm surge inundations. Where available, the lowest critical elevations are listed, indicating the elevation at which water will inundate a portion or all of a given structure if storm surge waters reach it. Water damage at these elevations is likely to occur and operation will be impeded. Lowest critical elevations are given in feet and are referenced to NAVD 1988.

The case study assumes that no adaptation or protection measures are taken now or in the future, unless indicated. Implementation of any structural or protective adaptation options or, in some cases, operational protective emergency measures, could diminish to various degrees the extent and impact of flooding depicted here.

Tunnels and Underground Structures

For tunnels and other underground structures, once storm waters reach the lowest critical elevation, water will flow down into the tunnel or underground structure. If the floodwaters stay above this critical elevation for sufficiently long, the tunnel and connected structures can fill completely to at or below the lowest critical elevation.[34]

The flood potential of the transportation systems listed below can be inferred by comparing the flood scenarios for the respective waterways listed in column 4 of **Table 9.13** (Appendix B) with the lowest critical elevations given, to determine whether the base flood elevation (2 and 4 feet, respectively) exceeds the lowest critical elevation (see **Table 9.13**, Appendix B).

Note that all elevations are uniformly relative to the NAVD 1988 datum. For all listings below, it can be inferred in conjunction with the data from **Table 9.13** whether:

- the lowest critical elevation is at or below the area-weighted average (Z_i) (or below Z_i for the 2-foot or 4-foot sea level rise scenarios, respectively), implying that the structure is within the 100-year flood zone for the given sea level, now or in the future; or
- the lowest critical elevation is above the area-weighted average (Z_i) (or above Z_i for the 2-foot or 4-foot sea level rise scenarios), implying that there is no 100-year flood hazard for the structure under the given sea level scenarios.

When the flood potential of a structure located outside New York City and outside the mapped waterways is assessed where no area-weighted average value Z_i was computed, the current FEMA 100-year base flood elevation is used directly (corrected for NAVD 1988 datum where needed) to allow similar inferences.

These methods were used to assess the flood potential for each of the structures discussed below.

New York City Transit Subway System

Most of the tunnel flooding analysis focused on the following three areas (**Figure 9.10**):

- Downtown Manhattan, with tunnels connecting below the East River to Brooklyn (six river crossings) (**Figure 9.11**)
- Midtown East Side Manhattan, with four tunnels crossing below the East River to Queens (Long Island City) with one nearby additional river-crossing tunnel segment (**Figure 9.12**) across the Newtown Creek at the boundary between Brooklyn and Queens
- Uptown Manhattan, with three tunnels crossing beneath the Harlem River into the Bronx (**Figure 9.13**)

The ClimAID subway flood study was cross-checked against an MTA-internal flood mapping effort,[35] which was carried out in 2006 for developing storm emergency plans. The study modeled the effects of the storm surge heights for "worst track" (i.e., direct hit) hurricanes of categories 1, 2, 3, and 4 as given by USACE (1995) based on NOAA's SLOSH (Sea, Lake, and Overland Surges from Hurricanes) computations then available. We reproduce here only the MTA map for the category-1 hurricane scenario (**Figure 9.14**).

This storm scenario has coastal storm surge elevations roughly comparable to the 100-year coastal storm

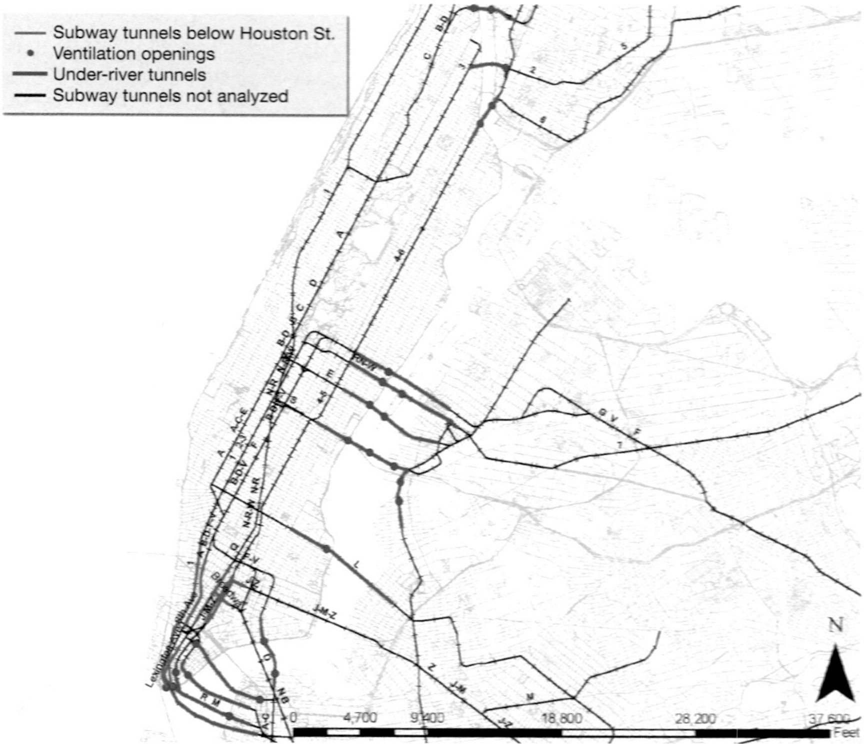

Note: Only the colored segments were considered in the ClimAID flood analysis. Purple lines are river-crossing tunnels and adjacent segments to the nearest land stations; green lines are analyzed tunnel segments on land, near or beneath flood zones. Subway stations are not shown, but non-station openings subject to potential flooding are indicated by circles. Source: LDEO/Civil Engineering, Columbia University

Figure 9.10 NYCT subway lines analyzed in the case study

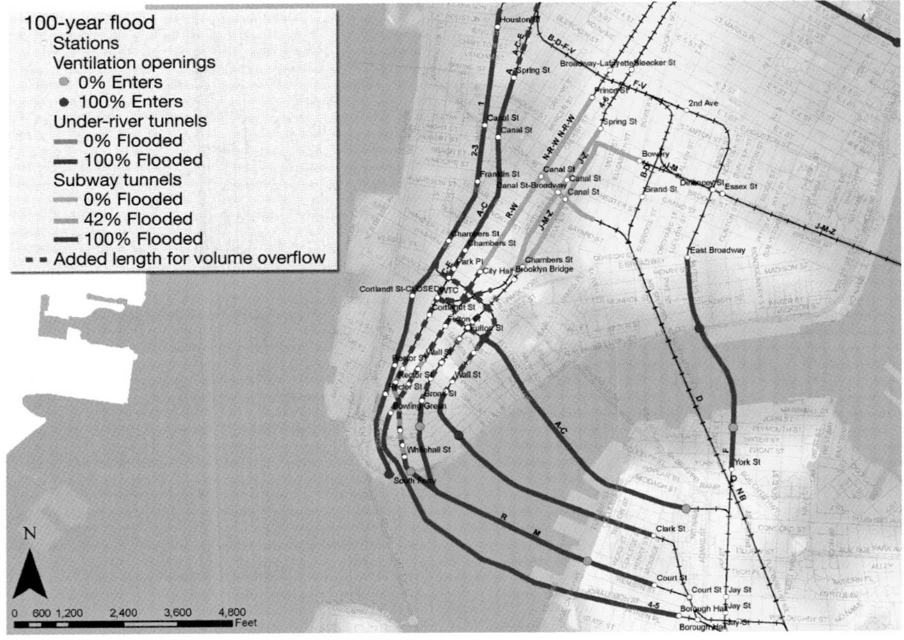

Source: LDEO/Civil Engineering, Columbia University

Figure 9.11A 100-year flooding without sea level rise of Lower Manhattan subways and adjacent East River tunnels crossing to Brooklyn; the heavy blue lines indicate fully flooded tunnels, and broken lines show overflow into tunnels located in areas that are not flooded above-ground; background colors show topographic surface elevations (yellow≥30ft)

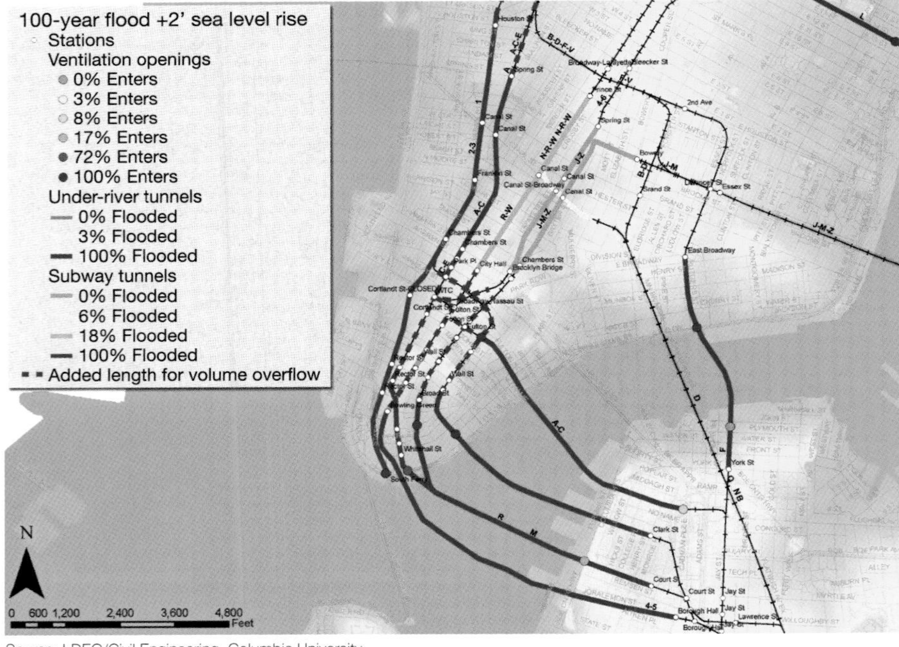

Source: LDEO/Civil Engineering, Columbia University

Figure 9.11B Same as A, but with 2-ft sea level rise; light blue lines are partially flooded

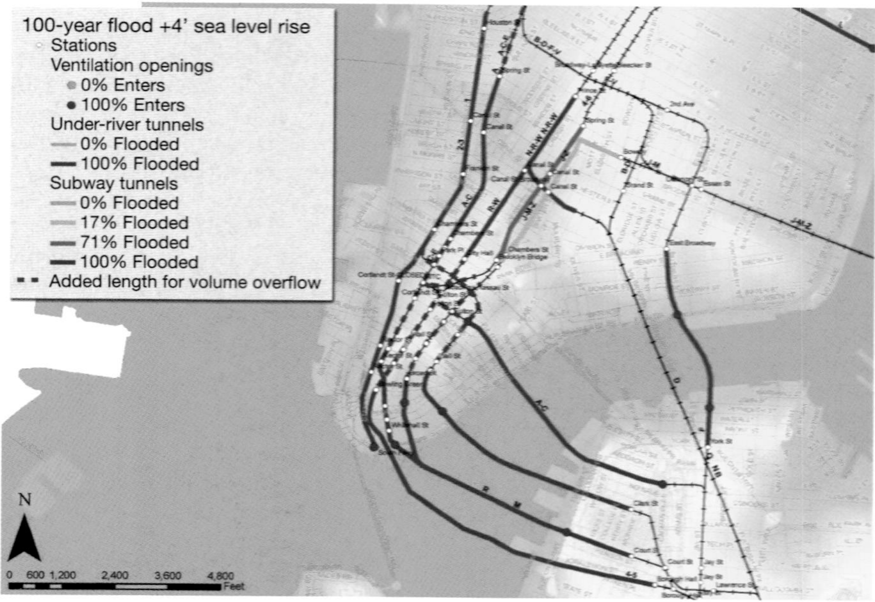

Source: LDEO/Civil Engineering, Columbia University

Figure 9.11C Same as A, but with 4-ft sea level rise; blue lines show additional partial or full flooding near Canal Street (Lines 4-6, J, M, Z); in all three cases East River tunnels for the 4, 5, R, M, 2, 3, and F lines are fully flooded

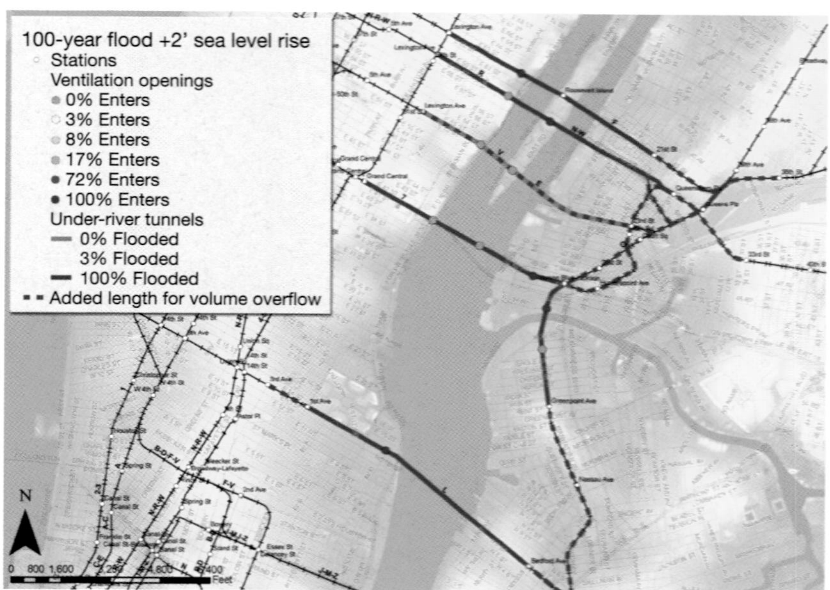

Note: Symbols as in Figure 9.11. The ClimAID team also performed hydraulic computations for S1 and S3 scenarios (no and 4ft SLR). They are omitted for brevity. Source: LDEO/Civil Engineering, Columbia University

Figure 9.12 100-year flooding with 2-ft sea level rise of Midtown Manhattan subways and tunnels across the East River to Brooklyn (L line) and Queens (F, N-W, V-E, and 7 lines), and across the Newtown Creek between Queens and Brooklyn (G line).

surge of the ClimAID case study S1, without sea level rise, but the MTA study assumed that the maximum flood height would be sustained sufficiently long to fill the tunnels to the full surge elevation. Therefore the map shows the maximum extent of tunnel flooding possible for the MTA category-1 hurricane scenario. Nevertheless, the map (**Figure 9.14**) shows a striking similarity in tunnel flooding extent to the ClimAID maps (**Figures 9.11** to **9.13**), despite the different coastal storm surge elevation patterns for this hurricane-1, storm-track-specific scenario used in the MTA study, and the more elaborate, time-dependent hydraulic flooding computations by the ClimAID team. The findings of very similar results of the two studies using different storm surge patterns and methodologies support two important points:

- It provides some validation of the results of either study carried out entirely independently from each other.
- It shows that, to a first order, the subway system in certain low-lying areas is flooded or not flooded depending on whether the flood surge height exceeds the critical ground elevations of 8 to 9 ft (NAVD, 1988). Any additional flood elevations somewhat extend the underground reach of the tunnel flooding, but not by very much. The reasons for this similarity of outcomes are twofold: 1) the effect of topography (discussed in more detail in Appendix B; extreme flood heights such as from category-3 or -4 direct-hit hurricanes would, however, extend the flooding considerably, especially on lines with modest tunnel climbing slopes); and 2) flooding of the tunnels occurs very fast to virtually the full height that the time-dependent storm surge elevations allow (see Appendix B).

One major difference between the MTA and ClimAID flood analyses is that ClimAID calculated, using hydraulic equations, the water entering the subway system as a function of time-dependent surge behavior. This approach tells how fast the tunnels are flooded, and how fast and far the flooding will spread, dependent on the amount of water that can enter the system, as long as the surge height is above the tunnel opening's lowest critical elevation (LCE). The LCE can be a station entrance, emergency exit, ventilation

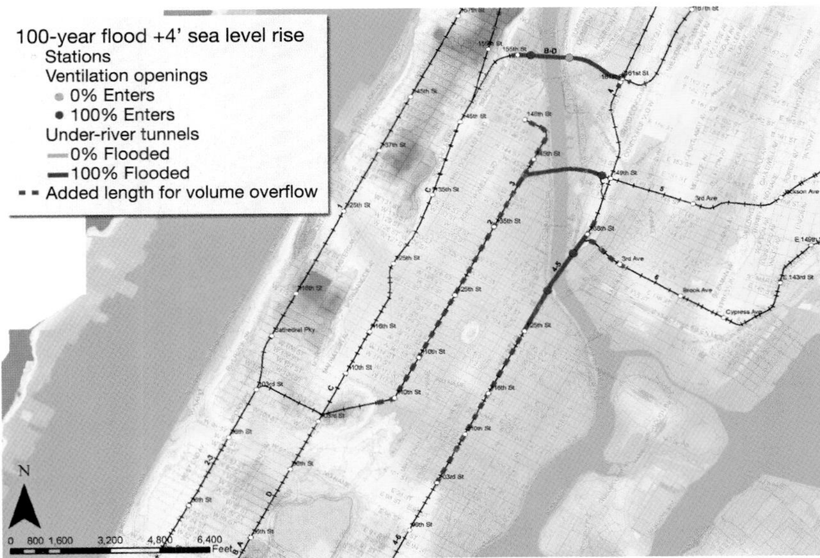

Note: The most northern tunnel (B-D lines), flooded for the 4-ft SLR scenario shown here, does not flood without SLR, and only partially floods with 2-ft SLR. Also the overflow extending towards midtown Manhattan is entirely absent along the #2/3 lines for 0 and 2-ft SLR, but extends for all 3 storm scenarios to 103rd Street along the #4, 5, and 6 lines. Subway tunnel and track elevations tend to follow the surface topography (track elevations are typically about 20ft below street grade; but many exceptions exist). It is therefore not surprising that where surface topography reaches an elevation of about ≥30ft (yellow shading), it creates a "dam" against underground flooding to proceed further, at least for these three storm scenarios [30 ft (topography) – 20ft (track depth) = 10 ft (approximate surge height)]. The ClimAID team also performed hydraulic computations for S1 and S2 scenarios (0 and 2-ft SLR). Blue shadings: ≥60-ft elevations. Source: LDEO/Civil Engineering, Columbia University

Figure 9.13 100-year flooding with 4-ft sea level rise of Uptown Manhattan subways and tunnels crossing into the Bronx beneath the Harlem River

shaft, or string of street-level ventilation grates or, as in most cases, combinations thereof.

The ClimAID hydraulic calculations show that *in most instances the tunnels fill up in less than 1 hour* as long as outside flood heights exceed the LCE, almost regardless of by how much. The total volume of water that needs to be pumped from the tunnels is discussed below.

The MTA analysis (**Figure 9.14**) for the category-1 Hurricane Flooding Scenario assumes that the flood surge and the corresponding high water level takes place for several hours; in other words, there is ample time for the subway flooding to occur, at least without any prevention response (e.g., possible sandbagging, or covering of vulnerable entry points such as entrances, vents, emergency exits, etc.). Thus, the extent of flooding depicted in **Figure 9.14** could be considered the "worst case" scenario for NYCT's system flooding for a category-1 hurricane.

Neither the MTA nor the ClimAID flood analyses take into account, however, recent ameliorative measures begun by the MTA, on a location-by-location basis, to address the propensity for storm-related flooding. For instance, planning is currently under way within the MTA to raise the Harlem River seawall along the 148th Street and Lenox Avenue subway train yard to protect the subway portal to the tunnels at that location. A program to raise ventilation grates to prevent water entry is also under way at some locations subject to recurrent flooding from high precipitation events.

Highway and Non-Subway Rail Tunnels

Discussed in this section are the potential flooding impacts to highway and non-subway rail tunnels. Critical parts of the road and rail system are vulnerable to flooding from sea level rise.

Highway Tunnels

There are four major highway tunnels connecting Manhattan with two other NYC boroughs: the Brooklyn-Battery (LCE=7.5 feet, Z1=9 feet) and Queens-Midtown (LCE=9.5 feet, Z2=11 feet) tunnels across the East River and its extension into the NY Inner Harbor (for locations see **Figures 9.2** and **9.3**); and two highway tunnels that connect Manhattan with New Jersey beneath the Hudson River, i.e., the Holland (LCE=12.1 feet*,[36] Z5=9 feet) and Lincoln (LCE= 22.6 feet*, Z5=9 feet) tunnels. The Lincoln tunnel has three tubes; all others have two tubes, with two lanes per tube.

Railroad Tunnels

In addition to the subway tunnels, the following river-crossing railroad tunnels exit from Manhattan and are used by Amtrak, Long Island Rail Road (MTA-LIRR), Metro-North (MTA-MNR), Port Authority Trans-Hudson (PATH), and NJ TRANSIT:

- **North River (Hudson) Railroad Tunnel**—The North River railroad tunnel has two tubes from New Jersey into Penn Station used by Amtrak and NJ TRANSIT. The tubes are connected into Penn Station, and therefore flooding could also potentially affect LIRR facilities in Penn Station and the West Side Rail Yard (LCE = 8.9 feet, Z5=9 feet). The top-of-rail (track) elevation in Penn Station is below sea level (LCE = -7.4 feet).
- **Two Pairs of PATH Tunnels**—Two pairs of PATH tunnels cross beneath the Hudson River with LCE=9.9 feet*, and Z5=9 feet. The critical elevations are located in New Jersey and imply closing the installed floodgates at the Hoboken station (without the Hoboken station flood gates, LCE would be 6.5 feet). Parts of the PATH system, both in Manhattan, and the much longer, also entirely below-ground system in New Jersey, are in their current configuration nominally flood prone, once the surge exceeds the LCE at various locations. Also note that PATH stations have internal passages that connect to NYCT subway stations along 6th Avenue at 14th, 23rd, and 33rd Streets, Manhattan.

  All PATH tunnels are interconnected in New Jersey and extend below grade into the Hackensack/Passaic River basin subject to tides and coastal storm surges. Several projects are currently under design to locally raise LCEs for some of the system openings (e.g., Washington Street Powerhouse and 15th Street Shaft, both in New Jersey). Until *all* lowest critical elevations within the system are raised above the respective base flood elevations plus sea level rise, the system may still remain vulnerable to floods, albeit may flood more slowly and hence potentially with less water to pump out.

* All LCE with an asterisk* attached are dependent on emergency operational measures (e.g., by sealing ventilation shaft doors etc.).

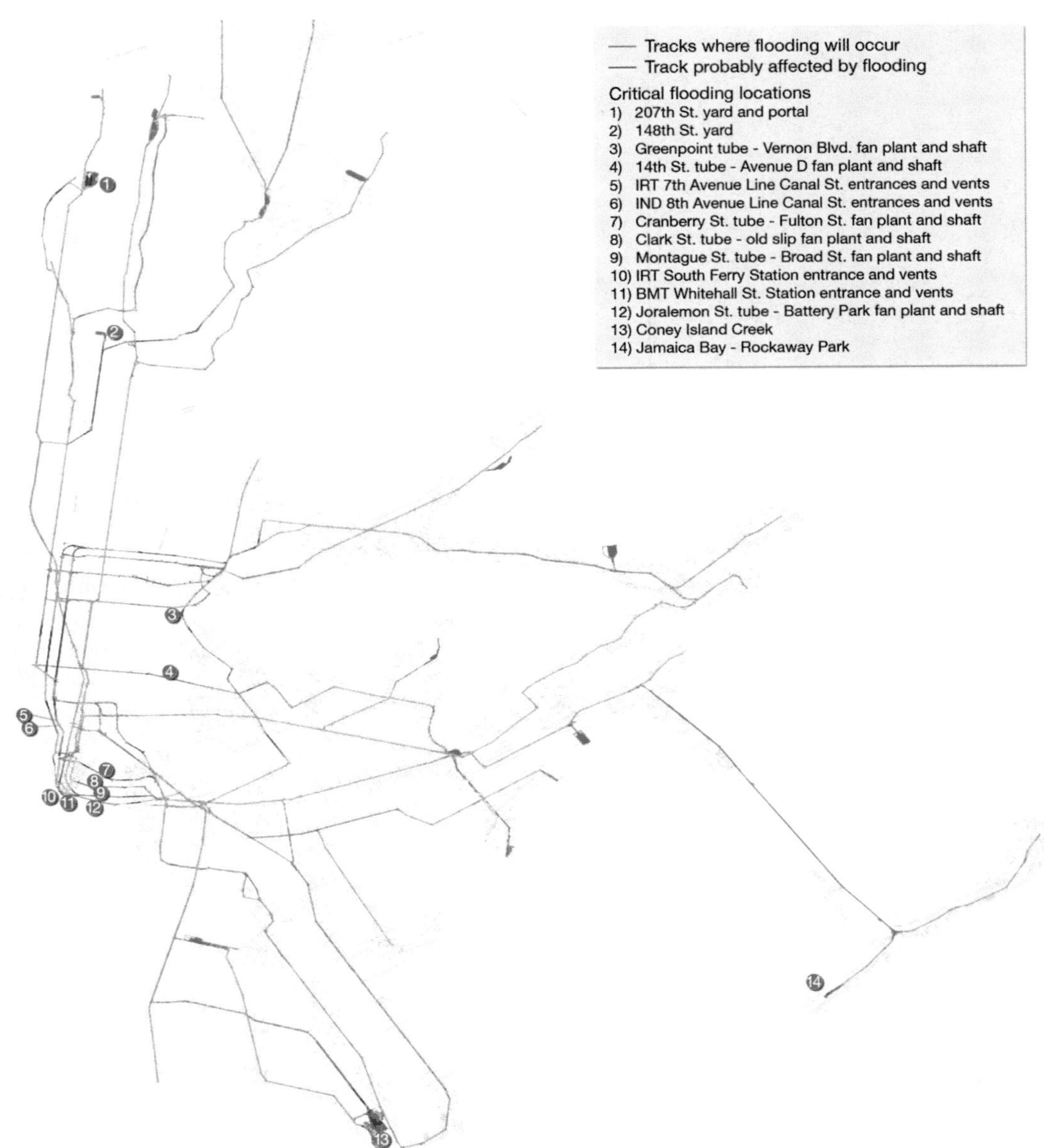

Note: Red symbols indicate definite flooding, blue ones potential flooding, the latter corresponding closely to the "overflow" segments in Figures 9.11–13. Green areas near the line tracks indicate locations of yards. Two yards along the Harlem River (upper left) are flooded, and so are facilities in Coney Island and Rockaway (bottom center and right). Source: MTA—NYCT, 2006, A. Cabrera (see Endnote 35)

Figure 9.14 MTA subway flooding map for a category-1 hurricane based on surge heights listed in USACE (1995)

The ClimAID team did not have proper terrain data (detailed digital elevation model data) to verify the flood potential of all New Jersey-based PATH stations and other potential entry points. While the Port Authority provided the lowest critical elevations for all stations, the topography by which the floodwaters may reach these potential entry points needs further investigation to fully assess their flood potential under the 100-year storm height and the 2-foot and 4-foot sea level rise scenarios. According to FEMA's Flood Insurance Rate Maps, several PATH system entry openings appear to be flood-prone. But it should be noted that during the December 1992 nor'easter storm, only the Hoboken station was flooded (then LCE=6.7 feet), while the next lowest entry point, Exchange Place (LCE=7.6 feet), was not. The World Trade Center PATH station, in New York, and adjacent terrain are currently in a state of reconstruction and, therefore, their current and future flood potentials are highly uncertain at this time. Based on the previous lowest critical elevation of the PATH system (USACE, 1995), this system appears to be flood-prone at the various states of sea level rise without additional protective measures. A more detailed analysis of all PATH entry points with updated digital elevation models and floodways is needed to better understand the flood vulnerability under current sea level and future sea level rise scenarios.

Flood vulnerability in tunnels varies, depending on whether adaptive or preventive structural (or even just operational emergency) measures are undertaken. They can best be implemented where only a limited number of openings provide flood access to the underground structures and systems. Such engineering measures are the prototype model for effective, albeit perhaps temporary, adaptation to sea level rise for tunnel systems with closed ventilation. This is unlike the ventilation system of the New York City subway, which is largely open to, and connected with, the street grade

Other Tunnel Systems

- **East River Tunnel**—This railroad tunnel is used by Amtrak and LIRR. It has a lowest critical elevation of 7.9 feet (Z2=11 feet), located in Long Island City, Queens. The tunnel provides an access route westward across midtown Manhattan into Penn Station and could potentially lead to flooding there into LIRR, Amtrak, and NJ TRANSIT facilities. The North River (Hudson) and East River tunnels and Penn Station have sump and ejector pump systems. Penn is also protected from flooded river tunnels by floodgates at the east and west ends of the station.
- **Access to the Region's Core (ARC) Mass Transit Tunnel**—A new tunnel system, the ARC Mass Transit Tunnel across the Hudson River, is currently under construction.[37] It will increase the capacity for NJ TRANSIT commuter trains and more than double commuter rail capacity between New Jersey and New York. The ARC project also includes a new expansion to Penn Station, New York. There is no interconnection between the ARC Mass Transit Tunnel tracks and the existing New York Penn Station tracks. However, NJ TRANSIT is building a pedestrian connection between the expansion and the existing Penn Station (the LCE of the pedestrian connection is 9.7 feet, while Z=9.0 feet, both relative to NAVD 88). Therefore the new Penn Station extension may become vulnerable to flooding via the pedestrian connector to the existing Penn Station for the scenario that assumes a 2-foot sea level rise (S2), or whenever sea level rise exceeds 0.65 feet. The ARC rail tunnel itself has the same LCEs (11.553 feet at both its Hoboken and the 12th Avenue, NYC, shafts). It therefore may become directly vulnerable to flooding from either end for a 100-year flood for scenario S3 (which assumes a sea level rise of 4 feet) or whenever sea level rise exceeds about 2.5 feet. Modifications to the Hoboken and 12th Avenue shaft designs and to the pedestrian connector design may have to be made to avoid future flooding on either side of the Hudson.
- **The 63rd Street Tunnel**—Another new railroad tunnel under construction is the MTA-LIRR's 63rd Street Tunnel. It crosses the East River as part of the East Side Access Project. Construction began in 1969 and the tubes making up the river-crossing tunnel were in place in 1972. The tunnel runs from the intersection of 63rd Street and 2nd Avenue in Manhattan to the intersection of 41st Avenue and 28th Street in Queens. The tunnel can accommodate four tracks on two levels (two for the subway on the upper level and two for the LIRR on the lower level). The MTA connected subway lines to the tunnel in 1989. The current East Side Access Project will build new tunnels in Manhattan to connect the LIRR portion of the 63rd Street Tunnel to Grand Central Terminal and the LIRR tracks in Queens. This connection brings the LIRR into Grand Central Terminal. The original 63rd Street Tunnel (used only for the B&Q subway lines) has an LCE of 11.6 feet (Z2=11 feet) on the Queens side. The new LCE is unknown at this time. The new LIRR train platforms in Grand Central Terminal will be at levels below the Metro-North track. Grand Central Terminal's current flooding potential is via the Steinway subway tunnel across the East River (42nd Street, No. 7 Line) (LCE=9.9 feet; Z2=11 feet).

Notes: The red arrow shows the lowest critical elevation, LCE=6.6 feet, of the rail tracks located in waterway zone Z3=9 feet. The elevated concrete structures are the passenger platforms at an elevation near 11 feet. Note the low-lying parking lot in background.
Source: http://en.wikipedia.org/wiki/Spuyten_Duyvil_(Metro-North_station)

Figure 9.15 Lowest critical elevation of the MTA-Metro-North Railroad Spuyten Duyvil Station, Bronx, next to the Harlem River

At- and Above-Grade Railroads (Commuter, Passenger, and Freight)

Outside Manhattan, many of the NJ TRANSIT and (below-ground) PATH tracks in the Hudson, Hackensack, and Passaic River Basins are flood prone, as demonstrated by the December 1992 nor'easter (USACE, 1995). MTA Metro-North trains can encounter flood-prone segments. Examples are near Spuyten Duyvil on the Harlem River (Bronx; LCE=6.6 feet; Z4=8 feet, see **Figure 9.15**) and Croton on Hudson (Westchester County, LCE=5.2 feet; 1%BFE=5.9 feet in 2000) for the Hudson Line. The LIRR may encounter flooding in Oceanside (Nassau County; LCE=8.5 feet; near Z7=8 feet; 1%BFE=6 feet) along the Long Beach Line; at Flushing (Queens, LCE=8.1 feet; Z2=11 feet) for the Port Washington Line; at low points along the Far Rockaway Line (LCE=8.1 feet; Z7=8 feet); and at the Oyster Bay Station (Nassau County, LCE=8.4 feet; near Z1=14 feet).

Hell Gate is a massive railroad bridge over the East River, connecting Astoria (Queens) with the now-joined

Notes: For Manhattan and parts of the Bronx and Queens (red=100-year base flood elevation at pre-2000 sea level; yellow=2-foot sea level rise scenario; and green=4-foot sea level rise scenario). The red-colored water-flooded areas in the Hudson River represent 9-foot sea levels (all measured in NAVD, 1988). The black lines represent railroads; the colored lines indicate various subway lines. Note that many of the railroads and subways traverse the outlined flood zones or natural bodies of water. For the details of their lowest critical elevations relative to the flood elevations, grouped by waterways in Table 9.13, see text. Source: Image from Google Earth (©2009 Google; ©2009 Tele Atlas; Image ©2009 DigitalGlobe; Image ©2009 Sanborn; Image ©2009 Bluesky); added data by ClimAID team

Figure 9.16 The Hudson, East, and Harlem Rivers and adjacent flood zones

at the same time is logistically possible. Therefore, five days is the minimum amount of time it would take under a best-case scenario; one week per tunnel is, perhaps, more realistic. The river subway tunnel operations alone would require 56 powered mobile pumps (four in each of the 14 tunnels) (see subsequent sections in this case study).

Assuming that the land-based tunnels can be pumped out more or less during the same time as the generally deeper river-crossing tunnels, the operation may need something in the order of 100 such pumps if pumping is to be achieved within one week. A smaller number of pumps, or not pumping all tunnels simultaneously, would lengthen the pumping time required.[42]

Rigorous, engineering-based assessments, combined with logistic management plans of how to procure such pumping capacity simultaneously, are urgently needed that can determine more precise estimates of the pumping system needs for New York City metropolitan-area tunnels.

The environmental impacts on the waters in the New York Harbor estuary from the simultaneous pumping activities could be significant and would be in addition to those from the debris and spills from surface sources, including toxic sites that were reached by the floodwaters. It is assumed that environmental emergency permits for disposing of the pumped tunnel waters are pre-event approved and would require no extra processing times. If pre-event approved permits do not exist, then additional delays may need to be assumed.

Such a storm as analyzed in the ClimAID assessment not only damages flooded tunnels, but also affects external support systems (power, communication, logistic preparations) needed for the pumping operations, subsequent inspection of damage in the tunnels, and to make the necessary repairs. The total projected outage times for transportation systems are summarized in **Table 9.5**.

Scenario	Flooded Tunnel Volume	Flooded Tunnel Length
S1 1%/y BFE*	400 million gallons	60,000 ft
S2 +2ft SLR	408 million gallons	60,600 ft
S3 +4ft SLR	411 million gallons	61,000 ft

* BFE = base flood elevation

Note: Flooded tunnel volume and flooded tunnel length for each of the S1, S2, and S3 sea level scenarios.

Table 9.6 Estimated total volume of flood-prone subway tunnels

The estimates of recovery times given in **Table 9.5** remain highly uncertain and may change substantially when the necessary engineering vulnerability and risk assessments of complex systems are performed in sufficient detail and when the emergency response capability of transportation operators can be quantified. Such assessments may take years for some of the more complex and older transportation systems, where the as-built or current state of repair information is not always readily available. Each operating agency will need to make these assessments in years to come before a more realistic picture will emerge for the expected damage and costs to the operating agencies and of the economic impact to the public (see Section 9.5.7).

For instance, there are likely to be other significant restraints on the ability of the NYCT subway system to recover from flooding that have not been incorporated into this analysis. Even if emergency pumping can be implemented, the impact of salt, brackish, and/or turbid water will last long after the water itself is removed. Deposits will need to be cleaned from signal equipment and controls, which may need to be replaced either in total or by component, and only very limited service could be provided after pumping is completed until signals are restored. Much of the equipment in the subways is of a specialized nature that requires orders from manufacturers with long lead times, especially for significant quantities. There probably are not enough personnel trained to rebuild and refurbish equipment simultaneously in multiple subway lines even if the equipment could be procured. There is some existing equipment that, if damaged, cannot be replaced because it is obsolete and is no longer manufactured, nor are there replacement parts for it. Such equipment would have to be redesigned and then installed—a process that can take a long time.

Finally, if significant soil movement or washouts occur, it is likely that structures throughout the system may experience some settlement, and there could be structural failure of stairs, vent bays, columns, etc.

Together, such conditions could easily extend the time it takes to restore to a 90-percent functionality of the subway system (**Table 9.5**) by three to six months (and perhaps longer). It is estimated that permanent restoration of the system to the full revenue service that was previously available could take more than two years.

In general, adaptation options (see sections 9.4, 9.6.2, and subsequent sections of this case study) will need to be carefully evaluated to arrive at a better understanding of the resources that will be needed to make the coastal and estuarine New York State transportation systems resilient to all types of climate change impacts, and to sea level rise in particular.

Methods for Calculating Restoration Time to 90 Percent of Functionality (T90, measured in days)

Table 9.5 represents ClimAID's best effort to combine stakeholder-provided information and publicly available data into outage/restoration time estimates. It is the basis for the case study, and contains key information, in compact numeric form.

The restoration time T90, after which a transportation system regains 90 percent of its pre-storm functional capacity, is computed for various transport systems as follows (see red numbers in columns 4, 5, and 6 in **Table 9.5**):

Equation 2. T90 (days) = Max{D, E, L|P>0} + Max{P, A, R} ≥ 1

All units are in days. The operator Max$\{x_1, x_2, x_3\}$ chooses the largest value of the values xi , where **D** is the surge duration; **E** is the electric grid restoration time; **L** is logistic set-up time (note that L|P>0 means that L is only counted when there is a finite pumping time P>0; otherwise L=0 since there is no logistic set-up time when pumping is not needed); **P** is pumping time; **A** is damage assessment time; and **R** is repair time. The maximum (largest value) rather than the sum of **D, E, L** is chosen since it is assumed that these times run largely in parallel, rather than being additive, although this choice may lead to underestimation of outage times from these causes.

A similar parallel set of activities is assumed between **P**, **A**, and **R**, although that may be even more optimistic. A minimum of T90≥1day is imposed on all facilities, assuming that even if all six variables were close to zero, the public would avoid using transport for general economic activity (businesses may be closed) on the day of the storm, and mass transit would largely be reserved for emergency evacuation according to NYC's emergency plans. For road tunnels the time for accessibility by emergency and essential traffic (repair crews, utilities, etc.) may be shorter than those shown, which are meant to indicate when the facility becomes operational for the general public. In **Table 9.5**, rows 1–4 address the first term, and rows 5–27 the second term of equation 2.

There are large uncertainties with each of these variables, and also for the functional relationships between them. It is possible to devise alternatives to equation 2. **D** is in most cases less than one day, but a stalled nor'easter storm could extend **D** from one to a few tidal cycles (roughly 12 hours apart) to as much as a few days. **E**, electricity restoration time, has been discussed in conjunction with **Figure 9.20**, but could range, for transportation priority customers, between zero and perhaps two days; for certain functions, it can be shortened by the availability of emergency generators. **L** is essentially the time to bring the pumps into place, ready for operation; with proper pre-storm planning it could be almost zero; if no preparations at all have been made, it may easily take a week to get so many pumps from across the nation to New York, especially if adjacent coastal communities have similar demands. **P** and **A** have been discussed above, and **R**, repair time, is highly uncertain and system-specific.

If, for instance in the case of subways, repairs need to be performed on existing relay, signal, and switching gear of older vintage (such as electric controls, pumps, and ventilation systems, which may need to be disassembled, cleaned, dried, reassembled, installed, and operationally tested because replacement by new spares are not an option), **R** may contribute the largest term and associated uncertainty in equation 2. For a new transport system, or a much simpler road tunnel, the **R** time may be shorter than, or comparable to **P**.

All numbers in column 3 are elevations in feet. All numbers in columns 4–6 are time estimates in days. Rows 1–4 are region-wide, generic (not structure-specific) estimations of days, i.e., **D, E, L** contributing to the service outage (except L is coupled to a facility by the operator |>P to whether pumping is needed, P>0; or is not needed (P=0) at any facility listed in Rows 5–27; the |>P operator determines whether L is accounted for when selecting Max{**D,E,L**}. The parentheses {**P,A,R**} in columns 4–6, rows 5–27, contain the days assigned to the delays caused by pumping P, assessing damage A, and repairs R, respectively. The maximum value of the triplet {**P,A,R**} is then added, for each scenario, to the resulting Max{**D,E,L**|P>0} listed in row 4 (for each scenario, columns 3–5; note that the upper bound is listed; for less complicated transport systems lesser values were chosen). This sum is then entered as the bold number T=... in columns 3–5, rows 5–27. This value T constitutes the estimated T90 (days) for each facility and storm surge/sea level rise scenario. Row 30, columns 3–5 list the range of T90 values obtained. These are assigned to T90min and T90max, respectively, as used for economic estimates in this chapter's case study, Appendix C, and Equation 4 therein.

The color code (see **Table 9.5**, footnote) indicates for which coastal storm surge scenario the respective facility becomes flooded (i.e., red for LCE≤ Z_i, orange for LCE≤Z_i+2 feet, green LCE≤Z_i+4 feet); or never becomes flooded (dark grey, LCE>Z_i+4 feet) for the modeled 100-year storms and sea level rise assumptions. The color scheme signals how readily a system/facility floods, from red as most vulnerable to grey as quite safe with orange and green in between.

Table 9.5 displays the results assuming no adaptation or protective measures are undertaken other than those indicated.

In specific cases, adaptation measures can drastically reduce the vulnerability of the systems and facilities. As such, the outage time and resulting economic impact, including fare/toll revenue losses to a system's operator, can be greatly reduced by taking preventative measures. Such protective measures also would avoid some of the damage and limit repair costs.

Economic Impact of the Vulnerability of New York City's Transportation Systems to Sea Level Rise and Coastal Storm Surges: Case Study Results vs. Losses from Hurricane Katrina

The social and economic impacts of a coastal storm with storm-surge flooding can be significant and in some instances long lasting. This has been vividly demonstrated by the extreme case of the effects of Hurricane Katrina on New Orleans in 2005, which cost in excess of $100 billion in losses, social disruptions, and displacements.

However, there are many differences between this ClimAID 100-year storm case study for the New York City metropolitan area and Hurricane Katrina in New Orleans. Portions of New Orleans are as much as 8 feet permanently below the average current sea level. So, once the levees were breached during Katrina, quasi-permanent flooding prevailed. Virtually all of the New York metropolitan area is above, albeit close to, sea level, with the important exception of some underground portions of the transportation and other infrastructure and of some excavated basement structures. Once the lowest critical elevations and/or the pumping capacities are exceeded by the floodwaters, then the physical circumstances simulate those of any inundated below-sea-level community.

Another difference is that Katrina was a hurricane of Saffir-Simpson category 3. As pointed out earlier, the 100-year storm used in this case study is closer to a non-direct but nearby hit of a hurricane of category 1 to 2.

On the other hand, the asset concentration in the New York City metropolitan region (some outside of New York State) is approaching $3 trillion—much larger than that of New Orleans. About half the assets are in buildings and half in infrastructure of all types. The metropolitan region's gross regional product is in excess of $1.466 trillion per year,[43] corresponding to a daily gross metropolitan product (DGMP) of nearly $4 billion per day.[44]

To assess the economic impact of such a storm on New York City, the ClimAID assessment made a number of assumptions. For example, after such an extreme event it is assumed that electricity and the economy come back not suddenly but gradually. The cost of a storm event depends on how quickly the economic activity can be restored. The analysis considers a range of how long this might take under current conditions and the two sea level scenarios, from a minimum restoration time to a maximum. The cost of a storm event must also consider the physical damage to the infrastructure. (For a complete list of assumptions and how the analysis was conducted, see Appendix C).

The procedure, described in Appendix C, yields a "time-integrated economic loss for the entire metropolitan" region (TIELEM), in dollars. Based on this analysis, the economic losses, due to failure of infrastructure systems in the entire New York City metropolitan region, range from $48 billion (current sea level) to $57 billion (2-foot rise) to $68 billon (4-foot rise). Economic recovery times would range from 1 to 29 days (**Table 9.5**). The results of this economic loss analysis are summarized in **Table 9.7**.

To these time-integrated economic losses (TIELEM), one must add the cost of the direct physical damages resulting from the storm. Then the total costs become even greater (**Table 9.5**). Physical damages alone are valued from $10 billion (current sea level scenario) to $13 billion (2-foot rise) to $16 billion (4-foot rise). For details on how the physical damage losses were derived, see Appendix C. Total losses, including both economic activity and physical damages, range from $58 billion (current), to $70 billion (2-foot rise), to $84 billion (4-foot rise) (**Table 9.8**).

Within these estimates there may be unaccounted for numerous other significant constraints on the ability of the transportation systems to recover from climate change-induced incidents. Such constraints include the age of equipment, the availability of replacement parts/equipment, and the need for these in appropriate quantities. These and other currently unknown and/or not-quantified factors could significantly increase climate change impacts in time, labor, and dollars.

Scenario	T90min (days)	T90max (days)	TIELEM ($Billion)
S1 (current sea level)	1	21	48
S2 (2-foot rise in sea level)	1	25	57
S3 (4-foot rise in sea level)	2	29	68

Note: T90min is the minimum amount of time (number of days) needed for the transportation system to regain 90 percent of its pre-storm functional capacity. T90max is maximum amount of time (number of days) needed for the transportation system to regain 90 percent of its pre-storm functional capacity. TIELEM is the time-integrated economic loss for the entire metropolitan region. 2010 assets and 2010-dollar valuation

Table 9.7 Economic losses for the New York City metropolitan region due to current 1/100 year coastal storms and future 1/100 year storms with 2 and 4 feet sea level rise

Scenario	Combined Economic ($ billion)	Physical Damage ($ billion)	Total Loss ($ billion)
S1 (current sea level)	48	10	$58
S2 (2-foot rise in sea level)	57	13	$70
S3 (4-foot rise in sea level)	68	16	$84

Note: 2010 assets and 2010-dollar valuation

Table 9.8 Combined economic and physical damage losses for the New York City metropolitan region for a 100-year storm surge under current conditions and two sea level rise scenarios

The losses summarized in **Table 9.8** do not include any monetary value for any lives lost. There are several reasons for excluding them: 1) it is very difficult to forecast loss of lives since such losses depend on the quality of storm forecasts, emergency planning, warnings, and readiness of the population to follow evacuation instructions and other behavior; 2) given that the New York City Office of Emergency Management and emergency services in the nearby counties in coordination with the New York State Emergency Management Office have extensive coastal storm evacuation plans in place, the loss of lives should be modest; and 3) it is difficult to assess the value of a human life.

The economic losses of Hurricane Katrina on New Orleans illustrate the significant economic impacts a coastal storm and associated storm surge can have. The economic impacts from the storm surge and sea level rise scenarios analyzed in this case study for the New York City area would be comparable with significant impacts and losses to transportation infrastructure.

Vulnerability and Social Equity

The social and economic effects of a 100-year storm would not be distributed evenly. Certain regions would be more likely to cope and recover quickly, while other regions might suffer to a greater degree and over a longer period of time. In general, underlying differences in patterns of poverty, income, levels of housing ownership, and demographics can give some indication of the resilience of an area. These effects are explored in more detail in the Chapter 5, "Coastal Zones", case study. This section builds upon that analysis by delving more deeply into the role of transportation access in mediating the effects of a storm along New York City and Long Island, both in the evacuation prior to landfall and during the resulting stages of relief and recovery.

This analysis illustrates existing transport disadvantages and the types of vulnerabilities that could be experienced with a storm event of this magnitude. It is important to note that, compared to other cities across the country, New York City has addressed these issues extensively as part of comprehensive evacuation plans. The New York City Office of Emergency Management and the MTA have incorporated income statistics and private-vehicle access into estimates of people who would need evacuation. Public information on the evacuation plans has been distributed in 11 different languages (Milligan, 2007).

Nevertheless, evacuation planning in the New York metropolitan region is very much a work in progress as it relates to transport-disadvantaged and special-needs populations (TRB, 2008b). To some degree, this is a result of intrinsic difficulties in managing an urban area as complicated as the New York metropolitan area, with three states and numerous agencies. While the Department of Homeland Security has been forthcoming with emergency planning funds, it has been less so for funding regional evacuation plans. These efforts are evolving slowly (TRB, 2008b).

Fully addressing transport disadvantage is also hampered by the structure of existing service delivery and the nature of the evacuation plans. The New York City Office of Emergency Management has conducted basic mapping of special-needs populations and made this information publicly available, but it does not have a complete picture of the location or needs of these populations and the resources available to them (TRB, 2008b). Furthermore, strategies that have worked well in places like Tampa, Florida, such as a special-needs registry, have not been attempted in New York City, largely because of the size and complexity of the city. The dominant strategy, therefore, is communicating preparedness through social networks, community groups, and community emergency-response teams, an approach that will not reach the many special-needs individuals who are isolated from consistent outreach services (Renne et al., 2009). As a last-resort option for those unable to arrange their own transport, the city offers "311" emergency services that would link individuals with the city's paratransport vehicles or, in critical situations, with fire and police. Still, there are lingering concerns that the paratransport fleet may be too small during any large evacuation (Renne et al., 2009) and that private-sector drivers might not report to work (TRB, 2008b). Further complicating the approach, there may be a conflict of priorities as public services (e.g., emergency personnel, buses) could be pulled away from the epicenter of evacuation to serve piecemeal needs.

The following section describes the broad climate change impacts, transport disadvantages, and transport resiliencies that extend along the coast of New York City and Long Island. Based on estimates generated for the ClimAID case study (and for current sea level), 90-percent-recovery times for specific parts of the New York City metropolitan transport system would vary from a few days to almost a month. This range in recovery would condition the relative regional severity of indirect economic impacts of a coastal storm surge. Those populations and areas dependent on less-resilient parts of the transport system would more likely suffer extended periods of lost wages and curtailed commercial operations. Some of those hardest hit by systemic failures would likely include populations dependent on the New York subway and those commuting to Manhattan by rail from New Jersey (via NJ TRANSIT) and Long Island (via LIRR), and the commuters of the northern suburbs relying on Metro-North Railroad (MNR).

In general, populations and regions with diverse and redundant transport options would more easily cope and recover from transport systems failure. Further hardship would confront transport-disadvantaged populations and regions, including communities constrained by geography to limited transport options, low-income households dependent on public transport, and individuals with limited mobility.

A recent study of environmental inequalities in Tampa Bay, Florida, suggests three census variables as proxies for transport disadvantage: households with no car, households with disabled residents, and households with residents 65 years or older (Chakraborty, 2009). The ClimAID analysis examines the distribution of these variables across the 100-year floodplain of New York City and Long Island to evaluate vulnerabilities and equity effects in the case of a 100-year storm. **Table 9.9** presents a regional comparison of these indicators.

	In Floodplain	Out of Floodplain
New York Coastal Zone		
% older than 65	14.3	11.9
% physically disabled, age 16-64	5.2	5.9
% households without a car	16.3	10.1
New York City		
% older than 65	13.1	11.1
% physically disabled, age 16-64	6.8	6.7
% households without a car	20.8	23.2
Long Island		
% older than 65	15.2	13.6
% physically disabled, age 16-64	4.1	4.4
% households without a car	2.4	2.1

Source U.S. Census 2000; authors' calculations

Table 9.9 Characteristics of transport-disadvantaged populations living in census block groups: New York Coastal Zone and the case study area

	In Floodplain	Out of Floodplain
New York Coastal Zone		
total workers using public transport	63,819	1,764250
total workers using public transport – bus	14,989	372,028
New York City		
total workers using public transport	48,943	1,635,907
total workers using public transport – bus	13,473	350,935
Long Island		
total workers using public transport	14,875	128,344
total workers using public transport – bus	1,515	21,094

Source U.S. Census 2000; authors' calculations

Table 9.10 Total workers living in New York Coastal Zone and using public transport as primary means of getting to work

	In Floodplain	Out of Floodplain
New York Coastal Zone		
% workers using public transport	27.8	42.1
% workers using public transport - bus	6.4	8.9
New York City		
% workers using public transport	44.9	52.7
% workers using public transport - bus	11.8	11.5
Long Island		
% workers using public transport	11.8	11.7
% workers using public transport - bus	1.3	1.8

Source U.S. Census 2000; authors' calculations

Table 9.11 Characteristics of transport-disadvantaged populations living in census block groups: New York Coastal Zone

Mirroring the statewide disparity in vehicle ownership between urban and rural areas, car access in ClimAID Region 4 (Chapter 1, "Climate Risks") heavily favors suburban areas of Long Island (**Figure 9.21**). In the urban centers of New York, rates of households with no car are nearly double those for the state as whole, a fact that would condition evacuation before and during a storm. Lower rates of car ownership partly reflect better access to public transportation (such as the New York subway and other trains). On average, working residents in floodplains in New York City are four times more likely than those on Long Island to use public transportation as their primary means of commuting.

In total, nearly 50,000 people live in the floodplain in New York City (**Tables 9.10** and **9.11**).

Evacuation from Long Island, on the other hand, would benefit from the flexibility offered by high vehicle access, but over-reliance could trigger potential delays and disruption from the clogging of highway systems. Despite a more equitable attempt at evacuation for Hurricane Rita following Hurricane Katrina later in 2005, the over-reliance on evacuation by car created a 100-mile long traffic jam, which generated its own vulnerabilities (Litman, 2005). The most critically vulnerable car-dependent populations include those with limited vehicle exit routes for evacuation, such as some populations along choke points in Suffolk County or those in Manhattan who depend on tunnel or bridge access to leave the city.

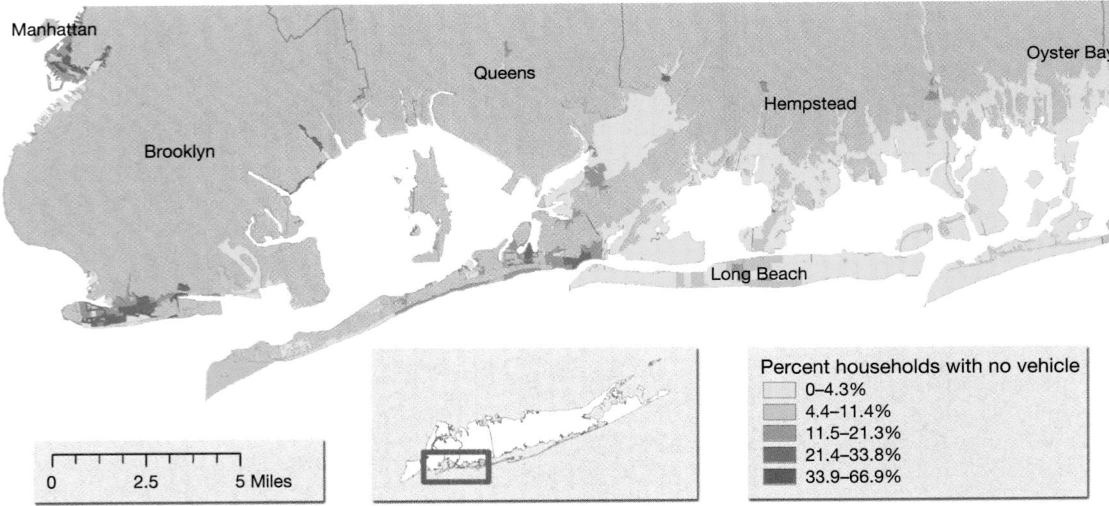

Source: US Census Data 2000, FEMA FIRM base map, with authors' computations and GIS graphics

Figure 9.21 Variations in access to a vehicle within the 100-year floodplain

Across census block groups, the percentage of people with access to a car ranges from less than 5 percent to more than 60 percent. Despite generally high rates of car ownership on Long Island, small pockets of low ownership are interspersed largely within Nassau County. A look at the demographic and socioeconomic makeup of a few of these census block groups underscores that car ownership is partly a function of underlying socioeconomic conditions. For example, a few such areas in Hempstead also have higher rates of poverty and lower average educational attainment compared to regional means. These conditions would act together as a group of stresses during a storm event, reinforcing the vulnerability of a person with no car. Put simply, not having vehicle access is a problem for anyone when it is time to prepare for a storm or evacuate, but if that person is elderly with existing mobility challenges or is living below the poverty line as a single mother with two children, then having no car can have a multiplier effect.

The mapping analysis builds on the basic methods used by New York City and Long Island transportation agencies as part of their compliance with requirements set out by Federal Executive Order 12898 on Environmental Justice. For example, the New York Metropolitan Transportation Council identifies the communities in **Table 9.12** as "communities of concern" on Long Island based on socioeconomic and racial status.

Social Justice and Adaptation

Securing transport systems for regional connectivity and mass commuter patterns are critical foci of hazards and adaptation planning. At the same time, successfully integrating equity into system-wide adaptations will require taking seriously the wide range of transport capacities mentioned in the previous section, including constraints on physical mobility, limited access to transportation options, and localized transport dependencies.

A frequently considered short-term adaptation is the selective "hardening" (i.e., protective measures such as buildings seawalls, raising road beds, and improving drainage) of transport infrastructure, but an important question remains: Hardening for whom? Will certain populations and regions benefit from secured commuting and mobility while others do not? For example, in and around New York City, populations reliant on specific local bus routes for commuting—often lower income—may be at a relative disadvantage

Source: US Census Data 2000, FEMA FIRM, and authors' computations and GIS graphics

Figure 9.22 Clustered poverty along the Long Island coast (Great South Bay)

Nassau County		Suffolk County	
Town	Village/Hamlet	Town	Village/Hamlet
Glen Cove	Glen Cove	Huntington	Huntington Station
Hempstead	East Garden City		Wyandanch
	Uniondale		Wheatley Heights
	Hempstead		N. Amityville
	Roosevelt		Copiague
	Freeport	Islip	Brentwood
	Elmont		Central Islip
	Inwood		Oakdale
	N. Valley Stream	Islip/Brookhaven	Holbrook
	Valley Stream		Holtsville
North Hempstead	New Cassel	Brookhaven	Patchogue
	Westbury		Stony Brook
Oyster Bay	East Massapequa		Centereach
			Selden
			Coram
			Middle Island

Source: NYMTC 2007

Table 9.12 Environmental justice communities of concern on Long Island

if hardening infrastructure is aimed at the short-term protection of arterial commuter rail lines and regional business connectivity to Manhattan. In New York City, bus commuters constitute 11.8 percent of the population in the floodplain (**Table 9.11**), many of whom are commuting within boroughs. On the other hand, bus systems are less vulnerable to storm surge flooding, since they generally can resume their function shortly after the floods retreat. Fixed rail lines, and especially those depending on tunnels, may require much longer recovery times after a storm as described in this case study.

A longer-term adaptation strategy is managed retreat, consisting of coastal buyout and relocations. Low-income regions and populations could be particularly sensitive to indirect effects of such interventions. For example, a protracted program could incrementally change land use and regional perception in ways that devalue communities prior to buyouts. There is also a risk that social support and monetary compensation are inadequate for successfully moving and reintegrating migrants. As **Figure 9.22** suggests, wealth and poverty tend to cluster in localized areas along the coast of Long Island and New York City. This uneven distribution would condition the response and sensitivity of different communities to a buyout program. Transport-specific issues include the exacerbation of spatial mismatches between jobs and housing centers as migrants put new pressures on local job and housing markets. This is a recurring challenge for planners on Long Island, where New York City's gravitational pull on the transport system exacerbates a mobility gap for those trying to commute north to south across the island rather than east to west (see, for example, http://www.longislandindex.org/).

Coastal Storm Surge Adaptation Options, Strategies, and Policy Implications

Options and time scales for adaptation measures vary over the short, medium, and long terms:

1) *Short-term Measures (over the next 5 to 20 years)*
   - Short-term measures (individual floodgates, berms, local levees, pumps, etc.) can be effective for a few decades for high-to-moderate probability events, i.e., surges with annual probabilities with low-to-moderate recurrence periods of 100 years or less (storms up to or weaker than the 100-year storm). These "concrete and steel" or "hard" engineering measures may be preceded by or combined with interim measures that improve a system's operational resiliency (e.g., those mentioned for the Lincoln and Holland Tunnel ventilation shaft doors, see footnotes to **Table 9.5** and **Table 9.4**). MTA NYCT is currently undertaking one such short-term measure by raising floodwalls at its 148th Street Yard along the Harlem River. This measure avoids the repeat of flooding already experienced in the past.

2) *Medium-term Engineering Hard Measures (over the next 30 to 100 years)*
   - System or site-specific (i.e., each station, rail track segment, substation, etc.) measures are needed to protect each site individually, such as by raising some structures or track segments.
   - Region-wide protective measures, such as constructing estuary-wide storm barriers, have been proposed (Aerts et al., 2009). These have been discussed in NPCC 2010.

3) *Long-term Sustainable Strategies (any time from now to beyond 100 years)*
   - Long-term measures include changing land use and providing more retreat options. These measures can be combined with the short- and

medium-term strategies indicated above. When sea level rise combined with coastal storm surges exceeds the design elevations of barriers and levees, these long-term strategies require comprehensive, sustainable plans that include time-dependent decision paths and "exit strategies."

To determine the optimal climate change adaptation for the transportation system in the coastal zone of New York State with the highest benefit-cost ratios, the time-dependent assessments listed below for current and projected future conditions need to be performed. Depending on the structure or system, these assessments may need to be projected out 100 or 150 years:

- Make probabilistic time- and sea level rise-dependent coastal storm surge hazard projections on a regular basis.
- Conduct a vulnerability assessment of transportation infrastructure systems given the hazard projections.
- Develop time-dependent transportation infrastructure asset-value estimation methodology and databases.
- Combine the above three items into regular time-dependent risk (loss) assessments.
- Assess costs and benefits of various adaptation options as a function of time.
- Conduct policy and finance assessments.
- Develop decision making and implementation strategies based on all of the items above.

Case Study Knowledge Gaps

The following major knowledge gaps for the transportation sector of the New York State Coastal Zone have been identified from the case study:

- High-resolution digital elevation models for terrains with infrastructure
- The as-built infrastructure elevations, geometry and volumes of the above- and below-grade structures, openings, hydrodynamics, flow rates, filling times
- Vulnerabilities (fragility curves) for coastal storm surge hazards for items listed in the prior bullet, especially when saltwater comes in contact with sensitive equipment
- Realistic estimation techniques for outage times, costs, and reduced losses versus benefits from adaptation measures
- Better economic models for the relationship of transport system outage to over-all economic losses
- Institutional and policy issues related to: How to foster strategic long-term planning at agencies? What is the legal/regulatory framework, and how can professional codes (engineering codes, FEMA's National Flood Insurance Program regulations, enforcement, etc.) be updated to take projected sea level rise and increased coastal storm damage into account ?

Case Study Conclusions

This detailed case study of 100-year coastal storm surges for current sea level and two sea level rise scenarios has provided insights into the technical, economic, and social consequences of climate change. They demonstrate, by example, the potential severity of climate change impacts on the state's transportation sector. Timing of adaptation paths, institutional transformations needed to embed adaptation measures into decision making, and allocation of funding present serious challenges. There is a broad range of policy options and measures that can be implemented to avoid future climate-related losses and to provide the state with a sustainable, climate-resilient transportation system.

Hazards, risks, and potential future losses from climate change—and especially sea level rise—to the region's transportation systems and general economy are increasing steadily. Costs, when annualized, may amount initially to an average of only about $1 billion per year over the next decade. By the end of the century, these costs will probably rise to tens of billions of dollars per year, on average. Note that these are long-term annualized averages. Individual storms may cost much more, as described above in the ClimAID scenario analysis.

Benefits versus Costs

Several thorough studies have shown, based on empirical data from the last 30 years, that there is an approximate 4-to-1 benefit-to-cost ratio of investing in protective measures to keep losses from disasters low

(MMC, 2005; CBO, 2007; GAO, 2007). If the 4-to-1 benefit-cost ratio for protective and other mitigation actions applies, then up to one-quarter of the expected annual losses should be invested every year. This approach provides rough guidance for the needed investments towards protective measures that can be considered cost-beneficial, if based on sound engineering and planning.

Based on the loss estimates given in **Table 9.8** for the 100-year storm,[45] this implies that hundreds of million dollars per year initially may be needed for protective adaptation measures, rising to billions per year at latest by mid-century. Such investment be needed by mid-century because of the long lead-times for infrastructure projects, and to ensure that adequate protections are in place before the end of century. Institutions must plan for the long term, sometimes as much as one to two centuries into the future, for instance when considering right-of-way and land-use decisions, especially in coastal areas. Such major climate change adaptation measures need to be integrated into the overall infrastructure upgrade and rejuvenation projects during the coming decades.

It is important to act before systems become inundated and damaged beyond easy repair.

Long-term Sea Level Rise

Decision-makers need to engage with scientists to monitor the Greenland Icesheet and the West Antarctic Ice Shield, which have the potential to contribute multiple feet to sea level rise this century. These impacts may need to be considered even when planning short- or medium-term adaptation strategies, in order to ensure their long-term sustainability.

In Europe, researchers have analyzed what to do under a scenario in which sea level rose by about 15 feet over the course of one century. The desktop exercise, named *Atlantis* (Tol et al., 2005), has been performed for three regions in Europe. The study areas included the Thames Estuary/London, the Rhine Delta/Netherlands/Rotterdam, and the Rhone Delta/South France. While the hypothetical scenario has a low probability, its high consequences put the larger societal issues into perspective for what, in reality, may turn out to be incremental solutions that are socially acceptable.

Indicators and Monitoring

The establishment of a climate indicators and monitoring network will enable the tracking of climate change science and impacts. Recording the changes in the physical climate (sea level rise), climate change impacts (flood events), and adaptation actions can provide critical information to decision-makers (Jacob et al., 2010).

References

Aerts, J., D.C. Major, M.J. Bowman, P. Dircke, and M.A. Marfai. 2009. Connecting Delta Cities. Coastal Cities, Flood Risk Management and Adaptation to Climate Change. VU University Press.

Bowman, M.J., B. Colle, R. Flood, D. Hill, R.E. Wilson, F. Buonaiuto, P. Cheng and Y. Zheng. 2005. "Hydrologic Feasibility of Storm Surge Barriers to Protect the Metropolitan New York - New Jersey Region, Final Report. Marine Sciences Research Center Technical Report." New York: Stony Brook University, March 2005.http://stormy.msrc.sunysb.edu/phase1/ Phase%201%20final%20Report%20Main%20Text.pdf

Bowman, M.J., D. Hill, F. Buonaiuto, B. Colle, R. Flood, R. Wilson, R. Hunter and J. Wang. 2008. "Threats and Responses Associated with Rapid Climate Change in Metropolitan New York." In *Sudden and Disruptive Climate Change*, edited by M. MacCracken, F. Moore and J.C. Topping, 327. London: Earthscan.

Bullard, R.D., ed. 2007. *Growing Smarter: Achieving Livable Communities, Environmental Justice, and Regional Equity*. Cambridge, MA: MIT Press.

Byron, J. 2008. "Bus Rapid Transit for New York?" *Gotham Gazette*, April 21. http://www.gothamgazette.com/article/sustain/20080421/210/2498

Chakraborty, J. 2009. "Automobiles, Air Toxics, and Adverse Health Risks: Environmental Inequities in Tampa Bay, Florida." *Annals of the Association of American Geographers* 99(4): 674-697.

Chen, D. 2007. "Linking Transportation Equity and Environmental Justice with Smart Growth." In *Growing Smarter: Achieving Livable Communities, Environmental Justice, and Regional Equity*, edited by R.D. Bullard. Cambridge, MA: MIT Press.

City of New York. 2011. PlaNYC: A green, greater New York. Update April 2011. Office of the Mayor, http://www.nyc.gov/html/planyc2030/

Colle, B.A., F. Buonaiuto, M.J. Bowman, R.E. Wilson, R. Flood, R. Hunter, A. Mintz and D. Hill. 2008. "New York City's Vulnerability to Coastal Flooding: Storm Surge Modeling of Past Cyclones." *Bulletin of the American Meteorological Society* 89(6):829-841.

DeGaetano, A.T. 2000. "Climatic perspective and impacts of the 1998 northern New York and New England ice storm." *Bulletin of the American Meteorological Society* 81:237–254.

Freed, A. 2009. New York City Office of Long-Term Planning and Sustainability, personal communication.

Godschalk, D.R., T. Beatley, P. Berke, D.J. Brower and E.J. Kaiser. 1999. *Natural hazard mitigation*. Washington D.C.: Island Press.

Hansen, J.E. 2007. "Scientific reticence and sea level rise." *Open Access E-Prints in Physics, Mathematics, Computer Science, Quantitative Biology and Statistics*, Cornell University. http://arxiv.org/pdf/physics/0703220v1.

Hess, D.B. and J.C. Gotham 2007. "Multi-Modal Mass Evacuation in Upstate New York: A Review of Disaster Plans." *Journal of Homeland Security and Emergency Management* 4(3): Article 11. http://www.bepress.com/jhsem/vol4/iss3/11.

Hill, D. 2008. "Must New York City Have Its Own Katrina?" American Society of Civil Engineers *Journal of Leadership and Management in Engineering* 8(3):132-138.

Horton, R., C. Herweijer, C. Rosenzweig, J.P. Liu, V. Gornitz and A.C. Ruane. 2008. "Sea level rise projections for current generation CGCMs based on the semi-empirical method." *Geophysical Research Letters* 35:L02715.

Horton, R., and C. Rosenzweig, 2010: Climate Risk Information. Climate change adaptation in New York City: Building a Risk Management Response, C. Rosenzweig, and W. Solecki, Eds., New York Academy of Sciences.

Jacob, K.H., N. Edelblum, and J. Arnold. 2000. Risk increase to infrastructure due to sea level rise. http://metroeast_climate.ciesin.columbia.edu/reports/infrastructure.pdf. Last accessed March 9, 2011. An abbreviated version, Jacob et al. (2001), is included as Chapter 4 in Rosenzweig and Solecki (2001).

Jacob, K.H., V. Gornitz and C. Rosenzweig. 2007. "Vulnerability of the New York City Metropolitan Area to Coastal Hazards, Including Sea level rise - Inferences for Urban Coastal Risk Management and Adaptation Policies." In *Managing Coastal Vulnerability: Global, Regional, Local*, edited by L. McFadden, R.J. Nicholls and E. Penning-Rowsell, 141-158. Amsterdam, London, New York: Elsevier.

Jacob, K.H., et al. 2008. "Climate Change Adaptation - a Categorical Imperative." New York Metropolitan Transportation Authority (MTA) White Paper. http://www.mta.info/sustainability/pdf/Jacob_et%20al_MTA_Adaptation_Final_0309.pdf

Jacob, K.H. 2009." Metropolitan Transportation Authority (MTA) Blue Ribbon Commission on Sustainability." In *Adaptation to Climate Change*.

Jacob, K., R. Blake, R. Horton, D. Bader, and M. O'Grady. 2010. Indicators and monitoring. Annals of the New York Academy of Sciences, 1196: 127–142. doi: 10.1111/j.1749-6632.2009.05321.x

Jones, K.F. and N.D. Mulherin. 1998. "An Evaluation of the Severity of the January 1998 Ice Storm in Northern New England." USA CoE Report for FEMA Region 1. http://www.crrel.usace.army.mil/techpub/CRREL_Reports/reports/IceStorm98.pdf.

Litman, T. 2005. *Lessons From Katrina and Rita: What Major Disasters Can Teach Transportation Planners*. Victoria Transportation Policy Institute (VTPI).

McGuire, R.K. 2004. "Seismic Hazard and Risk Analysis." Earthquake Engineering Research Institute. Monograph EERI-MNO-10. ISBN #0-943198-01-1.

Metropolitan Transportation Authority (MTA). 2007. "August 8, 2007 Storm Report." September 20. http://www.mta.info/mta/pdf/storm_report_2007.pdf.

Metropolitan Transportation Authority (MTA). 2008a. "Greening Mass Transit & Metro Regions: The Final Report of the Blue Ribbon Commission on Sustainability and the MTA." http://www.mta.info/sustainability/pdf/SustRptFinal.pdf.

Metropolitan Transportation Authority (MTA). 2008b. "MTA Adaptations to Climate Change: A Categorical Imperative." http://www.mta.info/sustainability/pdf/Jacob_et%20al_MTA_Adaptation_Final_0309.pdf.

Metropolitan Transportation Authority (MTA). 2009. http://www.mta.info/sustainability/pdf/SustRptFinal.pdf.

Metropolitan Transportation Authority (MTA). 2009. "Greening Mass Transit & Metro Regions." Blue Ribbon Commission on Sustainability. http://www.mta.info/sustainability/pdf/SustRptFinal.pdf.

Mileti, D.S. 1999. *Disasters by design*. Washington D.C.: Joseph Henry Press.

Millerd, F. 2007. *Global Climate Change and Great Lakes International Shipping*. Transportation Research Board (TRB) Special Report 291. 2007. http://onlinepubs.trb.org/onlinepubs/sr/ sr291_millerd.pdf.

Milligan and Company. 2007. *Transportation Equity in Emergencies: A Review of the Practices of State Departments of Transportation, Metropolitan Planning Organizations, and Transit Agencies in 20 Metropolitan Areas*. Washington D.C.: Federal Transit Administration.

Morrow, B. H. 1999. "Identifying and Mapping Community Vulnerability." *Disasters 23(1):1-18*.

Multihazard Mitigation Council (MMC). 2005. *Mitigation Saves*. National Institute of Building Sciences. http://www.nibs.org/index.php/mmc/projects/nhms.

National Infrastructure Simulation and Analysis Center (NISAC). 2006. *Hurricane Scenario Analysis for the New York City Area*. National Infrastructure Simulation and Analysis Center at Sandia National Laboratory of the Department of Homeland Security. Draft dated June 16.

National Oceanic and Atmospheric Administration (NOAA). 2004. "Aviation Climate Assessment Report." http://www.srh.noaa.gov/abq/avclimate/docs/Aviation_Climatology_Assessment_Report.pdf.

New York City Office of Emergency Management (NYC OEM). 2006. "Coastal Storm Plan. Version 2006." http://www.nyc-arecs.org/CoastalStormPlan_summary_06.pdf.

New York City Office of Emergency Management (NYC OEM). 2009. "Natural Hazard Mitigation Plan." http://nyc.gov/html/oem/html/about/planning_hazard_mitigation.shtml.

New York City Panel on Climate Change (NPCC). 2010. *Climate Change Adaptation in New York City: Building a Risk Management Response*. C. Rosenzweig and W. Solecki, Eds. Prepared for use by the New York City Climate Change Adaptation Task Force. *Annals of the New York Academy of Sciences* 1196. New York, NY.

New York Metropolitan Transportation Council (NYMTC). 2007. *Access to Transportation on Long Island: A Technical Report*. Accessed February 18, 2010. http://www.nymtc.org/project/LIS_access/documents/Final_TechRpt.pdf

New York Metropolitan Transportation Council (NYMTC). 2009. *A Coordinated Public Transit-Human Services Transportation Plan*. Accessed February 19, 2010. http://www.nymtc.org/project/PTHSP/PTHSP_documents.html.

New York State Emergency Management Office (NYSEMO). 2006. *SLOSH Model Hurricane Inundation Zones Revised*. Polygon Coverage. UTM NAD83. http://www.nysgis.state.ny.us/gisdata/inventories/details.cfm?DSID=1043

New York State Sea Level Rise Task Force (NYS SLRTF). 2010. http://www.dec.ny.gov/energy/45202.html.

Northeast Climate Impacts Assessment (NECIA). 2006. *Climate Change in the U.S. Northeast*. http://www.climatechoices.org/assets/documents/climatechoices/NECIA_climate_report_final.pdf

Northeast Climate Impacts Assessment (NECIA). 2007. *Confronting Climate Change in the U.S. Northeast*. http://www.climatechoices.org/assets/documents/climatechoices/confronting-climate-change-in-the-u-s-northeast.pdf.

Port Authority of New York and New Jersey (PANYNJ). 2008. "The Port Authority 2008 Annual Report." http://www.panynj.gov/corporate-information/pdf/annual-report-2008.pdf.

Press, E. 2007. "The new suburban poverty." *The Nation*, April 23. 18–24.

Pucher, J. and J.L. Renne. 2003. "Socioeconomics of Urban Travel: Evidence from the 2001 NHTS." *Transportation Quarterly* 57(3).

Renne, J.L., T.W. Sanchez, P. Jenkins and R. Peterson. 2009. "Challenge of Evacuating the Carless in Five Major U.S. Cities." *Transportation Research Record* 2119:36-44.

Rahmstorf, S. 2007. "A semi-empirical approach to projecting future sea level rise." *Science* 315:368-70. http://assets.panda.org/downloads/final_climateimpact_22apr08.pdf.

Root, A., L. Schintler and Button, K.J. 2000. "Women, travel and the idea of 'sustainable transport'." *Transport Reviews* 20(3):369-383.

Rosenzweig, C., R. Horton, D.C. Major, V. Gornitz and K. Jacob. 2007a. "Appendix 2: Climate." In "August 8, 2007 Storm Report." New York City Metropolitan Transportation Authority (MTA). September 20. http://www.mta.info/mta/pdf/storm_report_2007.pdf.

Rosenzweig, C. and W. Solecki, eds. 2001. *Climate Change and Global City: The Potential Consequences of Climate Variability and Change — Metro East Coast.* Report for the U.S. Global Change Research Program, National Assessment of the Potential Consequences of Climate Variability and Change for the United States. New York: Columbia Earth Institute.

Sanchez, T., R. Stolz and J. Ma. 2004. "Inequitable Effects of Transportation Policies on Minorities." *Transportation Research Record* 1885:104–110

Sanderson, A. August 2009. NYS Department of Transportation, personal communication.

Stedinger, J. 2010. State University of New York, Stony Brook. http://stormy.msrc.sunysb.edu.

Suro, R. and A. Singer. 2002 "Latino Growth in Metropolitan America: Changing Patterns, New Locations" *Census 2000 Survey Series.* Washington DC: The Brookings Institution.

Sze, J., and J.K. London. 2008. "Environmental justice at the crossroads." *Sociology Compass* 2(4):1331-1354.

Tol, X., et al. 2006. "Adaptation to 5 meters of sea level rise." *Journal of Risk Research* 9(5): 467–482.

Transportation Research Board (TRB). 2008a. "Potential Impacts of Climate Change on U.S. Transportation." *Special Report* 290. *National Research Council.*

Transportation Research Board (TRB). 2008b. "The Role of Transit in Emergency Evacuation." *Special Report* 294. *National Research Council.*

URS Corporation. 2008. "Appendix E: History and Projection of Traffic, Toll Revenues and Expenses. and Review of Physical Conditions of the Facilities of the Triborough Bridge and Tunnel Authority." http://www.mta.info/mta/investor/pdf/2008/TBTAReport2008Final.pdf.

United States Army Corps of Engineers, et al. 1995. *Metro New York Hurricane Transportation Study*. Interim Technical Data Report.

United States Bureau of Labor Statistics. 2010. Consumer Expenditure Survey.

United States Congressional Budget Office (CBO). 2007. *Potential Cost Savings from the Pre-Disaster Mitigation Program.* http://cbo.gov/ftpdocs/86xx/doc8653/09-28-Disaster.pdf.

United States Climate Change Science Program (CCSP). 2008a. "Impacts of Climate Change and Variability on Transportation Systems and Infrastructure: Gulf Coast Study, Phase I." A Report by the U.S. Climate Change Science Program and the Subcommittee on Global Change Research, edited by M.J. Savonis, V.R. Burkett and J.R. Potter. Department of Transportation, Washington, DC. http://www.climatescience.gov/ Library/sap/sap4-7/final-report/.

United States Climate Change Science Program (CCSP). 2008b. "Final Synthesis and Assessment Report 3-4: Abrupt Climate Change." http://www.climatescience.gov/Library/sap/sap3-4/final-report/default.htm#finalreport.

Untied States Government Accounting Office (GAO). 2007. *Natural Hazard Mitigation.* August. http://www.gao.gov/new.items/d07403.pdf.

Viscusi, W.K. and J.E. Aldy. 2002. "The value of a statistical life: a critical review of market estimates throughout the world." *Discussion Paper 392.* Harvard Law School. ISSN 1045-6333. http://www.law.harvard.edu/programs/olin_center/papers/pdf/392.pdf.

Appendix A. Stakeholder Interactions

Stakeholders of the New York State Transportation Sector cannot be easily differentiated by modes of transportation (air, water, ground), but are more readily described by their public, semi-public, and private institutional status, with considerable overlap across modes in these three classes of ownership.

The New York State Department of Transportation (NYSDOT) has the broadest statewide oversight function, in close coordination with U.S. federal transportation programs and guidelines. On a regional basis, government-established transportation authorities with a quasi-corporate administrative structure have the mandate to serve the public's transportation needs (examples include Metropolitan Transportation Authority (MTA), Port Authority of New York and New Jersey (Port Authority), New York State Thruway Authority, New York State Bridge Authority, etc.). In addition, there are many private transportation operators, including airlines, ferries, maritime and river barge operators, bus companies, rail freight companies, individual trucking operators and—last but not least—private truck and car owners, cyclists, and pedestrians. The ClimAID stakeholder process focused primarily on ground transportation, and on the public and semi-public transportation sector. Stakeholders of the ClimAID transportation sector thus included NYSDOT, MTA, the Port Authority of New York/New Jersey, Amtrak, CSX, New Jersey Transit, and others.

Stakeholders were invited to ClimAID meetings at the beginning of the project. Survey forms were sent to stakeholders early in the project asking for information related to a self-assessment of their vulnerabilities to climate change. In the New York City metropolitan area, ClimAID greatly benefited from the process that the NYC Climate Change Adaptation Task Force had undertaken to collect climate change vulnerability information and systematically order it in a risk matrix for importance/severity and adaptation feasibility (Adam Freed, personal communication, 2009; NPCC, 2010). The ClimAID stakeholder process also benefited greatly from close cooperation and coordination with the New York State Sea Level Rise Task Force on all matters related to sea level rise.

ClimAID transportation focus group meetings were held with individual agencies (MTA, the Port Authority of New York/New Jersey, and others) and by numerous conference-call working sessions to clarify survey questions and address security issues. The focus was previously on detailed technical issues regarding climate change vulnerabilities and protective measures.

Contributions to the chapter topics were solicited from the stakeholders. A total of at least three drafts of the chapter at various stages, and for some stakeholders several more, were provided for comment and input. Numerous comments, corrections, and improvements were received. This extensive iterative process led to the final version, which incorporated as many of these improvements as possible. But the responsibility for the final version rests with the ClimAID transportation sector research team.

Stakeholder Participants

- Amtrak
- CSX
- Federal Highway Administration
- Florida State University
- Long Island Railroad
- Metropolitan Transportation Authority
- New Jersey Transit
- New York City Office of Emergency Management
- New York City Office of Long-Term Planning and Sustainability
- New York City Transit
- New York State Department of Environmental Conservation
- New York State Department of Transportation
- New York State Office of Emergency Management
- New York University
- Port Authority of New York and New Jersey
- US Department of Homeland Security
- US Geological Survey

Appendix B. Method of Computation of Area-Weighted Average Flood Elevations for Nine Distinct Waterways in New York City

As stated in the main body of this chapter, the 2- and 4-foot sea level rise values are similar to the rapid ice-melt sea level rise scenario forecasts for the 2050s (2 feet) and 2080s (4 feet), described in Chapter 1, "Climate Risks," and by the New York City Panel on Climate Change (NPCC, 2010). Both sources provide more highly resolved sea level rise ranges: 19 to 29 inches by the 2050s and 41 to 59 inches by the 2080s, with central values of 24 inches and 50 inches. Within the integer-foot resolution (rounded whole number values) adopted for this case study, the investigators have approximated these two measures as 2 feet (2050s) and 4 feet (2080s). When in the course of this case study any maps or tables refer to 2-foot and 4-foot sea level rise, then this represents an approximation of the more precise sea level rise estimates and their range of uncertainties as given originally in the New York City Panel on Climate Change study for the rapid ice-melt model.

To analyze the risk that flooding poses to transportation infrastructure, the elevations of the structures relative to the elevation of the floodwaters according to FEMA's 100-year flood maps are analyzed. New flood zones that account for the anticipated 2- and 4-foot sea level rise are then also analyzed with respect to their impact on transportation structures.

When the effects of flooding on extended transportation networks are analyzed, then the relative elevation of the floodwaters to the transport system's critical elevations must be measured at many locations along the transport network's geographical extent. To achieve this task within the timeframe and resources available for this study, the ClimAID team used an approximation. FEMA's Flood Insurance Rate Maps (FIRMs) provide 100-year base flood elevations at a finite number of points along a waterway. The actual base flood elevations vary slightly from location to location within the flood zones mapped by FEMA that are shown, without alteration, as the red zones in **Figure 9.7**. The variations in flood elevations occur for hydrodynamic reasons related to bathymetry, topography, wave and wind exposure, etc.

When adding 2 and 4 feet of sea level rise, new flood zones of an indeterminate shape on their landward side result. That shape does not exactly follow terrain contours of constant elevations, just as the flood zone boundaries of FEMA's 100-year base flood elevations cross contours of constant elevations, according to hydrodynamic factors. To minimize the effort to determine the relative height of a transportation system versus flood elevations that vary slightly from location to location, the entire New York City water and land area was subdivided into nine waterways, based on their tidal and coastal storm surge characteristics (**Figure 9.8**).

Using the discrete FEMA-provided 100-year base flood elevation control points along the shores of each waterway, averaged base flood elevation control heights were computed for each of the nine zones. The arithmetic mean (simple average; **Table 9.13**, column 3) of the base flood elevation control points for each zone was, however, *not* applied. Instead, an area-weighted mean (Z_i, or area-weighted base flood elevation, column 4) was used. The weights were assigned proportional to the areas that the control points represent along the shorelines of each waterway. This weighting minimizes the undue influence of shore segments with unusually high density of control points that may skew the average base flood elevation for each waterway. **Table 9.13** (column 6) shows the number of control points for each zone (waterway) and the standard deviation (column 5) around the weighted mean for each area-weighted mean value.

Note that the original base flood elevations from FEMA's Flood Insurance Rate Maps are generally (at least for New York) referenced to the National Geodetic Vertical Datum of 1929 (NGVD 1929). The investigators, however, chose the new, averaged sea level rise-dependent flood zone elevations to reference to the more recent, and now generally more commonly used, North American Vertical Datum (NAVD 1988). Note that in contrast to FEMA maps in New York, FEMA maps for New Jersey use the NAVD 1988 datum. A constant difference of 1.1 feet between the two datums was used throughout the New York City area such that the numerical elevations above the two vertical datums relate to each other by Equation 3:

Equation 3. Elevation(ft) above NAVD'88 = Elevation(ft) above NGVD'29 - 1.1 ft

The so-derived, area-weighted average base flood elevations or area-weighted average (in the NAVD'88 reference frame) are rounded to the nearest integer foot for assessing the flood and sea level rise impact on transport in the region.

Once the area-weighted and integer-rounded average base flood elevations (or area-weighted averages) were obtained for the nine waterways, the 2- and 4-foot sea level rise estimates were added to these values. This allows the elevations of transport structures to be easily compared to the flood zone elevations.

In the regions outside New York City, including Long Island (Nassau and Suffolk counties), Westchester County, and the Lower Hudson Valley, much cruder approaches were used for a number of reasons. First, no high-resolution digital elevation model with a 1-foot vertical resolution was uniformly available for these regions outside of New York City. Additionally, for these areas, the lowest critical elevations are not known for many of the transportation systems and related structures as well as they are known within New York City. The New York City estimates were largely obtained from the

Zone (i)	Waterway	Rounded, Average Base Flood Elevation (feet) NGVD 88	Rounded, Area-Weighted Average Base Flood Elevation in NGVD 88, Z_i (feet)	Standard Deviation (feet)	Number of Points on FEMA Flood Map per Zone (n)	Relevant Boroughs
1	Long Island Sound	14	14	1.45	31	Bx, Q
2	East River	13	11	1.06	53	Bx, Q, M
3	Harlem River	9	9	1	3	Bx, M
4	Hudson River	8	8	0.71	2	Bx, M
5	Inner harbor	9	9	0.97	13	M, Bk, SI, (Q)
5A	Kill Van Kull	8	8	0.63	6	SI
6	Outer Harbor	10	10	1.20	48	SI, Bk
7	Jamaica Bay	7	8	0.72	32	Bk, Q
8	Rockaway (Atlantic and Jamaica Bay)	8	9	1.13	22	Q

Note: Bk=Brooklyn, Bx=Bronx, M=Manhattan, Q=Queens, SI=Staten Island

Table 9.13 New York City waterway zones and their rounded average values for obtained area-weighted base flood elevations

Hurricane Transportation Study (USACE, 1995), and the metropolitan east coast (MEC) climate change infrastructure study (Jacob et al., 2000 and 2007).[46]

This lack of elevation information points to the need for accurate, accessible digital elevation models in the storm-surge-prone coastal zones of New York State. These models need vertical resolutions of less than 1 foot. There is also a need for accurate as-built elevations of the transport structures. The digital elevation model resolution is technically achievable with carefully executed remote sensing technology (LIDAR surveys) and careful post-processing after acquiring the raw data. Some coverage with this technology exists in New York State, but needs to be undertaken systematically, at least for all flood-prone zones across the state that are affected by sea level rise and coastal storm surges. The collection of reliable elevations of transport structures in these critical areas is in the best interest of the operating agencies, but needs to be performed in the public interest as part of a concerted statewide flood-risk management plan.

Appendix C. Method to Compute Economic Losses (Appended to Case Study A, 100-Year Coastal Storm Surge with Sea Level Rise)

To estimate the economic losses from the ClimAID case study storm scenario, using the values summarized in **Table 9.5**, these assumptions were made:

- The economic activity is essentially zero from day zero to the lowest value of T90, for each scenario, listed in Row 30 of **Table 9.5**.
- The economic activity recovers gradually (assuming a linear relation) from day T90min to T90max, where the latter is the upper bound of the T90 value (in days) listed in Row 30 of **Table 9.5**, for each scenario.
- The recovery from 90 percent functionality to 100 percent functionality (on day T100) occurs with the same slope as between 0 and 90 percent functionality.

This concept of a gradual recovery of the economy (rather than coming to a total halt and then suddenly jumping back into full gear) is important for fully appreciating how the information in **Table 9.5** is used. TheT90 values in row 30, columns 3, 4, and 5, are not the times by which the economy is assumed to start recovering; these values are intended to mark the times by which the economy has recovered to 90 percent of its pre-disaster level, i.e., they mark the time by which the recovery has come almost to an end, and had made progress for the entire period in the days between T90min and T90max after the onset of the disaster.

All of these assumptions and approximations are highly uncertain, but can be justified by comparing them to the electric grid recovery curve shown in **Figure 9.20**, except the slightly upward convex curve of this figure is replaced with a linear relation. The basic concept is that electricity and economy come back not suddenly but gradually after such an event. Even if some transport modes do not work, commuters may find a way to substitute, work at home, or pay for and/or share a taxi (for caveats, see Vulnerability and Social Justice sections of the case study).

With these assumptions, the time-integrated economic losses for the entire metropolitan region (TIELEM) from the 100-year storm of the case study can be computed by integrating (summing up) over time the gradually (i.e., with time linearly) decreasing daily economic productivity losses from day zero to day T100. Using this concept of decreasing daily losses and increasing recovery of the economy yields Equation 4:

Equation 4: $TIELEM = DGMP\ [T90min + \frac{1}{2}\ (T90max - T90min)\ 100/90]$

Using the daily gross metropolitan product, DGMP = \$4 billion/day and the T90min and T90max values of **Table 9.5** for the three SLR scenarios S1 to S3, yields the TIELEM values summarized in **Table 9.7**.

Forward-Projection of Losses to 2050 and 2090

Note it has been assumed that all three SLR scenarios are applied to the 2010-DGMP. But the three scenarios require time for sea level to rise. The study assumes that the three scenarios occur in S1=2010, and that S2 occurs in the 2050s and S3 before 2090. Therefore, the study must account for what the economic trends for the next 40 and 80 years could be (a) by accounting for inflation and/or discount rates; and (b) by accounting for economic growth, expressed by increasing DGPM and/or increasing asset values. These trends can be formally treated in the same way as compounding interest for an interest rate of **r** % (say for inflation or economic growth rate), while adding a certain *fixed amount* of dollars **p** to every 100 dollars of built assets, say, *at the end of each year* (note that this means a steadily *decreasing percentage* addition of assets, since the dollar amount **p** stays constant while the initial asset value increases by compounding in relation to **r**).

Using, for example, the assumption that scenario S2 occurs around 2050, i.e., 40 years from now, and that scenario S3 occurs 80 years from now; and that for every $1 trillion/year in economic activity, another (constant) $20 billion per year (i.e., **p=2**) is added over the next 40 years or 80 years, respectively, then the multipliers for the S2-TIELEM of $57 billion, and for the S3-TIELEM of $68 billion, respectively, as a function of an effective economic growth rate **r** will be as indicated in **Table 9.14**.

Added to the economic losses (TIELEM) must be the direct *physical damage* **D** ($), incurred by the affected infrastructure during the storm. Since no vulnerability or fragility curves for the transportation systems, nor a realistic aggregate asset value of the transportation infrastructure, are known with any degree of accuracy or confidence at this time, proxies are used with uncertain validity. For a first-order approximation, we make the following working assumptions for estimating the direct damage **D** for this case study, and using several different approaches:

a) The regional combined transportation assets are on the order of $1 trillion (2010 dollars). The physical damage rates, based on typical flood scenario computation with the tool HAZUS-MH, are taken to be on the order on the order of 1.00, 1.25, and 1.50 percent of the asset values, respectively, for the three scenarios S1 to S3, respectively. This yields direct physical damage losses of **D=$10, $12.5, and $15 billion** (for 2010 assets) for the three scenarios, assuming they all were to occur in the year 2010. Since they do not, multipliers shown in **Table 9.14** would apply for S2 and S3 occurring in 2050 and before 2090, respectively, and assuming all other conditions would apply when the **Table 9.14** multipliers were computed (i.e., constant **p=2** or **$20 billion** annual infrastructure asset additions to the initial [2010] $1 trillion assets).
b) Based on limited observations, a finding is that losses for infrastructure assets during natural disasters in urban settings are typically of the same order of magnitude as for the building-related losses in the same area (e.g., Jacob et al., 2000). NYSEMO periodically computes losses (using the FEMA-sponsored HAZUS-MH software) associated with various storm scenarios for emergency exercises. One of these is a storm scenario in which a category 3 hurricane named "Eli" traverses Long Island making landfall near the boundary between Nassau and Suffolk county (D. O'Brien, NYSEMO, personal communication, October 2009). While this scenario is excessive for Nassau and Suffolk, it produced wind speeds and coastal storm surges for the five NYC boroughs and for Westchester County that are comparable to our 100-year storm scenarios. The building-related losses from the storm surge flooding in the five boroughs amounted to slightly over $20 billion, while in Westchester County it was just below $0.6 billion (for comparison, the wind damage in the five boroughs was only about $110 million and in Westchester $16 million). Moreover, an interesting observation is that the ratio of storm-surge flood- to wind-related losses was 3 to 1 for all counties in New York State affected by scenario "Eli."

Effective Economic Growth Rate *r* (%/year)	0	1.5	1.75	2.0
S2-TIELEM Multiplier for 40 Years:	1.8	2.91	3.16	3.44
S3-TIELEM Multiplier for 80 Years:	2.6	6.39	7.50	8.83

Table 9.14 Multipliers for 40- and 80-year time horizons as a function of growth rate *r* when p=2

If the results from the two approaches are combined, the conclusion is that the physical losses for all infrastructure systems for the entire scenario region due to coastal storm surge flooding is on the order of a few tens of billions of dollars; i.e., in the range of $10 to $20 billion. How much of it is attributable to damage to transportation versus other infrastructure? While at the moment there are no hard data to affirm this, the ClimAID Transportation study suggests, largely because so much of the transportation infrastructure assets are located at or below sea level and are therefore the most vulnerable, that at least half and perhaps as much as three-quarters of this total amount is attributable to damage to the transportation infrastructure.

If the physical damage and the economic losses are compared from the scenario event that are, directly or indirectly by its effect on the general economy, attributable to losses of functionality of the transportation infrastructure, then first-order approximation estimates of total losses from the three storm scenarios (all in 2010 dollars and for 2010 assets) can be obtained and are summarized in **Table 9.5** of the case study.

When reviewing these estimates, the ClimAID team again caution (as stated in the Case Study, in the paragraphs near equation 2) that there may be numerous other significant constraints on the ability of the transportation systems to recover from climate change-induced incidents. Such may include, for example, the age of equipment, the availability of replacement parts/equipment, and the need for such in appropriate quantities. These and other currently unknown and/or not quantified factors could significantly increase climate change impacts in time, labor, and dollars.

Note that **Table 9.14** multipliers for the losses associated with the scenarios S2 and S3 are applicable throughout to modify all losses; they transform them from their current 2010 time base to what they may be during the 2050s and the end-of-2080s, respectively, for the different economic projections and other assumptions stated.

---

1 http://www.bts.gov/publications/freight_shipments_in_america/html/table_03.html.
2 https://www.nysdot.gov/about-nysdot/history/past-present.
3 https://www.nysdot.gov/about-nysdot/responsibilities-and-functions.
4 https://www.nysdot.gov/about-nysdot/history/past-present.
5 http://www.nysba.state.ny.us/Index.html.
6 http://www.countyhwys.org/.
7 https://www.nysdot.gov/divisions/operating/opdm/passenger-rail/freight-rail-service-in-new-york-state.
8 http://www.aar.org/Homepage.aspx and foot note above.
9 Class I railroads are those with operating revenue of at least $272 million in 2002. http://www.nationalatlas.gov/articles/transportation/a_freightrr.html.
10 http://www.guardian.co.uk/environment/2006/nov/01/society.climatechange/print.
11 MEC infrastructure report (Jacob et al. 2000, 2001); FEMA FIRM flood zone maps; and http://www2.sunysuffolk.edu/mandias/38hurricane/.
12 MEC infrastructure report (Jacob et al. 2000, 2001); NPCC-CRI (2010).
13 See Chapter 1: "Climate Risks"; and New York City Panel on Climate Change "Climate Risk Information" (2010).
14 TRB (2008a).
15 CCSP, 2008a: Gulf Coast Study, Phase I. http://www.climatescience.gov/Library/sap/sap4-7/final-report/.
16 USACE, 1995; MEC, 2001; and MTA, 2007. The 08/08/07 Storm Report; NPCC, 2009, 2010 and NYCCATF (in preparation).
17 CCSP, 2008b; http://www.pogo.org/investigations/contract-oversight/katrina/katrina-gao.html.
18 DeGaetano 2000; Jones and Mulherin 1998.
19 The Tappan Zee Bridge is expected to be replaced with a new structure, but timing is uncertain.
20 Stedinger (2010).
21 TRB (2008a).

[22] The New York State Constitution provides for democratically elected legislative bodies for counties, cities, towns and villages. These legislative bodies are granted the power to enact local laws as needed in order to provide services to their citizens and fulfill their various obligations.

[23] E.g., for MTA see Jacob et al. 2009; Jacob, 2009; NYS SLRTF, 2010; NYC CCATF, 2010; NPCC-CRI, 2009; NYS CAC, 2010; and stakeholder cooperation with this ClimAID project.

[24] http://www.dec.ny.gov/energy/45202.html.

[25] http://www.nyc.gov/html/oem/html/hazards/storms_hurricaneevac.shtml.

[26] *New York Times*, January 28, 2010: "S.E.C. Adds Climate Risk to Disclosure List" http://www.nytimes.com/2010/01/28/business/28sec.html?sq=sec&st=cse&scp=2&pagewanted=p.

[27] http://www.fta.dot.gov/about/offices/about_FTA_927.html#Mission and file:///Downloads/Post%209_11regional_offices_4154.html.

[28] http://www.gothamgazette.com/graphics/2008/04/DotDensityLowIncomeCommute.jpg.

[29] For a purely random occurrence of storms in time, statistics indicate that the probability that a 100-year storm does occur within the 100-year time period is only 63 percent. This is because the 100-year period is an average; thus, there are periods between such storm events that are longer than 100 years. These longer periods make up for occasional shorter recurrence intervals.

[30] Based on the Poisson Distribution, the probability for an event with average recurrence period T to occur in the time interval t is: p = 1 - e ^ -(t/T). When t equals T, in this case 100 years, the result turns out to be ~63%.

[31] The technical term of the average flood elevations for the waterways is: "area-weighted base flood elevations (AW BFE). These are later labeled, for simplicity, the Z_i values. For details and listing of the Z_i values in Table 9.13, see Appendix B.

[32] More could be added when maps of Long Island (Suffolk and Nassau County) for base flood elevations (BFE) of 1% per year and 2 and 4-ft sea level rise become available.

[33] The numbers in this figure were derived using a standard GIS intersection operation applied to the New York City street grid and to the three flood zones shown in Figure 9.7.

[34] A nearly complete and more detailed listing of lowest critical elevations of transportation systems in the New York City metropolitan region can be found in USACE (1995), with the caveats that (i) the lowest critical elevations in that reference are given with respect to NGVD, 1929; and (ii) that some modifications to structures or the terrain may have been made since the 1995 report was issued. Where we provide new information not contained in USACE (1995), the source is indicated where identifiable.

[35] MTA, 2006, courtesy A. Cabrera; communication of December 2009.

[36] The Port Authority has an emergency operational plan for Holland and Lincoln Tunnel and for part of its PATH system that will be activated prior to the arrival of a storm. LCE without such measures would be lower (e.g., Holland Tunnel vent shaft: LCE=7.6 feet; and Lincoln Tunnel vent shaft: LCE=10.6 feet).

[37] The ARC project was put on halt in 2010 to explore less costly options.

[38] Each step in this procedure is associated with large uncertainties. The procedure outlined here is site- and system-dependent, especially in the absence of a complete engineering risk and vulnerability assessment. Such an assessment is urgently needed to perform this task rigorously. The stakeholders provided physical data regarding tunnel volumes and pumping capacity of the most essential transport systems, but were unable to provide estimates of system vulnerability, repair, and restoration times and/or associated costs because there are too many unknown variables. Another large uncertainty is whether grid power will remain uninterrupted and, if interrupted, how long it will take power providers to restore it.

[39] ClimAID uses the hydraulic calculations for estimating the total floodwater volume in the tunnels.

[40] These numbers are preliminary and may change subject to more detailed engineering analyses.

[41] In contrast, the pumps installed in the NYCT subway tunnels are of older vintage and their purpose is not pumping out a flooded tunnel but draining the tunnels under normal operational conditions. NYCT's more than 750 pumps in 300 pump stations drain about 8 to 13 million gallons of water per day from the subway system, depending on whether it is a dry or wet day. Using 13 million gallons per day and 750 pumps yields 17,000 gallons/pump/day or just 12 gallons per pump per minute. If the total available pumping capacity after the scenario storm were 17,000 gallons per day (though the actual capacity is higher), it would take nearly 80 days to drain the system. However, not all of the 750 pumps are installed in the sections that would be flooded and, therefore, the process could take even longer. Note that the 12 gallons per minute value does not constitute the pumping capacity available during an extreme event. It is the pumping capacity used during a typical rainy day.

[42] If N is the number of pumps working in parallel at any given time, then the time required would be 1 week x (100/N).

[43] Based on Price Waterhouse Cooper (PWC) data for 2008.

[44] This daily gross regional product for the metropolitan region (DGMP), when used with the outage times listed in Table 9.5, allows the study to estimate the order of magnitude of the economic impact of outages. While the focus of this chapter is on transportation, the highly simplified assumption is used that the economic productivity is a direct function of the operational functionality of the transportation sector. In reality it reflects the functionality of all types of infrastructure (electricity, gas, water, waste, communication, etc.). But because most of these systems are so tightly coupled, the time estimates for transportation (Table 9.5) are, to a first-order approximation, a seemingly rational choice for a proxy for the functioning of all economic activity.

[45] And forward-projected to 2050 and 2090 by the multipliers of Table 9.14 in Appendix C.

[46] More could be added when maps of Long Island (Suffolk and Nassau County) for BFE of 1% per year and 2 and 4-ft sea level rise will become available.

Chapter 10

Telecommunications

Authors: Klaus Jacob,[1] Nicholas Maxemchuk,[2] George Deodatis,[2] Aurelie Morla,[2] Ellen Schlossberg,[2] Imin Paung,[2] Madeleine Lopeman,[2] Radley Horton,[4] Daniel Bader,[4] Robin Leichenko,[3] Peter Vancura,[3] and Yehuda Klein[5]

[1] Sector Lead (formerly at Columbia University)
[2] Columbia University
[3] Rutgers University, Department of Geography
[4] Columbia University Earth Institute, Center for Climate Systems Research
[5] City University of New York, Brooklyn College, Department of Economics

Contents

Introduction....364
10.1 Sector Description....365
10.1.1 Economic Value....366
10.1.2 Non-Climate Stressors....366
10.2 Climate Hazards....367
10.2.1 Temperature....367
10.2.2 Precipitation....368
10.2.3 Sea Level Rise, Coastal Floods, and Storms....368
10.2.4 Other Extreme Events....369
10.3 Vulnerabilities and Opportunities....369
10.3.1 Ice Storms....370
10.3.2 Hurricanes....372
10.3.3 Rain, Wind, and Thunderstorms....374
10.3.4 Extreme Heat and Heat Waves....374
10.3.5 Snowstorms....374
10.3.6 Electric Power Blackouts....375
10.3.7 Causes of Telecommunications Outages....375
10.4 Adaptation Strategies....377
10.4.1 Key Technical Adaptation Strategies....377
10.4.2 Larger-Scale Adaptations....379
10.5 Equity and Environmental Justice Considerations....380
10.5.1 Landline Dependency and Adaptation Decisions....381
10.5.2 Cascading Inequities and Challenges....381
10.5.3 Digital Divide....381
10.5.4 Deploying Rural Broadband as an Adaptation Strategy....383
10.5.5 Equity and Equity-Governance....383
10.5.6 Information and Telecommunication Technology Adaptation Strategies and Climate Change Mitigation....384
10.6 Conclusions....384
10.6.1 Key Vulnerabilities....385
10.6.2 Adaptation Options....385
10.6.3 Knowledge Gaps....386
Case Study A. Winter Storm in Central, Western, and Northern New York....386
References....392
Sources of Information....393
Appendix A. Stakeholder Interactions....394

Introduction

The telecommunications and broadcasting industries are vital elements of New York State's economy. Their combined direct economic contributions to the state's gross domestic product are on the order of $44 billion.[1] Telecommunications capacity and reliability are essential to the effective functioning of global commerce and of the state's main economic drivers, including the finance, insurance, information, entertainment, health, education, transportation, tourism, and service-based industries. It is essential to the daily life of every business, farmer, and citizen across the state, from rural to urban regions, and is especially vital during emergencies. Reduction in communication capacity for an extended period results in commercial and economic losses. This is a critical concern especially in the financial-service markets concentrated in and around the New York City area (The New York City Partnership, 1990).

The communications industry, perhaps more than any other sector, has undergone and continues to undergo a perpetual rapid technological revolution. It has experienced major deregulation and institutional diversification and functions in a state of fierce internal competition. In large part due to rapid technological changes, the planning horizons and lifespans for much of its infrastructure are at best on the order of a decade. This is a very short time horizon relative to the significant climate changes taking place over the scale of multiple decades to centuries. It is also short compared to that for other sectors, for example the public transportation sector, in which some rights of way, bridges, and tunnels have useful lifespans of 100 years or more. That is not to say that some parts of the communication infrastructure cannot be quite old. There are oilpaper-wrapped copper cables hung from poles or in the ground in some places, including New York City, many of which are older than 50 years.

The rapid technological turnover of communication infrastructure versus the pace of climate change gives rise to several inferences and issues:

In the context of the industry's vulnerability to weather and climate, it is essential to focus on its present uninterrupted service under extreme conditions. This, however, depends on the extent to which the market is willing to pay for such reliability and/or the extent to which the State and society at large demand and support higher reliability, including resilience to extreme events. The key questions are: What is the tolerable balance between reliability and cost? And who will bear the costs?

If service reliability and continuity are achievable at an acceptable cost for current weather extremes and if service disruptions can be better decoupled from electric grid power failures, there is good reason to expect that the industry could maintain high reliability vis-à-vis the additional hazards caused by climate change and be able to adapt to such changes with the help of new technologies.

Therefore, unlike many of the other sectors in the ClimAID report, addressing future climate change is arguably less important than addressing the communication industry's vulnerability to the current climate extremes. Additional hazards are expected from climate change in the sense that the frequency and severity of some extreme events are more likely to increase than not. Such events include excessive wind and lake effect snow in the coming decades, bringing down power and communication lines and even some wireless facilities. Some recent events have caused extensive and prolonged service failures with substantial economic and social impacts. Also, where centralized communications infrastructure is located at low elevations near the coast or near rivers and urban flood zones, climate change will pose additional risks that need to be managed comprehensively (see Chapter 5, "Coastal Zones," and Chapter 4, "Water Resources"). The areas at risk of flooding are expected to become larger, increasing the extent of flood zones as well as extending to higher elevations at the currently designated flood zones. In other words, the risk will increase in frequency and severity because of sea level rise and more extreme precipitation events. But these additional climate-change-induced risks are likely to be manageable in the future if currently existing vulnerabilities can be reduced.

There are a number of factors that make reducing

- The industry is experiencing strong internal competition and market pressures, which tend to limit redundancy to what dynamic free markets and profit motives are willing to pay for—on both the customer's and service provider's side. Market pressures and the short lifespan of certain telecom technologies result in an industry tendency to replace infrastructure as it becomes damaged, rather than to "harden" existing facilities. This would appear to be a reasonable response to lesser climate threats but it leaves critical components of the network vulnerable to rare but catastrophic events.
- Regulation and related mandatory reporting of service outages are limited and unequal among the different service modes and technologies.
- Customers have little accessible data to make choices based on reliability and built-in redundancy of services; instead, decisions are based largely on convenience, accessibility, marketing, and price.

Reducing current vulnerability while these factors prevail requires balance of policies between providing incentives to and regulation of the telecommunications industry. It can be argued whether it is valid to compare the risk-taking and aversion to regulation that has prevailed in the financial services sector to that of the technology-intensive communications sector. But such a comparative assessment may yield insight into changes to both business and public governance and policies that can guarantee the industry's reliable and continuous delivery of services—even during external shocks from climate-related (and other) extreme events. This could be for the benefit of the sustained economic health of the industry itself as well as of its customers and society at large.

A focus of the telecommunications infrastructure sector—including that of the service providers, the government, and the customer—is on how to ensure that the ongoing introduction of new technologies enhances the reliability and uninterrupted access to services, rather than degrading the reliability of these services. Such a focus is essential both now and in the future, when the impacts from climate change may increase.

The ClimAID telecommunications sector research team interacted with stakeholders from industry and government. A description of this process and the list of stakeholders are contained in Appendix A.

10.1 Sector Description

Telecommunications is one of the fundamental infrastructure systems on which any modern society depends. Its technological sophistication, availability, accessibility, broadband capacity, redundancy, security, and reliability of services for the private and public sectors are telling indicators of a region's economic development and internal social equity.

According to a report by the Federal Communications Commission (2009), the penetration rate for telephone service (land and cell combined) for all New York households was 91.4 percent in 1984, 96.1 percent in 2000 and 93.7 percent in 2008. Nationwide, the penetration rate was 95.2 percent in 2008, 1.5 percent higher than that of New York State. Demographic factors and level of aid to low-income households contribute to the differences in telephone service penetration among states. There is also considerable variance for income groups around the average of 93.7 percent within New York State.

At present, the telecommunications infrastructure sector comprises point-to-point public switched telephone service; networked computer (Internet) services, including voice over Internet protocol (VoIP), with information flow guided by software-controlled protocols; designated broadband data services; cable TV; satellite TV; wireless phone services; wireless broadcasting (radio, TV); and public wireless communication (e.g., government, first responders, special data transmissions) on reserved radio frequency bands.

The various domains are highly interconnected, overlapping, and networked. The boundaries between the different media are fluid and shift rapidly, often in concert with changes in technologies. Increasingly, the boundaries between technology providers versus content providers are also in flux.

Ongoing telecommunications innovations include the transition from analog to digital communication, introduction of networked computers, the Internet, broadband services, satellites, fiber optics, and the rapid expansion of wireless communication (including mobile phones and hand-held devices). Fourth-generation (4G) wireless technologies, such as Long Term Evolution (LTE) and WiMAX (Worldwide Interoperability for Microwave Access), provide an advanced IP-based

(Internet protocol) wireless platform for telephony, broadband Internet access, and multimedia services. These are some of the technologies that have transformed telecommunications in the last few decades. Some of these technologies have the potential to expand wireless voice and broadband coverage in unserved and underserved areas of the state.

In concert with technology, the institutional landscape of the industry has changed radically. Telecommunications giants, operating as regulated utilities with quasi-monopolies, were broken up in the United States in the mid-1980s to foster competition and innovation. The breakup was paired with considerable deregulation fostering robust intermodal competition followed by more deregulation. Among all types of service infrastructure on which society has come to rely, the telecommunications industry is almost entirely privately owned. It functions more competitively than most basic services that require large infrastructure, including electric power distribution (but not generation), transportation, and water and waste.

10.1.1 Economic Value

Telecommunications is an important sector in New York State's economy. Its total annual revenues contribute some $20 billion to the state's economy, about 2 percent of New York's entire gross state product of about $1.1 trillion (2007 dollars). Telecommunications is critical to the success of many of New York's largest industries and to many of the industries that will drive the state's growth in the future. New York City's status as a global financial center, for example, is heavily dependent on the capacity and reliability of its telecommunications networks. The New York Clearing House processes as many as 26 million financial transactions per day, at an average value of $1.5 trillion per day, for 1,600 financial institutions in the United States and around the world (NYCEDC et al., 2005).

10.1.2 Non-Climate Stressors

Not all areas of New York State have equal access to broadband wire services. **Figure 10.1** (top) shows a map of central offices (where subscriber lines are connected on a local loop), differentiating between those that are DSL-capable (digital subscriber line) and those that are not. **Figure 10.1** (bottom) shows the cable-modem

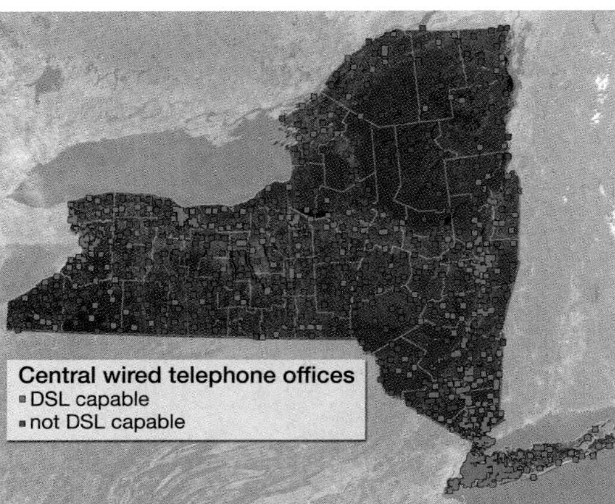

Distribution in 2010 of central offices for wired telephone in New York State. Those in green are capable of providing digital subscriber lines (DSL, 2009). Source: http://www.dslreports.com/comap/st/NY; basemap NASA

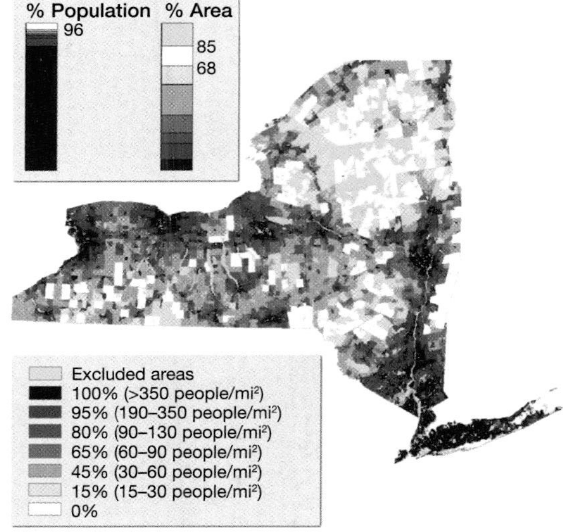

Source: Redrawn from NYS GIS, 2009

Figure 10.1 Distribution of central offices for landline telephone in New York State, 2010 (top); Predicted cable modem broadband availability, 2009 (bottom)

availability for 2009 as determined by the New York State Office of Cyber Security and Critical Infrastructure Coordination (CSCIC). Note that these are CSCIC's own projections and not based on data provided by service providers.[2]

The Federal Communications Commission (FCC) has oversight of the industry on the federal level, and the New York State Public Service Commission (PSC) exercises oversight on the state level. The stated

restored until December 19, over a week after the storm began (**Figure 10.7**).

The American Red Cross of Northeastern New York opened multiple shelters around Albany to give residents a warm place to stay and eat. At least four deaths were attributed to the storm. Three of the deaths (two in New York) were caused by carbon monoxide poisoning, the sources of which were gas-powered generators used indoors.

Hotels, hardware stores, malls, and restaurants that either had power or had a generator saw a boom in business during that weekend, as many residents finished holiday shopping, ate, and sought warmth. Most schools closed on Friday, December 12, and some colleges ended the semester early due to the severity of the storm.

Federal disaster aid topped $2 million for the nine New York counties that suffered damages from the December 2008 ice storm. Aid distributed to these counties and the State of New York is listed in **Table 10.1**.

Several weeks after the New England storm, a similar ice storm struck the midwestern United States, knocking out power to a million people and leading to at least 38 deaths.

Of note is that most outage reports cover the failure of power. Only some of these outages lead to telecommunications failures, which more commonly are experienced by consumers and less often by service providers. No consistent data for the failures of telecommunications services are in the public domain for the 2008 ice storm nor are such data available for many of the other storms described below, unless otherwise indicated.

Western New York State: April 3–4, 2003

During this ice storm, 10,800 telecommunications outages were reported. It took 15 days from the beginning of the storm to return conditions to normal. More than $25 million in federal aid was provided to help in the recovery (FEMA, 2003).

Northeast United States and Canada: January 4–10, 1998

The extent, thickness of accumulated ice, duration, and overall impact of the January 4–10, 1998, ice storm are

County	Federal Aid
Albany County	$295,675
Columbia County	$123,745
Delaware County	$324,199
Greene County	$203,941
Rensselaer County	$203,079
Saratoga County	$166,134
Schenectady County	$300,599
Schoharie County	$324,569
Washington County	$173,393

Table 10.1 Federal aid distributed to New York Counties as a result of the December 2008 ice storm

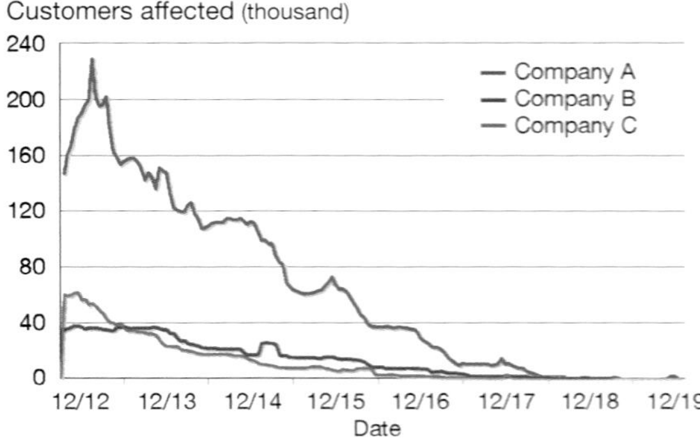

Source: NYS DPS 2009a

Figure 10.6 Number of reported customers with power outages versus time during the December 12–19, 2008 ice storm

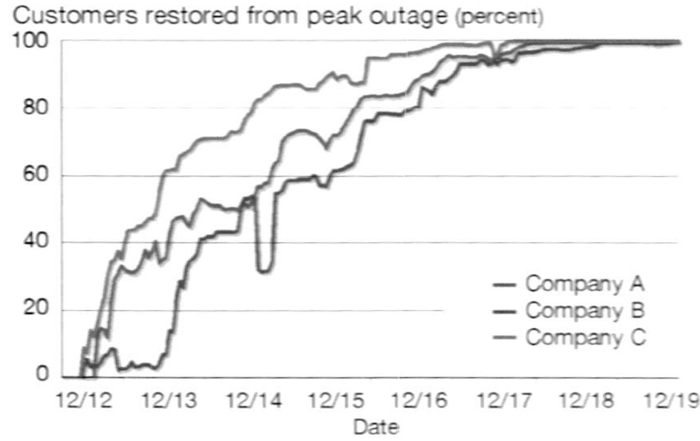

Source: NYS DPS 2009a

Figure 10.7 Percentage of customers with restored power versus restoration time during the December 12–19, 2008 ice storm

considered the most severe of any ice storm to hit eastern North America in recent history (DeGaetano, 2000). The storm affected both Canada and the United States (**Figure 10.8**).

In northern New York, tens of thousands of people living in isolated rural areas lost power and/or telephone service. Power was not restored in all parts of Jefferson County until 25 days after the start of the storm. It took another two to three weeks for services to be fully restored. Approximately 129,000 telecommunications problems were reported to one company (Jones and Mulherin, 1998; NYS PSC, 2007).

Emergency communications systems became stretched beyond capacity as a result of the ice storm. There was a sudden increase in emergency radio communications, and a number of calls were blocked because of overload of lines (**Figure 10.9**).

Pre-1998 Ice Storms Affecting New York State

Between 1927 and 1991, at least seven severe ice storms affected New York and/or New England states. Descriptions of their effects are given in USACE (1998). **Figure 10.10** depicts one of these storms, which devastated western and northern New York, Vermont, New Hampshire, and Massachusetts in 1991.

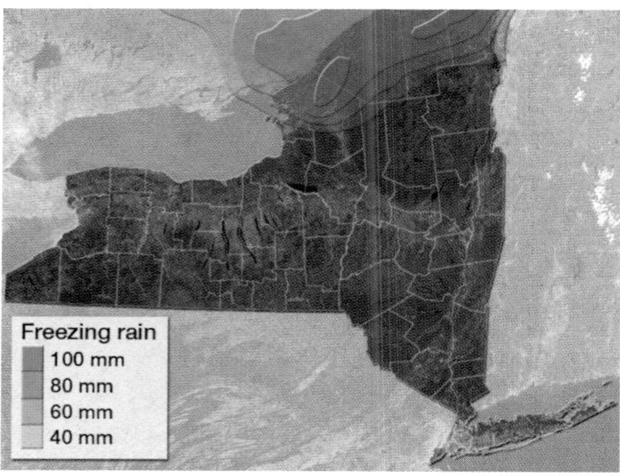

The blue-shaded areas represent freezing rain accumulations of more than 1.5 to nearly 4 inches (40–100 millimeters; 20-millimeter gradient). Affected areas reached from Lake Ontario to Nova Scotia, including four U.S. states (New York, Vermont, New Hampshire, and Maine) and four Canadian provinces (Ontario, Quebec, New Brunswick, and Nova Scotia). Source: Redrawn from Federal Communications Commission Spectrum Policy Task Force: Report of the Spectrum Efficiency Working Group. November 15, 2002; basemap NASA, based on data from Environment Canada

Figure 10.8 Distribution of ice accumulations between January 4 and 10, 1998

Six additional reported severe ice storms during this period occurred on the following dates:

- February 14–15, 1986
- January 8–25, 1979
- March 2–5, 1976
- December 22, 1969–January 17, 1970
- December 4–11, 1964
- December 29–30, 1942
- December 17–20, 1929

10.3.2 Hurricanes

To have maximum effect on the New York City metropolitan area, a hurricane would have to make landfall on the New Jersey coast, between Atlantic City and Sandy Hook. Since New York has not been directly impacted by a serious hurricane for the past several decades, this analysis uses hurricanes that have hit in the Gulf States as examples of the potential impact such a hurricane could have on telecommunications infrastructure in New York.

In 1938, the highest-category storm New York State has experienced made landfall in central Long Island, east of New York City (Hurricane Saffir Simpson 3). New York City was spared from the storm's worst effects, because the eastern side of the storm did not directly hit the city. (In the Northern Hemisphere, the eastern side is associated with the highest wind speeds and storm surges.)

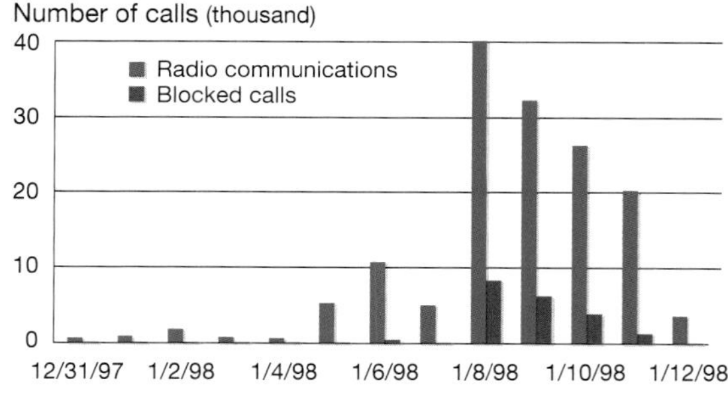

The first five days show normal background traffic, prior to when the storm hit. Source: http://www.stanford.edu/~rjohari/roundtable/sewg.pdf

Figure 10.9 Number of emergency radio communications per day and blocked calls because of overload in a single New York State county during the 1998 ice storm

Hurricane Katrina: August–September 2005

An excellent source of information on telecommunications vulnerabilities that became apparent with Hurricane Katrina, which made landfall as a category 3 storm, is FCC (2006). Hurricane Katrina struck the Gulf Coast in August 2005 and caused widespread flooding and wind damage, both of which affected telecommunications infrastructure. The duration of power outages during Hurricane Katrina exceeded the length of time that back-up batteries and fuel to power generators could supply communications. There were no means nor any plans and too many obstacles to restock fuel and batteries. Fuel to power the base stations lasted 24–48 hours, and batteries for portable radios lasted 8–10 hours. Thirty-eight 911 call centers went down and lacked an advance plan for rerouting calls. Most call centers in the low-impact areas took 10 days to restore. More than 3 million customer telephone lines lost phone service due to damage to switching centers and the fiber network and lack of sufficient diversity in the call-routing system.

Figure 10.11 shows the spatial distribution of causes of wired telephone system failure; lack of fuel supply for standby power features prominently. **Figure 10.12** indicates the failure mode for wireless services. In the area that experienced the largest service loss, diesel fuel ran out for back-up generators and supplies could not be replenished in time. It took 10 days to restore 90 percent of phone service.

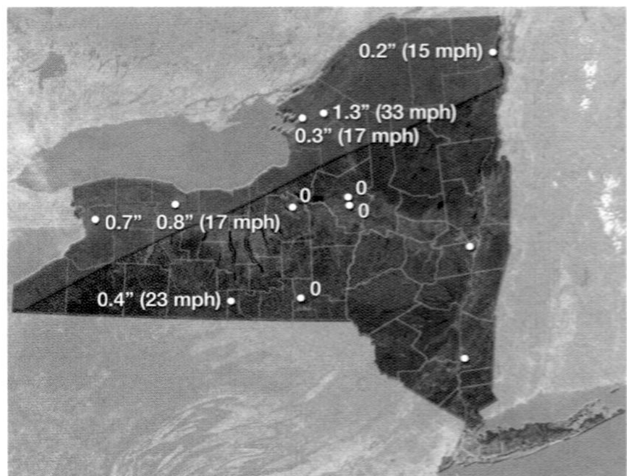

Source: Redrawn from (USACE, 1998), basemap NASA

Figure 10.10 Ice loads (inches) and wind speeds (mph) reported for the March 3-6, 1991 ice storm

In all, 35 broadcast radio stations failed, and only 4 stations worked during the storm. Also, 28 percent of television stations experienced downtime in the storm zone.

Hurricane Ike, September 2008

Hurricane Ike made landfall as a strong category 2 hurricane on September 13, 2008, near Galveston, Texas. On September 15, 2008, 75 percent of one company's customers in coastal Texas did not have service. Service was restored over the following days, with 60 percent lacking service on September 17, 48 percent on September 23, 30 percent on September 24, and 20 percent on September 26. As much as seven weeks later, some TV channels were not operative in severely hit areas. Most satellite TV customers also lost service. In the greater-Houston region, the functionality of cell phone services, on average, ranged between 60 and 85 percent in the days immediately following the storm in September 2008.

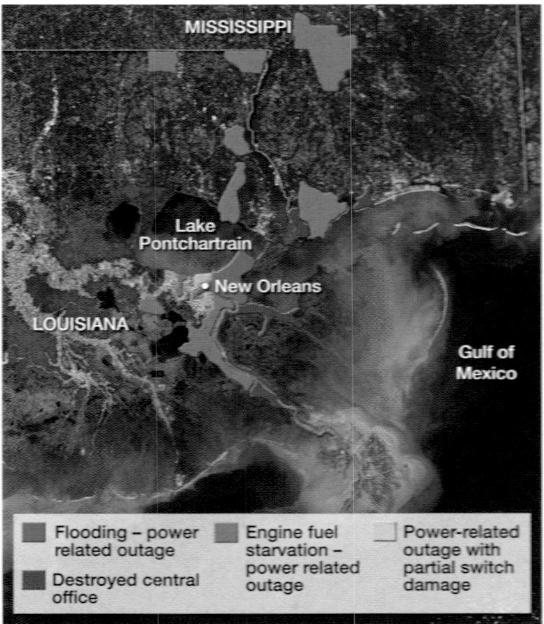

Note: Central office is where subscriber lines are connected to a local service loop. Source: Redrawn from: https://netfiles.uiuc.edu/akwasins/www/ Intelec06_Katrina.pdf; basemap: Jeff Schmaltz, MODIS Rapid Response Team, NASA/GSFC

Figure 10.11 Failure modes of the wired telephone systems after Hurricane Katrina

10.3.3 Rain, Wind, and Thunderstorms

Rain is generally of little consequence for communications facilities, except when buried facilities or central offices are flooded during urban flash floods or by overflow from nearby flooding rivers. Wind and thunderstorms are more substantial hazards to above-ground communications facilities, in part from falling trees and downed wires.

Nationally, an example was a windstorm in Washington State on December 16, 2006. Approximately 15,000 customers lost high-speed Internet for up to 48 hours. Rural areas in Kitsap and east King Counties experienced service disruptions. More than 46,000 customers lost telephone service between December 16 and 22; distribution-plant and power problems interrupted service for another 100,000 telephone customers, 400,000 Internet customers, and 700,000 television customers.

Closer to home, New York State experienced, for instance, the 1998 Labor Day thunderstorm affecting the Rochester to Syracuse and Utica regions. Approximately 37,000 telecommunications trouble reports were filed. It took 16 days from the start of the storm for service to return to normal.

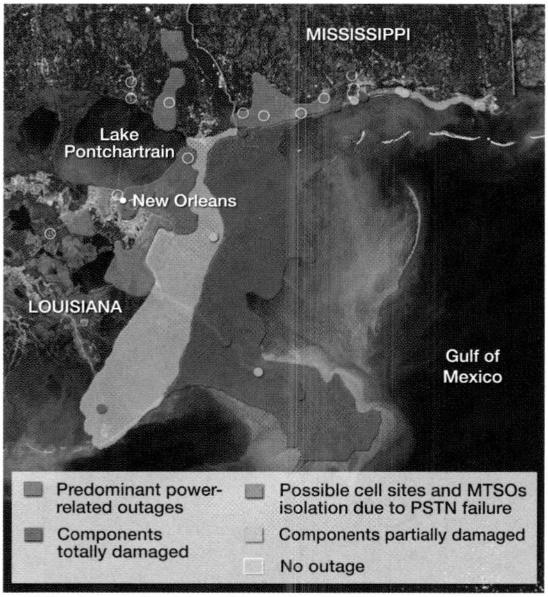

Circles show the locations (cell towers) included in the sample. MTSO stands for mobile-telephone switching office (which connects all individual cell towers to the central office); PSTN for public switched telephone network (which connects landline services).
Source: Redrawn from https://netfiles.uiuc.edu/akwasins/www/Intelec06_Katrina.pdf; basemap credit: Jeff Schmaltz, MODIS Rapid Response Team, NASA/GSFC

Figure 10.12 Zones of predominant failure type of wireless phone services

10.3.4 Extreme Heat and Heat Waves

Most heat-wave-related outages for the telecommunications sector are related to power outages that, in turn, are related to unmet peak power demands for air conditioning. Because of these similarities, see the example discussed below in Section 10.3.6, "Electric Power Blackouts."

10.3.5 Snowstorms

Several recent noteworthy snowstorms that affected either power or telecommunications systems, or both, in New York revealed considerable vulnerabilities of the telecommunications systems, often in connection with power failures.

Western New York: October 2006

Wet snow fell on October 13, when there was still foliage on the trees and many of them snapped under the heavy load (NYSDPS, 2007). From October 13 to November 10 (29 days), there were 93,000 reported disruptions to telephone service affecting one company's customers out of the roughly 475,000 access lines (i.e., an outage rate of about 19.6 percent) in the area affected by the storm. The company replaced about 350 downed poles and about the same number of distribution and feeder cables, and it repaired about 46,000 drop wires (i.e., wires connecting poles to homes or other buildings). **Figure 10.13** shows customer-reported service disruptions and the service restorations over the 29-day period that it took to fully restore wired phone services.

Power failures on Friday, October 13, affected approximately 400,000 customers as a result of the storm. The power companies completed restorations to full electrical service in 10 days. It took almost three times as long to complete restoration of wired telephone and cable TV services. From October 13 to November 10 (29 days), one company reported 149,000 cable television outages and repaired 46,000 lines. Most of

networking technologies that may or may not add diversity or robustness include:

- **Free-space optics** (FSO), an optical communication technology that uses light propagating in free space to transmit data between two points. The technology is useful where the physical connections by means of fiber optic cables are impractical due to high costs or other considerations. Free-space optics is only good for a few hundred yards to maintain high reliability (i.e., better than 0.999 or 0.9999). Any longer distances will produce circuit errors in heavy rain or fog.
- Commercial versions of ***ad hoc* networking** techniques typically relying on wireless communication. *Ad hoc* networks lack a designed infrastructure and form cooperative links between users to forward data. The structure of the network reflects the bandwidth requirements of the users in an area and the availability of access to the network infrastructure. However, ultimately they depend on the connection to the backbone wired network infrastructure, except in some relatively localized settings, which may be limited to urban environments.
- **Transmission via power lines**, which would reduce redundancy and couple power and communication failures more than they are currently.
- **Delay-tolerant networking techniques**. These networks can provide emergency communications during weather-related disasters, but are limited in data rate and quality. They include, for instance, those being proposed to provide communications to nomadic reindeer herders in Arctic latitudes. They are typically applicable to e-mails and text messages that are delay-tolerant.
- **Satellite phones and ham radio operators**, which have played important roles in emergency situations. The United Nations regularly distributes satellite phones in disaster regions internationally. These phones were in high demand during Hurricane Katrina. Satellite phones continued to operate following Hurricane Katrina and more than 20,000 satellite phones were used in the Gulf Coast region in the days following Katrina. Amateur ham operators have been the lifeline in many disasters and, perhaps, should be better organized. Not only should first responders be tied to them (some local emergency offices have such arrangements), but utilities should be organized to link with them as well.

10.4.2 Larger-Scale Adaptations

This section focuses on broader adaptation strategies for the telecommunications sector.

Diversification of Communications Media

Cable television and telephone distribution networks were originally different. Telephone systems used twisted wire pairs to connect to a central office, while cable television used coaxial-cable-based tree topology. A major difference between the cable company hybrid fiber-coax networks and the traditional telephone networks is that the former are more reliant on commercial power in the field and on electronic relays and amplifiers that have no back-up capability. They are not designed to operate in a power loss or blackout. Traditional telephone networks are designed to work even after a loss of commercial power. This critical reliability difference still exists today.

To some degree, the technologies in both networks have become more similar. They both use a fiber-optic network from a central location that connects to a customer's neighborhood with a short coax (cable television), twisted pair of wires, or a fiber connection (telephone systems) from the neighborhood node to a customer's premises. Both systems provide the same services to the end users (voice communications, high-speed data, and video distribution). The more recent technologies are more power-dependent, which affects reliability, resiliency, and recovery, although some use passive optical fiber technology requiring no power for "the last mile" (i.e., the last segment of telecommunications delivery from provider to customer).

It is possible that separate cable and telephone networks may evolve into a single monopoly distribution network that may be provided by a separate private or public utility company. Companies similar to the current cable and telephone companies may compete as service providers. If this occurs, a redundancy that currently exists in the multiple distribution networks may disappear, and the network may become more susceptible to failures caused by weather-related events. However, telephone and cable lines, while separate, are not really redundant in the sense that they are located on the same poles; if the poles are damaged in a storm, both cable and telephone lines may fail.

The Hurricane Katrina communications panel recommended more diversity of call routing in wireline networks to avoid reliance on a single route. The Public Service Commission instituted such diversity requirements following the September 11, 2001, outages that largely affected New York City (discussed further below) (NYSDPS, 2002; Case 03-C-0922). This approach is useful for routing traffic between switches, but does not help when the problem is in "the last mile," near the end customer. Also, the increasing use of Internet protocol for telephone services will provide routing diversity, because the information processing system will automatically search for any surviving physical routes. On the other hand, Internet-based networks often experience more widespread outages than a traditional network does when a major node or other centralized critical function location or equipment fails. This is common because these providers must leverage economies of scale to compete with bigger traditional companies and have fewer distributed facilities and less redundancy.

Natural Competition: Wired versus Wireless Networks

Wired networks provide point-to-point links that are more secure and private and can currently support much higher total data rates in a given geographic area. Improving antenna technologies, such as multiple-input and multiple-output (MIMO),[8] will continue to change this imbalance, but it is unlikely that the data rates provided by wireless technologies will exceed the rates provided by wired networks.

While wireless networks are in general dependent on wireline networks in order to backhaul data from cell sites to the backbone network, they do provide seamless communications to mobile, untethered users. They transfer information that is broadcast to a large set of receivers more naturally than wireline systems.

The current federal and state broadband initiatives could potentially encourage competition between wired and wireless media by developing both. However, major wireline companies own large portions of the wireless companies with major market shares in New York State. The development of either technology is likely to occur naturally by consumer choice, desired data rates, and considerations of quality versus price. Whether wired communications are more likely to prevail in densely populated, disadvantaged areas, while wireless communications prevail in sparsely populated rural areas, is questionable. In either case—wireline or wireless networks— in a competitive free-market telecommunications environment, commercial operators need a customer base to support the cost of infrastructure. Rural areas will continue to have more difficulty in obtaining access to high-speed broadband than urban areas, unless it is publicly supported, or prices may tend to be higher in the rural areas that often are least able to afford them.

Prior Adaptation Policy Recommendations

It is instructive to revisit what kind of measures and actions New York State agencies have already recommended vis-à-vis experiences from past extreme events, whether of natural or manmade origins. A review of these assessments reveals that nearly all proposed policy options and recommendations for reducing communications vulnerability to extreme events, made without particular reference to climate change, are directly relevant to the kind of extreme weather events discussed in the ClimAID report.

In the context of telecommunications, there is a comprehensive document that combines many of the findings, options, and conclusions for this important infrastructure sector: *Network Reliability After 9/11*, a white paper issued by the New York State Department of Public Service (NYSDPS, 2002). While it was originally inspired by the lessons learned from the September 11 events in 2001, it looked far beyond this single event and addressed fundamental systemic telecommunications vulnerability and reliability issues.

10.5 Equity and Environmental Justice Considerations

The rapid rate of innovation in telecommunications technology and the relative impermanence of the infrastructure mean the sector is potentially in a relatively good position to respond to climate change, signaled either by perceived physical risk or price changes. Yet flexibility and mobility present some challenges to enhancing social equity and ensuring that these technologies facilitate wide-ranging social resilience rather than exacerbate isolation and lack of access to information among more vulnerable people.

Because of the rapid changes taking place in the sector, monitoring equity involves examining the distribution of and access to old technology as well as rates of adoption and use of new technology. As climate risks affect decisions about types of infrastructure to deploy and where it can be built, a number of questions stand out: Are there specific regions, communities, or demographic groups that are likely to lose out? Which types of telecommunications technology and infrastructure are inherently more resilient? Will some adaptation decisions create new vulnerabilities for those using less resilient and obsolete infrastructure?

10.5.1 Landline Dependency and Adaptation Decisions

Because of enormous growth in new technologies, telecommunication companies are increasingly losing landline subscriptions. As of mid-2008, landline subscribers in the state had declined 55 percent since 2000. This is, in part, due to competition from increasing mobile phone penetration (which, in the context of storm vulnerability, may provide higher reliability where mobile services are available). The New York landline loss rate is comparable to that of the decline in landlines in New Jersey (50 percent), but surpasses the lowest rates in Connecticut (10 percent), Texas (20 percent), and California (21 percent) (Cauley, 2008). In the last year alone, one company lost 12 percent of its landlines. At the same time, the cost of maintaining the lines is increasing, and there are reports that some companies are pulling back on the upkeep of lines (Hansell, 2009; NYS DPS, 2009b).

Amid these changes, 14 percent of Americans are neither cell phone nor Internet users (Horrigan, 2009). Some of these customers are simply late adopters, but many others are households in isolated rural areas where new technologies have simply not yet penetrated. This leaves them dependent on landlines for lifeline services in emergency situations.[9] Adaptation strategies that focus disproportionately on the use of newer technologies and on implementation in areas with opportunities for greatest cost recovery may exacerbate the relative vulnerability of those reliant on landlines in more remote locations. Natural progression of technology can have a profound and beneficial impact on the reliability of networks if combined with responsible and realistic policies to address these concerns.

10.5.2 Cascading Inequities and Challenges

Similar to the way localized energy problems can ripple through the grid, a relatively localized disturbance to telecommunications infrastructure can create cascading impacts across regions and cripple widespread economic operations. For example, commercial transactions are increasingly reliant on credit card authorization, ATM withdrawals, and computer networks, services that are incapacitated with power and telecommunications outages (Quarantelli, 2007). Coping capacity reflects the underlying social and financial capital as well as the degree of isolation and service repair capacity. Rural and low-income communities are likely to be at a disadvantage.

On the other hand, it is possible that a progressive policy of universal service offers an opportunity to expand newer (wireless) technologies to the outer reaches of the network. This is comparable to developing nations "skipping" legacy telecommunication technologies. Cellular expansion in rural areas could make disaster recovery less burdensome (e.g., fewer drops to fix); allow utilities to pursue more efficient, centralized recovery strategies; and allow the severity of long-term power outages to be mitigated more easily. For example, rural customers are more likely to be able to use and recharge cell phones using car batteries, because vehicle ownership is more prevalent in rural areas. In contrast, modern fiber and cable networks are heavily dependent on the availability of commercial power.

10.5.3 Digital Divide

According to the 2008 State New Economy Index, New York ranks within the third quartile in terms of digital economy competitiveness (NYS Council for Universal Broadband, 2009), i.e., use of digital communication is widespread. At the same time, disparities in access to technologies and different rates of adopting them ensure that some areas and groups within New York State will benefit more than others from the potential of new information and communications technology to drive social and economic development and wellbeing. Sustainable development is an important tool for building local and regional resilience to climate stresses and shocks. Technology disparities are discussed in the next section as well as how infrastructure deployment aimed at

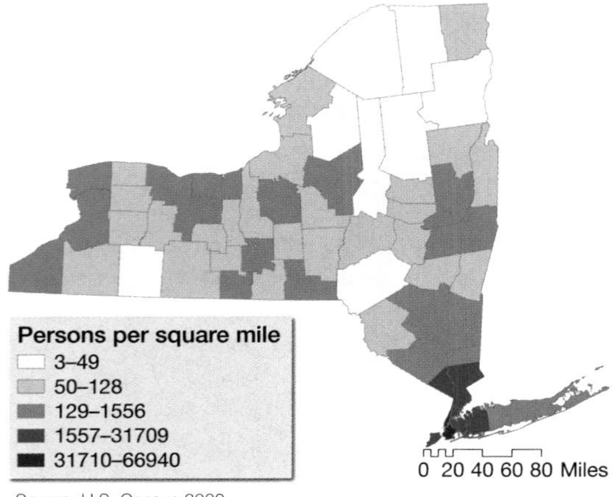

Source: U.S. Census 2000

Figure 10.17 Variation in population density in New York counties

minimizing these disparities could be part of a broad adaptation strategy.

Since the 1990s, the term "digital divide" has been employed to describe persistent differences in access to digital technology based on race, gender, age, geography, and socioeconomic condition (Light, 2001). For example, in a recent survey, low-income households adopted broadband at less than half the rate of higher-income households, and a wide gap was noted between white adults and African American adults (Horrigan, 2007 and 2008).

Demographic differences in rates of adopting technologies are compounded by regional differences in access to technologies. A national survey found that 24 percent of Internet users did not have broadband access because it was unavailable in their area (NYS Council for Universal Broadband, 2009a). Similarly, throughout New York State, there are communities where broadband is neither available nor affordable. The most sparsely populated counties are clustered in the Adirondack region and in Delaware and Allegany Counties (**Figure 10.17**). These areas also tend to have limited access to broadband. Notably, large parts of Franklin, Essex, and St. Lawrence have no availability at all. Compare this to the near-universal access in and around most of the state's urban centers (**Figure 10.18**). Perhaps most striking is the variation within counties. In Albany County, a noticeable division exists between urban centers such as the city of Albany, with coverage rates of 95 to 100 percent, and surrounding towns with less than 50 percent availability (**Figure 10.19**).

Access to wireless services (cell phones) is also limited in rural areas with low population densities. The same applies to the expansion of competitive wired networks, such as digital cable. Unfortunately, this is the reality of a non-regulated competitive industry. If there are not enough people to break even (much less turn a profit) on

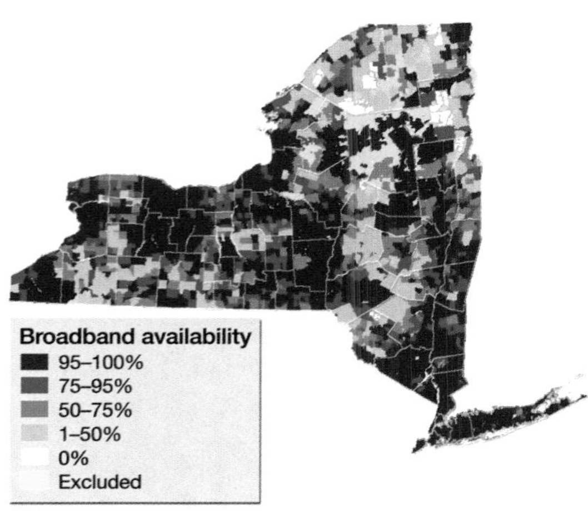

Source: Redrawn from NYS Council for Universal Broadband 2009a

Figure 10.18 Variations in wired broadband availability (cable-modem and DSL) in New York State, February 2009

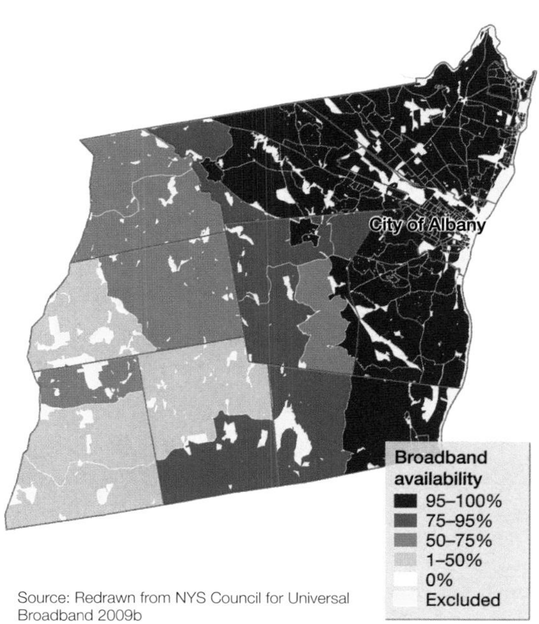

Source: Redrawn from NYS Council for Universal Broadband 2009b

Figure 10.19 Wired broadband availability (cable-modem and DSL) within Albany County, February 2009

the infrastructure required to deliver the service, it is very difficult for service providers to make that investment when other areas with higher population densities are in a similar need for additional capacity and speed. Some rural cell towers, unless they are on a highway corridor, operate at a loss. With continued downward pressure on wireless service prices, equitable distribution will continue to be a difficult problem to solve.

Introducing new technologies and maintaining equitable and reliable access are often conflicting. New technologies are introduced where they are most profitable, i.e., in high-density population areas. Noting this reality, short-term goals then could be to preserve service and access so that customers and critical services are not abandoned. The long-term solution should be to deploy a more reliable and equitable technology network that can be sustained by viable operators.

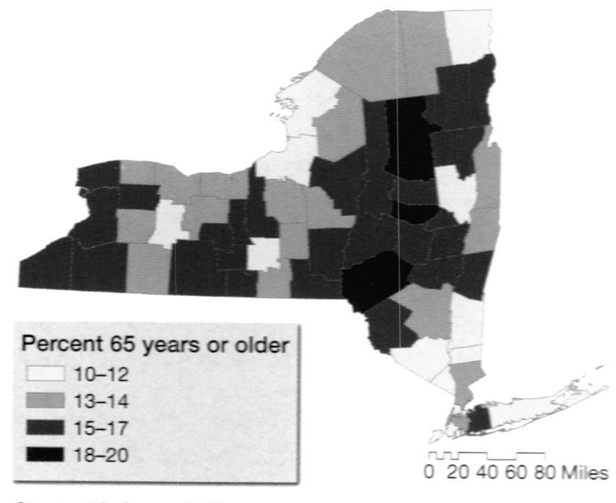

Figure 10.20 Regional variations in concentrations of population 65 years and older

Another demographic trend is that lower-income groups drop landlines faster than higher-income groups and use wireless as their sole means of communication.[10] On the one hand, this reduces redundancy in emergency situations, but on the other, because wireless is less vulnerable to extreme weather events, it implies more continuity of services during extreme events as long as customers find a way to recharge their mobile batteries (e.g., via charges from cars).

10.5.4 Deploying Rural Broadband as an Adaptation Strategy

Broadening the penetration and use of affordable and fast information and telecommunications technology can help strengthen the types and degree of connectivity between lower-income rural communities and economic centers, educational options, business services, and health infrastructure.

As part of a comprehensive development strategy aimed at employment and business diversity, for example, deploying broadband could help build social and economic resilience in regions dependent on climate-sensitive industries such as agriculture and natural resources (see Figure 3.4 of Chapter 3, "Equity and Economics"). It also could help increase citizen capacity to respond to climate-related disasters via better communication of risks and preparedness strategies. Recently, the federal National Telecommunications and Information Administration awarded a $40-million grant for the ION Upstate New York Rural Initiative to deploy a 1,300-mile fiber-optic network in northern New York State as part of the federal government's broadband stimulus program.

Rural deployment of broadband would tend to target regions with higher-than-average rates of aggregate population vulnerabilities. For example, Delaware County, one of the state's most sparsely populated counties, is located within the high-risk zone for ice storms and was hard hit by flooding in 2006. On top of this, it is also among those counties with the highest rates of poverty outside of New York City (see Figure 3.2 of Chapter 3, "Equity and Economics") and the highest proportion of elderly people(**Figure 10.20**). In the current recession, lower-income rural, elderly populations are especially vulnerable to additional climate extremes. These extremes could multiply the burden of regional economic decline on the elderly and also could cause the state to roll back the social supports that serve them (see e.g., *New York Times*, 2009).

10.5.5 Equity and Equity-Governance

Focusing on the use of information and telecommunications technologies as part of a broader strategy of inclusive community participation and sustainable development opens a range of possible strategies for equitable social, economic, and environmental gains in communities that might otherwise be exposed and sensitive to a variety of climate stressors.

Following the framework identified by a 2008 report (MacLean, 2008), information and telecommunications technologies can be coordinated for first-, second-, and third-order effects. Applied to adaptation, first-order effects include using innovative forms of technology to monitor and research climate change and adaptation, as well as to disseminate information on best practices and critical vulnerabilities. Second-order effects include using social networking and emergent forms of cooperative dialogue that build adaptive capacity and enable modes of debating and evaluating potential adaptations and risks. Finally, third-order effects encompass a whole suite of networked government measures related to equity, ranging from those that facilitate access to and coordination across branches of government to those that increase procedural justice by encouraging active executive participation among isolated or disengaged stakeholders.

To adopt these strategies, citizens must have equitable access to affordable information and telecommunications technology networks and knowledge of how to use these resources. Equally important is equitable access for local governments, where wide disparities in technological infrastructure exist across local planning departments in New York State (for an example, see Gross, 2003).

On a more sophisticated level, governance strategy to enhance equity requires building local capacity (e.g., through education, new management practices, behavioral changes) so that communities and governments have the means to creatively use technology for information gathering, dialogue, or participation. However, no amount of access can overcome persistent ignorance about how and when to use technology. Situations in which people do not know how to use technology may generate a false sense of security or control. In some cases, this can even increase vulnerability when the equipment malfunctions at a critical stage.

Telecommunication systems are designed so that the installed capacity can handle the typical daily peak traffic load. Add in a disaster, and the system will likely be overwhelmed. As long as telecommunications companies running the networks have to pay to operate and maintain the infrastructure on a competitive basis, change is unlikely. Wireless phone technology (and, to some extent, landlines) can augment capacity fairly quickly when needed in emergency situations. Some capacity-enhancing measures can be implemented immediately, trading off voice quality for additional traffic. Adding radios and backhaul capacity can take a few days, depending on the situation.

A useful adaptation strategy is to educate people about the impacts their behavior will have on a network during a disaster. To educate customers to send a text message about the tornado, as opposed to taking a picture and sending it from their cell phone (which uses more network capacity), is one example.

10.5.6 Information and Telecommunication Technology Adaptation Strategies and Climate Change Mitigation

Any significant expansion of information and telecommunications technology services needs to be evaluated with respect to the impact of increased energy use on household budgets. The expansion also needs to be evaluated with regard to its wider impact on greenhouse gas emissions. Cooling and operating more information and telecommunications technology servers and applications will result in increased energy demands. These processes already account for 1.5 percent of the energy consumption in the United States, and it is a percentage that is growing quickly (*The Economist*, 2008). Evaluating the efficiency gains of new technologies relative to this increased energy usage is a critical area for further research.

10.6 Conclusions

As discussed in this ClimAID chapter, telecommunications is an essential sector that is vital to New York State's economy and welfare. It is largely privately operated but has important public functions. Because of rapidly changing telecommunications technology and deregulated, fiercely competitive markets, some service providers tend to focus on short-term market share and profitability rather than pursuing long-term strategies to achieve reliability and redundancy. Business planning horizons are at most five to ten years, which is short compared to projected climate change trends over many decades. Even under current climate conditions, there are serious vulnerabilities that prevent the telecommunications sector from uniformly delivering reliable services to the public during extreme events. New York State can proactively engage industry to help prepare for more severe and more frequent extreme climate events in the future.

10.6.1 Key Vulnerabilities

The telecommunications sector is vulnerable to several climate hazards, many of which are projected to change in the future with climate change. The sector's key vulnerabilities include the following:

- Telecommunication service delivery is vulnerable to severe wind, icing, snow, hurricanes, lightning, floods, and other extreme weather events, some of which are projected to increase in frequency and intensity.
- In coastal and near-coastal areas, sea level rise in combination with coastal storm-surge flooding will be a considerable threat during this century to some central offices and underground installations. This risk extends up the tide-controlled Hudson River to Albany and Troy.
- The delivery of telecommunications services is sensitive to power outages, some of which result from increased energy demands during heat waves. Heat waves are expected to increase in frequency and duration.
- Telecommunication lines and other infrastructure are vulnerable to the observed and projected increase in heavy precipitation events resulting in floods or icing during freezing rain.
- Populations in underserved areas, especially in remote rural areas, often have only one type of service and hence lack redundancy. They may have difficulty reporting outages during extreme events and potentially life-threatening emergencies. For instance, during ice or snow storms, mobility can be severely hindered.

10.6.2 Adaptation Options

There are adaptation options and opportunities that can help the telecommunications sector prepare for the impacts of climate change. Key adaptation options and strategies include the following:

- Make the backbone network redundant for most if not all service areas, and resilient to all types of extreme weather events; provide reliable backup power with sufficient fuel supply for extended grid power outages.
- Decouple communication infrastructure from electric grid infrastructure to the extent possible, and make both more robust, resilient, and redundant.
- Minimize the effects of power outages on telecommunications services by providing backup power at cell towers, such as generators, solar-powered battery banks, and "cells on wheels" that can replace disabled towers. Extend the fuel storage capacity needed to run backup generators for longer times.
- Protect against outages by trimming trees near power and communication lines, maintaining backup supplies of poles and wires to be able to replace expediently those that are damaged, and having emergency restoration crews at the ready ahead of the storm's arrival.
- Place telecommunication cables underground where technically and economically feasible.
- Replace segments of the wired network most susceptible to weather (e.g., customer drop wires) with low-power wireless solutions.
- Relocate central offices that house telecommunication infrastructure, critical infrastructure in remote terminals, cell towers, etc., and power facilities out of future floodplains, including in coastal areas increasingly threatened by sea level rise combined with coastal storm surges.
- Further develop backup cell phone charging options at the customer's end, such as car chargers, and create a standardized charging interface that allows any phone to be recharged by any charger.
- Assess, develop, and expand alternative telecommunication technologies if they promise to increase redundancy and/or reliability, including free-space optics (which transmits data with light rather than physical connections), power line communications (which transmits data over electric power lines), satellite phones, and ham radio.
- Reassess industry performance standards combined with appropriate, more uniform regulation across all types of telecommunication services, and uniformly enforce regulations, including mandatory instead of partially voluntary outage reporting to the regulatory agencies.

- Develop high-speed broadband and wireless services in low-density rural areas to increase redundancy and diversity in vulnerable remote regions.

10.6.3 Knowledge Gaps

The industry generally lacks computerized databases that readily show the location and elevations of installed telecommunication facilities and lifelines and their operational capacity. Such data can be crucial in extreme weather events to make rapid damage, loss, and consequence assessments in potential hazard and damage zones. For security reasons, such databases need to be fully protected to allow only restricted, authorized accessibility.

The public lacks standardized easy access to information on service outages and expected restoration times. This information can be crucial in response actions taken during emergencies, by public first responders, businesses, and private households. Some consideration must be given to what kind of information is publicly accessible and what additional information is only accessible to authorized parties (government, first responders, etc.), because of security reasons. But these concerns must not prevent the public from having ready access to information in order to minimize the potential impact of emergencies.

A sound financial model is needed for telecommunications companies to implement costly reliability and resiliency measures and to remain competitively viable, since these companies 1) have obligations to serve high-cost rural customers, and 2) provide backbone services for all other communication modes described in this report.

The ClimAID assessment suggests both technical and policy options for effective adaptation strategies and reducing vulnerability/improving resilience. The following potential responses emerge from this assessment:

- Overcome the lack of and unevenness in transparency with respect to reporting and assessing vulnerabilities to climate-related hazards for both the current and future communication infrastructure systems and operations. Attune state actions to balancing the competing needs for public safety versus concerns for free-market competition and cyber security.
- Perform a comprehensive assessment of the entire telecommunications sector's current resiliency to existing climate perils, in all of their complexities. Extend this assessment to future climate projections and likely technology advances in the telecommunications sector. This includes the assessment of co-dependency between the telecommunications and power sectors' relative vulnerabilities. Provide options and incentives to decouple one from the other while improving resiliency of each.
- Implement measures to improve public safety and continuity of communications services during extreme events. Any such actions need to be risk-informed and need to consider the benefits versus costs to both the public and the industry for increased resilience to extreme events. They need to foster security for both the public and the industry and simultaneously advance competition, technological innovation, and equitable and affordable customer access across the state.

Case Study A. Winter Storm in Central, Western, and Northern New York

This ClimAID case study analyzes the impacts of a severe winter storm in central, western, and northern New York State, concentrating on two specific climate hazards based on geographic location in the state. For central New York, the focus is on an ice storm that produces freezing rain and ice accumulation. Snow accumulation is the focus for western and northern New York.

The case study's primary focus for the societal impacts of the winter storm is on the telecommunications infrastructure. However, a secondary area of examination is the effects of the winter storm on the electric power grid.

Ice Storm Scenario

Severe winter storms in New York generally follow a certain pattern, as described in section 10.2. A low-pressure system moves up the Atlantic Coast bringing warm moist air that encounters cold dry air in a high-pressure system over Canada and extends into the northern parts of New York. The northward movement of the counterclockwise-rotating storm system causes warm air to overrun the cold air mass. This typically forms three moving bands of precipitation (**Figure 10.21**):

- a southwest-northeast band of heavy rain closest to the coast
- parallel to it but farther inland, a band of freezing rain (ice)
- farther toward the northwest, another parallel band of precipitation that gradually grades from snow pellets into snow

The jet stream's position, strength, and persistence, as well as other meteorological factors, determine how large the storm system is; where and how fast or slowly it moves; how much total precipitation it will produce as rain, freezing rain/ice, and snow; how wide and long the three bands of precipitation stretch; and how the bands move in time and, hence, how long each phase of precipitation lasts at any location. Any given location may go through more than one precipitation phase (from rain to freezing rain to snow pellets to snow), while other locations may be affected only by a single precipitation band.

In this case study, a hypothetical composite of historical extreme winter storms is assumed. While the three precipitation categories (rain, freezing rain, and snow) would not necessarily be expected to occur concurrently in these proportions, each of these types of extreme winter precipitation is currently expected to occur on average at least once per century:

- up to 8 inches of rain falling in the rain band in near-coastal New York over a period of 36 hours
- up to 4 inches of freezing rain precipitating in the ice band in central New York, of which between 1 and 2 inches (radial, i.e., the thickness of accumulated ice as measured outward from the collection surface, such as a twig) accumulates as ice, over a period of 24 hours
- up to 2 feet of snow accumulating in the snow band in northern and western New York over a period of 48 hours

Figure 10.22 shows the three precipitation bands of the scenario storm system in relation to county boundaries within the state. The center of the ice band covers the cities of Binghamton, Albany/Troy, and Schenectady, and several rural areas in between and in their vicinity. The snow band covers Buffalo, Rochester, Syracuse, Utica, Plattsburg, and the Adirondacks. The rain precipitates over Long Island, New York City, and the mid-Hudson Valley counties to halfway between New York City and Albany.

Of New York State's 62 counties, 12 are assumed to be dominated by rain and about 20 by snow; about 30 are subjected to freezing rain. The county population density varies significantly from extreme urban (65,000

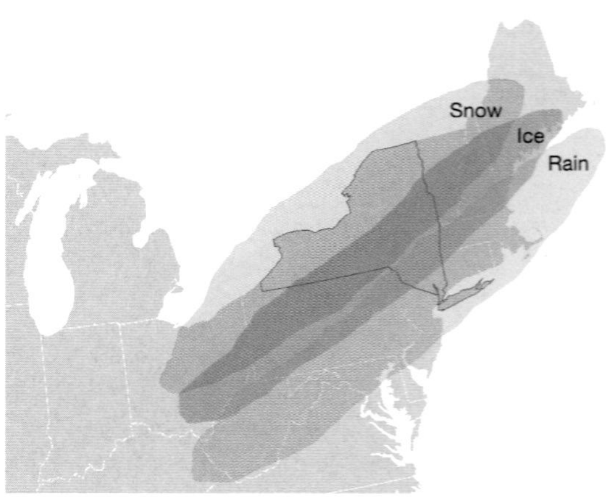

Note: The ice band includes a zone in New York State stretching from Binghamton through Albany into the Berkshires.

Figure 10.21 Typical pattern of severe winter storms in New York State

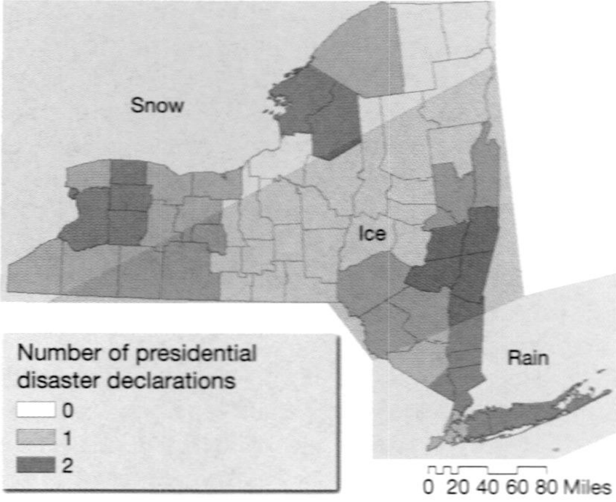

Source: Redrawn from NYSEMO historic map of presidential disaster declarations of winter storms in New York State for 1953 to 2007

Figure 10.22 Approximate overlay of the precipitation bands for the winter storm analyzed in the case study

people per square mile in Manhattan) to very rural (three people per square mile in Hamilton). Of the nearly 20 million people living in New York State, about 12 million are assumed to be affected largely by heavy rains, 4 million by freezing rain and ice, and about 4 million by snow. This weather-affected population (individuals) translates into about half of the above-quoted numbers as electric grid customers (households or businesses), with 6 million electric grid customers affected by heavy rain, 2 million affected by freezing rain and ice, and about 2 million affected by snow. About 95 percent of these customers in each of the precipitation categories are connected by wire (cable), wireless services, or both.

While there may be some urban flooding in the rain band, this assessment focuses on electric grid and telecommunications outages. Thus, the analysis largely examines the approximately 2 million New York customers in the ice band and the approximately 2 million customers in the snow band.

There are an estimated 4.1 million utility poles along about 145,000 pole miles in New York State,[11] i.e., an average of about 28 poles per pole-mile. Nearly one-third (almost 1.4 million poles) would fall into each of the three precipitation zones. This implies, on average, about 0.7 poles per customer in the less populated ice and snow bands and only slightly more than 0.2 poles per customer in the metropolitan area of the rain band, which, at least in New York City, has a large portion of the electric wires and phone lines running underground. These are average numbers, and the local values of poles per customer may vary in inverse relation to the population density, with more poles per person in less densely populated areas. Therefore, on average, rural customers have a higher chance of wire line problems from snow and ice loads than do city dwellers. Of course, if an urban area is struck by power outages, each outage can affect a much larger number of customers.

But because of the much longer average wireline per rural customer, and the assumed rate of ice and snow load failure is proportional to wire length (although other factors, such as proximity to trees and wind exposure, play a considerable role), rural customers can expect longer restoration times. Another factor is that utilities may decide to bring back the largest possible number of customers at the earliest possible time with the finite number of repair crews available. For this reason, there is a tendency to make restoring lines with a high customer density a higher priority. This may leave rural areas at a lower priority, not by intent but for technical reasons. The pattern of restoration often starts from the core of the network and radiates outward from there. Also, telecommunications companies generally follow the electric grid restoration, and hence the pace and pattern of electric grid restoration largely controls the pace and pattern of telecommunications restoration.

The Public Service Commission monitors restoration plans on a regular basis and works with utility companies via post-storm reviews to improve restoration planning and performance. This information is also important for updating emergency response and assistance readiness.

The electric grid outage rate during the 2008 ice storm left about 12.4 percent of customers without power (see section 10.3.1). The percentage varied from county to county and from township to township, affecting between a few percent of customers up to almost 60 percent of customers (with the largest outages in rural Otsego County, which has a population density of only five people per square mile). The 2008 ice storm was centered on Albany County. There, it had a (radial) ice thickness that rarely exceeded 1 inch.

This analysis considers an ice storm with 1 to 2 inches of radial ice accumulation, which raises the average outage to 25 percent of customers, notwithstanding the possible strong local deviations from this average. This would imply that within the ice band a total of some 500,000 New York State customers would be without power. Fewer customers would probably be without power in the snow zone. Most customers without electricity are likely to lose communication services sooner or later due to dropped wirelines placed on the same poles as electric lines; from the inability to sustain back-up power at central phone offices when they run out of fuel; from drained batteries that cannot be recharged in customers' wireless home sets or in their wireless phones; or from drained batteries, inside the customers' homes, located at the end of fiber-optic drop lines.

Exhausted batteries in fiber loop converters that serve wireless cell sites could also contribute significantly to the loss of wireless communication. Typically, a single fiber loop converter serves all the wireless carriers at a tower. If one of the carriers cannot get generator power to the fiber loop converter, the sites of all carriers go down at the tower.

Restoration Times

Estimates of likely restorations for power and communication services are based on the recent storms described in Section 10.3.1 of this chapter regarding reported power failure and restoration times, including those times given for the 1998 Canada/United States ice storm and the December 2008 New York ice storm centered on Albany. This scenario also assumes that the ice thickness is greater than the ice thickness in two out of the three ice storms described, and that adjacent states are also affected by the scenario ice storm and, thus, need some of their utility repair crews to restore their own outages.

Restoration Time Estimates

Based on the assumptions above, the estimated restoration times for the central ice band are as follows:

- Ten percent of customers who lost power will have their electricity restored within 24 hours after the ice stops accumulating (i.e., the first 50,000 of the half million customers in the band of freezing rain/ice).
- Fifty percent of customers will have electricity restored after 10 days (i.e., 250,000 customers).
- Ninety percent of customers will have their power restored after three weeks (i.e., 450,000 of the half million customers in the band of freezing rain/ice).
- Full restoration of power will take about five weeks (i.e., for the remaining 10 percent, or 50,000 customers, who are most likely located in remote, rural locations).

The restoration times in the snow zone may be slightly shorter than in the ice band. From the trends and historic cases described earlier, it is likely that the majority of customers in most of the larger cities (e.g., Albany, Binghamton, and the Schenectady area in the freezing-rain zone, and Buffalo, Rochester, Syracuse, Ithaca, and Utica in the snow zone) will be part of the first 50 percent of customers who lost power to have it restored, i.e., within the first 10 days.

However, large uncertainties exist, and local restoration times may depend, in part, on how well prepared a utility is to cope with the consequences of the storm. Preventive tree trimming, stocking poles and wires, and arranging for outside crews to assist in the restoration can all make a difference, either by reducing the failure rate or by shortening restoration times. Tree trimming is unpopular with many homeowners, and in some areas utilities have succumbed to political pressure and reduced the clearance they ordinarily would maintain.

Economic and Social Impacts: Productivity Losses, Damage, and Equity and Environmental Justice Issues

To estimate economic productivity and damage losses, the case study uses the number of people affected and the number of customers restored per number of days until restoration from the previous section. It also uses New York State's average per-person contribution to the state's gross domestic product ($1.445 trillion per year per 19.55 million people equals about $58,600 per person per year, which is equal to $160.50 per person per day).

Loss Estimates

Based on these assumptions, the losses to the state's economy are about $600 million in the first 10 days, $240 million between days 10 and 20, and $60 million in the remaining time from days 20 to 35. In total, this amounts to about $900 million ($0.9 billion) from productivity losses alone.

In addition to costs associated with lost productivity, costs associated with direct damages must be included as well (e.g., spoiled food; damaged orchards, timber, and other crops; replacement of downed poles and electric and phone/cable wires; medical costs; emergency shelter costs). These costs are likely to be of the same order as those of the productivity losses, which would imply a total ice storm cost of about $2 billion in New York State. This estimate does not include the snow effects on the state's economy and potential economic losses in the areas covered by snow. The loss estimate of $2 billion is probably on the low side, given that the 1998 ice storm resulted in losses of about U.S. $5.4 billion in Canada alone.

Equity and Environmental Justice Issues

The equity and environmental justice analysis uses the October 2006 snow storm in western New York as a historical analogue for illustrating potential social vulnerabilities during the recovery and restoration phase. The case considers rural areas and particular segments of the population who might be especially vulnerable during a protracted recovery. A primary advantage of analyzing this event instead of the 1998 ice storm is that the 2006 storm reflects a more current state of telecommunications technology. Its similarity to other severe ice storms is confirmed by one company's report that the degree of infrastructure damage and the magnitude of the company's response for the 2006 storm were comparable to those of the historic 1998 ice storm. Also, the 2006 storm triggered a recovery lasting nearly a month (NYSDPS, 2007), which is comparable with the estimates for restoration in this case study.

Following the 2006 storm event, the New York State Public Service Commission published a report detailing the steps leading up to the infrastructure failures and the subsequent difficulties in diagnosing problems and restoring service (NYSDPS, 2007). The report did not explicitly address population vulnerabilities, but it does reveal the limits of one communication company's capacity to respond, and it suggests a number of areas where these limits could be differentially experienced across regions and groups.

Date	Opening Trouble Load	Incoming Troubles	Troubles Cleared	Repair Technicians
10/14/2006	7,004	6,539	1,305	278
10/15/2006	10,811	6,274	2,467	372
10/16/2006	11,774	4,155	3,192	453
10/17/2006	15,699	7,196	3,271	497
10/18/2006	17,373	5,473	3,799	497
10/19/2006	18,263	4,791	3,901	535
10/20/2006	19,947	4,479	2,795	509
10/21/2006	19,604	4,015	4,358	514
10/22/2006	19,100	2,896	3,400	519
10/23/2006	19,700	2,068	1,468	568
10/24/2006	20,368	5,307	1,639	589
10/25/2006	21,218	3,830	2,980	599
10/26/2006	20,674	3,191	3,735	617
10/27/2006	20,157	3,213	3,730	608
10/28/2006	18,965	2,726	3,918	606
10/29/2006	17,361	1,986	3,590	607
10/30/2006	15,397	2,064	4,028	614
10/31/2006	14,884	3,164	3,677	606
11/01/2006	14,121	2,713	3,476	603
11/02/2006	13,055	2,358	3,424	649
11/03/2006	11,652	1,844	3,247	772
11/04/2006	10,085	1,801	3,368	776
11/05/2006	8,290	1,009	2,804	758
11/06/2006	6,113	934	3,111	732
11/07/2006	3,995	1,826	3,944	675
11/08/2006	2,540	1,747	3,202	636
11/09/2006	1,779	2,133	2,894	629
11/10/2006	1,388	1,306	1,697	448
11/11/2006	1,034	968	1,322	287

Table 10.2 Daily opening trouble reports, incoming troubles, troubles cleared, and staffing levels for October 2006 snow storm

The majority of damage in 2006 (and large amounts in the 1998 ice storm) was to tens of thousands of drop wires to individual building units. Nearly 93,000 trouble reports (not all may indicate that customers are out of service) were registered over a three-week period, with the peak report load being reached nearly two weeks after the storm (**Table 10.2**). These reports are a guide to restoration activities, with extended lag times on customer response complicating such efforts. As the report notes, one reason for the widespread delays was that customers were unaware that they were responsible for reporting the outage or assumed that service would be restored in time with power. One could expect that customers with better access to communications and information or who were socially and geographically more connected would be in a better position to understand their personal responsibility and act on the situation. On the other hand, isolated or impaired individuals or those who were in disconnected households in rural areas would be at higher risk of lengthened hardship.

The New York State Department of Public Service (2007) report notes another key variable in delays to restoring service: Large numbers of affected customers may have lost the incentive to promptly report outages because they simply switched to cell phones or left their homes. Whether these individual cases of non-reporting might contribute to aggregate, systemic, communitywide misdiagnoses and delays is unclear. But it does raise the prospect of one group's coping strategies potentially exacerbating the vulnerability of less mobile or otherwise isolated individuals who are located within the same communities. The report found it credible, for instance, that use of cellular phones likely contributed to delays in the company's initial damage assessment, which is key to the above suggestion that it delayed the restoration of more vulnerable customers.

In all such emergencies, there remains one big issue: How do households in rural communities report a telephone outage when the telephone services are out?

Coping during Service Restoration

Initial concentration on centralized and reported infrastructure failures is a technically logical reaction to the magnitude of the problem, but one that inevitably favors more densely populated areas. In more general terms, restoration after an ice storm would happen first in urban areas and then in rural areas, with smaller, remote communities likely to be restored last. This pattern is reinforced by the relative inaccessibility of remote areas in the aftermath of a storm, which prevents service technicians from safely restoring lines, particularly when the latter are in unapproachable areas in backs of houses, as was noted in the 2006 storm. Both of these issues are pertinent since central New York is marked by wide variations in population density and rapid transitions between accessible urban areas and more isolated rural areas.

The ability to cope through the lifecycle of a power and telecommunications outage partly reflects access to diverse telecommunications and transport options. In the 2006 ice storm, large numbers of households did cope by leaving their homes or switching primarily to cell phones. (The cell phone network relies, however, entirely on the landline network, except for the wireless link from the tower to the mobile phone. The tower is typically connected to the network over landline facilities, so cell phone service can fail when the lines feeding the towers are damaged.) Both of these strategies (leaving homes and cell phone use) rely on physical mobility, wealth, and geographic integration. More wealthy, urban populations with access to public transportation, adaptive vehicles (e.g., sport utility vehicles, all-terrain vehicles), or affordable temporary housing are substantively more resilient than elderly, low-income, disabled, rural, or otherwise transport-disadvantaged populations.

Under some conditions, cell phones can become a coping mechanism even when other parts of the communication network are down. However, cell phone coverage varies across providers and regions, and most major companies have dead zones within parts of rural New York State. Furthermore, during localized power outages, rural households with access to power exclusively from the electric grid will be—for as long as the latter is down—unable to recharge their cell phones without supplemental solar or car phone chargers.

Special Considerations and Communication Needs

Individuals with cognitive and physical impairments are less likely to receive emergency messages and to correctly interpret the recommended actions. This vulnerability could be compounded by mismanaged or misleading information disseminated by telecom providers (or other institutions).

In 2006, providers struggled to communicate critical information regarding service restoration promptly and consistently to the local media. At times, communication with public institutions bypassed local officials on the town and village level, officials who arguably would have been best placed to spread emergency communications (NYSDPS, 2007).

Case Study Conclusions

In summary, the case study shows that with the current state of vulnerability of power and telecommunications systems to winter storms, interruption of these services in New York State can affect hundreds of thousands of customers for many weeks from a single event. The resulting business interruptions and direct losses combined tend to produce losses in the hundreds of millions of dollars. Services for remote rural customers are typically the last to be restored and pose social injustice and inequities, and in some cases life-threatening emergency conditions.

References

Belinfante, A. 2009. "Telephone Subscribership in the United States (Data through March 2009)." Report for the Industry Analysis and Technology Division, Wireline Competition Bureau, Federal Communications Commission (FCC), August. http://hraunfoss.fcc.gov/edocs_public/attachmatch/DOC-292759A1.pdf.

Bennett, J. 2002. *State of the U.S. Telecommunication Networks and Root Cause Analysis: 2001 Annual Report*. ATIS Report, May.

Cauley, L. 2008. "Consumers ditching land-line phones." *USA Today*, May 14. http://www.usatoday.com/money/industries/telecom/2008-05-13-landlines_N.htm

Changnon, S.A. 2003. "Characteristics of ice storms in the United States." *Journal of Applied Meteorology* 42:630-639.

Changnon, S.A. and T.R. Karl. 2003. "Temporal and spatial variations of freezing rain in the contiguous United States 1948-2000." *Journal of Applied Meteorology* 42:1302-1315.

Congress of the United States Congressional Budget Office (CBO). 2007. *Potential Cost Savings from the Pre-Disaster Mitigation Program*. http://cbo.gov/ftpdocs/86xx/doc8653/09-28-Disaster.pdf.

DeGaetano, A.T. 2000. "Climatic Perspective and Impacts of the 1998 Northern New York and New England Ice Storm." *Bulletin of the American Meteorological Society* 81(2): 237-254.

DSL Reports. *DSL-ready map*. http://www.dslreports.com/comap/st/NY/.

The Economist. 2008, May 22. "Buy our stuff, save the planet." http://www.economist.com/opinion/displaystory.cfm?story_id=11412495

Federal Communications Commission (FCC). 2002.

Federal Communications Commission (FCC). 2008.

Federal Emergency Management Agency (FEMA). 2003. Ice Storm Disaster Aid Tops $25 Million [Press release]. Retrieved from http://www.fema.gov/news/newsrelease.fema?id=3947

Fox, S. 2005. *Digital Divisions*. Washington, DC: Pew Internet & American Life Project.

Global Climate Change Impacts in the United States. 2009. Edited by T.R. Karl, J.M. Melillo and T.C. Peterson. U.S. Global Change Research Program. http://downloads.globalchange.gov/usimpacts/pdfs/climate-impacts-report.pdf

Gross, D.L. 2003. "Assessing Local Government Capacity for Ecologically Sound Management of Flowing Water in the Landscapes of the Great Lakes Basin." Unpublished M.S. thesis, Cornell University, Department of Natural Resources.

Hansell, S. 2009. "Will the Phone Industry Need a Bailout, Too?" *New York Times*, May 8. Accessed May 20, 2009. http://bits.blogs.nytimes.com/2009/05/08/will-the-phone-industry-need-a-bailout-too/.

Horrigan, J. 2009. *The Mobile Difference*. Washington, DC: Pew Internet & American Life Project.

Horrigan, J. 2007. *Why it will Be Hard to Close the Broadband Divide*. Washington, DC: Pew Internet & American Life Project.

Horrigan, J. 2008. *Home Broadband Adoption*. Washington, DC: Pew Internet & American Life Project.

Johnson, K. 2009. "For Rural Elderly, Times are Distinctly Harder." *New York Times*. December 10.

Jones, K.F. 1996. "Ice accretion in freezing rain." U.S. Army Corps of Engineers Cold Regions Research and Engineering Laboratory. Rep. 96-2.

Jones, K.F. and N.D. Mulherin. 1998. "An evaluation of the severity of the January 1998 ice storm in northern New England." U.S. Army Corps of Engineers Cold Regions Research and Engineering Laboratory report.

Karl, T.R., J.M. Melillo, and T.C. Peterson (eds.). 2009. Global Climate Change Impacts in the United States. Cambridge University Press.

Light, J. 2001. "Rethinking the digital divide." *Harvard Educational Review* 71(4):709-735.

MacLean. 2008. "ICTs, Adaptation to Climate Change, and Sustainable Development at the Edges." *An International Institute for Sustainable Development Commentary*.

Manuta, L. 2009. "New FCC Telephone Penetration Statistics Reveal Problems in New York." *The Public Utility Law Project of New York (PULP) Network blog*, August 14. http://pulpnetwork.blogspot.com/2009/08/new-fcc-telephone-penetration.html.

McKenna, C. 2009. "New York Statewide Wireless Interoperable Communications Network Refocused on Regional Systems." *Government Technology*. http://www.govtech.com/gt/635218?id=635218&full=1&story_pg=1.

Multihazard Mitigation Council (MMC). 2005. "Natural Hazard Mitigation Saves: An Independent Study to Assess the Future Savings from Mitigation Activities." National Institute of Building Sciences. http://www.nibs.org/index.php/mmc/projects/nhms and http://www.nibs.org/client/assets/files/mmc/Part1_final.pdf.

National Oceanic and Atmospheric Administration (NOAA) National Weather Service (NWS). 2008. "The Historical 11-12 December 2008 Ice Storm." http://cstar.cestm.albany.edu/PostMortems/CSTARPostMortems/2008/12dec2008icestorm/dec2008icestorm.htm.

New York City Economic Development Corporation (NYC EDC), New York City Department of Information Technology and Telecommunications (NYC DoITT), New York City Department of Small Business Services (NYC DSBS). 2005. *Telecommunications and Economic Development in New York City: A Plan for Action*. A Report to Mayor Michael R. Bloomberg. March.

New York City Partnership (NYCP). 1990. "The $1-Trillion Gamble: Telecommunications and New York's Economic Future." A Report by the New York City Partnership in Collaboration with Booz, Allen & Hamilton, Inc., June.

New York State Council for Universal Broadband. 2009b. *Broadband Map, Albany County* http://www.nysbroadband.ny.gov/maps/counties/albany.htm.

New York State Geographic Information Systems (NYS GIS). 2009. *Clearinghouse*. http://www.nysgis.state.ny.us/coordinationprogram/reports/presentations/gisconf09/2009_State_of_State_FINAL.

New York State Office for Technology (NYS OFT). 2008. "Going from Good to Great." http://www.oft.state.ny.us/News/FinalNYS2008GoalsandStrategies.pdf.

New York State Office for Technology (NYS OFT). 2009a. "Connecting New York to the World for Sustainable Adoption: New York State Universal Broadband Strategic Roadmap." New York State Council for Universal Broadband, June. http://www.nysbroadband.ny.gov/assets/documents/Final_Broadband_Strategy_June2009.pdf.

New York State Office of Cyber Security and Critical Infrastructure Coordination. 2009. "2009 State of the State in GIS." http://www.nysgis.state.ny.us/coordinationprogram/reports/presentations/gisconf09/2009_State_of_State_FINAL.pdf

New York State Public Service Commission (NYS PSC). 2002. "Network Reliability After 9/11". A Staff White Paper on Local Telephone Exchange Network Reliability. NYS DPS Office of Communications. November 2.

New York State Public Service Commission (NYS PSC). 2003. "Case 03-C-0922: Network Reliability Proceeding." http://www.dps.state.ny.us/03C0922.html.

New York State Public Service Commission (NYS PSC). 2004. "Initial Report on the August 14, 2003 Blackout." February. http://www3.dps.state.ny.us/pscweb/WebFileRoom.nsf/Web/5FA2EC9B01FE415885256E69004D4C9E/$File/doc14463.pdf?OpenElement.

New York State Public Service Commission (NYS PSC). 2005. "Second Report on the August 14, 2003 Blackout." October. http://www3.dps.state.ny.us/pscweb/WebFileRoom.nsf/Web/DA5944111F5033A4852570B4005E778A/$File/FinalSecondBlackoutRpt-4-03-06-errata.pdf?OpenElement.

New York State Public Service Commission (NYS PSC). 2006. *Mission Statement*. http://www.dps.state.ny.us/mission.html.

New York State Public Service Commission (NYS PSC). 2007. "October 2006 Western New York Major Snowstorm: A Report on Utility Performance" May. http://www3.dps.state.ny.us/pscweb/WebFileRoom.nsf/Web/2D5910CDFE0696CA852572DD005F9D80/$File/WesternNewYorkOctober2006SnowstormFinalReport.pdf.

New York State Public Service Commission (NYS PSC). 2009a. "October and December 2008 Winter Storms. A Report on Utility Performance," June. http://documents.dps.state.ny.us/public/MatterManagement/CaseMaster.aspx?MatterSeq=30915.

New York State Public Service Commission (NYS PSC). 2009b. Case 09-M-0527. http://documents.dps.state.ny.us/public/MatterManagement/CaseMaster.aspx?MatterSeq=31654.

Quarantelli, E.L. 2007. "Problematical Aspects of the Computer Based Information/ Communication Revolution with Respect to Disaster Planning and Crisis Managing." Working Paper 358, University of Delaware Disaster Research Center. Accessed May 29, 2009. http://dspace.udel.edu:8080/dspace/bitstream/19716/3055/1/PP%20358.pdf.

Sullivan, B. 2006. "Why cell phone outage reports are secret." *MSNBC The Red Tape Chronicles*, December 15. http://redtape.msnbc.com/2006/12/why_cell_phone_.html.

United States Government Accounting Office (US GAO). 2007. *Natural Hazard Mitigation* August. http://www.gao.gov/new.items/d07403.pdf.

Untied States Army Corps of Engineers (USACE). 1998. "Ice storms in New England and New York."

Victory, N. 2006. "Report and Recommendations of the Independent Panel Reviewing the Impact of Hurricane Katrina on Communications Networks." Memo to the Federal Communications Commission (FCC), June 12. http://www.fcc.gov/pshs/docs/advisory/hkip/karrp.pdf.

Sources of Information

http://www.dps.state.ny.us/mission.html

http://www.govtech.com/gt/635218?id=635218&full=1&story_pg=1

http://www.oft.state.ny.us/News/FinalNYS2008GoalsandStrategies.pdf

http://redtape.msnbc.com/2006/12/why_cell_phone_.html

Case 09-M-0527 brought before the NYS PSC re the Universal Service Fund http://documents.dps.state.ny.us/public/MatterManagement/CaseMaster.aspx?MatterSeq=31654

http://www3.dps.state.ny.us/pscweb/WebFileRoom.nsf/web/B3C502B29ECB84FD85256DF100756D51/$File/doc4618.pdf?OpenElement

Variations in Broadband Availability; NYS Broadband Council. http://www.nysbroadband.ny.gov/assets/documents/Final_Broadband_Strategy_June2009.pdf

http://www.nielsenmobile.com/documents/WirelessSubstitution.pdf

http://www.cdc.gov/nchs/data/nhis/earlyrelease/wireless200905.htm

http://www.ntia.doc.gov/broadbandgrants/BTOPAward_IONHoldCoLLC_121709.pdf

http://en.wikipedia.org/wiki/December_2008_Northeast_ice_storm

http://www.fema.gov/news/newsrelease.fema?id=48163

http://www.associatedcontent.com/article/1428697/kentucky_storms_weather_cause_phone.html and http://www.kentucky.com/news/state/story/682579.html

Kentucky power outages 1/28/09: http://vielmetti.typepad.com/.a/6a00d8341c4f1a53ef010536ff9773970c-800wi]

http://www.aimclearblog.com/2008/04/11/massive-spring-blizzard-takes-out-duluth-internet-services-electricity/

http://www.fema.gov/news/newsrelease.fema?id=3538

http://raincoaster.com/2006/06/08/the-ice-storm-quebec-1998/

http://www.nytimes.com/1998/01/11/nyregion/us-declares-five-counties-disaster-area.html

http://www.greenspun.com/bboard/q-and-a-fetch msg.tcl?msg_id=001HBW

http://www.stanford.edu/~rjohari/roundtable/sewg.pdf

Bell South/Katrina: https://netfiles.uiuc.edu/akwasins/www/Intelec06_Katrina.pdf

http://news.vzw.com/news/2008/09/pr2008-09-16s.html

NYS Public Workshop on Utility Preparation, Response, and Recovery from the December 2006 Wind Storm (Docket No. U-070067). http://www.governor.wa.gov/news/2007-03-09_storm_workshop_summary.pdf

Annual frequency of outages vs. customers affected for US 1984 to 1997
https://reports.energy.gov/B-F-Web-Part3.pdf.
Open File Report on Black Out 2003: http://www3.dps.state.ny.us/pscweb/WebFileRoom.nsf/Web/5FA2EC9B01FE415885256E69004D4C9E/$File/doc14463.pdf?
http://www.beyondpesticides.org/wood/pubs/poisonpoles/tables/table2.html
http://www.pcworld.com/article/159630/universal_chargers_to_finally_become_a_reality.html and http://reviews.cnet.com/8301-13970_7-10165603-78.html
http://www.arrl.org/
http://tsp.ncs.gov
http://www.iec.org/online/tutorials/ss7/index.asp

Appendix A. Stakeholder Interactions

The first ClimAID project stakeholder meeting for the Telecommunications sector was held in conjunction with the Transportation sector stakeholders on February 12, 2009. Following this initial meeting, a questionnaire was developed and sent to the stakeholders. The questionnaire highlighted information that would allow an assessment of the most important challenges posed by climate change.

ClimAID telecommunications infrastructure stakeholders were invited to comment on a chapter draft dated January 8, 2010. We acknowledge the thorough reviews by several stakeholders.

Stakeholder Questionnaire

NYS ClimAID: Telecommunications Survey for Information Covering the Entire State of New York (4/07/2009)

A. Commercial Power

1) How many a) office facilities (central offices, head-ends, mobile switch centers) and b) outside plant facilities (cell towers, controlled environmental vaults, fiber nodes, etc.) have back-up power generation? (Give both percentage and actual number for both a. and b.)
2) What portion of facilities with back-up power generation is provided by a) battery and b) generator, or c) some other type of back-up generation?
3) How long can facilities operate on back-up generation types identified in question 2?
4) What arrangements are in place to replenish backup generation fuel and supplies for extended commercial power outages?

B. Wireless Networks

5) How many transmitters/repeaters are a) singularly located on towers, and b) co-located on towers with other service providers? (Give both percentage and actual number for both a. and b.)
6) Do you expect the arrangements in question 5 to change significantly over the next 5 years? 10 years?
7) What portion of the backbone network interconnecting transmitters/repeaters to the mobile switching offices are comprised of the following facilities: a) wireless, b) telephone company, c) cable company, d) other service provider?
8) What portions of cable facilities are a) aerial and b) underground?

C. Wireline (cable TV, telephone) Networks

9) How much of the outside cable plant is a) aerial cable, and b) underground cable?
10) How much of the outside cable plant is a) copper cable, and b) fiber optic cable? (Give both percentage and actual miles for both 9. and 10.)

D. Climate Hazard Thresholds

11) Do outside plant facilities (towers, antennas, aerial cables) meet or exceed industry recommended standards for surviving maximum wind velocities (mph) and ice loading? What are these maximum limits?
12) How many a) office facilities (central offices, head-ends, mobile switch centers) and b) outside plant facilities (cell towers, controlled environmental vaults, fiber nodes, etc.) are located in FEMA-designated flood zones (according to FIRM maps)?
13) What restoration/contingency plans are in place to prevent or mitigate service interruptions if these facilities become inundated? Note: FIRM maps are web accessible by state/county from: http://msc.fema.gov/

In an effort to improve healthcare provision, in 1996 New York State initiated a data and knowledge communication program linking a wide range of partners, including hospitals, local health departments, nursing homes, diagnostic centers, laboratories, insurance provider networks, and federal agencies. Current communication networks—the Health Alert Network (state and city levels), the Health Provider Network, and the Health Information Network—are viewed as "both very helpful and very underutilized" by the Public Health Association of New York City (PHANYC, 2001). However, as a result of non-standardized data systems, the value of these networks across user groups is often compromised (PHANYC, 2001). These would be appropriate organizations to target for climate-health educational outreach and to evaluate climate-health interventions.

11.1.2 New York City Public Health System

New York City has been at the forefront of public health programming and policy since the founding of the Board of Health in 1866, the first such agency in the United States. More recently, New York City conducted the nation's first regional Health and Nutrition Examination Survey (NYC HANES), modeled after the CDC's National Health and Nutrition Examination Survey, providing policymakers and public health professionals with invaluable population-based health information (NYC DOHMH, 2007).

In 1995, the New York City Department of Health and Mental Hygiene (DOHMH) instituted a system of syndrome-based surveillance to locate potential disease outbreaks through ongoing monitoring of public health service use patterns and analysis for time- and location-related deviations. What started primarily as a means to detect waterborne illnesses that cause diarrhea through tracking influenza-like symptoms has evolved into electronic reporting of diverse health-related data. It now incorporates city emergency departments, pharmacy and over-the-counter medication purchases, employee absenteeism, and ambulance dispatch calls (Heffernan et al., 2004). With 39 city emergency departments participating, the electronic surveillance system covered about 75 percent of annual emergency department visits in its first year of operation alone (Heffernan and Mostashari et al., 2004).

11.1.3 Public Health Funding: Sources and Targets

Local health departments are funded by a combination of federal and state income streams and grants, complemented by fees levied through the local tax base and distributed by the State in proportion to county population. According to the Public Health Association of New York City (PHANYC), in 2001, New York City accounted for 46 percent of State aid, with the next six largest counties (Suffolk, Nassau, Erie, Westchester, Monroe, and Onondaga) receiving an additional 22 percent. Together these most-populous counties, which contain 72 percent of the state's population, accounted for 70 percent of the State aid to local health departments (PHANYC, 2001). In the 2001 fiscal year, the New York City Department of Health and Mental Hygiene budget drew 62 percent of funding from city tax revenues (PHANYC, 2001).

There is growing concern among public health practitioners that the confluence of State budget tightening with increasing needs of emerging chronic illnesses and emergency programming may threaten provision of basic healthcare services—both climate and non-climate related (NYS ACHO, 2008). While post-September 11 federal funding for emergency preparedness programming has benefitted the entire state and many aspects of surveillance and programming, the sufficiency and security of these funds into the future is a matter of serious concern (NYSPHC, 2003). It is also important to note that the federal health care landscape is evolving in significant ways as a result of the recent passage of health care reform legislation.

11.1.4 Emergency Preparedness

Projected changes in frequency and severity of extreme weather events will call upon the emergency preparedness plans within New York State. The New York State Disaster Preparedness Commission, made up of 23 State agencies and the American Red Cross, is responsible for disaster planning as well as communications with all levels of local, state, and federal-related bodies. The attacks of September 11 highlighted both strengths and gaps in New York City's public health infrastructure and underscored the importance of preparedness for the state in general. Immediate responses demonstrated the coordination of multiple health agencies to quickly and effectively react to threats to the public health of the city (Rosenfield, 2002). Transfer of the Office of Emergency Management command center from the World Trade Center (a high-profile, vulnerable location) to its current location in Brooklyn was one of the lessons learned. Most important, the events made clear that investments in preparedness infrastructure benefit the daily operations and effectiveness of the public health system.

In 2002, Congress designated Centers for Disease Control and Prevention funding for nationwide capacity building and emergency response training initiatives and research through the Academic Centers for Public Health Preparedness program (Rosenfield, 2002). Columbia University in New York City was one of these centers and continues to provide valuable contributions in research and training to public health professionals through its National Center for Disaster Preparedness.

11.1.5 Current Health Status for Climate-sensitive Diseases

People whose health is already compromised by pre-existing disease are likely to be among the most vulnerable to emerging climate impacts. This is likely to be the case for a wide range of disease types. We can also identify a subset of diseases that may be particularly climate sensitive, either because the existing burden of disease is especially high or because climate change could directly impact the incidence or severity of the disease. Here we highlight three broad disease categories—asthma, cardiovascular, and infectious diseases—that are likely to be particularly climate sensitive in New York State. These were selected based on the limited evidence that currently exists on climate change and health. However, we do not mean to imply that these are the only disease categories for which climate change is or will be relevant in New York State. Ongoing research and reassessment will be critical to identify and target emerging health risks.

Asthma

Asthma is potentially a climate-sensitive disease. It is already well established that asthma is exacerbated by certain weather patterns, pollen and mold seasons, and air pollution, and also is affected by indoor allergens like dust mites. Asthma can have allergic (such as pollen) or non-allergic (such as ozone) triggers, with the majority being of the allergic type. Many asthmatics are considered of mixed type, i.e., they are potentially sensitive to both types of triggers.

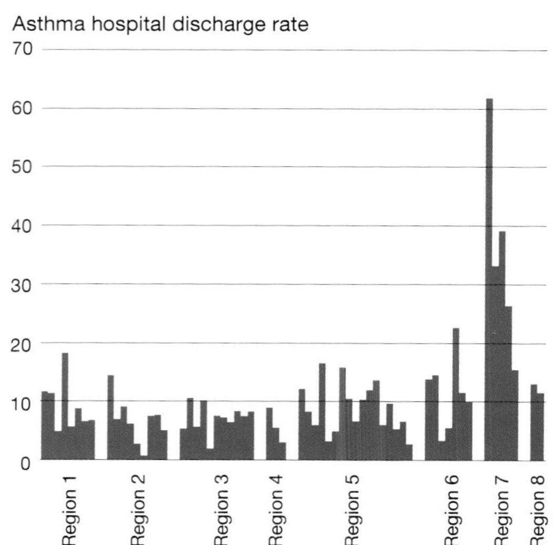

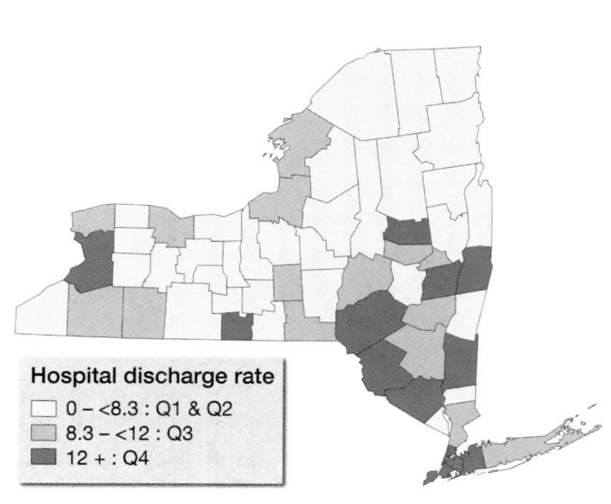

Note: Counties are shaded based on quartile distribution. Source: Adapted from Figure 7-13 of New York State Asthma Surveillance Report, October 2007, accessed March 18, 2009 at http://www.health.state.ny.us/diseases/asthma/

Figure 11.2 Hospital discharge rate for asthma per 10,000 population age 5 to14, 2005–2007 for (left) ClimAID regions (see Chapter 1, "Climate Risks," for definition of regions) and (right) for New York State counties

Childhood asthma is an important current health challenge in many parts of New York State—especially in the five counties that comprise New York City. Asthma events can be severe enough to require hospital admission (see **Figures 11.2** and **11.3**). However, the threshold of severity that triggers a hospital visit and admission likely differs by socioeconomic status. Wealthier individuals with health insurance, under doctor supervision, and with access to controller medications are less likely to have asthma attacks and are less likely to go to the hospital for care than are lower-income individuals lacking these resources.

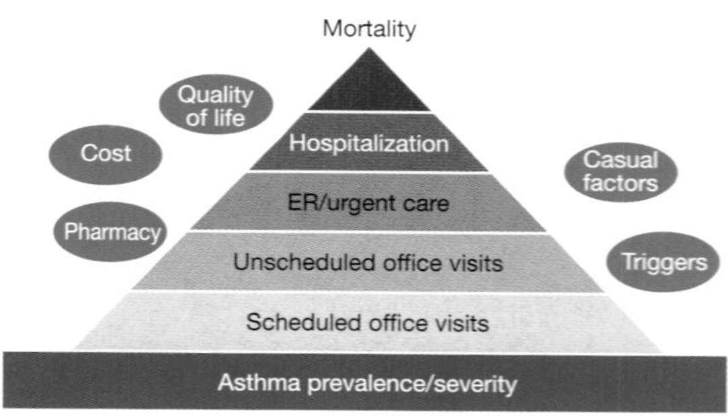

Source: Figure 3-1 of New York State Asthma Surveillance Report, October 2007, accessed March 18, 2009 at http://www.health.state.ny.us/diseases/asthma/

Figure 11.3 Asthma surveillance pyramid

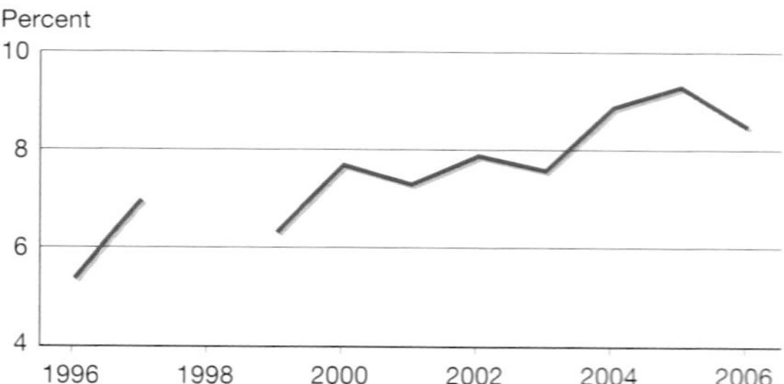

Source: Figure 5-1 of New York State Asthma Surveillance Report, October 2007, accessed March 18, 2009 at http://www.health.state.ny.us/diseases/asthma/

Figure 11.4 Prevalence of current asthma among adults: 1996-2006 in New York State

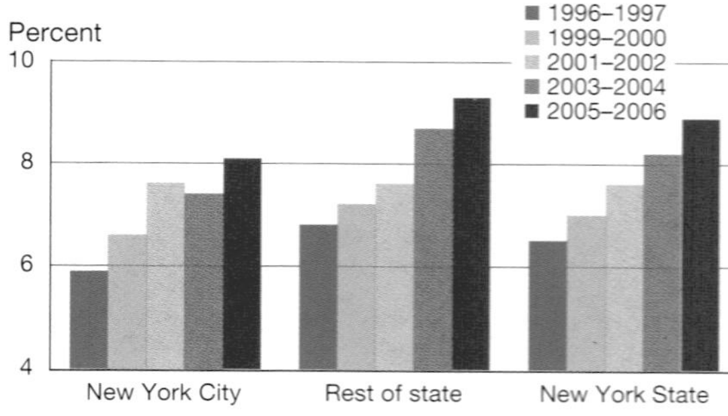

Source: Figure 5-2 of New York State Asthma Surveillance Report, October 2007, accessed March 18, 2009 at http://www.health.state.ny.us/diseases/ asthma/

Figure 11.5 Prevalence of current asthma among adults, by region

Figure 11.4 shows that the percentage of New York State adults reporting that they currently have asthma that was diagnosed by a physician (based on survey methods) has trended generally upward between 1996 and 2006. In terms of prevalence as opposed to hospital admissions, New York City shows similar trends to the remainder of New York State (**Figure 11.5**).

Cardiovascular Disease

Cardiovascular disease is the leading cause of death in New York State (**Figure 11.6**). Underlying cardiovascular disease can interfere with a body's ability to regulate temperature in response to heat stress and, thus, can be an important predisposing factor for vulnerability to heat-related deaths. In addition, air pollution is a risk factor for cardiovascular disease (Kheirbek et al., 2011).

Cardiovascular disease is composed of several disease conditions, the most prevalent of which is coronary heart disease. Coronary heart disease, which is the single-greatest killer of New York State residents, occurs

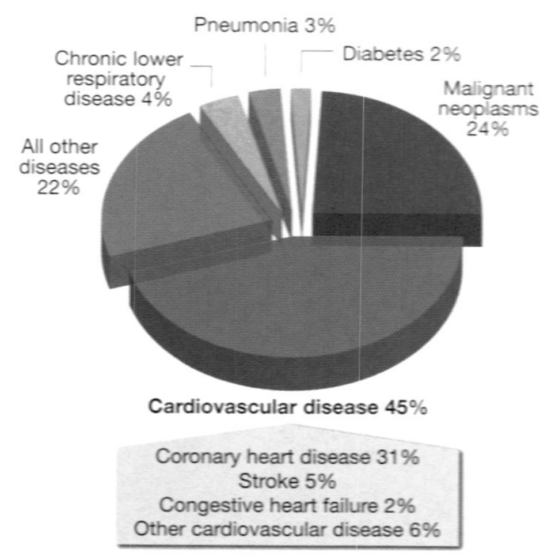

Source: New York State Vital Statistics, 1999

Figure 11.6 New York State causes of death

due to thickening and hardening of arteries, resulting in insufficient blood supply and potentially severe damage to heart tissue and other organ systems in the body. Age-adjusted coronary heart disease mortality for persons aged 35 and older in New York State is the highest in the nation, mostly due to coronary heart disease in persons 65 and older. Fortunately, however, there has been a steady reduction in cardiovascular death rates in the state, from the 1979 level of about 600 per 100,000 residents to the 1999 level of less than 400 per 100,000 residents (Fisher et al., 2000).

Infectious Diseases

Infectious diseases were the most important health challenge in New York City during the 1800s and were the prime focus of the New York City Department of Health activities starting in 1866. The advent of antimicrobial drugs in the 1900s strongly reduced the burden of infectious disease. However, the end of last century and the early part of this century have seen the emergence and re-emergence of infectious pathogens in New York State and globally. Climate-sensitive infectious diseases include those spread by contaminated food (**Figure 11.7**) and water as well as those transmitted by insects and other vectors.

New York State has experienced the emergence of several vector-borne diseases in the past few decades. For instance, the state leads the nation in numbers of Lyme disease cases. Between 2002 and 2006, the top two counties in the United States for number of cases, and four of the top 10 counties in Lyme disease incidence rate (cases per 100,000 people) were in New York State. Illness caused by West Nile virus in the state peaked in 2002 at 82 cases, and the state has had the highest numbers of cases on the East Coast since 2005. Both Lyme disease and West Nile virus tend to be most prevalent in the Hudson Valley, Long Island, and New York City areas with dense and growing human populations. The factors responsible for the concentration of Lyme disease and West Nile fever in the southeastern region of the state are not well understood. Similar southeastern concentrations of *Borrelia burgdorferi*-infected blacklegged ticks, as well as of West Nile virus in mosquitoes and wild birds, suggest that ecological conditions, possibly including warmer climate, might be important.

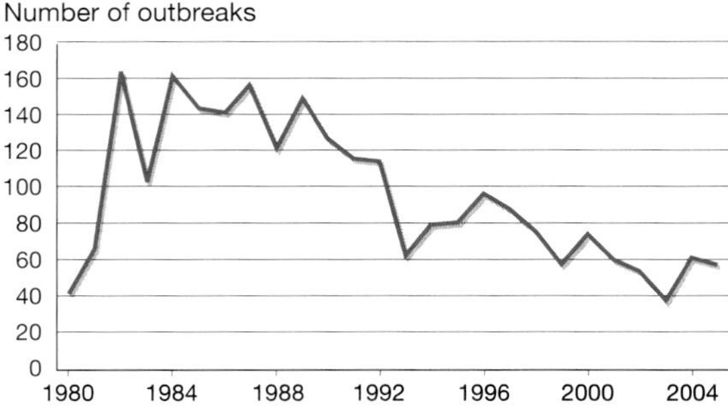

Source: Adapted from Figure 1 of Foodborne Disease Outbreaks in New York State, 2005, NYS DOH, Bureau of Community Environmental Health and Food Protection, January 2005; accessed March 19 2009 at http://www.health.state.ny.us/statistics/diseases/foodborne/outbreaks/2005/ report.htm

Figure 11.7 Reported food-borne disease outbreaks in New York State, 1980–2005

11.1.6 Economic Value

The size of the public health sector is roughly reported in the official State GDP figures issued by the U.S. Bureau of Economic Analysis. The New York State full- and part-time employment in health care and social assistance for 2008 was 1,486,598 (New York State Department of Labor, 2008). The 2008 current dollar state GDP was $1.144 trillion; of this total, more than $82 billion was in the public health sector (U.S. Department of Commerce Bureau of Economic Analysis, 2009). (See also the ClimAID economic analysis in Annex III to the full report.)

11.2 Climate Hazards

Climate factors and measures that are particularly relevant to the health of New Yorkers are highlighted and briefly introduced below. Some of these factors are discussed in more detail in Vulnerabilities and Opportunities (Section 11.3) and in case studies at the end of the chapter.

11.2.1 Temperature

Historical observations over the past 40 years provide clear evidence of increasing average temperatures in New York State. Projected increases in average temperatures in the coming decades will also be associated with increases in other temperature measures, such as the minimum and maximum temperature and the minimum, average, and maximum daily apparent temperature (perceived outdoor temperature, including factors such as wind and humidity, as well as air temperature). Other temperature measures of relevance to public health include the number of days with temperature exceeding 85, 90, and 95°F, all of which are projected to increase. Consequently, heat-related mortality could increase, and persons with heat-sensitive conditions are at particular risk.

As temperature increases, and with potential increases in the frequency of stagnant air events over New York State, conditions favoring high ozone days could increase. Daily maximum 8-hour ozone concentrations and the number of days with 8-hour ozone concentrations above 60–70 parts per billion (ppb) represent useful measures of changing ozone-related risks for respiratory irritation and damage. These risks are particularly relevant for people working or exercising outdoors, including children and those with respiratory disease.

11.2.2 Precipitation

Extreme precipitation and flooding events can have significant direct health impacts due to injury and drowning, and can have a wide range of indirect impacts such as diminished water and food supply and quality, interruption of healthcare service delivery, mental health consequences, and respiratory responses to indoor mold. The most relevant precipitation metrics are not yet known and will likely vary for different health-related outcomes. Research is needed to elucidate the links between precipitation metrics and health in New York State.

11.2.3 Changing Patterns of Monthly Temperatures and Precipitation

Average temperature and precipitation pattern shifts can impact ecosystems (see Chapter 6, "Ecosystems") and can affect vector habitats and prevalence. West Nile virus as well as other diseases carried by mosquitoes, ticks, or other vectors could change their distribution or pattern of occurrence. In addition, allergy triggers such as pollen and molds could change in timing and intensity.

11.3 Vulnerabilities and Opportunities

Climate change vulnerabilities in the public health sector are, to a large extent, ones in which public health and environmental agencies are already engaged. However, climate change places an additional burden on public health agencies that are already burdened by low levels of staffing and funding. Climate-related risk factors include heat events, extreme storms, disruptions of water supply and quality, decreased air quality, changes in timing and intensity of pollen and mold seasons, and alterations in patterns of infectious disease vectors and organisms. Climate-sensitive health vulnerabilities include heat-related mortality (death) and morbidity (illness), respiratory disorders stemming from aeroallergen and/or air pollution exposures, trauma and complex downstream effects related to storm events, and a range of infectious diseases.

In later sections of this chapter, we present case studies to highlight a subset of health vulnerabilities for New York State over coming decades for which adequate information and expertise currently exists to make qualitative or in some cases quantitative assessments. The case studies examine health impacts related to heat, ozone, extreme storms, and West Nile virus. These were chosen as examples based on the current (albeit limited) knowledge base, and should not be viewed as a complete list of future health vulnerabilities for New York State. Evolving science and experience will continue to clarify the picture of health vulnerabilities in coming years. In the present section, our goal is to provide a broad sense of the range of potential health vulnerabilities.

Information on public health vulnerabilities to climate variability and change in New York State is available from a series of assessments carried out over the past decade, including the Metropolitan East Coast Climate Impact Assessment (Rosenzweig and Solecki, 2001), the New York Climate and Health Project (www.globalhealth.columbia.edu/projects/RES0716289.html), and the Northeast Climate Impact Assessment (Frumhoff et al., 2007). Based on an assessment of this and subsequent work, a review of current health challenges in New York State, and on our engagement with stakeholders, several climate-related health vulnerabilities emerged. These include increased risk for all natural-cause mortality associated with more frequent and severe heat waves (Knowlton et al., 2007; Kinney et al., 2008), asthma exacerbations and mortality associated with ozone air pollution (Knowlton et al., 2004), allergy and asthma associated with altered pollen and mold seasons, water- and food-borne diseases, emergence and/or changing distributions of vector-borne diseases, and impacts of extreme storm events, especially coastal storms in the New York City metropolitan area and Long Island.

These vulnerabilities span a range from the relatively direct, data-rich, and well-understood to more complex, multi-factorial systems for which both data and models are currently underdeveloped. Even the direct and relatively well-studied effects of heat waves on mortality among the urban elderly and those with low incomes require further work to assess potential future impacts of climate change against a backdrop of changing economics, energy constraints, demographics, and adaptation responses (Kinney et al., 2008).

Uncertainties pervade any effort to predict either direct or indirect health impacts of climate change. These uncertainties relate to projections of site-specific climate change itself, due to uncertain future pathways of global greenhouse gas emissions and the behavior of the climate system in response. This complicates future projections of climate metrics, including temperature, sea level rise, and the effects of changing temperature and humidity on health outcomes like communicable and vector-borne diseases. Additional uncertainties arise in projecting future health impacts due to potential future pathways of population demographics, economic development, and adaptation measures. These multiple uncertainties increase the importance of building resilience into the public health system to cope with inevitable surprises to come. Vulnerability assessments combined with a full accounting of uncertainties will help in prioritizing climate-health preparedness plans, informing communities on which actions should be taken first, and which information gaps are most critical to fill.

11.3.1 Temperature-Related Mortality

Extreme temperature events have been linked with higher mortality rates and premature death, in particular among vulnerable populations (elderly, young children, or those suffering from cardiovascular or respiratory conditions) (WHO, 2004; Basu and Ostro, 2009). More than 70,000 deaths were associated with the heat wave in Western Europe during the summer of 2003 (Robine et al., 2008). In the United States, mortality rates from higher than normal temperatures have also been documented, with approximately 10,000 deaths during the summer of 1980 (Ross and Lott, 2003). Large metropolitan areas where the heat-island effect is prevalent are particularly affected. It has been estimated that in Chicago, between 600 (Dematte, 1998) and 739 (Klinengberg, 2002) people died during the July 1995 heat wave, and an additional 80 cases were attributed to a second extreme heat episode during the summer of 1999. Similarly, 118 died in Philadelphia during the July 6–14, 1993 heat wave. Moreover, the combined effects of extreme temperature and air pollution have been seen to increase morbidity and mortality cases during heat waves (Cheng, 2005).

There is also emerging evidence for effects of heat on hospital admissions for respiratory and cardiovascular diseases. For example, in a study of summertime hospital admissions in New York City during the period from 1991 to 2004, Lin and colleagues (2009) from the NYSDOH found significant associations between high temperatures and increased risk of both respiratory and cardiovascular admissions. While effects were seen throughout the population, elderly and Hispanic residents appeared to be especially vulnerable.

Those at higher risk for heat-related health effects are among the most vulnerable urban residents: the elderly, those with low incomes, those with limited mobility and social contact, those with pre-existing health conditions and belonging to nonwhite racial/ethnic groups, and those lacking access to public facilities and public transportation or otherwise lacking air conditioning. Children, urban residents, and communities in the northern parts of the state that are not adapted to heat may also be vulnerable subgroups for temperature-related mortality (death) and morbidity (illness). As stated earlier, cardiovascular disease can impair a body's ability to regulate temperature in response to heat stress and thus can be an important predisposing factor for vulnerability to heat-related deaths. Further, persons with cardiovascular disease are often under close medical supervision and care, and thus may be especially vulnerable to disruptions of health care access following extreme storm and flood events. Since physical activity reduces the risk of cardiovascular disease, changing patterns of physical activity due to climate change could impact disease in either positive or negative directions.

As a result of climate change, New York State will experience increased temperatures that could have significant health consequences. Climate change is shifting the overall temperature distribution in the United States such that extreme high temperatures will become hotter. This will change the timing of heat waves and also increase their frequency. Urban areas are especially vulnerable because of the high concentrations of susceptible populations and the influence of the urban heat island effect. Thus, preparing for and preventing heat-related health problems is likely to be of growing importance in urban areas. Health departments, city planners, and emergency response agencies all can benefit from assessments aimed at determining future heat/health vulnerabilities under a changing climate. While the largest changes may lie 50 to 100 years in the future, smaller but still health-relevant changes are likely to occur over time horizons of interest to planners, e.g., 20 to 30 years. However, to be useful, future projections should take account not only of climate change, but also changes in population characteristics, infrastructure, and adaptive measures.

In a relevant recent study, Knowlton et al. (2007) examined potential climate change impacts on heat-related mortality in the New York City metropolitan area. Current and future climates were simulated at a 36-kilometer grid scale over the northeastern U.S. with a global-to-regional climate modeling system. Summer heat-related premature deaths in the 1990s and 2050s were estimated using a range of scenarios and approaches to modeling acclimatization. Acclimatization describes physiological adaptation in the human body that allows for maintenance of normal body temperature range during heat exposure through increased evaporative cooling (sweating), thereby mitigating cardiovascular system stress. Projected regional increases in heat-related premature mortality by the 2050s ranged from 47 to 95 percent, with a mean 70 percent increase as compared to the 1990s. Acclimatization reduced regional increases in summer heat-related premature mortality by about 25 percent. Local impacts varied considerably across the region, with urban counties showing greater numbers of deaths and smaller percentage increases than less urbanized counties. While considerable uncertainty exists in climate forecasts and future health vulnerability, the range of projections developed suggested that by mid-century acclimatization may not completely mitigate the effects of temperature change in the New York metropolitan region, resulting in an overall net increase in heat-related premature mortality.

It is important to note that more people die on average in winter than in summer in New York State and in the United States as a whole. However, winter mortality is heavily influenced by influenza and other viral infections, which are more prevalent during the winter season, likely due to low indoor and outdoor humidity and activity patterns. Temperature per se appears to play a minor role. Thus, it appears unlikely that climate warming will significantly reduce winter mortality in the foreseeable future. To examine this issue further, we present below a new case study of the impacts of daily temperature throughout the year on daily mortality due to all natural causes in New York County (i.e., Manhattan). We first fitted the U-shaped exposure-response function linking temperature with mortality over the full year using an 18-year record of daily observations. The analysis controlled for seasonal and day-of-week cycles in the data. We then used the fitted function to compute future mortality under the alternative climate models and scenarios included in the ClimAID project. While temperature-related mortality was projected to diminish slightly in winter under climate change, increases in warm-season mortality far outweighed this benefit in all cases. Further, we noted that, on a percentage basis, future mortality increases will be most prominent in the spring and fall seasons.

11.3.2 Air Pollution and Aeroallergens

Climate variables such as temperature, humidity, wind speed and direction, and mixing height (the vertical height of mixing in the atmosphere) play important roles in determining patterns of air quality over multiple scales in time and space. These linkages can operate through changes in air pollution emissions, transport, dilution, chemical transformation, and eventual deposition of air pollutants. Policies to improve air quality and human health take meteorologic variables into account in determining when, where, and how to control pollution emissions, usually assuming that weather observed in the past is a good proxy for weather that will occur in the future, when control policies are fully implemented. However, policymakers now face the unprecedented challenge presented by changing climate baselines. Air quality planning is a very important function of the New York State Department of Environmental Conservation, which is charged with the difficult task of developing and implementing strategies to achieve air quality standards despite being downwind of several states that host major emission sources.

There is growing recognition that development of optimal control strategies to control future levels of key health-relevant pollutants like ozone and fine particles ($PM_{2.5}$)* should incorporate assessment of potential future climate conditions and their possible influence on the attainment of air quality objectives. Given the significant health burdens associated with ambient air pollution, this is critical for designing policies that maximize future health protection. Although not regulated as air pollutants, naturally occurring air contaminants of relevance to human health, including smoke from wildfires and airborne pollens and molds, also may be influenced by climate change. Thus there is a range of air contaminants, both anthropogenic and natural, for which climate change impacts are of potential importance.

In spite of the substantial successes achieved since the 1970s in improving air quality, many New Yorkers continue to live in areas that do not meet the health-based National Ambient Air Quality Standards for ozone and $PM_{2.5}$ (www.epa.gov/air/criteria.html). Ozone is formed in the troposphere mainly by reactions that occur in polluted air in the presence of sunlight. The key precursor pollutants for ozone formation are nitrogen oxides (emitted mainly by burning of fuels) and volatile organic compounds (VOCs, emitted both by burning of fuels and evaporation from stored fuels and vegetation). Because ozone formation increases with greater sunlight and higher temperatures, it reaches unhealthy levels primarily during the warm half of the year. Daily peaks occur near midday in urban areas, and in the afternoon or early evening in downwind areas. It has been firmly established that breathing ozone can cause inflammation in the deep lung as well as short-term, reversible decreases in lung function. In addition, epidemiologic studies of people living in polluted areas have suggested that ozone can increase the risk of asthma-related hospital visits and premature mortality (Peel et al., 2005; Peel et al., 2007; Kinney et al., 1991; Levy et al., 2005). Vulnerability to ozone effects on the lungs is greater for people who spend time outdoors during ozone periods, especially those who engage in physical exertion, which results in a higher cumulative dose to the lungs. Thus, children, outdoor laborers, and athletes all may be at greater risk than people who spend more time indoors and who are less active. Asthmatics are also a potentially vulnerable subgroup.

$PM_{2.5}$ is a complex mixture of solid and liquid particles that share the property of being less than 2.5 μm (millionths of a meter) in aerodynamic diameter. Because of its complex nature, $PM_{2.5}$ has complicated origins, including primary particles emitted directly from a variety of sources and secondary particles that form via atmospheric reactions of precursor gases. $PM_{2.5}$ is emitted in large quantities by combustion of fuels by motor vehicles, furnaces and power plants, wildfires, and, in arid regions, windblown dust (Prospero et al., 2003). Because of their small size, $PM_{2.5}$ particles have relatively long atmospheric residence times (on the order of days) and may be carried long distances from their source regions (Prospero et al., 2003; Sapkota et al., 2005). For example, using satellite imagery and ground-based measurements, Sapkota and colleagues tracked a wildfire plume over 621 miles (1,000 km) from northern Quebec, Canada, to the city of Baltimore, Maryland, on the East Coast of the U.S. (Sapkota et al., 2005). Research on health effects in urban areas has demonstrated associations between both short-term and long-term average ambient $PM_{2.5}$ concentrations and a variety of adverse health outcomes, including premature deaths related to heart and lung diseases (Samet et al., 2000; Pope et al., 2002; Schwartz, 1994). In addition, smoke from wildfires has been associated with increased hospital visits for respiratory problems in affected communities (Hoyt and Gerhart, 2004; Johnston et al., 2002; Moore et al., 2006). In a study of acute asthma emergency room visits in NYC, the pollutants most associated were ozone, sulfur dioxide and one-hour $PM_{2.5}$. A more robust health impact was observed for the daily maximum $PM_{2.5}$ concentration than the 24-hour mean, suggesting peak exposure may have larger health impacts (NYSERDA, 2006).

Airborne allergens (aeroallergens) are substances present in the air that, upon inhalation, stimulate an allergic response in sensitized individuals. Aeroallergens can be broadly classified into pollens (e.g., from trees, grasses, and/or weeds), molds (both indoor and outdoor), and a variety of indoor proteins associated with dust mites, animal dander, and cockroaches. Pollens are released by plants at specific times of the year that depend to varying degrees on temperature,

* PM2.5 is a complex mixture of solid and liquid particles that are less than 2.5 μm (millionths of a meter) in diameter.

sunlight, moisture, and CO_2. Allergy is assessed in humans either by skin prick testing or by a blood test, both of which involve assessing reactions to standard allergen preparations. A nationally representative survey of allergen sensitization spanning the years 1988–1994 found that 40 percent of Americans are sensitized to one or more outdoor allergens, and that prevalence of sensitization had increased compared with data collected in 1976–1980 (Arbes et al., 2005).

Allergic diseases include allergic asthma, hay fever, and atopic dermatitis. More than 50 million Americans suffer from allergic diseases, costing the U.S. healthcare system over $18 billion annually (American Academy of Allergy, Asthma, and Immunology, 2000). For reasons that remain unexplained, the prevalence of allergic diseases has increased markedly over the past three to four decades. Asthma is the major chronic disease of childhood, with almost 4.8 million U.S. residents affected. It is also the principal cause for school absenteeism and hospitalizations among children (O'Connell, 2004). Mold and pollen exposures and home dampness have been associated with exacerbation of allergy and asthma, as has air pollution (Gilmour et al., 2006; IOM, 2000; IOM, 2004; Jaakkola and Jaakkola, 2004).

The influence of climate on air quality is substantial and well established (Jacob, 2005), giving rise to the expectation that changes in climate are likely to alter patterns of air pollution concentrations. Higher temperatures hasten the chemical reactions that lead to ozone and secondary particle formation. Higher temperatures, and perhaps elevated carbon dioxide (CO_2) concentrations, also lead to increased emissions of ozone-relevant VOC precursors by vegetation (Hogrefe et al., 2005).

Weather patterns influence the movement and dispersion of all pollutants in the atmosphere through the action of winds, vertical mixing, and rainfall. Air pollution episodes can occur with atmospheric conditions that limit both vertical and horizontal dispersion. For example, calm winds and cool air aloft limits dispersion of traffic emissions during morning rush hour in winter. Emissions from power plants increase substantially during heat waves, when air conditioning use peaks. Weekday emissions of nitrogen oxides (NOx) from selected power plants in California more than doubled on days when daily maximum temperatures climbed from 75°F to 95°F in July, August, and September of 2004 (Drechsler et al., 2006). Changes in temperature, precipitation, and wind affect windblown dust, as well as the initiation and movement of forest fires.

Finally, the production and distribution of airborne allergens such as pollens and molds are highly influenced by weather phenomena, and also have been shown to be sensitive to atmospheric CO_2 levels (Ziska et al., 2003). The timing of phenologic events such as flowering and pollen release is closely linked with temperature.

Human-induced climate change is likely to alter the distributions over both time and space of the meterologic factors described above. There is little question that air quality will be influenced by these changes. The challenge is to understand these influences better and to quantify the direction and magnitude of resulting air quality and health impacts.

Hogrefe and colleagues were the first to report results of a local-scale analysis of air pollution impacts of future climate changes using an integrated modeling approach (Hogrefe et al., 2004a; Hogrefe et al., 2004b). In this work, a global climate model was used to simulate hourly meteorologic data from the 1990s through the 2080s based on two different greenhouse gas emissions scenarios, one representing high emissions and the other representing moderate emissions. The global climate outputs were downscaled to a 36-kilometer (22-mile) grid over the eastern U.S. using regional climate and air quality models. When future ozone projections were examined, summer-season daily maximum 8-hour concentrations averaged over the modeling domain increased by 2.7, 4.2, and 5.0 ppb in the 2020s, 2050s, and 2080s, respectively, as compared to the 1990s, due to climate change alone. The impact of climate on mean ozone values was similar in magnitude to the influence of rising global background ozone by the 2050s, but climate had a dominant impact on hourly peaks. Climate change shifted the distribution of ozone concentrations toward higher values, with larger relative increases in future decades occurring at higher ozone concentrations.

The finding of larger climate impacts on extreme ozone values was confirmed in a study in Germany (Forkel and Knoche, 2006) that compared ozone in the 2030s and the 1990s using a downscaled integrated modeling system. Daily maximum ozone concentrations increased by 2–6 ppb (6–10 percent) across the study region. However, the number of cases where daily maximum ozone exceeded 90 ppb increased by nearly four-fold, from 99 to 384.

More recently, the influence of climate change on $PM_{2.5}$ and its component species have been examined in the northeastern U.S., including New York State, using an integrated modeling system (Hogrefe et al., 2006). Results showed that $PM_{2.5}$ concentrations increased with climate change, but that the effects differed by component species, with sulfates and primary particulate matter increasing markedly but with organic and nitrated components decreasing, mainly due to transformation of these volatile species from the particulate to the gaseous phase.

The health implications of wildfire smoke have been tragically demonstrated by events in Russia during the summer of 2010. Because the risk of wildfire initiation and spread is enhanced with higher temperatures, decreased soil moisture, and extended periods of drought, it is possible that climate change could increase the impact of wildfires in terms of frequency and area affected (IPCC, 2007a; Westerling et al., 2006). Among the numerous health and economic impacts brought about by these more frequent and larger fires, increases in fine particulate air pollution are a key concern, both in the immediate vicinity of fires as well as in areas downwind of the source regions. Several studies have been published examining trends in wildfire frequency and area burned in Canada and the U.S. Most such studies report upward trends in the latter half of the 20th century that are consistent with changes in relevant climatic variables (Westerling et al., 2006; Gillett et al., 2004; Podur et al., 2002). Interpretation of trends in relation to climate change is complicated by concurrent changes in land cover and in fire surveillance and control. However, similar trends were seen in areas not affected by human interference (Westerling, et al., 2006) or under consistent levels of surveillance over the follow-up period (Podur et al., 2002). Several studies have looked at wildfire risk in relation to climate change (Lemmen and Warren, 2004; Williams et al., 2001; Flannigan et al., 2005; Bergeron et al., 2004).

Aeroallergens that may respond to climate change include outdoor pollens generated by trees, grasses, and weeds, and spores released by outdoor or indoor molds. Historical trends in the onset and duration of pollen seasons have been examined extensively in recent studies, mainly in Europe. Nearly all species and regions analyzed have shown significant advances in seasonal onset that are consistent with warming trends (Root et al., 2003; Beggs, 2004; Beggs and Bambrick, 2005; Clot, 2003; Emberlin et al., 2002; Galan et al., 2005; Rasmussen, 2002; Teranishi et al., 2000; van Vliet et al., 2002; World Health Organization, 2003). There is more limited evidence for longer pollen seasons or increases in seasonal pollen loads for birch (Rasmussen, 2002) and Japanese cedar tree pollen (Teranishi et al., 2000). Grass pollen season severity has been shown to be greater with higher pre-season temperatures and precipitation (Gonzalez et al., 1998). What remains unknown is whether and to what extent recent trends in pollen seasons may be linked with upward trends in allergic diseases (e.g., hay fever, asthma) that have been seen in recent decades.

In addition to earlier onset of the pollen season and possibly enhanced seasonal pollen loads in response to higher temperatures and resulting longer growing seasons, there is evidence that CO_2 rise itself may cause increases in pollen levels. Experimental studies have shown that elevated CO_2 concentrations stimulate greater vigor, pollen production, and allergen potency in ragweed (Ziska et al., 2003; Ziska and Caufield, 2000; Singer et al., 2005). Ragweed is arguably the most important pollen species in the U.S., with up to 75 percent of hay fever sufferers sensitized (American Academy of Allergy, Asthma, and Immunology, 2000). Significant differences in allergenic pollen protein were observed in comparing plants grown under historical CO_2 concentrations of 280 ppm, recent concentrations of 370 ppm, and potential future concentrations of 600 ppm (Singer et al., 2005). Interestingly, significant differences in ragweed productivity were observed in outdoor plots situated in urban, suburban, and rural locales where measurable gradients were observed in both CO_2 concentrations and temperatures. Cities are not only heat islands but also CO_2 islands, and thus to some extent represent proxies for a future warmer, high-CO_2 world (Ziska et al., 2003).

With warming over the longer term, changing patterns of plant habitat and species density are likely, with gradual movement northward of cool-climate species like maple and birch, as well as northern spruce (IPCC, 2007a). Although these shifts are likely to result in altered pollen patterns, to date they have not been assessed quantitatively.

As compared with pollens, molds have been much less studied (Beggs, 2004). This may reflect in part the relative paucity of routine mold monitoring data from which trends might be analyzed, as well as the complex relationships between climate factors, mold growth, and spore release (Katial et al., 1997). One study examining the trends in Alternaria spore counts between 1970 and 1998 in Derby, U.K., observed significant increases in seasonal onset, peak concentrations, and season length. These trends parallel gradual warming observed over that period.

In addition to potential effects on outdoor mold growth and allergen release related to changing climate variables, there is also concern about indoor mold growth in association with rising air moisture and especially after extreme storms, which can cause widespread indoor moisture problems from flooding and leaks in the building envelope. Molds need high levels of surface moisture to become established and flourish (Burge, 2002). In the aftermath of Hurricane Katrina, very substantial mold problems were noted, causing unknown but likely significant impacts on respiratory morbidity (Ratard, 2006). There is growing evidence for increases in both the number and intensity of tropical cyclones in the north Atlantic since 1970, associated with unprecedented warming of sea surface temperatures in that region (IPCC, 2007a; Emanuel, 2005).

Taken as a whole, the emerging evidence from studies looking at historic or potential future impacts of climate change on aeroallergens led Beggs to state (Beggs, 2004):

> [This] suggests that the future aeroallergen characteristics of our environment may change considerably as a result of climate change, with the potential for more pollen (and mold spores), more allergenic pollen, an earlier start to the pollen (and mold spore) season, and changes in pollen distribution.

11.3.3 Infectious Diseases

Infectious diseases that are transmitted by arthropod vectors, such as mosquitoes and ticks, are highly sensitive to climate change. Effects of even small increases in average temperatures can increase rates of population growth and average population densities of mosquitoes and other vectors (Harvell et al., 2002; Epstein, 2005). In addition, both the biting rates of mosquitoes and the replication rates of the parasites and pathogens they transmit increase with increasing temperatures (Harvell et al., 2002). Nevertheless, the degree to which recent and future climate change affects the distribution and intensity of vector-borne diseases remains controversial (Harvell et al., 2002; Ostfeld, 2009). One common criticism of the contention that climate warming will cause vector-borne diseases to spread geographically is that, just as some areas that are below the suitable temperature range will move into this range, others that are currently suitable might become too warm. Evidence to support this contention, however, is scant (Ostfeld, 2009). Moreover, because the overall climate of New York State appears to be well below any detectable upper thresholds for vector-borne disease, it seems that climate warming is more likely to increase, rather than decrease, the burden of vector-borne disease in the state.

In the case of Lyme disease, a climate-based spatial model (Brownstein, et al., 2005) suggested that the conditions under which blacklegged tick populations can be supported will expand northward into Canada as the climate warms. However, this model assumed that ticks currently occupy the entire state of New York and therefore was unable to make predictions relevant to the expansion of Lyme disease within the state. Other models (Ogden et al., 2005) also predict northward expansion of blacklegged ticks into areas currently assumed to be too cold to support them. These models are based on assumed, rather than empirically verified, relationships between temperature and tick demography (Killilea et al., 2008). In contrast, the relationships between specific climatic parameters and cases of West Nile virus illness or mosquito vector demography are better established. Therefore, this chapter focuses on West Nile virus in Case Study D.

11.4 Adaptation Strategies

Climate is often considered a factor that will change the frequency and severity of existing health problems more than create entirely new ones. From this point of view, the challenge is more about integrating specific information about climate-related vulnerabilities into ongoing programs of public health surveillance, prevention, and response than developing new programs to deal with unique challenges. While largely valid, this view misses the mark in one important way, namely that changing climate brings the possibility of entirely new health risks, for example from new infectious diseases or coastal storm events of unprecedented magnitude.

Here we briefly review a range of adaptation options that should be considered in addressing climate-related health risks in New York State.

11.4.1 Key Adaptation Strategies

Avoiding or reducing the health impacts of climate change will ultimately depend on public health preparedness. In the sections that follow, a number of adaptations, or preparedness strategies, are discussed.

Heat Adaptation

Heat-related mortality has been recognized as an important public health challenge for many decades. As a result, heat warning and response systems have been implemented in many cities in the United States and Europe, including New York City. These warning systems include collaboration with local meteorologists for forecasting as well as coordination with multiple agencies and community groups. The goal is to maximize dissemination of actionable information for both immediate health protection and provision of additional services during the period of intense heat. Often the additional services include longer hours at community centers for seniors (called cooling centers during the time they are open during a heat wave) as well as reduced fare on public transportation or the implementation of neighborhood buddy systems. In addition, the NYSDOH distributes statewide a fact sheet entitled "Keep Your Cool During Summer Heat" that provides information on what to do before and during a heat event, how to recognize and act on heat-related illness, and who is most vulnerable. The NYSDOH also has worked with the State Environmental Health Collaborative Climate Workgroup to develop several climate indicators. These include indicators for the vulnerable population (elderly and people living in low-income neighborhoods), cardiovascular disease, hospital readmissions for respiratory diseases due to heat, maximum/minimum temperature, and air pollution change due to heat. One important priority with respect to these efforts is to evaluate their effectiveness in reducing morbidity and mortality.

Home air conditioning is a critical factor for prevention of heat-related illness and death (Bouchama et al., 2007). Air conditioning is especially important for elderly, very young, and health-compromised individuals, all of whom have a lower internal capacity to regulate body temperature (CDC, 2009).

Within New York City, approximately 84 percent of housing units had some form of indoor air conditioning in 2003. Air conditioning rates are not uniform across the city, however. Neighborhoods with higher poverty rates, including Central Harlem, Washington Heights, Fordham, the South Bronx, Greenpoint, Williamsburg, Bedford-Stuyvesant, and others, have lower rates of in-home air conditioning than more affluent parts of the city (**Figure 11.8**). These differences suggest that many residents living in lower-income neighborhoods of the city may be more vulnerable to heat-related illness and mortality.

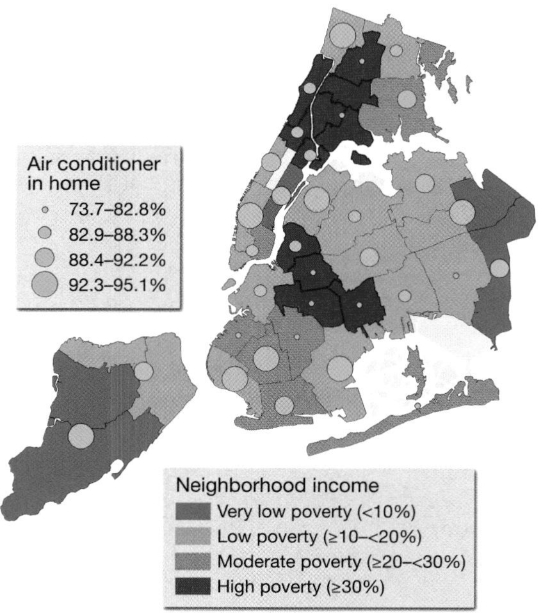

Note: Percentages are age adjusted. Poverty is categorized by the percent of residents in each neighborhood living below the federal poverty level.
Source: NYC Community Health Survey 2007; Bureau of Epidemiology Services, NYC DOHMH; U.S. Census 2000/NYC Department of City Planning

Figure 11.8 Air conditioning distribution and neighborhood-level poverty in New York City

The presence of an air conditioner does not necessarily equate to its effective use during a heat wave. Also, while fans can be helpful at moderate temperatures, Wolfe (2003) points out that their effectiveness diminishes at very high temperatures and humidity.

As noted in the Chapter 8 ("Energy"), energy costs associated with use of air conditioning are a major concern for lower-income households and particularly for lower-income elderly populations (Tonn and Eisenberg, 2007). Even during periods of extreme heat, low-income elderly residents, particularly those living alone, may be reluctant to use their air conditioners due to concerns about energy costs. While age and social isolation were key factors in predicting mortality in the 1995 Chicago heat wave (Semenza et al., 1996), presence of air conditioning in the home did not necessarily have a mitigating effect. Many of the Chicago heat wave's elderly victims had working air conditioners in their apartments, but the machines were not in use at the time of death (Klinenberg, 2003). Thus, to improve the effectiveness of air conditioning as an adaptive measure, it will be important to develop strategies to ensure energy access for low-income, vulnerable individuals, as well as ensure that functional, high-efficiency air conditioners are widely available and in use. Possible measures include monetary support of low-income populations to ensure the use of air-conditioning and programs for peak load and or voltage reduction (Warren and Riedel, 2004). The costs to implement such measures are not well documented.

In addition to these measures, infrastructure investments, particularly in vulnerable urban neighborhoods, could yield substantial health benefits. Urban greening programs, green roofs, and building codes requiring reflective exterior surfaces are among the options that should be considered.

Air Pollution

Implementation strategies addressing ozone and fine particles are well developed in New York State and are described on the New York State Department of Environmental Conservation website (www.dec.ny.gov/chemical/8403.html; see State Implementation Plan). However, integrating climate forecasts into ongoing planning for air quality is a challenge that must be addressed in collaboration with stakeholders at the New York State Department of Environmental Conservation and the U.S. Environmental Protection Agency.

11.4.2 Larger-scale Adaptations

Comparative health-risk assessments of climate change adaptation (and also mitigation) measures, such as the health effects of the combustion byproducts of biofuels and gases of varying ethanol blends, are important. Data gaps, such as the specifics of relationships between certain climate factors and some health outcomes and projections of climate impacts on multiple types of disease and vulnerable subpopulations, and the specific ongoing need for increased environmental monitoring linked to health outcome reporting, are also key to adaptation. Additionally, stakeholders have voiced the importance of public health communication. Alerts regarding known health risks should be tested and tailored to most effectively convey information and needed action to vulnerable communities. Cross-cutting environment and health initiatives that bridge the divide in legislation between ecosystems and human health should also be developed.

11.4.3 Co-benefits and Opportunities

This chapter has focused primarily on potential negative health impacts of a changing climate in New York State. However, it is possible that climate change may bring some positive impacts on health. For example, warmer winters may reduce the burden of some cold-related health effects (e.g., hypothermia among the homeless, snow-related accidents and injuries) and could encourage greater physical activity during extended periods of mild weather. In addition, policies enacted in New York State to reduce greenhouse gas (GHG) emissions by curtailing fossil fuel burning will reduce emissions of other pollutants, and may deliver health benefits as well. Furthermore, unlike climate benefits, these health co-benefits accrue locally in space and time, enhancing their value in economic analyses (Burtraw et al., 2003; Dessus and O'Connor, 2003; Proost and Van Regemorter, 2003; Wang and Smith, 1999; Bloomberg and Aggarwala, 2008). For 20 years at least, researchers have attempted to quantify co-benefits (Ayres and Walter, 1991; Viscusi, 1994). Most studies have found that the magnitude of the ancillary benefits are large, even relative to the large outlays required by GHG mitigation. Most of the literature to date emphasizes co-benefits that accrue from reductions in air pollution, particularly $PM_{2.5}$ and ozone precursors. However, GHG mitigation policies may improve health in other ways, e.g., via increased physical activity, decreased meat consumption, and reduced traffic accidents. For comprehensive reviews see Bell et al. (2008) and Nemet et al. (2010).

11.5 Equity and Environmental Justice Considerations

Climate change is an evolving problem for human health conditioned by unequal access to resources and differential exposure to unhealthy landscapes. The negative impacts of climate change on health may be particularly consequential for people living in poverty or communities segregated by race.

11.5.1 Vulnerability

There are two important pathways for climate-related health inequities. First, lower-income populations and communities of color may be concentrated in areas exposed to more climate-sensitive health risks. For example, compared to higher-income white populations, low-income segregated African-American and Hispanic communities tend to have greater exposure to allergens and smog, and live in homes that are less able to regulate temperature and humidity (Williams and Collins, 2001; Evans and Kantrowitz, 2002). Second, exposure may impose added burdens on pre-existing vulnerabilities of health, living conditions, and socioeconomic position. For example, low-income communities tend to have inferior public infrastructure, higher risk of underlying health conditions such as cardiovascular disease, and less access to quality, affordable health care (Williams and Collins, 2001; Evans and Kantrowitz, 2002). Other indicators of pre-existing vulnerabilities to climate-related health shocks include lower wages or unemployment, lack of insurance, occupational stresses, and poor nutrition.

Higher temperatures will likely increase the duration and intensity of heat waves and associated heat-related health stresses. Heat-related health stresses are felt disproportionately in inner-city urban areas, where a preponderance of heat-trapping surfaces and a scarcity of heat-reducing infrastructure (trees, parks, water) contribute to the urban "heat island" effect (Rosenzweig et al., 2006). The urban heat island effect has been implicated in past heat wave events (Kunkel et al., 1996). Because of residential segregation patterns, these inner-city neighborhoods also tend disproportionately to house low-income communities of color (Williams and Collins, 2001).

Health risks can be intrinsic or extrinsic. Intrinsic heat-related health risks include age, disability, and underlying medical conditions, such as depression or cardiovascular problems (Stafoggia, 2006; Worfolk, 2000). Some of these medical conditions are more prevalent in low-income communities or within communities of color. Extrinsic risks encompass contextual factors such as behavior, quality of housing, community integration, and access to cooling infrastructure and transportation (Kovats and Hajat, 2007; Epstein and Rogers, 2004; Klinenberg, 2003). Some of these risks are also associated with lower-income status, such as the higher probability of residing in heat-trapping buildings and lacking air conditioning (Klinenberg, 2003). All these risks generally interrelate to create unique, magnified vulnerabilities. For example, elderly persons may be medically sensitive to heat stress (intrinsic), while at the same time may lack coping strategies such as access to community support networks (extrinsic) (Worfolk, 2000; Klinenberg, 2003).

Heat-related morbidity also has its own suite of inequities (Lin et al., 2009). Those most likely to die from heat stress are not necessarily those who would suffer the contextual and indirect harms associated with heat morbidity, such as lost wages and productivity and health care expenses.

Air pollution and respiratory health is another area in which environmental justice concerns arise in the context of climate change. African Americans tend to live in urban centers that are more exposed to primary air pollutants. They also are significantly more likely to be hospitalized and die from asthma (Prakash, 2007). Rising temperatures and increasing emissions create conditions for ozone formation and further inequitably distributed health burdens.

Another climate impact is the probability of increased levels of mold and other allergens. This also contributes to respiratory health problems (Beggs, 2004). Dampness of households, a key variable for mold growth, is associated with socioeconomic status (Gold, 1992).

Environmental justice activists have become increasingly concerned about the contribution of mold to the high rates of hospitalization for asthma among African Americans in cities such as New York (NYS Department of Environmental Conservation, 2008). Tackling these high rates of urban asthma or home allergens through health adaptation programs is one way to reduce health disparities.

Securing access to affordable, good quality, nutritious food for lower-income urban communities of color is a priority area for environmental justice advocates in New York State (NYS Department of Environmental Conservation, 2008). Impacts of climate change on local agriculture could make this goal more challenging to achieve.

11.5.2 Adaptation

Some cities, such as New York, have begun developing adaptation programs because of existing health burdens related to heat stress (Rosenzweig et al., 2006). Other more northerly cities in the state may confront new emergent heat stress. They will need to be proactive to avoid any evolving health inequities related to differential coping capacities within their populations.

Since heat danger is frequently mediated by underlying vulnerabilities, one way to build equity into climate change adaptation mechanisms is a broad-based effort to improve health and reduce social isolation among vulnerable populations, including increasing access to health insurance and social support systems, broadening and diversifying economic activities, and improving education. More targeted adaptations include short-term social mechanisms such as warnings and outreach in conjunction with long-term technical design approaches that reduce ambient heat (Bernard and McGeehin, 2004; Rosenzweig et al., 2006). Ensuring equitable implementation of social prevention requires tailoring messages among and within groups. This means confronting language barriers in outreach and warning systems and targeting at-risk groups, such as elderly, disabled, or otherwise isolated persons. For example, the Phoenix heat wave in 2005 took a particular toll on homeless people (Epstein, 2005). Designing a warning for itinerants with tenuous access to information is a challenge for any outreach system. Through the CDC's Climate-Ready States and Cities Initiative, the New York State Department of Health is conducting an assessment that will examine a range of health outcomes related to extreme weather events, as well as waterborne, food-borne, and vector-borne diseases (www.cdc.gov/climatechange/climate_ready.htm).

One way to build social justice into heat adaptive design is to prioritize energy efficiency and retrofits of public housing, such as installing cooling surfaces and insulation. These synergistic approaches are also discussed in Chapter 8, "Energy." Other strategies that enforce climate-adaptive regulations, such as new building codes, might need to provide support mechanisms, funding incentives, or loans for low-income homeowners and small businesses.

11.6 Conclusions

This ClimAID assessment has identified a set of key existing and future climate risks for public health in New York State. Some health risks arise from increases in the frequency, duration, or intensity of weather events, such as diverse health consequences from more storms and flooding events, and from heat-related mortality and morbidity. Other risks may arise due to gradual shifts in weather patterns, such as changes in vector-borne disease prevalence and distribution, worsening air quality (smog, wildfires, pollen), and related cardiovascular and respiratory health impacts. Similarly, risks to water supply and food production may arise due to increased temperatures and shifting precipitation patterns. While the analyses presented here have been from the perspective of New York State, it is important to note that many of our findings can be generalized to other U.S. locations.

11.6.1 Main Findings on Vulnerability and Opportunities

- Climate will likely change the frequency and severity of existing health problems, while also bringing the possibility of entirely new health risks.
- Impacts of climate change will be particularly significant for people in New York State made more vulnerable because of age, preexisting illness, and/or poverty.
- Illness and death from heat will particularly impact low-income urban residents, the elderly, and those with pre-existing health conditions.

- Climate-related changes in air pollution patterns will be particularly significant for asthmatics and for persons who work, play, or exercise out of doors.

11.6.2 Adaptation Options

Adaptation to climate-related health vulnerabilities in New York State is an evolving process. Aside from heat wave warning and response planning, few climate-specific adaptation strategies yet exist in New York State. Climate impacts and adaptation strategies for the health sector build upon the existing public health system of New York State, which is already engaged to some extent with most of the health domains likely to be relevant to climate change. However, there is the possibility that future climate impacts in the health sector may fall outside of historical experience, presenting new challenges. Of particular concern is that information and capacity for integrating climate change into public health planning remains limited at the local level.

Future adaptations in the health sector should begin by enhancing capacity for climate planning within the existing public health system of New York State, and also by strengthening linkages between health and environmental initiatives.

One key objective is to expand ongoing surveillance of climate-sensitive environmental and health indicators. Surveillance is a central public health function that can inform periodic assessments of emerging risks and anticipated future impacts, and help to guide ongoing adaptation planning.

Another key area of focus should be the development of early warning systems and response plans for a broader range of climate risks, building on the experience with heat systems. Adaptation strategies and messaging should be particularly targeted at, and tailored for, protecting vulnerable populations.

Air quality control efforts will need to increasingly take climate change into account, as well as be integrated with greenhouse gas mitigation strategies, so that maximal health co-benefits are achieved.

A general point worth emphasizing is the importance of integrated health planning across multiple sectors, including environmental quality, parks and recreation, urban planning, food and water supply, and others.

With respect to equity and environmental justice, care is called for in designing both adaptation and mitigation strategies so that disparities can be reduced. Without making this an explicit goal, existing health disparities are likely to be worsened by climate change. People in northern parts of the state may be at particular risk for heat-related health impacts due to lack of adaptation to high temperatures. Mitigation and adaptation actions by New York State should ensure an equitable distribution of costs and benefits.

11.6.3 Knowledge Gaps

Future efforts to address health risks due to climate change will require ongoing, state-based research to inform periodic policy developments. Of particular importance is research to identify cross-sectoral interactions and win-win options for adaptation/mitigation, including extensive health co-benefits assessments.

It is also important to develop and analyze local health impact projections of climate factors and related disease outcomes. Information and capacity building for integrating climate change into public health planning at all levels of government is needed.

Examining the effectiveness of heat warning systems and related adaptive strategies, and translating these strategies to urban areas across the state, should be high priorities.

Enhanced environmental monitoring of climate-related factors linked to health outcome reporting, particularly of airborne allergens and infection vectors, is crucial for improving the knowledge foundation on which decisions are based.

Case Study A. Heat-related Mortality among People Age 65 and Older

As a result of climate change, New York State will experience increased temperatures that could have significant consequences for health, particularly for the most vulnerable members of the population: the elderly, those with low incomes, those with limited mobility and social contact, those with pre-existing health conditions and belonging to nonwhite racial/ethnic groups, and those lacking access to public facilities and transportation or otherwise lacking air conditioning. Urban areas are especially vulnerable because of the high concentrations of susceptible populations and the influence of the urban heat island effect. Thus, preparing for and preventing heat-related health problems is likely to be of growing importance in urban areas.

Projecting Temperature-related Mortality Impacts in New York City under a Changing Climate

Climate change has led to increasing temperatures in urban areas in recent decades, and these changes are likely to accelerate in the coming century. These changes may result in more heat-related mortality but also might alter winter mortality, and the net impact remains uncertain. Our objective was to explore a methodology for projecting future temperature-related mortality impacts over the full year in New York County across a range of climate change models and scenarios. The ClimAID climate team provided temperature projections for the 2020s, 2050s and 2080s over New York County, obtained from five different global climate models (GFDL, GISS, MIROC, CCSM and UKMO) that were run with the Intergovernmental Panel on Climate Change (IPCC) A2 and B1 greenhouse gas emissions scenarios (see Chapter 1, "Climate Risks," for details). Monthly differences between modeled future temperatures and those modeled for the climatological baseline period of 1970–1999 were used to adjust observed daily temperatures for 1970–1999 in Central Park, NY to the future time periods.

The association between maximum temperature and daily mortality in 1982–1999 was modeled using log-linear Poission regression analysis. Seasonal cycles were controlled using a natural spline function with 7 degrees of freedom per year. Day-of-week effects were also controlled. Temperature effects were fit using a natural spline with 2 degrees of freedom, yielding a U-shaped curvilinear relationship (**Figure 11.9**). Percentage changes in mortality in both winter and summer were calculated relative to the minimum point on **Figure 11.9**. This analytical approach is similar to those used extensively in the literature (for example, Curriero, Heiner, et al., 2002; Curriero, 2003; O'Neill, Zanobetti, et al., 2003; Anderson and Bell, 2009). We analyzed mortality in relation to maximum daily temperature observed on the same day as death (i.e., lag zero) for both heat and cold effects. This contrasts with the approach used by Anderson and Bell (2009) in which cold effects were modeled as a 25-day moving average. We avoided this approach because it might lead to confounding by winter season effects, that is, a tendency to mis-attribute seasonal effects to the cold slope. The heat- and cold-related deaths in the 1970s, 2020s, 2050s, and 2080s were estimated by integrating the results from the climate models and the empirical exposure-response relationship, with results shown in **Tables 11.1** and **11.2**, and **Figure 11.10**.

During the baseline period, 1970–1999, we estimated there were on average 604 mean annual temperature-related deaths. Under the A2 scenario, mean annual temperature-related deaths increased to 686 in the 2020s, 782 in the 2050s, and 920 in the 2080s. In the B1 scenario, the mean annual temperature-related deaths were 681 in 2020s, 741 in the 2050s, and 779 in the 2080s. Differences across models and scenarios were minimal early in the century but increased by mid-century (**Figure 11.10**). Warm season impacts on mortality expanded in both number and in annual extent (i.e., earlier in spring and later in fall) as the century progressed (**Table 11.2**). Additional sensitivity analyses using alternative lags of temperature and different reference temperatures are under way. However, these preliminary results suggest that, over a range of models and scenarios of future greenhouse gas

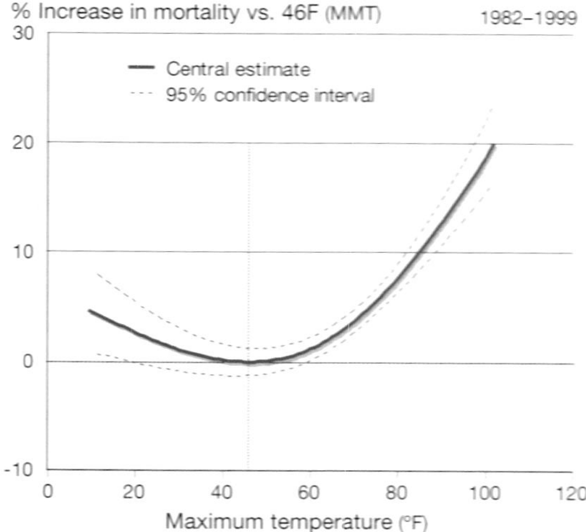

Figure 11.9 Predicted mortality vs. maximum temperature, based on analysis of daily observations from 1982 through 1999

emissions, increases in heat-related mortality could outweigh reductions in cold-related mortality. Further, while the two emissions scenarios produce similar mortality estimates through mid-century, the lower-emission B1 scenario could result in substantially smaller annual mortality impacts by the 2080s.

Economic Impacts of Mortality Due to Heat Waves

As noted above, climate projections can be used in assessing the impact of heat waves on the public health sector and society as well as the effectiveness of potential remedies. Measures to prevent increased mortality during extreme weather events may be

Climate Model	Scenario	Net Temperature Effect			Heat Effect			Cold Effect		
		T_{maxave}(°F)[a]	Deaths[b]	Percent Change[c]	Days Above MMT	Deaths[b]	Percent Change[c]	Days Below MMT	Deaths[b]	Percent Change[c]
Baseline[d]		62.7	604		287	586		78	18	
GFDL	2020s A2	64.4	676	11.92%	294	660	12.63%	72	16	-11.11%
	2020s B1	64.6	674	11.59%	297	659	12.46%	68	15	-16.67%
	2050s A2	66.6	763	26.32%	304	751	28.16%	61	12	-33.33%
	2050s B1	66.0	748	23.84%	299	735	25.43%	66	14	-22.22%
	2080s A2	69.5	902	49.34%	320	894	52.56%	46	8	-55.56%
	2080s B1	66.8	778	28.81%	304	765	30.55%	61	13	-27.78%
GISS	2020s A2	64.4	670	10.93%	295	655	11.77%	70	15	-16.67%
	2020s B1	64.9	679	12.42%	300	666	13.65%	65	13	-27.78%
	2050s A2	66.1	726	20.20%	306	716	22.18%	59	10	-44.44%
	2050s B1	65.2	694	14.90%	299	681	16.21%	65	13	-27.78%
	2080s A2	68.5	818	35.43%	320	812	38.57%	46	7	-61.11%
	2080s B1	65.5	715	18.38%	299	702	19.80%	64	12	-33.33%
MIROC	2020s A2	65.2	697	15.40%	300	685	16.89%	65	13	-27.78%
	2020s B1	65.3	696	15.23%	301	684	16.72%	64	12	-33.33%
	2050s A2	67.8	798	32.12%	314	790	34.81%	52	9	-50.00%
	2050s B1	67.0	765	26.66%	310	755	28.84%	55	10	-44.44%
	2080s A2	71.5	957	58.44%	333	953	62.63%	32	4	-77.78%
	2080s B1	68.3	819	35.60%	317	811	38.40%	47	7	-61.11%
CCSM	2020s A2	65.3	695	15.07%	300	683	16.55%	65	12	-33.33%
	2020s B1	65.6	700	15.89%	302	689	17.58%	62	11	-38.89%
	2050s A2	68.0	807	33.61%	314	798	36.18%	50	9	-50.00%
	2050s B1	66.6	728	20.53%	313	720	22.87%	52	9	-50.00%
	2080s A2	70.6	927	53.48%	326	922	57.34%	39	5	-72.22%
	2080s B1	66.4	735	21.69%	306	725	23.72%	59	10	-44.44%
UKMO	2020s A2	64.6	685	13.41%	294	673	14.85%	71	16	-11.11%
	2020s B1	64.0	658	8.94%	292	643	9.73%	72	15	-16.67%
	2050s A2	67.4	819	35.60%	306	805	37.37%	59	10	-44.44%
	2050s B1	66.5	768	27.15%	302	756	29.01%	63	12	-33.33%
	2080s A2	71.3	997	65.07%	323	991	69.11%	42	6	-66.67%
	2080s B1	68.6	850	40.73%	317	842	43.69%	48	8	-55.56%
Average Across Models	2020s A2	64.4	686	13.6%	297	671	14.5%	68	14	-22.2%
	2020s B1	64.6	681	12.7%	299	668	14.0%	66	13	-27.8%
	2050s A2	66.6	782	29.5%	309	772	31.7%	56	10	-44.4%
	2050s B1	66.0	741	22.7%	305	729	24.4%	60	11	-38.9%
	2080s A2	69.5	920	52.3%	324	914	56.0%	41	6	-66.7%
	2080s B1	66.8	779	29.0%	309	769	31.2%	56	10	-44.4%

[a] Mean daily maximum temperature (MMT) in °F for typical year, from observations for baseline period and from climate models simulations for 2020s, 2050s, 2080s.
[b] Central effect estimate for the net temperature, cold- and heat- related additional deaths in a typical year.
[c] Percentage change in central estimate of additional deaths in a typical year, relative to the baseline.
[d] Baseline refers to 1970-1999 reference period.

Table 11.1 Summary of projected annual mean daily maximum temperature and associated additional deaths in the 1970s versus the 2020s, 2050s, and the 2080s, in the A2 and B1 scenarios for 5 of the 16 global climate models used in ClimAID

Month	Base	A2			B1		
		2020s	2050s	2080s	2020s	2050s	2080s
1	9	8	18	19	8	19	19
2	7	7	16	19	7	17	17
3	10	11	35	52	12	31	38
4	27	35	105	130	34	103	99
5	63	73	206	251	73	198	210
6	99	108	305	354	111	291	305
7	135	151	418	476	148	401	420
8	124	139	394	454	137	369	390
9	79	90	260	297	88	241	257
10	34	42	130	160	41	121	125
11	12	16	48	66	16	45	50
12	6	6	19	24	7	18	19

These are 5 of the 16 GCMs used for ClimAID climate projections.

Table 11.2 Average (across five global climate models) projected monthly additional deaths in the 1970s versus the 2020s, 2050s, and 2080s, A2 and B1 scenarios

evaluated in terms of economic net effects. Most public policy decisions requiring economic assessments include estimating the costs of the proposed actions against those ensuing from inaction. The calculus of economic losses from increased mortality includes assigning monetary values to human life as well as estimating costs associated with services rendered before death (e.g., emergency/ ambulance services and/or hospital stay) and/or averting behavior (e.g., purchasing air conditioning units).

Some economics assessments measure mortality as the change in the probability of dying for a specific population due to a change in health status. This does not represent the "crude" mortality rate of the population, measured as the ratio of the total number of deaths divided by the total number of individuals in the population. Instead, some economics methods assume that individuals are able to rank other traded goods against the "value of a statistical life" (VSL) or the "value of a statistical death avoided" (Krupnick, 1996). In this perspective, death and illness are treated as probability rates and individuals as willing to pay to reduce marginal changes in the probability of death or incidence of illness. Thus, people are assumed to be making informed choices about the rate of substitution between small changes in the probability of death or illness, and other traded goods.

Based on such assumptions, various studies have developed coefficients to estimate the value of a statistical life in order to evaluate economic losses ensuing from premature mortality. Two methods may be

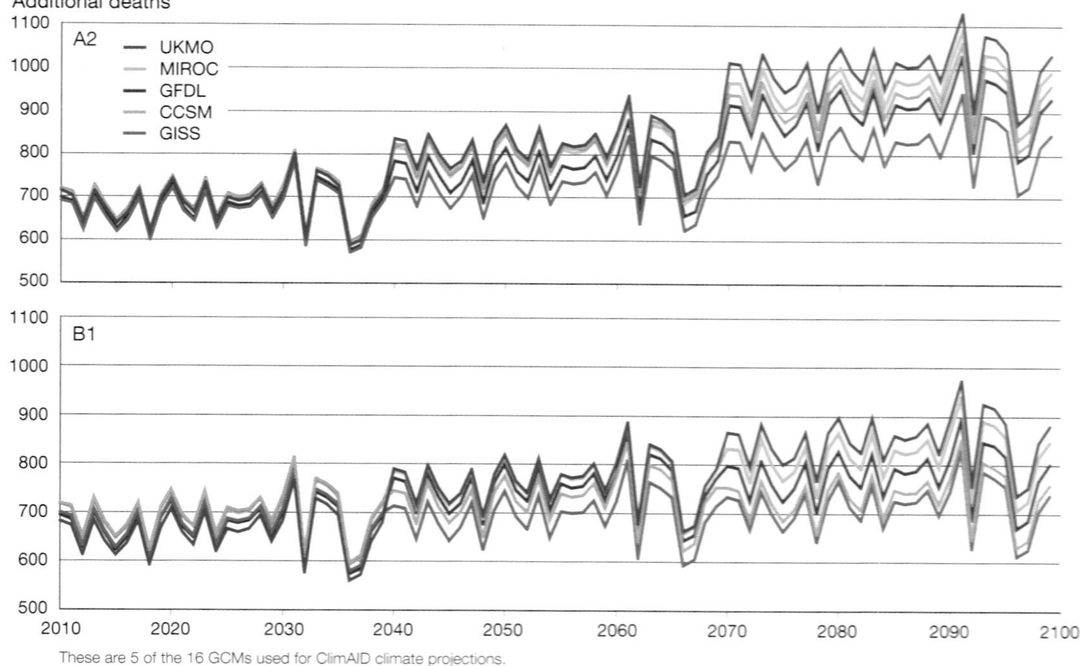

These are 5 of the 16 GCMs used for ClimAID climate projections.

Figure 11.10 Annual net additional deaths in the 21st century for five global climate models for A2 (top) and B1 (bottom) emissions scenarios

used to identify the VSL in relation to reduced mortality risks. The first is based on surveys that gather information on people's willingness to pay (WTP) to decrease mortality risks. The second one is based on the "revealed preferences" method and applies a "willingness to accept compensation" (WTA) approach to estimate VSLs by using hedonic wages or differential wage rates (Ebi et al., 2004). Hedonic wages are statistically based estimates of the wage rates of different types of jobs based on the characteristics of the jobs. Jobs that are more unpleasant or pose health and safety risks for workers typically pay higher wages than other types of jobs, and hedonic models can be used to estimate the value of these wage differences. In general, the results of both methodologies have been found to be similar (Ebi et al., 2004), with heterogeneity in age and income levels playing a role in explaining variations.

Most studies applying the above methodologies report VSL in dollars per life saved. For example, when evaluating the benefits of policies to reduce pollution, the U.S. Environmental Protection Agency reported VSLs ranging from $2.3M to 11.8M (Smith et al., 2001). Updated estimates provide a central VSL of $7.4 M (in 2006 dollars) (U.S. EPA, 2000, 2004, and 2010). Other recent studies have estimated the value of statistical life averaging $7 million (Viscusi & Hersh, 2008). Another study, which assessed VSL values for Ontario based on wage rates, placed the value of a statistical life as ranging from 0.92M to 4.54M (Krupnick et al., 2000).

Results from surveys assessing WTP to reduce mortality risks are expected, in theory, to reflect the individual characteristics of respondents. These results may be subject to a certain degree of heterogeneity, in particular because of differences in age and income levels of the population sample surveyed. With respect to age, VSLs are seen to increase up to age 50 and then decrease, with older people having the lowest values. For example, a WTP survey of Canadians found that individuals in a 70 to 75 year-old cohort were less willing to pay to reduce mortality risks than cohorts of younger adults (Krupnick et al., 2000). Nevertheless, this study found that the VSL did not decline (per age group) for people whose health is compromised, regardless of the health problem. Another VSL study explored the simultaneous effect that income levels and age have on WTP surveys, within the context of the hedonic wage model (Evans and Schaur, 2010). The authors found that the impact of age on the wage–risk tradeoff varies across the wage distribution. Results are shown in **Table 11.3**.

An alternative approach measures the VSL based on the years of potential life lost (YPLL). This approach has been advanced to consider younger age groups that may lack income streams by assigning heavier weights to premature mortality at younger ages (CDC, 1986). The YPLL approach has also been used to account for differential health status by ethnic background (CDC, 1989).

The economic burden to the health care system must also be taken into account when estimating losses from increased mortality due to heat waves. The elderly, children, and persons with certain medical conditions are at greatest risk for heat-related illness and death. Of particular concern are those individuals affected by cardiovascular disease (CVD), which accounts for more deaths in the United States than any other major cause, with roughly two-thirds related to coronary heart disease (CHD) and stroke (Yazdanyar, 2009). In 2009, costs associated with treating CVD and stroke in the United States were expected to exceed $475 billion, with direct costs, such as services at hospitals or nursing home facilities, professional fees, and medicines, estimated to reach over $313 billion. While not all such

Point in the Real Wage Distribution	Real Hourly Wage	Marginal Impact of Risk	VSL (million $)	Marginal Impact of Risk	VSL (million $)	Marginal Impact of Risk	VSL (million $)
		50-year-old		55-year-old		60-year-old	
10%	6.49	0.07	9.08	0.025	3.24	<0	<0
25%	8.85	0.089	15.75	0.049	8.67	0.009	1.59
50%	13.07	0.251	65.59	0.231	60.36	0.211	55.14
75%	19.49	0.156	60.81	0.141	54.97	0.126	49.12
Mean	15.97	0.046	14.69	0.016	5.11	<0	<0

The VSL estimates are given in 1998 dollars, and have been calculated as: Marginal Impact of Risk*Real Wage*x*y*z, where x=40, y=50, and z=10,000
Source: Evans & Schaur, 2010

Table 11.3 Estimated marginal impacts of risk on the real wage and associated value of statistical life estimates by age and real wage

Case Study D. West Nile Virus

In the U.S., more than 25,000 cases of human disease caused by West Nile virus have been reported since its introduction to North America in 1999, and hundreds of thousands of birds have been killed by the infection. The disease-causing pathogen replicates within some species of wild birds and is transmitted among birds and other hosts (including humans) via the bite of infected mosquitoes. Human risk of exposure to West Nile virus is correlated with both the abundance and infection prevalence of mosquitoes carrying the pathogen (Allan et al., 2009). Although the number of infected mosquitoes depends on the infection rate of the hosts upon which they feed, the number of mosquitoes is likely to increase with rising temperatures and a wetter climate. In New York State, the species of mosquitoes that are most likely to carry West Nile virus are those that breed in natural or artificial containers, such as ponds and discarded tires, respectively, including *Culex pipiens*, *Culvex restuans*, and *Aedes albopictus*. While West Nile virus infections in humans and birds have only been reported in a limited part of the state, the prevalence of West Nile virus in mosquitoes is more widespread throughout the state (**Figures 11.11a** and **11.11b**).

In the eastern United States, human incidence of disease caused by West Nile virus at the county level is correlated with above-average total precipitation in the previous year (Landesman et al., 2007). Higher total precipitation likely results in more immature mosquitoes surviving over the winter, which leads to a greater abundance of adults the following year. In Erie County, New York, a higher number of adult mosquitoes in the summer is correlated with cooling degree days base 63 and 65 (degree days above 63 to 65°F) seven to eight weeks earlier, with the product of cooling degree days base 63 and precipitation four weeks earlier, and with rates of evapotranspiration (the loss of water from soil evaporation and plant transpiration) five weeks earlier, although these relationships are complex and nonlinear (Trawinski and MacKay, 2008).

At the national level, higher incident rates of West Nile virus disease are associated with increased weekly maximum temperature, increased weekly average temperature, increased average weekly dew point temperature (the temperature at which water vapor condenses into water), and the occurrence of at least one day of heavy rainfall within a week (Soverow et al., 2009).

Climate change is expected to increase precipitation and summer temperatures in New York. Therefore, in general, risk of human exposure to West Nile virus is expected to increase in the state as the climate becomes warmer and wetter. Quantitative predictions about changes in risk that are specific to regions within the state will require more extensive site-specific data on the relationships between climate variables, the distribution of mosquitoes, the density of their populations and their behavior, and virus replication rates.

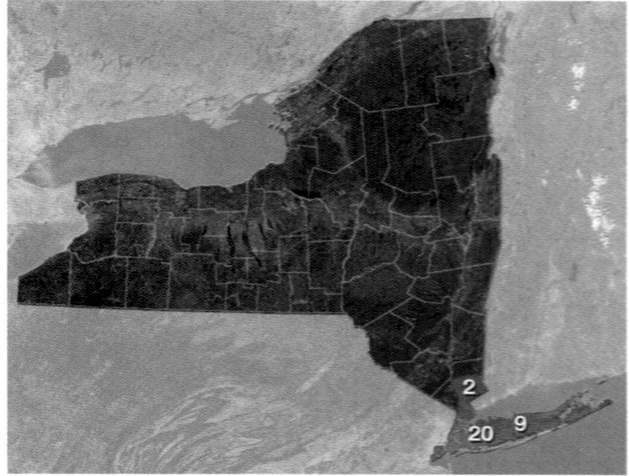

Figure 11.11a Numbers of cases of West Nile illness in humans, New York State, 2008

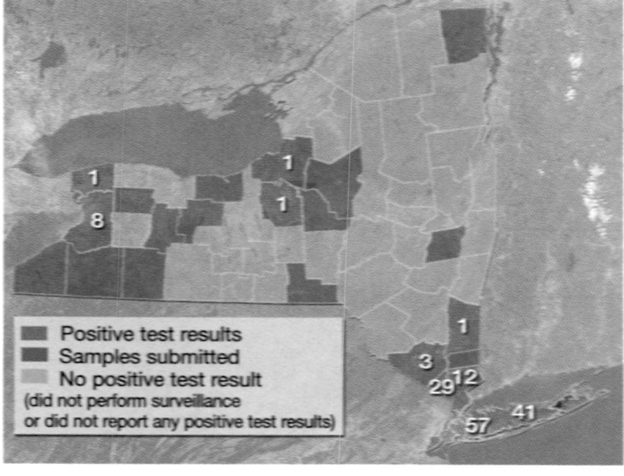

Figure 11.11b Numbers of mosquito samples testing positive for West Nile virus, New York State, 2008

References

Allan, B.F., W.A. Ryberg, R.B. Langerhans, W.J. Landesman, N.W. Griffin, R.S. Katz, K.N.Smyth, B.J. Oberle, M.R. Schultzenhofer, D.E. Hernandez, A. de St. Maurice, L. Clark, R.G. McLean, K.R. Crooks, R.S. Ostfeld, and J.M. Chase. 2009. "Ecological correlates of risk and incidence of West Nile Virus in the United States." *Oecologia* 158:699-708.

American Academy of Allergy, Asthma and Immunology. 2000. *The allergy report: science based findings on the diagnosis & treatment of allergic disorders, Volume 1*. Quoted on http://www.aaaai.org/patients/gallery/prevention.asp?item=1a Accessed: 20 May 2011.

American Heart Association. 2008. *Heart Disease and Stroke Statistics – 2008 Update at a Glance*. Accessed December 22, 2009. http://www.americanheart.org/downloadable/heart/1200078608862HS_Stats%202008.final.pdf.

Arbes S.J., Jr., P.J. Gergen, L. Elliott and D.C. Zeldin. 2005. "Prevalences of positive skin test responses to 10 common allergens in the U.S. population: results from the third National Health and Nutrition Examination Survey." *Journal of Allergy and Clinical Immunology* 116(2):377–83.

Auld, H., D. MacIver and J. Klaassen. 2004. "Heavy Rainfall and Waterborne Disease Outbreaks: The Walkerton Example." *Journal of Toxicology and Environmental Health* 67:1879-1887.

Ayres R.U. and J. Walter. 1991. "The greenhouse effect: Damages, costs and abatement." *Environmental and Resource Economics* 1:237-270.

Barney, L.D., H. White. 1992. "Hedonic wage studies and the value of a life." *Atlantic Economic Journal* September 1.

Basu, R. and B.D. Ostro. 2008. "A Multicounty Analysis Identifying the Populations Vulnerable to Mortality Associated with High Ambient Temperature in California." *American Journal of Epidemiology* 168(6):632-637. Accessed online on December 17, 2009. http://aje.oxfordjournals.org/cgi/content/abstract/ 168/6/632.

Beggs, P.J. 2004. "Impacts of climate change on aeroallergens: past and future." *Clinical and Experimental Allergy* 34:1507–1513.

Beggs, P.J. and H.J. Bambrick. 2005. "Is the global rise of asthma an early impact of anthropogenic climate change?" *Environmental Health Perspectives* 113(8):915–9.

Bell, M.L., D.L. Davis, L.A. Cifuentes, A.J. Krupnick, R.D. Morgenstern and G.D. Thurston. 2008. "Ancillary human health benefits of improved air quality resulting from climate change mitigation." *Environmental Health* 7:41.

Bell, M.L. and F. Dominici. 2008. "Effect modification by community characteristics on the short-term effects of ozone exposure and mortality in 98 US communities." *American Journal of Epidemiology* 167(8):986-997.

Bell, M.L., R. Goldberg, C. Hogrefe, P.L. Kinney, K. Knowlton, B. Lynn, et al. 2007. "Climate change, ambient ozone, and health in 50 U.S. cities." *Climatic Change* 82(1-2):61-76.

Bergeron, Y.M., M. Flannigan, S. Gauthier, A. Leduc and P. Lefort. 2004. "Past, current and future fire frequency in the Canadian boreal forest: implications for sustainable forest management." *Ambio* 6:356–60.

Bernard, S.M. and M.A. McGeehin. 2004. "Municipal heatwave response plans." *American Journal of Public Health* 94:1520-1521.

Blane, D. and F. Drever. 1998. "Inequality among men in standardized years of potential life lost, 1970-93." *British Medical Journal* 317:255-256.

Bloomberg, M.R. and R.T. Aggarwala. 2008. "Think Locally, Act Globally: How Curbing Global Warming Emissions Can Improve Local Public Health." *American Journal of Preventive Medicine* 35:414–423.

Bornehag C.G., G. Blomquist, F. Gyntelberg, B. Jarvholm, P. Malmberg, L. Nordvall, A. Nielsen, G. Pershagen and J. Sundell. 2001. "Dampness in buildings and health." Nordic interdisciplinary review of the scientific evidence on associations between exposure to "dampness" in buildings and health effects (NORDDAMP). *Indoor Air* 11:72-86.

Bouchama, A., M. Dehbi, G. Mohamed, F. Matthies, M. Shoukri and B. Menne. 2007. "Prognostic factors in heat wave related deaths: a meta-analysis." *Archives of Internal Medicine* 167(20):2170-6.

Brownstein, J.S., T.R. Holford and D. Fish. 2005. "Effect of Climate Change on Lyme Disease Risk in North America." *Ecohealth* 2:38-46.

Bullard, R. 1990. *Dumping in Dixie: Race, Class and Environmental Quality*. Boulder, CO: Westview.

Burge, H.A. 2002. "An update on pollen and fungal spore aerobiology." *Journal of Allergy and Clinical Immunology* 110(4):544–52.

Burtraw, D., A. Krupnick, K. Palmer, A. Paul, M. Toman and C. Bloyd. 2003. "Ancillary benefits of reduced air pollution in the US from moderate greenhouse gas mitigation policies in the electricity sector." *Journal of Environmental Economics and Management* 45:650–673.

CCSP (U.S. Climate Change Science Program). 2008. *Analyses of the effects of global change on human health and welfare and human systems*. A Report by the U.S. Climate Change Science Program and the Subcommittee on Global Change Research, edited by J.L. Gamble, written by K.L. Ebi, F.G. Sussman and T.J. Wilbanks. U.S. Environmental Protection Agency, Washington, DC, USA.

Centers for Disease Control and Prevention (CDC). 1986. "Premature mortality due to sudden infant death syndrome." *Morbidity and Mortality Weekly Report* 35:169-70.

Centers for Disease Control and Prevention (CDC). 1989. "Black/White Comparisons of Premature Mortality for Public Health Program Planning - District of Columbia." *Morbidity and Mortality Weekly Report* 38(3):33-37. Accessed online on December 23, 2009. http://www.cdc.gov/mmwr/preview/mmwrhtml/00001773.htm.

Centers for Disease Control and Prevention (CDC). 1995. "Heat-related illness and deaths—United States, 1994–95." *Morbidity and Mortality Weekly Report* 44: 577–579.

Centers for Disease Control and Prevention (CDC). 2003. "Heat-Related Deaths — Chicago, Illinois, 1996-2001, and United States, 1979-1999." *Morbidity and Mortality Weekly Report* 52(26):610-613. Accessed online on December 23, 2009. http://www.cdc.gov/mmwr/preview/mmwrhtml/mm5226a2.htm.

Centers for Disease Control and Prevention (CDC). 2006. "Health concerns associated with mold in water-damaged homes after Hurricanes Katrina and Rita – New Orleans area, Louisiana, October 2005." *Morbidity and Mortality Weekly Report* 55(2):41-4.

Centers for Disease Control and Prevention (CDC). 2009. "Extreme Heat: A Prevention Guide to Promote Your Personal Health and Safety." Accessed on December 21, 2009. http://www.bt.cdc.gov/disasters/extremeheat/heat_guide.asp

Cheng, C.S., M. Campbell, et al. 2005. "Differential and Combined Impacts of Winter and Summer Weather and Air Pollution due to Global Warming on Human Mortality in South-central Canada; Technical Report." Project Number of the Canada's Health Policy Research Program: 6795-15-2001/4400011. http://www.toronto.ca/health/hphe/pdf/weather_air_pollution_impacts.pdf.

Williams, D.R. and C.A. Collins. 2001. "Racial residential segregation: A of racial disparities in health." *Public Health Reports* 116:404–416.

Wolfe, R. 2003. "Death in heat waves: Beware of fans…" *British Medical Journal* 327(7425):1228.

Worfolk, J.B. 2000. "Heat waves: their impact on the health of elders." *Geriatric Nursing* 21 (2):70 –77.

World Health Organization (WHO). 2004. "Health aspects of air pollution. Results from the WHO project *Systematic review of health aspects of air pollution in Europe*". Report E83080. http://ec.europa.eu/environment/archives/cafe/activities/pdf/e83080.pdf

World Health Organization (WHO). 2003. *WHO briefing note for the fifty-third session*. WHO Regional Committee for Europe. Vienna, Austria: 8–11 September 2003.

World Health Organization (WHO). 2003. *Phenology and human health: allergic disorders*. Report on a WHO meeting. Rome. Italy.

Yazdanyar, A. and A.B. Newman. 2009. "*The Burden of Cardiovascular Disease in the Elderly: Morbidity, Mortality, and Costs.*" *Clinics in Geriatric Medicine* 25(4).

Ziska, L.H. and F.A. Caufield. 2000. "Rising carbon dioxide and pollen production of common ragweed, a known allergy-inducing species: implications for public health." *Australian Journal of Plant Physiology* 27:893–8.

Ziska, L.H., P.R. Epstein and W.H. Schlesinger. 2009. "Rising CO_2, climate change, and public health: exploring the links to plant biology." *Environmental Health Perspectives* 117:155-158.

Ziska, L.H., P.R. Epstein and C.A. Rogers. 2008. "Climate Change, Aerobiology, and Public Health in the Northeast United States." *Mitigation and Adaptation Strategies for Global Change* 13:607–613. DOI 10.1007/s11027-007-9134-1.

Ziska, L.H., D.E. Gebhard, D.A. Frenz, S. Faulkner, B.D. Singer and J.G. Straka. 2003. "Cities as harbingers of climate change: common ragweed, urbanization, and public health." *Journal of Allergy and Clinical Immunology* 111(2):290–5.

Appendix A. Stakeholder Interactions

A diverse network of stakeholders and partner organizations has been developed over the course of several assessments carried out by the ClimAID Public Health sector team since the late 1990s. The stakeholders include city, state, and federal governmental agencies in the areas of environment, health, planning, and emergency management; non-governmental environmental organizations; academic institutions with research interests in public health and climate change; environmental justice organizations; clinical health sector organizations; and community-based organizations targeting the elderly, youth, and low-income populations. Stakeholder engagement, involving approximately 100 stakeholders, included direct interviews, informal discussions, attendance at specially convened task forces, and an online survey administered to county health officials across the state.

Stakeholder Concerns

Our first approach involved phone interviews with a subset of key stakeholders at the following agencies and organizations: New York City Department of Health, New York City Mayor's Office of Long Term Planning and Sustainability, a national environmental non-governmental organization, and the New York State Department of Environmental Conservation. The climate-related health issues identified in these interviews included concerns about heat events; vector-borne illnesses such as West Nile virus (the first case in the United States occurred in New York City); other emerging infections; extreme storms (causing health risks from contaminated watersheds as a result of coastal storms, which cause flooding hazards, injury risks, and surface water quality issues that necessitate beach closures); waterborne illness; air pollution such as ground-level ozone, particulate matter and airborne allergens; and population displacement. Additional concerns expressed included the need for a full assessment of potential health effects of adaptation measures such as air pollutants from biofuels.

The stakeholders also identified needs for planning and adaptation. They reported that specific geographic variation of health impacts as well as specific population vulnerability information would be helpful in tailoring community-level adaptation projects and media messaging. Additionally, they reported that health cost-benefit analyses could assist policymakers in choosing between various planning options. Overall, there was strong consensus regarding the need for ongoing environmental and environmental health monitoring and for more data on the effectiveness of different adaptation measures. Evaluation research on the effectiveness of different adaptation measures was also identified as useful, e.g., heat-response plans, including cooling centers, public advisories about heat and the need for hydration, and buddy systems.

Some stakeholders raised concerns that transcended sectors. They questioned if the energy grid can provide continuous output during an extended heat wave and whether there is potential for failure of the power grid. Additional concerns involved energy and air quality feedbacks that could have potential health effects (i.e., power plants may burn dirtier fuels during heat waves to accommodate power demands). Also, as the risk of flooding increases, potential mold problems could increase. Lastly, concerns were raised over the impact of climate change effects on New York City's water supply. This relates to a more general area of interest voiced by our stakeholders: the increased risk of waterborne illness following high precipitation events. The importance of identifying vulnerable communities—by virtue of age, socioeconomic status, or underlying medical conditions, for example—and particular areas statewide that are more likely to be affected was emphasized.

Similar issues were raised in our informal group meetings with physicians, students, and community residents. There is a considerable amount of interest and concern about climate change and its potential health impacts. However, the knowledge base remains limited.

Emerging Adaptations

New York City has been proactive in developing climate-risk information processes for several health-relevant climate risks (NPCC, 2010). Climate-protection levels developed by an advisory group for 2050 and 2080, which include the projected number and severity of heat waves, sea level rise, and extreme rain events, are being used to guide infrastructure policy and codes. Infrastructure is broadly defined to include water, energy, and bridges. Additionally, there are efforts to increase the proportion of the vulnerable population with access to home air conditioning.

Additional adaptation measures that are within the purview of the New York City health and housing codes include beach closing after extreme rain events until water quality meets safety standards and wiring in buildings for energy efficiency and safety.

On the state level, there is a "Climate Smart Community" initiative (see www.dec.ny.gov/energy/50845.html). This initiative encourages municipalities and businesses to jointly form strategies for mitigation while also raising awareness of public health officials for coordinated effort to approach climate change.

Nongovernmental organizations are generating fact sheets and briefing reports on health preparedness for inevitable climate change. The goal is to inform policy discussions and to encourage win-win efforts. There are also efforts to transcend the artificial divide in much legislation between ecosystem and human health. The general perception by these stakeholders was that thinking about climate change and the future risks it poses provides an opportunity to improve our current level of preparedness.

Stakeholders

Natural Resources Defense Council
NYC Department of Health and Mental Hygiene
NYC Office of the Mayor NYS Department of Health
NYS Department of Environmental Conservation
New York State Association of County and City Health Officials (NYSACCHO)
U.S. Environmental Protection Agency Region II
WE ACT for Environmental Justice

Survey of City and County Health Department Directors across New York State

This part of ClimAID stakeholder engagement involved administering an online survey to New York State county health officials. This survey was adapted from the 2007 national survey of city and county health department directors—"Are We Ready?"—which revealed critical gaps between expected climate-related health impacts and local health department capacity to respond. The 2007 national survey results included evidence that 1) the majority of respondents believe that climate change already has and will continue to represent significant health threats in their jurisdiction; 2) a majority perceived lack of knowledge and expertise at all levels; 3) there is minimal incorporation of long-range weather and climate projections; and 4) a majority call for increased funding, staff and training (Maibach et al., 2008).

Climate-related health outcomes were included for specific questions pertaining to perceived current or future threats and adaptation capacity: heat-related illness, hurricanes and floods, droughts, vector-borne infectious disease, water- and food-borne disease, water supply and quality, mental health conditions, and services and infrastructure for populations affected by extreme events. While nearly all departments had some programmatic activity in one of the climate-health categories included, few indicated that they had new programming areas planned. General questions about programming activity levels, knowledge capacity, and resource needs were stratified by climate-related health driver, such as heat waves and disease vectors. Results of the New York State survey are comparable to the national survey and generate meaningful insights into local preparedness infrastructure and needs.

As part of the ClimAID project, city and county health department directors were invited to participate in a statewide replication of this national survey during the winter of 2009–2010. The "Are We Ready?" survey instrument was adapted for online administration and distributed to all department directors. The survey questions are included at the end of this section. A letter of support from the National Association of County and City Health Officials (NACCHO) and the New York State Association of County and the City Health Officials encouraged officials to participate. Responses were anonymous and have no geographic identifiers.

The survey had an overall participation rate of 39 percent. While 57 percent of respondents agreed that climate change would affect their local area in the next 20 years, only 39 percent thought that climate change would cause health problems during that same time period. However, the majority (79 percent) disagreed or strongly disagreed that their local health department had "ample" expertise to assess the impacts of climate change in their jurisdiction. And over 70 percent of respondents reported no use of long-range weather or climate information in their departments' planning.

Among respondents who believed that climate-sensitive health impacts would stay the same or increase over the next 20 years, the following were cited as areas of perceived threat:

- heat waves and heat-related illnesses
- storms, including hurricanes and floods
- droughts, forest fires, or brush fires
- vector-borne infectious diseases
- water- and food-borne diseases
- anxiety, depression, or other mental health conditions
- quality or quantity of freshwater available
- quality of the air, including air pollution
- unsafe or ineffective sewage and septic system operation
- housing for residents displaced by extreme weather events
- healthcare services for people with chronic conditions during service disruptions, such as extreme weather events
- food security
- shoreline damage/loss of shoreline/wetlands/groundwater and saltwater interaction
- severe cold and ice

As permitted by the survey, respondents could choose more than one area of concern regarding the health impacts of climate change. Heat-related health impacts were selected by 30 percent of respondents and storms by 33 percent, vector-borne disease by 56 percent, and air quality changes by 22 percent. Planned and active adaptation programming for these same four areas were reported as heat-related health programs in 33 percent of jurisdictions, storms in 54 percent of jurisdictions, vector-borne disease in 63 percent, and air quality adaptation programming in 25 percent. Of note, these percentages were all less than when respondents simply reported on current program activity in these same four areas. Of those that had a planned or active program in one of these areas, 5 percent deemed the allocated budget insufficient.

Two quotes from survey respondents that speak to the constraints regarding some of these issues:

> "With the current fiscal crisis in our region we are challenged to achieve basic health department mandated functions. We also do not have the

expertise to address this issue nor the funds to expand the programs we currently run."

"The local health department has not traditionally had a primary response role to environmentally related issues although we do support the emergency services department. While we understand that this is a role that public health should have, current fiscal restraints prevent us from being able to address climate change health effects in a suitable manner. Issues with food, water, etc. are covered by New York State Dept. of Health."

Overall, the New York State respondents showed a similar variety of concerns as the national sample though a smaller percentage deemed climate change a current or future threat to the health of residents in their jurisdiction. A non-respondent analysis is currently being explored to address the potential for generalizing these findings.

Survey Questions

Background

1. What is your position at your health department?
2. What is the approximate annual budget for your health department?
3. Approximately how many staff members in full-time equivalents does your health department have?

Climate change

4. People have different ideas about what climate change is. In your own words, what do you think the term "climate change" means?

Knowledge

5a. I am knowledgeable about the potential public health impacts of climate change.
○ Strongly disagree ○ Disagree ○ Agree ○ Strongly agree ○ Don't know

5b. The other relevant senior managers in my health department are knowledgeable about the potential public health impacts of climate change.
○ Strongly disagree ○ Disagree ○ Agree ○ Strongly agree ○ Don't know

5c. Many of the other relevant appointed officials in my jurisdiction outside of the public health system—such as environmental, agricultural, forestry and wildlife, energy and transportation officials—are knowledgeable about the potential public health impacts of climate change.
○ Strongly disagree ○ Disagree ○ Agree ○ Strongly agree ○ Don't know

5d. Many of the relevant elected officials in my jurisdiction are knowledgeable about the potential public health impacts of climate change.
○ Strongly disagree ○ Disagree ○ Agree ○ Strongly agree ○ Don't know

5e. Many of the business leaders in my jurisdiction are knowledgeable about the potential public health impacts of climate change.
○ Strongly disagree ○ Disagree ○ Agree ○ Strongly agree ○ Don't know

5f. Many of the leaders of the health care delivery system in my jurisdiction— including the hospitals and medical groups—are knowledgeable about the potential public health impacts of climate change.
○ Strongly disagree ○ Disagree ○ Agree ○ Strongly agree ○ Don't know

Perception

6a. My jurisdiction has experienced climate change in the past 20 years.
○ Strongly disagree ○ Disagree ○ Agree ○ Strongly agree ○ Don't know

6b. My jurisdiction will experience climate change in the next 20 years.
○ Strongly disagree ○ Disagree ○ Agree ○ Strongly agree ○ Don't know

6c. In the next 20 years, it is likely that my jurisdiction will experience one or more serious public health problems as a result of climate change.
○ Strongly disagree ○ Disagree ○ Agree ○ Strongly agree ○ Don't know

6d. My health department currently has ample expertise to assess the potential public health impacts associated with climate change that could occur in my jurisdiction.
○ Strongly disagree ○ Disagree ○ Agree ○ Strongly agree ○ Don't know

6e. Preparing to deal with the public health effects of climate change is an important priority for my health department.
○ Strongly disagree ○ Disagree ○ Agree ○ Strongly agree ○ Don't know

7a. Would you say that preventing or preparing for the public health consequences of climate change is among your health department's top ten current priorities?
○ Yes ○ No ○ Don't know

7b. (If Yes for Q7a) Which number—from one to ten, with one being the highest priority—would you say best characterizes the priority given to climate change currently in your health department?

Programmatic activity

8. Are the following health issues currently areas of programmatic activity for your health department?

a. Heatwaves and heat-related illnesses?	○ Yes	○ No	○ Don't know
b. Storms, including hurricanes and floods?	○ Yes	○ No	○ Don't know
c. Droughts, forest fires or brush fires?	○ Yes	○ No	○ Don't know
d. Vector-borne infectious diseases?	○ Yes	○ No	○ Don't know
e. Water- and food-borne diseases?	○ Yes	○ No	○ Don't know
f. Anxiety, depression or other mental health conditions?	○ Yes	○ No	○ Don't know
g. Quality or quantity of fresh water available to your jurisdiction?	○ Yes	○ No	○ Don't know
h. Quality of the air, including air pollution, in your jurisdiction?	○ Yes	○ No	○ Don't know
i. Unsafe or ineffective sewage and septic system operation?	○ Yes	○ No	○ Don't know
j. Food safety and security?	○ Yes	○ No	○ Don't know
k. Housing for residents displaced by extreme weather events?	○ Yes	○ No	○ Don't know

l. Health care services for people with chronic conditions during service disruptions, such as extreme weather events?
○ Yes ○ No ○ Don't know

9a. Are there other possible health effects associated with climate change in your jurisdiction that I have not mentioned?
○ Yes ○ No ○ Don't know

9b. (If Yes for Q9a) What are those health effects?

9c. (If Yes for Q9a) Is this health issue currently an area of programmatic activity for your department?
○ Yes ○ No ○ Don't know

10a. Does your health department use long-range weather or climate information in planning or implementing any programmatic activities?

○ Yes ○ No ○ Don't know

10b. (If Yes for Q10a) Do you use long-range weather or climate information in your planning or implementation of (each of the health issues a–l listed above)?

○ Yes ○ No ○ Don't know

11. Do you think climate change has already affected (each of the health issues a–l listed above) in your jurisdiction?

○ Yes ○ No ○ Don't know

12. Do you think that over the next 20 years climate change will likely make (each of the health issues a–l listed above) more common or severe, less common or severe, or that the problem will remain the same in your jurisdiction over the next 20 years?

○ More common or severe ○ Less common or severe ○ Remain the same ○ Don't know

13. Which of the potential health impacts of climate change that we have discussed, if any, are of greatest concern to you as a public health official? Feel free to name up to three outcomes.

14. Which of these three is your greatest concern? And which is your second greatest concern?

Adaptation expertise

15a. My health department currently has ample expertise to create an effective climate change adaptation plan.

○ Strongly disagree ○ Disagree ○ Agree ○ Strongly agree ○ Don't know

15b. My state health department currently has ample expertise to help us create an effective climate change adaptation plan in this jurisdiction.

○ Strongly disagree ○ Disagree ○ Agree ○ Strongly agree ○ Don't know

15c. The Centers for Disease Control and Prevention currently has ample expertise to help us create an effective climate change adaptation plan in this jurisdiction.

○ Strongly disagree ○ Disagree ○ Agree ○ Strongly agree ○ Don't know

15d. The health care delivery system in my jurisdiction—including the hospitals and medical groups—has ample expertise to create an effective climate change adaptation plan.

○ Strongly disagree ○ Disagree ○ Agree ○ Strongly agree ○ Don't know

Adaptation plans

16. Is your health department currently incorporating, planning to incorporate or not planning to incorporate adaptation into your programs for (each of the health issues a–l listed above)?

○ Currently incorporating ○ Planning to incorporate ○ Neither currently nor planning to incorporate ○ Don't know

17. How many staff members—in full-time equivalents—does/will this program have?

18. What is/will be the annual budget for this program?

19. In your opinion, is this an adequate level of funding for the program?

○ Yes ○ No ○ Don't know

The following question only asked if the response to Q16 was "currently":

20. Next year, will the annual budget for this program increase, decrease or remain about the same?

○ Increase ○ Decrease ○ Remain the same ○ Don't know

Mitigation expertise

21a. My health department currently has ample expertise to create an effective climate change mitigation plan.

○ Strongly disagree ○ Disagree ○ Agree ○ Strongly agree ○ Don't know

21b. My state's health department currently has ample expertise to help us create an effective climate change mitigation plan in this jurisdiction.

○ Strongly disagree ○ Disagree ○ Agree ○ Strongly agree ○ Don't know

21c. The Centers for Disease Control and Prevention currently has ample expertise to help us create an effective climate change mitigation plan in this jurisdiction.

○ Strongly disagree ○ Disagree ○ Agree ○ Strongly agree ○ Don't know

Mitigation plans

22. Does your department currently have, plan to have, or not have nor plan to have programs focused on the following activities?

a. Mitigating climate change by reducing greenhouse gas emissions from the health department?

○ Currently have ○ Plan to have ○ Neither currently nor plan to have ○ Don't know

b. Helping residents of your jurisdiction reduce their greenhouse gas emissions?

○ Currently have ○ Plan to have ○ Neither currently nor plan to have ○ Don't know

c. Reducing fossil fuel use or conserving energy in the operation of the health department?

○ Currently have ○ Plan to have ○ Neither currently nor plan to have ○ Don't know

d. Helping residents of your jurisdiction reduce their fossil fuel use or conserve energy?

○ Currently have ○ Plan to have ○ Neither currently nor plan to have ○ Don't know

e. Encouraging or helping people to use active transportation such as walking or cycling?

○ Currently have ○ Plan to have ○ Neither currently nor plan to have ○ Don't know

f. Encouraging or helping people to use mass transportation?

○ Currently have ○ Plan to have ○ Neither currently nor plan to have ○ Don't know

g. Encouraging or helping people to change the way they purchase foods such as buying locally grown foods, organic foods or plant-based foods?

○ Currently have ○ Plan to have ○ Neither currently nor plan to have ○ Don't know

h. Educating the public about climate change and its potential impact on health?

○ Currently have ○ Plan to have ○ Neither currently nor plan to have ○ Don't know

23a. Are there other activities associated with climate change mitigation in your jurisdiction that I have not mentioned?

○ Yes ○ No ○ Don't know

23b. (If Yes for Q23a) What are those activities?

23c. (If Yes for Q23a) Is this a current, future or not an area of programmatic activity for your department?

○ Yes ○ No ○ Don't know

The following questions only asked if the response to Q22 was "currently" or "planning":

24. How many staff members—in full-time equivalents—does/will this program have?

25. What is/will be the annual budget for this program?

26. In your opinion, is this an adequate level of funding for the program?

○ Yes ○ No ○ Don't know

The following question was only asked if the response to Q22 was "currently":

27. Next year, will the annual budget for this program increase, decrease or remain about the same?

○ Increase ○ Decrease ○ Remain the same ○ Don't know

Regulatory role

28. Does your health department have any regulatory responsibility for the following functions?

a. Water supply and quality?	○ Yes	○ No	○ Don't know
b. Air quality?	○ Yes	○ No	○ Don't know
c. Food safety and security?	○ Yes	○ No	○ Don't know
d. Sewage or septic systems?	○ Yes	○ No	○ Don't know
e. Health care services?	○ Yes	○ No	○ Don't know
f. Mental health services?	○ Yes	○ No	○ Don't know
g. Housing code?	○ Yes	○ No	○ Don't know

Resources

29a. Are there resources that your department does not currently have that, if made available, would significantly improve its ability to deal with climate change as a public health issue?

○ Yes ○ No ○ Don't know

29b. (If Yes for Q29a) What are those resources?

○ Additional Staff ○ Staff Training ○ Equipment ○ Budget/Money/Funding ○ Other

Respondents were asked to describe their answers in further detail:

a. How many additional staff and what would they do?

b. What kind of training?

c. What kind of equipment?

d. How much money and what would you use it for?

Conclusion

Is there anything else that will help us understand the public health response to climate change in your jurisdiction?

Appendix B. Technical Information on Heat Wave Cost

Year	Event Type	Region Affected	Sector(s) Most Affected	Total Costs / Damage Costs (billion $)	Deaths
2000	Severe drought & persistent heat	South-central & southeastern states	agriculture and related industries	$4.2	140
1998	Severe drought & persistent heat	TX / OK eastward to the Carolinas	agriculture and ranching	$6.6–9.9	200
1993	Heat wave/drought	Southeast US	agriculture	$1.3	16
1988	Heat wave/drought	Central & Eastern US	agriculture & related industries	$6.6	5000–10,000
1986	Heat wave/drought	Southeast US	agriculture & related industries	$1.8–2.6	100
1980	Heat wave/drought	Central & Eastern US	unspecified	$48.4	10,000

Source: Ross and Lott, 2003

Table 11.5 Costs for major heat waves in the United States

Appendix C. Annotated Heat-Mortality, Wildfires, and Air Pollution Methods References

Anderson, B. G. and M. L. Bell. 2009. "Weather-related mortality: how heat, cold, and heat waves affect mortality in the United States." *Epidemiology* 20(2):205–213.

Background:

Many studies have linked weather to mortality; however, the role of such critical factors as regional variation, susceptible populations, and acclimatization remain unresolved.

Methods:

They applied time-series models to 107 U.S. communities allowing a nonlinear relationship between temperature and mortality by using a 14-year dataset. Second-stage analysis was used to relate cold, heat, and heat-wave effect estimates to community-specific variables. They considered exposure timeframe, susceptibility, age, cause of death, and confounding effects of pollutants. Heat waves were modeled with varying intensity and duration.

Results:

Heat-related mortality was most associated with a shorter lag (average of same day and previous day), with an overall increase of 3.0 percent (95 percent posterior interval: 2.4 percent–3.6 percent) in mortality risk comparing the 99th and 90th percentile temperatures for the community. Cold-related mortality was most associated with a longer lag (average of current day up to 25 days previous), with a 4.2 percent (3.2 percent–5.3 percent) increase in risk comparing the first and 10th percentile temperatures for the community. Mortality risk increased with the intensity or duration of heat waves. Spatial heterogeneity in effects indicates that weather-mortality relationships from one community may not be applicable in another. Larger spatial heterogeneity for absolute temperature estimates (comparing risk at specific temperatures) than for relative temperature estimates (comparing risk at community-specific temperature percentiles) provides evidence for acclimatization. They identified susceptibility based on age, socioeconomic conditions, urbanicity, and central air conditioning.

Conclusions:

Acclimatization, individual susceptibility, and community characteristics all affect heat-related effects on mortality.

Hoyt, K. S. and A. E. Gerhart. 2004. "The San Diego County wildfires: perspectives of healthcare providers [corrected]." *Disaster Management and Response* 2(2):46–52.

The wildfires of October 2003 burned a total of 10 percent of the county of San Diego, California. Poor air quality contributed to an increased number of patients seeking emergency services, including healthcare providers affected by smoke and ash in hospital ventilation systems. Two large hospitals with special patient populations were threatened by rapidly approaching fires and had to plan for total evacuations in a very short time frame. A number of medical professionals were forced to prioritize responding to the hospital's call for increased staff during the disaster and the need to evacuate their own homes.

Johnston, F. H., A. M. Kavanagh, et al. 2002. "Exposure to bushfire smoke and asthma: an ecological study." *Medical Journal of Australia* 176(11):535–538.

Objective:

To examine the relationship between the mean daily concentration of respirable particles arising from bushfire smoke and hospital presentations for asthma.

Design and Setting

An ecological study conducted in Darwin (Northern Territory, Australia) from 1 April–31 October 2000, a period characterised by minimal rainfall and almost continuous bushfire activity in the proximate bushland. The exposure variable was the mean atmospheric concentration of particles of 10 microns or less in aerodynamic diameter (PM_{10}) per cubic metre per 24-hour period.

Outcome Measure:

The daily number of presentations for asthma to the Emergency Department of Royal Darwin Hospital.

Results:

There was a significant increase in asthma presentations with each $10 \mu g/m^3$ increase in PM_{10} concentration, even after adjusting for weekly rates of influenza and for weekend or weekday (adjusted rate ratio, 1.20; 95 percent confidence interval (CI), 1.09–1.34; $P < 0.001$). The strongest effect was seen on days when the PM_{10} was above $40 \mu g/m^3$ (adjusted rate ratio, 2.39; 95 percent confidence interval (CI), 1.46–3.90), compared with days when PM_{10} levels were less than $10 \mu g/m^3$.

Conclusions:

Airborne particulates from bushfires should be considered as injurious to human health as those from other sources. Thus, the control of smoke pollution from bushfires in urban areas presents an additional challenge for managers of fireprone landscapes.

Kinney, P. L. and H. Ozkaynak. 1991. "Associations of daily mortality and air pollution in Los Angeles County." *Environmental Research* 54(2):99–120.

They report results of a multiple regression analysis examining associations between aggregate daily mortality counts and environmental variables in Los Angeles County, California for the period 1970 to 1979.

Methods:

Mortality variable included total deaths not due to accidents and violence (M), deaths due to cardiovascular causes (CV), and deaths due to respiratory causes (Resp). The environmental variables included five pollutants averaged over Los Angeles County: total oxidants (Ox), sulfur dioxide (SO_2), nitrogen dioxide (NO_2), carbon monoxide (CO), and KM (a measure of particulate optical reflectance). Also included were three metereological variables measured at the Los Angeles International Airport: temperature (Temp), relative humidity (RH), and extinction coefficient (B_{ext}), the latter estimated from noontime visual range. To reduce the possibility of spurious correlations arising from the shared seasonal cycles of mortality and environmental variables, seasonal cycles were removed from the data by applying a high-pass filter. Cross-correlation functions were examined to determine the lag structure of the data prior to specifying and fitting the multiple regression models relating mortality and the environmental variables.

Results:

The results demonstrated significant associations of M (or CV) with Ox at lag 1, temperature, and NO_2, CO, or KM. Each of the latter three variables was strongly associated with daily mortality but all were also highly correlated with one another in the high-frequency band, making it impossible to uniquely estimate their separate relationships to mortality

Conclusions:

The results of this study show that small but significant associations exist in Los Angeles County between daily mortality and three separate environmental factors: temperature, primary motor vehicle-related pollutants (e.g., CO, KM, NO_2), and photochemical oxidants.

Levy, J. I., S. M. Chemerynski, et al. 2005. "Ozone exposure and mortality: an empiric Bayes metaregression analysis." *Epidemiology* 16(4): 458–468.

Background:

Results from time-series epidemiologic studies evaluating the relationship between ambient ozone concentrations and premature mortality vary in their conclusions about the magnitude of this relationship, if any, making it difficult to estimate public health benefits of air pollution control measures. Authors conducted an empiric Bayes metaregression to estimate the ozone effect on mortality, and to assess whether this effect varies as a function of hypothesized confounders or effect modifiers.

Methods:

They gathered 71 time-series studies relating ozone to all-cause mortality, and they selected 48 estimates from 28 studies for the metaregression. Metaregression covariates included the relationship between ozone concentrations and concentrations of other air pollutants, proxies for personal exposure-ambient concentration relationships, and the statistical methods used in the studies. For the metaregression, they applied a hierarchical linear model with known level-1 variances.

Results:

They estimated a grand mean of a 0.21 percent increase (95 percent confidence interval = 0.16–0.26 percent) in mortality per 10-$\mu g/m^3$ increase of 1-hour maximum ozone (0.41 percent increase per 10 ppb) without controlling for other air pollutants. In the metaregression, air-conditioning prevalence and lag time were the strongest predictors of between-study variability. Air pollution covariates yielded inconsistent findings in regression models, although correlation analyses indicated a potential influence of summertime $PM_{2.5}$.

Conclusions:

These findings, coupled with a greater relative risk of ozone in the summer versus the winter, demonstrate that geographic and seasonal heterogeneity in ozone relative risk should be anticipated, but that the observed relationship between ozone and mortality should be considered for future regulatory impact analyses.

O'Neill, M. S., A. Zanobetti, et al. 2003. "Modifiers of the temperature and mortality association in seven US cities." *American Journal of Epidemiology*. 157(12):1074–1082.

This paper examines effect modification of heat- and cold-related mortality in seven U.S. cities in 1986–1993.

Methods:

City-specific Poisson regression analyses of daily noninjury mortality were fit with predictors of mean daily apparent temperature (a construct reflecting physiologic effects of temperature and humidity), time, barometric pressure, day of the week, and particulate matter less than 10 micro m in aerodynamic diameter. Percentage change in mortality was calculated at 29°C apparent temperature (lag 0) and at -5°C (mean of lags 1, 2, and 3) relative to 15°C. Separate models were fit to death counts stratified by age, race, gender, education, and place of death. Effect estimates were combined across cities, treating city as a random effect.

Results:

Deaths among Blacks compared with Whites, deaths among the less educated, and deaths outside a hospital were more strongly associated with hot and cold temperatures, but gender made no difference. Stronger cold associations were found for those less than age 65 years, but heat effects did not vary by age. The strongest effect modifier was place of death for heat, with out-of-hospital effects more than five times greater than in-hospital deaths, supporting the biologic plausibility of the associations.

Conclusions:

Place of death, race, and educational attainment indicate vulnerability to temperature-related mortality, reflecting inequities in health impacts related to climate change.

Peel, J. L., K. B. Metzger, et al. 2007. "Ambient air pollution and cardiovascular emergency department visits in potentially sensitive groups." *American Journal of Epidemiology* 165(6):625-633.

Limited evidence suggests that persons with conditions such as diabetes, hypertension, congestive heart failure, and respiratory conditions may be at increased risk of adverse cardiovascular morbidity and mortality associated with ambient air pollution.

Methods:

The authors collected data on over four million emergency department visits from 31 hospitals in Atlanta, Georgia, between January 1993 and August 2000. Visits for cardiovascular disease were examined in relation to levels of ambient pollutants by use of a case-crossover framework. Heterogeneity of risk was examined for several comorbid conditions.

Results:

The results included evidence of stronger associations of dysrhythmia and congestive heart failure visits with comorbid hypertension in relation to increased air pollution levels compared with visits without comorbid hypertension; similar evidence of effect modification by diabetes and chronic obstructive pulmonary disease (COPD) was observed for dysrhythmia and peripheral and cerebrovascular disease visits, respectively. Evidence of effect modification by comorbid hypertension and diabetes was observed in relation to particulate matter less than 10 microm in aerodynamic diameter, nitrogen dioxide, and carbon monoxide, while evidence of effect modification by comorbid COPD was also observed in response to ozone levels.

Conclusions:

These findings provide further evidence of increased susceptibility to adverse cardiovascular events associated with ambient air pollution among persons with hypertension, diabetes, and COPD.

Peel, J. L., P. E. Tolbert, et al. 2005. "Ambient air pollution and respiratory emergency department visits." *Epidemiology* 16(2):164-174.

Background:

A number of emergency department studies have corroborated findings from mortality and hospital admission studies regarding an association of ambient air pollution and respiratory outcomes. More refined assessment has been limited by study size and available air quality data.

Methods:

Measurements of five pollutants (particulate matter [PM_{10}], ozone, nitrogen dioxide [NO_2], carbon monoxide [CO], and sulfur dioxide [SO_2]) were available for the entire study period (1 January 1993 to 31 August 2000); detailed measurements of particulate matter were available for 25 months. Authors obtained data on four million emergency department visits from 31 hospitals in Atlanta. Visits for asthma, chronic obstructive pulmonary disease, upper respiratory infection (URI), and pneumonia were assessed in relation to air pollutants using Poisson generalized estimating equations.

Results:

In single-pollutant models examining three-day moving averages of pollutants (lags 0, 1, and 2): standard deviation increases of ozone, NO_2, CO, and PM_{10} were associated with 1–3 percent increases in URI visits; a 2 μg/m increase of $PM_{2.5}$ organic carbon was associated with a 3 percent increase in pneumonia visits; and standard deviation increases of NO_2 and CO were associated with 2–3 percent increases in chronic obstructive pulmonary disease visits. Positive associations persisted beyond three days for several of the outcomes, and over a week for asthma.

Conclusions:

The results of this study contribute to the evidence of an association of several correlated gaseous and particulate pollutants, including ozone, NO_2, CO, PM, and organic carbon, with specific respiratory conditions.

Westerling, A. L., H. G. Hidalgo, et al. 2006. "Warming and earlier spring increase western U.S. forest wildfire activity." *Science* 313(5789): 940–943.

Background:

Western United States forest wildfire activity is widely thought to have increased in recent decades, yet neither the extent of recent changes nor the degree to which climate may be driving regional changes in wildfires has been systematically documented. Much of the public and scientific discussion of changes in western United States wildfires has focused instead on the effects of 19th- and 20th-century land-use history.

Methods:

They compiled a comprehensive database of large wildfires in western United States forests since 1970 and compared it with hydroclimatic and land-surface data.

Results:

Here, the authors show that large wildfire activity increased suddenly and markedly in the mid-1980s, with higher large-wildfire frequency, longer wildfire durations, and longer wildfire seasons. The greatest increases occurred in mid-elevation, Northern Rockies forests, where land-use histories have relatively little effect on fire risks and are strongly associated with increased spring and summer temperatures and an earlier spring snowmelt.

Chapter 12

Conclusions and Recommendations

Contents

Climate Change and New York State 440
Integrating Themes 440
  Climate 440
  Vulnerability 442
  Adaptation 454
  Equity and Environmental Justice 457
  Economics 457
Recommendations 457
  Policy 457
  Management 458
  Knowledge Gaps and Information Needs 458
  Science and Research 460
Responding to Future Climate Challenges 460

Climate Change and New York State

Adapting to a changing climate is challenging in New York State due to its diverse nature geographically, economically, and socially. The main drivers of climate change impacts—higher temperature, sea level rise and its potential to increase coastal flooding, and changes in precipitation—will have a wide variety of effects on the sectors and regions across the state and will engender a wide range of adaptation strategies. Climate change will bring opportunities as well as constraints, and interactions of climate change with other stresses, such as population growth, will create new challenges.

While New York State ranks 27th among the states in area (54,556 square miles, including 7,342 square miles of inland water), it is subject to a much wider range of climate impacts than its size in square miles would suggest. The north-to-south distance from the Canadian border to the tip of Staten Island is over 300 miles; from east to west (from the longitude of the eastern tip of Long Island to the longitude of the western border of New York State at Lake Erie), the distance is over 400 miles. Further diversity stems from the presence of the densely populated New York City, while much of the state is rural in character. Thus, climate hazards are likely to produce a range of impacts on the rural and urban fabric of New York State in the coming decades.

The adaptation strategies described in the ClimAID Assessment could be useful in preparing for and responding to climate risks now and in the future. Such adaptation strategies are also likely to produce benefits today, since they will help to lessen impacts of climate extremes that currently cause damage. However, given the scientific uncertainties in projecting future climate change, monitoring of climate and impacts indicators is critical so that flexible adaptation pathways for the region can be achieved over time.

This chapter summarizes the overall conclusions and recommendations of the ClimAID assessment. They focus on the five integrating themes (climate, vulnerability, adaptation, equity and environmental justice, and economics) and the eight sectors (Water Resources, Coastal Zones, Ecosystems, Agriculture, Energy, Transportation, Telecommunications, and Public Health). The conclusions and recommendations highlight sectoral, geographical, and temporal dimensions in responding to the risks posted by climate change in New York State.

Integrating Themes

This section highlights the conclusions focused on the five integrating themes.

Climate

The humid continental climate of New York State varies from warmer to cooler and from wetter to dryer regions. The weather that New York State has experienced historically provides a context for assessing climate changes that are projected for the rest of the century. The ClimAID Assessment found that much of the state is already warming and that projected climate changes in temperature and other variables could bring significant impacts.

Observed Climate Trends

Observed climate trends include the following:

- Annual temperatures have been rising throughout the state since the start of the 20th century. State-average temperatures have increased by approximately 0.6°F per decade since 1970, with winter warming exceeding 1.1°F per decade.
- Since 1900, there has been no discernable trend in annual precipitation, which is characterized by large interannual and interdecadal variability.
- Sea level along New York's coastline has risen by approximately 1 foot since 1900.
- Intense precipitation events (heavy downpours) have increased in recent decades.

As a whole, New York State has experienced a significant warming trend over the past three to four decades. Sea level along New York's coastline has increased approximately 12 inches over the past century. Given these trends and projections of future changes, past climate will likely be a less consistent predictor of future climate, and, in turn, reliance on past climate records may not suffice as benchmarks for forecasting.

Water Resources (continued)

Main Climate Variable	Specific Climate Variable	Probability of Specific Climate Variable	Climate Variable Notes	Impact on Resource	Likelihood of Impact	Consequence without Adaptation	Magnitude of Con-sequence*
Water Quality							
Temperature	Increase in mean annual temperature	Very likely	Increase in mean temperatures may be greater 1) in the north than south, and 2) in winter than in summer in the north	Favorable corn-based ethanol production	Medium	May lead to increased agricultural land use in NYS	Medium
				Greater pathogen survivability in waters	High	Increased potential for disease in aquatic life	High
				Increased algal growth in water bodies as well as increased dissolved organic matter exported from soils and wetlands	High	Impairs recreational use and normal ecosystem function; increased organic matter may increase the concentration of disinfection by-products (DBP) in drinking water (potentially harmful chemicals that form when chlorine added to kill pathogens reacts with organic matter)	High
	Increase in water temperature of streams and rivers	Likely/ very likely	Depends on many factors besides air temperature, such as precipitation, water demand, and land cover	Warmer water holds less dissolved oxygen (DO), so warmer waters will increase strain on streams that already experience oxygen depletion	Medium	High DO levels are detrimental to aquatic organisms	Medium
Precipitation	Increase in mean annual precipitation	More likely than not	N/A	Expanded agriculture in water-rich areas	Medium	Increased nutrient (nitrogen and phosphorus) loading, which leads to degraded water quality and ecosystem health	Medium
	Increase in extreme precipitation events	More likely than not	N/A	Increased runoff and reduced infiltration of rain into natural ground cover and soils	High	Greater potential for CSOs	High

Notes: N/A = Not Applicable
CSO = Combined sewer overflow
* Factors that are considered when determining the magnitude of consequence, defined as the combined impact of the occurrence should a given hazard occur, include: effects on internal operations, capital and operating costs, public health, the economy, and the environment, as well as the number of people affected. (see Annex II to the full report, "Adaptation Guidebook")

Coastal Zones

Main Climate Variable	Specific Climate Variable	Probability of Specific Climate Variable	Climate Variable Notes	Impact on Resource	Likelihood of Impact	Consequence without Adaptation	Magnitude of Consequence*
Infrastructure and Coastal Property							
Sea level rise	Permanent inundation of coastal areas	N/A	By 2050, only a small increase in the area permanently inundated is expected	Entrances to bridges, tunnels, segments of highways, wastewater treatment plants, and sewer outfall systems permanently under sea water	High	Failure of systems	High
				Coastal properties permanently under sea water	High	Abandonment of waterfront structures and residences (ground floor or potentially altogether)	Medium
				Increase in salinity of influent into wastewater pollution control plants	Medium	Corrosion of materials and equipment, failure of systems	High
	Increased frequency, intensity, and duration of storm surge and coastal flooding	Likely/very likely	Will depend both on sea level rise and on uncertain changes in tropical cyclones and nor'easters	Coastal property damage	High	Potential loss of life	High
						Economic impact	High
						Complications to evacuation routes	Medium
						Failure of systems	Medium
				Increased wear and tear on equipment not designed for salt-water exposure	Medium	More frequent delays and service interruptions on public transportation and low-lying highways	Medium
Ecosystems							
Temperature	Warmer coastal sea surface temperatures	Likely	N/A	Heightened disease, harmful algae blooms, and increased competition over resources	High	Ecosystem vulnerability	Medium
				Northward shift in range of habitat for many commercially important fish and shellfish species	High	Decline in fishing industry	High
Precipitation	Increased mean precipitation	More likely than not	N/A	Affect rates of groundwater recharge lake levels	Medium	Potential shortages of drinking water availability	High
				Increased or reduced stream flow	Medium	Affect the delivery of nutrients and pollutants to coastal waters potentially leading to poorer water quality	Medium
Sea level rise	Permanent inundation of coastal areas	N/A	By 2050, only a small increase in the area permanently inundated is expected	Permanent inundation of wetlands	High	Loss of critical wetland habitat	High
	Increased storm surge and coastal flooding	Likely/very likely	Will depend both on sea level rise and on uncertain changes in tropical cyclones and nor'easters	Increased beach erosion	High	Barrier migrations and loss of barrier islands resulting in exposure of the bay and mainland shoreline to more oceanic conditions	High
	Increased wave action	Likely	Will depend both on sea level rise and on uncertain changes in tropical cyclones and nor'easters	Erosion and reshaping of shorelines	Medium	Affect the location and extent of storm surge inundation	High

Notes: N/A = Not Applicable
* Factors that are considered when determining the magnitude of consequence, defined as the combined impact of the occurrence should a given hazard occur, include: effects on internal operations, capital and operating costs, public health, the economy, and the environment, as well as the number of people affected. (see Annex II to the full report, "Adaptation Guidebook")

Ecosystems							
Main Climate Variable	Specific Climate Variable	Probability of Specific Climate Variable	Climate Variable Notes	Impact on Resource	Likelihood of Impact	Consequence without Adaptation	Magnitude of Consequence*
Plants							
Temperature	Increase in mean annual temperature	Very likely	Increase in mean temperatures may be greater 1) in the north than south, and 2) in winter than in summer in the north	Potential increase in plant growth with large differences between species	Medium	Altered plant community structure and potential for invasives	Low
	Warmer winters	Very likely	N/A	Longer growing season	Medium	Shift in ecosystems	High
				Earlier blooming of perennials	High	Potential to throw off symbiotic relationships	High
				Potential changes in sap flow	Medium	Negative effects on maple syrup production requiring some regions to increasingly rely on more expensive technology	High
Animals and Insects							
Temperature	Increase in mean annual temperature	Very likely	Increase in mean temperatures may be greater 1) in the north than south, and 2) in winter than in summer in the north	Insects see more generations per season	Medium	Rate of invasive and pest species rises	High
	Warming waters	Likely/very likely	Depends on air temperature, precipitation, water demand, and land cover	Decline in coldwater fish species such as brook trout and other native species	High	Changes in coldwater ecosystems	High
						Decline in fishing industry for coldwater species	Medium
	Warmer winters	Very likely	N/A	Northward shift in range of many species, including undesirable pests, diseases and vectors of disease, invasives	High	Changes in ecosystems, decline of native species	High
				Increased winter survival of deer populations	High	Increasing deer inflicted damage to plants	Medium
				Increased survival of marginally over-wintering insect pests	Medium	Increased pest threat to ecosystems	Medium
				Earlier arrival of migratory birds	High	Potential to throw off symbiotic relationships	High
	Reduction in snow cover	Unknown	Earlier snowmelt is likely/very likely	Negative effects on survival of animals and insects who depend on snow for insulation and protective habitat	High	Changes in ecosystems, decline of native species	High
				Increased winter deer feeding	High	Increased vegetation damage	Medium
Recreation							
Temperature	Reduction in snow cover	Unknown	Earlier snowmelt is likely/very likely	Less natural snow for ski industry	High	Smaller, more southerly or lower altitude ski operations may have more difficulty keeping up with increasing demands on artificial snowmaking capacity	Medium

Notes: N/A = Not Applicable
* Factors that are considered when determining the magnitude of consequence, defined as the combined impact of the occurrence should a given hazard occur, include: effects on internal operations, capital and operating costs, public health, the economy, and the environment, as well as the number of people affected. (see Annex II to the full report, "Adaptation Guidebook")

Agriculture							
Main Climate Variable	Specific Climate Variable	Probability of Specific Climate Variable	Climate Variable Notes	Impact on Resource	Likelihood of Impact	Consequence without Adaptation	Magnitude of Consequence*
Crops							
Temperature	Increase in mean temperatures	Very likely	Warming may be greater 1) in the north than south, and 2) in winter than in summer in the north	Longer growing season for certain crops	High	Potentially increased crop yield and may expand market opportunity for some crops, but also prices go down	Medium
						Weeds will grow faster and will have to be controlled for longer periods	Medium
						Increased seasonal water and nutrient requirements	Medium
				Increased weed, disease, and insect pressure	Medium	Lower native crop survival, increase in prices	High
				Increased relative risk of freeze or frost damage and/or reduced winter chill-hour accumulation required for normal spring development	High	Lower survival of perennial fruit crops	High
				Weed species more resistant to herbicides	Low	Change in species composition potentially not favoring native crops	Medium
				Northward expansion of disease range and weeds (plants that have not built immunity to new pathogens are more susceptible to disease and larger populations of pathogens survive to initially infect plants)	High	Lower crop survival	High
				Crop damage due to sudden changes, such as increased freeze damage of woody plants due to loss of winter hardiness or premature leaf-out and frost damage	Medium	Decrease in crop yield	Medium
	Warmer winters	Very likely	N/A	Lengthened growing season	Medium	Could increase productivity or quality of some woody perennials (e.g., European wine grapes)	High
				Not enough freeze days for certain crops	Medium	By mid to late century, negatively affect crops adapted to current climate (e.g., Concord grape, some apple varieties)	Medium
				More winter cover crop options; depending on variability of winter temps, can lead to increased freeze or frost damage of woody perennials	Medium	Decrease in crop yield	Medium
	Increase in extreme heat events	Likely	N/A	Stress on crops, especially if extreme events occur in clusters	Medium to High	Major crop and profit loss	Medium to High
				Heat stress effects	High	Negatively affect yield or quality of many cool-season crops that currently dominate the ag economy, such as apple, potato, cabbage, and other cold crops	High

Notes: N/A = Not Applicable

* Factors that are considered when determining the magnitude of consequence, defined as the combined impact of the occurrence should a given hazard occur, include: effects on internal operations, capital and operating costs, public health, the economy, and the environment, as well as the number of people affected. (see Annex II to the full report, "Adaptation Guidebook")

Agriculture (continued)							
Main Climate Variable	Specific Climate Variable	Probability of Specific Climate Variable	Climate Variable Notes	Impact on Resource	Likelihood of Impact	Consequence without Adaptation	Magnitude of Consequence*
Crops (continued)							
Precipitation	Increase in mean precipitation	More likely than not	N/A	Increased flooding resulting in inability to access field during critical times	Medium	Direct crop damage, increased chemical contamination of waterways and harvested crops	Medium
				Increased flooding risk could delay spring planting and harvest	High	Negatively affect market prices; reduction in the high-value early season production of vegetable crops	High
				Increased soil compaction because of tractor use on wet soils	High	Increased vulnerability to future flooding and drought; increasing runoff and erosion; plants have difficulty in compacted soil because the mineral grains are pressed together leaving little space for air and water, which are essential for root growth	High
				Increased crop root disease and anoxia	High	Decrease in crop productivity and yield	High
				Wash-off of applied chemicals	Medium	Decrease in crop productivity and yield	High
	Increase in droughts	Uncertain	N/A	Decrease the duration of leaf wetness and reduce forms of pathogen attack on leaves	High	Decrease in crop productivity	High
				Increased stress on plants	High	Reduced yields and crop losses, particularly for rain-fed agriculture	Medium
				Inadequate irrigation capacity for some high value crop growers	High	Decrease in crop yield	Medium
				Dry streams or wells	Medium	Increased pumping costs from wells	Medium
	Increase in intense precipitation events	More likely than not	N/A	Stress on crops, especially if extreme events occur in clusters	Medium to High	Major crop and profit loss	Medium to High
	Changes in cloud cover and radiation	Uncertain	N/A	Cloudy periods during critical development stages impacts plant growth	High	Affect plant growth, yields, and crop water use	High
Livestock (Dairy)							
Temp-erature	Increase in extreme heat events	Likely	N/A	Increased stress to livestock	High	Decrease in milk production; reduced calving rates	Medium
Insects and Weed Pests							
Temperature	Increase mean temperatures	Very likely	Warming may be greater 1) in the north than south, and 2) in winter than in summer in the north	More generations per season; shifts in species range	High	Increased vulnerability of crops to pests	High
	Warmer winters	Very likely	N/A	Increased spring populations of marginally overwintering insects	High	Increased vulnerability of crops to pests and invasives	High
				Northward range expansion of invasive weeds			

Notes: N/A = Not Applicable

* Factors that are considered when determining the magnitude of consequence, defined as the combined impact of the occurrence should a given hazard occur, include: effects on internal operations, capital and operating costs, public health, the economy, and the environment, as well as the number of people affected. (see Annex II to the full report, "Adaptation Guidebook")

Energy							
Main Climate Variable	Specific Climate Variable	Probability of Specific Climate Variable	Climate Variable Notes	Impact on Resource	Likeli-hood of Impact	Consequence without Adaptation	Magnitude of Con-sequence*
Energy Resources							
Temp-erature	Increased mean temperatures	Very Likely	Warming may be greater 1) in the north than south, and 2) in winter than in summer in the north	Changes in biomass available for energy generation	Medium	Decreased reliability of biomass as an alternative energy source	Low
Precipitation	Increases in mean precipitation	More likely than not	N/A	Availability of hydropower reduced	Medium	Decreased reliability of hydropower as an alternative energy source	Low
	Cloud cover	Uncertain	N/A	Changes in solar exposure	High	Decreased reliability of solar power as an alternative energy source	Low
Extreme events	Wind	Uncertain	N/A	Availability and predictability is reduced with variation in wind	High	Decreased reliability of wind energy as an alternative energy source	Low
Generation Assets							
Temperature	Increase in mean temperatures	Very Likely	Warming may be greater 1) in the north than south, and 2) in winter than in summer in the north	Reduced water cooling capacity	Medium	Water-cooled nuclear power plants become more at risk for overheating and failure of equipment; the thermal efficiency of power generation is reduced	High
Sea level rise	Increased frequency, intensity, and duration of storm surge and coastal flooding	Likely/very likely	Will depend both on sea level rise and on uncertain changes in tropical cyclones and Nor'easters	Damage to coastal power plants	High	Reduced generation	Medium
Transmission and Distribution Assets							
Temp-erature	Increase in mean temperatures	Very Likely	Warming may be greater 1) in the north than south, and 2) in winter than in summer in the north	Sagging power lines	Medium	More frequent power outages	Medium
				Wear on transformers	Medium	Transformers rated for particular temperatures may fail more frequently	Medium
Precipitation	Snow storms	Uncertain	N/A	Transmission infrastructure damage	Low	Changes in power outage frequency	Medium
	Ice storms	Uncertain	N/A	Transmission lines sagging due to freezing/collecting ice	Low	Changes in power outage frequency	Medium
Electricity Demand							
Temperature	Increase in mean annual temperatures	Very Likely	Warming may be greater 1) in the north than south, and 2) in winter than in summer in the north	Increased energy demand	High	Increase in number of instances of peak load during summer, winter, and shoulder season	Medium
	Increase in extreme heat events; decrease in extreme cold events	Likely	N/A	Overwhelmed power supply system	Low	Increased frequency of blackouts and brownouts and reduced availability and reliability of power for downstate regions	High
Buildings							
Extreme events	Hurricanes and nor'easters	Uncertain	N/A	Heightened storm regime may reveal weaknesses in building envelopes	Medium	Increased chance of structural failure	Low
	Extreme wind events	Uncertain	N/A				
	Increased intense precipitation events	More likely than not	N/A	Low lying areas susceptible to more frequent flooding	High	Potential for structural damage to boilers	High

Notes: N/A = Not Applicable
* Factors that are considered when determining the magnitude of consequence, defined as the combined impact of the occurrence should a given hazard occur, include: effects on internal operations, capital and operating costs, public health, the economy, and the environment, as well as the number of people affected. (see Annex II to the full report, "Adaptation Guidebook")

Transportation

Main Climate Variable	Specific Climate Variable	Probability of Specific Climate Variable	Climate Variable Notes	Impact on Resource	Likelihood of Impact	Consequence without Adaptation	Magnitude of Con-sequence*
Physical Assets							
Temperature	Increase in mean temperature	Very likely	Warming may be greater 1) in the north than south, and 2) in winter than in summer in the north	Freezing and thawing more common than steady below-freezing temperatures	Medium	Increased strain on road surface materials and potential for cracks and potholes in roads	Low
				Increased strain on A/C capacity	Medium	Increased strain on electricity grid	Medium
				Increased strain on runway material	Low	More frequent flight delays or cancellations	Medium
				Rail buckling	High	Delays in railroad schedules	Medium
				Increased strain on bridge materials	High	Sagging of large bridges	High
Precipitation	Increase in mean precipitation	More likely than not	N/A	Increased street flooding	Medium	Traffic delays	Low
	Amplified stream flow	More likely than not	N/A	Increased scour potential for bridge foundations	Medium	Delays in public transportation systems	Medium
						Reduced lifespan of current structures, potential need for new regulations	High
				Damage to road and rail embankments	Medium	Increased traffic and public transportation delays and rerouting	Medium
	Mudslides and landslides	Uncertain	N/A	Road and rail closures	Medium	Increased traffic and public transportation delays and rerouting, potential threat to lives	High
Temperature/ precipitation	Increase in droughts	Uncertain	Towards the end of the century, warm season droughts will more likely than not increase	Lower water level of lakes and canals due to higher rates of evaporation	Medium	Reduction in shipping capacity and increased costs of shipping due to required additional trips	Medium
Sea level rise	Increased storm surge and coastal flooding	Likely/very likely	Will depend both on sea level rise and on uncertain changes in tropical cyclones and nor'easters	Clearances of some bridges across waterways diminished below the limits set by the U.S. Coast Guard or other jurisdictions	High	Closure of bridges	High
				Flooding of bridge access ramps, tunnel entrances and ventilation shafts, and general highway beds	High	Traffic delays due to inundation	Low
				Reduced effectiveness of collision fenders on bridge foundations	High	Increase in impacts of ships or barges	Medium
				Flooding of roadways, railways, fuel storage farms and terminals, or maintenance facilities	Medium	Potential for equipment failure	High

Notes: N/A = Not Applicable
* Factors that are considered when determining the magnitude of consequence, defined as the combined impact of the occurrence should a given hazard occur, include: effects on internal operations, capital and operating costs, public health, the economy, and the environment, as well as the number of people affected. (see Annex II to the full report, "Adaptation Guidebook")

Telecommunications

Main Climate Variable	Specific Climate Variable	Probability of Specific Climate Variable	Climate Variable Notes	Impact on Resource	Likelihood of Impact	Consequence without Adaptation	Magnitude of Consequence*
Transmission and Distribution Assets							
Temperature	Increase in extreme heat events	Likely	N/A	Increase energy demand causing power failures	High	Reduction in telephone and cable services	High
Sea level rise	Increased frequency, intensity, and duration of storm surge and coastal flooding	Likely/very likely	Will depend on both sea level rise and on uncertain changes in tropical cyclones and nor'easters	Flooded central offices and underground installations	Medium	Reduced service	Medium
Extreme events	Extreme wind events	Uncertain	N/A	Fallen trees and downed wires	Low	Increased disruption of telephone and video service	Medium
	Snow storms	Uncertain	N/A	Strain on trees and utility lines from wet snow	Low	Reduction and delays in wired and cellular telephone service, as well as cable services	Medium
	Hurricanes	Uncertain	N/A	Power failures caused by high winds and storm surge	Medium	Increased strain on rerouting abilities of emergency calling centers	High
	Ice storms	Uncertain	N/A	Damage to utility lines and electrical equipment	Medium	Increased emergency communications and reduction in cable-provided services	High

Notes: N/A = Not Applicable
* Factors that are considered when determining the magnitude of consequence, defined as the combined impact of the occurrence should a given hazard occur, include: effects on internal operations, capital and operating costs, public health, the economy, and the environment, as well as the number of people affected. (see Annex II to the full report, "Adaptation Guidebook")

Public Health

Main Climate Variable	Specific Climate Variable	Probability of Specific Climate Variable	Climate Variable Notes	Impact on Resource	Likelihood of Impact	Consequence without Adaptation	Magnitude of Consequence*
Air Quality							
Temperature	Increase in mean temperature	Very likely	Warming may be greater 1) in the north than south, and 2) in winter than in summer in the north	Extension of pollen and mold seasons	High	Asthma, which exhibits strong seasonal patterns related to pollen and mold seasons, is exacerbated	High
				Dust mites and cockroaches thrive at high temperatures and especially high absolute air humidity, which they depend upon for hydration	High	Asthma exacerbations triggered by greater presence of indoor allergens	High
				Increase in emission of volatile organic compounds	Medium	Increase in the amount of ozone being ingested results in short-term, reversible decreases in lung function and inflammation in the deep lung; also, epidemiology studies of people living in polluted areas have suggested that ozone can increase the risk of asthma-related hospital visits, and premature mortality	High
	Increase in extreme heat events	Likely	N/A	Peak in air conditioning use	High	Greater amount of emissions and resulting pollution from power plants	Medium
				Loss of on-site electricity	Low	Increase CO poisoning as a result of non-evacuated residents without back-up power	High
Precipitation	Increase in mean precipitation	More likely than not	N/A	Weather patterns influence the movement and dispersion of all pollutants in the atmosphere	Medium	Potential increase in severe ozone episodes	High
Disease/Contamination							
Temperature	Increase in mean temperature	Very likely	Warming may be greater 1) in the north than south, and 2) in winter than in summer in the north	Increased population density and increase in biting rates of mosquitoes and ticks	Medium	Increase in infectious diseases spread by contaminated foods and water as well as those transmitted by insects	Medium
				Greater rates of overwinter survival of immature mosquitoes	High	Greater abundance of adults the following year that could potentially spread WNv	Medium
Precipitation	Increase in mean precipitation	More likely than not	N/A	Increased runoff from brownfields and industrial contaminated sites	Medium	Increased exposure to toxins creates health problems in respiratory and gastrointestinal tracts	High
				Receding floodwaters release molds and fungi that proliferate and release spores	High	Inhaled spores can cause respiratory irritation and allergic sensitization	High
Sea level rise	Increased storm surge and coastal flooding	Likely/Very likely	Will depend both on sea level rise and uncertain changes in tropical cyclones and nor'easters	Greater frequency of flooding events	High	Greater potential for drowning, delayed health service delivery	High
Mental Health							
Sea level rise	Increased storm surge and coastal flooding	Likely/Very likely	Will depend both on sea level rise and uncertain changes in tropical cyclones and nor'easters	Increased property damage (e.g., loss), displacement/family separation, violence, stress effects	High	Increase in anxiety, depression, PTSD as a result of low resilience capacity, lack of access to evac transportation, low SES	High

Notes: N/A = Not Applicable
CO = Carbon Monoxide
PTSD = Post traumatic stress disorder
* Factors that are considered when determining the magnitude of consequence, defined as the combined impact of the occurrence should a given hazard occur, include: effects on internal operations, capital and operating costs, public health, the economy, and the environment, as well as the number of people affected. (see Annex II to the full report, "Adaptation Guidebook")

Adaptation

New York State has significant resources and capacity for effective adaptation responses, which are characterized by a wide range of types, actors, levels of effort, timing, and scales (**Table 12.3**). A critical resource for the state are the existing codes, standards, and regulations that could be enhanced in a comprehensive adaptation approach. Developing climate change adaptation plans requires input from a breadth of academic disciplines as well as stakeholder experience to ensure that recommendations are both scientifically valid and practically sound (see Annex II to the full report).

Identifying the co-benefits of adaptation strategies is important, since they are positive effects that adaptation actions can have on mitigating climate change (i.e., reduction of greenhouse gas emissions) or on improving other aspects of the lives of New York State citizens. An example of a mitigation co-benefit is the establishment of green roofs that keep residents cooler while reducing the use of air conditioners, thereby reducing fossil fuel emissions at power plants. An example of a co-benefit with other aspects is the upgrading of combined sewer and stormwater systems to reduce current water pollution, while helping to prepare for future climate change impacts.

Some adaptation options may either complement or negatively affect mitigation efforts to reduce greenhouse gas emissions. For example, avoiding adverse public health impacts related to heat waves may result in increased reliance on air conditioning. This could counteract mitigation options designed to reduce energy consumption and could potentially result in increased energy demand during summer peak-load conditions.

Key Sector Adaptations

Potential adaptation strategies for the identified climate vulnerabilities are summarized in **Table 12.4**. These are to be considered as options for adaptive measures and should not be considered as an exhaustive list. For each sector, selected adaptation strategies that respond to key climate risks are presented in terms of short-, medium-, and long-term time scales and by operations/management, capital investment, and policy categories. The three categories are presented as a way of illustrating the varying range and focus of potential adaptation strategies. It is recognized that in many cases there will be significant overlap among the categories when the strategies are operationalized.

The key adaptations are broken into time groups: 0 to 10 years (i.e., to 2020), 10 to 40 years (i.e., to 2050), and more than 40 years (i.e., beyond 2050) (see **Table 12.4**). The short-term adaptations that are identified in the tables will often be continued into the medium and long terms, but to facilitate a focused overview, they are not necessarily repeated in each column of the table. Thus, while a short-term operations/management strategy—one involving small adjustments to everyday practices—will probably be continued throughout the longer period, it is listed as short-term to indicate its earliest use/implementation. "Ongoing" refers to work that is taking place at present and expected to continue over time.

Adaptation Mechanism	Definitions
Type	Behavior, management/operations, infrastructure/physical components, risk-sharing, and policy (including institutional and legal)
Administrative group	Private vs. public; governance scale – local/municipal, county, state, national
Level of effort	Incremental action, paradigm shift
Timing	Years to implementation, speed of implementation (near-term/long-term)
Scale	Widespread, clustered, isolated/unique

Table 12.3 Adaptation categories

Table 12.4 Selected Adaptation Strategies by Sector

Selected adaptation strategies by sector responding to key climate risks	Type*	Timing**
Water Resources		
Build on the existing capacity of water managers to handle large variability	O/M	O
Expand basin-level commissions to provide better oversight of water supplies in systems with multiple users, address water quality issues, and take leadership on basin-level monitoring, conservation, and coordination of emergency response	CI, P	S
Update and enlarge stockpiles of emergency equipment, including mobile pumps, water tanks, and filters, to help small water supply systems and to assist during emergencies	CI	S
Establish streamflow regulations that mimic natural seasonal flow requirements to protect aquatic and ecosystem health	O/M, P	S
Increase water use efficiency through leak detection programs, low-flow devices, rainwater harvesting, and equitable water-pricing programs	O/M, P	S
Develop more comprehensive drought management programs that include improved monitoring of water supply storage levels and that institute specific conservation measures when supplies decline below set thresholds	O/M, P	S to M
Explore new economic opportunities for New York State's relative wealth of water resources	P	M
Upgrade combined sewer and stormwater systems to reduce pollution and mitigate climate change impacts	CI	M
Adopt stormwater management infrastructure and practices to reduce the rapid release of stormwater to water bodies	O/M, P	M to L
Relocate and rebuild aging infrastructure out of high-risk flood-prone areas; construct levees and berms where necessary to remain in the flood plain	CI	L
Coastal Zones		
Site new developments outside of future floodplains, taking into consideration the effects of sea level rise, barrier island and coastline erosion, and wetland inundation	P	O
Improve building codes to promote storm-resistant structures and increase shoreline setbacks	O/M, P	S
Use rolling easements to protect coastal wetlands (recognize nature's right-of-way to advance inland as sea level rises)	P	M
Use engineering-based and bio-engineered strategies to protect coastal communities from floods or to restore wetlands	O/M	M
Maintain and expand beach renourishment and wetland restoration programs	O/M, P	M
Relocate coastal infrastructure and small, rural developments to higher elevations	CI, P	L
Buy out land or perform land swaps to encourage people to move out of flood-prone areas	CI, P	L
Ecosystems		
Minimize stressors such as pollution, invasive species, sprawl, and other habitat-destroying forces	O/M	O
Develop reliable indicators of climate change impacts on biodiversity and ecosystem services, and cost-effective strategies for assessing climate change impacts	O/M	O
Manage primarily for important ecosystem services and biodiversity rather than attempting to maintain the current mix of species present today	O/M	O
Facilitate natural adaptation to climate change by protecting stream (riparian) zones and migration corridors for species adjusting to changes in the climate	O/M, P	S
Institutionalize a comprehensive monitoring effort to track species range shifts and to track indicators of ecosystem response to climate change	O/M, P	M
Develop cost-effective management interventions to reduce vulnerability of high-priority species and communities, and determine minimum area needed to maintain boreal or other threatened ecosystems	O/M, P	M
Agriculture		
Change planting dates, varieties, or crops grown; increase farm diversification	O/M	S
Develop strategic adaptation decision tools to assist farmers in determining the optimum timing and magnitude of investments to cope with climate change	CI, P	S
Increase control of pests, pathogens, and weeds and use of new approaches to minimize chemical inputs	O/M	S
Improve cooling capacity and use of fans and sprinklers in dairy barns	CI	M
Invest in irrigation and/or drainage systems	CI	M
Develop new crop varieties for projected New York State climate and market opportunities	CI	M
Build supplemental irrigation with good drainage capacity for high-value crops	CI	M

Note: The key adaptations are broken into time groups: 0 to 10 years (i.e., to 2020), 10 to 40 years (i.e., to 2050), and more than 40 years (i.e., beyond 2050). The short-term adaptations that are identified will often be continued into the medium and long terms, but to facilitate a focused overview, they are not necessarily repeated in each column of the table. Thus, while a short-term operations/management strategy—one involving small adjustments to everyday practices—will probably be continued throughout the longer period, it is listed as short term to indicate its earliest use/implementation.

* O/M = Operations/Management, CI = Capital Investment, P = Policy,

** S = Short-term, M = Medium-term, L = Long-term, 0 = Ongoing

Selected adaptation strategies by sector responding to key climate risks	Type*	Timing**
Energy		
Balance the need to make energy systems more resilient with the cost of such investments and changes	O/M	O
Improve system resiliency with the replacement cycle of energy system assets	CI	O
Use transformers and wiring that function efficiently at higher temperatures	CI	S
Maintain and expand tree trimming programs next to power lines	O/M	S
Adjust reservoir release policies to ensure sufficient summer hydropower capacity	O/M	S
Prioritize demand-side management, which encourages consumers to use energy more efficiently	P	S
Shade buildings and windows or use highly reflective roof paints and surfaces to reduce warming in buildings from sun exposure	O/M	S
Improve energy efficiency in areas likely to have the largest increases in demand, to reduce strain on electrical equipment during heat waves	O/M, P	S
Construct berms and levees to protect infrastructure from flooding; install saltwater-resistant transformers to protect against sea level rise and saltwater intrusion	CI	M to L
Transportation		
Adopt operational measures to cope with high wind speeds, such as allowing bridge traffic only at reduced speeds or, for higher wind speeds, suspending traffic	O/M, P	S
Form alliances among agencies to set performance standards and work together to reduce risks, such as through mutual insurance pools that spread risks across time, space, and type	O/M	S
Perform engineering-based risk assessments of assets and operations and complete adaptation plans based on these assessments	CI, P	S to M
Relocate critical systems to higher ground out of future flood zones	CI	M
Create strategies to protect against heat hazards, including increasing the seat length of expansion joints on bridges, lengthening airport runways, and increasing and upgrading air conditioning on trains, subways, and buses	CI	M to L
Devise engineering-based solutions to protect against coastal hazards, including constructing levees, sea walls, and pumping facilities; elevating infrastructure, including bridge landings, roads, railroads, and collision fenders on bridge foundations; and designing innovative gates at subway, rail, and road entrances	CI	M to L
Develop engineering-based solutions to protect against heavy-precipitation hazards, including increasing the capacity of culverts and other drainage systems; raising and/or strengthening road and rail embankments to make them more resistant to flood-related erosion and river scour; and creating more permeable surfaces or regrading slopes to direct runoff away from critical transportation infrastructure	CI	L
Telecommunications		
Reassess industry performance standards combined with more uniform regulation across all types of communication services; provide better enforcement of regulations, including uniform mandatory reporting of outages to regulatory agencies	O/M, P	S
Further develop backup cell phone charging options, such as car chargers, and create a charging interface that allows any phone to be recharged by any charger	CI	S
Develop high-speed broadband and wireless services in low-density rural areas to increase redundancy and diversity in vulnerable remote regions	CI	S
Trim trees near power and communication lines, maintain backup supplies of poles and wires to replace those that are damaged, and have emergency restoration crews at the ready to protect against outages	O/M	S, O
Assess, develop, and expand alternative communication technologies with the goal of increasing redundancy and/or reliability, including free-space optics (which transmits data with light rather than physical connections), power line communications (which transmits data over electric power lines), satellite phones, and ham radio	CI	M
Place communication cables underground where technically and economically feasible	CI	M
Decouple communication facilities from electric grid infrastructure to the extent possible, and/or make these infrastructures more robust, resilient, and redundant	CI	M
Minimize the effects of power outages on communications services by providing backup power at cell towers, such as with generators, solar-powered battery banks, and "cells on wheels" that can replace disabled towers; extend the fuel storage capacity needed to run backup generators for extended times	CI	M
Relocate central offices that house communications infrastructure out of future floodplains	CI, P	L
Public Health		
Integrate adaptation strategies into existing surveillance, prevention, and response programs	O/M	S
Better coordinate environment and health initiatives so they address both human health and ecosystem health and avoid the legislative divide that often exists between them	O/M, P	S
Increase use of air conditioning during heat waves for vulnerable individuals, but use alternative energy sources to avoid increased greenhouse gas emissions	O/M	S
Provide alerts regarding potential health risks, such as those from extreme heat events, which convey information and needed actions to vulnerable communities	O/M, P	S
Implement extreme-heat response plans, such as longer opening hours for air-conditioned community centers for seniors, reduced fares on public transportation, and neighborhood buddy systems to check on those most vulnerable	O/M, P	S
Plant low-pollen trees in cities to reduce urban heat without increasing allergenic pollen	CI	M
Invest in structural adaptations to reduce heat vulnerability, including tree planting, green roofs, and high-reflectivity building materials	CI	M to L

Note: See previous page

Equity and Environmental Justice

Certain groups, types of communities, and regions within the state are better able to respond to climate risk and vulnerabilities than others. Communities, groups, and locations currently at risk because of limited response capacity and resilience to climate hazards (e.g., those who are economically marginal) are, in most cases, those that will be most vulnerable to future climate change impacts. Such groups include the elderly and disabled, as well as people with low incomes and the underprivileged.

Elderly and health-compromised individuals are more vulnerable to climate hazards, including floods and heat waves. Low-income groups have limited ability to meet higher energy costs, making them more vulnerable to the effects of heat waves. Those who lack affordable healthcare are more vulnerable to climate-related illnesses such as asthma. Those who depend on public transportation to get to work, and lack private cars for evacuating during emergencies, are also vulnerable. Farm workers may be exposed to more chemicals if pesticide use increases in response to higher pest infestations brought about by a warming climate.

It is not clear at this time how the costs of adaptation will be distributed. In general, groups with more limited means to respond to increased risks or to provide funds for adaptation, such as smaller businesses, may be less able to cope. This condition extends across both the public and the private sectors.

Economics

The costs of climate change impacts will vary across and within sectors (see Annex III to the full report). Overall costs of impacts within the energy, transportation, and coastal zone sectors will be most significant, likely by many-fold, but impacts within each sector will be significant depending on the structure of that sector. This is well illustrated in the agriculture and ecosystem sectors, where particular components such as specific crops and modes of production or rare and endangered ecosystems and species could be significantly affected by climate change in comparison to other parts of the sectors.

There are several types of costs associated with climate impacts and adaptation. Direct costs include costs that are incurred as the direct economic outcomes of a specific climate event or aspect of climate change. Indirect costs are those incurred as secondary outcomes of the direct costs of a specific event or facet of climate change. Impact costs are direct costs associated with the impacts of climate change, and adaptation costs include the direct costs associated with adapting to those impacts. The direct costs of impacts that cannot be adapted to are the costs of residual damage.

The costs of adapting to climate change are already occurring and will grow over time. Adaptation response costs and benefits will not be evenly distributed throughout the state. For example, a significant amount of the benefits of adaptation to sea level rise will be experienced only by communities and property owners in the coastal zone.

Recommendations

This section presents recommendations for policy and management that arise from the ClimAID Assessment. Policy recommendations are aimed at statewide decision-makers, and management recommendations are associated with everyday operations within stakeholder agencies and organizations, as they respond to the challenge of climate change. Sector-specific knowledge gaps and information needs are identified, as well as recommended directions for further science and research activities.

Policy

Key policy recommendations, targeted for New York State decision-makers, are discussed in this section.

- Promote adaptation strategies that enable incremental and flexible adaptations within sectors, among communities, and across time.
- Analyze environmental justice issues related to climate change and adaptation on a regular basis.
- Evaluate design standards and policy regulations based on up-to-date climate projections.
- Consider regional, federal, and international climate-related approaches when exploring climate adaptation options. This is crucial because it is clear that New York State's adaptation potential (and

mitigation potential as well) will be affected by national and international policies and regulations as well as state-level policies.

- Improve public and private stakeholder and general public education and awareness about all aspects of climate change. This could encourage the formation of new partnerships for developing climate change adaptations, especially given limited financial and human resources, and the advantages of shared knowledge.
- Identify synergies between mitigation and adaptation. Taking steps to mitigate climate change now will help to reduce hazards and enhance opportunities for co-benefits. Conversely, many potential adaptation strategies present significant mitigation opportunities.
- Develop standardized, statewide climate change mitigation and adaptation tools, including a central database of climate risk and adaptation information resulting from ongoing partnerships between scientists and stakeholders.

Management

Management recommendations associated with everyday operations in stakeholder agencies and organizations are described here.

- Integrate climate adaptation responses into the everyday practices of organizations and agencies, with the potential for synergistic or unintended consequences of adaptation strategies taken into account.
- Take climate change into account in planning and development efforts.
- Identify opportunities for climate adaptation partnerships among organizations and agencies.

Knowledge Gaps and Information Needs

There has been great advancement in knowledge surrounding climate change, impacts, and adaptation over the past few decades. However, there are still areas where further research would complement and further the understanding, help to reduce uncertainties, and aid in better decision-making. Key areas of knowledge gaps and information needs for each sector are outlined in **Table 12.5**.

Table 12.5 Knowledge Gaps and Information Needs by Sector

Sector-specific and statewide knowledge gaps and information needs	Type (Climate science, impact, adaptation)
Water Resources	
Identification of critical pollutant-contributing areas and processes	Impact
More in-depth assessment of how fundamental hydrologic processes, such as groundwater recharge, stream low-flows, evaporation, and flooding, might be altered by a changing climate	Impact
Refinement of existing monitoring networks	Climate science
Updated estimates of streamflow and water temperature scenarios based on future climate changes	Climate science
Models of the impacts on the quality of water bodies receiving effluent	Impact
Coastal Zones	
Research on the response of barrier islands to accelerated rates of sea level rise	Climate science
Improved understanding of regional sediment transport processes along the coast and continental shelf	Climate science
Quantified and monitored land use and coastal water quality	Impact
Assessment of ecosystem services for natural and engineered shorelines	Impact
Monitoring program for submarine groundwater discharge	Impact
Systematic mapping (every two to five years) and standardized mapping protocols for all New York State coastal regions	Climate science
GIS-based data repository to facilitate interagency collaboration and future assessments	Impact
Improved hydrodynamic modeling capability for the Hudson River	Climate science
Ecosystems	
Reliable indicators of climate change impacts on biodiversity and ecosystem functions, and cost-effective strategies for monitoring these impacts	Climate science/impact
Cost-effective management interventions to reduce vulnerability of high-priority species and communities, and determination of the minimum area needed to maintain boreal and other threatened ecosystems.	Impact
Evaluation techniques for rapid and reliable assessment of vertebrate abundance at the landscape scale	Climate science
Improvements in techniques used to identify and target invasive species likely to benefit from climate change	Climate science
Development of citizen-science programs that can provide accurate and reliable data on change in species distributions and movements	Impact
Agriculture	
Non-chemical control strategies for weed and pest threats	Impact
New economic decision tools for farmers	Impact
Sophisticated real-time weather-based systems for monitoring and forecasting crop stress	Climate science
Crops with increased tolerance to climate stresses	Impact
Energy	
Review of thermoelectric power intake or discharge rules in light of a changing climate	Impact
Identification of temperature tipping points related to failure of the energy supply system	Impact
Potential impacts of climate change on wind patterns and speeds in selected areas currently used or proposed for wind farm development	Climate science/impact
Potential impacts of climate change on biomass-based heat production (either at a large central station or co-firing facilities)	Climate science/impact
Assessment of potential impacts of climate change on hydropower availability in different parts of the state	Climate science/impact
Evaluation of potential climate impacts on the demand for natural gas and other heating fuels given anticipated decreases in heating degree-days over the coming decades	Impact
Better understanding of the impact of extreme events on electricity demand	Climate science
Transportation	
Accurate, high-resolution LIDAR surveys to facilitate the development of digital elevation models (DEM) of sufficiently high vertical and horizontal resolution to perform forward-looking flood risk assessments and regional planning of sustainable developments	Impact
Development of updated climate information that includes climate change projections for standards and regulations	Climate science
Comprehensive program of research and technological development for advancing innovative, cost-effective, and climate-resilient urban and inter-urban transportation infrastructure	Impact
Telecommunications	
Creation of computerized (proprietary) databases that show the location and elevations of installed communication facilities and lifelines and their operational capacity and other details	Impact
Improved knowledge-sharing tools to disseminate information about service outages and expected restoration times to the public	Impact
Public Health	
Ongoing, state-based research to inform periodic policy developments, especially that which identifies cross-sectoral interactions and win-win options for adaptation/mitigation, including extensive health co-benefits assessments	Impact
Development and analysis of local health impact projections of climate factors and related disease outcomes	Impact
Information and capacity-building for integrating climate change into public health planning at all levels of government	Impact

Science and Research

This section presents recommendations for future science and research.

- Refine climate change scenarios for New York State on an ongoing basis, as results from new climate models and downscaled products become available.
- Conduct research on understanding climate variability, including stakeholder-identified variables, such as ice storms, extreme precipitation events, wind patterns, etc.
- Conduct targeted impacts research in conjunction with regional stakeholders.
- Implement and institutionalize an indicators and monitoring program focused on climate, impacts, and adaptation strategies.
- Improve spatial analysis and mapping to help present new data.
- Focus studies on specific systems that may enter into a phase change or similar shifts in process, known as "tipping points." Work should be encouraged to understand the potential for tipping points associated with climate change impacts on natural and social systems.
- Develop a better understanding of the economic costs of climate change and benefits of adaptations.

Responding to Future Climate Challenges

New York State is highly diverse, with simultaneous and intersecting challenges and opportunities presented by a changing climate. Among the people, sectors, and regions of the state, those that are already facing significant stress will likely be placed most at risk by the effects of future climate change. Responding to these challenges and opportunities will depend on how stakeholders develop effective adaptation strategies by connecting climate change with ongoing proactive management and policy initiatives within the state and beyond.

The adaptation strategies suggest several important perspectives: First, there is a wide range of adaptation needs across sectors. Second, there are many adaptation needs that can be undertaken or reviewed in the near term, in most cases at relatively modest cost. Third, there are some potential infrastructure investments—especially relating to the transportation sector and coastal zones—that could be needed in the long term and that may be expensive. These perspectives also suggest the need for increased interactions between scientists and policy-makers, and consideration of methods for ensuring that science better informs policy, as well as increased scientific and technical capabilities. The overall goal is the development of equitable and efficient climate resilience throughout New York State in the decades to come.

Annex I
Expert Reviewers for the ClimAID Assessment

Chapter	Reviewer	Affiliation
Introduction	Virginia Burkett	U.S. Geological Survey
	Paul Fleming	Seattle Public Utilities
Climate	David Yates	National Center for Atmospheric Research
	Ron Stouffer	Geophysical Fluid Dynamics Laboratory
	Kathy Jacobs	White House Office of Science and Technology Policy
	Paul Fleming	Seattle Public Utilities
Vulnerability and Adaptation	Kirstin Dow	University of South Carolina
	Lynne M. Carter	Adaptation Network
	Stewart Cohen	Environment Canada / University of British Columbia
Equity and Economics	Rae Zimmerman	New York University
	Mike Beck	The Nature Conservancy
	Karen O'Brien	University of Oslo
	Vicki Arroyo	Georgetown University
Water Resources	Doug Burns	U.S. Geological Survey, New York Water Science Center
	Brad Udall	University of Colorado
Coastal Zones	Robert Deyle	Florida State University
	Paul Kirshen	Battelle Duxbury MA
Ecosystems	Jerry Jenkins	Wildlife Conservation Society Forest Issues
	Gary Lovett	Cary Institute of Ecosystem Studies
	Nicholas Rodenhouse	Wellesley College
Agriculture	Greg Albrecht	NYS Department of Agriculture and Markets / Empire Soil and Water Conservation Society
	David Abler	Pennsylvania State University
Energy	Edward Vine	Lawrence Berkeley National Laboratory
	Stanley Bull	National Renewable Energy Laboratory
	Vatsal Bhatt	Brookhaven National Laboratory
	Matthias Ruth	University of Maryland
Transportation	Mark Horner	Florida State University
	Michael J. Savonis	U.S. Department of Transportation
Telecommunications	Mike Hainzl	Ericsson Inc.
	Craig Faris	Accenture
Public Health	Christine Rogers	University of Massachusetts, Amherst
	Paul Epstein	Harvard Medical School Center for Health and the Global Environment

ClimAID Annex II

Climate Adaptation Guidebook for New York State

Authors: Cynthia Rosenzweig, Arthur DeGaetano, William Solecki, Radley Horton, Megan O'Grady, Daniel Bader

New York State Energy Research and Development Authority

Climate Adaptation Guidebook for New York State

NYSERDA
November 2011

NYSERDA's Promise to New Yorkers:

New Yorkers can count on NYSERDA for objective, reliable, energy-related solutions delivered by accessible, dedicated professionals.

Our Mission: Advance innovative energy solutions in ways that improve New York's economy and environment.

Our Vision: Serve as a catalyst—advancing energy innovation and technology, transforming New York's economy, and empowering people to choose clean and efficient energy as part of their everyday lives.

Our Core Values: Objectivity, integrity, public service, and innovation.

Our Portfolios

NYSERDA programs are organized into five portfolios, each representing a complementary group of offerings with common areas of energy-related focus and similar objectives.

Energy Efficiency & Renewable Programs

Helping New York to achieve its aggressive clean energy goals – including programs for consumers (commercial, municipal, institutional, industrial, residential, and transportation), renewable power suppliers, and programs designed to support market transformation.

Energy Technology Innovation & Business Development

Helping to stimulate a vibrant innovation ecosystem and a clean energy economy in New York – including programs to support product research, development, and demonstrations, clean-energy business development, and the knowledge-based community at the Saratoga Technology + Energy Park.

Energy Education and Workforce Development

Helping to build a generation of New Yorkers ready to lead and work in a clean energy economy – including consumer behavior, K-12 energy education programs, and workforce development and training programs for existing and emerging technologies.

Energy and the Environment

Helping to assess and mitigate the environmental impacts of energy production and use – including environmental research and development, regional initiatives to improve environmental sustainability, and West Valley Site Management.

Energy Data, Planning and Policy

Helping to ensure that policy-makers and consumers have objective and reliable information to make informed energy decisions – including State Energy Planning, policy analysis to support the Low-Carbon Fuel Standard and Regional Greenhouse Gas Initiative, nuclear policy coordination, and a range of energy data reporting including *Patterns and Trends*.

New York State Energy Research and Development Authority

17 Columbia Circle
Albany, New York 12203-6399

toll free: 1 (866) NYSERDA
local: (518) 862-1090
fax: (518) 862-1091

info@nyserda.org
www.nyserda.ny.gov

Climate Adaptation Guidebook for New York State

Annex II to the ClimAID Integrated Assessment for Effective Climate Change Adaptation Strategies in New York State

Authors: Cynthia Rosenzweig, Arthur DeGaetano, William Solecki, Radley Horton, Megan O'Grady, Daniel Bader

NYSERDA
November 2011

Responding to Climate Change in New York State

Climate change is already beginning to affect the people and resources of New York State, and these impacts are projected to grow. At the same time, the state has the potential capacity to address many climate-related risks, thereby reducing negative impacts and taking advantage of possible opportunities.

ClimAID: The Integrated Assessment for Effective Climate Change Adaptation Strategies in New York State was undertaken to provide decision-makers with cutting-edge information on the state's vulnerability to climate change and to facilitate the development of adaptation strategies informed by both local experience and scientific knowledge.

This state-level assessment of climate change impacts is specifically geared to assist in the development of adaptation strategies. It acknowledges the need to plan for and adapt to climate change impacts in a range of sectors: Water Resources, Coastal Zones, Ecosystems, Agriculture, Energy, Transportation, Telecommunications, and Public Health.

The author team for the report is composed of university and research scientists who are specialists in climate change science, impacts, and adaptation. To ensure that the information provided would be relevant to decisions made by public and private sector practitioners, stakeholders from state and local agencies, non-profit organizations, and the business community participated in the process as well.

This Guidebook will help develop climate change adaptation strategies using a risk management approach. The larger technical report provides useful information to decision-makers, such as state officials, city planners, water and energy managers, farmers, business owners, and others as they begin responding to climate change in New York State.

Table of Contents

I. Climate Change and New York State 6

II. Framing Adaptation 8

III. Current Climate and Climate Change Projections 9

IV. Adaptation Assessment Steps 19

V. Other Adaptation Tools 27

VI. Summary 29

Rosenzweig, Cynthia, Arthur DeGaetano, William Solecki, Radley Horton, Megan O'Grady, Daniel Bader. 2011. ClimAID Adaptation Guidebook for New York State. Annex II of *Responding to Climate Change in New York State: The ClimAID Integrated Assessment for Effective Climate Change Adaptation Strategies in New York State*. New York State Energy Research and Development Authority (NYSERDA). Albany, NY.

I. Climate Change and New York State

Over the last century, global mean temperatures and sea levels have been increasing and the Earth's climate has been changing. As these trends continue, climate change is increasingly being recognized as a major global concern. In 1988, the World Meteorological Organization (WMO) and the United Nations Environment Programme (UNEP) formed an international panel of leading climate scientists, coined the Intergovernmental Panel on Climate Change (IPCC), to provide objective and up-to-date information regarding the changing climate. In its 2007 Fourth Assessment Report (AR4), the IPCC states that there is a greater than 90 percent chance that rising global temperatures, as observed since 1750, are primarily the result of human activities.

As predicted in the 19th century, the principal driver of climate change over the past century has been the increase in levels of atmospheric greenhouse gases (GHGs) associated with fossil-fuel combustion, changing land-use practices, and other human activities. The atmospheric concentrations of the major GHG carbon dioxide (CO_2) are now more than one-third higher than in pre-industrial times. The concentrations of other important GHGs, including methane (CH_4), ozone (O_3), and nitrous oxide (N_2O), have increased as well. Largely resulting from work performed by the IPCC and the United Nations Framework Convention on Climate Change (UNFCCC), global efforts to mitigate the severity of climate change by limiting levels of GHG emissions are now underway.

Because some of the added GHGs will remain in the atmosphere for centuries, and some parts of the climate system respond in a gradual manner, awareness is growing that some climate changes are inevitable. Responses to climate change have evolved from focusing on *mitigating* or reducing the amount of GHGs released into the atmosphere to including *adaptation* measures in an effort to both minimize the impacts and prepare for unavoidable future changes. In some cases, climate change may bring opportunities. (For more information, see the full ClimAID Technical Report.)

New York State possesses a wide range of vulnerabilities to a changing climate and, at the same time, has great potential to adapt to its impacts. From the Great Lakes to Long Island Sound, from the Adirondacks to the Susquehanna Valley, climate change will affect the people and resources of New York State. Risks associated with climate change include higher temperatures leading to greater incidence of heat stress caused by more frequent and intense heat waves; increased summer droughts and extreme rainfall affecting food production, natural ecosystems, and water resources; and sea level rise causing exacerbated flooding in coastal areas.

Climate change—and associated uncertainties in future climate projections, as well as complex linkages among climate change, physical systems, biological systems, and socioeconomic factors—poses special challenges for New York State decision-makers. However, there is a knowledge base that decision-makers can use to make progress in reducing vulnerability to climate change and building adaptive capacity needed to respond to extremes in the current climate, as well as increased climate risks in the future.

This *Climate Adaptation Guidebook for New York State* describes a risk management approach to developing climate change adaptation strategies. The climate change adaptation process involves understanding climate trends and projections, identifying vulnerabilities, assessing the risk levels, and developing and prioritizing strategies. The guidebook discusses these key aspects in the context of New York State. By developing climate change adaptation strategies following a risk management approach, New York State can effectively respond to future climate impacts.

Key Definitions for Responding to Climate Change

Adaptation – Actions that take place in response to a changing climate. Actions can create opportunities or challenges.

Adaptive capacity – Ability of a system to adjust to actual or expected climate stresses, or to cope with the consequences.

Adaptation strategies – Operational, managerial, budgetary, or infrastructure changes that will result in reducing risk and/or taking advantage of potential opportunities associated with climate change. A strategy is usually developed for a key vulnerability. Adaptation strategies do not directly include actions that reduce the likelihood of climate change occurring.

Climate resilience – A state in which climate risk information, vulnerability, and adaptation knowledge are taken into account in order to reduce the level of physical, social, or economic impact of climate variability and change.

Climate risks – Generally, risk is a product of the likelihood of an event occurring (typically expressed as a probability) and the magnitude of consequences should that event occur. For climate change impacts, risk can be thought to have three dimensions: the probability of a climate hazard occurring; the likelihood of impacts associated with that hazard; and the magnitude of consequence, should that impact occur. These risk estimates can be adapted and improved as additional information becomes available.

Impacts – The natural or potential effects a change in climate has or could have on natural or human systems.

Mitigation – Direct actions that reduce the concentrations of greenhouse gases in the atmosphere and other factors that are currently altering, or have the potential to alter, the earth's climate system.

Prioritization – Methods to assess and evaluate a set of adaptation strategies to determine those that are more pressing or suitable to undertake. Various prioritization criteria can be used.

Vulnerability – The degree to which geophysical, biological, and socio-economic systems are susceptible to, and unable to cope with, adverse impacts of climate change.

Sources: IPCC (2007) and New York City Panel on Climate Change (NPCC) (2010)

II. Framing Adaptation

Developing climate change adaptation involves understanding how the *climate* in New York State might change; identifying potential *vulnerabilities* a change in climate might create; assessing *risk* levels of those vulnerabilities; developing *adaptation strategies* that will help to minimize those risks; and prioritizing those strategies. This process helps to distill the complexities involved in considering climate change, its impacts, and how to adapt. The outcome of the process involves enhancing the overall *adaptive capacity* of a particular region, jurisdiction, or organization. *Adaptive capacity* is defined as the ability of a system to adjust to actual or expected climate stresses, or to cope with the consequences (see Figure 1).

Risk Management

Climate adaptation strategies and actions have a direct connection to risk and hazards management. Individuals and organizations reduce their vulnerability and exposure to threats through risk management as they develop protocols to avert and manage hazards and promote disaster risk reduction, especially around areas of uncertainty. Stakeholders can modify risk management tools, such as a risk matrix, for climate change adaptation, especially as a way to deal with the uncertainties surrounding climate hazards and associated impacts. Other uncertainties that may affect climate change adaptation include changes in technologies and social dynamics. The exact need and context in which stakeholders develop adaptation strategies reflect both the history and emerging understanding of the amount and significance of ongoing climate change.

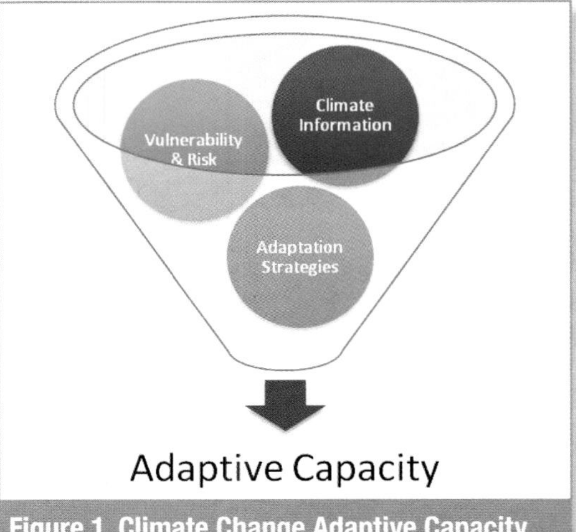

Figure 1. Climate Change Adaptive Capacity

Climate Resilience and Flexible Adaptation Pathways

To build climate resilience, climate change adaptation should allow for flexible responses to changing climate conditions. Flexible adaptation consists of implementing actions or infrastructure that stakeholders can adjust or shift over time in response to new climate science and evidence from ongoing monitoring, as well as implementing shifts in policies and strategies to better respond to emerging climate threats and opportunities (see Figure 2).

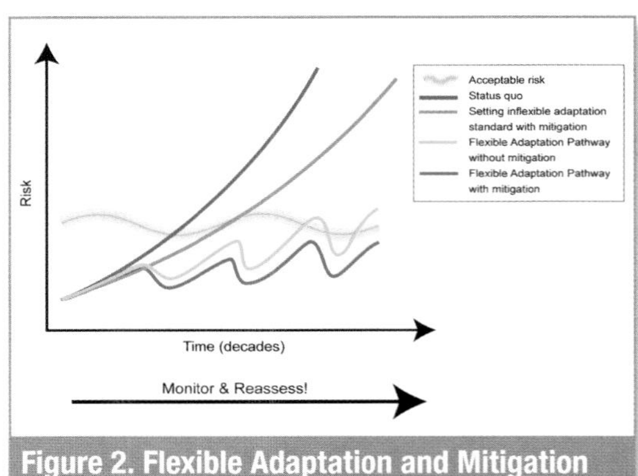

Figure 2. Flexible Adaptation and Mitigation Pathways

Graphic adapted from Lowe (2009)

An acceptable level of risk, as determined by society, is likely to change over time; for instance, the acceptable level of risk is likely to be lower after an extreme event, such as a hurricane. A one-time static or inflexible adaptation is better than maintaining the status quo, but such actions would still eventually result in crossing into an unacceptable level of risk, when climate conditions change beyond what the action was designed to withstand. *Flexible adaptation pathways* that include both adaptation and mitigation allow policymakers, stakeholders, and experts to develop and implement strategies that evolve as climate change progresses. The process of adaptation assessment can be summarized in an eight-step process (see Section IV) and adjusted as needed, depending on varying circumstances.

III. Current Climate and Climate Change Projections

This section provides an overview of the current climate in New York State and summarizes the climate change projections for New York. Understanding the climate is the first step in developing adaptation strategies for New York State (see Section IV).

New York State's Climate

The following components are key features of New York State's climate:

- Average annual temperature varies from 40°F in the Adirondacks to near 55°F in the New York City metropolitan region.
- Average annual precipitation ranges from approximately 30 inches in Western New York to close to 50 inches in the New York City region, Tug Hill Plateau, and Adirondacks.
- The state experiences a variety of extreme events:
  - **Heat waves** are common in urban areas, especially in the southern parts of the state.
  - **Short-duration flooding,** which can result from heavy rainfall and/or runoff from snowmelt, affects the entire state.
  - **Lake-effect snow** is a major climate hazard in western and central New York State.
  - **Coastal storms** along the Atlantic coast and Hudson River Valley bring heavy precipitation, high winds, and flooding.

Because New York State's climate is varied, climate impacts and effective adaptation strategies will be varied as well.

New York State Climate Regions

The climate of New York State varies from the Great Lakes to Long Island Sound. To help in developing adaptation strategies, the ClimAID assessment divided New York State into seven regions, as shown in Figure 3.

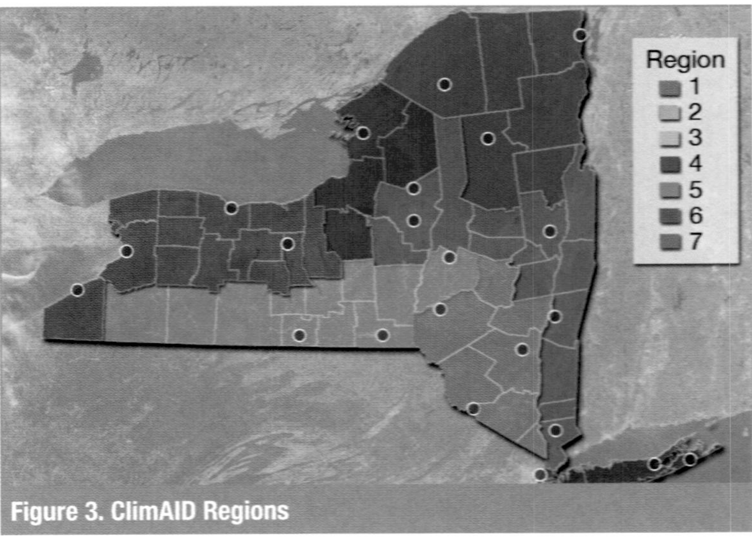

Figure 3. ClimAID Regions

Observed Climate Trends

Temperatures in New York State have risen over the course of the 20th century, with the greatest warming occurring in recent decades. New York State has experienced an increase in extreme hot days (days at or above 90°F) and a decrease in cold days (days at or below 32°F). In addition, the sea level has steadily risen in the coastal areas of the state. Figure 4 shows observed 20th century trends in temperature, precipitation, and sea level rise for New York City (ClimAID Region 4); these trends serve as an example of how the climate has already begun to change in different parts of the state.

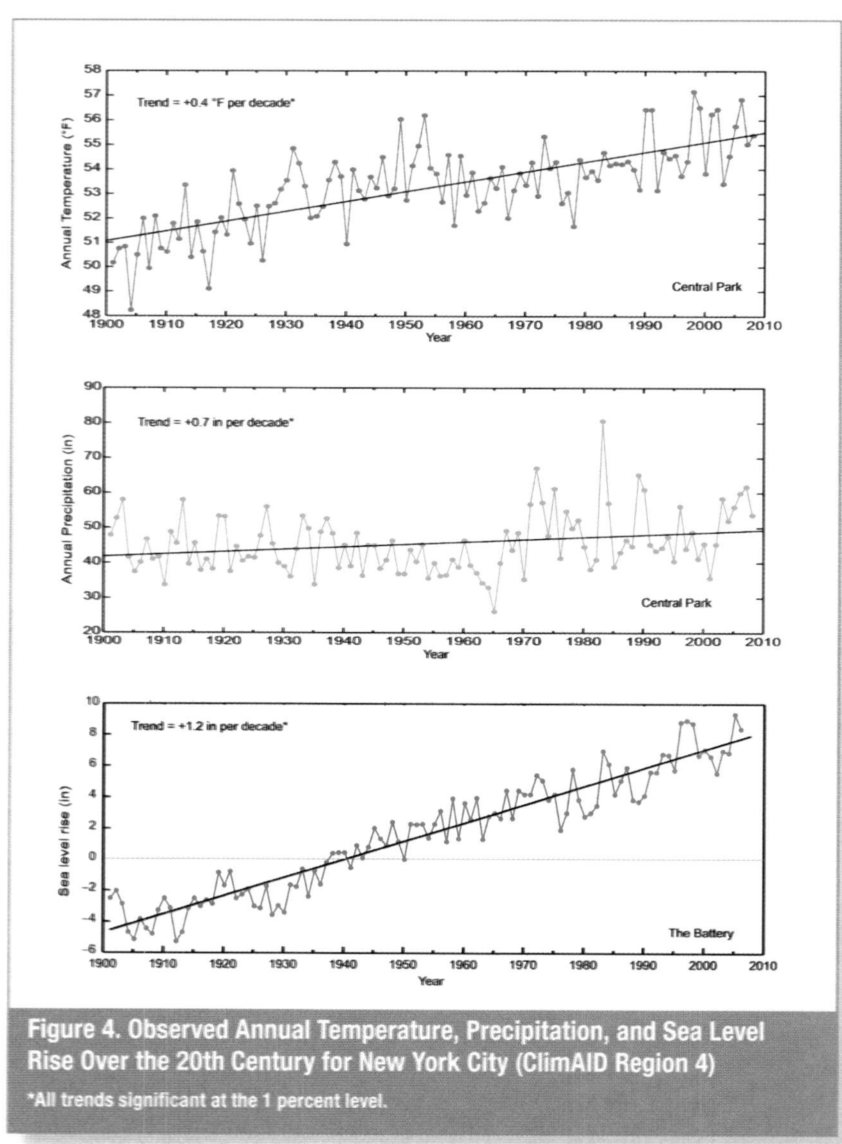

Figure 4. Observed Annual Temperature, Precipitation, and Sea Level Rise Over the 20th Century for New York City (ClimAID Region 4)

*All trends significant at the 1 percent level.

Future Projections

To produce future climate scenarios, experts use global climate models with a number of possible GHG emissions scenarios. Each emissions scenario represents a set of different demographic, social, economic, technological, and environmental assumptions about the future, called "storylines" (IPCC, 2000). The ClimAID team used three GHG emissions scenarios, as shown in Figure 5. The three scenarios and the storylines the team used in the ClimAID Assessment are described in Table 1.

Scenario	Storyline
A2	Relatively rapid population growth and limited sharing of technological change combine to produce high GHG levels by the end of the 21st century, with emissions growing throughout the entire century.
A1B	Effects of economic growth are partially offset by the introduction of new technologies and decreases in global population after 2050. This trajectory is associated with relatively rapid increases in GHG emissions and the highest overall CO2 levels for the first half of the 21st century, followed by a gradual decrease in emissions after 2050.
B1	This scenario combines the A1/A1B population trajectory with societal changes tending to reduce GHG emissions growth. The net result is the lowest GHG emissions of the three scenarios, with emissions starting to decrease by 2040.

Table 1. Greenhouse Gas Emissions Scenarios and Storylines

Other emissions scenarios yield different GHG concentrations by the end of the 21st century as compared to the three scenarios ClimAID used. The IPCC's "A1FI" scenario, for example, projects even higher CO_2 concentrations than those shown in Figure 5. The A1FI scenario was not included, however, because very few global climate model results are available for the scenario. However, experts should continue to reassess high-end climate change scenarios such as this over time.

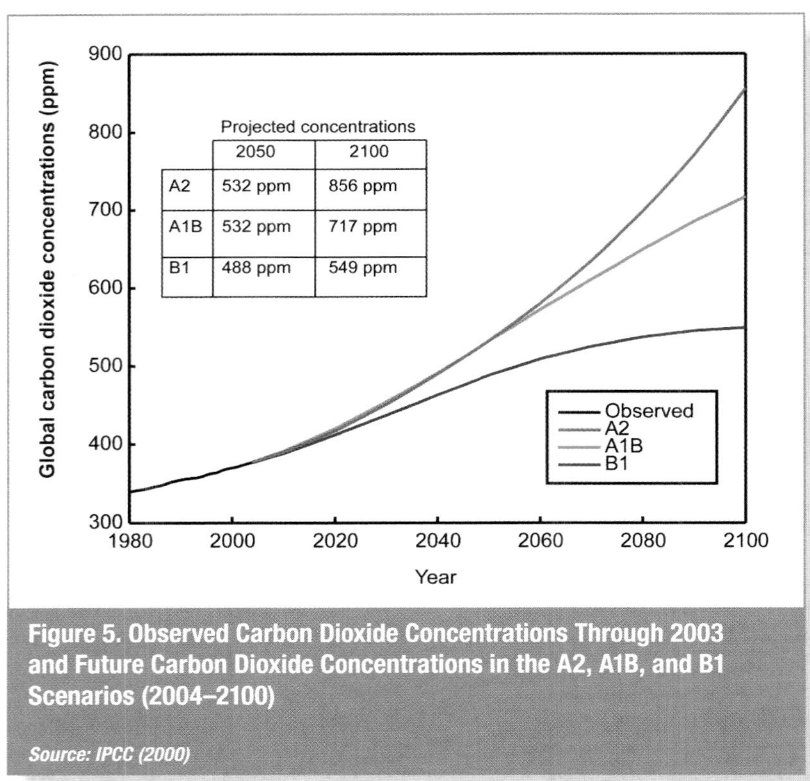

Figure 5. Observed Carbon Dioxide Concentrations Through 2003 and Future Carbon Dioxide Concentrations in the A2, A1B, and B1 Scenarios (2004–2100)

Source: IPCC (2000)

The ClimAID team divided the projections produced from the global climate models into two categories: mean annual changes and changes in extreme events. For the ClimAID Assessment, the team produced projections for each of the seven regions shown in Figure 3. The sections below present projections for a few of the regions, as examples. For the full suite of the ClimAID Assessment projections, please see the full Technical Report.

Mean Annual Changes

The maps and graphs shown in Figures 6 and 7 display temperature, precipitation, and sea level rise projections, based on a range of climate models and scenarios of possible future GHG concentrations. Table 2 and Figure 8 display both the global climate model-based sea level rise projections and a second set of higher projections (the rapid ice-melt scenario) based on the possibility of accelerated melting of land-based ice sheets and glaciers.

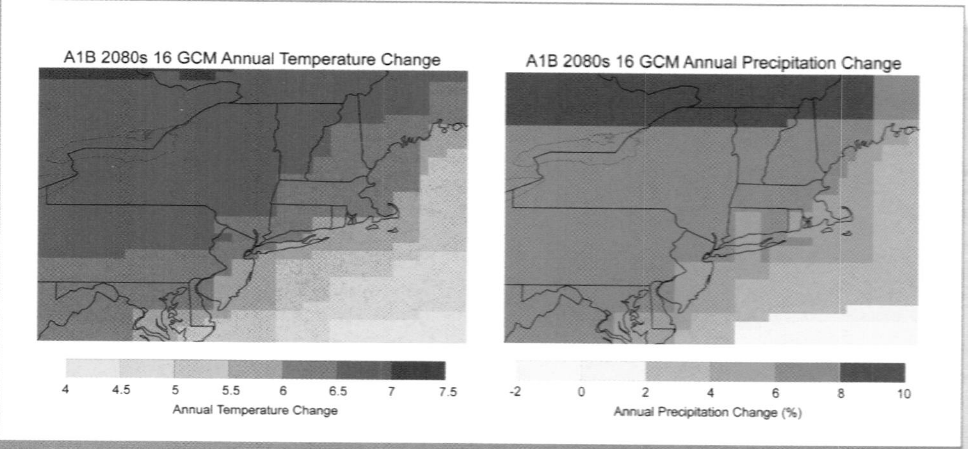

Figure 6. Projected Change in Annual Temperature and Precipitation in the Northeast for the 2080s, Relative to the 1980s Baseline (Under the A1B Emissions Scenario)

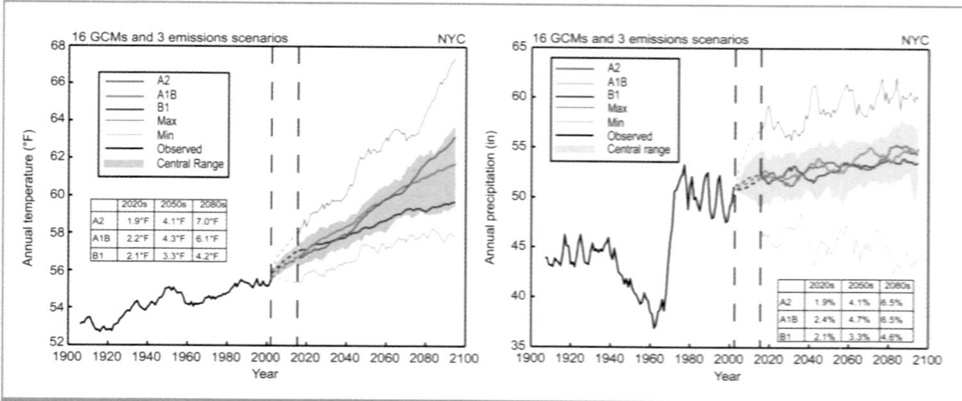

	2020s	2050s	2080s
A2	1.9°F	4.1°F	7.0°F
A1B	2.2°F	4.3°F	6.1°F
B1	2.1°F	3.3°F	4.2°F

	2020s	2050s	2080s
A2	1.9%	4.1%	6.5%
A1B	2.4%	4.7%	6.5%
B1	2.1%	3.3%	4.6%

Figure 7. Temperature and Precipitation Observations and Projections for the New York City Area (ClimAID Region 4)

Projected model changes through time are applied to the observed historical data. The three thick lines (green, red, and blue) show the average for each emissions scenario across the 16 GCMs. Shading shows the central range (middle 67%). The bottom and top lines, respectively, show each year's minimum and maximum projections across the suite of simulations. A ten-year filter has been applied to the observed data and model output. The dotted area between 2003 and 2015 represents the period that is not covered due to the smoothing procedure.

Region 4: Lower Hudson Valley & Long Island	Baseline (1971–2000)	2020s	2050s	2080s
Sea level rise[1] GCM-based	NA*	+ 2 to 5 in**	+ 7 to 12 in	+ 12 to 23 in
Sea level rise[2] Rapid ice-melt	NA	~ 5 to 10 in	~ 19 to 29 in	~ 41 to 55 in
Region 5: Mid Hudson Valley & Capital Region	**Baseline (1971 – 2000)**	**2020s**	**2050s**	**2080s**
Sea level rise[1] GCM-based	NA	+ 1 to 4 in	+ 5 to 9 in	+ 8 to 18 in
Seal level rise[2] Rapid ice-melt	NA	~4 to 9 in	~ 17 to 26 in	~ 37 to 50 in

Table 2. Sea Level Rise Projections

***NA: not applicable**
****in: inch**

[1] The central range (middle 67 percent) of values from GCM-based probabilities rounded to the nearest inch is shown.
[2] The rapid ice-melt scenario is based on acceleration of recent rates of ice melt in the Greenland and West Antarctic ice sheets and paleoclimate studies.

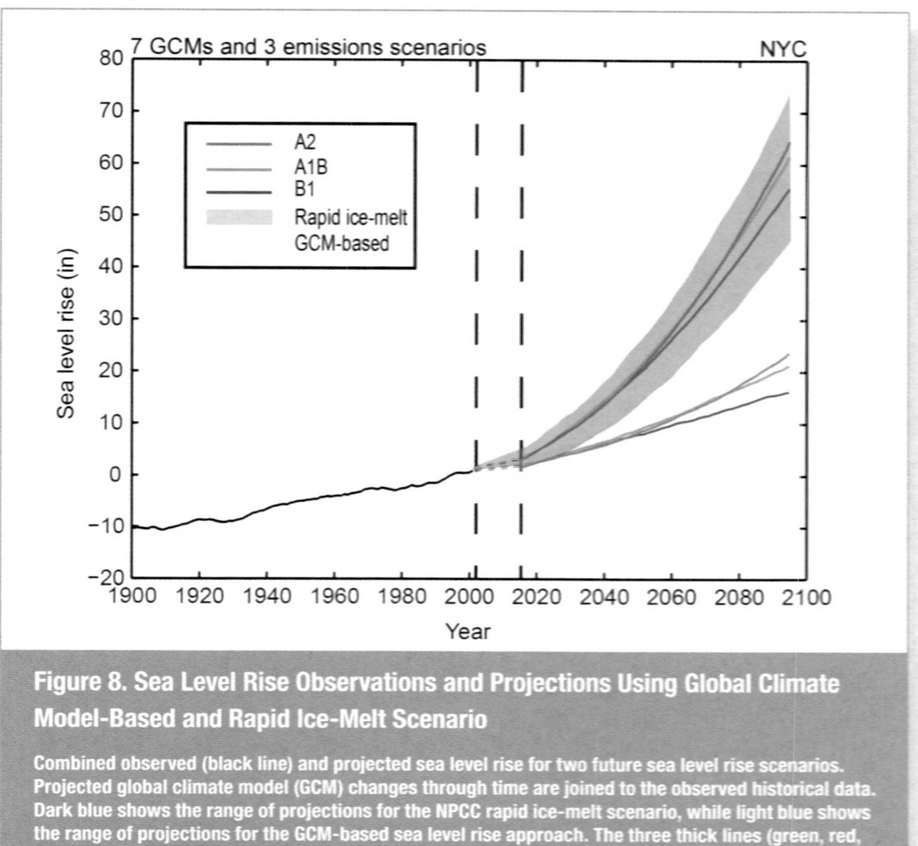

Figure 8. Sea Level Rise Observations and Projections Using Global Climate Model-Based and Rapid Ice-Melt Scenario

Combined observed (black line) and projected sea level rise for two future sea level rise scenarios. Projected global climate model (GCM) changes through time are joined to the observed historical data. Dark blue shows the range of projections for the NPCC rapid ice-melt scenario, while light blue shows the range of projections for the GCM-based sea level rise approach. The three thick lines (green, red, and blue) within each sea level rise scenario show the average for each emissions scenario across 7 GCMs. A ten-year filter has been applied to the observed data and modeled output.

Higher temperatures and sea level rise are extremely likely for New York State in the future. All global climate models project continuing temperature and sea level rise increases over the century, with the central range (the middle 67 percent of all projections) projecting more rapid temperature and sea level rise than what occurred over the 20th century. Although most projections indicate small increases in precipitation, some do not, and decade-by-decade precipitation variability is large; therefore, precipitation projections are less certain than temperature projections.

Region-specific projections of mean changes in temperature and precipitation are provided in Table 3. Figure 9 shows seasonal projections for the Adirondacks (ClimAID Region 7).

		Baseline[1] 1971–2000	2020s	2050s	2080s
Region 1					
Stations used for Region 1 are Buffalo, Rochester, Geneva and Fredonia.	Air temperature[2]	48ºF	+1.5 to 3.0ºF	+3.0 to 5.5ºF	+4.5 to 8.5ºF
	Precipitation	37 in	0 to +5%	0 to +10%	0 to 15%
Region 2					
Stations used for Region 2 are Mohonk Lake, Port Jervis, and Walton.	Air temperature[2]	48ºF	+1.5 to 3.0ºF	+3.0 to 5.0ºF	+4.0 to 8.0ºF
	Precipitation	48 in	0 to +5%	0 to +10%	+5 to 10%
Region 3					
Stations used for Region 3 are Elmira, Cooperstown, and Binghamton.	Air temperature[2]	46ºF	2.0 to 3.0ºF	+3.5 to 5.5ºF	+4.5 to 8.5ºF
	Precipitation	38 in	0 to +5%	0 to +10%	+5 to 10%
Region 4					
Stations used for Region 4 are New York City (Central Park and LaGuardia Airport), Riverhead, and Bridgehampton.	Air temperature[2]	53ºF	+1.5 to 3.0ºF	+3.0 to 5.0ºF	+4.0 to 7.5ºF
	Precipitation	47 in	0 to +5%	0 to +10%	+5 to 10%
Region 5					
Stations used for Region 5 are Utica, Yorktown Heights, Saratoga Springs, and the Hudson Correctional Facility.	Air temperature[2]	50ºF	+1.5 to 3.0ºF	+3.0 to 5.5ºF	+4.0 to 8.0ºF
	Precipitation	51 in	0 to +5%	0 to +5%	+5 to 10%
Region 6					
Stations used for Region 6 are Boonville and Watertown.	Air temperature[2]	44ºF	+1.5 to 3.0ºF	+ 3.5 to 5.5ºF	+4.5 to 9.0ºF
	Precipitation	51 in	0 to +5%	0 to +10%	+5 to 15%
Region 7					
Stations used for Region 7 are Wanakena, Indian Lake, and Peru.	Air temperature[2]	42ºF	+1.5 to 3.0ºF	+3.0 to 5.5ºF	+4.0 to 9.0ºF
	Precipitation	39 in	0 to +5%	0 to +5%	+5 to 15%

Table 3. Projections of Mean Annual Changes in Air Temperature and Precipitation for New York State Climate Regions

[1] The baselines for each region are the average of the values across all the stations in the region.
[2] The central range (middle 67 percent) of values from model-based probabilities is shown; temperature ranges are rounded to the nearest half-degree and precipitation to the nearest 5 percent.

Source: Columbia University Center for Climate Systems Research. Data are from USHCN and PCMDI.

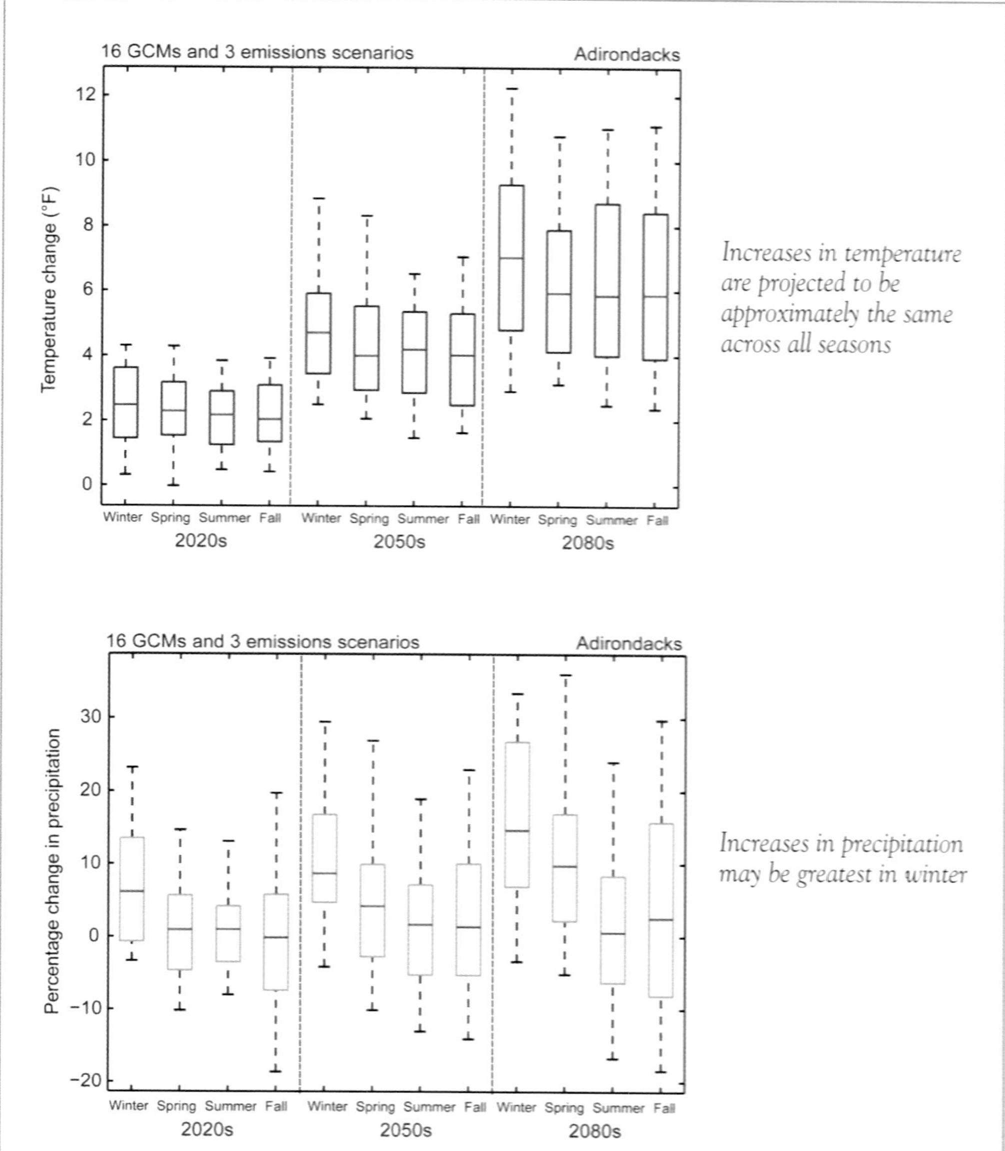

Figure 9. Seasonal Temperature Projections for the Adirondacks (ClimAID Region 7)

The full range of values across the 16 GCMs and three emissions scenarios and key points in the distribution are shown here. The central 67 percent of values are shown in the boxed areas; the median is indicated by the red line. Winter runs from December to February, while Spring runs from March through May, Summer from June through August, and Fall from September through November.

Extreme Events

Extreme events can have disproportionate effects on both urban and rural systems throughout New York State. During the 21st century:

- Heat waves are expected to become more frequent and intense
- Intense precipitation events are expected to become more frequent
- Storm-related coastal flooding is expected to increase due to rising sea levels

Table 4 presents projections for some extreme events for the Southern Tier (ClimAID Region 3).

Elmira (Region 3): Full range of changes in extreme events: minimum, (central range*), and maximum

	Extreme event	Baseline	2020s	2050s	2080s
Heat Waves & Cold Events	Number of days per year with maximum temperature exceeding				
	90°F	10	11 (14 to 19) 25	15 (21 to 33) 45	19 (26 to 56) 70
	95°F	1	2 (2 to 4) 7	2 (4 to 10) 18	4 (7 to 24) 38
	Number of heat waves per year[2]	1	1 (2 to 3) 3	2 (3 to 4) 6	2 (3 to 8) 9
	average duration	4	4 (4 to 5) 5	4 (4 to 5) 5	4 (5 to 5) 7
	Number of days per year with min. temp. at or below 32°F	152	116 (122 to 124) 145	86 (106 to 122) 168	68 (87 to 114) 124
Intense Precipitation	Number of days per year with rainfall exceeding:				
	1 inch	6	5 (6 to 7) 8	5 (6 to 7) 8	5 (6 to 8) 10
	2 inches	0.6	0.5 (0.6 to 0.9) 1	0.5 (0.6 to 1) 1	0.4 (0.7 to 1) 2

Table 4. Extreme Event Projections for the Southern Tier

The minimum, central range (middle 67 percent), and maximum of values from global climate model-based probabilities across the GCMs and GHG emissions scenarios are shown.

[1] Decimal places are shown for values less than 1, although this does not indicate higher precision/certainty. The high precision and narrow range reflect model-based results. Due to multiple uncertainties, actual values and range are not known to the level of precision shown in this table.
[2] Defined as three or more consecutive days with maximum temperature exceeding 90°F.

Extreme Event	Probable Direction Throughout 21st Century	Likelihood[1]
Heat Index[2]	^	Very likely
Ice storms/Freezing rain	^	About as likely as not
Snowfall frequency & amount	v	Likely
Downpours (precipitation rate/hour)	^	Likely
Lightning	Unknown	
Intense hurricanes	^	More likely than not
Nor'easters	Unknown	
Extreme winds	^	More likely than not

Figure 10. Qualitative Changes in Extreme Events for New York City/Long Island (ClimAID Region 4)

Potential for changes in other variables are described in a more qualitative manner, as quantitative information is either unavailable or considered less reliable. Figure 10 shows the likelihood of each of these changes occurring in New York City/Long Island.

[1] Likelihood definitions: Very likely = >90 percent probability of occurrence; Likely = >66 percent probability of occurrence; More likely than not = >50 percent probability of occurrence.
[2] The National Weather Service uses a heat index related to temperature and humidity to define the likelihood of harm after prolonged exposure or strenuous activity (http://www.weather.gov/om/heat/index.shtml).

IV. Adaptation Assessment Steps

Adaptation to climate change focuses on actions that stakeholders take in response to a changing climate. Adaptation strategies do not directly include actions that reduce the likelihood of climate change from occurring (i.e., climate change *mitigation*) but instead present actions to lessen the impact of climate change or take advantage of changes unleashed by a shifting climate. In the context of the ClimAID assessment, the ClimAID team examined the following two categories of adaptation strategies:

- Those that reduce the level of physical, social, or economic impact of climate change and variability
- Those that take advantage of new opportunities emerging from climate change

The process of adaptation assessment can be summarized in an eight-step process (see Figure 11), which can be adjusted as needed depending on varying circumstances.

1. Identify current and future climate hazards
2. Inventory vulnerabilities and opportunities
3. Prioritize vulnerabilities
4. Identify and categorize adaptation strategies
5. Evaluate and prioritize adaptation strategies
6. Link strategies to capital and rehabilitation cycles
7. Create an adaptation plan
8. Monitor and reassess

Developing adaptation strategies starts with learning about current climate and how climate is projected to change in the future (see Section III). After understanding how the climate in New York State is projected to change, the next step in developing adaptation strategies is identifying the vulnerabilities a change in climate might create, as well as assessing risk levels. Vulnerabilities and risks can then be prioritized based on several criteria. The risk ratings resulting from the process of prioritizing vulnerabilities can help in the development of adaptation strategies. Several different types of adaptation strategies can be developed in response to a particular climate risk, and a set of factors can be used to evaluate and prioritize these strategies. The final step of the adaptation process is monitoring and reassessing climate changes, impacts, and adaptation strategies (see Figure 11).

These adaptation assessment steps are intended to be general enough to be useful for a range of jurisdictions and infrastructure sectors, yet specific enough to serve as the template for developing and implementing a sector's adaptation efforts. These steps may be used to develop climate change adaptation in any urban area, with region-specific adjustments related to climate risk information, critical infrastructure, and protection levels.

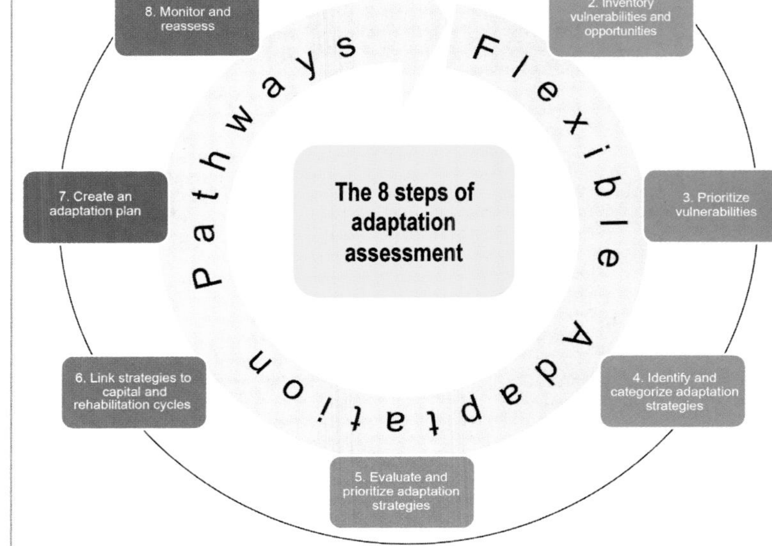

Figure 11. Adaptation Assessment Steps

Step 1: Identify Current and Future Climate Hazards

The first step in developing adaptation strategies is learning about current climate and how climate is projected to change in the future. For more information on the climate of New York State and future projections, see Section III.

Step 2: Inventory Vulnerabilities and Opportunities

A focus on key vulnerabilities is necessary to help policymakers and stakeholders assess the level of risk, prioritize, and design pertinent response strategies. In most instances, inventories of vulnerabilities will be qualitative, based on expert knowledge and relevant climate hazards. Factors that help characterize vulnerability include:

- Magnitude
- Timing
- Persistence and reversibility
- Likelihood
- Distributional aspects
- Importance of the at-risk systems
- Potential for adaptation
- Thresholds or trigger points that could exacerbate the change

Based on these factors, the ClimAID team developed an inventory of key vulnerabilities for New York State; examples of key vulnerabilities for New York State by climate factor, for each of the ClimAID sectors, are shown in Table 5.

Climate Hazard \ Sector	Water Resources	Coastal Zones	Ecosystems	Agriculture	Energy	Transportation	Telecommunications	Public Health
Temperature and Heat Waves	Increased wear and tear on materials Potential changes in drinking supply	Shifts in marine species due to warmer waters	Increased frequency of summer heat stress on plants Potential changes in pest populations and habits Changes in species composition due to warmer winters	Changes in distribution of primary crops such as apples, cabbage, and potatoes Decline in dairy milk production	Increased demand on energy supply Increased vulnerability of energy infrastructure	Increased wear and tear on infrastructure Extreme event-related delays and hazards	Increased wear and tear on materials	More heat-related deaths Decline in air quality
Precipitation, Extreme Precipitation, and Drought	Increased vulnerability of infrastructure Potential changes in drinking supply Increased risk of changes in river flooding	Potential permanent inundation of coastal lands, including critical wetland habitat	Potential changes in pest populations and habits	Changes in distribution of primary crops such as apples, cabbage, and potatoes	Increased vulnerability of energy infrastructure Greater uncertainty around future availability of alternative energy sources	Flooding of key rail lines, roadways, and hubs Increased wear and tear of materials Extreme event-related delays and hazards	Flooding of central facilities Increased wear and tear on materials	Outbreaks of illness related to water-borne pathogens
Sea Level Rise and Coastal Flooding	Saltwater intrusion into freshwater aquifers	Increased risk of storm surge-related flooding Potential permanent inundation of coastal lands, including critical wetland habitat	Effects on marine and freshwater species	Salinization of coastal agriculture areas	Increased vulnerability of energy infrastructure	Episodic and permanent inundation of key rail lines, roadways, and hubs Extreme event-related delays and hazards	Flooding of central facilities Increased wear and tear on materials	Direct physical harm and trauma

Table 5. Examples of Key Vulnerabilities for New York State by Climate Factor

Step 3: Prioritize Vulnerabilities

Vulnerabilities are prioritized depending upon those systems or regions whose failure or reduction in function is likely to carry the most significant consequences. One tool used in risk assessment is a matrix that assesses the magnitude of consequence of an event against the likelihood of the event occurring. For climate adaptation assessment, there are at least three layers of uncertainty that need to be considered to yield an approximate overall risk of a particular climate hazard and a particular impact (see Figure 12). The overall risk rating can then assist in the creation of adaptation strategies. Risk categories to be considered include:

Probability of a given climate hazard – The general probability for change in a climate hazard (such as temperatures or extreme precipitation events) occurring. Using climate risk information as a guide, these can be defined as:

- **High** probability of the climate hazard occurring
- **Medium** probability of the climate hazard occurring
- **Low** probability of the climate hazard occurring

Likelihood of impact occurrence – The likelihood that a change in a given climate hazard (e.g., temperature rise) will result in a particular impact (e.g., material failure). Examples of likelihood categories include:

- **Virtually certain/already occurring** – Nearly certain likelihood of the impact occurring over the useful life of the infrastructure, and/or the climate hazard may already be impacting infrastructure
- **High** likelihood of the impact occurring over the useful life of the infrastructure
- **Moderate** likelihood of the impact occurring over the useful life of the infrastructure.
- **Low** likelihood of the impact occurring over the useful life of the infrastructure.

Magnitude of consequence – The combined impacts should a given hazard occur, taking into account such factors as:

- **Internal operations**, including the scope and duration of service interruptions, reputational risk, and the potential to encounter regulatory problems
- **Capital and operating costs**, including all capital and operating costs to the stakeholder and revenue implications caused by the climate change impact
- **Number of people impacted**, including considerations related to any impacts on vulnerable populations (including, but not limited to seniors, low-income communities, mentally or physically disabled citizens, homebound residents, and children).
- **Public health**, including worker safety
- **Economy**, including any impacts to the city's economy, the price of services to customers, and clean-up costs incurred by the public
- **Environment**, including the release of toxic materials and impacts on biodiversity, the state's ecosystems, and historic sites

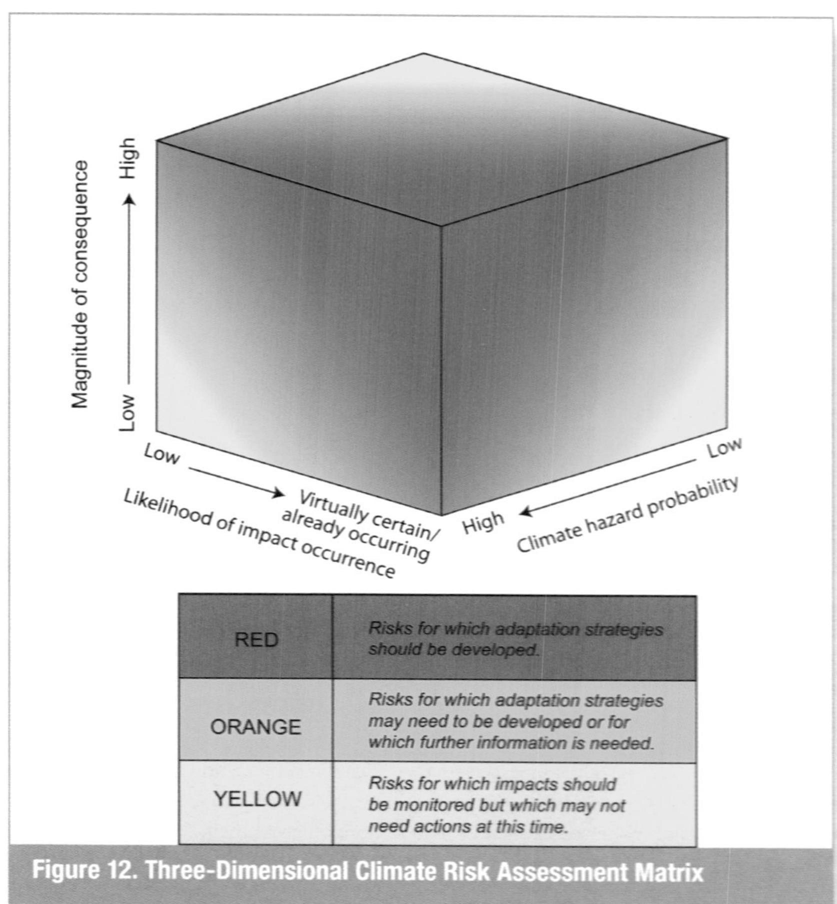

RED	Risks for which adaptation strategies should be developed.
ORANGE	Risks for which adaptation strategies may need to be developed or for which further information is needed.
YELLOW	Risks for which impacts should be monitored but which may not need actions at this time.

Figure 12. Three-Dimensional Climate Risk Assessment Matrix

Step 4: Identify and Categorize Adaptation Strategies

Building on internal risk-management and assessment policies, stakeholders can begin to brainstorm strategies for those infrastructure classes that fall into the red and orange categories of the risk matrix (Figure 13). Adaptation strategies may be divided into a set of categories, including:

- The **type** of adaptation strategy depends on whether the strategy is focused on management and operations, infrastructural change (particularly with the physical component of the sector), or policy adjustments.
- The **administration** element of adaptation strategies defines the strategy as either emerging from the public or private sectors, and from which level of government (i.e., local/municipal, county, state, or national).
- **Condition** is defined by whether an adaptation strategy is an incremental action or a larger-scale paradigm shift.
- **Timing** highlights the period during which the adaptation strategy will be implemented. Given what is understood about the rate of climate change and the sensitivities of the system, a primary question is whether the adaptation should take place in the short term (less than 5 years), medium term (5 to 15 years), or long term (more than 15 years). A crucial consideration regarding the issue of timing is whether there are tipping points associated with dramatic shifts

in the level of impacts and/or vulnerabilities and whether these tipping points become triggers for new policies and regulations.

- **Geography** relates to the overall spatiality of the adaptation impacts, specifically, cataloging if the adaptation strategy is widespread, clustered, or isolated/unique (e.g., if the impact is associated with a specific site or location) throughout the state.

Potential adaptation strategies can be further defined within a range of elements including economics and institutional organization. Economic issues include the costs and benefits of adaptation, and the relative distribution of both. A critical economic issue is the overall cost-to-benefit ratio and how much economic advantage there is to taking a specified action. It is also important to determine potential opportunity costs, as well as the capacity (e.g., human and capital resources) and capability (e.g., regulatory mandate, legal ability) of the entity considering the adaptation.

Step 5: Evaluate and Prioritize Adaptation Strategies

Prioritization of which adaptation to undertake is a critical component of developing an adaptation strategy. Prioritization criteria include considerations of climate risk levels, vulnerability and exposure, maximum benefit-cost ratio, cost effectiveness, distributional and equity concerns, and institutional capacity and capability. Other criteria include the spatial and temporal character of a strategy's impact and the potential for flexible adaptation.

There may be multiple strategies to consider during adaptation planning. Once stakeholders have an initial list of adaptation strategies, they can evaluate these strategies in order to determine an order in which they should be implemented, and begin to create a broader agency- or organization-wide adaptation plan. There are a variety of available methods and perspectives to aid in evaluating individual actions and strategies (see example in Table 6). Elements to consider as part of evaluating adaptation strategies could include:

- **Cost** – What are the general costs of the proposed strategy, including human and other resources? General costs can yield a rough measure of benefits and costs to the extent that the consequences are measured in economic terms. There will also be important non-economic consequences in most decision problems.
- **Timing** – Timing of implementation should be considered relative to the timing of impact. Specifically, if the impact will occur in a time frame comparable to the time required for implementation, there is need for immediate consideration.
- **Feasibility** – How feasible is the strategy for implementation both within an organization and from perspectives such as engineering, policy, legal, and insurance? Are there expected technological changes that would impact future feasibility?
- **Efficacy** – To what extent will the strategy, if successfully implemented, reduce the risk?
- **Robustness** – Is there the potential to install equipment or upgrade infrastructure that is designed to withstand a range of climate hazards? Are there opportunities for flexible adaptation pathways?
- **Co-benefits** – Will strategies have a negative or positive impact on other stakeholders or sectors? Is there potential for cost sharing? Are there impacts on mitigation of greenhouse gases? Are there impacts on the environment or a vulnerable population?

Other factors to consider include equity, social justice, sustainability, institutional context, and unique circumstances.

Adaptation Strategy	Strategy Cost (1=low to 3=high)	Strategy Feasibility (1=low to 3=high)	Timing of implentation (1=low to 3=high)	Efficacy (1=low to 3=high)	Resiliency rating (1=low to 3=high)	Co-benefits (1=low to 3=high)	Average*	Notes & institutional considerations
Clean drains	1	1	1	2	2	2	1.8	
Build flood walls	3	2	2	1	3	2	2.2	

Table 6. Strategy Prioritization Framework with Adaptation Strategy Examples

***1=high priority strategy, 2=medium priority strategy, 3=low priority strategy**

Source: NPCC (2010)

Step 6: Link Strategies to Capital and Rehabilitation Cycles

Stakeholders have capital budgets that extend over a variety of time periods; in some cases, budgets extend over decades. Stakeholders should review these budgets to determine which adaptation strategies can be undertaken within existing funding constraints and what additional resources need to be identified. Linking adaptation strategies to planned projects or other non-adaptation efforts can result in significant cost savings. In turn, stakeholders are advised to put priority on exploring low-cost adaptation strategies, especially in times of fiscal austerity.

Step 7: Create an Adaptation Plan

The conclusion of the climate adaptation assessment process is really just the beginning. Stakeholders can combine and distill the knowledge gained from the assessment into an adaptation plan, which, in turn, can help operationalize adaptation planning.

An adaptation plan could include the following components:

- Discussion of key climate vulnerabilities
- List of prioritized adaptation strategies
- Consideration of other adaptation tools
- Plan for establishing indicators and monitoring
- Timeline to reassess strategies as new information comes to light

An adaptation plan should be seen as a living document and be revisited on a semi-regular basis to ensure that it incorporates the latest research and knowledge. By doing so, stakeholders can develop flexible adaptation pathways that lead to an ongoing adaptive capacity for systems, sectors, regions, and groups.

Step 8: Monitor and Reassess

Monitoring climate change on a regular basis, as well as other factors that might directly or indirectly influence climate change risks, will help development of flexible adaptation pathways. Consistent monitoring protocols are needed for climate change indicators, particularly those related to changes in the climate, climate science updates, climate impacts, and adaptation activities. Monitoring of key indicators can help stakeholders initiate course corrections in adaptation policies and/or changes in timing of their implementation. These indicators need to be developed and tracked over time to provide targeted quantitative measures of climate change, its impacts, and adaptation. This will provide useful information to decision-makers regarding the timing and extent of needed adaptation actions.

V. Other Adaptation Tools

There are other climate change adaptation tools to consider that include regulatory, design, and engineering standards; legal structures; and insurance opportunities.

Climate Protection Levels

Climate protection levels (CPLs) refer to building and construction codes and regulations, design standards, and best practices that pertain to climate, as adopted by the professional engineering community and various government entities.

The general framework for the development of CPLs and/or recommendations for future study are summarized in the following steps:

1. Develop regional/local-specific climate change projections.
2. Select climate hazards of focus (e.g., coastal flooding and storm surge, inland flooding, heat waves, and extreme events).
3. Solicit feedback from operators and regulators of infrastructure through questionnaires to identify potential impacts of climate change hazards on infrastructure.
4. Identify existing design and/or performance standards relevant to critical infrastructure
5. Review and reassess these standards in light of the climate change projections.
6. Highlight those standards that may be compromised by climate change and/or need further study to determine if revised CPLs are necessary to facilitate climate resiliency.

To meet the criteria for development of a recommended CPL, a regulation, policy, or practice needs to:

- Guide the formation or maintenance of critical infrastructure at risk to climate-related hazards.
- Dictate action in order to maintain acceptable risk levels with respect to climate-related hazards.
- Allow for adjustments that will enable a stable level of risk protection in response to a changing climate.

CPL recommendations can take multiple forms and offer content that is broad-based, design-specific, measurable/quantifiable, policy relevant, or suggestive of future studies. The following examples illustrate the types of recommendations for CPLs:

- **Quantitative statements** – Statements that emerge from the interplay between quantitative design, performance standards, and quantitative climate risk information.
- **General statements** – Narrative comments on the relevance of climate risk information to existing design standards.
- **Infrastructure analysis** – Recommendations for further analysis of critical parts of the infrastructure for which more information is needed to create CPLs. For example, more specific information on the existing design standards of street catch basins for inland street level flooding is required to determine if a CPL is needed to address the issue.
- **Engineering-based studies** – Suggestions for engineering studies such as hydrologic studies that need to be performed in order to determine if and/or how current standards need to be changed. These are necessary in situations where there are limitations in the knowledge of the system/material-level response to climate change and variability (e.g., responses of materials to increased heat).
- **Policy and planning issues** – Evaluation of system-wide processes such as the distribution of impervious surfaces, land-use changes, and public health alerts.

Legal Framework

Another climate change adaptation tool is the updating of laws and legal frameworks that guide planning, zoning, building codes, health codes, and materials usage. In many cases, the addition of a climate change component to an Environmental Impact Statement or equivalent regulation could be an efficient way to encourage the consideration of climate change impacts. Current federal, state, and local laws could be reassessed; new regulations should incorporate climate change into their formulations.

Insurance

Insurance can be a powerful risk-sharing tool for climate adaptation. Insurance companies are now being brought into discussions about climate change adaptation. As an example, insurance companies influence the level of development in coastal areas. If potential future changes in sea level rise are taken into account, insurance companies could factor these risks into their hazard models and help to disperse certain risks associated with climate change.

VI. Summary

The risk-management adaptation strategies described in this guidebook will be useful in helping stakeholders reduce climate impacts in the future. Climate change is extremely likely to bring warmer temperatures to New York State, while climate hazards are likely to produce a range of impacts on the urban and rural fabric of the state in the coming decades. Heat waves are very likely to become more frequent, intense, and longer in duration. An increase in total annual precipitation is more likely than not; brief, intense rainstorms are also likely to increase. Additionally, rising sea levels are extremely likely, and are very likely to lead to more frequent and damaging flooding related to coastal storm events in the future.

It is important to note that adaptation strategies are also likely to produce benefits today, as such strategies will help to lessen impacts of climate extremes that cause current damage. Given the scientific uncertainties in projecting future climate change, however, monitoring of climate and impacts indicators is critical so that flexible adaptation pathways for the region can be achieved.

Climate variables should be monitored and assessed on a regular basis. Indirect climate change impacts, such as those caused by climate change in other regions, should also be taken into consideration. By evaluating this evolving information, New York State can be well positioned to develop robust and flexible adaptation pathways that maximize climate and societal benefits while minimizing climate hazards and costs.

References

Intergovernmental Panel on Climate Change (IPCC). 2000. *Emissions Scenarios: A Special Report of IPCC Working Group III*. Cambridge University Press.

Intergovernmental Panel on Climate Change (IPCC). 2007. *The Physical Science Basis, Contribution of Working Group I to the Fourth Assessment Report*. Cambridge University Press.

Lowe, J., T. Reeder, K. Horsburgh, and V. Bell. 2009. "Using the new TE2100 science scenarios." United Kingdom Environment Agency.

New York State Climate Action Council. November 2010. *Climate Action Plan Interim Report*. http://www.nyclimatechange.us.

New York State Sea Level Rise Task Force. December 2010. *Report to the Legislature*. http://www.dec.ny.gov/docs/administration_pdf/slrtffinalrep.pdf.

New York City Panel on Climate Change. (NPCC). 2010. *Climate Change Adaptation in New York City: Building a Risk Management Respons* C. Rosenzweig and W. Solecki, Eds. Prepared for use by the New York City Climate Change Adaptation Task Force. Annals of the New York Academy of Science: New York, NY.

United States Global Change Research Program. 2009. *Global Climate Change Impacts in the United States*. T, R. Karl, J.M. Melillo, and T.C. Peterson, Eds. Cambridge University Press.

NYSERDA, a public benefit corporation, offers objective information and analysis, innovative programs, technical expertise and funding to help New Yorkers increase energy efficiency, save money, use renewable energy, and reduce their reliance on fossil fuels. NYSERDA professionals work to protect our environment and create clean-energy jobs. NYSERDA has been developing partnerships to advance innovative energy solutions in New York since 1975.

To learn more about NYSERDA programs and funding opportunities visit www.nyserda.org.

New York State
Energy Research and
Development Authority

17 Columbia Circle
Albany, New York 12203-6399

toll free: 1 (866) NYSERDA
local: (518) 862-1090
fax: (518) 862-1091

info@nyserda.org
www.nyserda.ny.gov

Climate Adaptation Guidebook for New York State

November 2011

State of New York
Andrew M. Cuomo, Governor

New York State Energy Research and Development Authority
Vincent A. DeIorio, Esq., Chairman | Francis J. Murray, Jr., President and CEO

ClimAID Annex III

An Economic Analysis of Climate Change Impacts and Adaptations in New York State

Authors: Robin Leichenko,[1] David C. Major, Katie Johnson, Lesley Patrick, and Megan O'Grady

[1] Rutgers University

Table of Contents

Executive Summary 3
1 Introduction 7
2 Water Resources 19
3 Ocean and Coastal Zones 30
4 Ecosystems 43
5 Agriculture 58
6 Energy 76
7 Transportation 93
8 Telecommunications 103
9 Public Health 112
10 Conclusions 131
11 References 134

Executive Summary

This study provides an overview assessment of the potential economic costs of climate change impacts and adaptations to climate change in eight major economic sectors in New York State. These sectors, all of which are included in the ClimAID report are: water resources, ocean and coastal zones, ecosystems, agriculture, energy, transportation, communications, and public health. Without adaptation, climate change costs in New York State for the sectors analyzed in this report may approach $10 billion annually by midcentury. However, there is also a wide range of adaptations that, if skillfully chosen and scheduled, can markedly reduce the impacts of climate change by amounts in excess of their costs. This is likely to be even more true when non-economic objectives such as environment and equity are taken into account. New York State as a whole has significant resources and capacity for effective adaptation responses; however, given the costs of climate impacts and adaptations, it is important that the adaptation planning efforts that are now underway are continued and expanded.

Methods

The methodology for the study entails a six-step process that utilizes available economic data, interviews, and risk-based assessment to identify and where possible to assign costs of key sectoral vulnerabilities and adaptation options for climate change for eight economic sectors. The study draws conceptually from the general framework of benefit-cost analysis (recognizing its significant limitations in evaluating adaptation to climate change) to provide an overview assessment of the potential costs of key impacts and adaptation options. For all sectors, key economic components with significant potential impact and adaption costs are highlighted.

Sector Assessments

All of the eight sectors examined will have impacts from climate change, and for all sectors a range of adaptations is available. Because New York State is a coastal state and is highly developed, the largest direct impacts and costs are likely to be associated with coastal areas. Among the sectors in this study, these include the ocean coastal zone, transportation, energy and part of the water sector. However, impacts and costs will be significant throughout the state in sectors such as public health, transportation and agriculture. Impacts must be judged not only on the basis of direct economic costs, but also on the overall importance of sector elements to society. In terms of adaptation costs, the largest costs may be in the transportation sector, with significant adaptation costs for water, ocean coastal zones, energy, agriculture and ecosystems. The largest positive differences between benefits and costs among the sectors are likely to be in ecosystems and public health.

In addition to the overall analysis of the report, illustrative cost and benefit projections were made for one or more elements of the sectors. The results in terms of mid-century (2050s) annual costs (in $2010) of impacts are: water resources, $116-203 million; ocean coastal zones, $44-77 million; ecosystems, $375-525 million; agriculture, $140-289 million; energy, $36-73 million; transportation, $100-170 million; communications, $15-30 million, and public health

$2,998-6,098 million. These figures understate the aggregate expected costs, especially for heavily developed coastal areas, because they are for selected elements of the sectors for which extrapolations relating to climate data could be made. (Because of differences in method and data availability and the extent of coverage within sectors, these numbers are not directly comparable. For example, the high annual costs in public health are partly a function of the U.S. Environmental Protection Agency's estimate of the value of a statistical life (USEPA 2000; 2010.) The extent to which explicit public planning for adaptation will be required will differ among sectors: energy, communications and agriculture are sectors with regular reinvestment that has the effect of improving the resilience of the sector for present and future climate variability and other factors, and so climate adaptation will be more easily fit into the regular processes of these sectors. For the other sectors, much more public evaluation and planning will be required.

Overview assessments by sector are:

Water Resources. Water supply and wastewater treatment systems will be impacted throughout the state. Inland supplies will see more droughts and floods, and wastewater treatment plants located in coastal areas and riverine flood plains will have high potential costs of impacts and adaptations. Adaptations are available that will have sizable benefits in relation to their costs.

Coastal Zones. Coastal areas In New York State have the potential to incur very high economic damages from a changing climate due to the enhanced coastal flooding due to sea level rise and the development in the area with residential and commercial zones, transportation infrastructure (treated separately in this study), and other facilities. Adaptation costs for coastal areas are expected to be significant, but relatively low as compared to the potential benefits.

Transportation. The transportation sector may have the highest climate change impacts in New York State among the sectors studied, and also the highest adaptation costs. There will be effects throughout the state, but the primary impacts and costs will be in coastal areas where a significant amount of transportation infrastructure is located at or below the current sea level. Much of this infrastructure floods already, and rising sea levels and storm surge will introduce unacceptable levels of flooding and service outages in the future. The costs of adaptation are likely to be very large and continuing.

Agriculture. For the agriculture sector, appropriate adaptation measures can be expected to offset declines in milk production and crop yields. Although the costs of such measures will not be insignificant, they are likely to be manageable, particularly for larger farms that produce higher value agricultural products. Smaller farms, with less available capital, may have more difficulty with adaptation and may require some form of adaptation assistance. Expansion of agricultural extension services and additional monitoring of new pests, weeds and diseases will be necessary in order to facilitate adaptation in this sector.

Ecosystems. Climate change will have substantial impacts on ecosystems in New York State. For revenue-generating aspects of the sector, including winter tourism and recreational fishing, climate change may impose significant economic costs. For other facets of the sector, such as forest-related ecosystems services, heritage value of alpine forests, and habitat for endangered species, economic costs associated with climate change are more difficult to quantify. Options for adaptation are currently limited within the ecosystems sector and costs of adaptation are only beginning to be explored. Development of effective adaptation strategies for the ecosystems sector is an important priority.

Energy. The energy sector, like communications, is one in which there could be large costs from climate change if ongoing improvements in system reliability are not implemented as part of regular and substantial reinvestment. However, it is expected that regular investments in system reliability will be made, so that the incremental costs of adaptation for climate change will be moderate. Even with regular reinvestments there may be increased costs from climate change. Moreover, the energy sector is subject to game-changing policies and impacts such as changes in demand from a carbon tax (either directly or via cap and trade) and large investments in stability that could be undertaken to deal with the potential impacts of electromagnetic storms.

Communications. The communications sector is one in which there could be large costs from climate change if ongoing adaptations are not implemented as part of regular reinvestment in the sector or if storms are unexpectedly severe. However, it is expected that regular adaptations will be made, so that additional costs of adaptation for climate change will be relatively small.

Public Health. Public health will be impacted by climate change to the extent that costs could be large if ongoing adaptations to extreme events are not implemented. Costs could also be large if appropriate adaptations are not implemented in other sectors that directly affect public health, particularly water resources and energy. The costs associated with additional adaptations within the public health sector need further study.

The Future

This study is an important starting point for assessing the costs of climate change impacts and adaptations in New York. Much further work needs to be done in order to provide the extensive, detailed estimates of comprehensive costs and benefits associated with climate change required for planning. This work will have to deal with challenges such as the lack of climate-focused data sets and the fact that the feasibility of many potential adaptations has not been adequately analyzed. However, the basic conceptual approaches to future work have been identified, and even initial benefit-cost analyses of major impacts and corresponding adaptation options can help to illustrate the economic benefits of adaptation and thus to shape policy. This study therefore provides an important source of information for policy makers as to the relative size of climate impacts across major sectors of state activities and the adaptations that might be undertaken to deal with them. Because of the extensive impact and adaptation costs facing New York State, planning for adaptation to climate change must continue. With effective planning and implementation, the benefits from adaptation are likely to be significant because there are many opportunities for development of resilience in all sectors and regions.

1 Introduction

This study provides an overview assessment of the potential economic costs of impacts and adaptation to climate change in eight major economic sectors in New York State in the ClimAID report. The goal of the study is to provide information on the economic impacts of climate change and adaptation for use by public officials, policy makers, and members of the general public. The study is also intended to provide information that will assist the New York State Climate Action Council with identification and prioritization of adaptation areas for the state. While this study, because of limitations of data, case studies, methods and time, does not achieve the detail of the highly specific project evaluation that should be undertaken in the future in New York State, it nonetheless provides an important source of information for policy makers as to the relative size of climate impacts across major sectors of state activities and the adaptations that might be undertaken to deal with them. The state of the art of assessing the economic costs of climate impacts and adaptations is still nascent, so that this and other contemporary studies (cited throughout this report) perform important functions but cannot yet be considered as comprehensive.

The study draws from the information provided in the eight ClimAID sectors, supplemented by interviews with the sector leaders and other experts and by information from other studies of the costs of impacts and adaptation in New York State and elsewhere in the US and other countries. All these data sources are used to develop the information and assessments in the eight sector chapters in the report. Based on the study results, climate change costs, without adaptation, may approach $10 billion annually by mid-century for the sectors studied. However, there are a wide range of adaptations that, if skillfully chosen and scheduled, can markedly reduce the impacts of climate change in excess of their costs. This is likely to be even more true when non-economic objectives, such as the environment and equity, are taken into account.

This introductory chapter describes the framing approaches and methods of the study. Section 1.1 provides an overview of methods and some main results. Section 1.2 provides an overview of methodological concepts used in the study, including key terms and concepts, benefit-cost analysis, interest rates, the use of analogs, and the classification of impacts and adaptations. Section 1.3 describes the six steps used to develop the sectoral chapters and their results; and Section 1.4 is a summary of the methods used for the illustrative benefit-cost analyses.

Each of the eight sectoral chapters is organized according to the following pattern. The first part describes key economic risks and vulnerabilities and the illustrative benefit-cost analysis done for the sector. In the second part, the economic importance of the sector in New York State is described followed by a discussion of key climate sensitivities. Impact costs and adaptation costs are then examined from available information and additional information developed for the study, followed by a list of knowledge gaps for the sector. Technical notes describing the methods used in the benefit-cost analysis conclude each chapter. Consolidated references for the entire study follow the Conclusions chapter. Throughout the report, an attempt has been made to utilize stakeholder input of data, language and presentation, and to harmonize the work with the ClimAID chapters.

1.1 Summary of Methods and Main Results

The methodology for the study entails a six-step process that utilizes available economic data, interviews, and risk-based assessment (New York City Panel on Climate Change [NPCC] 2010) to identify and where possible to assign costs of key sectoral vulnerabilities and adaptation options for climate change in New York State. The study draws conceptually from the general framework of cost benefit analysis (recognizing its significant limitations in evaluating adaptation to climate change [Weitzman, 2009]) to provide an overview assessment of the potential costs of key impacts and adaptation options.

As part of the overall assessments for each sector, key economic components with significant potential costs were identified based on economic evaluation of the findings from the ClimAID sectors and the analyses of this study. Due to data limitations, costs could not be estimated for every component in each sector at this time. Table 1.1 presents a summary of the expected annual climate change impact costs at midcentury (i.e., for the 2050s) and the expected costs of adaptation options for the specified components of each sector, for which both impact and adaptation costs could be estimated. Details on the methods used to develop these extrapolations, and their limitations, are given in each specific sector chapter for the three study benchmark periods of the 2020s, 2050s, and 2080s.

A key issue for assigning costs of climate change is whether to focus on the effects of changes in the most damaging extreme events, such as coastal storms, or to focus on the changes in average climatic conditions. This study considers both of these types of climate changes. Estimates are made for costs and benefits with changes in extreme events for wastewater treatment plants, insured value for coastal zones, the transportation sector, energy, and health. The climate hazards include sea level rise, large coastal storms and heat waves. For agriculture and ecosystems, changes in the mean (average) value of climate variables are used. However, in all sectors broadly considered, both means and extremes matter.

Table 1.1 Available Estimated Annual Incremental Impact and Adaptation Costs of Climate Change at Mid-century for specified components of the ClimAID sectors. (Values in $2010 US.)

Sector	Component	Cost of annual incremental climate change impacts at mid-century for selected components, without adaptation	Costs and benefits of annual incremental climate change adaptations at mid-century for selected components
Water Resources	Flooding at Coastal Wastewater Treatment	$116-203 million	Costs: $47 million Benefits: $186 million
Coastal Zones	Insured losses	$44-77 million	Costs: $29 million Benefits: $116 million
Ecosystems	Recreation, tourism, and ecosystem service losses	$375-525 million	Costs: $32 million Benefits: $127 million
Agriculture	Dairy and crop losses	$140-289 million	Costs: $78 million Benefits: $347 million
Energy	Outages	$36-73 million	Costs: $19 million Benefits: $76 million
Transportation	Damage from 100 year storm	$100-170 million	Costs: $290 million Benefits: $1.16 billion
Communications	Damage from 100 year storm	$15-30 million	Costs: $12 million Benefits: $47 million
Public Health	Heat mortality and asthma hospitalization	$2.99-6.10 billion	Costs: $6 million Benefits: $1.64 billion
All Sectors	Total of Available Estimated Components	$3.8 – 7.5 billion/yr	Costs: $513 million/yr Benefits: $3.7 billion/yr

Note: see chapters for definitions of the selected components, and details of the estimation methods used. All values in $2010 US. The figures are not strictly additive because of the different methods used in each case

In each of the sector chapters, impacts and adaptations are evaluated according to four classes:

Level 1. Detailed assessment of costs for 2020s, 2050s, and 2080 where data permit (these are the components of the sectors that are represented in Table 1.1);

Level 2. Generalized estimates where data are limited. These estimates are based on literature and expert judgment;

Level 3. Qualitative discussion where cost data are lacking but there is general knowledge of impact and adaptation types;

Level 4. Identification of areas where costs are unknown because impacts and/or adaptation options are unknown or cannot be assigned.

An important strength of this and the ClimAID study is that the identification of economic risks and sensitivities to climate change is based on detailed, stakeholder-based investigation of specific sectors. Prior studies of the economic costs associated with climate change have generally entailed either top-down global assessments of impact costs (e.g., Stern 2007; Parry et al 2009), or highly generalized regional assessments for specific U.S. states that contain limited information on adaptation options (e.g. Niemi et al. 2009). This study of New York State provides an overview assessment of the costs of climate change impacts and adaptation that is grounded in empirical knowledge of key vulnerabilities and adaptation options.

The study of the economics of climate impacts and adaptations is relatively recent, so there are not enough examples of detailed studies, whether in New York State or elsewhere, to provide a wide assessment of costs. Further work needs to be done in order to fully estimate the comprehensive costs and benefits associated with climate change. This work will have to deal with challenges such as the lack of climate-focused data sets and the fact that the feasibility of many potential adaptations has not been adequately analyzed. On the other hand, the basic conceptual approaches to future work have been identified, and initial cost-benefit analyses of major impacts and corresponding adaptation options illustrate the economic benefits of adaptation.

1.2 Assessing the Economic Costs of Climate Change Impacts and Adaptation

The economic costs associated with both mitigation and adaptation to climate change are a topic of growing concern for national, state, and local governments throughout the world. Major research efforts to date, however, have primarily emphasized assessment of the aggregate costs of climate change impacts and adaptation at the global level across major country categories (e.g., developing countries), major world regions (e.g., Africa; South Asia), or specific sectors or countries, (e.g., World Bank 2006; Stern 2007; United Nations Framework Convention on Climate Change [UNFCCC] 2007; UNDP 2007; Cline 2007; Parry et al 2009). The estimates for the total costs of adaptation to the impacts of climate change are highly variable among these studies (see Agrawala and Fankhauser 2008). For example, estimates of the annual costs of adaptation in developing countries range from $10 to 40 billion/year (World Bank 2006) to $86 billion/year (UNDP 2007). The UNFCCC (2007) estimates of the annual global costs of adaptation in 2030 range between $44 billion and $166 billion. Reasons for this wide range of estimates include differences in how adaptation is defined, whether residual damages (see Table 1.2) are included in the estimates, and the comprehensiveness of the studies. A recent evaluation of the current state of knowledge for global adaptation cost estimates concluded that such estimates are preliminary and incomplete, and that important gaps and omissions remain (Fankhauser 2010, p. 25). Similar shortcomings are noted by Fankhauser (2010, p. 22) in studies conducted at the country level, particularly for estimates associated with National Adaptation Programmes of Action (see UNFCCC, n.d.), which also vary in scope, quality, and coverage. Despite limitations of both global and national studies, these studies nonetheless provide general guidance on the types of adaptations that may be needed within various sectors, as well as rough estimates of the types of costs that may be associated with these measures. A recent World Bank (2010) study uses an extrapolation framework similar to that used for the examples in Table 1.1.

While most prior work on adaptation costs has emphasized the global and national levels, several recent assessments of the costs associated with the impacts of climate change have been conducted for states including Washington, Maryland, and New Jersey (e.g., Niemi et al. 2009; CIER 2008; Solecki et al. 2011). These studies provide useful estimates of the general range of costs that may be associated with climate change impacts at a regional level. An important limitation of the existing state studies, however, is that these studies are not based on detailed climate hazard and vulnerability assessments, as have been conducted for the ClimAID project for each of eight major sectors. Many of the prior studies also lack detailed stakeholder-based considerations of adaptation options in the cost-benefit estimates.

In a few cases, estimates of the overall benefits of adaptation to climate change have been made. A leading example is in Parry et al. (2009, Ch. 8). Using runs of a simulation model, and the assumptions of the Stern Review (2007), the benefits of an invested dollar are estimated at $58. A more moderate estimate for adaptations to current variability in the United States (Multihazard Mitigation Council, 2005a) gives an overall estimate of $4 in benefits for each dollar invested in adaptation to current hazards. It can be expected that the benefits from adaptation will be significant in New York State. This is for two reasons: first, New York State is a coastal state, with enormous assets in the coastal counties that are at risk from sea level rise and storm surge; and, second, throughout the state, and not just in coastal areas, relatively little has been done by way of adaptation, so many favorable opportunities for adaptations with significant returns can be expected.

A third category of economic cost studies entails highly detailed analysis of one type of impact or adaptation option for a particular sector within a specific region. For example, a study by Scott et al. (2008) explores the potential costs associated with loss of snowpack in the Adirondacks for snow-dependent tourism industries in the region. These types of detailed studies, which are relatively scarce for New York State, help to inform estimates of the costs associated with specific impacts and adaptations in each sector.

Key terms and concepts

In discussing costs associated with impacts and adaptation to climate change, there are several types of costs that may be considered, as listed in Table 1.2. This study focuses primarily on identification of direct impact costs and direct adaptation costs (and benefits) (see Table 1.2).

Table 1.2. Defining different types of costs

Direct costs. The costs that are incurred as the direct economic outcome of a specific climate event or facet of change. Direct costs can be measured as by standard methods of national income accounting, including lost production and loss of value to consumers.
Indirect costs. The costs that are incurred as secondary outcomes of the direct costs of a specific event or facet of climate. For example, jobs lost in firms that provide inputs to a firm that is directly harmed by climate change.
Impact costs. The direct costs associated with the impacts of climate change (e.g., the reduction in milk produced by dairy cows due to heat stress higher mean temperatures and humidity under climate change.)
Adaptation costs. The direct costs associated with adapting to the impacts of climate change (e.g., the cost of cooling dairy barn to reduce heat stress on dairy cows).
Costs of residual damage. The direct costs of impacts that cannot be avoided through adaptation measures (e.g., reductions in milk production due to heat stress that may occur if cooling capacity is exceeded).

A discussion of adaptation costs, avoided damages, and residual damages both at a single point in time and over time is in Parry et al. (2009). In their discussion, these authors suggest that the costs of avoiding damage tend to increase in a non-linear fashion, becoming substantially higher depending on how much damage is avoided. Adaptation to the first 10% of damage will likely be disproportionately cheaper than adaptation to 90% of damage (Parry et al. 2009, p. 12). It is also important to recognize that while adaptation can reduce some damage, it is likely that damage will occur even with adaptation measures in place. This is particularly true over the long term, as both impacts and costs of adaptation increase.

Benefit-cost analysis, the statewide assessment and public policy

This study draws some insights from the approach of benefit-cost analysis, which has been developed over many years. The first use of the approach that required that project benefits exceed costs was embodied in the Flood Control Act of 1936 (United States Congress, 1936). Following World War II, standard economic benefit-cost analysis methods were developed and, by the early 1960s were widely accepted (Krutilla and Eckstein, 1958; Eckstein, 1958). This was followed by the development of methods for assessing non-economic as well as economic objectives (Maass et al., 1962; Marglin, 1967; Dasgupta et al., 1972; Major, 1977).

At the project level, benefit-cost analysis consists of identifying the stream of benefits and costs over time for each configuration of a project (such as a dam to control flooding), bringing these back to present value by means of an interest rate (discounting), and then choosing the project configuration that yields the maximum net benefits. This approach, widely used by the World Bank and other agencies for project analysis (Gittinger, 1972 is a classic World Bank example), embodies a range of (sometimes debatable) assumptions about the meaning of economic costs and benefits and the value of these over time (see Dasgupta et al., 1972 for an excellent evaluation of these issues). The benefit-cost approach has proven its utility as a framing method, and where benefit and cost estimates are good, relatively robust conclusions can be drawn about optimal project configuration, or, more specifically for the subject of this report, optimal adaptation design. On the other hand, the approach can be misused or used ineffectively; the quality of the work must be judged on a case-by-case basis. A further issue with benefit-cost analysis as usually employed is that it does not typically capture the sometimes extensive delays in design and implementation of measures in the public sector, which can lead to inappropriate choice of designs because projects are designed for the wrong level of climate change. Benefit-cost analysis has two roles in this study. First, the relatively few available benefit-cost studies are described in each of the chapters to help develop an overview of climate change impacts and adaptations in each sector. Second, the method is used as a framing device for the sectoral elements for which general estimates of future benefits and costs over the planning horizon can be made.

A more general issue is whether economic benefit-cost analysis should serve as the basis for public decisions in circumstances such as climate change in which potentially extreme outcomes are not captured by the method. Stern (2009, ch. 5) presents a carefully argued case for using ethical values beyond the market when dealing with climate change. Weitzman (2009) suggests (in response to Nordhaus 2009) that standard cost-benefit analyses of climate change are limited as guides for public policy because deep structural uncertainties about climate extremes render the technique inappropriate for decision-making. These uncertainties include: the implications of GHG concentrations of CO_2 outside of the long ice core record; the uncertainty of climate (temperature) sensitivity to unprecedented increases in CO_2; potential feedbacks exacerbating warming (e.g., release of methane in permafrost); and the uncertainty in extrapolating damages from warming from current information. Taken together, these factors suggest that although formal benefit-cost analysis can be helpful in some respects, it brings with it the danger of "undue reliance on subjective judgments about the probabilities and welfare impacts of extreme events" (Weitzman 2009, p. 15). While these arguments have typically been made at the global level, they are relevant for jurisdictions such as New York State that face potentially very large impacts from climate change; public decision-making efforts must go beyond the information presented in standard economic benefit-cost analysis.

At the same time, agencies should make use of the conceptual framework of benefit-cost analysis (for example in detailed studies comparing the cost of adaptations during the rehabilitation cycle with later stand-alone adaptations) where this approach is helpful. An example of adaptation relevant to New York State is the implementation of adaptations for wastewater treatment plants during rehabilitation, rather than the more expensive attempt to add on adaptations when climate change occurs. Appropriate studies for other issues can help substantially in determining how to schedule adaptations intended to achieve broad public policy goals; many such studies are needed.

Interest rates

In detailed studies, the interest rate is a key element in assessing future benefits and costs from climate change, because the present value of such effects can change greatly depending on the value of the interest rate. (The limitations of standard cost-benefit analysis for climate change have been addressed in significant part through discussions of the interest rate, i.e., the intertemporal weighting assigned to future events). There are advocates for low social rates of discount, most notably Stern (2007) as well as more standard opportunity cost rates (Nordhaus, 2007). Higher interest rates have the effect of postponing action on climate change, as future benefits are more heavily discounted. Stern (2009) argues persuasively that the risks of inaction are quite high (and largely uncertain or unknown), when compared to the costs of action (about 1-2% of GDP for several decades; Stern (2009, p. 90). The use of higher interest rates carries the implicit assumption that actions are reversible, which they are likely not to be in transformative conditions such as climate change.

A practical alternative for the interest rate currently available is for decision-makers to consider the consequences for decisions of using a range of interest rates from low to high. The Stern report uses very low interest rates—a range of 1-2%; market rates can range upward from 8% (Stern, 2007). In this report, interest rates are embodied in many of the available case studies. The estimates for elements of sectors use estimates of GDP growth rates, as discussed below in Section 1.4, but are not discounted back to the present. (The actual estimated values per benchmark year are given instead.) A recent report on the economics of adaptation to climate change suggests the use of sensitivity analysis on the interest rate (Margulis et al. 2008, p. 9). It is also important to note that while methods for integrating a social rate of discount (i.e. a socially-determined interest rate, rather than a market rate) with shadow pricing (an estimate of true opportunity cost) for private sector investments foregone have long been available (Dasgupta et al., 1972), shadow pricing has not been developed to confront the significant uncertainty of climate change.

Use of analogs

Ideally, a study such as this could provide a broad assessment of the costs of climate change impacts and adaptations based only on detailed studies in New York State. In fact, some examples of the economic costs of climate impacts and adaptations are available from cases in New York State, including a few cases in the main ClimAID report, and these are used where possible. However, because the detailed study of the economics of climate impacts and adaptations is relatively recent, there are not enough examples from New York State alone to provide a wide assessment of costs. Nonetheless, a larger range of examples of the economic costs of climate impacts and adaptations is available from other states, cities and countries. Some of these examples are relevant, and often quite analogous to, the types of climate change costs and adaptations that might be expected in New York State. Cost estimates from such cases are used in this study. In addition, there is another group of cases, both from New York State and elsewhere, that relate to adaptations to current climate variability rather than to climate change. These can often also be used to estimate costs for the same or analogous adaptations to climate change, and they are so used in this study as well. Both of these cases are representative of the "Value Transfer Method" (Costanza et al., 2006), in which values from other studies that are deemed appropriate are used for a new study. A further point is that processes for planning infrastructure are broadly the same across many sectors (Goodman and Hastak, 2006). By extension, information on planning climate change adaptations from one sector can be helpful in considering some elements of adaptation in other sectors.

Classifying impacts and adaptations

Thus, as part of the basis of the study, several classes of impacts and adaptations were reviewed and extended to the extent possible.

Impacts.

1. Impacts where good cost estimates exist, either in New York State or elsewhere;

2. Impacts where cost estimates can be obtained or extended within the resources of the project;

3. Impacts where cost estimates could be obtained with a reasonable expenditure of additional resources for new empirical analysis beyond the scope of this project. In such cases it is sometimes possible to describe the general size of costs; and

4. Impacts where it would be very difficult to estimate costs even with large expenditures of resources.

For some impacts, estimates can be made about the time period during which they will be felt, and thus some information is provided about the potential effects of discounting on these costs.

Adaptations. These can be specifically for climate change, but also can be for existing extreme events while being applicable to climate change.

1. Adaptations where good cost estimates exist, either in New York State or elsewhere. In some cases, benefits will be available as well;

2. Adaptations where cost estimates can be obtained within the resources of the project; in some cases benefit estimates can also be obtained;

3. Adaptations where cost estimates may be obtained with reasonable expenditure of resources for new analysis beyond the scope of this project. In such cases it is possible that the general size of costs can be described. This can sometimes also be true for benefits; and

4. Adaptations where it would be very difficult to estimate costs even with large expenditures of resources.

Adaptations can occur at any point over the time horizon of a project, and therefore their costs will also be subject to discounting. However, in many cases, adaptations will occur in the near term and therefore the effect of discounting will be relatively small, especially if low rates of interest are used.

As noted above, for each of the ClimAID sectors, a specific benefit-cost analysis is applied to a major sector element and a related adaptation strategy. For other impacts and adaptations, the extent to which examples of the eight cases described above have been found and analyzed is described in the chapter texts; where possible generalizations are made about the overall level of impact and adaptation costs and benefits for each sector.

1.3 Study Methods and Data Sources

The study design entailed six interrelated tasks. Each of these tasks was performed for each of the eight ClimAID sectors. The tasks entailed the following general sequence of activities:

Step 1: Identification of Key Economic Components

Drawing upon the sectoral knowledge and expertise of the ClimAID sector leaders and teams and recent studies of the economic costs of climate change (e.g., CIER 2007; Parry et al. 2009, Agrawala and Fankhauser 2008), this step entailed description of the major economic components of each ClimAID sector that are potentially vulnerable to the impacts of climate change (e.g., the built environment in the Ocean Coastal Zones sector). The information developed in this step is used to guide the remainder of the analysis for each sector.

Methods for this step included review of existing New York State economic data, compilation of data on economic value of the key components in each sector, and the use of a survey instrument developed for the research group's related study in New Jersey (Solecki et al., forthcoming) as the basis for interviews with sector leaders. The survey instrument includes questions about the key economic components of each sector and, for Steps 2-4 below, the sensitivity of those components to climate change and the potential costs associated with those sensitivities. Estimates of the value of production, employment, and/or assets in each sector were developed based on review of existing New York State economic data from the U.S. Economic Census, the Census of Agriculture, the Bureau of Economic Analysis, and other sources specific to each sector.

Step 2: Identification of Climate Impacts

Drawing upon on knowledge developed by the ClimAID sector team and other New York State experts, as well as current literature on the sectoral impacts of climate change (e.g., NPCC 2010 for infrastructure; Kirshen et al. (2006) and Kirshen (2007) for the Water Sector), the second step entailed identification of the facets of climate change (e.g., flood frequency, heat waves, sea level rise) that are likely to have significant impacts on the key economic components of each sector (as identified in Step 1). Methods used include developing a climate sensitivity list for each sector based on review of existing sectoral literature, New York State documents, ClimAID materials, results of interviews with ClimAID Sector Leaders (SLs), and consultation with ClimAID team members and other New York State experts.

Step 3: Assessment of Climate and Economic Sensitivity

The third step entailed further refinement of the climate sensitivity matrix developed for each sector in order to specify which climate-related changes identified in Step 2 will have the most significant potential costs for the key economic components of each sector. The step draws from the risk-based approach used in the NPCC (see Yohe and Leichenko 2010) to identify which economic components in each sector are most at risk from climate change (i.e., which components have highest value and/or largest probability of impact). In addition to results of the interviews as discussed above, this step also draws from the findings of NPCC (2010) and other relevant studies of the costs of adaptation to climate change (e.g., Parry et al. 2009; Agrawala and Fankhauser 2008).

Step 4: Assessment of Economic Impacts

This step entailed estimation, to the extent permitted by the available data, of the range and value of possible economic impacts based on the definition of the most important economic components and potential climate-related changes (Steps 1-3). Impacts are defined as direct costs that will be incurred as the result of climate change, assuming that the sector is operating in a "business as usual" frame and is not taking specific steps to adapt to climate change. Methods include evaluation of "bottom-up" results from ClimAID case study data where available, New York State economic data, and other economic data, and analysis of "top-down" data from the interviews with SLs and other experts. The estimates are quantitative where possible and qualitative where the data do not permit suitable quantitative estimates. The aim in both cases is to provide the best available information to decision makers. For each sector, available data is assessed for quality and comprehensiveness, supplemented where possible, and extended on an estimated basis to future time periods. In each case, costs for sector components are estimated and checked against other sources where possible. The uncertainties relating to the estimates are also discussed.

Step 5: Assessment of Adaptation Costs and Benefits

The next step entailed estimation of the costs and benefits of a range of adaptations based on the ClimAID sector reports and available case studies. The costs of adaptation are defined as the direct costs associated with implementing specific adaptation measures. Once adaptation measures are put into place, it is expected that some sectors will still incur some direct costs associated with climate change (i.e., residual damage). These costs are defined as the costs of impacts after adaptation measures have been implemented (see Table 1.2). The work in this step is framed using the standard concepts of benefit-cost analysis, with full recognition of the limitations of these techniques under the uncertainties inherent in climate change (Weitzman, 2009). This framework is combined with ideas of flexible adaptation pathways to emphasize the range of policy options available. Methods for this step include combining extrapolated case study information (see the next section) and results from interviews with SLs and other experts and identifying and assessing the relevance of other adaptation cost and benefit studies.

Step 6: Identification of Knowledge Gaps

The final step entails identification of gaps in knowledge and recommends further economic analyses, based on assessments of work in Steps 1-5.

1.4 Benefit-Cost Analyses Methods Summary

This study emerged based on a recognized need for additional information on the economic costs associated with climate change both in terms of the costs of the potential impacts and the costs and benefits of various adaptation strategies. The process described here provides a specific estimate of benefits and costs for a major component of each ClimAID sector as well as the broader-scale overview of economic impacts and costs of adaptations in each chapter. With the information from Steps 1-6, the general method to extrapolate costs and benefits used was first to identify current climate impact costs for a key component of each sector, and then to project these into the future, generally using a real growth rate for GDP of 2.4%. This value is a conservative estimate of the future long-term growth rate of the U.S. economy, which was 2.5% between 1990 and 2010 (see United States Department of Commerce, Bureau of Economic Analysis, n.d.). The estimate of 2.4% can be taken as a central tendency around which sensitivity analyses could be performed. It should be noted that this procedure does not capture possible climate feedbacks on GDP growth, nor does it take into account the potential impacts of climate change adaptation and mitigation efforts. Rather the approach provides general estimates of future costs without climate change based on reasonable assumptions applicable to each sector. Next, specific climate scenario elements from ClimAID are applied to estimate costs with climate change. Then, estimates of adaptation costs based on information in the text are made, as well as estimates of costs avoided (benefits).

This assessment takes into account in a broad way the with and without principle—identifying those sectors in which climate change adaptations are likely to be made as part of general sector reinvestment, whether or not there are specific adaptation programs in effect. Benefit estimates are from available literature on adaptation. The results are plausible scenarios that yield information on the magnitude of the figures involved, and that are reasonably resilient to changes in input assumptions. To illustrate the potential range of variation, key elements of the input assumptions have been varied, and the results are described in each chapter text.

While the economic costs estimates for impacts and adaptations are approximate, both because of data uncertainties and because they deal with future events, they nonetheless provide a useful starting point for prioritization of adaptation options in the state. The approach used represents a generalized framework that could be applied in a more comprehensive analysis. It should be recognized that the further out in time that the forecasts or extrapolations go, the less reliable they are. Other issues that impinge on the usefulness of these types of analytic tools in climate impact assessment include irreversibility, uncertainty (noted above in the discussion of benefit-cost analysis), and the associated possibility of non-linear or catastrophic changes. A further point is that the procedures used, tailored to each sector, differ, and thus the benefit and cost estimates for the various sectors are not strictly additive. Taken together, however, they give a general picture of the potential impacts and adaptation costs that New York State faces over the next century.

2 Water Resources

The water resources sector in New York State is an essential part of the economy and culture of the state. With its many outputs, such as water supply and flood control, and organizations both public and private, it is a complex sector. The principal impacts expected from climate change will be on various types of infrastructure that will be subject to increased risks from flooding as sea levels rise as well as significant impacts from droughts and inland flooding. These impacts, without adaptation, are likely to be at least in the tens of billions of dollars. There is a wide range of adaptations that is available in the water sector, including many that are contemplated now for current variability and dependability. The largest adaptation costs are likely to be those for wastewater treatment, water supply, and sewer systems.

PART I. KEY ECONOMIC RISKS AND VULNERABILITIES AND BENEFIT-COST ANALYSIS FOR WATER RESOURCES

Key Economic Risks and Vulnerabilities

Of the many risks and vulnerabilities, the most economically important include the risks to coastal infrastructure, including wastewater treatment plants and water supply systems (ground and surface) from rising sea levels and associated storm surges. Inland flooding statewide is also an important economic risk; Figure 2.1 shows the location of some of the state's wastewater treatment plants within the current 100 year flood zone. Other economically important risks and vulnerabilities include the costs of droughts of potentially increased size and frequency, losses in hydropower production, and increased costs of water quality treatment. A loss of power can be costly in both economic and regulatory terms to water supply and wastewater treatment plants; on August 14, 2003, the blackout covering much of the Northeast caused shutdowns in the New York City Department of Environmental Protection's (NYCDEP) Red Hook and North River wastewater treatment plants, resulting in the discharge of untreated waters into New York Harbor. The resulting violations brought legal action by the United States Attorney's Office for the Southern District of New York (New York City Municipal Water Finance Authority [NYCMWFA], 2009, p. 54).

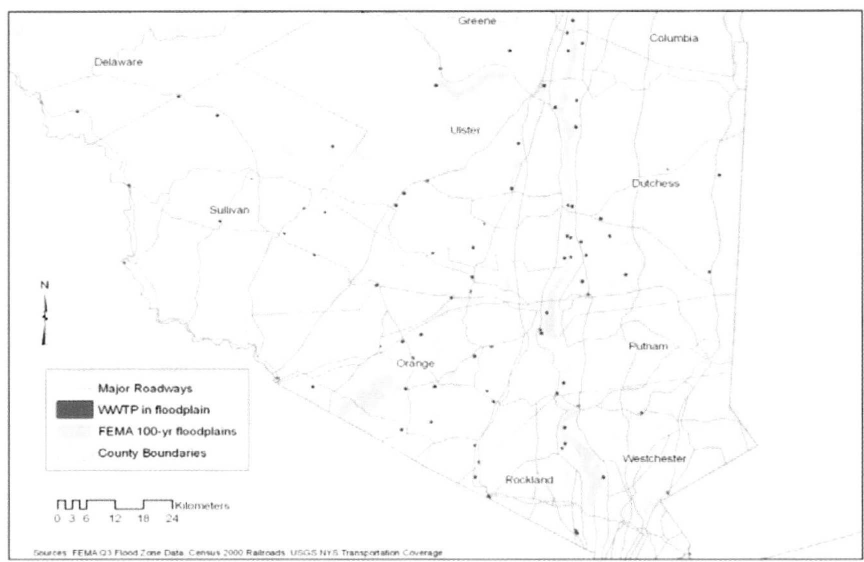

Figure 2.1. WWTPs in close proximity to floodplains in the Hudson Valley and Catskill Region. WWTPs along the Hudson are at risk from sea level rise and accompanying storm search.

One challenge in estimating future damages resulting from climate change is that the recurrence intervals of serious floods and droughts will become more difficult to estimate (Milly et al., 2008), and historical records will no longer be suitable as the sole basis for planning. The expected changes in the non-hydrologic drivers of floods and drought (e.g., development, population increases, and income growth) must also be taken into account.

The main relationships of climate and economic sensitivity in the water sector in New York State are shown in Table 2.1.

Table 2.1. Climate and Economic Sensitivity Matrix: Water Resources Sector (Values in $2010 US.)

Element	Main climate variables: Temperature	Main climate variables: Precipitation	Main climate variables: Extreme events: heat	Main climate variables: Sea level rise & storm surge	Economic risks and opportunities: – is Risk + is Opportunity	Annual incremental impact costs of climate change at mid-century, without adaptation	Annual incremental adaptation costs and benefits of climate change at mid-century
Coastal flooding		●		●	– Damage to wastewater treatment plants – Blockage from SLR of system outfalls – Salt water intrusion into aquifers	Coastal flooding of WWTPs $116-203M	Costs: $47M Benefits: $186M
Inland flooding	●	●			– Increased runoff leading to water quality problems – Damage in inland infrastructure	High direct costs Statewide estimated $237M in 2010.	Restore natural flood area; decrease permeable surfaces; possible use of levees; control turbidity
Urban flooding		●			– Drainage system capacity exceeded; CSOs – Damage to infrastructure	Violation of standards	Very high costs of restructuring drainage systems
Droughts	●	●			– Reduction in available supplies to consumers – Loss of hydroelectric generation – Impacts on agricultural productivity	1960s drought in NYC system reduced surface safe yield from 1800 mgd to 1290 mgd	Increased redundancy and interconnected-ness costs for irrigation equipment
Power outages	●	●	●		– Loss of functionality of wastewater treatment plants and other facilities	Violation of standards	Flood walls
Total estimated costs of key elements						$353-440M	Costs: $47M Benefits: $186M

(See Technical Notes at end of chapter for details. Total flooding costs are calculated minus an allowance for WWTP costs.)

Key for color-coding:

(dark gray)	**Analyzed example**
(light gray)	**From literature**
(white)	**Qualitative information**
(white)	**Unknown**

The costs of climate change are expected to be substantial in the water sector, both for upland systems and for those parts of the system, such as drainage and wastewater treatment plants

(WWTPs), located near coastal area. An estimate for climate change impacts resulting from increased flooding of coastal WWTPs is given in Table 2.2; details of the calculation are in the technical notes at the end of this chapter. While these costs are expected to be significant, they will be just a part of total impacts costs for the water sector, which will be quite high. These costs will include the cost of infrastructure for improving system resilience and intersystem linkages, the costs of drought (both to consumers and water agencies), and the increased costs of maintaining water quality standards with changing temperature and precipitation patterns. Adaptation costs for the sector will also be higher than what is presented in the table and will include costs for adaptation of urban drainage and sewer systems, the costs of managing droughts, and the costs of preventing inland flooding. However, it is important to note that much of the drainage, wastewater and water supply infrastructure in New York is antiquated and inadequately maintained, with an estimated cost for upgrades of tens of billions of dollars. An important policy opportunity would be to use the need for infrastructure improvement as a simultaneous chance to adapt to anticipated climate change impacts, thereby reducing future risk and saving water currently lost through leaks or inefficient operations.

Table 2.2. Illustrative Key Impacts and Adaptations: Water Resources Sector (Values in $2010 US.)

Element	Timeslice	Annual costs of current and future climate hazards without climate change ($M)[1]	Annual incremental costs of climate change impacts, without adaptation ($M)[2]	Annual costs of adaptation ($M)[3]	Annual benefits of adaptation ($M)[4]
All New York State wastewater treatment plant damages from 100 year coastal event	Baseline	$100	-	-	-
	2020s	$143	$14-$43	$23	$91
	2050s	$291	$116-$203	$47	$186
	2080s	$592	$415-$533	$95	$379

[1] *Based on the most recent approximate 100 year WWTP flooding event (Nashville) and estimated repair costs, scaled up by population for New York City, Nassau, Suffolk, and 10% of Westchester (to represent lessened flooding risks there and up the Hudson). Growth in cost is scaled by US long term GDP growth of 2.4%.*

[2] *Ranges are based on changing flood recurrence intervals from NPCC (2010) p. 172.*

[3] *Costs are based on Rockaway WWTP total retrofit estimate, annualized and scaled up for New York City capacity and scaled up by Nassau, Suffolk and Westchester (10%) population.*

[4] *Benefits are based on the empirically-grounded benefit to cost ratio of 4:1 from Multihazard Mitigation Council (2005a) and the reference in Jacob et al. (forthcoming-a).*

Results

As the example of Table 2.2 indicates, costs of impacts may be large; adaptations are available, and their benefits may be substantial. While the numbers in the example depend on the input assumptions, within a fairly wide set of assumptions the magnitude will be in the same range. As other examples in the sector where climate change impacts are expected to be substantial, upstate WWTPs will be subject to flooding, and water supply systems will be subject to increased droughts as climate change progresses.

PART II. BACKGROUND

2.1 Water Resources in New York State

The water resource systems of New York State are many and complex, with a range of system outputs. These resources are abundant: New York State averages almost 40 inches of rain per year, and it is bordered by large fresh water lakes: Erie, Ontario, and Champlain. The outputs of New York State water systems include public water supply; industrial self-supply; cooling water for power plants; hydroelectric energy production; irrigation for agricultural and non-agricultural uses; dams for flood control; water-based recreation; flood control; water quality; wastewater treatment; instream flows for ecological systems preservation; and navigation. The sector has many components, reflecting the diversity of outputs: water supply utilities; wastewater treatment plants; agricultural and industry self-supply systems; hydroelectric generating stations; water-based recreation facilities; canals and navigable rivers; and wetlands and other ecological sites affected by water systems. The most important element of the sector to most citizens is probably public water supply. Schneider et al. (forthcoming) deals primarily with flooding, drinking water supply, water for commercial uses (mainly agriculture and hydropower), and water quality. This chapter uses examples from these and other system outputs.

Because of the number and variety of outputs of water systems, "water" is not a category in the North American Industrial Classification System (NAICS) (United States Bureau of Economic Analysis, n.d.); rather, the values of water system outputs are distributed among industries, utilities, government, transportation and others. Despite this diversity, the water sector has, particularly with regard to projects with Federal participation, a unifying factor: the application of multipurpose economic benefit-cost analysis. The water resources sector was among the first in which benefit-cost analysis was required (United States Congress, 1936), and relatively standard economic benefit-cost analysis methods had been developed by the early 1960s (Krutilla and Eckstein, 1958; Eckstein, 1958), followed by the development of methods for assessing non-economic as well as economic objectives (Maass et al., 1962). With this background, and because water systems deal with natural variability, there is a base of information that can be used to estimate more fully the impact and adaptation costs in the water sector brought about by a changing climate.

To focus just on water supply in the state's large and complex water sector, the state's water utilities vary widely in sources, public/private operations, and size. The largest in the state, the New York City Water Supply System (Figure 2.1), serves a population of more than 9 million people in New York City and upstate counties, nearly half of the state's population. The sources of supply are upland reservoirs in the Croton, Catskill, and Delaware Systems. The NYCDEP has already embarked on significant climate change activities (Rosenzweig et al., 2007b; NYCDEP, 2008). Other New York State utilities use a wide variety of sources: Poughkeepsie, drawing from the Hudson, Long Island utilities using groundwater; and Buffalo, drawing from Lake Erie. There are also many small suppliers in New York State, for which the New York Rural Water Association provides an umbrella organization. Some suppliers are public entities; others are private, and some public utilities have contracts with private water firms to manage their facilities. These New York State utilities face a wide variety of climate challenges, as exemplified in NPCC (2010). For all these reasons, New York State water utilities provide a range of challenges and opportunities in climate risk management. It is of interest that water resource utilities were among the first industries to be concerned with the impacts of climate change (Miller and Yates, 2005).

In addition to considerations of planning and management within the state, there are interstate and international institutional considerations affecting water supply in New York State, such as the Delaware River Basin Commission (DRBC) and the Great Lakes Basin Commission. Water utilities are regulated by a variety of laws and rules (Sussman and Major, 2010), including the Clean Water Act. While it is challenging to estimate the capital value of water utility infrastructure throughout the state, an idea of the size of this part of the sector can be gathered by considering that the NYCDEP's capital program for 2010 through 2019 is just over $14 billion (NYCMWFA, 2009, p. 24).

2.2 Key Climate Change Sensitivities

There is a very large range of potential impacts of climate change on the state's water resources from the principal climate drivers of rising temperatures, rising sea levels, higher storm surges, changing precipitation patterns, and changes in extreme events such as floods and droughts. These are described in detail in Schneider et al. (forthcoming); a comprehensive list for the nation as a whole is in Lettenmaier et al. (2008). Some of the most significant are presented in Table 2.3.

Table 2.3. Key Climate Change Sensitivities: Water Resources Sector

Impacts of rising sea levels, and the associated storm surges and flooding, on the water resources and water resources infrastructure in the state in coastal areas, including aquifers, wastewater treatment plants, and distribution systems.
Potentially more frequent and intense precipitation leading to inland flooding and more runoff and potential water quality problems in reservoirs.
Rising temperatures and potential changes in the distribution of precipitation leading to increases in the frequency and severity of droughts.
Potentially more intense precipitation events leading to increased urban flooding.
An intersectoral vulnerability is the loss of power, which shuts down pumping stations and wastewater treatment plants that do not have adequate back-up generation facilities.

2.3 Impact Costs

In estimating the costs of climate change in the water sector in New York State, relatively standard methods can be applied; however, data are often inadequate and the uncertainties in the future climate are large, compounded by uncertainties in other drivers such as population and real income growth. Nevertheless, in many cases costs or level of magnitude of costs have been estimated or could be obtained with reasonable additional effort.

As an example, the costs of sea level rise and storm surge on the water supply and wastewater treatment systems of Charlottetown, Prince Edward Island, have been estimated (McCulloch et al., 2002). Charlottetown, the provincial capital, has a population of some 32,000, and is therefore similar in size to many New York State coastal towns and smaller cities. A storm that generated a maximum height of 4.23 m above Chart Datum was used for the study. (The Chart Datum is the lowest theoretical astronomical tide at a site.) Under the hypothesized conditions, the replacement costs of the water, sanitary, and storm pipes, lift stations, sewage treatment plant and related infrastructure impacted were estimated to be $13.5 million Canadian (about $26 million US adjusted for inflation and exchange rates) (McCulloch et al., 2002). Because smaller coastal cities in New York State have similar infrastructure at low elevations, this suggests large climate impacts in the aggregate for coastal municipal water supply systems in New York State, bolstering the example in Table 2.2.

There are potential impacts of climate change on water resources in New York State that could be substantially larger. Very significant cost impacts on wastewater treatment plants and sewer system outfalls can be expected as sea level rises. Sea level rise will cause the salt water front in the Hudson to move northward; under some scenarios, this would require the repositioning of the intakes for the City's Chelsea Pump Station and the Poughkeepsie water supply system. (Cost estimates for these impacts are not available.) In the Delaware, there could be substantial institutional and operating costs relating to the integrated operation of the river with the New York City water supply system, which releases specified flows to the river from its Delaware watershed reservoirs (Major and Goldberg, 2001) which might have to be modified over time as new infrastructure came on line for Philadelphia. (This could potentially include complex legal issues, as flows are currently regulated by U.S. Supreme Court rulings.)

Other impact costs will relate to precipitation changes and increased evapotranspiration that can lead both to more intense precipitation and more droughts. More intense precipitation could bring about increased turbidity in New York City's watersheds. In this case, turbidity control measures could be brought to bear, for example utilizing the Croton System more effectively to minimize use of the Catskill System during turbidity events. With respect to droughts, should droughts increase in frequency and intensity toward the end of the century, as is widely expected, costs could reach significant amounts both for losses to water system consumers and for emergency measures. Estimating the current value of such impacts is challenging. The recurrence intervals of the drought of record and more serious droughts are difficult to estimate, given both the loss of stationarity incumbent upon climate change, and the expected changes in the non-hydrologic drivers of population and income growth. Droughts will impact the availability of water for a variety of sectors including household supply, including irrigation for agriculture.

Another impact of precipitation changes could be increased inland flooding of towns, cities, and other areas. Considering just the issue of wastewater treatment, many of the state's wastewater treatment plants are located in areas subject to inland flooding (Figure 2.1). As for damages to all sectors in one basin, flooding in 2006 in the Susquehanna Basin caused estimated damages of $54 million (Schneider et al. (forthcoming). Interpreting this figure, the estimate may be too low for future storms if these become more frequent and/or intense; the additional costs would be attributable to climate change. In addition, asset values may increase over time, which will increase the costs of such climate-related precipitation changes.

A cost estimate for flooding in a neighboring state is of interest in this regard. In 1999, there was an estimated $80 million in damages from flooding in the Green Brook sub-basin of the Raritan. This sub-basin is continually subject to severe and sometimes devastating flood damage (United States Army Corps of Engineers [USACE], n.d.). If there are more frequent and intense rainfall events with climate change, as many observers expect, such damages will be larger and/or occur more frequently and will therefore be an economic consequence of climate change. While the aggregate future dollar values have not been estimated, is seems clear that flooding impacts from climate change in New York, as in its neighbors, could be quite large.

2.4 Adaptation Costs

There is a wide range of potential adaptations to the impacts of climate change on water resource systems; these can be divided into adaptations for: management and operations; infrastructure investment; and policy. Adaptations can also be classified as short-, medium- and long-term. Costs vary substantially among different types of adaptations; and the adaptations need to be staged, and integrated with the capital replacement and rehabilitation cycles (Major and O'Grady, 2010). There has begun to be a substantial number of studies of estimating the costs of adaptations, and in some cases, cost estimates (Parry et al. 2009; Agrawala and Fankhauser, eds., 2008). Several adaptations have been estimated that relate to climate change. As one example relating to planning and research as components of adaptation to climate change, the NYCDEP's study of the impacts of climate change on its facilities (NYCDEP, 2008b) is expected to cost less than $4 million but at least several million dollars. A second research adaptation to climate that is already in place in NYCDEP is the use of future climate scenarios to study potential needed changes in system operation, using the Department's reservoir operating models (NPCC, 2010, App. B). The costs of a series of model runs over an extended period can be approximated by the cost of a single post-doc employee at NYCDEP hired through a major research university for one year. In 2010, such an employee would be paid $55K, and with benefits and overhead at typical levels the total would be $92K.

Costs for capital adaptations are of course much greater than costs for research and planning. The costs of raising key equipment at the Rockaway Wastewater Treatment Plant are estimated at $30 million; this is an adaptation that will help both with current variability and future sea level rise. Total adaptation costs for coastal wastewater treatment plants and low-lying parts of the water supply and sewer systems are likely to be very large. In addition to the climate change study referenced above, which has not yet begun, the NYCDEP has underway its Dependability Study (NYCDEP, 2008a), which is designed to provide for continuity of service in the event of outage of any component, is considering among other possibilities interconnections with other jurisdictions; increased use of groundwater supplies; increased storage at existing reservoirs; withdrawals and treatment from other surface waters; hydraulic improvement to existing aqueducts and additional tunnels (NYCMWFA, 2009, p. 48). All of these measures, for many of which costs are in process of being estimated, would also be suitable candidate adaptations to climate change. The climate change and Dependability studies together will provide a good basis for estimates of adaptation to climate change in the New York City Water Supply System.

A drought emergency measure for which costs could be re-estimated is the cost of the pipe laid across the George Washington Bridge in 1981 to allow New York City to meet some of its Delaware obligations from its east-of-Hudson watershed (Major and Goldberg, 2001). (A recent search of NYCDEP records was unsuccessful in finding the original costs.) This drought adaptation was explicitly authorized by the Delaware River Basin Commission, and although never used, could be replicated today in appropriate conditions. There is a range of other actual and potential adaptations for which costs have not yet been estimated but for which costs could be estimated from existing information and reasonable forecasts; this is work that should be undertaken in the near future.

The proposed costs for adaptation to current conditions in the Green Brook NJ case are of interest to New York State because the Green Brook area is highly developed, as is the case with some New York State inland riverine areas, and therefore flood characteristics are partly human-created. The United States Army Corps of Engineers (USACE) is planning to spend, including local contributions, $362 million over 10 years to build levees/floodwalls, bridge/road modifications, channel modifications, closure structures, dry detention basins, flood proofing and pump stations in Green Brook (USACE, n.d.). The estimated benefit-cost ratio for this work is 1.2:1. The plan is designed to deal with floods up to the current 150 year recurrence interval in the lower basin and the current 25 year recurrence interval in the upper basin, so that expected damages from floods within these recurrence intervals would be expected to decrease (USACE, n.d.). However, the recurrence intervals of the given floods may be reduced (the floods became more frequent) with climate change, and their intensity may also increase, thus offsetting some of the effects of the proposed adaptations.

2.5 Summary and Knowledge Gaps

From the standpoint of improving the ability of planners to do economic analysis of the costs of impacts and adaptations in the transportation sector, there are many knowledge gaps to which resources can be directed. These include:

- A comprehensive data set in GIS or CAD form of as-located elevations of water system infrastructure

- Updating of FEMA and other flood maps to reflect the impacts of rising sea levels.

- Undertaking of a series of comprehensive benefit-cost analysis of potential adaptations to aid in long term planning, building upon current studies of the NYC system and other systems.

- Developing a comprehensive data base, GIS referenced, on the condition of water infrastructure projects across the state, including wastewater treatment plants, CSOs, and water supply systems which could be used to prioritize and allocate climate adaptation funding as it becomes available.

- Integration of population projections into climate change planning.

- More advanced planning for power outages and their impacts on wastewater treatment plants and other facilities.

Technical Notes – Water Resources Sector

Water extrapolation methods for the text example:

1. The initial annual cost is based on the most recent approximately 100 year event that flooded a WWTP, in Nashville in 2010. The estimated repair costs for the Dry Creek plant are $100 million; the population served by the Dry Creek plant is 112,000 (Nashville Water Services Department, personal communication).

2. These costs were scaled up by population for NYC, Nassau, Suffolk and 10% of Westchester. This gives total costs of 10$B, or annual costs of $100 million over 100 years. Scaling by population rather than number of plants gives a more general estimate of costs.

3. This figure is then extrapolated assuming a US GDP real growth rate of 2.4%.

4. The range of flood recurrence with SLR is then applied to yield the increase in damages; these ranges are based on NPCC (2010), p. 177. Flood damages (because of SLR) become about 10% more frequent in the 2020s, 40% more frequent in the 2050s, and 70% more frequent in the 2080s (NPCC 2010) for the low estimate of SLR, and become about 30% more frequent in the 2020s, 70% more frequent in the 2050s, and 90% more frequent in the 2080s (NPCC 2010) for the low estimate of SLR.

5. To prepare for climate change—and growth—NYC is spending $30 million to raise pumps and other electrical equipment at the Rockaway WWTP plant well above sea level. These costs are used for adaptation costs in the example, annualized and scaled up by capacity for NYC and by population for Nassau and Suffolk and 10% of Westchester.

6. Reductions in impacts (benefits from adaptations) are estimated using the empirically determined 4:1 benefit to cost estimate (from the references in Jacob et al. (forthcoming-a), which is appropriate for infrastructure-intensive sectors.

7. For Table 3.1, the estimated total flooding in the state, estimated at $100 million in $US 2009, is assumed to grow at an annual rate of GDP (2.4%). It is assumed conservatively that 80% of this is unrelated to WWTP flooding, and thus the figures are assumed to be additive.

3 Ocean Coastal Zones

The ocean coastal zone in New York State is an essential part of the economy and culture of the state; with its many economic and natural outputs and governing organizations, it is a complex system. Total losses from climate change on coastal areas (without further adaptation, and excepting transportation, discussed in the Transportation chapter of this report), over the next century will be in the hundreds of billions of dollars, primarily from rising sea levels and the associated higher storm surges and flooding. Adaptations are available to reduce some of these impacts; their costs may be in the tens of billions of dollars, and they will need to be carefully scheduled over the course of the century for maximum effectiveness and efficiency.

PART I. KEY ECONOMIC RISKS AND VULNERABILITIES AND BENEFIT-COST ANALYSIS FOR COASTAL ZONES

Key Economic Risks and Vulnerabilities

Of the many risks and vulnerabilities, the most economically important are the multifaceted risks to coastal zones from higher sea levels and consequent higher storm surges. Substantial economic losses can be expected in buildings, infrastructure, natural areas, and recreation sites. Other impacts from precipitation changes, higher temperatures, higher ocean temperatures and ocean acidification will also have significant impacts. Table 3.1 provides a summary of climate and economic impact categories. The negatives shown substantially outweigh the positives.

Table 3.1. Climate and Economic Sensitivity Matrix: Ocean Coastal Zones Sector (Values in $2010 US.)

Element	Main Climate Variables: Temperature	Main Climate Variables: Precipitation	Main Climate Variables: Sea Level Rise & Storm Surge	Economic risks and opportunities: – is Risk + is Opportunity	Annual incremental impact costs of climate change at mid-century, without adaptation	Annual incremental adaptation costs and benefits of climate change at mid-century
Coastal Flooding (Insured damages)			•	– Significant damage to buildings, transportation, other infrastructure and natural and recreation areas	$44-77M	Costs: $29M Benefits: $116M
Inland flooding and wind damage in coastal areas		•		– Damage from more intense and frequent precipitation events	Comparable to coastal flooding	Emergency evacuation procedures
Salt front			•	– Salt front moving further up the Hudson – Impacts on water intakes – Impacts on natural areas	Moderate costs for water supply; significant impacts on natural areas	Relocation of intakes
Marine ecosystems	•	•	•	– Impacts from higher ocean temperatures – Impacts from increased ocean acidity	Unknown	Need for additional research; global mitigation efforts required
Recreation	•		•	– Loss of some recreation areas + Longer warm season for some types of recreation	Annual cost of loss of 10% of beach area in Nassau/Suffolk estimated as $345M	Beach nourishment
Freshwater sources	•	•	•	– Potential salt water intrusion into aquifers – Water quality problems from heat and turbidity	Unknown	Turbidity management measures
Natural areas	•	•	•	– Recession of wetlands from sea level rise – Damage from more intense storms – Ecosystem changes from heat – Beach and bluff erosion	$49M annually for loss of 10% of natural areas	Mitigation and retreat
Total costs of estimated elements					$416-449	Costs: $29M Benefits: $116M

(See technical notes at the end of the chapter for details of calculations)

Key for color-coding:

	Analyzed example
	From literature
	Qualitative information
	Unknown

The expected costs of climate change on coastal zones in New York State are expected to be very large. An estimate based on extrapolation of insured damages for New York State coastal zone is presented in Table 3.2, with details on methods in the technical notes included in this section. While there are other significant damages, including damages from winds and inland floods, uninsured damages, and damages to self-insured public infrastructure, insured damages are a substantial element in total sector damages.

Table 3.2. Illustrative Key Impacts and Adaptations: Ocean and Coastal Zones Sector (Values in $2010 US.)

Element	Timeslice	Annual costs of current and future climate hazards without climate change ($M)[1]	Annual incremental costs of climate change impacts without adaptation ($M)[2]	Annual costs of adaptation ($M)[3]	Annual benefits of adaptation ($M)[4]
Coastal flooding insured damages[5]	Baseline	$38	-	$10	-
	2020s	$54	$5-$16	$14	$57
	2050s	$110	$44-$77	$29	$116
	2080s	$225	$157-$202	$59	$237

[1] *See the technical notes for the estimation of the baseline and future impacts from insured damages information*

[2] *Based on increased frequency of coastal floods (NPCC,2010, p. 177) for range of climate scenarios*

[3] *Based on potential annual expenditures for building elevation, sea walls, emergency planning, beach nourishment and wetlands management estimated from case studies in the Coastal Zone text, especially Tables 3.6 and 3.7. The total of $10 million is based on the following figures (in millions): building elevation, 2; sea walls 2; emergency management 1; beach nourishment 2; and wetlands management 1. The total assumes no surge barrier construction within the scenario time frame.*

[4] *Based on the empirical 4:1 benefit to cost relationship from Jacob et al. (forthcoming-a) references. Rounding in the calculations results in this relationship being approximate in the table.*

[5] *Insured damages in the example include losses to property from coastal flooding, and in some cases, business interruption losses.*

Results

As the example in Table 3.2 indicates, costs of impacts may be large; adaptations are available, and their benefits may be substantial. While the numbers in the example depend on the input assumptions, within a fairly wide set of assumptions, the magnitude will be in the same range. Furthermore, most public infrastructure, such as the New York City subway system, bridges, and tunnels, is self-insured, so that while it is not included in the insured estimates used for the example the loss potential is large. In addition, although smaller in dollar terms, impacts on natural areas will be substantial.

PART II. BACKGROUND

3.1 The Ocean Coastal Zone in New York State

The ocean coastal zone of New York State comprises parts of the 5 counties of New York City, Nassau, Suffolk and Westchester counties, as well as the counties bordering the Hudson River to Troy Dam, since these too will be impacted by sea level rise. The characteristics of the coastal zone in New York State are very varied. The most striking element is the high level of urban development along the coast in New York City, but there are also many natural coastal features, including coastal and marine ecosystems, beaches, and bluffs. Most of these areas are open to the ocean; in the Hudson Valley, much of the original shoreline has been engineered for railways and other purposes (Buonaiuto et al., forthcoming). Because of the wide range of coastal systems, both impacts and adaptations will vary geographically in the New York State coastal zone. Due to the number and variety of elements in the ocean coastal zone, this sector of ClimAID is not a category in the North American Industrial Classification System (NAICS) (U.S. Bureau of Economic Analysis, n.d.). The values produced by economic activity in the ocean and coastal sector are distributed among a wide variety of industry, government, commercial and private activities. However, a simple metric of economic worth is the total insured value in coastal counties in New York State in 2004. This was nearly 2 trillion dollars: $1,901.6 billion, or 61% of the total insured value in New York State of $3123.6 billion (AIR Worldwide Corporation, 2005). (AIR (2007) reported and estimated $2,378.9 billion of insured coastal exposure in New York State.)

3.2 Key Climate Change Sensitivities

There is a very large range of potential impacts of climate change on the state's ocean coastal zone from the principal climate drivers of rising sea levels, higher storm surges, rising temperatures, changing precipitation patterns, and changes in extreme events such as floods and droughts. Some of the most significant are presented in Table 3.3.

Table 3.3. Key Climate Change Sensitivities: Ocean Coastal Zones Sector

Rising sea levels and the associated storm surges and flooding will impact all coastal areas, including buildings, transportation and other infrastructure, recreation sites and natural areas.
Potentially more frequent and intense precipitation events will cause more inland flooding in coastal areas.
Rising temperatures and potential changes in the distribution of precipitation will impact natural areas.
Higher temperatures will change the use and seasons of recreation areas.
Movement of the salt front up the Hudson as a result of sea level rise will impact both natural areas and water intakes.
Sea level rise may degrade freshwater sources, infrastructure and other facilities through salt water intrusion.
Sea level rise and storm surge will cause beach erosion.
Sea level rise and storm surge will cause bluff and wetland recession.
Rising ocean temperatures will impact marine ecosystems.
Increased ocean acidity will impact marine life.

3.3 Impact Costs

In estimating the costs of climate change on the ocean coastal zone in New York State, relatively standard methods can be applied; however, data are often inadequate and the uncertainties in the future climate are large, compounded by uncertainties in other drivers such as population and real income growth. Nevertheless, in many cases costs or level of magnitude of costs have been estimated.

One approach to estimating the size of impacts of climate change on coastal counties, largely relating to the built environment, is to consider insured losses from storms in New York State. Insured losses for all natural and man-made catastrophic events in the United States are available from Property Claims Services (PCS), a division of Insurance Services Offices, located in Jersey City, NJ. The PCS database covers from 1950 to present day, and insured market losses are available by state, by event and by year. Available in event-year dollars, the insured losses are brought to as-if estimates by assuming a compound annual growth rate of 6.75%.

The three weather perils which drive insured losses in New York State are winter storms (both lake-effect events and nor'easters are included in this category), hurricanes and severe thunderstorms. Nor'easters and hurricanes have the largest impact on coastal regions, while other winter storms and thunderstorms are prevalent throughout the state. Nor'easters/winter storms contribute the most to both annual aggregate losses and event-based losses in New York State; nor'easters can cripple the NYC metro area and significant lake-effect snow events can be highly problematic for Syracuse, Buffalo and Rochester. Due to their infrequent occurrence, hurricanes do not contribute significantly to annual aggregate losses, but do have high event-based losses. The opposite is true with severe thunderstorms; the event-based insured losses caused by severe thunderstorms are not often substantial, but the losses can accrue to a significant amount on an annual basis.

Since 1990, ten years have seen annual aggregate as-if losses in excess of $500 million US. With over $1 billion dollars (2010 as-if) in insured losses, 1992, which featured the December '92 nor'easter, was the costliest year in terms of natural catastrophe loss. Future losses can certainly exceed the historical losses of the most recent 20 years. For example, Pielke et al. (2008, p. 35) adjusted the losses from the 1938 hurricane to account for inflation, changes in population density (and thus exposures) and asset value, and estimated that the 1938 storm, if it occurred today, would cause $39.2 billion (2005 $US) in economic damages.

This information gives insight into the magnitude of potential insured losses from climate events without further adaptation measures. As sea level rises, the probability of any given amount of flooding rises. For example, the same event that causes a 25-year flood today might produce a 10-year flood later in the 20th century when the storm surge impacts are compounded by increased sea level. The incremental increases in flooding and damages at each level (adjusted for population and development changes unrelated to climate change) are therefore attributable to climate change. For example, if the flooding levels from the 1992 storm were replicated once over the coming century, the amount attributable to climate change would be the damages from that storm minus the damages that would have occurred absent SLR. When summed over all storms, this number will be quite large during the coming century, almost certainly in the tens of billion dollars and quite possibly in the hundreds of billion dollars. This number is an estimate of the impacts of storm flooding, and does not consider permanent losses from sea level rise, which will also be very significant.

This approach is useful for the general size of impacts. However, the use of insured loss figures has some limitations that prevent their use as complete estimates of impact. Primarily, the insured loss figures understate total losses because of the substantial amount of uninsured properties and self-insured facilities such as subways, bridges, tunnels, recreation areas, and natural areas. There are also institutional complications that will affect the values of insured property in the future. For example, the federally mandated U.S. National Flood Insurance Program is active in New York. Any residence with a mortgage backed by a federally regulated or insured lender located a in high-risk flood area, defined as an area within the 100 year flood plain, is required to have flood insurance. Homes and businesses located outside the 100-year flood plain are typically not required to have insurance (http://www.floodsmart.gov). The average flood insurance policy costs less than $570/year (http://www.floodsmart.gov), which is regarded as well below a true actuarially based risk premium. Many analysts feel that NFIP (due for reauthorization on September 30, 2011) is unsustainable over the long run, and in the event of a large loss, many insured parties will not be able to receive a payout and the financial burden is then transferred to the tax payers. Many private insurers do not offer personal line flood insurance because they are not able to charge the true rate that would be required.

Another approach to the size of impacts of climate change in the New York State ocean coastal sector relates to ecosystem services, focusing more on natural areas or human-affected natural systems, rather than on the built environment. (This is a subject that overlaps with the analysis of Chapter 4, Ecosystems.) A range of estimates for per-acre annual ecosystem services for different types of ecosystems has been developed for New Jersey (Costanza et al., 2006). Several different approaches to valuation were used; the figures cited here are the so-called "Value Transfer Method" figures, which are essentially figures from existing studies of some relevance to New Jersey. They are relevant to New York also because of the similarity of many coastal zone ecosystems in the two states. The figures used here are from "Type A" studies, the best attested, from either peer-reviewed journal articles or book chapters. Each type of ecosystem has different services. Beaches, for example, are credited with disturbance regulation (buffering from wave action and other effects), esthetic and recreational values, and a smaller component of spiritual and cultural value. For the sum of these services, in $2004, the study gives an annual value of $42,127 per acre per year averaged over the available Type A studies. Salt water wetlands, with services including disturbance regulation, waste treatment, habitat/refugia, esthetic and recreational, and cultural and spiritual, have an average estimated value per acre per year of $6,527. These values should be reasonably applicable to New York State coastal zones, although in order to make firm estimates a wide range of assumptions would have to be examined. To examine impacts (losses of ecosystems and their services) from climate change, the total number of acres estimated to be lost in each category over the coming century would be estimated using flood mapping and other techniques. These and other coastal ecosystem estimates per acre per year are given in Table 3.4 (from Costanza et al. (2006, p. 17).

Table 3.4. Summary of average annual value of ecosystem services per acre for New Jersey, $2004

Coastal Shelf	$620
Beach	$42,147
Estuary	$715
Saltwater Wetland	$6,527

Source: Costanza et al. 2006

The totals for beach losses would be expected to be quite high for New York State coastal zones over the coming century. While of course not all acres would be affected, it is of interest that in 2006 it was estimated that there were 24,320 acres of beach and dune in Nassau and Suffolk Counties, and, from the only available but outdated (and thus probably high) estimates, 23,578 acres of tidal marsh in these two counties (Table 3.4). The estimated costs of losing 10% of each type of ocean landscape using the Costanza et al. (2006) estimates are $102.5 million (2004) year and $15.4 million (2004) year. A project underway by The Nature Conservancy (www.coastalresilience.org) has developed and is now applying a coastal mapping tool that will enable the detailed estimation of losses of coastal landscapes from sea level rise and storm surge over the course of the century for southern Long Island and Long Island Sound.

Table 3.5. Estimated Beach/Dune and Tidal Marsh Acres in Nassau and Suffolk Counties and Impacts of Loss of 10% of Acres

County	Est. Beach/Dune Acres 2006	Est. Tidal Marsh Acres 1974
Nassau	3,420	9,655
Suffolk	20,900	13,923
Totals	24,320	23,578
Annual $2004 impact of losing 10% of estimated acreage	$102.5 million	$15.4 million

Sources: 2006 Beach/Dune, The Nature Conservancy, n.d.; 1974 Tidal Marsh, New York State Department of Environmental Conservation, 1974; loss estimates/acre/year Costanza et al., 2006.

3.4 Adaptation Costs

There is a wide range of potential adaptations to the impacts of climate change on the New York State coastal zone; these can be divided into adaptations for: management and operations; infrastructure investment; and policy. Adaptations can also be classified as short-, medium- and long-term. Costs vary substantially among different types of adaptations; the adaptations need to be staged, and integrated with the capital replacement and rehabilitation cycles (Major and O'Grady, 2010). There has begun to be a substantial number of studies about how to estimate the costs of adaptations, and in some cases, cost estimates (Parry et al. 2009; Agrawala and Fankhauser, eds., 2008). Several adaptations have been estimated that relate to climate change. For coastal zone climate impacts, there will be some losses (e.g. some natural areas) that are essentially unpreventable; for many other losses, some appropriate menu of adaptations that varies over time can be developed. Some of these adaptations for either or both of climate change and current variability are given here, with the figures summarized in Table 3.6.

- Emergency evacuation planning is an emergency management/operations measure that is already in place for current climate variability. The costs of improving this program over time as SLR rises will be relatively small, although they have not yet been estimated, and the benefits are potentially large.

- Some infrastructure costs can be modest. As an example of an adaptation to a long-standing problem with a salt marsh, the separation of a salt marsh on the Connecticut shore of Long Island Sound from the Sound by development is presented in Zentner et al. (2003). The estimated costs/acre for a 10 acre salt marsh where a dike has been breached range from $6,000 to $14,100 depending on the nature of the levees that are constructed to improve the flow of salt water from the sound to the marsh (Zentner et al., 2003, p. 169). This is an example of a type of adjustment for a marsh that could be relevant to some marshes as the sea rises, and is directly relevant to New York State salt marshes, at least those on LI Sound.

- On the other hand, estimates for some wetlands restoration are substantially higher. Like beach nourishment (below), such costs may be more appropriate for the earlier part of the century than later, especially for wetlands that have no retreat route. Estimates from a personal communication (Frank Buonaiuto), suggest a wide variation. In the mid range is the cost of recreating the marsh islands of Jamaica Bay-Elders West, about $10 million for 40 acres ($250,000/acre); for a project at Soundview, including excavation costs, the total would be about $5 million for 4 acres, or $1.25 million/acre.

- An example of adjustment to storms that involves a moderately expensive capital investment for sea walls and other facilities is the proposal for Roosevelt Island in New York City set out by the USACE in its Roosevelt Island Seawall Study and announced by Congresswoman Maloney (Maloney, 2001). The study advocated wall repair (rather than wall replacement that could cost 10 times as much) for the existing seawall, noting particular concern for the northwest shoreline and the eastern sections adjacent to an underground steam tunnel. The estimated cost for this repair work was $2,582,000. Besides repair work, the USACE recommended further testing of the walls and the establishment of a design/maintenance standard for the seawall. To protect the southern shoreline from storms and erosion, the study finds a vinyl sheet pile (a wall of hard plastic anchored into the ground) to be the most cost-effective and environmentally desirable. The estimated cost is $3,640,000, bringing the total cost for seawall maintenance and shore stabilization to $6,222,000.

- More expensive is a common current adaptation to climate variability in coastal zones, beach nourishment. Beach nourishment costs for projects in New York State as well as all coastal states on the East and Gulf coasts are given in NOAA (n.d.). Among projects in New York State in the 1990s are Coney Island (1995), with an estimated project cost of $9 million and a length of 18,340 feet; and Westhampton Beach in Suffolk County (1996), with an estimated cost of $30.7 million and a length of 12,000 ft. Beach nourishment provides a good example of how appropriate adaptations will vary with time. With increasing SLR, beach nourishment is likely to become less attractive, especially in areas with no retreat room for beaches. In addition, as sea level rises beach nourishment can be counterproductive if it encourages increased coastal construction

- An example of large-scale adaptation measures for the coastal zone is the set of surge barriers that have been suggested as a possible protective measure for parts of New York City. These would consist of barriers on the upper East River, the Arthur Kill, and the Narrows, or alternative a larger Gateway system. The hydrologic feasibility of such barriers is studied in Bowman et al. (2005). Preliminary estimates for the NY Harbor barriers given by the designers were $1.5 billion for the upper East River site, $1.1 billion for the Arthur Kill, $6.5 billion for the Narrows barrier, and $5.9 billion for the Gateway barrier system (American Society of Civil Engineers [ASCE], 2009). These options are described in Aerts et al. (2009). According to those authors, "These options are at present only conceptual, and would require very extensive study of feasibility, costs, and environmental and social impacts before being regarded as appropriate for implementation. New York City has high ground in all of the boroughs and could protect against some levels of surge with a combination of local measures (such as flood walls) and evaluation plans; and barriers would not protect against the substantial inland damages from wind and rain that often accompany hurricanes in the New York City region" (Aerts et al., 2009, p. 75). Thus, the barrier costs cannot be directly compared to insured losses of property, because they would only protect against a subset of the surge impacts that will be expected; further detailed study would be required for a full benefit-cost analysis. Moreover, there is no obvious barrier system for Long Island short of Dutch-style dikes protecting large stretches of the region.

Table 3.6. Adaptations to Climate Change/Current Variability, with Locations and Costs

Adaptation	Climate (current or future) and/or other variables	Location	Estimated Cost
Reconnecting a salt marsh	Adapt to development	LI Sound (CT shoreline)	Total cost $60,000 to $141,000 for 10 acres
Wetlands restoration	Sea level, storm surge	Jamaica Bay-Elders West	$10 million for 40 acres
Wetlands restoration	Sea level, storm surge	Soundview	$5 million for 4 acres
Sea wall repair	Sea level, storm surge	Roosevelt Island	$6,222,000
Beach nourishment	Sea level, storm surge	Coney Island (1995)	$9,000,000
Beach nourishment	Sea level, storm surge	Westhampton Beach (1996)	$30,700,000
Storm surge barriers	Sea level, storm surge	New York Harbor	$9.1 billion for 3-barrier system

In considering this set of adaptation examples, it becomes clear that the menu of adaptations for the coastal zone will vary over time and space. There are some adaptations that are reasonable in cost (evacuation planning, sea walls) that are likely to avoid some impact costs in the next few decades. There are other adaptations that are likely to become less appropriate later in the century as beaches and salt marshes are lost; and there may be large-scale infrastructure investment that would be appropriate later in the century and that need to be studied more intensively.

The Multihazard Mitigation Study (2005a) presented a full benefit-cost analysis of FEMA Hazard Mitigation grants, including one set of grants to raise streets and structures in Freeport, NY (pp. 63-64 and 107) to prevent flooding under existing conditions. The analysis for housing elevation is presented here (the street analysis is in the transportation chapter). The total costs were $2.36 million; the grants for raising private structures required local matching funds of 25 %; the match for raising private buildings was paid by the owners. The study examined a wide range of parameter values of benefits and costs, and concluded that the total Freeport benefit-cost ratio best estimate for this adaptation to coastal flooding was 5.7, with a range of 0.18-16.3 (Table 3.7). This provides some sense of what might be required in the future in coastal areas such as Freeport, which of course do not have underground transit lines as does the inner core of the NYMA.

Table 3.7. Costs, Benefits, benefit-cost ratios and ranges for HMGP grant activities in Freeport, NY.

Activity in Freeport, NY	**Total Costs (2002 $M)**	**FEMA Costs (2002 $M)**	**Best Estimate Benefits (2002 $M)**	**Best Estimate Benefit-Cost Ratio**	**BCR Range**
Building Elevation	$2.36	$1.77	$13.5	5.7	0.18-16.3

Source: adapted from: Multihazard Mitigation Council, 2005b, vol. 2

3.5 Summary and Knowledge Gaps

From the standpoint of improving the ability of planners to do economic analysis of the costs of impacts and adaptations in the ocean and coastal sector, there are many knowledge gaps to which resources can be directed. Some of these are similar to recommendations for the transportation sector.

- A comprehensive data set in GIS or CAD form of as-located elevations of coastal infrastructure
- Updating of FEMA and other flood maps for rising sea levels
- A new Department of Environmental Conservation (NYSDEC) study of the amounts of coastal wetland remaining in New York State
- Studies of marsh and beach retreat areas, and the development of a typology of such areas that indicates which are most likely to be protectable with available adaptations
- Evaluation of the relationship of insured property to total property values
- Undertaking of a series of comprehensive benefit-cost analysis of potential adaptations to aid in long term planning.

- Review of local and state planning and environmental regulations to insure that, to the extent possible, they are compatible with and act as drivers of coastal adaptation measures.

Technical Notes – Ocean Coastal Zones Sector

Method for extrapolation of insured damages:

1. To consider plausible future damage figures from coastal flooding, the average insured damages figure for New York State is a starting point. This figure was $440 million (2010 $) for the period from 1990 to 2009. Insured damages in the example include losses to property from coastal flooding, and in some cases, business interruption losses.
2. To estimate 2010 damages, the average was taken at the midpoint (1999) and increased by 2.4% annually, to $545 million.
3. Of insured damages in New York State, about 46% are in coastal counties (2004 figures). Of those damages, 61% are from winter storms and hurricanes, and perhaps one quarter of this is from flooding (the rest is from winds); the damages from flooding and winds are not calculated separately in the data.
4. Applying these factors to the starting point of $545 million in insured damages, the figure applicable to coastal flooding is $38 million.
5. This figure will grow (at 2.4%) as shown in Table 3.2. These are damages without the impact of sea level rise and the consequent increase in flooding at each level.
6. Floods (because of SLR) become about 10% more frequent in the 2020s, 40% more frequent in the 2050s, and 70% more frequent in the 2080s (NPCC 2010) for the low estimate of SLR, and become about 30% more frequent in the 2020s, 70% more frequent in the 2050s, and 90% more frequent in the 2080s (NPCC 2010) for the low estimate of SLR.
7. These factors were applied to the damages in order to yield estimates of the additional flooding damages brought about by climate change. These figures, which are approximations because of topographical considerations for the specified years are given in the table. From these figure for 3 separate years, it will become apparent that total increased damages from coastal flooding over the forecast year will be in the many billions of $US. This conclusion will hold even with sensitivity on the assumptions.
8. Estimated adaptation costs are based on examples in the text for building elevation, sea walls, emergency planning, beach nourishment, and wetlands management.
9. Reductions in impacts (benefits from adaptations) are estimated using the empirically determined 4:1 benefit to cost estimate (references in the ClimAID transportation chapter), which is appropriate for infrastructure-intensive sectors.
10. For Table 3.1, beach and natural area losses are increased by GDP growth (2.4%) annually. These losses and the losses from the insured sector have some overlap, so that the figures are not strictly additive.
11. The insurance industry, which compiles the insured value data cited here, has long been concerned with climate change, as evidenced by the participation of one large company, Swiss Re, in the Economics of Climate Change Working Group (2009).

4 Ecosystems

The ecosystems sector in New York State includes the plants, fish, wildlife, and resources of all natural and managed landscapes in the state. Ecosystem services provided by New York's landscapes include preservation of freshwater quality, flood control, soil conservation and carbon sequestration, biodiversity support, and outdoor recreation (Wolfe and Comstock, forthcoming-a). Climate change is likely to have substantial impacts on the state's ecosystems, yet knowledge about both the precise nature of these impacts and options for adaptation is extremely limited. A further difficulty with economic cost estimates arises because ecosystems have intrinsic, non-market value associated with provision of habitat for many species, and preservation of wild places and heritage sites. Monitoring of the effects of climate change on ecosystem health, including threats from invasive species, and identification of viable adaptation options will be essential for protection of the state's ecosystems. Preservation of critical ecosystem services will also be an important step for minimizing some of the costly impacts of climate change in other sectors in New York State including water resources, agriculture, and public health.

PART I: KEY ECONOMIC RISKS AND VULNERABILITIES AND BENEFIT-COST ANALYSIS FOR ECOSYSTEMS

Key Economic Risks and Vulnerabilities

Climate change will alter baseline environmental conditions in New York State, affecting both ecosystem composition and ecosystem functions. The most economically important components of the ecosystem sector that are at risk from various facets of climate change include impacts on tourism and recreation, forestry and timber, and riparian and wetland areas. While it is possible to estimate the costs associated with climate change impacts for some of the key, revenue-generating facets of the ecosystem sector, such as snow-related recreation, fishing, and timber and forestry production, the impacts of climate change on many other types of ecosystem services, particularly forest-related ecosystem services are presently unknown. Viable options for adaptation within the ecosystems sector and the costs associated with these options are only beginning to be explored.

Information on key economic risks associated with climate change in the ecosystems sector is summarized in the climate and economic sensitivity matrix presented in Table 4.1. Table 4.1 presents mid-century estimates of the impact costs for three illustrative components of the sector including skiing (currently valued at approximately $1 billion/year), snowmobiling (currently valued at approximately $500 million/year), timber (currently valued at $300 million/year), trout fishing (currently valued at $60.5 million/year). Table 4.1 also includes a rough estimate of the impacts of climate change on freshwater wetland ecosystems services (currently valued at $27.7 billion/year).

Table 4.1. Climate and Economic Sensitivity Matrix: Ecosystems Sector (Values in $2010 US.)

Element	Main Climate Variables: Temperature	Precipitation	Extreme Events: rainfall	Sea Level Rise	Atmospheric CO_2	Economic risks and opportunities: – is Risk + is Opportunity	Annual incremental impact costs of climate change at mid-century, without adaptation	Annual incremental adaptation costs and benefits of climate change at mid-century
Outdoor recreation and tourism	●	●				+ Summer tourism with longer season – Winter ski tourism with reduced snowpack – Winter snowmobile tourism with reduced snowpack	$694-844M/yr (winter snowmobiling and skiing loss)	Costs: $54M/yr Benefits: $73M/yr
Freshwater Wetlands and riparian areas			●	●		– Sea level rise and extreme rainfall events threaten viability of coastal riparian areas – Inland wetlands threatened by drought and extreme rainfall events	$358 M/yr (estimated value of the loss of 5 % of ecosystem services)	Unknown
Recreational fishing	●					+ Warm water fishing with higher water temperatures – Cold water fishing with higher lake temperatures	$46 M/yr (trout fishing loss)	Costs: $2M/yr Benefits: $9M/yr
Timber industry	●	●			●	+ Longer growing season + Increase growth with higher levels of CO2 – Increased damage from pests and invasive species	+$15 M/yr (timber harvest gain)	Costs: $12M/yr Benefits: $45M/yr
Forest ecosystem services	●	●	●		●	+ Longer growing season + Increase growth with higher levels of CO2 – Increased damage from precipitation variability and extreme events – Loss of high alpine forests	Unknown	Unknown
Total estimated costs of key elements							$1083-1233M/year	Costs: $68M/yr Benefits: $127M/yr

Key for color-coding:

	Analyzed example
	From literature
	Qualitative information
	Unknown

Together, the components included in table 4.1 are estimated to account for roughly one half of the total value of the ecosystems sector in the state. Important values that are not included in the impact cost numbers include new revenue that may be associated with expansion of summer recreational opportunities and expansion of warm-water recreational fishing. Although precise estimates of adaptation costs are presently unavailable, these costs are provisionally estimated to be approximately 1 to 3 percent of the projected economic value of each sector by 2050, and are expected to increase thereafter. It is also important to recognize that some adaptations (e.g. snowmaking to preserve skiing), may not be feasible later in the century due to substantially altered baseline climatic conditions.

Illustrative Key Costs and Benefits

Although the costs associated with climate change for some of the major ecosystem service components of the sector are uncertain or unknown, it is nonetheless possible to develop estimates of the costs of climate change impacts for critical, revenue-generating facets of the ecosystems sector. In Table 4.2 below, detailed estimates of the costs of climate change impacts on the state's snowmobiling, trout fishing, and timber industries are presented. Estimation of climate change impact costs for all revenue-generating facets of the ecosystems sector was beyond the scope of this study, however the three components selected for detailed analysis are illustrative of a range of revenue-generating ecosystem services which may be affected by climate change. Because the feasibility and costs of a range of adaptation measures for these three facets of the ecosystem sector have not been fully assessed, all estimates for adaptation costs and benefits should be regarded as provisional.

Results

Results (see Table 4.2) suggest that the impacts of climate change are likely to be highly varied across these three facets of the ecosystems sector. Substantial negative impacts are projected for both trout fishing and snowmobiling, both of which may be largely eliminated in New York State by the 2080s as the result of climate change. By the 2080s, annual losses associated with reductions in snowmobiling are expected to range from over $600 million to more than one billion dollars. Annual losses associated with the elimination of trout fishing are estimated to be in the range of $150 million. By contrast, climate change is expected to have positive effects for the state's future timber harvests due to both longer growing seasons and increased levels of atmospheric CO_2. By the 2080s, gains in timber harvesting as the result of climate change are expected total more than $40 million per year.

Table 4.2. Illustrative key impacts and adaptations: Ecosystems Sector (Values in $2010 US.)

Element	Timeslice	Annual costs of current and future climate hazards without climate change ($M)	Annual incremental costs of climate change impacts, without adaptation ($M)	Annual costs of adaptation ($M) [6]	Annual benefits of adaptation ($M) [7]
Snowmobiling and reduced snowpack[1]	Baseline	$25[2]	-	-	-
	2020s	$29	$139-$140[3]	$11	$46
	2050s	$45	$344-$494[3]	$18	$73
	2080s	$71	$649-$1068[3]	$28	$113
Trout fishing and impacts of higher water temperatures[1]	Baseline	$3[2]	-	-	-
	2020s	$7	$7[4]	$1	$6
	2050s	$12	$46[4]	$2	$9
	2080s	$18	$162[4]	$3	$15
Timber industry and impacts of longer growing season[1]	Baseline	$3[2]	-	-	-
	2020s	$3	$ -3[5]	$7	$28
	2050s	$5	$ -15[5]	$12	$45
	2080s	$8	$ -45[5]	$18	$71
TOTAL[8]	Baseline	$31	-	-	-
	2020s	$39	$144	$19	$80
	2050s	$62	$375-$525	$32	$127
	2080s	$97	$760 - $1180	$49	$199

[1] *Value of sector is projected to increase between 1.0 and 2.0 percent per year in New York State. Average increases of 1.5 percent per year are shown in the table. Climate change impact and adaptation cost estimates in the table are estimated based on a growth rate of 1.5 percent.*

[2] *Baseline losses are assumed to be 5% per year for snowmobiling, 5% per year for trout fishing and 1% per year for timber harvesting.*

[3] *Based on Scott et al. (2008) estimates of reductions in snowmobile days for four New York snowmobile regions using low (B1) and high (A1fi) emissions scenarios.*

[4] *As the result of climate change impacts, trout fishing is expected to be eliminated in unstratified lakes by 2050 and in stratified lakes by 2080 (Wolfe and Comstock, forthcoming-a, trout fishing case study).*

[5] *Climate change is expected to have positive impact on timber harvests in New York State due to longer growing season and increased CO_2. Impacts are estimated for a range of values: .5 to 1.5 percent in 2020, 2 to 3 percent in 2050, and 4 to 6 percent in 2080. Midpoint values are shown in the table.*

[6] *Estimates of the costs of climate change adaptation are assumed to be approximately 1 to 3 percent of the total economic value each sector. Midpoint values are shown in the table. It should be noted that these estimates are provisional. Further analysis of adaptation options, feasibility and costs is needed.*

[7] *Benefits of adaptations are assumed to total four times the value of each dollar spent on adaptation. These estimates are preliminary and provisional. Further analysis of adaptation options, feasibility and costs is needed.*

[8] *Totals are based on mid-point values, expect in cases where multiple climate change scenarios are available.*

Overall, development of options for adaptation to climate change in the ecosystem sector is still in a preliminary stage. We assume for illustrative purposes that adaptation costs will range from approximately 1 to 3 percent of annual revenue in the three sectors. By the 2080s, midpoint estimates of annual adaptation costs for all three components are approximately $49 million per year.

PART II: BACKGROUND

4.1 Ecosystems in New York State

The state's terrestrial ecosystems include forests, meadows, grasslands and wetlands. Coastal ecosystems include coastal wetlands, beaches and dune areas, and Hudson River tidal processes. Sixty one percent of New York's land area, or 18.5 million acres, is covered by forest canopy, 40 percent of which (7.4 million acres) is occupied by Northern hardwoods. Tree species with important functional roles include spruce and fir, which are key components of the unique and highly cherished high-elevation forests of the Adirondacks, and hemlocks, which provide shade to stream banks (essential for coldwater fish species) and habitat for many other species. New York's inland aquatic ecosystems depend upon the state's rich abundance of water resources including seventy thousand miles of streams and rivers and 4,000 lakes and ponds (Wolfe and Comstock, forthcoming-a; NYSDEC 2010a).

New York's terrestrial and aquatic ecosystems provide habitat for 165 freshwater fish species, 32 amphibians, 39 reptiles, 450 birds, including many important migratory bird species, 70 species of mammals, and a variety of insects and other invertebrates. Three mammal species - the New England cottontail (*Sylvilagus transitionalis*), the small-footed bat (*Myotis leibii*) and the harbor porpoise (*Phocoena phocoena*) - are state species of concern and one species, the Indiana bat (*Myotis sodalis*) is federally endangered. The Hudson River Valley is globally significant for its diversity of turtles (Wolfe and Comstock, forthcoming-a).

The vast majority of New York's forests and other natural landscapes are privately owned (e.g., over 90 percent of the state's 15.8 million acres of potential timber land). The state also contains over 2.4 million acres of freshwater wetlands, 1.2 million of which are legally protected and administered by the DEC and 0.8 million by the Adirondack Park Agency (NYSDEC 2010b). The Army Corps of Engineers also has jurisdiction over some wetlands in New York State. The economic value of goods and services provided by New York's ecosystems includes recreational and tourism value, the value of commodities such as timber and maple system, and the value of wide array of ecosystem functions including such as: carbon sequestration; water storage and water quality maintenance; flood control; soil erosion prevention; nutrient cycling and storage; species habitat and biodiversity; migration corridors for birds and other wildlife. These functions have substantial economic value, but quantifying them is complex. Also difficult to quantify are the "existence" or "non-use" values, associated with concepts such as preservation of cultural heritage, resources for future generations, charismatic species, and "wild" places (Wolfe and Comstock, forthcoming-a).

A useful illustration of the economic value of ecosystems services in New York is the example of New York City's decision in 1997 to invest in the protection of Catskills watersheds in order to avoid the cost of constructing and operating a large-scale water filtration system for the city's upstate water supplies. The new, larger filtration system was estimated to cost between $2 billion to $6 billion (National Research Council 2004) with operation costs estimated to be $300 million annually for a total estimate of $6 to $8 billion (Chichilnisky and Heal, 1998). By contrast the cost estimates of the city's watershed protection efforts within the Catskills are in the range of $1 billion to $1.5 billion over 10 years, therefore preservation of the ecosystem services provided by the Catskills watersheds has saved the city between $4.5 and $7 billion in avoided costs.

A recent study of the value of ecosystems services in New Jersey also provides some useful estimates for the per acre value of a range of other ecosystem services. The New Jersey study identified a broad spectrum of services that are provided by the state's beaches, wetlands, forests, grasslands, rivers, estuaries, including regulation of climate and atmospheric gas, disturbance prevention (e.g., flood and storm surge protection), freshwater regulation and supply, waste assimilation, nutrient regulation, species habitat, soil retention and formation, recreation, aesthetic value, pollination. The study provided estimates of the average per acre and total values of these services within the state based on value transfer methods, hedonic analysis and spatial modeling (Costanza et al. 2006). The study found that some of the highest per acre value ecosystems are provided by beaches ($42,147/acre-year), followed by estuaries ($11,653/acre-year), freshwater wetlands ($11,568/acre-year), saltwater wetlands ($6,131/acre-year), and forests ($1,476/acre-year). In total, the report estimates that New Jersey's ecosystem services provide economic value for the state of between $11.4 and $19.4 billion per year (Costanza et al. 2006, p. 18). Given New York's vastly greater land area (New Jersey's land area is 5.5 million acres compared to more than 30 million acres in New York), the value of ecosystem services in New York would be expected to be substantially larger. New York's 18.5 million acres of forest canopy alone would have an estimated value of more than $27 billion, based on the estimate of $1,476 annual value per acre used in the New Jersey study.

While ecosystem service values can be difficult to quantify, values associated with human recreational usage of ecosystems are somewhat more straightforward. Outdoor recreation and tourism directly contributes over $4.5 billion to the state's economy. Over 4.6 million state residents and nonresidents fish, hunt, or wildlife watch in New York State (USFWS 2006), spending $3.5 billion, including equipment, trip-related expenditures, licenses, contributions, land ownership and leasing, and other items. The 2007 New York State Freshwater Angler Survey indicated over 7 million visitor-days fishing for warm water game fish (predominantly smallmouth & largemouth bass, walleye and yellow perch), and nearly 6 million days in pursuit of coldwater gamefish (predominantly brook, brown, or rainbow trout) (NYSDEC 2009). Total annual fishing expenditure at the fishing site was $331 million in 2007 (Connelly and Brown 2009a, p. 77). Trout fishing (brook, brown, and rainbow) accounted for 18.3 percent of estimated angler days in the state in 2007 (estimated based on Connelly and Brown, 2009a, p. 16), and the annual value of trout fishing for the state's economy is estimated to be $60.5 million/year.

The state's ski areas host an average of 4 million visitors each year, contributing $1 billion to the state's economy and employing 10,000 people (Scott et al. 2008). New York is also part of a six-state network of snowmobile trails that totals 40,500 miles and contributes $3 billion a year to the Northeast regional economy. Assuming New York accounts for one-sixth of this economic impact, it is estimated that snowmobiling currently brings $500 million to the state's economy overall. The local economies of the Adirondacks, Catskills, Chautauqua-Allegheny, and the Finger Lakes areas are especially dependent on outdoor tourism and recreation, including skiing, hiking, boating and fishing. Table 4.3 provides 2008 data on the economic impact of tourism in these regions. In total, visiting spending in these five regions surpassed $5.3 billion and generated more than $353 million in state tax revenue and $336 million in local tax revenue.

Table 4.3. Economic Impact of Tourism in Selected Regions of New York State.

Region	Visitor Spending ($ millions)	Total employment in tourism and recreation	Share of regional employment in tourism and recreation	State Tax Revenue associated with tourism ($ millions)	State Tax Revenue associated with tourism ($ millions)
Adirondacks	$1,128	20,015	17%	$78	$74
Catskills	$988	17,411	15%	$64	$64
Chautauqua-Allegheny	$500	11,101	11%	$33	$32
Finger Lakes	$2,606	57,083	6%	$180	$166
Total	$5,223	105,610		$354	$337

Source: Tourism Economics 2009. Total figures calculated by authors.

Timber and non-timber forest products such as maple syrup are also significant for the state's economy. In 2005, the estimated value of timber harvested in the state exceeded $300 million (North East Foresters Association [NEFA], 2007). The manufactured conversion of these raw timber components into wood products such as commercial grade lumber, paper and finished wood products adds considerably to the value of this industry to the state. The total forest-based manufacturing value of shipments in 2005 was $6.9 billion (NEFA 2007). Each 1000 acres of forestland in New York is estimated to support 3 forest-based manufacturing, forestry and logging jobs. In 2007, the state's wood products industry employed 9,991 people with an annual payroll of $331 million (United States Census Bureau 2010a). The state's paper manufacturing industries employed 16,868 people with an annual payroll of $748 million (United States Census Bureau 2010a). These industries are particularly important to the regional economies of areas like the Adirondacks, where wood- and paper-product companies employ about 10,000 local residents (Jenkins 2008). In 2007, New York produced 224,000 gallons of maple syrup (2nd in the US, after Vermont) at a value of $7.5 million (USDA NYSS 2009). The Northeast State Foresters Association, using US Forest Service statistics for 2005, found that forest-based recreation and tourism provided employment for 57,202 people and generated a payroll of $300 million in the region (NEFA 2007).

4.2 Key Climate Change Sensitivities

Climate change is likely to have substantial effects of the composition and function of New York State's ecosystems. While this report emphasizes climate change related impacts, it is important to recognize that effects of climate change cannot be viewed in isolation, as other stressors such as urbanization and land use change, acid rain, and invasive species are also affecting ecosystems and will affect vulnerability and capacity to adapt to climate change. Key climate related ecosystem sensitivities are summarized in Table 4.4:

Table 4.4. Climate change sensitivities: Ecosystems Sector (See Wolfe and Comstock, forthcoming-a, for further details).

Higher atmospheric carbon dioxide can increase growth of many plant species. Higher levels of CO_2 are likely to alter species composition in some New York State ecosystems, favoring some species over others. Fast-growing invasive plants and aggressive weed species tend benefit most from higher levels of CO_2.
Warmer summers and longer growing seasons will affect species composition, benefitting some plant and animals species, but harming others. Insects and insect disease vectors will benefit in multiple ways, such as higher food quality of stressed plants, more generations per season and increased over-winter survival. In aquatic systems, warmer waters will tend to be more productive, but are also more prone to nuisance algal blooms and other forms of eutrophication.
Higher temperatures and increased frequency of summer heat stress affects many plant and animal species, constraining their habitable range and influencing species interactions. Temperature increases will drive changes in species composition and ecosystem structure, most notably leading to eventual loss or severe degradation of high elevation spruce-fir, krumholz, and alpine bog and tundra habitats.
Warmer, more variable winters, with less snow cover will have substantial effects on species composition. The habitable ranges of many plant, animal, and insect species that are currently located south of New York may shift north.
Increasing frequency of high rainfall events and associated short-term flooding is currently an issue and is projected to continue. This leads to increased run off from agricultural and urban landscapes into waterways with possible pollution or eutrophication effects, erosion and damage to riparian zones, flood damage to plants, and disturbance to aquatic ecosystems. Extreme events from climate change can cause radical to ecosystem composition. Ecosystems that are already under stress (e.g. forested areas that have been subject to drought or insect invasion) are less resilient to extreme events.
Summer soil water deficits are projected to become more common by mid- to late-century, and the impacts on ecosystems will include reduced primary productivity, and reduced food and water availability for terrestrial animals. Summer water deficits could lead to a reduction of total wetland area, reduced hydroperiods of shallow wetlands, conversion of some headwater streams from constant to seasonal flow, reduced summer flow rates in larger rivers and streams, and a drop in the level of many lakes.

4.3 Impact Costs

Existing efforts to assess the impact costs of climate change for ecosystems are quite limited and typically focus on impacts associated with specific facets of ecosystem services such as snow-dependent tourism in Northeast U.S. (Scott et al. 2008). Broad-based global assessments of ecosystems costs of climate change are also limited (e.g., Tol 2002; Nordhaus and Boyer 2000). More typically, ecosystem studies include qualitative discussion of potential costs associated with climate change (e.g. Parry et al. 2007). For New York State, it is possible to identify a number of areas where impact costs are likely to be incurred. It is important to note, however, that the climate change impacts to New York State's ecosystems are likely to be substantial, regardless of our ability to assign a dollar amount to each impact.

Winter and summer recreation. Under climate change, higher temperatures, reduced snowfall and more variable winter temperatures will have a detrimental effect on the state's $1.5 billion snow-dependent recreational industries including skiing and snowmobiling. While substantial losses in the ski industry are unlikely until much later in the century due to the snowmaking capacities of many resort areas, conditions will become less favorable for skiing within the next several decades. Snowmobiling – which is more dependent on natural snow – is likely to decline substantially in western, northeastern, and southeastern New York within the next several decades (Scott et al. 2008, p. 586). By the mid-21st century, annual economic losses for snowmobiling alone could total $420 million/year (see Tables 4.1 and 4.2). By mid-century expected annual reductions of ski-season length for three major ski regions in New York (Western, Northeastern and Southeastern) are expected to be in the range of 12 to 28 percent. The lower estimates are based on the B1 (lower) emissions scenario while the higher estimates are based on the A1Fi (higher) emissions scenario. Excluding the costs associated with snowmaking, the direct costs associated with these reductions in the ski season range from approximately $200 million per year to more than $500 million per year. A midpoint loss estimate of $350 million is used in Table 4.1 above. Addition of snowmaking costs would substantially increase the total cost estimates.

Summer recreational opportunities such as hiking, swimming and surface water sports are likely to expand with earlier onset of spring weather and higher average summer temperatures. Outdoor tourism and recreation is especially important for rural counties in the Adirondacks, Catskills, and Finger Lakes regions. It is possible that a large share of winter recreation losses could be offset by increases in summer recreational activities.

Recreational fishing. Rising temperatures are likely to have a deleterious effect on cold-water recreational fish species, including brook and lake trout, which currently add more than $60 million per year to the state's economy from on-site fishing-related expenditures (see Table 4.2). Although warm-water species such as bass are likely to benefit from climate change, cold-water recreational species are more desirable for many angler tourists from other regions where these species are less plentiful. Within the Adirondacks, total fishing-related expenditures within the local region were estimated at approximately $74.5 million in 2007, and expenditures by anglers from other regions of New York and out-of-state represented more than 85 percent of this total (Wolfe and Comstock, forthcoming-a; Connelly 2010; Connelly and Brown 2009a, 2009b). Loss of revenue associated with those anglers from other regions or states who are specifically coming for trout and other cold-water species would represent a significant economic blow to the area's tourism-related industries such as hotels, gas stations, and restaurants. For the state as whole, annual trout-fishing losses are estimated to be more than $40 million/year by mid-century (see Tables 4.1 and 4.2).

Timber Industry. Climate change presents both opportunities and challenges for the state's timber industry. Climate change is expected to enhance hardwood production in the state as the result of higher levels of atmospheric CO_2 and a longer growing season. By mid-century the estimated additional value to the timber industry is estimated to be $14 million/year (see Tables 4.1 and 4.2). However, it is also possible that the state's forested areas could become less ecologically diverse as climate changes. Moreover, the transition to a warmer climate may create stresses for some tree species making them less able to withstand normal climatic shocks, leading to dramatic shifts in species composition following extreme events. The timber industry will also face additional costs to manage greater populations of deer and other invasive species that threaten tree survival and timber quality.

Maple syrup production. Maple syrup production may increase under climate change. However, syrup production in lower cost regions such as Quebec may also increase, potentially affecting the competitiveness of the industry.

Heritage value of spruce forests. Spruce forests in New York State have aesthetic and heritage value for state residents, and are also an attraction for summer recreational tourists. These forest ecosystems are not expected to survive under climate change.

Impacts on Riparian Areas. Water quality and flood protection are key ecosystem services provided by riparian areas. These areas also provide critical avenues for species dispersal. Within New York State, the ecosystem services associated with freshwater wetlands are currently valued at more than $27 billion. Although the direct impacts of climate change on wetland and riparian areas are unknown, these areas are already under considerable stress due to land use changes, particularly urban development. New development in and around riparian areas often undermines the water quality and flood protection services associated with these areas.

Costs of invasive species. Invasive plant and animal species have profound ecological and economic impacts and climate change is expected to exacerbate invasive species threats. Within New York State, invasive species pose serious economic threats to agriculture, forestry, maple sugar production, and recreation (Wolfe and Comstock, forthcoming-a). For the U.S. as a whole, invasive species have been estimated to cost the U.S. $120 billion per year in damage and control expenditures (Pimentel et al. 2005). A single species, the emerald ash borer (***Agrilus planipennis*** Fairmaire), which is now established in 13 states including New York, is estimated to cost $10.7 billion from urban tree mortality alone over the next 10 years (Kovacs et al. 2010). Within New York State, Hemlock is currently threatened by infestations of the insect pest, hemlock wooly adelgid (Paradis et al 2008), and grassland ecosystems are also threatened by a number of fast-growing invasive species.

4.4 Adaptation Costs

Assessments of the adaptation costs of climate change for ecosystems are also limited and tend to be focused on specific ecosystem subsectors, such as forestry, within particular regions or countries. With the exception of the United Nations Framework Convention on Climate Change (UNFCCC 2007), recent comprehensive studies of adaptation costs such as that of Stern (2007) do not explicitly include ecosystem adaptation cost estimates. Furthermore, many proposed options for specific adaptations are based largely on ecological theory and have not been tested for their practical effectiveness (Berry 2009). The UNFCCC adaptation costs estimates, which are based primarily on enhancement of the global terrestrial protected areas network, indicate that additional annual expenditures of $12 to $22 billion are needed. Because these estimates do not include marine protected areas or adaptation for non-protected landscapes, they are likely to underestimate the full costs of ecosystem adaptation (Berry 2009).

Despite the lack of generally knowledge about the true costs associated with ecosystem adaptation and the effectiveness of ecosystems adaptation measures, there is nonetheless a consensus within the literature that human intervention will be needed in order to enhance ecosystem adaptation and protect ecosystem integrity and ecosystem services (Berry 2009).

Monitoring and responding to climate change threats to ecosystem functions. A key adaptation entails institutionalizing a comprehensive ecosystems database and monitoring effort. This could potentially entail a state government position with an agency such as the Department of Environmental Conservation. Monitoring and development of indicators for species movement are critical for the management of climate change adaptation by species. In many cases, the need to monitor invasive species and to react quickly, perhaps even with chemical intervention. Costs associated with responding to insect pests can be substantial. For example, since 1996, the annual cost of controlling Asian longhorned beetles in New York City and Long Island has ranged between $13 million and $40 million (New York Invasive Species Clearinghouse 2010).

The costs associated with monitoring efforts for invasive species would likely be similar to the costs associated with the Integrated Pest Management (IPM) program for agriculture. That program, budgeted at $1 million/year entails monitoring of insect pests in New York State and development of responses that can be implemented by farmers while minimizing use of chemical insecticides (NYSIPM 2010). An effort that is similar in scope to the IPM program would monitor indicators of climate change and identify threats to ecosystem services associated with climate change. In particular, the monitoring program would need to: identify good indicators of ecosystem function; monitor these indicators; monitor native species and species interaction – e.g. presences of correct food at correct time of year for migrating birds; monitor invasive species, with a focus on tracking devastating species that may be entering New York State. The annual cost of such a program would be on at least on par with the $1 million/year IPM program budget. The broader goal of such a monitoring program would be to help maintain ecosystem functions under climate change, including management of transitions to new climate conditions.

Adapting outdoor tourism to new climatic conditions. While outdoor tourism will likely continue to be a robust sector in New York State, adaptation to climate change will require new investment on the part of tourism operators in order to maintain profitability and take advantage of opportunities associated with a warmer climate. Within the skiing industry, for example, potential strategies may include expansion of snowmaking capacity and addition of summer season offerings at ski resorts such as hiking and mountain biking or development of new ski resorts at higher altitude and in more northern areas. Managers of state parks and forests will also need to prepare for changes in patterns or seasonality of tourism and demand for recreational services, such as greater use of campgrounds during the fall and spring seasons.

Protection of Forests, Riparian and Wetland Areas. Intact forests, particularly in riparian areas, provide critical ecosystems services including flood control and maintenance of water quality. Forest related ecosystems services are also critical for meeting the state's climate change mitigation goals. Planned mitigation programs that entail incentives for private landowners to leave forests intact could potentially dovetail with the goals of adaptation. Protection of natural corridors in forested riparian areas may provide other ecosystem benefits such as facilitating adaptation of species to climate change. Protection and/or restoration of wetlands in both inland and coastal areas is also critical for flood control, maintenance of water quality, and preservation of habitat for many species.

The benefits associated with protection of wetlands are illustrated in Table 4.5, based on the estimates of Costanza et al. (2006) on the per acre value of wetlands. Once a wetland has been lost or destroyed, the costs of restoration can be very high on a per acre basis. Table 4.5 provides per acre cost estimates for both coastal and inland restoration in New York State. The coastal costs per acre are based on the costs of restoration for two areas on Long Island, while the inland costs are based on costs associated with restoration of wetlands around the Peconic River. For the state as a whole, freshwater wetlands provide ecosystem service benefits valued at more than $27 billion per year. Costs of freshwater restoration of wetlands can range from $3,500 to $80,000 per acre and may entail activities ranging from simple preparation of soils and planting new vegetation to replacement of soils, grading, and planting trees (Brookhaven National Laboratory [BLN] 2001).

Table 5.1. Climate and Economic Sensitivity Matrix: Agriculture Sector (Values in $2010 US.)

Category	Main Climate Variables: Temperature	Precipitation	Extreme Events: rainfall	Sea Level Rise	Atmospheric CO_2	Economic risks and opportunities – is Risk + is Opportunity	Annual incremental impact costs of climate change at mid-century, without adaptation	Annual incremental adaptation costs and benefits of climate change at mid-century
Dairy and livestock	●			●		– Increased stress to livestock – Reduced milk production due to heat	$110M/yr (cost heat stress on dairy production)	Costs: $5M/yr (cooling dairy barns) Benefits: $79M/yr
Field Crops	●	●	●		●	+ Longer growing season + Increase growth with higher levels of CO2 – Increased weed and pest pressures – Higher risk of crop damage from drought	$20-102M/yr (cost extreme events and drought)	Costs: $42M/yr (pesticides, weed control, cropping changes) Benefits: $153M/yr
Perennial fruit crops, vegetables, nursery crops	●	●	●	●	●	+ Longer growing season + New crops and new varieties possible with warmer climate – Increased weed and pest pressures – Higher risk of crop damage from drought	$10-77M/yr (cost of extreme events and drought	Costs: $31M/yr (irrigation, pesticides, weed control, changes in crops varieties) Benefits: $115M/yr
Total estimated costs of key elements							$ 140-289M	Costs: $78M/yr Benefits: $347M/yr

Key for color-coding:

(dark shading)	Analyzed example
(light shading)	From literature
(no shading)	Qualitative information
(no shading)	Unknown

Illustrative Key Costs and Benefits

As described in Table 5.1, the impacts of climate change on the state's agricultural sector are likely to be mixed. While higher temperatures and increased pest pressures will impose strains on dairy and crop production, a longer growing season with more frost free days is likely to have a beneficial effect for many crops, particularly if irrigation capacity is expanded. Table 5.2 presents rough estimates of the costs associated with climate change for the three main facets of the state's agricultural sector. Baseline climate impacts for each facet are based on either empirical documentation of historical losses or extrapolation of losses associated with past events. The costs of impacts of climate change entail estimation of the incremental increase in losses as the result of climate change, beyond the baseline estimates. For dairy production, these loss estimates are based on modeled scenarios of the impacts of climate change on milk production (see Wolfe and Comstock, forthcoming-b, Dairy case study). Estimates of the costs and benefits of adaptation are based on modeling results for the dairy sector (see Wolfe and Comstock, forthcoming-b, Dairy case study), and research suggesting that, with adaptation, most of the impacts of climate change could be substantially reduced or eliminated for agriculture within the Northeast U.S. (see Cline 2007).

For the other components of the sector, the climate change loss estimates are based on the assumption that, without adaptation, average climate change losses for agriculture will increase as the climate changes. Estimated losses in the range of 1% to 5% in 2020 and 2050, and 5% to 10% 2080, respectively, are used as illustrative estimates of the potential magnitude of the impacts of climate change. These estimates may be regarded as provisional pending a more detailed assessment of the effects of climate change on crop production under a range of climate scenarios, which was beyond the scope of this study.

Results

Results indicate that without adaptation, climate change will have substantial costs for the state's agricultural sector, potentially leading to losses of between $766 and $1047 million by the 2080s. However, with the implementation of adaptation strategies including cooling systems for dairy barns, expanded irrigation of crops, and expanded efforts at weed and pest control, future climate change impacts can be minimized. The gains with adaptation are expected to more than offset anticipated losses associated with climate change, leading to net gains in total crop production. By 2050, for example, crop production losses (i.e., losses of fruit, vegetables, nursery, and field crops) due to climate change are estimated to total as much as $179 million, while gains from adaptation measures are expected to total more than $268 million. Annual adaptation costs for the agricultural sector are expected to increase over time, totaling over $300 million/year by the 2080s.

Table 5.2. Illustrative Key Impacts and Adaptations: Agriculture Sector (Values in $2010 US.)

Element	Timeslice	Annual costs of current and future climate hazards without climate change ($M)	Annual incremental costs of climate change impacts, without adaptation ($M)	Annual costs of adaptation ($M)	Annual benefits of adaptation ($M)
Dairy Production and heat stress[1]	Baseline	$25[9]	-	-	-
	2020s	$29	$20[4]	$3[5]	$25[6]
	2050s	$45	$110[4]	$5[5]	$79[6]
	2080s	$71	$488[4]	$12[5]	$252[6]
Fruit, Vegetable and Nursery Crop Production and extreme events, drought, and higher temps[1]	Baseline	$13[10]	-	-	-
	2020s	$17	$9 - $49	$9[3]	$20[8]
	2050s	$27	$10 - $77[2]	$31[3]	$115[8]
	2080s	$43	$120 - $240[2]	$126[3]	$360[8]
Field Crop Production extreme events, drought, and higher temps[1]	Baseline	$33[10]	-	-	-
	2020s	$39	$13 - $55[2]	$14[3]	$26[8]
	2050s	$61	$20 - $102[2]	$42[3]	$153[8]
	2080s	$96	$158 - $319[2]	$167[3]	$479[8]
TOTAL	Baseline	$71	-	-	-
	2020s	$85	$42 - $124	$26	$71[7]
	2050s	$133	$140 - $289	$78	$347[7]
	2080s	$210	$766 - $1047	$305	$1091[7]

[1]*The baseline value of agricultural production is projected to increase between 1.0 and 2.0 % per year in New York State, based recent growth rates of GDP in this sector. Average values of 1.5 % per year are shown in the table.*
[2]*As the result of climate change impacts without adaptation, projected value is assumed to decline by between 1 and 5 percent in both 2020 and 2050, and 5 to 10% in 2080.*
[3]*Estimated costs of adaptation including additional irrigation, pest and weed control, and shifts in crop varieties. These estimated costs are provisionally estimated to range from .5 to 1.5% of value of baseline production in 2020, 1 to 3% percent of baseline production in 2050 and 4 to 6% percent in 2080. Average values are used in the table.*
[4] *Based on Wolfe and Comstock, forthcoming-b, estimates of costs of heat stress on milk production under the A2 climate change scenario and assuming changes in diet but no additional cooling capacity in dairy barns (see Wolfe and Comstock, forthcoming-b, Table 7.5)*
[5]*Estimated costs of adaptation based on costs of addition and operation of cooling systems for dairy barns, assuming costs per cow range from $10 to $110 (see Wolfe and Comstock, forthcoming-b, Dairy case study). Midpoint values are used in the table.*
[6]*With adaptation, the negative effects of heat stress on dairy production are estimated to be reduced by 50%.*
[7]*With adaptation, the total net effect of climate change on New York agriculture is expected to be positive with gains in crop production offsetting losses in dairy production.*
[8]*With adaptation, the net effect of climate change on crop production is expected to be positive due to both longer growing season and on-farm adaptations (e.g. irrigation, changing crop varieties, pest control). Gains of 1% in 2020, 2.5% in 2050, 5.0% in 2080, are projected based on Cline's (2007) estimates of 5% gain by 2080 without assuming CO_2 fertilization; values for 2020 and 2050 were extrapolated.*
[9] *Estimated current annual heat-related losses in dairy and livestock sector (see Wolfe and Comstock, forthcoming-b).*

[10]Current annual climate-related losses for fruit, vegetables and nursery products and field crops are assumed to range from approximately 1.0 to 2.5 percent/year of the total value.

PART II. BACKGROUND

5.1 Agriculture in New York State

New York State's agricultural sector contributes approximately $4.5 billion to the state's economy (USDA 2009). Table 5.3 summarizes some of the most recent (2007) New York agriculture statistics (www.nass.usda.gov/ny). Some of the largest commodities in terms of value include dairy ($2.4 billion), hay ($322 million), grain corn ($300 million), silage corn ($262 million), apples ($286 million), floriculture ($199 million), and cabbage ($100 million). New York is the dominant agricultural state in the Northeast, and typically ranks within the top five in the U.S. for production of apples, grapes, fresh market sweet corn, snap beans, cabbage, milk, cottage cheese, and several other commodities (see Table 5.4) (Wolfe and Comstock, forthcoming-b).

Table 5.3. 2007 NY Agriculture Value

Commodity	2007 Value (thousands)	2007 Harvested Acres (thousands)
Dairy and Livestock	2,727,299	N/A
Total Fruit Crops	368,267	84.25
Total Vegetable Crops	422,000	109.1
Total Field Crops	1,070,873	2769.5
Total Floriculture, Nursery, Greenhouse	357,661	
Total Livestock & Crops	4,454,294	

Source: USDA Nat Ag Stat Service: www.nass.usda.gov/ny
From Wolfe and Comstock, forthcoming-b, p. 36-37.

The agriculture sector plays a particularly important role in many of the state's rural regions. Although dairy farms occur throughout the state, they are the dominant component of the agricultural economy of many counties in the northern, central, and southern regions (Figure 5.1). In some of these more rural regions, a large fraction of the total economy is affected by the fate of the dairy sector. Many dairy farms also produce hay, corn (for grain and silage), and maintain some pasture land to support their own livestock, and for sale of hay. A large fraction of the state's high-value fruit and vegetable crops are grown in western New York, where cash receipts for these crops are highest. Long Island and the Hudson Valley region are also important fruit and vegetable crop areas (see Wolfe and Comstock, forthcoming-b). Small farms throughout the state are also vital to the economy of many rural areas, and fill an important market niche for fresh, high quality, affordable local produce (Wolfe and Comstock, forthcoming-b). About half of New York's 34,000 farms have sales below $10,000 (www.nass.usda.gov/ny), while 18 percent have sales exceeding $100,000. (Table 5.5).

Table 5.4. 2007 NY Agricultural Commodities: Significant Crops in Total Value for NY State and/or Crops with Top 5 National Rank

Product	2007 Total value (thousands)	NY State Rank	National Rank
Dairy products	2,377,987	1	1 (cottage cheese) 3 (milk)
Cattle, hogs, sheep	118,742		2 (calves) 6 (lambs & sheep)
Apples (total)	286,000	4	2
Grapes (total)	49,222		3
Tart cherries	4,369		4
Pears	5,120		4
Cabbage (fresh)	101,190		2
Sweet corn (fresh)	72,600		4
Snap bean (fresh)	49,749		4
Pumpkins (fresh)	22,694		4
Onions (fresh)	94,182		5
Potatoes (TOTAL)	64,372		11
Grain corn	300,355	3	22
Silage corn	262,548	5	3
All hay	322,128	2	22

Source: USDA Nat Ag Stat Service: www.nass.usda.gov/ny
From Wolfe and Comstock, forthcoming-b, p. 36-37.

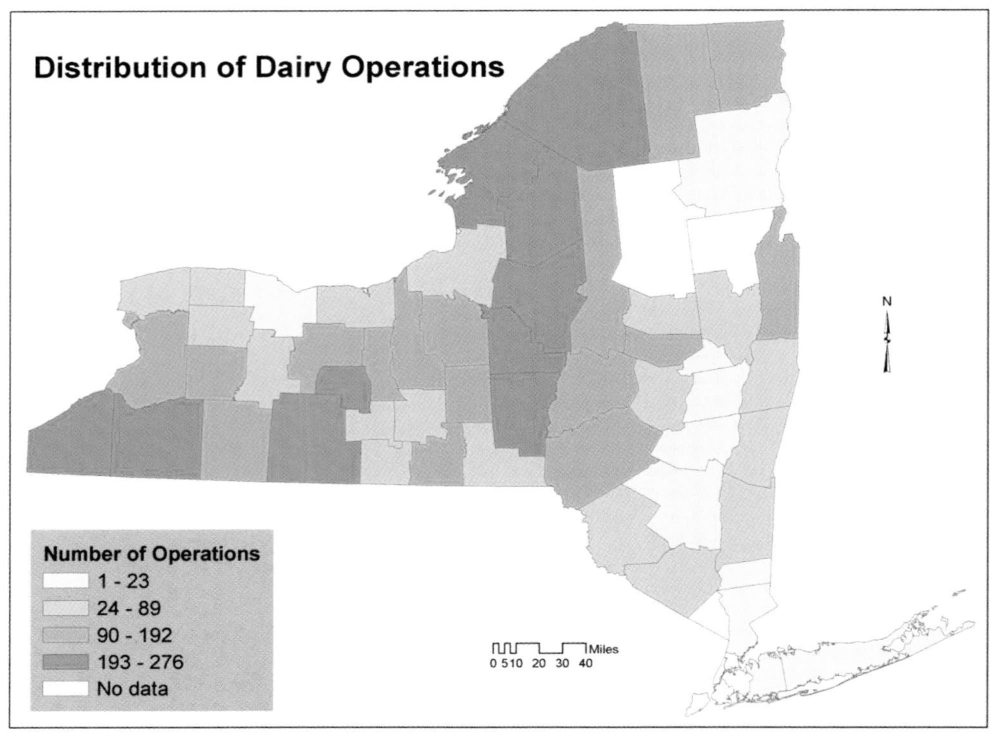

Figure 5.1. Locations of dairy operations in New York State.
Source: USDA 2009.

Approximately 56,900 people in New York State were involved in farming and ranching in 2007 as key farm operators, and almost 60,000 farm laborers were hired statewide (New York Office of the State Comptroller 2010). Within the state's food processing sector, much of which is directly tied to the state's agricultural output for activities such as canning and preserving of fruit and vegetables and dairy product manufacturing, total employment was 48,815 in 2007. Payroll in the state's food processing sector totaled more than $1.7 billion in 2007 (United States Census Bureau 2010a).

Table 5.5. Changes in NY Farm Characteristics

	1997	2002	2007
Approximate total land area (acres)	30,196,361	30,216,824	30,162,489
Total farmland (acres)	7,788,241	7,660,969	7,174,743
Cropland (acres)	4,961,538	4,841,367	4,314,954
Harvested Cropland (acres)	3,855,732	3,846,368	3,651,278
Woodland (acres)	1,655,185	1,649,585	1,559,522
Pastureland (acres)	520,150	550,225	714,615
Land in house lots, ponds, roads, wasteland, etc. (acres)	651,368	619,792	585,652
Farmland in conservation or wetlands reserve programs (acres)	97,617	211,996	115,546
Average farm size (acres)	204	206	197
Farms by size (percent)			
1 to 99 acres	45.9	47.9	51.2
100 to 499 acres	45.1	42.8	40.4
500 to 999 acres	6.7	6.6	5.5
1000 to 1,999 acres	1.9	2.2	2.1
2,000 or more acres	0.4	0.6	0.8
Farms by sales (percent)			
Less than $9,999	51.6	55.9	54.6
$10,000 to $49,999	20.7	18.5	20.4
$50,000 to $99,999	9.1	8.2	6.2
$100,000 to $499,999	15.9	14.4	14.0
More than $500,000	2.6	2.9	4.8
Farm organization			
Individuals/family, sole proprietorship (farms)	32,813	32,654	30,621
Family-held corporations (farms)	1,593	1,388	1,885
Partnerships (farms)	3,465	2,846	3,347
Non-family corporations (farms)	178	193	225
Others - cooperative, estate or trust, institutional, etc. (farms)	215	174	274

Data Source: USDA 2010 (,U.S. Census of Agriculture: 1997, 2002, 2007. More information on farm characteristics available from the Census of Agriculture.

The value of agriculture to the state extends beyond farming and food processing. For example, New York is the second-largest producer of wine in the nation behind California, with wine sales in excess of $420 million in 2007. In 2008, the state's 208 wineries employed approximately 3,000 workers (NY State Office of the Comptroller, 2010). An analysis of the total value of the New York grape and wine industry that included multipliers such as regional tourism and supporting industries estimated that the total economic impact of this industry in 2004 was over $6 billion (MKF Research 2005).

Agricultural areas encompass about one quarter of the state's land area (over 7.5 million acres). Reduction of pollution as the result of farming practices continues to be a priority for New York State farmers. Farm landscapes also provide important and economically valuable ecosystem services such as preservation of soil and water resources, habitat to enhance biodiversity, and carbon sequestration to mitigate climate change (Bennet and Balvanera 2007) (Wolfe and Comstock, forthcoming-b). The state also has an active Farmland Protection Program. As of 2009, the state had awarded over $173 million to assist municipal and county governments and local project partners on projects in 29 counties. Upon completion, these projects will permanently protect over 72,000 acres of agricultural land (USDA NASS 2010). To date, more than 160 farmland protection projects have been completed in the state, protecting over 31,000 acres with a state investment of more than $84 million (USDA NASS 2010).

The response of New York agriculture to climate change will occur in the context of numerous economic and other forces that will be shaping its future, including pricing pressures, trends toward farm consolidation, rising energy and production costs, and increasing competition for water resources (Wolfe and Comstock, forthcoming-b). As illustrated in Table 5.5, the state's agricultural sector has undergone a number of changes over the past decade including a decline in total acres of farmland from 7.78 million in 1997 to 7.17 million in 2007, a decline in average farm size, from 204 acres in 1997 to 195 acres in 2007, and increases in the number of very small farms (under 99 acres) and very large farms (over 2000 acres). Although examination of how climate change may intersect or influence these trends is beyond the scope of the present study, it important to recognize that these broader trends will condition the impacts of climate change and the adaptation strategies available.

5.2 Key Climate Change Sensitivities

Climatic conditions are a critical driver of agricultural activity and production worldwide. A number of aspects of climate change are particularly relevant to the agriculture sector in New York State. These factors are summarized in Table 5.6 and described in detail in Wolfe and Comstock, forthcoming-b.

Table 5.6. Climate change sensitivities: Agriculture sector (See Wolfe and Comstock, forthcoming-b, for further details)

Higher atmospheric carbon dioxide (CO_2) levels can potentially increase growth and yield of many crops under optimum conditions. However, research has shown that many aggressive weed species benefit more than cash crops, and weeds also become more resistant to herbicides at higher CO_2.
Warmer summer temperatures and longer growing seasons may increase yields and expand market opportunities for some crops. Some insect pests, insect disease vectors, and pathogens will benefit in multiple ways, such more generations per season, and for leaf-feeding insects, an increase in food quantity or quality.
Increased frequency of summer heat stress will negatively affect yield and quality of many crops, and negatively affect health and productivity of dairy cows and other livestock.
Warmer winters will affect suitability of various perennial fruit crops and ornamentals for New York. The habitable range of some invasive plants, weeds, insect and disease pests will have the potential to expand into New York, and warmer winters will increase survival and spring populations of some insects and other pests that currently marginally overwinter in this area.
Less snow cover insulation in winter will affect soil temperatures and depth of freezing, with complex effects on root biology, soil microbial activity, nutrient retention (Rich 2008) and winter survival of some insects, weed seeds, and pathogens. Snow cover also will affect spring thaw dynamics, levels of spring flooding, regional hydrology and water availability.
Increased frequency of late summer droughts will negatively affect productivity and quality, and increase the need for irrigation.
Increased frequency of high rainfall events is already being observed with negative consequences such as direct crop flood damage, non-point source losses of nutrients, sediment via runoff and flood events and costly delays in field access.

5.3. Impact costs

This section discusses the potential costs associated with impacts of climate change across the major components of the state's agricultural sector. Numerous assessments of the costs of climate change on agriculture and food production have been conducted on a global level and for specific countries including the United States (e.g., Cline 2007; McCarl 2007; Parry et al. 2004). These studies typically employ methods that include either modeling of the impact of climate change on crop yields and agricultural output or estimation of how land values vary as a function of climatic conditions. In recent years, crop model assessments have also incorporated different future development scenarios based on the IPCC Special Report on Emissions Scenarios (SRES) which allow for variations in projected population, income levels, and emissions (e.g., Parry et al. 2004).

Results of these types of studies provide a 'top down' gauge of the potential costs of climate change both for the U.S. as a whole and for major subregions. A widely cited study by Cline (2007), for example, finds increases in agricultural output for the U.S. Lakes and Northeast region as the result of climate change, despite overall losses for the United States as a whole. Under a scenario that does not assume crop fertilization from CO_2, the study finds that climate change will lead to an increase in agricultural production of 5.0 percent for the Great Lakes and Northeast region by the 2080s, but that the U.S. as a whole will experience a net loss of 5.9 percent, largely due to reduced production in the Southeast and Southwest regions (Cline, 2007, p. 71).

Although these types of aggregate studies provide an indication of the direction and general magnitude of the impacts of climate change, they provide little information that is specific to key economic components of the New York's agricultural sector. As described below, climate change may have significant costs for various facets of New York State's sector, particularly if appropriate adaptation measures are not taken. Such costs, as described below, include declining yields in the dairy sector, declines in yield and quality of perennial fruit crops, and crop losses associated with drought, weeds and pests (see also Tables 5.1 and 5.2).

Heat Stress and Milk Production. Dairy is the largest component of New York State's agricultural sector. Higher temperatures and summer heat stress on dairy cattle may result in lower milk production, decreased calving, and increased risk of other health disorders – all of which impact costs and profitability. The negative economic impacts of climate change on the dairy sector are likely to be substantial without significant adaptation (Wolfe and Comstock, forthcoming-b).

Heat stress has an especially significant effect on milk production and calving rates for dairy cows. Historical economic losses due to heat stress for dairy and other livestock industries in New York have been estimated to be $24.9 million per year (St. Pierre et al. 2003, p. E70). Under climate change, higher temperature and humidity indices (THI) are likely to have a significant negative effect on total milk production. High-producing dairy cows (85lb/day) are especially sensitive to the effects of heat stress, and even small declines in dairy milk production (e.g. 2 pounds per day), translate into large losses of milk (400-500 lbs) over a lactation period. At current milk prices of $12/100 lbs, a 400-500 lbs loss would amount to $48-$60/cow (Wolfe and Comstock, forthcoming-b, Dairy case study). As average THI increases over the next century, losses are expected to increase substantially, potentially approaching 8 to 10 pounds per day during the hottest days for regular (65lb/day) and high (85lb/day) cows, respectively (Wolfe and Comstock, forthcoming-b, dairy case study).

By the 2080s, the projected annual economic losses under climate change could approach 248 lbs per year for regular cows and 437 lbs per day for high-producing cows. These losses, which represent a 6-fold increase over the historical average, would lead to economic losses of approximately $37 and $66 per cow for regular and high producing cows, respectively (Wolfe and Comstock, forthcoming-b). Assuming the total number of cows in the state in the future is relatively constant -- in 2006 there were approximately 640,000 dairy cows in New York State

(New York State, Department of Agriculture and Markets, 2007) - the value of these types of economic losses by 2080 would total more than $400 million for the dairy sector (see Table 5.2).

Climate change stresses on fruit, vegetable, and nursery crops. New York State's fruit, vegetable and nursery crops are worth approximately $807 million/year (USDA NASS 2009). Among fruit crops, perennial fruits such as apples and grapes are especially at risk from climate change. For apples, reduced winter chill periods are likely to reduce apple harvests and negatively affect fruit quality, possibly necessitating changes in apple varieties grown. Over the long term, apples may be substituted for other perennial crops, such as peaches, that are better suited to shorter winters and higher summer temperatures. In the short term, climate change is likely to have negative impact on the profitability of apple production. By contrast, grape producers in New York State are likely to benefit from climate change because warmer temperatures are more conducive to grape production. Over time, climate change may allow producers to shift to more desirable and profitable varieties for use in wine production.

Vegetable production is also vulnerable to climate change. New York currently specializes in cold-weather adapted crops such as cabbage and potatoes. Production of these types of crops is likely to decline as temperatures warm. Over time, it is likely that producers will substitute cold-weather crops with crops that are more suited to warmer growing conditions. A major economic cost for vegetable producers will entail identification of more suitable crops, purchase of seeds and capital needed to produce these new crops, and marketing of the new crops (Wolfe and Comstock, forthcoming-b).

Nursery crops are also a major industry in New York State. These high-value crops are especially vulnerable to heat stress and drought. In order to reduce present-day climate risks, the state's nursery industries are increasingly making use of controlled environments. Under climate change, the need for such environments may expand in order to cope with insects, disease, weeds, drought and heat stress.

A key climate-change related uncertainty for crop production entails changes in the frequency, timing, and magnitude of extreme events. Fruit, vegetable and outdoor nursery crop production are all highly sensitive to extreme climate events. Hail, heavy rain, and high-wind events can damage many types of crops, especially if such events occur during the growing season, and particularly near harvest time (Wolfe and Comstock, forthcoming-b). A single event during or near the harvest period, such as a brief hail storm, can virtually wipe out an entire crop in an affected region. Increased variability of temperatures during winter months is a particularly threat for perennial fruit crops. For example, during the winter of 2003-2004, mid-winter freeze damage led to substantial production losses in the Finger Lakes wine growing region. For the state as a whole, grape production declined from 198,000 tons in 2003 to 142,000 tons in 2004, with an associated loss of value of more than $6 million (USDA NASS New York Office, 2009, p. 35). These losses were primarily due to "dehardening" of the vines during an unusually warm December, which increased the susceptibility of the vines to cold damage during a subsequent hard freeze that occurred in January. (Wolfe and Comstock, forthcoming-b). Drought is also a threat to fruit and vegetable crops, the majority of which are not currently irrigated. Without adaptation, climate change-related economic losses for fruit, vegetable, and nursery crops are estimated to be nearly $230 million per year by 2080 (see Table 5.2).

Field crops and drought. Field crops such as grain and silage corn and soybeans provide a critical source of feed for the dairy and livestock sector (Wolfe and Comstock, forthcoming-b). Worth approximately $1.1 billion per year, field crops are particularly vulnerable to drought, and farmers currently incur substantial economic losses when field crops harvests are reduced or lost during drought periods. Drought related losses are likely to increase under climate change due to increased variability of summer precipitation and higher temperatures. Estimates of annual field crop losses under climate change and the benefits of adaptation, as presented in Table 5.2 above, suggest that losses under climate change may total more than $300 million by 2080 without appropriate adaptation. Such losses will directly affect feed costs for dairy and livestock farmers.

Insect damage and weeds. Higher temperatures and more CO_2 are conducive to insect reproduction and weed growth. Crop losses due to insects and weeds have been substantial in the past, and are likely to increase under climate change, without appropriate adaptations. Insect and weed pressures affect all types of crop production in New York State and costs for control of these pressures are likely to increase with climate change.

5.4 Adaptation Costs

Planning for adaptation is a critical step for New York's agricultural sector, not only in preparation for challenges such as new invasive species, but also to take advantage of warmer climates and longer growing seasons. The literature regarding the costs of adaptation within the agricultural sector generally suggests that within advanced economies such as the United States, the incremental costs of adaptation measures are likely to be relatively small in comparison with the amount that is already being invested in research and development within the sector (Wheeler and Tiffin 2009). The current literature also indicates that the need for additional, adaptation-related capital investment in the near term is likely to be less pressing than in the middle to longer term because most agricultural capital has a 10-20 year lifespan and is likely to replaced before significant climatic change impacts occur (UNFCCC, 2007, pp. 101-102). A recent top down global assessment of the total costs of climate change for agriculture estimates that adaptation in the agricultural sector will require a ten percent increase in research and development expenditure and a two percent increase in capital formation, beyond what would be spent without climate change (McCarl 2007). The costs of these additional expenditures will in the range of $11.3 to $12.6 billion globally in the year 2030, with mitigation (SRES B1) and without mitigation (SRES A1B1), respectively (Wheeler and Tiffin 2009). Another recent study, which took a "bottom up" approach by focusing on the costs for a specific type of adaptation, estimates a cost of $8 billion per year globally in 2030 for increased irrigation capacity in order to adapt climate change, under a scenario that includes mitigation (SRES B1) (Fischer et al. 2007).

Within New York State, numerous adaptations are possible in order to mitigate the impacts of climate change within the agricultural sector. While some adaptations may have negligible costs (e.g., shifting to earlier planting dates), most will entail some type of financial outlays on the part of farm operators, and some will require significant new investment. In addition to new investments will be needed, above and beyond the normal investments that would be made anyway. There is a related need for decision support tools to help farmers decide when to make investments in appropriate adaptation technologies. This section discusses costs and benefits associated with some key adaptation options for the sector. Many of these adaptations are steps that individual farmers may take, while others would require state-level involvement and coordination.

Reduction of heat stress for dairy cows. Adjustment of diet and feeding management can reduce some of the impacts of heat stress with minimal impacts on production costs. However, as temperatures increase under climate change, improvement of cooling capacities and dairy barns will be a critical adaptation in order to reduce heat stress and maintain productivity. Farmers can enhance cooling via increased use of existing fans, sprinklers, and other cooling systems (Wolfe and Comstock, forthcoming-b). The major costs for these types of adaptations would include additional energy usage and additional labor. Improvement in the cooling capacity of housing facilities is also likely to be needed, especially as average THI increase under climate change. While such systems represent added costs, these investments have a high likelihood of paying for themselves, through increased milk production, over a short time span (1 to 3 years depending on the numbers of days that the system is in operation) (Turner, 1997). For example, installation of a tunnel ventilation system for a small, 70-cow herd producing 75 lb per cow is estimated to cost $7,694 ($110/cow), including both operational costs and interest on a 5-year loan (Wolfe and Comstock, forthcoming-b). For the sector as a whole, the costs of addition and operation of cooling systems for the dairy sector are estimated to total approximately $5 million/year by the 2050s (see Tables 5.1 and 5.2).

Diversification of fruit crops and vegetable crops. Near term adaptations to climate change for fruit and vegetable producers will entail adjustments to planting or harvesting dates to coincide with early onset of spring or later occurrence of the first frost. While such steps have minimal cost, availability of labor and market demand will be critical limiting factors. As climate change progresses, farmers will need to consider new crop varieties that are more heat or drought tolerant, and may also shift to different crops that are more suitable to new climatic conditions. The costs associated with shifting crops typically include new planting or harvesting equipment and new crop storage facilities. In the case of fruit trees, it typically takes several years for a new tree to bear fruit, which also adds to the costs of adaptation.

Insect and weed control. Increase use of chemical inputs and non-chemical techniques will be a necessary adaptation in order to control increased insect, pathogen, and weed pressures under climate change. For crops such as sweet corn, the number of insecticide applications that are needed could double or even quadruple. Current climate conditions in New York require 0 to 5 insecticide applications against a key sweet corn pest (lepidopteran insects), while states with warmer climates such as Maryland and Delaware require 4-8 applications and Florida requires

15-32 applications (Wolfe and Comstock, forthcoming-b). Because chemical use is expensive and harmful to human and ecosystem health (e.g., New York potato farmers currently spend between $250 and $500 per acre for a total of $5 to $10 million statewide on fungicides to prevent late blight, [Wolfe and Comstock, forthcoming-b]), other means of adaptation to control insects and weeds will also be needed. Integrated pest management techniques are an effective means of controlling insects that minimize the use of chemical inputs. Within New York, the annual budget for state's Integrated Pest Management Program is approximately $1,000,000 (NYSIPM 2010). Such a program would likely need to be continued and substantially expanded in order to facilitate adaptation to climate change.

Irrigation and/or drainage systems. Expansion of irrigation capacity and drainage systems may be necessary in order to maintain productivity and allow farmers to take advantage of new opportunities under warmer climatic conditions. While expanded use of existing irrigation systems is possible for some farmers, installation of new systems requires significant capital investment. These systems currently draw water from local streams, but it also possible that they may require more extensive and costly infrastructure to enable water transfers between basins. The fixed capital costs associated with adding an overhead moveable pipe irrigation system within New York state are estimated to be on the order of $1000 per ha or $405 per acre (Wilks and Wolfe, 1998) (1 ha = 2.47 acres), a figure slightly higher than the nationwide estimate of approximately $290/hectare or $117/acre (Fischer et al. 2007). This type of system also requires labor costs to move the pipes with each irrigation, as well as energy costs for pumping the water. The estimated annual irrigation and annual labor costs associated with energy use are estimated to be approximately $12.50/ha ($5.06/A) and $32.50/ha ($13.16/A) respectively (not adjusted into constant dollars; Wilks and Wolfe, 1998).

Given the relatively high cost of irrigation, it is expected that such systems would only be put into place as an adaptation to climate change for production of high value fruit, vegetable, and horticulture crops. In 2007, approximately 1.5 percent of New York State's million acres were irrigated (U.S Department of Agriculture, 2009). This translates into approximately 68,000 irrigated acres (USDA 2009). During 2008, approximately half of the state's total irrigated acreage was irrigated including approximately 20,158 acres of fruit, vegetables, and other food crops and 8,765 acres of non-food horticultural crops (USDA 2010). A key reason for reduced irrigation in 2008 was adequate soil moisture (USDA 2010).

If we assume total irrigated acreage capacity in New York State would need to double for high value crops in order to adapt climate change, we can estimate both the fixed costs and variable costs associated with adding this new capacity as well as the added benefits. Table 5.7 presents estimates of both the fixed and variable costs associated with a doubling of irrigation capacity for vegetables, orchards and berries, and nursery stock, as well as the benefits associated within increased crop yields. Benefits associated with increase in yields are based on the results of Wilkes and Wolfe (1998). Wilkes and Wolfe (1998) found that addition of irrigation increases the annual per hectare value of lettuce production in New York State by more than 50 percent, from $8000/hectare to $12,500/hectare. In addition to benefits associated with increased drought resilience, which might entail preservation of much of the value of a particular crop during a drought year, added benefits from irrigation of fruits and vegetables include higher total yields and improved quality. Results indicate that fixed costs associated with the doubling of irrigation capacity for these three crop categories would be approximately $19.6 million and the labor, energy and interests costs assuming a five year loan would be an additional $1,861,000 annually. Benefits of the adding irrigation capacity for these three crop categories are estimated to be approximately $33.2 million per year in added value of crop production.

Table 5.7. Benefit Cost Analysis of Potential Climate Change Adaptation: Expansion of irrigation

Crop	Total Acres (2007)	Irrigated Acres (2007)	Percent irrigated	Annual value of crop (2007) ($M)	Fixed costs to double total acres irrigated ($M)	Annual labor, energy and interest cost of additional irrigation ($M)	Increased annual value with added irrigation ($M)
Vegetables	160,146	34,170	21.3	$338	$13.8	$1.4	$18.0
Orchards and berries	104,349	11,038	11.0	$368	$4.5	$0.4	$9.7
Nursery stock (open)	14,638	3,161*	21.6	$101	$1.3	$0.1	$5.5
Total				$807	$19.5	$1.9	$33.2

**2008 data*

Data sources: USDA 2010; U.S. Census of Agriculture, Farmer and Ranch Irrigation Survey 2008; Authors' calculations.

Research, monitoring, extension, and decision support tools. Within the agriculture sector, effective adaptation to climate change will require monitoring of new threats (e.g., new pathogens or invasive species) and extension assistance to facilitate successful transitions to new crop varieties and new crops. These types of monitoring and extension efforts can also be accompanied by development and dissemination of decision support tools. Such tools can assist farmers in making strategic adaptation choices, particularly with respect to the timing of new capital investments in adaptation such as new cooling facilities for dairy farms.

5.5 Summary and Knowledge Gaps

The broad findings for New York State agriculture echo the general findings from the literature regarding the costs of impacts and adaptation within the agricultural sector, which suggest that appropriate adaptation measures can be expected to offset declines in projected yields for the next several decades (e.g., McCarl 2007; Agrawala et al, 2008; Parry et al. 2009). Although the costs of such measures will not be insignificant, they are likely to be manageable, particularly for larger farms that produce higher value agricultural products. Smaller farms, with less available capital, may require adaptation assistance in the forms of grants or loans, in order to facilitate adaptation. Expansion of agricultural extension services will also be necessary in order to assist farmers with adaptation to new climatic conditions.

In order to facilitate adaptation in New York State, key areas for additional investment in research and extension include:

- Monitoring of new pests, weeds and other disease threats to agricultural crops;
- Improvement of techniques for integrated pest management to deal with these new threats, while minimizing use of pesticides, herbicides and other hazardous materials;
- Improvement of techniques for integrated pest management to deal with these new threats, while minimizing use of pesticides, herbicides and other hazardous materials;
- Investigation of alternative irrigation technologies that are less water and energy intensive; and
- Development of decision support tools to help farmers select and time new capital investments in order take advantage of opportunities associated with climate change, while minimizing risks.

Technical Notes – Agriculture Sector

1. Current value of production, based on the Census of Agriculture, 2007, is \$2.4 billion in the dairy and livestock sector, \$807 million in fruits, vegetables and nursery crops, and \$1.1 billion in field crops (most of which are used as feed for dairy and livestock). Agricultural value in New York State is projected to grow by a rate of between 1.0 and 2.0 percent per year (all calculations above are based on an average growth rate of 1.5%/year). A lower rate of growth is used in this sector as compared to the state overall because the agriculture sector has been growing more slowly than other facets of the state's economy and limits on land availability are likely to constrain future growth.

2. Dairy sector estimates are based on costs of heat stress on milk production assuming changes in diet but no additional cooling capacity in dairy barns (see Wolfe and Comstock, forthcoming-b, Table 7.5). The estimated cost of adaptation are based on costs of addition and operation of cooling systems for dairy barns, assuming costs per cow range from \$10 to \$110 (see Wolfe and Comstock, forthcoming-b, Dairy case study). With adaptation, the effects of heat stress on dairy production are expected to be reduced by 50%. (This is the assumed benefit of adaptation.)

3. Current annual climate-related losses for fruit, vegetables and nursery products are assumed to range from approximately 1.0 to 2.5 percent/year of the total value. Without adaptation, projected values are assumed to decline by 1.0% in 2020, 5% in 2050 and 10% in 2080. With adaptation, the net effects of climate change are expected to be positive due to both longer growing season and on-farm adaptations (e.g. irrigation, changing crop varieties, pest control). Gains of 1% in 2020, 2.5% in 2050, 5.0% in 2080, are based Cline (2007). Cline (2007) estimates of 5% gain by 2080 in agricultural productivity for the U.S. Northeast, without assuming CO_2 fertilization. Values for 2020 and 2050 were estimated based on extrapolation. The benefits of adaptation are calculated by subtracting the total value of production under climate change without adaptation from the total value of production with adaptation.

4. Current annual climate-related losses for field crop products are assumed to range from approximately 1.0 to 5.0 percent/year of the total value. Projected values are assumed to decline between 1% and 5% in 2020 and 2050, and between 5% and 10% in 2080 without adaptation. With adaptation, the net effects of climate change are expected to be positive due to both longer growing season and on-farm adaptations (e.g., changing crop varieties, pest control). Gains of 1% in 2020, 2.5% in 2050, 5.0% in 2080, are based Cline (2007), as described above. The net benefits of adaptation are calculated by subtracting the total value of production under climate change without adaptation from the total value of production with adaptation.

6 Energy

New York State's electricity and gas supply and distribution systems are highly reliable; they are designed to operate under a wide range of temperature and weather conditions – from 0 to 100°F, in direct sunlight or under the weight of snow and ice. The system is deliberately robust and resilient because utility companies are risk averse. When designing energy supply and distribution systems companies use conservative engineering estimates (industry standards plus 30%) and typically look 20 years into the future. In some cases, threshold conditions (as opposed to the mean or standard conditions), or shifts in the threshold caused by climate change can create vulnerability within the energy sector (Hammer, 2010) and substantially increase the cost of maintaining reliability.

PART I. KEY ECONOMIC RISKS AND VULNERABILITIES AND BENEFIT-COST ANALYSIS FOR ENERGY SECTOR

Key Economic Vulnerabilities

This section provides estimates of the extent to which climate related changes will affect economic components of the energy sector. Table 1 identifies the climate variables that are likely to impact the sector along with the project economic outcome. Note that economic risks significantly outweigh opportunities.

Table 6.1. Climate and Economic Sensitivity Matrix: Energy Sector (Values in $2010 US.)

Element	Main Climate Variables						Economic risks and opportunities: – is Risk + is Opportunity	Annual incremental impact costs of climate change at mid-century, without adaptation	Annual incremental adaptation costs and benefits of climate change at mid-century
	Temperature	Precipitation	Sea Level Rise	Extreme Events: Heat	Extreme Events: Intense Precipitation	Extreme Events: Hurricanes, Nor'easters, & Wind			
Energy Supply	●	●	●				– Changes in biomass available for generation – Availability of hydropower reduced – Potential Changes in solar exposure – Availability and predictability is reduced with variation in wind – Reduced water cooling capacity – Damage to coastal power plants – Sagging power lines – Wear on transformers – Transmission infrastructure damage – Transmissions lines sagging due to freezing/collecting ice	$36-73M	Costs: $19M Benefits: $76M
Electricity Demand	●		●	●			– Increased energy demand for cooling – Increased demand for pumping at coastal energy producing locations – Potential increases in pumping for industrial cooling water – Decreased demand for winter heating	Increased supply costs	Net total of increased air conditioning use in summer and heat in winter and pumping demands
Buildings				●	●	●	– Heightened storm regime may reveal weaknesses in building envelopes – Low-lying areas susceptible to more frequent flooding + Installation of green roofs	Structural damage from extreme events; Increased insurance costs	Cost for repairs and upgrades
Total estimated costs of key elements								$37-73M	Costs: $19M Benefits: $76M

Key for color-coding:

	Analyzed example
	From literature
	Qualitative information
	Unknown

For the energy sector, climate change will affect both energy supply and energy demand.

Energy Supply

Milder winter weather may help alleviate some of the stresses on the supply chain of New York State's energy system, however it is more commonly projected that climate change will adversely affect system operations, increase the difficulty of ensuring supply adequacy during peak demand periods, and exacerbate problematic conditions, such as the urban heat island effect (Rosenzweig and Solecki, 2001). The following climate impacts pose the greatest economic risks and vulnerabilities to energy supply:

Impacts on thermoelectric power generation and power distribution due to floods and droughts, increases in air and water temperatures, and ice and snow storms. The threat of ice storms affecting upstate energy infrastructure is potentially large (Hammer, 2010). Additionally, sea level rise and storm surges will threaten coastal power plants.

Impacts on natural gas distribution infrastructure due to the flood risk associated with extreme weather events (Associated Press 1986, New York Times 1994), and frost heaves (Williams and Wallis, 1995) (although the effect that climate change will have on frost heaves is still unclear). These potential impacts would be alleviated to some extent because natural gas supplies adequate to provide some level of insurance against natural disasters that may disrupt production and delivery systems are stored in underground facilities in western New York and Pennsylvania (Hammer and Parshall, forthcoming).

Impacts on renewable power generation due to changes in the timing and quantity of the natural resource available for power generation (Hammer and Parshall, forthcoming). For example, the lost capacity for inexpensive hydropower may be replaced by more expensive forms of power generation, creating significant cost repercussions for the state (Morris et al., 1996).

Energy Demand

The following climate impacts pose the greatest economic risks and vulnerabilities to energy demand:

Shifts in the number of heating degree-days and cooling degree-days (i.e. demand space for heating and cooling) will occur due to changes in mean and extreme temperatures. The direction and magnitude of changes in energy demand depend on changes in heating and cooling degree-days, other climate shifts, and the sensitivity of demand to climate factors

(Hammer and Parshall, forthcoming). As electricity consumption climbs and peak demand grows in summer months, the current energy supply and demand equilibria will be disrupted. With higher mean temperatures and increased numbers of extremely hot days, the cost of maintaining a reliable supply of electricity is likely to increase in all parts of the state. For New York City in particularly, where the system is already taxed during very hot summer days, climate change will place additional pressures. Meeting the demand for electricity may also become more expensive due to extreme weather events (The Center for Integrated Environmental Research, 2008, p. 4). There may also be increases in demand for industrial uses due to changing climate, for example increases in pumping cooling water for industrial uses. Changes in incomes, technology, law and population will probably result in greater impacts on energy demand than climate change. The energy sector, among the ClimAID sectors, is perhaps the most likely to see game-changing policies in the next decade. For example, a carbon tax in any form (either directly, or indirectly through cap-and-trade) could radically alter demand and supply conditions in the energy sector.

To the extent that climate change causes additional economic impacts on the sector, these are likely to be for increased capacity and smarter grids. There is also the possibility of increased climate-related blackouts due to increased demand. This possibility depends on the level of investment within the energy sector. There are regular, ongoing new investments in the sector that will continue to be undertaken even without specific new programs for adaptation to climate change; to the extent that these contribute to a more stable system under both present and future climate conditions, blackouts will be reduced. (If the electrical system becomes hardened against electromagnetic storms, that will go even further to accommodate the impacts of climate change.) However, the potential uncertainty in the pattern and extent of extreme heat events could increase outages, although fewer than would be expected absent the ongoing improvements in system reliability that can be assumed. Even with regularly improved systems, therefore, the probability is that some additional adaptations will be needed that specifically take climate change into account, particularly to handle extreme heat; some utilities are already beginning to incorporate climate change into their planning processes. The possibility of a slightly increased incidence of blackouts can be used to illustrate the costs of climate change in the energy sector if such adaptation measures are not undertaken.

As the likelihood of a blackout is exacerbated by heat waves and associated thunderstorms (as well as other extreme storm events), and as heat waves are likely to increase in the future, it is likely that blackouts may occur somewhat more frequently, although to an extent reduced by the regular, ongoing investment of the electricity industry. A study by the Wharton School (2003) indicates that the energy system is designed for a 1-in-10 year blackout, over the past thirty years New York City has experienced four major events in 1977, 1999, 2003 and 2006. Climate change could, without ongoing investment, increase the number of blackouts above that for which the system is designed. Cost estimates vary widely from these events, as it can be difficult to ascertain exact expenses directly related to the blackout. However, using a range of estimates, it is possible to calculate an average cost per event. From this estimate, based on the assumption that a blackout occurs once every ten years, an annual cost can be obtained. Using the heatwave projections given in Horton et al. (forthcoming) future cost of impact estimates can be estimated based on these assumptions and the impacts of regular upgrades in investment.

One key adaptation put forward to reduce the likelihood of heat-related blackouts is the installation of a smart grid, as discussed in the adaptation section of this chapter. Additionally, the Multi-hazard Mitigation Council has estimated that every $1 spent in public disaster mitigation results in a $4 savings. Based on these findings an approximate adaptation cost and benefit calculation can be estimated. These calculations are shown in Table 6.2.

Table 6.2. Energy sector illustrative key impacts and adaptations (Values in $2010 US.)

Element	Timeslice	Annual costs of current and future climate hazards without climate change ($M)	Annual incremental costs of climate change impacts, without adaptation ($M)	Annual costs of adaptation ($M)	Annual benefits of adaptation ($M)
Heat related blackout	Baseline[1]	$18	-	-7	-26
	2020s	$21	$10 - $22	$9	$37[2]
	2050s	$36	$36 - $73	$19	$76
	2080s	$62	$92 - $206	$38	$154

Notes: The relationship in the tables is not exact due to rounding in calculations. See Technical Notes at the end of the chapter for complete methodology.

[1] *The baseline is based on the cost estimates from blackouts that occurred during the 30-year period from 1966 to 2006, where blackouts occurred in 1977, 1999, 2003, and 2006. All costs were indexed to 2006 values. Blackout costs based on New York City blackouts; scaled up by 3 to produce a state-wide estimate.*

[2] *Based on the findings by the Multi-hazard Mitigation Council (2005a) that every $1 spent in public disaster mitigation results in a $4 savings in non-incurred disaster losses (see also the references in Jacob et al., forthcoming-a).*

Results

Based on the range of estimates from the previous four major blackouts in New York City, indexed to current value and scaled up to New York State, a baseline annual cost of historic heat-related blackouts was found to be $16 million. Assuming no changes in the current climate, this estimate was scaled up with a 2.4% GDP growth rate to find estimates for the midpoints of the 2020s, 2050s, and 2080s. These results were $27 million for the 2020s, $54 million for the 2050s and $111 million for the 2080s. The costs from impacts assuming a change in current climate were then imposed on these values based on the projections of the increase in heatwaves from the Horton et al. (forthcoming). Without adaptation, the estimated annual incremental costs of heat-related blackouts above the baseline estimates were estimated at $13 to 27 million for the 2020s, $54 to 110 million for the 2050s and $161 to 332 million for the 2080s. As explained in the Technical Notes, both the extrapolated without climate change and extrapolated with climate change figures are reduced because of assumed regular, ongoing investment by the energy sector, so that the number of blackouts per heatwave declines over time. In any event, better climate projections will assist the utilities in their planning both for climate and other drivers of energy demand.

If, however, a smart grid system is installed and maintained in New York State, these costs are reduced significantly. For the calculations, it is assumed that one-half of the cost of the smart grid is for climate change; the other half is assumed to be part of regular investment by the energy sector. Additionally, better climate projections will assist utilities in incorporating the changing climate into their planning processes.

PART II. BACKGROUND

6.1 Energy in New York State

This section describes the most important economic components of the energy sector with respect to value at risk to climate change. Energy supply and demand projections for a twenty-year time frame are emphasized in the discussion below. For longer time frames, there are substantial uncertainties associated with the pace of technological change and the development of alternative forms of energy, as well as shifts in the policy and regulatory environment. While this report assumes a GDP growth rate of 2.4 percent for New York State over the next century, is also important to realize that rates of population and economic growth are also uncertain and will have substantial impacts on both energy supply and demand. Taken together, technological changes, policy changes, and rates of growth in demand are likely to be more significant drivers or change of the energy sector than climate change.

The energy sector is generally very risk averse, utilizing a short term planning horizon, conservative engineering estimates, and acting only on reliable information. The risk and probability divisions within utility companies handle climate change, and they are essentially making a bet on the level of climate change that might occur. Utilities hesitant to make investments in this area are concerned with recovering adaptation costs and realize that customers might not want to bear the costs to create a more responsive energy system that would protect against threshold climate conditions (Hammer, 2010).

State GDP and Employment

The size of the energy sector is reported almost exactly in the official State GDP figures issued by the U.S. Bureau of Economic Analysis. The main NAICS classification for energy is Utilities, and the subsidiary parts are: Electric Power Generation, Transmission, and Distribution, Natural Gas Distribution, and Water, Sewage, and Other Systems. (The ClimAID energy sector does not include Water, Sewage, and Other Systems.) New York State has substantial components in each of these. For the 2008 current dollar State GDP figures, New York State GDP was $1.144 trillion; of this total, $20.914 billion was in the utilities sector.

6.2 Key Climate Change Sensitivities

Changes in temperature, precipitation, extreme events, and sea level are anticipated to have adverse effects on energy resources, generation assets, transmission and distribution assets, electricity demand, and buildings. "Weather-related stressors can damage equipment, disrupt fuel supply chains, reduce power plant output levels, or increase demand beyond operational capacity," (Hammer and Parshall, forthcoming). This section specifies which facets of climate change will impact the key economic components of the energy sector (Table 6.3). See also Summary of climate risks to New York energy system; Hammer and Parshall, forthcoming.

Table 6.3. Climate Change Sensitivities: Energy Sector

Increases in mean temperature will affect the thermal efficiency of power generation, change the amount of biomass available for energy generation, alter the water-cooling capacity at power plants, lead to a rise in energy demand, and cause power lines to sag and wear on the transformers. Electrical lines and transformers will fail more often as energy demands exceed the equipments rated capacity.
Increases in extreme heat events and decreases in cold events will change electricity demand patterns and may overwhelm the power supply system in times of summer peak energy demand.
Increases in mean precipitation will reduce the availability and reliability of hydropower generation, as they are dependent upon the timing and quantity of precipitation and snowmelt.
Increases in intense precipitation events will make building and homes more susceptible to flooding, creating the potential of structural damage to boilers.
Snow and ice will damage transmission lines, causing them to sag.
Hurricanes, nor'easters, and extreme winds will damage buildings and energy infrastructure and cause power outages. Extreme weather events may also change energy demand patterns.
Sea level rise will damage coastal power plants.

6.3 Impact Costs

Climate change is anticipated to impact the energy sector in two ways: first, energy demand will change due to a different combination of heating and cooling needs, and second, the physical structures (power plants, electrical lines, etc.) will be affected by changing climate conditions (Dore & Burton, 2000, p. 78). Additional indirect impacts on the energy sector, such as the financial impacts on investors or insurance companies linked to vulnerable energy system assets or on customers forced to grapple with changing energy prices resulting from changing climate conditions, should not be forgotten as they may even be greater than the direct impacts (Hammer and Parshall, forthcoming). The following section presents the costs of climate change impacts for New York State, which are primarily incurred through outages, power prices, loss of income to the utility companies, benefit transferred to the consumer, and additional research.

Power Outages

Economic losses from electric service interruptions are not trivial, as indicated by estimates of damage costs ensuing from major power outages, which may occur during periods of increased energy demand, such as heat waves. The economic impact of the 25-hour blackout that affected most of New York City in July 1977 was assessed at $60 million (estimate may include costs of riots and looting), while the cascading blackout of August 14, 2003 has been estimated to affect approximately 22,000 restaurants, which lost from $75 million to $100 million in foregone business and wasted food. In addition, the City of New York reported losses of $40 million in lost tax revenue and $10 million in overtime payments to city workers (Wharton School 2003).

Other localized service outages in New York City include the July 3-9, 1999 blackout that affected 170,000 Con Edison customers, including 70,000 in Washington Heights (New York State Public Service Commission, 2000); as well as the nine-day blackout that started on July 16, 2006 in Long Island City, Queens, which affected 174,000 residents (Chan 2007). Total claims paid by Con Edison in 2006 amounted to $17 million ($350 to compensate residents and $7,000 to business customers); and an additional $100 million was estimated to be spent by the utility on recovery costs to repair and replace damaged equipment (Office of the Attorney General, 2007). Preventing the losses described above, as well as the number of mortality cases due to heat stress, will require further strengthening of the reliability of the electric grid in order to decrease the number of power outages (paragraph based on Leichenko et al. forthcoming).

Additional analogous impact costs for the energy sector outside NY include:

- In 1998, a massive multi-day ice storm resulted in more than $1 billion in damage across the northeastern United States and eastern Canada. In New York State alone, dozens of high-voltage transmission towers, 12,500 distribution poles, 3,000 pole-top transformers and more than 500 miles of wire conductor required replacement, affecting 100,000 customers from Watertown to Plattsburgh. Most of the repairs were completed within two months, although some areas were not completely repaired for four months (Hammer and Parshall, forthcoming).

- A 2001 survey report found that the estimated cost to US consumers of business losses was between $119 billion to $188 billion per year due to poor power quality, outages and other disruptions (referred to collectively as "reliability events"). The Pacific Gas & Electric Company used direct costs of reliability events to assess that such power disruptions cost its customers approximately $79 billion per year. A 2004 Berkeley National Laboratory comprehensive study of end-users focusing on just power outages, estimated annual losses to the national economy of approximately $80 billion. The figures provided by these studies coincide with estimates by the US Department of Energy, ranging from $25 billion to $180 billion per year (Hammer and Parshall, forthcoming).

- A 2006 IJC report examining alternatives to the 1958-D Order of Approval estimated that the economic impact on hydropower production at NYPA's St. Lawrence/FDR project could vary from -$28.5 million to $5.86 million, depending on which GCM is employed. (The "not-so-warm/wet" scenario was the only one of the four models to produce a positive impact.) The NYPA has developed its own internal estimate, however, that a 1 meter decrease in the elevation of Lake Ontario would result in a loss of 280,000 MWh of power production at the St. Lawrence/FED project (Hammer and Parshall, forthcoming)

The information summarized in the tables below shows the impact costs of power outages and disruptions. Large commercial and industrial customers will experience losses averaging $20,000 and $8,166 for a 1-hour power interruption during a winter afternoon and summer afternoon, respectively. As the power outage increases in duration, so do costs – sharply during the winter and significantly in the summer (Hammer and Parshall, forthcoming).

The total economic cost of a blackout can be estimated by multiplying the affected customers' average value of electricity by data on the magnitude and duration of the power outage. Based on previous analyses, ICF Consulting estimated that the value assigned by consumers to electric power service reliability is on average 100 times its retail price (or a range from 80 to 120 times the retail price). In the case of the 2003 blackout, and assuming a total outage period of 72 hours and using the average electricity price for the region of $93/MWh, the economic cost to the national economy was estimated to be between $7 and $10 billion (Hammer and Parshall, forthcoming).

Table 6.4. Estimated Average Electric Customer Interruption Costs Per Event US 2008$ by Customer Type, Duration and Time of Day

	Interruption Duration				
Interruption Cost	Momentary	30 minutes	1 hour	4 hours	8 hours
Medium and Large C&I					
Morning	$8,133	$11,035	$14,488	$43,954	$70,190
Afternoon	$11,756	$15,709	$20,360	$59,188	$93,890
Evening	$9,276	$12,844	$17,162	$55,278	$89,145
Small C&I					
Morning	$346	$492	$673	$2,389	$4,348
Afternoon	$439	$610	$818	$2,696	$4,768
Evening	$199	$299	$431	$1,881	$3,734
Residential					
Morning	$3.7	$4.4	$5.2	$9.9	$13.6
Afternoon	$2.7	$3.3	$3.9	$7.8	$10.7
Evening	$2.4	$3.0	$3.7	$8.4	$11.9

Source: (Hammer and Parshall, forthcoming).

Table 6.5. Estimated Average Electric Customer Interruption Costs Per Event US 2008$ by Duration and Business Type (Summer Weekday Afternoon)

Interruption Cost	Interruption Duration				
	Momentary	30 minutes	1 hour	4 hours	8 hours
Medium and Large C&I					
Agriculture	$4,382	$6,044	$8,049	$25,628	$41,250
Mining	$9,874	$12,883	$16,366	$44,708	$70,281
Construction	$27,048	$36,097	$46,733	$135,383	$214,644
Manufacturing	$22,106	$29,098	$37,238	$104,019	$164,033
Telecommunications & Utilities	$11,243	$15,249	$20,015	$60,663	$96,857
Trade & Retail	$7,625	$10,113	$13,025	$37,112	$58,694
Fin., Ins. & Real Estate	$17,451	$23,573	$30,834	$92,375	$147,219
Services	$8,283	$11,254	$14,793	$45,057	$71,997
Public Administration	$9,360	$12,670	$16,601	$50,022	$79,793
Small C&I					
Agriculture	$293	$434	$615	$2,521	$4,868
Mining	$935	$1,285	$1,707	$5,424	$9,465
Construction	$1,052	$1,436	$1,895	$5,881	$10,177
Manufacturing	$609	$836	$1,110	$3,515	$6,127
Telecommunications & Utilities	$583	$810	$1,085	$3,560	$6,286
Trade & Retail	$420	$575	$760	$2,383	$4,138
Fin., Ins. & Real Estate	$597	$831	$1,115	$3,685	$6,525
Services	$333	$465	$625	$2,080	$3,691
Public Administration	$230	$332	$461	$1,724	$3,205

Source: (Hammer and Parshall, forthcoming).

Table 6.6. Estimated Average Electric Customer Interruption Costs Per Event US 2008$ by Customer Type, Duration, Season and Day Type

Outage Cost	Outage Duration				
	Momentary	30 minutes	1 hour	4 hours	8 hours
Medium and Large C&I					
Summer Weekday	$11,756	$15,709	$20,360	$59,188	$93,890
Summer Weekend	$8,363	$11,318	$14,828	$44,656	$71,228
Winter Weekday	$9,306	$12,963	$17,411	$57,097	$92,361
Winter Weekend	$6,347	$8,977	$12,220	$42,025	$68,543
Small C&I					
Summer Weekday	$439	$610	$818	$2,696	$4,768
Summer Weekend	$265	$378	$519	$1,866	$3,414
Winter Weekday	$592	$846	$1,164	$4,223	$7,753
Winter Weekend	$343	$504	$711	$2,846	$5,443
Residential					
Summer Weekday	$2.7	$3.3	$3.9	$7.8	$10.7
Summer Weekend	$3.2	$3.9	$4.6	$9.1	$12.6
Winter Weekday	$1.7	$2.1	$2.6	$6.0	$8.5
Winter Weekend	$2.0	$2.5	$3.1	$7.1	$10.0

Source: (Hammer and Parshall, forthcoming).

Table 6.7. Value of Service Direct Cost Estimation

Facility Outage Impacts			Annual Outages		Annual Cost	
Power Quality Disruptions	Outage Duration per Occurrence	Facility Disruption per Occurrence	Occurrences per Year	Total Annual Facility Disruption	Outage Cost per Hour*	Total Annual Costs
Momentary Interruptions	5.3 Seconds	0.5 Hours	2.5	1.3 Hours	$45,000	$56,250
Long-Duration Interruptions	60 Minutes	5.0 Hours	0.5	2.5 Hours	$45,000	$112,500
Total			3	3.8 Hours		$168,750
Unserved kWh per hour (based on 1,500 kW average demand)			1,500 kWh			
Customer's Estimated Value of Service (VOS), $/unserved kWh			$30 /unserved KWh			
Normalized Annual Outage Costs, $/kW-year			$113 $/kW-year			

Source: (*Hammer and Parshall, forthcoming*).

6.4 Adaptation Costs

Adaptation costs in the energy sector are positively correlated with the level of temperature increases and economic growth (Dore & Burton, 2000, p. 79). In addition to temperature change, other important factors that influence economic costs in the energy sector include population growth projections, fuel price changes, and the GDP (Dore & Burton, 2000, p. 80). However, current literature on adaptation costs is primarily focused on increases in energy demand for cooling in the summer and reduced heating in the winter (Agrawala et al, 2008, p. 56). Many studies have concluded that for the United States the adaptation costs of increased cooling will be greater than the benefits of reduced heating demands (Agrawala et al, 2008, p. 57-58). An overview of adaptation possibilities in the energy sector is in AAC (2010), pp. 88-91. Some estimates of the costs of climate change adaptation strategies relevant to New York State are given in the following paragraphs.

The existing power system infrastructure in the US was recently valued at $800 billion (Hammer and Parshall, forthcoming). Because this system requires constant refurbishment and eventual replacement over long timescales, it will make sense to align implementation of adaptation measures into the natural replacement cycle of vulnerable system assets.

Adaptation strategies generally target either supply or demand. Supply related measures often emphasize physical improvements to enhance the capacity of power generation, transmission, and distribution to better operate under a range of future climate conditions. Demand related measures target all types of energy consumption, from taxes to public education programs (Hammer and Parshall, forthcoming).

Out of the numerous adaptation strategies presented, Hammer and Parshall (forthcoming) have identified NYSERDA as a stakeholder in the position to implement the following measures:

Energy Supply

- Install solar PV technology to reduce effects of peak demand
- Develop non-hydro power generation resources to reduce need for hydropower generation during winter

Energy Demand

- Design new buildings with improved flow-through ventilation to reduce air conditioning use
- Increase use of insulation in new buildings and retrofit existing buildings with more insulation and efficient cooling systems
- Improve information availability on climate change impacts to decision makers and public
- Plant trees for shading and use reflective roof surfaces on new and existing buildings
- Install power management devices on office equipment
- Upgrade building interior and lighting efficiency
- Improve domestic hot water generation and use
- Improve HVAC controls
- Upgrade elevator motors and controls
- HVAC design improvements
- More efficient HVAC equipment
- Improved steam distribution
- Weatherize low income households

The costs of several adaptations are as follows:

Saltwater Resistant Transformers

Con Edison voluntarily launched a 10-year plan beginning in 2007 to replace 186 underground transformers located in Category 1 floodplains around NYC for a cost of $7 million. New saltwater submersible transformers can better handle storm surge intrusion than the equipment currently in place (Hammer and Parshall, forthcoming; New York State Department of Public Service, 2007). However, utility companies can be reluctant to install more of these transformers if they think that they will be unable to recover the costs through higher rates.

Back-up Generators

The energy grid may change over time to more distributive power (Hammer, 2010). Gridpoint's Connect Series unit, a battery back-up system for houses, is a step in this direction. The unit costs around $10,000 and is the size of a refrigerator. It has the capacity to store 12kWh of usable AC electricity and helps electricity utilities and customers manage energy more intelligently. Telecommunication grade lead acid batteries are used in the unit, which last for five years and cost about $185 per usable kilowatt-hour of AC current.

The benefits of distributive storage include reliable constant power, even during power outages, because stored electricity can be discharged back into the grid beyond the break line. Also, electricity can be stored during low off peak rates and discharged when rates are higher in markets where energy pricing is tiered. Distributive power can even flatten the electricity load and relieve congestion on the grid by pushing power into the grid during peak hours of demand from distributed sources. Distributed renewable energy sources, i.e. wind and solar, can be captured by the storage system during their limited hours of collection and utilized at any time (EcoWorld, http://www.ecoworld.com/technology/gridpoints-storage.html).

Smart grid. Smart grid technology provides operators with the information necessary to properly manage power flows and transmission systems by creating a clearer metric of potential risk to avoid major power outages. A recent study proposed installing sensors every ten miles over the existing 157,000 miles of transmission lines nationwide at a cost of $25,000 per sensor, amounting to $100,000,000 if the sensors are replaced every five years. Average residential monthly utility bills would increase by 0.004 cents per kilowatt-hour. The total cost for the proposed service would be about one tenth of the estimated annual cost of blackouts (Hammer and Parshall, forthcoming). Other components of smart grids include two-way communication systems between producers and consumer, and can include the possibility of integrating renewable energy generated by consumers into the system.

Costs for additional adaptation strategies include:

- The Energy Department expects that electricity use and production will increase by 20% over the next decade; however the nation's high-voltage electric network will only increase by 6% in the same time period. After the major blackout of 2003 many have been calling for investments ranging from $50 billion to $100 billion to reduce severe transmission bottlenecks and increase capacity (Hammer and Parshall, forthcoming).

- In some places adaptation cost incentive programs can be used to prevent power outages. Customers participating in voluntary options such as the "Distribution Load Relief" program must be reduced at least 50kW or 100kW, for individuals or aggregators respectively to receive compensation of at least $0.50 per kWh after each event (Hammer and Parshall, forthcoming).

6.5 Summary and Knowledge Gaps

- Research is needed to better understand how climate change may affect markets for gas and oil, as well as how climate change may affect the breakdown of demand for natural gas for building heat versus power generation (Hammer and Parshall, forthcoming).

- There is a need for additional research analyzing trends in a wider range of climate variables, including how seasonal and extreme trends may affect electricity demand (Hammer and Parshall, forthcoming).

- Research is also necessary to better understand how upstate utility companies will be monetarily affected by a decreased heating demand in the future (Hammer, 2010).

- An initial assessment of the relationship of a carbon tax (or cap and trade) on the energy sector is needed as a foundation for a range of policy choices, including the impacts or climate change and adaptations on the sector.

- A more extensive analysis of how substantial investments not now planned, such as making the electric grid resilient against electromagnetic storm will impact policies for climate adaptation is needed.

- Both supply and demand adaptation strategies often serve a dual role as climate change mitigation strategies, depending on the temporal scale, cost level, target audience, technology and policy decisions, and decision rules emphasized and more should be learned about these dual roles (Hammer and Parshall, forthcoming).

Technical Notes – Energy Sector

Impact: Heat-related blackouts
Adaptation: Smartgrid

Assumptions

- 2.4% GDP growth rate *(= to the long term US GDP growth rate)*
- Heat-related blackouts can also serve as a proxy for heat waves and thunderstorms.
- The baseline is based on the 30-year period from 1966 to 2006, where blackouts occurred in 1977, 1999, 2003, and 2006.
- All costs were indexed to 2006 values.
- Blackout costs based on New York City blackouts; scaled up by 1.3 to produce a state-wide estimate.
- Based on the findings by the Multihazard Mitigation Council that every $1 spent in public disaster mitigation results in a $4 savings in non-incurred disaster losses (Jacob et al., forthcoming-a).
- Based on a report finding the cost to install a $25,000 sensor every 10 miles over the existing US transmission line system that would cost $100M per year if the sensors are replaced every 5 years (Apt et al, 2004, http://www.issues.org/20.4/apt.html).
- Electricity customer and consumption information from http://www.eia.doe.gov/cneaf/electricity/esr/table5.html.

Baseline:

1. To find the baseline impact cost of blackouts in NYC, estimates of impacts were taken from available literature and studies, including Hammer and Parshall (forthcoming), to create a potential range of impact costs for each previous blackout (1977, 1999, 2003, and 2006).
    a. For the 1977 New York City-wide blackout, the ClimAID Energy chapter notes that the impact cost estimates for the blackout are roughly around $60M (low range). Another estimate from a 1978 report prepared for the Department of Energy by Systems Control Incorporated estimated the total cost of the blackout to be $290M (http://blackout.gmu.edu/archive/pdf/impact_77.pdf) (high range).
    b. To calculate the 1999 costs estimate for the heat wave that affected 170,000 Con Edison customers, the literature reported that ConEd compensated individuals $100 for spoilage of food and medicine and businesses $2,000. The low estimate assumption is that all 170,000 affected were residents while the high estimate assumes that all customers were businesses. Therefore, the total costs range from $17M to $340M.
    c. For the 2003 city-wide storm, estimates range from $125M (estimates from Hammer and Parshall [forthcoming]: $75-100M lost by restaurants, $40 in lost tax revenue, and $10M in overtime payments to city workers) to $1B (given by NYC's Comptroller William Thompson).

   d. The 2006 Queens blackout low cost estimate of $117M includes the Con Edison total claims amount, plus the estimated spending on recovery costs to repair and replace damaged equipment ($17M + $100M). The high end of the range is $188M, found in a study done by the Pace Energy and Climate Center (http://www.crainsnewyork.com/article/20100716/FREE/100719876).

2. Average the range of costs for each blackout. The averages are: $175M in 1977, $179M in 1999, $563M in 2003, and $153M in 2006.
3. Index these costs to $2006. All values were indexed using the CPI Inflation Calculator on the US BLS website: http://www.bls.gov/data/inflation_calculator.htm. The indexed averages are: $582M in 1977, $217M in 1999, $617M in 2003, and $153M in 2006.
4. Take the average of the indexed values (=$392M).
5. To calculate the annual costs, divide the average of indexed values by the number of years (30) over which these blackouts occurred (1966-2006). The annual blackout cost over a 30-year period is $13M.
6. To scale up the annual cost from New York City to New York State, multiply by 1.3 (based on the assumption that, on average, annual state-wide costs would be 30% of those for a New York City blackout). The total is $17M.
7. Project the baseline cost into the future using a 2.4% GDP. To find the total cost per blackout (for use in later calculations), multiply the annual blackout cost by 10 (based on the assumption of a 1-in-10 year blackout).

Annual incremental cost of climate change impacts, without adaptation:

8. Based on the ClimAID heat wave observations and projections, there are currently 2 heat waves per year (defined as 3 or more consecutive days with a maximum temperature exceeding 90°F). Assuming blackouts occur once in every 10 years (Wharton School 2003), it can be estimated that 1 out of every 20 heat waves results in a blackout. However, it can be assumed that the energy sector's continued investment for general purposes (rather than specifically for climate change)—the "without" investment--will reduce this incidence, perhaps substantially, as the industry routinely operates in a warmer environment.
9. Following the climate change heat wave projections in ClimAID, the projected increase in heatwaves per year is 3 to 4 per year in the 2020s, 4 to 6 per year in the 2050s and 5 to 8 year in the 2080s. Based on this information, and if blackouts were to continue to occur once in every 20 heatwaves, then blackout occurrences would increase to 1 blackout every 6.7 to 5 years in the 2020s, 1 blackout every 5 to 3.3 years in the 2050s, and 1 blackout every 4 to 2.5 years in the 2080s. However, it would be more realistic to assume a lower incidence of blackouts/heatwaves, as noted above. Instead, for this extrapolation, it is assumed that in the 2020s blackouts will occur once in every 25 heatwaves (instead of the one in 20 now; the estimates for the 2050s and 2080s are one in every 30 heatwaves, and one in every 35. This secular improvement in system reliability is assumed to reflect constant improvements in the industry.
10. Using the total cost per blackout found in step 7, estimate projected annual blackout costs by dividing the new yearly occurrence interval into the total cost per blackout for the respective timeslice. These annual costs were then subtracted from the annual

average baseline costs without climate change for the respective timeslices . All of the costs calculated in this way, both with and without climate change, were reduced by the factors of 20/25, 20/30, and 20/35, respectively, for the 2020s, 2050s, and 2080s, reflecting the secular improvement in system efficiency.

Annual costs of adaptation:

11. The annual estimated cost to install and maintain a smart grid system in the US (with 1 sensor every 10 miles over 157,000 miles of transmission wire, where sensors cost $25,000 and need to be replaced every 5 years) is $100M per year (Apt et al, 2004). It can then be assumed that the cost to New York State is proportional to its energy consumption when compared to the national level, which is 4%. Therefore, the estimated cost of a smart grid system for New York State is $4M per year. It was assumed that this was one of 5 adaptation options of the same cost, and that 0.3 of the total was due to adaptation and the remainder to other pressures., so that adaptation costs in the first year of the example are $6.

Annual benefits of adaptation:

12. Based on the Multihazard Mitigation Council finding that "for every $1 spent in public disaster mitigation there is a savings of $4 in non-incurred disaster losses" (Jacob et al., forthcoming-b), multiply the total annual adaptation cost of $4M by 4. This results in an annual benefit of $16M.
13. Project out the annual future benefit ($16M) at a 2.4% GDP growth rate, adjusted for the 50% element that is not for climate adaptation.

Incremental costs of climate change impacts with adaptation:

Subtract the findings from step 13 from the incremental annual costs without adaptation found in step 10.

$US 2010 adjustment:

All of the figures in the example were adjusted to $US2010 using the United States Bureau of Labor Statistics CPI Inflation Calculator, http://data.bls.gov/cgi-bin/cpicalc.pl to yield the final calculations. This calculator was also used for other adjustments throughout the report.

7 Transportation

The transportation sector in New York State is an essential part of the economy and culture of the state; with its many modes and organizations, it is a complex system. There is a very large range of potential impacts of climate change on the state's transportation sector from the principal climate drivers of rising temperatures, rising sea levels, higher storm surges, changing precipitation patterns, and changes in extreme events such as floods and droughts. This analysis estimates that total impacts without adaptation could be in the hundreds of billions of dollars. Adaptations are available that would be cost-effective. Planning for these should begin as soon as possible.

PART I. KEY ECONOMIC RISKS AND VULNERABILITIES AND BENEFIT-COST ANALYSIS FOR TRANSPORTATION SECTOR

Key Economic Risks and Vulnerabilities

Of the many vulnerabilities, the most economically important include first the impacts on infrastructure investment and management of rising sea levels and the accompanying increase in storm surges for coastal areas. These effects will impact all forms of transportation in coastal areas, where a large proportion of fixed investment is close to the present sea level (roads, airports, surface rail) and a significant fraction (tunnels, subways) is below sea level (Jacob et al., forthcoming–a). One of many examples of low-lying infrastructure is the Corona/Shea yards in Queens, NYC (Rosenzweig et al., 2007a). These yards are used to store subway and LIRR cars, respectively, for rush hour and other use. They flood under current conditions, and will be still more vulnerable as sea level rises. In addition to coastal flooding from sea level rise and storm surges inland flooding and urban flooding from intense storms create other important vulnerabilities in the transportation sector.

Another important vulnerability economically is increased transportation outages attributable to climate change. To the extent that extreme events increase in frequency (floods, droughts, ice storms, wind) these will impact all forms of transportation throughout New York State. The August 8, 2007 storm, for example, had severe impacts on transportation throughout the NYC area; these are detailed by mode in Metropolitan Transportation Authority (MTA) 2007. The main climate and economic sensitivities are shown in Table 7.1.

The expected impacts of climate change on transportation in New York State are very great. An example for the 100-year hurricane, based on the detailed example in Jacob et al. (forthcoming-a) and potential adaptation costs are given in Table 7.2. An increment for upstate storms is included also. In this sector, the stated storm (100-year hurricane) essentially covers all transportation for the given storm. However, this will be an understatement of damages, as many other storms will also take place, including contributions from both smaller and some greater than the 100-year storm; and from non-storm related climate factors (e.g. heat waves).

Table 7.1. Climate and Economic Sensitivity Matrix: Transportation Infrastructure Sector (Values in $2010 US.)

Element	Main Climate Variables				Economic risks and opportunities: – is Risk + is Opportunity	Annual incremental impact costs of climate change at mid-century, without adaptation	Annual incremental adaptation costs and benefits of climate change at mid-century
	Temperature	Precipitation	Sea Level Rise & Storm Surge	Atmospheric CO_2			
Permanent and temporary coastal flooding from SLR and storm surge			●	●	-Damage to all modes of transportation in low-lying areas, including increased transportation outages	$100-170M for 100-year hurricane and some upstate losses	Costs: $290M Benefits: $1,160M
Inland flooding		●			-Damages to all modes of transportation in flood plains, including increased transportation outages	Substantial costs to be estimated	Improved culvert design, flood walls
Track and other fixed investment	●		●		-Potential buckling of tracks -Damage to road surfaces + Longer season for maintenance and repairs	Monitoring of climate change required	Revised design standards
Power Outages	●	●	●		-Impacts on subway and train power -Impacts on signals on highways an local streets -Impacts on airport operation	Significant economic and social impacts	Smart grid and other investment costs
Total estimated costs of key elements						$100-$170M	Costs: $290M Benefits: $1,160M

Note that the damages are annualized, although the incident is a single storm.

Key for color-coding:

(dark gray)	**Analyzed example**
(light gray)	**From literature**
(white)	**Qualitative information**
(white)	**Unknown**

Table 7.2. Illustrative key impacts and adaptations: Transportation Infrastructure Sector (Values in $2010 US.)

Element	Timeslice	Annual costs of current and future climate hazards without climate change ($M)[1]	Annual incremental costs of climate change impacts, without adaptation ($M)[2]	Annual costs of adaptation ($M)[3]	Annual benefits of adaptation ($M)[4]
Outages from 100 year hurricane and upstate intense rainfall	Baseline	$520	-	-	-
	2020s	$740	$10 - $40	$140	$570
	2050s	$1510	$100 - $170	$290	$1160
	2080s	$3080	$320 - $410	$590	$2370

[1] *Based on the 100-year hurricane study in the Transportation chapter, adjusted to remove the estimated New Jersey portion of the NY Metro area, and increased by 5% to reflect upstate intense rainfall events, and annualized.*
[2] *Based on the growth of damages given in Jacob et al (forthcoming-a). between the present sea level and a SLR of 2 feet, using the range of SLR scenarios in NPCC (2010) SLR scenarios, p. 172, and scaled up for growth in damages.*
[3] *Taken as beginning in 2010 with $100m in annual investment, the low end of the range of figures given in Jacob et al. (forthcoming-a) (100s of $millions to $billions annually).*
[4] *Based on the estimate in Multihazard Mitigation Council (2005a) of a 4:1 benefit cost ratio for hazard mitigation investments (see also the references in Jacob et al. (forthcoming).*

Results

The costs of climate change are expected to be substantial in the transportation sector, with its heavy fixed capital investment, much of it at or below sea level and subject to large impacts from sea level rise and storm surges. As the example in Table 7.2 indicates, costs of impacts are expected to be very large; adaptations are available, and their benefits may be substantial. While the numbers in the example depend on the input assumptions, within a fairly wide set of assumptions the estimates will be very large. As other examples in the sector where climate change impacts are expected to be substantial, all modes of upstate transportation systems will be affected by more intense storms, inland flooding, winds and heat.

PART II. BACKGROUND

7.1 Transportation in New York State

Transportation is an essential element of New York State's economy and society. The state not only has a full complement of roads and road traffic, but also possesses, in the New York metropolitan area, the major share of the largest public transportation complex in the United States. Further, the Port of New York and New Jersey is one of the largest in the nation; there are 3 high-traffic airports in the New York City area, and many smaller commercial and private airports. There is also an extensive rail network. These systems are quite dense, most of all in the New York Metropolitan Area (see Figure 7.1 for subways and rail lines), but also in terms of the highway and rail networks of New York State as a whole. As fully described in Jacob et al. (forthcoming-a), these systems are operated by a multitude of public and private entities.

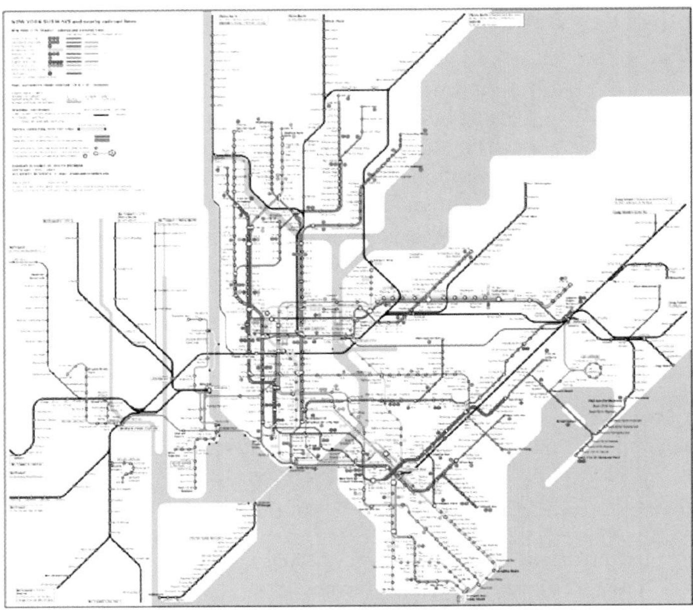

Figure 7.1. Schematic map of rail systems of the NYMA.
Source: http://www.columbia.edu/~brennan/subway/Subwaymap.gif

The transportation sector is one of those in ClimAID in which the size of the sector is reported almost exactly in the official state GDP figures issued by the U.S. Bureau of Economic Analysis. Industries are divided into North American Industry Classification System (NAICS), (U.S. Bureau of Economic Analysis, n.d.) covering Canada, the U.S. and Mexico; these replace the former Standard Industrial Classification codes used in the US. The main NAICS classification for transportation is transportation and warehousing, excluding Postal Service, and the subsidiary parts are: Air transportation; Rail Transportation; Water transportation; Truck transportation; Transit and ground passenger transportation; Pipeline transportation; and Other transportation and support activities. New York State has substantial components in each of these. For the 2008 current dollar state GDP figures, New York State GDP was $1,144,481,000,000; of this total, $19,490,000,000 was in the transportation sector. (The state figures do not break down the subcomponents.) It is also of interest that total 2008 current dollar GDP for the NY-Northern NJ-Long Island NY-NJ-Pa Metropolitan Statistical Area (MSA) was $1,264,896,000,000; the transportation sector figure is not provided to avoid disclosure of confidential information. This MSA includes 1 county in PA (Pike) and none in CT.

These figures, while of great interest in comparing current output of different sectors, are flow figures, that is, output per period of time (in this case, one year). They thus understate the immense importance of transportation to the state, which is perhaps better defined in terms of the way in which transportation activities are intertwined in nearly every action of government, businesses, and private citizens. This importance is also emphasized by the enormous capital investments in the transportation sector in New York State. As examples, Jacob et al.

(forthcoming-a) cites asset values of $10 billion for Metro North, $19 Billion for the Long Island Rail Road, and $25 billion for MTA bridges and tunnels.

7.2 Key Climate Change Sensitivities

Climate sensitivities in the transportation sector are described in detail in Jacob et al. (forthcoming-a); a comprehensive list for the nation as a whole is given in the Annexes to Chapter 5 in National Research Council (2008). Another comprehensive source is Canadian Council of Professional Engineers (2008). The most significant impacts are shown in Table 7.3:

Table 7.3. Key climate changes sensitivities: Transportation Infrastructure Sector

Rising sea levels and the associated storm surges will cause flooding of the large transportation systems in the state in coastal areas, including road, rail, aviation and maritime transport facilities.
Potentially more frequent and intense precipitation will cause inland flooding from events on roads, public transit systems and railroads, leading to more frequent outages.
Increased ice storms, especially in Central and Northern New York State, will impact all forms of transportation.
Weather-related power failures will impact all forms of transportation.
Higher temperatures and more frequent heat waves may adversely impact rail tracks and other fixed investment.

7.3 Impact costs

In estimating the costs of climate change in the transportation sector in New York State, relatively standard methods can be applied; however, data are often inadequate and the uncertainties in the climate sector are large, compounded by uncertainties in other drivers such as population and real income growth. In many cases, however, an assessment of magnitude can be obtained. Such is the result of the case study in Jacob et al. (forthcoming-a), in which a moderately strong storm's flooding impacts on the New York Metropolitan region are estimated, and then sea level rise is added to indicate the impact of climate change. The selected storm is a hurricane that would produce coastal flooding equivalent to the 100 year flood (as currently calculated). Then, sea level rises of 2 and 4 feet are added, and the flooding from the same storm is estimated. Impacts on the relevant transportation structures are calculated, and then estimates are made of the extent of transportation outages. These damages include both above-ground and below-ground systems that will require repair (Jacob et al., forthcoming-a). (In addition, hurricanes result in flooding damages to non-transportation infrastructure below street level, and much of this infrastructure is needed for a fully functioning transportation system.) Using the simplifying assumption that the overall economic impact would be a direct result of the relative functionality of the transportation systems, an estimate is made of the economic loss per day until nearly full functionality is restored. In addition to the economic losses, direct damages to physical transportation infrastructure are estimated. The results are given in Jacob et al. (forthcoming-a) Table 4, adapted here as Table

7.4, where estimates of combined economic costs and physical infrastructure damage are given for the 3 scenarios. These are given for 2010 asset values and 2010 dollar valuation.

Table 7.4. Combined Economic Production and Physical Damage Losses, in Billions, for the Metropolitan Region for a 100-year Storm Surge for three SLR Scenarios (for 2010-Assets and 2010-Dollar Valuation).

Scenario	Economic Production ($Billion)	Physical Damage ($Billion)	Total Loss ($ Billion)
S1	$48	$10	$58
S2	$57	$13	$70
S3	$68	$16	$84

S1=current sea level; S2 = S1 + 2 ft; S3=S1 + 4 ft.

Interpreting the results, the climate change costs of the impacts are the initial scenario costs subtracted from the larger future costs due to sea level rise, or $12 billion and $26 billion respectively for the chosen storm. These costs are underestimates, because asset values will rise over time; and they may be underestimates also because storm frequency and intensity may increase.

In the Jacob et al. (forthcoming-a) study, the possibility of lives being lost is acknowledged but not included. The most recent northeast hurricane that caused significant loss of life was Floyd (1999), a Category 2 hurricane. Blake et al. (2007) give the number of lives lost as 62 for that event. For the future, the possibility of deaths from hurricanes in the New York State coastal region depends on several factors. The coastal counties have well-developed evacuation plans (Jacob et al., forthcoming-a), with most residents living within a relatively short distance of higher ground. At the same time, it can be expected that hurricane tracking systems will improve continuously, so that the available time for evacuation will tend to grow over the years. However, there are some possible scenarios where there could be extensive loss of life, from wind damage as well as flooding, and this should be taken into account in adaptation planning. As a monetary measure of lives lost (not of course a full basis for decision-making), the Public Health chapter of this report gives an estimate of $7.4 million ($2006) per life.

For a full accounting of sea level rise and associated storm surge damages in the NYMA, the costs from all storms with different recurrence intervals or annual probabilities would have to be examined and the results summed, an effort that would be difficult to accomplish with current data; however, the case study shown, by indicating the magnitude of damages from a moderate storm, suggests very much higher damages if all storm probabilities and their related costs are considered. It should also be noted that one reason that impacts on transportation are high in the NYMA is that much of the fixed investment is underground, at or below sea level and is currently not well protected. It should be noted that these are the costs of impacts without adaptation measures—there will undoubtedly be adaptations that would reduce these impacts.

In summary, while there are many assumptions that go into such a calculation, the overall level of magnitude indicates that losses from climate change in the NYMA from SLR and storm surge will be substantial without suitable adaptation. These costs, without adaptation, for the transportation sector could be in the hundreds of $billions. The reductions in such costs that are attributable to adaptation measures constitute the benefits of the adaptations. Many available adaptations to climate change in this sector will be both worthwhile and essential. These will have to be planned and implemented in a carefully staged manner to stay ahead of the worst of the impacts.

7.4 Adaptation Costs

There is a wide range of potential adaptations to the impacts of climate change on transportation systems; these can be divided into adaptations for: management and operations; infrastructure investment; and policy. Adaptations can also be classified as short-, medium- and long-term; examples of these are in Jacob et al. (forthcoming-a). Costs vary substantially among different types of adaptations; and the adaptations need to be staged, and integrated with the capital replacement and rehabilitation cycles (Major and O'Grady, 2010). There has begun to be a substantial number of studies about how to estimate the costs of adaptations, and in some cases, cost estimates (Parry et al. 2009; Agrawala, and Fankhauser, eds., 2008).

Among adaptations for New York State transportation systems will be changes to cope with rising sea levels and the accompanying higher storm surges, and climate-related transportation and power outages throughout New York State. While costs for adaptations, as opposed to discussions of methods, are not widely available as yet, some sense of the magnitude can be obtained by considering available information on hazard reduction. The Multihazard Mitigation Study (2005b) examined the benefits and costs of FEMA Hazard Mitigation grants, including one set of grants to raise streets in Freeport, NY (pp. 63-64 and 107) to prevent flooding under existing conditions. (A companion effort to raise buildings is described in the OCZ chapter.) These totaled about $2.76 million, including a 25% local matching contribution. The study examined a wide range of parameter values of benefits and costs, and concluded that the total Freeport benefit-cost ratio best estimate was 2.4; the range is shown Table 7.5. This provides some sense of what might be required in the future in coastal areas such as Freeport, which of course do not have underground transit lines as does the inner core of the NYMA.

Table 7.5. Benefit Cost Analysis of Potential Climate Change Adaptation: Raising Local Streets Subject to Flooding

Activity in Freeport, NY	Total Costs (2002 $M)	FEMA Costs (2002 $m)	Best Estimate Benefits (2002 $M)	Best Estimate Benefit-Cost Ratio	BCR Range
Street grading/elevation	$2.76	$2.07	$6.52	2.4	0.19-9.6

Source: adapted from: Multihazard Mitigation Council, 2005b, vol. 2, p.107,Table 5-14.

An example of larger costs for adaptation of transportation systems comes from Louisiana, which is in the process of upgrading and elevating portions of Louisiana Highway 1, which in its current configuration floods even in low-level storms. The project has several phases and includes a four-lane elevated highway between Golden Meadow, Leeville, and Fourchon to be elevated above the 500-year flood level and a bridge at Leeville with 22.3-m (73-ft) clearance over Bayou LaFourche and Boudreaux Canal. Construction has begun on both the bridge project and a segment of the road south of Leeville to Port Fourchon. The bridge project has a value of $161 million, and while this might be taken as an adaptation to current conditions and risks rather than climate change, it is indicative of the level of costs for large infrastructure projects subject to coastal storms, the impact of which will increase substantially with rising sea levels. (Savonis et al., 2008, p. 4-55).

A second example of estimating the costs of actual design for climate change adaptation of a transportation project is in Asian Development Bank (2005). This case study examined a road building development plan for Kosrae in the Federated States of Micronesia, specifically a 9.8-km unbuilt portion of the circumferential road north of the Yela Valley. This route is subject to flooding; the specific design climate driver was chosen in this case is the hourly rainfall estimated with a 25 year return interval. This was forecast to rise from 190 mm to 254 mm in 2050. There is a detailed climate-proofed design plan for the road design, including construction, maintenance and repair costs for the built and unbuilt sections of the road. The estimated marginal cost for climate-proofing is $500,000; the study further concludes that would be more costly to climate proof retroactively. As of the report date, the Kosrae state government decided not to proceed with construction of the road until additional funds were available for climate proofing. This example, although in a tropical area with higher rainfall than New York State, presents a typical problem in road design that is relevant to the state—adaptation of designs to more intense rainfall.

A pioneering large infrastructure decision actually made on the basis of adaptation to sea level rise is in Canada: "…the designers of the new causeway to Prince Edward Island made it one meter higher than it would otherwise have been" (Titus, 2002, p. 141). This structure, completed in 1996, is called the Confederation Bridge. Because the adaptation to sea level rise was included in the initial designs, the marginal cost of the adaptation was not estimated. (This might, however, be possible with a detailed examination of the design documents.)

A very large-scale adaptation relevant to the reduction of climate change impacts on transportation is a set of surge barriers for New York Harbor; these are described in the OCZ chapter. However, such a regional solution needs a thorough analysis of its long-term sustainability for the scenarios under which sea level rise continues beyond the height and useful lifetime of such barriers (say, for example, 100 years)--an exit strategy. Benefit-to-cost ratios can change with time, and the question arises what is the proper time horizon for making decisions, and how can adaptation (and its cost) be adjusted to uncertain future long-term conditions of climate, economics and demographics.

For still other adaptations, on a much shorter time scale, costs have not yet been estimated but could be estimated from existing information and reasonable forecasts. For example, the New York State Department of Transportation has a 24/7 emergency command center in Albany to deal with road blockages and outages from extreme events. The NYSDOT is able to move resources among its divisions fairly quickly because of this information center. If extreme events increase due to climate change, it would be expected that the budget for this operation and the associated costs of resource movement would increase gradually over time; these budget increases would be costs of adaptation.

7.5 Summary and Knowledge Gaps

From the standpoint of improving the ability of planners to do economic analysis of the costs of impacts and adaptations in the transportation sector, there are many knowledge gaps to which resources can be directed. These include:

- A comprehensive data set in GIS or CAD form of as-located elevations of transportation infrastructure relative to current and future storm surge inundation zones and elevations.
- Increased staffing of planning and risk management units in transportation agencies
- Updating of FEMA and other flood maps to reflect the impacts of rising sea levels.
- Undertaking of a series of comprehensive benefit-cost analysis of potential adaptations to aid in long term planning.
- Integration of population projections into climate change planning.
- More advanced planning for power outages and their impacts on transportation.
- Forecasts of improvements in information technology, such as hurricane models, which should be able to provide improved real-time forecasts to enable more efficient evacuation planning.

Technical Notes – Transportation Infrastructure Sector

Methods for estimating transportation impact and adaptation costs for 100-year hurricane:

1. This extrapolation is based on the transportation case study in Jacob et al. (forthcoming-a).

2. The total loss for the baseline is $58 billion for the reference study, or $.580 billion annually.

3. This is for the NY Metro area. This includes 1 county in PA (Pike), 10 in NJ, and none in CT.

4. The total loss was reduced by 15% to exclude the transportation-related losses for NJ, and was then increased by 5% to include transportation related intense rainfall outages in New York State. This yields $.520 billion annually. The growth in annual costs was projected with the long term US GDP growth rate of 2.4%. This was used because the example in the transportation chapter is for current asset values.

5. Then, the incremental losses were estimated by using the range of SLR in inches for benchmark years, times the increased loss per inch. The increased loss per inch is $.5 billion, taken linearly from the increase of 12 billion for an increase of 24 inches. The annualized incremental loss is 5 million.

6. Adaptation costs were reduced by judgment to the low end of the ranges given in the ClimAID Transportation chapter, which go upward into the billions of dollars per year. The lower range was chosen because the ClimAID figures include not only adaptations to future climate but also needed infrastructure spending for general purposes.

7. Benefits (reduction in costs) were based on empirically derived 4:1 figure in the Transportation chapter. Because so many important adaptations have not been made, annual benefits may be higher than the conservative estimate used here.

8 Telecommunications

The capacity and reliability of New York State's communication infrastructure are essential to its economy and consequently to the effective functioning of global commerce (Jacob et al., forthcoming-b). The communications sector includes point-to-point switched phone (voice) services; networked computer (Internet services, with information flow guided by software-controlled protocols; designated broadband data services; cable TV; satellite TV; wireless phone services; wireless broadcasting (radio, TV); and public wireless communication (e.g. government, first responders, special data transmissions) on reserved radio frequency bands (Jacob et al., forthcoming-b). The sector poses special challenges to climate change analysis. Businesses in the sector are reluctant to disclose some classes of information that would be relevant to climate change assessments, due to competitive pressures and also concerns about potential additional regulation (Jacob et al., forthcoming-b). Thus, as compared to some other ClimAID sectors, it is relatively difficult to quantify the costs of climate change impacts on capacity and reliability and adaptation strategies to protect these assets. Adaptation costs can be minimized if adaptations to climate change are incorporated into the existing short-term planning schedule. Adaptation costs could then become standard equipment update/upgrade costs rather than additional replacement costs.

PART I. KEY ECONOMIC RISKS AND VULNERABILITIES AND BENEFIT-COST ANALYSIS FOR TELECOMMUNICATIONS SECTOR

Key Economic Risks and Vulnerabilities

By affecting systems operations and equipment lifespan, more intense precipitation events, hurricanes, icing and lightning strikes, and higher ambient air temperatures (Connecticut Climate Change Infrastructure Workgroup of the Adaptation Subcommittee, 2010) will impact the capacity and reliability of the communications infrastructure sector. Table 8.1 identifies the climate variables that are likely to impact the sector along with the project economic outcome. Note that economic risks significantly outweigh opportunities. Furthermore, this sector integrates and overlaps with each of the other sectors and impacts in the communication sector will likely have secondary or tertiary effects throughout the economy.

Table 8.1. Climate and Economic Sensitivity Matrix: Telecommunications Sector (Values in $2010 US.)

Elements	Main Climate Variables				Economic risks and opportunities: – is Risk + is Opportunity	Annual incremental impact costs of climate change at mid-century, without adaptation	Annual incremental adaptation costs and benefits of climate change at mid-century
	Extreme Events: Heat	Extreme Events: Ice and Snow Storms	Extreme Events: Hurricanes, Rain, Wind & Thunderstorms	Electric Power Blackout			
Equipment Damage System Failure	●	●	●	●	– Damaged power and communication lines and poles – Infrastructure damage – Unmet peak energy demands (i.e. for AC) will cause power outages and incidentally communication outages	$15-30M	Costs: $12M Benefits: $47M
Total estimated costs of key elements						$15-30M	Costs: $12M Benefits: $47M

Key for Color-Coding:

(dark gray)	**Analyzed example**
(light gray)	**Analogous number or order of magnitude**
(very light)	**Qualitative information**
(white)	**Unknown**

Winter storms can result in outages in communications systems, a key concern for the sector relating to climate change. Past storms have resulted in communications outages, which have translated to several million dollars of lost revenue and damage. One advantage in the communications sector is that, due to the frequently updated technology, the equipment is often replaced on a short time cycle. This allows for the opportunity to include climate change into the new design or life-cycle replacement of equipment. However, because the costs of a communication outage can be so significant, it is still important to consider the investment of adaptations to minimize the impacts from climate change. Table 8.2, below, illustrates the estimation of costs from a communication outage due to a severe winter storm and the benefits that two different types of backup systems could bring. For complete methodology, see technical note at the end of this chapter.

Table 8.2. Illustrative key impacts and adaptations (Values in $2010 US.)

Element	Timeslice	Annual costs of current and future climate hazards without climate change ($M)[1]	Annual incremental costs of climate change impacts, without adaptation ($M)[2]	Annual costs of adaptation ($M)[3]	Annual benefits of adaptation ($M)[4]
Outages from a 1-in-50 yr storm[1, 2]	Baseline	$40	-	-	-
	2020s	$72	$7 - $14[3]	$6	$23[5,6]
	2050s	$147	$15 - $30[3]	$12	$47[5,6]
	2080s	$300	$30 - $60[3]	$24	$95[5,6]

[1] *From the case study in Jacob et al , forthcoming-b), "Communications outage from a 1-in-50 year winter storm in Central, Western and Northern New York"*

[2] *The values presented are based on a growth rate for GDP of 2.4%.*

[3] *Based on the findings by the Multi-hazard Mitigation Council that every $1 spent in public disaster mitigation results in a $4 savings in non-incurred disaster losses (Jacob et al., forthcoming-a).*

[4] *Future changes in winter storms are highly uncertain, however, because it is more likely than not that severe coastal storms will become more frequent, 10% and 20% increases in storm damage are estimated here to serve as a sensitivity test, but should be used for illustrative purposes only.*

[5] *Based on the findings that it would cost $10 million to develop a rooftop wireless backup network in lower Manhattan (Department of Information Technology and Telecommunications, & Department of Small Business Services [NYCEDC, DoITT, & DSBS] 2005, p.37) and the assumption that this network would have a 10-year lifespan. Additionally, it is assumed that annual NYC-wide costs for a wireless backup network system would be 3 times the costs of Lower Manhattan (based on the 2 other concentrated building locations in midtown Manhattan and downtown Brooklyn).*

[6] *Based on the annual estimated costs for fiber optic network from Jacob et al. (forthcoming-b) and the assumption that this network would have a 40-year lifespan. The fiber optic network was not scaled down to include NYC based on the assumption that there is already a fiber optic network in place there.*

Results

Based on the economic impact estimate of $2 billion from the ClimAID Telecommunications chapter of the damage and lost revenue from a severe winter storm, calculations were made taking into consideration the potential future impacts that may result from climate change. The baseline costs can be estimated to increase at the rate of GDP growth in the future. Based on an estimate of a 2.4 % GDP growth rate, the annual costs from a communications outage without climate change were estimated to between $72 million in the 2020s, $147 million in the 2050s and $300 million by the 2080s. Since the climate information regarding changes in winter storms is not certain enough to give a precise predication regarding the increased frequency of winter storms in the future, an estimate of a 10% increase and 20% in these types of storms during each time period was used to serve as a sensitivity test. In this case, the incremental annual cost of a communications outage above the baseline was estimated to be $7 to $14 million for the 2020s, $15 to $30 million for the 2050s, and $30 to 60 million for the 2080s.

In order to reduce the impacts of climate on the communications sector, there are a number of adaptation options. The two illustrative examples chosen in this case study were the development of a rooftop wireless backup network for New York City with a lifespan of 10 years and the development of a fiber optic network for upstate with a lifespan of 40 years. These two examples were selected because they are feasible with current technology. If these kinds of adaptations were put in place, the result would be annual incremental benefits through the end of the century of $33 million for the 2020s, $40 for the 2050s, and $98 for the 2080s. The annual benefits of adaptation can then be calculated to be $25 million for the 2020s, $61 for the 2050s and $147 for the 2080s. These costs can be compared to the annual costs of adaptation for these systems of $4 million.

PART II. BACKGROUND

8.1 Telecommunication Infrastructure in New York State

Because communications infrastructure is replaced on approximately a 10-year cycle, adaptation to climate change can be more of an ongoing, integrated process in this sector than in sectors with longer-lasting infrastructure.

State GDP and Employment

The size of the Communications sector is roughly reported in the official state GDP figures issued by the U.S. Bureau of Economic Analysis. The NAICS classification for Communications is Broadcast and Telecommunications. For the 2007 (2008 n/a) current dollar state GDP figures, New York State GDP was $1.144 trillion; of this total, $43.763 billion was in the Broadcast and Telecommunications sector. This NAICS includes a wider range of industries than are discussed in the telecommunications sector included in ClimAID. The total annual revenue for telecommunications is $20 billion, contributing approximately 2% of the $1.1 trillion gross state product (GSP) (Jacob et al., forthcoming-b).

More than 43,000 people are employed by telecommunications, cable, and Internet service companies in New York City, earning an average salary of $79,600. In 2003, these telecommunications, cable, and internet service companies produced a combined output of over $23 billion, totaling more than three percent of the city's economy (New York City Economic Development Corporation, Department of Information Technology and Telecommunications, & Department of Small Business Services [NYCEDC, DoITT, & DSBS], 2005, p. 9).

8.2 Key Climate Change Sensitivities

Communications in New York State are interconnected, overlapping, and networked, and boundaries are constantly in flux (Jacob et al., forthcoming-b). Due to network complexity, communications infrastructure is vulnerable to many different failure modes. The primary cause of failure for communication networks is commercial grid and service provider back-up power failures due to communications interdependence with power (Jacob et al., forthcoming-b). This section identifies the facets of climate change that will cause broadcast, telecommunication, and power outages and thereby affect the key economic components of the sector.

Table 8.3. Climate Change Sensitivities: Telecommunications Sector

Ice storms will damage power and telecommunication lines and poles. In December 2008, federal disaster aid totaled more than $2 million for nine New York counties that suffered damage from an ice storm.
Hurricanes. A slight increase in the intensity of hurricanes or storm surges will likely cause a substantial increase in infrastructure damage (Stern, (2007) Communications in coastal areas will be vulnerable to coastal flooding intensified by sea level rise.
Rain, wind, and thunderstorms will damage power and telecommunication lines and poles. Riverine and inland flooding caused by intense precipitation will also threaten low-lying Communications.
Heat. Unmet peak energy demands for air conditioning will cause power outages. This will indirectly lead to communication outages.
Snowstorms will damage power and telecommunication lines and poles.
Electric power blackouts. Power outages are often weather related and are a leading cause for communication outages. Risks are becoming increasingly significant as the proportion electric grid disturbances caused by weather related phenomena has more than tripled from about 20% in the 1990s to about 65% more recently.

8.4 Impact Costs

The costs of climate change impacts in the communications infrastructure sector are incurred through direct damage of equipment and productivity losses (Jacob et al., forthcoming-b). Telecommunication companies generally consider the economic data that is relevant to the ClimAID study as proprietary information. This, coupled with the limited and often voluntary requirements for communications operators to report service outages to the New York Public Service Commission (Jacob et al., forthcoming-b), combined with the fact that some of this information is not publicly accessible, makes it nearly impossible to determine the total costs of climate impacts on infrastructure. This section presents the available costs of climate change impacts for New York State.

Loss Estimates

Damage costs are fairly straightforward and include things such as the replacement of downed poles and wires, etc.

Ice and Snow Storms. The ClimAID communications case study found that the total estimated cost of a major winter storm in NY is nearly $2 billion dollars, of which nearly $900 million comprises productivity losses (due to service interruption) and $900 million comprises direct damage (spoiled food, damaged orchards, replacement of downed poles and electric and phone/cable wires, medical costs, emergency shelter costs etc.) To estimate damage and economic productivity losses, the case study used the number of people affected and the number of customers restored per number of days until restoration. It also used New York State's average-per-person contribution to the state's gross domestic product ($1.445 trillion per year per 19.55 million people equals about $58,600 per person per year, which is equal to $160.50 per person per day). Losses to the state's economy were approximated at about $600 million in the first 10 days, $240 million between days 10 and 20, and $60 million in the remaining time from days 20 to 35. In total, this amounts to about $900 million ($0.9 billion) from productivity losses alone (Jacob et al., forthcoming-b, Economic Impacts of a Blackout Case Study).

Federal aid for New York State ice storms: During an April 3-4, 2003 ice storm affecting western New York State, 10,800 telecommunications outages were reported. It took 15 days from the beginning of the storm to return conditions to normal. More than $15 million in federal aid was provided to help in the recovery (Jacob et al., forthcoming-b).

Federal disaster aid topped $2 million for the nine New York counties that suffered damages from the December 2008 ice storm. The aid for these counties and to the State of New York was (Jacob et al., forthcoming-b):

- Albany County - $295,675
- Columbia County - $123,745
- Delaware County - $324,199
- Greene County - $203,941
- Rensselaer County - $203,079
- Saratoga County - $166,134
- Schenectady County - $300,599
- Schoharie County - $324,569
- Washington County - $173,393
- State of New York - $ 10,070

Additional impact costs of ice storm events outside New York State include:

- Between 1949 to 2000, freezing rain caused more than $16.3 billion in total property losses in the United States (Changnon 2003; Jacob et al., forthcoming-b).

- The estimated cost of the 1998 ice storm that hit Northeastern US and Canada caused damages in Canada alone totaling (U.S.) $5.4 billion. In Quebec, telephone service was cut off to more than 158,500 customers. Several thousand kilometers of power lines and telephone cables were rendered useless; more than 1,000 electric high-voltage transmission towers, of which 130 were major structures worth $100,000 each, were toppled; and more than 30,000 wooden utility poles, valued at $3,000 each, were brought down. 28 people died in Canada, many from hypothermia, and 945 people were injured (Environment Canada). More than 4 million people in Ontario, Quebec and New Brunswick lost power. About 600,000 people had to leave their homes. By June 1998, about 600,000 insurance claims were filed totaling more than $1 billion (Jacob et al., forthcoming-b).

Productivity loss is slightly more complicated but can be estimated in terms of potential business that would have been done under normal circumstances. For example, the *New York Clearing House* processes up to 26 million transactions per day for an average value of $1.5 trillion (NYCEDC, DoITT, & DSBS, 2005); if the communications infrastructure is down then this business productivity loss is an impact cost of climate change.

8.4 Adaptation Costs

There are two types of adaptations in infrastructure: (1) modifications in the operations of infrastructure that is directly affected by climate change, and (2) changes in infrastructure needed to support activities that cope with climate sensitive resources (UNFCCC, 2007, p. 121). This section deals with the latter and presents the costs of climate change adaptation strategies for communications infrastructure in New York State.

Rapid changes in technology and intra-industry competition drive the constantly evolving communications sector, allowing for a planning horizon of only 10 to 20 years. Therefore adaptation to climate change will not bear significant costs if it is incorporated into the existing communications plans. It has been determined that for every $1 spent in public disaster mitigation there is a savings of $4 in non-incurred disaster losses (Jacob et al., forthcoming-b). Following this reasoning, proactively modifying communications infrastructure to adapt to climate change will benefit the sector.

Proposed adaptations to ensure a higher level of reliability in the sector include the following (Jacob et al., forthcoming-b):

- Move wired communications from overhead poles to buried facilities
- Emergency power generators and strategies for refueling generators
- Standardization of power systems for consumer communication devices
- Diversification of communication media
- Natural competition between wired and wireless networks
- Develop alternate technologies (free space optics, power line communications, etc.)

Costs are available for several specific adaptations proposed in NYC's telecommunications Action Plan:

- It will cost an average of $250,000 per building in lower Manhattan to bolster resiliency by having (1) two or more physically separate telecommunication cable entrances, (2) carrier-neutral dual risers within buildings, and (3) rooftop wireless backup systems (NYCEDC, DoITT, & DSBS, 2005, p. 33).

- It will cost approximately $10 million to develop a rooftop wireless backup network in lower Manhattan to ensure that the building's tenants could move data in the event that landline communications are disrupted (NYCEDC, DoITT, & DSBS, 2005, p. 37).

Some additional examples of adaptation costs in NY include:

- Recently, the federal National Telecommunications and Information Administration awarded a $40-million grant for the ION Upstate New York Rural Initiative to deploy a 1,300-mile fiber optic network in upstate regions as part of the federal government's broadband stimulus program (Jacob et al., forthcoming-b).

Initial analysis determined that 62 percent of telephone central offices in New York State have geographic diversity (the ability to transmit/receive signals from one location to another via two distinct and separate cable routes), while 38 percent of do not. Company estimates determined that the cost to provide geographic diversity to all remaining offices was approximately $174 million. The Public Service Commission performed a critical-needs analysis, which concluded that 40 percent of the non-diverse central offices could be equipped with geographic route diversity at a significantly lower total cost of about $13.3 million. Following this recommendation, 77 percent of central offices have now achieved geographic route diversity, covering 98 percent of the total lines in New York. This enhanced route diversity of outside cable facilities substantially increases access to emergency services, overall network reliability and the resiliency of telephone service during emergency situations.

8.5. Summary and Knowledge Gaps

From the standpoint of improving the ability of planners to do economic analysis of the costs of climate change impacts and adaptations in the communications sector, there are many knowledge gaps to which resources can be directed. These include:

- There is a need for comprehensive data bases showing the locations and elevation of installed communications facilities as well as other details. These data bases will have to be secure, but accessible to qualified researchers.
- From locational data as above, assessment need to be completed of vulnerability of infrastructure components to coastal and inland flooding.
- Within the monitoring systems that should be developed for climate analysis, wind records in relation to communications systems should be included.
- As climate changes, the important of public access to outage information will increase.
- Public health aspects of communications infrastructure should continue to be monitored.

Technical Notes – Telecommunications Sector

Impact: Communications outage from a 1-in-50 year winter storm

Adaptations: Develop a wireless backup network in New York City and construct a fiber optic broadband network in Upstate New York

Annual costs of current and future climate hazards without climate change:

1. Annualize the total storm cost given by ClimAID Telecommunications Chapter 10 based on the 1-in-50 year storm ($2,000M/50=$40M).
2. Project out annualized $40M baseline cost to 2100 accounting for the 2.4% growth in GDP (Baseline: $40M, 2020s: $72M, 2050s: $147M, 2080s: $300M).

Annual incremental costs of climate change impacts without adaptation:

3. Assume a 10% and 20% increase in baseline costs associated with an increase in storm frequency due to climate change.

Annual costs of Adaptation:

4. Estimate from the annual cost for a rooftop wireless backup network assuming 10-year lifespan ($10M/10 = $1M). Multiply this cost by 3 to scale up to the city level (representing two other concentrated areas in the city, Midtown Manhattan and Downtown Brooklyn).
5. Estimate the annual cost for fiber optic network assuming 40-year lifespan ($40M/40 = is $1M).
6. Add the totals from steps 4 and 5 for a total annual adaptation cost of $4M.
7. Projected out the costs of adaptation ($4M) to 2080 based on 2.4% GDP growth (2020s: $6M; 2050s: $12M; 2080s: $24M)

Annual benefits of adaptation:

8. Based on the Multi-hazard Mitigation Council finding that "for every $1 spent in public disaster mitigation there is a savings of $4 in non-incurred disaster losses" (Multihazard Mitigation Council 2005a; Jacob et al., forthcoming-a), take the annual adaptation cost of $4M and multiply it by 4 to find the savings in non-incurred disaster losses (=$16M).
9. Projected out the savings from adaptation ($16M) to 2100 based on 2.4% GDP growth are as follows: 2020s: $23M; 2050s: $47M; 2080s: $95M

9 Public Health

Climate change is anticipated to have widespread and diverse impacts on public health. On the whole these impacts will be negative, with the exception of a potential reduction in cold-related health outcomes (Parry et al, 2009, p.108). Maintenance of public health is critically linked with other sectors, particularly water resources and energy. In many cases, adaptation to climate change within other sectors is as important as the enhancement of conventional public health programs for reducing the health impacts of climate change. Appropriate adaptation in these other sectors will insure that the public health costs of climate change will be manageable (Kinney, 2010). Taking steps to prepare for climate related hazard events, to maintain grid reliability during heat waves, to secure food and water supplies, and to implement infrastructure improvements will significantly reduce the impacts of climate change on public health (Parry et al, 2009, p.52).

PART I. KEY ECONOMIC RISKS AND VULNERABILITIES AND BENEFIT-COST ANALYSIS FOR PUBLIC HEALTH

Key Economic Risks and Vulnerabilities

This section identifies climate-related changes that will have significant potential costs for the public health sector. Table 9.1 identifies the climate variables that are likely to impact some of the key facets of the public health sector with the projected economic impact by mid-century. Based on existing data, it is possible to develop rough, provisional estimates of the direct climate-change related costs for some facets of the public health sector, including costs associated with loss of life due to extreme heat and hospitalizations due to asthma. For other types of impacts including the potential costs associated with emergent, vector-borne diseases and water-borne illnesses, costs are currently unknown. The mid-century estimate of total impact costs of between roughly $3 and $6 billion dollars is an estimate of some of the critical, potential costs associated with mortality and hospitalization as the result of climate change (without adaptation). Other types of impacts may amount to several hundred million or more per year in additional costs.

Many climate change related threats to public health can be substantially reduced or even eliminated with preventative measures and adaptations such as heat wave warning programs, asthma awareness and treatment programs, and development of new vaccines for emergent vector-borne diseases. Other impacts can be reduced via appropriate adaptations action within other sectors such as maintenance of water quality to protect residents from water-borne illness. Table 9.1 provides mid-century estimates of costs associated with heat warning systems and asthma prevention programs, and also describes qualitatively a number of other types of potential adaptation costs that may be incurred with climate change.

Table 9.1. Climate and Economic Sensitivity Matrix: Public Health Sector (Values in $2010 US)

Element	Main Climate Variables: Temperature	Main Climate Variables: Precipitation	Main Climate Variables: Extreme Events: Heat	Main Climate Variables: Sea Level Rise & Storm Surge	Economic risks and opportunities: – is Risk + is Opportunity	Annual incremental impact costs of climate change at mid-century, without adaptation	Annual incremental adaptation costs of climate change at mid-century
Temperature related deaths	●		●		+ Fewer cold related deaths – More heat related deaths – Loss of life and productivity – Hospitalization costs	$ 2,988M-$6,040M (value of loss of life from heat-related deaths using VSL of $7.4M ($2006, indexed to $2010)	Costs: $ .6M heat wave warning system; Benefits: $1,636M
Air quality and respiratory health	●	●	●		– Extension of pollen and mold seasons – More suitable environment for dust mites and cockroaches – Increased ozone concentrations, due in part to higher emission of VOCs – Peak in AC use, potentially leading to loss of electricity – Change in the dispersion of pollutants in the atmosphere	$10M – $58M additional asthma hospitalization costs	Costs: $5M asthma prevention Benefits: $8M
Water supply and food production	●	●		●	– Water quality – Safety of food supply – Higher food prices + Longer growing season for local crops	Increase in water and food-borne illness; malnutrition	Increased water treatment and protection of food supply
Storms and flooding		●		●	– Loss of life from large storm event (e.g., hurricane) – Mental health issues caused by displacement and family separation, violence, or stress – Increased runoff from brownfields and industrial contaminated sites – Flooding favors indoor molds that can proliferate and release spores	Costs associated with loss of life, treatment of post-traumatic stress, and treatment of mold-related illnesses	Expansion of emergency preparedness
Vector borne and infectious disease	●	●		●	– Increased population and biting rate of mosquitoes and ticks – Greater rates of overwinter survival of immature mosquitoes	Doctor or hospital costs for treatment	Mosquitoes spraying, vaccination
Total estimated costs of key elements						$2,998 - $6,098M	Costs $6M: Benefits: $1,644M

Key for color-coding:

	Analyzed example
	Analogous number or order of magnitude
	Qualitative information
	Unknown

Table 9.2 provides more detailed estimates of the costs of climate change impacts associated with temperature-related deaths in New York City and asthma hospitalizations in New York State. Every year, several hundred deaths within New York City can be attributed to temperature-related causes, both from extreme heat and extreme cold. With a changing climate, heat-related deaths may increase due to more frequent heat waves and more days with extreme hot temperatures. A reduction in extreme cold days may mean a decrease in the number of deaths from cold. Extreme heat can also exacerbate other health problems such as cardiovascular disease and asthma, and individuals with these conditions are particularly vulnerable to heat-related illness (Kinney et al. 2008). Elderly populations and those with pre-existing health conditions are especially at risk. The number of state residents at risk for temperature-related illness is likely to increase in the future with an aging population.

Asthma is a major public health issue within New York State. Between 2005 and 2007, approximately 39,000 state residents were hospitalized annually due to asthma-related illness (New York State Department of Health [NYSDOH 2009]). In 2007, the total annual cost of these hospitalizations was approximately $535 million (NYSDOH 2009). Climate change may lead to an increase in asthma hospitalizations in New York State as the result of an increase in the frequency of high ozone days. Concentrations of ambient ozone are expected to increase in urbanized areas of the state as the climate changes due to both higher daily temperatures and increases in precursor emissions (Kinney et al. 2000; Kinney 2008; Knowlton et al., 2004, Bell et al. 2007).

Table 9.2. Illustrative key impacts and adaptations: Public Health Sector (Values in $2010 US)

Element	Timeslice	Annual costs of current and future climate hazards without climate change ($M)	Annual incremental costs of climate change impacts, without adaptation ($M)	Annual costs of adaptation ($M)	Annual benefits of adaptation ($M)
Heat-related deaths	Baseline 2050s	307 307	- 147 to 292	- NA	- 79[5]
Heat-related deaths – VSL ($7.4 M)[1, 2]	Baseline 2050s	$2,462 $6,358	- $2,988 - $6,040	- $ .622[4]	- $1,636
Cold-related deaths	Baseline 2050s	102 102	- -40 to -45	- NA NA	- NA NA
Cold-related deaths – VSL ($7.4M)[1, 2]	Baseline 2050s	$ 818 $2,112	- $-1,174 to $-1,291	- NA	- NA
Asthma (ozone)[3]	Baseline 2020s 2050s 2080s	$620 $786 $1,601 $3,262	- $2 to $11 $10 to $58 $32 to $193	- $3[6] $5 $11	- $2[7] $8 $27
TOTAL –	Baseline 2050s	$3,900 $10,071	- $1,824 to $4,807	- $ 6	- $1,644

[1] *Heat and cold baseline mortality projections from Kalkstein and Greene (1997). Climate change heat projections based on Knowlton et al. 2007. Climate change cold projections based on Kinney et al. (2010). Climate change scenario projections are only available for 2050 from Knowlton et al. (2007).*

[2] *Based on a 2.4% GDP growth rate (BEA) and using a VSL of $7.4 million (in 2006 $), as prescribed by the U.S. Department of Environmental Protection (USEPA) (USEPA 2010, 2000).*

[3] *Asthma hospitalization projections are based on Bell et al. (2007) of the impacts of climate change on asthma hospitalizations as the result of ambient ozone in U.S. cities.*

[4] *Estimates based on average number of lives saved and average costs to run the PWWS. Actual values vary from year-to-year.*

[5] *Calculated based on the findings of Ebi, et al.'s (2004) study of the Philadelphia Hot Weather – Health Watch/Warming System (PWWS) , which estimated the system saved 117 lives between 1995 and 1998*

[6] *Estimates based on annual costs to run New York State Health Neighborhoods program.*

[7] *Calculated based on the study of Lin et al. (2004), which found that the New York State Healthy Neighborhoods Program lead to a 24% decrease in asthma hospitalizations in eight participating counties between 1997 and 1999.*

Results

Results of the temperature and asthma analyses suggest that climate change may have substantial public health costs for New York State. New York State already incurs significant economic costs as the result of both extreme heat and extreme cold. Kalkstein and Greene (1997) estimate that there are presently 307 heat-related deaths and 102 cold-related deaths on an annual basis in New York City. We estimate the annual costs associated with temperature-related deaths in New York City using a standard VSL of $7.4 million (in $2006), as recommended by U.S. Department of Environmental Protection (USEPA) (USEPA 2010, 2000).

Even without climate change the costs of heat-related deaths in the state are substantial, approaching $2.5 billion annually. With climate change, the annual number of heat-related deaths could increase between 47 and 95 percent by the 2050s (Knowlton et al. 2007). These estimates are based on Knowlton et al.'s (2007) forecasts of increases in summer heat related deaths in the New York region under both low (B2) and high (A2) emissions scenarios. These additional temperature related deaths due to climate represent estimates of the number of lives that may be lost without appropriate adaptation. By contrast, cold related deaths are expected to decrease in New York State with climate change (Kinney et al. 2010). However, as illustrated in Table 9.2, the costs of heat-related mortality far outweigh the benefit of decreased cold-related mortality.

Heat-related deaths in the state could be considerably reduced with adaptation. Adaptation will also likely occur through expanded use of air conditioning in homes, schools and offices. Air conditioning prevalence in private dwellings has increased steadily in recent decades, and this trend is likely to continue. However, affordability of the units and energy costs continues to be a major concern. New York City has initiated a program to provide free air conditioners to elderly residents who are unable to afford them. This program cost approximately $1.2 million for each year 2008 and 2009, and entailed distribution of approximately 3000 air conditioning units to residents over 60 years old (Sheffield, 2010). Substantial expansion of this type of program may be needed to foster adaptation to climate change, given the high number of at-risk seniors not only in New York City but throughout the state. Other on-going efforts to reduce heat related mortality in New York include development of a network of cooling centers to help residents cope with extreme heat. The capital, energy and pollution-related costs of air conditioning should be borne in mind.

In the example above, implementation of a heat wave warming system, similar to the one put into place in Philadelphia (see Ebi et al. 2004) would save an average of 79 lives per year and thus lower the annual incremental costs of temperature-related deaths by $1,636 million in the 2050s, assuming a VSL of $7.4 million (USEPA 2000, 2010). Based on data from the Philadelphia study (Ebi et al 2004) such a program is estimated to cost less than $1 million annually to establish and run. Even if such a program saved only one life, the benefits would exceed the costs.

Asthma-related hospitalizations may also be affected by climate change, due largely to increases in ozone concentrations absent more aggressive emissions controls of ozone precursors (Kinney 2008). The costs associated with such hospitalizations are estimated to exceed $600 million today. Without climate change, these costs will increase over the next century, approaching $3.2 billion by the 2080s. Climate change is expected to increase the number of asthma related hospitalizations due to increased levels of ambient ozone and an increase in the severity and length of the pollen season. The above analysis estimates costs associated with increased ozone-related hospitalizations in the state under climate change based on Bell et al. (2007). Results suggest that climate change will lead to additional annual costs in the ranges of $2 to $11 million in the 2020s, $10 to $58 million in the 2050s, and $32 to $193 million by the 2080s. Adaptation may reduce these costs somewhat. In Table 9.2, we estimate the benefits associated with implementation of an asthma intervention program similar to the New York State Healthy Neighborhoods Program, which was found to reduce asthma hospitalization rates by approximately 24 percent within eight counties in New York State (Lin et al. 2004). The benefits of adapting monetarily increase in the future and eventually outweigh the costs of asthma intervention programs.

PART II. BACKGROUND

9.1 Public Health in New York State

The public health sector in New York State encompasses disease prevention and the promotion of healthy lifestyles and environments, as well as clinical medicine and the treatment of sick people. Within the state, 99% of health care spending is currently allocated to medicine while approximately 1% is spent on the public health system (Kinney, 2010). The county-based public health system in New York State is highly decentralized with non-uniform provision of its core services. According to the New York State Public Health Council, this decentralization of the public health service delivery system is a key obstacle for climate health preparedness (Kinney et al., forthcoming).

State GDP and Employment

The size of the public health sector is roughly reported in the official state GDP figures issued by the U.S. Bureau of Economic Analysis. The NAICS classification for public health is Health Care and Social Assistance, excluding Social Assistance, and the subsidiary parts are: Ambulatory Health Care Services, and Hospitals and Nursing and Residential Care Facilities. Employing more than 1.3 million people, the Health Care and Social Assistance industry accounted for 7% of the total state GDP in 2008 (New York State Department of Labor, 2008). For the 2008 current dollar state GDP figures, New York State GDP was $1.144 trillion; of this total, $82.580 billion was in the Public Health sector (United States Department of Commerce Bureau of Economic Analysis, 2009). See Table 9.3.

Table 9.3. 2007 New York State Census Data for Health Care and Social Assistance

Type of care/assistance	# Of establishments	# Of paid employees	Receipts/ revenue ($1,000)	Annual payroll ($1,000)
Health care and social assistance	53,948	1,326,039	128,595,239	54,422,381
Ambulatory health care services	38,284	439,960	46,191,651	18,512,293
Offices of physicians	17,279	134,142	21,801,478	8,589,789
Offices of dentists	9,101	50,896	6,124,859	1,993,816
Offices of other health practitioners	8,071	34,808	3,037,320	1,080,660
Outpatient care centers	1,454	43,522	4,330,922	1,875,468
Medical and diagnostic laboratories	924	16,433	2,967,253	999,220
Home health care services	944	144,246	6,432,091	3,444,280
Other ambulatory health care services	511	15,913	1,497,728	529,060
Hospitals	278	416,273	54,026,089	23,216,717
General medical and surgical hospitals	216	368,682	48,395,169	20,465,979
Psychiatric and substance abuse hospitals	44	25,258	2,073,753	1,220,277
Other specialty hospitals	18	22,333	3,557,167	1,530,461
Nursing and residential care facilities	5,048	237,061	15,820,321	7,160,538
Nursing care facilities	651	128,310	9,432,676	4,263,973
Residential mental health facilities	3,316	64,872	3,627,477	1,737,770
Community care facilities for the elderly	655	26,992	1,703,565	619,091
Other residential care facilities	426	16,887	1,056,603	539,704
Social assistance	10,338	232,745	12,557,178	5,532,833
Individual and family services	4,122	131,331	7,005,336	3,275,727
Emergency and other relief services	1,059	18,401	2,164,252	563,746
Vocational rehabilitation services	492	21,184	1,052,240	484,654
Child day care services	4,665	61,829	2,335,350	1,208,706

Source: United States Census Bureau 2010b

Health Care Expenditures

Billions of dollars are spent each year on the prevention and treatment of mortality and morbidity. In 2004, health care expenditures in New York State totaled approximately $126 billion (The Kaiser Family Foundation, 2007). Hospital care and professional medical care services accounted for over 50% of these health care expenditures statewide. See Table 9.4.

Table 9.4. Distribution of Health Care Expenditures (in millions), in 2004

	NY %	**NY $**	**US %**	**US $**
Hospital Care	36.10%	**$45,569**	37.70%	$566,886
Physician and Other Professional Services	23.20%	**$29,230**	28.20%	$446,349
Drugs and Other Medical Nondurables	14.10%	**$17,722**	13.90%	$222,412
Nursing Home Care	10.60%	**$13,364**	7.40%	$115,015
Dental Services	4.30%	**$5,445**	5.20%	$81,476
Home Health Care	4.80%	**$6,021**	2.30%	$42,710
Medical Durables	1.30%	**$1,685**	1.50%	$23,128
Other Personal Health Care	5.60%	**$7,040**	4.00%	$53,278
Total	100.00%	**$126,076**	100.00%	$1,551,255

Source: The Kaiser Foundation, 2007

9.2 Key Climate Change Sensitivities

Climate change is compounding existing vulnerabilities within New York State's public health sector. Changes in temperature, precipitation and sea level are anticipated to have adverse effects on air quality, disease and contamination, and mental health. Table 9.5 specifies which facets of climate change will impact the key economic components of the public health sector. See Kinney et al., forthcoming, for additional details.

Table 9.5. Climate Change Sensitivities: Public Health Sector (see Kinney et al., forthcoming)

Increases in mean temperature will affect air quality and the spread of disease and contamination
Increases in extreme heat events will contribute to more heat related deaths and air quality problems
Increases in mean precipitation will impact air quality, the spread of disease and contamination, and food production
Increases in storm surges and coastal flooding will contribute to mental health issues and the spread of disease and contamination **Decrease in soil moisture** could lead to greater risk of wildfires, which place residents at risk.

9.3 Impact Costs

Impact and adaptation costs in the public health sector are heavily interrelated. The level of impact is dependent upon preparedness, and adaptation strategies undertaken are dependent upon the type and severity of the impact. The following section presents costs associated with most common health vulnerabilities within New York State: heat waves, asthma and allergies, storms and flood, vector borne and infectious diseases, and food and water supply. Impact costs can be divided into three categories: morbidity, mortality, and lost productivity.

Although many aspects of public health are not easily quantifiable, the Environmental Protection Agency has approximated the value of a statistical life to be $6.9 million (See Kinney et al., forthcoming, "Economic Impacts of Mortality due to Heat Waves" for more information on estimating the value of a statistical life.) Other studies use substantially lower values. For this study, we used a range of estimates from $1.0 million to $6.9 million for the value of a statistical life.

Temperature-Related Deaths

Heat Waves. Heat waves are the leading cause of weather related deaths in the US and are anticipated to increase in magnitude and duration in areas where they already occur (Kalkstein & Greene, 1997; Knowlton et al. 2007). Heat events also lead to an increase in hospital admissions for cardiovascular and respiratory diseases (Lin et al. 2009). Without adaptation in New York State, there will likely be a net increase in morbidity and mortality due to heat waves. Fewer cold days should lower the number of cold-related deaths; however, new heat related deaths would outnumber these lives saved. The heat wave threat however may be a near term problem as it is expected that most homes will be climate controlled by the second half of this century. Adaptation costs will include air conditioning, but there is also a trend of increased air conditioning use in New York State (Kinney, 2010). This section presents various impact costs for heat waves that have occurred in other areas. Table 9.2 above contains estimates for heat impact costs in New York City.

Table 9.6 provides a summary of the costs associated with major heat waves that occurred in the U.S. over the past 30 years. Costs per heat event range from $1.8 billion to $48.4 billion (Kinney et al., forthcoming).

Table 9.6. Costs for Major Heat Waves in the United States, 1980-2000

Year	Event Type	Region affected	Total Costs / Damage Costs	Deaths
2000	Severe drought & persistent heat	South-central & southeastern states	$4.2 B	140
1998	Severe drought & persistent heat	TX / OK eastward to the Carolinas	$6.6-9.9 B	200
1993	Heat wave/ drought	Southeast US	$1.3B	16
1988	Heat wave/ drought	Central & Eastern US	$6.6B	5000-10,000
1986	Heat wave/ drought	Southeast US	$1.8-2.6B	100
1980	Heat wave/ drought	Central & Eastern US	$48.4B	10,000

Additional impact costs of extreme heat events outside New York State include:

- The number of premature deaths linked with hot weather events in Canada has been reported as 121 in Montreal, 120 in Toronto, 41 in Ottawa, and 37 in Windsor. The value per premature death, based on lost earning potential, is estimated at $2.5 million. These cities are spending an additional $7 million per year on health care (Kinney et al., forthcoming).

Concerning hospital admissions and extreme heat, Lin et al. (2009) found increased rates of hospital admissions for both cardiovascular and respiratory disorders in New York City. These effects, which were investigated for summer months between 1991 and 2004 were especially severe among elderly and Hispanic residents. As discussed in the Energy chapter, extended heat events may also be associated with increased likelihood of blackouts, with compounding effects on public health. In a study of the health impacts in New York City of the 2003 blackout, Lin et al. (2010) found that the blackout event had a stronger negative effect on public health than comparable hot days. In particularly, the study found that mortality and respiratory hospital admissions increased significantly (2 to 8 fold) during the blackout event (Lin et al. 2010).

Cardiovascular Disease. Extreme temperature events have been linked to higher rates of premature death and mortality among vulnerable populations, including children, elderly, and people suffering from cardiovascular or respiratory conditions (Kinney et al., forthcoming). Cardiovascular disease is a predisposing factor for heat related deaths because it can interfere with the body's ability to thermoregulate in response to heat stress (Kinney et al., forthcoming). Table 9.7 includes information on the costs of treating and suffering from cardiovascular disease. Nearly $16 billion was spent on cardiovascular disease in New York State in 2002. This number will likely increase as temperatures continue to climb.

- The costs associated with treating CVD and stroke in the U.S. in 2009 were expected to exceed $475 billion, with estimates of direct costs reaching over $313 billion. Although not all such costs are related to extreme heat events, CVD prevalence is likely to be exacerbated during such periods, thereby putting additional strain on the Public Health System and its efforts to reduce CVD incidence. Costs are projected to increase in future decades, as the size of the elder population is also expected to grow. (Kinney et al., forthcoming). As noted earlier, nearly $16 billion was spent on cardiovascular in 2002 disease in New York State alone.

Table 9.7. New York State Costs for Cardiovascular Disease, 2002 (in Millions of dollars)

Type of Cost	Coronary Heart Disease	Stroke	Congestive Heart Failure	Total Cardiovascular Disease
Direct Costs				
Hospital/Nursing Home	$3,751.20	$1,189.20	$828.10	$6,120.90
Physicians/Other Professionals	$771.80	$116.50	$86.00	$1,451.40
Drugs/Other	$0.00	$0.00	$0.00	$0.00
Medical Durables	$556.40	$38.80	$107.60	$1,543.60
Home Health Care	$143.60	$150.50	$129.10	$567.90
Total direct expenditures	$5,223	$1,495.00	$1,150.80	$9,683.80
Indirect Costs				
Lost Productivity/Morbidity	$753.80	$271.80	NA	$1,499.90
Lost Productivity/Mortality	$4,056.30	$631.00	$96.80	$4,795.80
Total indirect expenditures	$4,810.20	$902.90	$96.80	$6,295.70
Grand Totals	$10,033.20	$2,397.90	$1,247.60	$15,979.50

Source: http://www.nyhealth.gov/diseases/cardiovascular/heart_disease/docs/burden_of_cvd_in_nys.pdf

Asthma and Allergies

The spending on asthma, allergies, and respiratory problems in New York State is anticipated to increase with climate change (Kinney, 2010). Current spending on asthma in the U.S. is on the order of $10 billion per year. Within New York State, spending on asthma-related hospitalizations exceeded $535 million in New York State in 2007 (NYSDOH 2009). As described in Table 9.2 and below, asthma hospitalization costs may increase as the result of higher levels of ambient ozone with climate change. Asthma-related spending is also likely to increase as heat, higher levels of CO_2, increased pollen production, and a potentially longer allergy season (or shift in the start date of the season) may increase cases of allergies and asthma in New York State (Kinney, 2010).

Vulnerable populations, including children and the elderly, poor, and those with predisposing health conditions, face the greatest threats and therefore costs. Consider, for example, the costs of childhood asthma. Children are among those most vulnerable to the public health impacts of climate change. One study found that the average per capita asthma-related expenditures totaled $171 per year for US children with asthma -- $34 for asthma prescriptions, $31 for ambulatory visits for asthma, $18 for asthma ED visits, and $87 for asthma hospitalizations. Average yearly health care expenditure for children with asthma were found to be $1129 per child compared with $468 for children without asthma, a 2.8-fold difference (Lozano et al, 1999). Within New York State, the cost for asthma hospitalizations for children

15 and under between 2005 and 2007 exceeded $317 million (NSYDOH, 2009). Such costs are likely to increase as the result of climate change.

Ambient Ozone
Many areas within New York State do not meet the health-based National Ambient Air Quality Standards for ozone. Surface ozone formation is anticipated to increase with climate change, as a result of changing airmass patterns and rising temperatures (the latter leads to an increase in the emissions of ozone relevant precursors from vegetation) (Kinney 2008). Unhealthy levels are reached primarily during the warm half of the year in the late afternoon and evening. Asthmatics and people who spend time outdoors with physical exertion during high ozone episodes (i.e. children, athletes, and outdoor laborers) are most vulnerable to ozone and respiratory disease because of increasing cumulative doses of ozone to the lungs (Kinney et al., forthcoming). Recent estimates by Knowlton et al. (2004) and Bell et al. (2007) indicate that climate change is likely to cause significant increases in both asthma hospitalizations and asthma mortality in New York City. Knowlton et al. (2004) project a median increase in asthma mortality of 4.5 percent for the New York Metropolitan region by 2050. Bell et al. (2007) project an increase of 2.1 percent average in asthma hospitalizations across all U.S. cities included in the study. At the 95 percent confidence level, Bell et al.'s (2007) estimates range from .6% to 3.6%. This range of values is used in Table 9.2 above.

Storms and Floods
Storms and coastal and inland flooding will result in the loss of lives and property, as well as cause physical injury, mental distress, and the spread of disease and contamination. More intense storms are anticipated to disrupt energy and communication infrastructure, which will adversely impact public health as the sector has recently become increasingly dependent on high-quality, high-speed telecommunications (NYCEDC, DoITT, & DSBS, 2005, p. 9).

Emergency preparedness and response are crucial components of the public health sector and its ability to forewarn and respond to extreme storms. More extreme events may require better and more extensive emergency response systems, particularly with respect to coastal storms and flooding and ice storms. There will be costs associated with protecting the public from injury and death as the result of more frequent extreme events. The state currently has emergency response systems in place, e.g. DOT, to keep sectors running smoothly during and after storms. These systems will need to be expanded to deal with more frequent and severe extreme events (Kinney, 2010).

Vector-Borne and Other Infectious Diseases
Changes in temperature and precipitation will affect the patterns of vector-borne and other infectious disease in New York State, likely increasing the incidence of West Nile and Lyme Disease. This may require more spending on pest management and vaccinations and enhancement of existing surveillance programs.

Arthropod vectors, transmitters of infectious disease, are extremely sensitive to climate change because population density and behavior are correlated with ambient air temperature, humidity, and precipitation. West Nile Virus and Lyme Disease are particularly prevalent in New York City, Long Island, and Hudson Valley due to favorable climate conditions for vectors (Kinney et al., forthcoming), and human exposure is generally expected to increase as New York State gets wetter and warmer (Kinney et al., forthcoming).

Water Supply and Food Production

The increased cost of water treatment to ensure public health safety in the face of more extreme storm events (e.g. cost of treating additional turbidity) will likely become one of the most significant economic costs within this sector (Kinney, 2010). See also Chapter 2: Water Resources and Chapter 5: Agriculture for a more complete discussion of the economic costs associated with maintaining a secure and reliable supply of water and food.

9.4 Adaptation Costs

Adaptations are wide-ranging and constantly evolving in the public health sector. Cost are incurred through measures to improve the health protection system to address climate change, introduce novel health interventions, meet environmental and health regulatory standards, improve health systems infrastructure, occupational health, research on reducing the impact of climate change, and the prevention of additional cases of disease due to climate change (Parry et al, 2009, p.53).

Because climate change in New York State will mainly alter the frequency of existing health care problems, public health and environmental agencies in New York State are already involved in activities that address climate change vulnerabilities. The most effective adaptation strategy will be to further integrate climate change information into ongoing public health surveillance, prevention, and response programs. Additional investment should be made in comparative health risk assessments, environmental monitoring and reporting, communication and information dissemination, and environment-health crosscutting initiatives. This section discusses potential costs of adaptation to climate change in the public health sector in New York State. While some of adaptation measures and costs described below are based on studies of New York State, others are based on studies conducted in other states in the Northeast or in other parts of the United States. Additional, detailed analysis of the feasibility and costs of these measures is needed to ensure that they would be appropriate and effective in New York State.

Temperature-Related Deaths

Heat Watch/Warning Systems. Early warning systems for extreme heat events are an effective method to reduce heat-related morbidity and mortality. One example of an effective program that may apply to New York is that The Philadelphia Hot Weather–Health Watch/Warning System (PWWS). PWWS was developed in 1995 to serve as an early warning system for extreme heat events. Ebi et al.'s 2004 study examined the costs and benefits of the system and concluded that if any lives are saved, then the system has significant benefits. The VSL for even one life is greater than the cost of running the system. These findings are based on the additional wages required to pay workers to run the system, totaling around \$10,000 per day.

Over a three-year period between 1995 and 1998, the City of Philadelphia issued 21 alerts, and costs for the system were estimated at $210,000. The value of 117 lives saved over the same time period were estimated to be $468 million; therefore the net benefits of the issued heat wave warnings were estimated to be nearly $468 million for the three-year period (Ebi et al, 2004; Kinney et al., forthcoming). In Table 9.2 above, results from the Ebi study are used to develop estimates of adaptation costs and benefits of a similar heat wave warning system for New York State.

Air Conditioning and Cooling Centers
Expanded use of air conditioning is another important adaptation to extreme heat. As described above, New York City has initiated a program to provide free air conditioners to elderly residents who are unable to afford them at a program cost of approximately $1.2 million for each year 2008 and 2009. The program entailed distribution of approximately 3000 air conditioning units to residents over 60 years old (Sheffield, 2010). Substantial expansion of this type of program may be needed to foster adaptation to climate change, given that high number of at-risk seniors not only in New York City but throughout the state. As noted, other on-going efforts to reduce heat related mortality in New York include development of a network of cooling centers to help residents cope with extreme heat.

Asthma Prevention
Prevention of asthma hospitalizations is a priority for New York State (New York State Department of Health 2005). One option for prevention of asthma hospitalizations entails implementation of a statewide program similar to the New York State Healthy Neighborhoods Program. In this program, which was implemented in eight New York counties between 1997 and 1999, outreach workers initiated home visits and also provided education about asthma, asthma triggers, and medical referrals. The program was found to reduce asthma hospitalization rates by approximately 24 percent within eight counties in New York State (Lin et al. 2004). Such a program may help reduce additional hospitalizations as the result of climate change.

Vector-Borne and Other Infectious Diseases
Vector Control. Without adaptation, cases of West Nile virus may increase in New York State. One potential adaptation option is aerial spraying to control mosquito populations. The benefits of this type of spraying have been found to outweigh the costs in other parts of the country. For example, 163 human cases of West Nile virus (WNV) disease were reported during an outbreak in Sacramento County, California in 2005. Emergency aerial spraying was conducted by the Sacramento-Yolo Mosquito and Vector Control District In response to WNV surveillance indicating increased WNV activity. The economic impact of the outbreak included both vector control costs and the medical cost to treat WNV disease. Approximately $2.28 million was spent on medical treatment and patients' productivity loss for both West Nile fever and West Nile neuroinvasive disease. Vector control costs totaled around $701,790 for spray procedures and worker's overtime hours. The total economic impact of WNV was $2.98 million. A cost-benefit analysis indicated that only 15 cases of West Nile neuroinvasive disease would need to be prevented to make the emergency spray cost-effective (Barber et al, 2010).

Vaccination. Another option for adapting to increased threats of vector-borne disease entails vaccination programs. Such programs can be a cost-effective means to reduce the public health impacts of climate change. An evaluation of the cost effectiveness of vaccinating against Lyme disease in Atlanta, GA revealed that there may be substantial economic benefits from vaccination. Within the study, a decision tree was used to examine the impact on society of six key components, including the cost per case averted. Assuming a 0.80 probability of diagnosing and treating early Lyme disease, a 0.005 probability of contracting Lyme disease, and a vaccination cost of $50 per year, the mean cost of vaccination per case averted was $4,466. Increasing the probability of contracting Lyme disease to 0.03 and the cost of vaccination to $100 per year, the mean net savings per case averted was found to be $3,377. Because most communities have average annual incidences of Lyme disease <0.005, economic benefits will be greatest when vaccination is used on the basis of individual risk, especially for those whose probability of contracting Lyme disease is ≥0.01 (Meltzer et al, 1999, p. 321-322).

In addition to known diseases such as West Nile virus, climate change may also bring emerging diseases to New York State, or lead to the introduction of diseases that are present in more tropical climates. There will be a need to monitor for new diseases as part of the public health system (Kinney, 2010). Options for treatment or prevention of these new diseases will be an important public health priority.

9.5 Summary and Knowledge Gaps

The public health system in New York State is highly decentralized and county-based, with non-uniform provision of its core services. According to the state's Public Health Council, this decentralization of the public health service delivery system is a key obstacle for climate health preparedness (Kinney et al., forthcoming). Adaptations within this sector will help lessen the impacts of climate change on resident's health and investment in preparedness infrastructure will also enhance the effectiveness of the day-to-day operations of the public health system (Kinney et al., forthcoming).

Knowledge gaps and areas for further action include:

- Additional monitoring of emergent diseases and development of effective options for treatment and vaccination;

- Additional monitoring of threats to food and water supplies and development of appropriate strategies to reduce these threats;

- Expansion of emergency preparedness planning throughout the state in order to prepare for more frequent and severe extreme climate events;

- Expansion of community-based public health warning systems for extreme heat; and

- Expansion of programs to reduce asthma-related hospitalizations.

Maintenance of public health is linked with other sectors and adaptation within other sectors is likely to be as important as the enhancement of conventional public health practices for reducing the health impacts of climate change. That is, if we take care of adaptation in these other sectors, then the public health costs of climate change will be manageable (Kinney, 2010). Particularly, disaster mitigation, food and water security, and infrastructure improvements will significantly reduce the impacts of climate change on public health (Parry et al, 2009, p.52).

Technical Notes – Public Health Sector

Impact: Heat-related deaths

Adaptation: Create a heat watch/warning system similar to Philadelphia

Assumptions

- From ClimAID Ch. 11 Case Study, "Projecting Temperature-Related Mortality Impacts in New York City under a Changing Climate"
- Based on a 2.4% GDP growth rate *(*United States Department of Commerce Bureau of Economic Analysis, nd.)
- $7.4 million ($2006), Environmental Protection Agency (EPA) Value of a Statistical Life (VSL) (USEPA 2000, 2010). (The use of the EPA value for VSL was suggested by the New York State Department of Health).
- 30X to 604 temperature-related deaths per year for New York County (Kinney et al., forthcoming; and Kalkstein and Greene 2007)
- Calculated based on the findings of Ebi, et al., 2004 study of the Philadelphia Hot Weather – Health Watch/Warming System (PWWS) that estimated the system saved 117 lives between 1995 and 1998
- Based on 2000 population data for New York County (Manhattan) (1,537,195) and Philadelphia County (1,517,542) (United States Census Bureau, 2000a)
- Based on average costs to run the PWWS. Actual expenses vary from year-to-year.

Annual costs of current and future climate hazards without climate change:

1. Project out the $7.4M VSL ($2006) to 2080 using a 2.4% GDP growth rate to find the VSL for 2020, 2050, and 2080.
2. Using these VSL projections, estimate future costs of lives lost by multiplying the respective values by the projected number of lives lost in New York State due to temperature-related deaths per year under both the low and high scenario to find the totals.

Annual incremental costs of climate change impacts without adaptation:

3. Multiply the heat-related mortality projections under climate change in the ClimAID chapter figures by the respective future VSL estimates to find the projected costs of climate change -related deaths.

Annual benefits of adaptation:

4. Based on the estimated number of lives saved from the Philadelphia Hot Weather-Health Watch/Warning System (PWWS) over a three-year period (117), find the annual lives saved by dividing by 3 (39). In order to ascertain what percentage of the population was saved by PWWS, divide number of lives saved per year (39) by the total population of Philadelphia County (1,517,542) (0.0026%).
5. Using this percentage, estimate the total number of New York City deaths that could be saved by a similar system. Assuming that twice the New York County population is

vulnerable to temperature-related deaths, multiply 0.0026% by twice the New York County population: (0.0026% x (2 x 1,537,195)) = 79.

6. To find economic benefit from the number of lives saved, multiply the future VSL estimate (step 1) by the estimated number of lives saved in New York City (79 from step 8).
7. Project this benefit out to 2080 using the 2.4% GDP growth rate.

Annual costs of adaptation:

8. The PWWS study that found it cost approximately $210,000 to run the system over 3 years. Therefore the average annual cost of the system is $70,000 (=$210,000/3). Find the per person annual cost of the PPWS by dividing the annual cost by the number of people in Philadelphia County ($70,000/1,517,542=$0.05).
9. Find the annual cost to NYC by multiplying the estimated vulnerable population (step 8) by the annual per person cost to run the system (step 12) (3,074,390 x $0.05=$141,813).

Impact: Cold-related deaths
Adaptation: None

Assumptions

- From Kinney et al. (forthcoming) Case Study, "Projecting Temperature-Related Mortality Impacts in New York City under a Changing Climate"
- Based on a 2.4% GDP growth rate.
- $7.4 million ($2006) Environmental Protection Agency (EPA) Value of a Statistical Life (VSL) (USEPA 2000, 2010).

Annual costs of current and future climate hazards without climate change:

10. Using the estimated cold-related deaths of 18 in New York County per year for the baseline period of 1970-1999) from Kinney et al. (forthcoming), calculate the current VSL costs of cold-related deaths.
11. Project out the VSL values to obtain values for 2020, 2050, and 2080.
12. Using these VSL projections, estimate future costs of lives lost by multiplying the respective values by the projected number of lives lost in New York State due to cold-related deaths per year.

Annual incremental costs of climate change impacts without adaptation:

13. Reduce the cold-related death projections given in Kinney et al. (forthcoming) for each timeslice to scale up to New York State.
14. Multiply these figures by the respective future VLS estimates to find the projected reductions in costs due to reduced temperature-related deaths.

Impact: Asthma
Adaptation:
Implementation of a statewide New York Health Neighborhoods program. This program was found to reduce asthma related hospitalizations by 24% between 1997 and 1999 in the eight counties where it was implemented (Lin et al. 2004).

Assumptions
- Based on a 2.4% GDP growth rate.

Annual costs of current and future climate hazards without climate change:

1. Asthma hospitalizations cost the state approximately $535 million in 2007 (New York State Department of Health (2009). In 2007, the average cost per asthma hospitalization in New York State was $14,107 (NYSDOH 2009).

2. These costs are each assumed to increase over time at a rate of 2.4% based on the midpoint growth rate of GDP.

Annual incremental costs of climate change impacts without adaptation:

3. Bell et al. (2007) provide estimates of the number of additional asthma hospitalizations U.S. cities as the result of the climate change in 2050. These values were extrapolated to obtain estimates for 2020 and 2080. Costs were estimated based on the cost of hospitalization in each year multiplied by the number of additional projected hospitalizations.

Annual costs of adaptation

4. Lin et al. (2004) provide data on the annual cost of the New York State Healthy Neighborhoods program in eight counties in New York State. These costs were assumed to increase at an average rate of 2.4% per year, and were extrapolated to the state as a whole to obtain estimates of the costs of adaptation in 2020, 2050 and 2080.

Annual benefits of adaptation:

5. Lin et al. (2004) found that the New York Healthy Neighborhoods program reduced asthma hospitalizations by 24 percent in New York State. A similar reduction rate was used for climate change-related hospitalizations in order to obtain estimates of the benefits of adaptation.

$US 2010 adjustment:

The final calculations in tables 9.1 and 9.2 were adjusted to $US2010 using the United States Bureau of Labor Statistics CPI Inflation Calculator, http://data.bls.gov/cgi-bin/cpicalc.pl to yield the final calculations.

10 Conclusions

This study has aimed to provide an overview assessment of the potential economic costs of impacts and adaptation to climate change in eight major sectors in New York State. It builds on the sectoral knowledge of climate change impacts and adaptation developed in the ClimAID Assessment Report as well as on economic data from New York State and analyses of the costs of impacts and adaptations that been have conducted elsewhere. This chapter presents the principal conclusions of the study.

Costs of impacts and adaptation are expected to vary across sectors in New York State, with some sectors more at risk to climate change than others and with some sectors potentially requiring more costly adaptations. Because New York is a coastal state, and because of the heavy concentrations of assets in coastal counties, the largest impacts in dollar terms will be felt in coastal areas, including impacts on transportation, other coastal infrastructure, and natural areas. There will be significant costs of climate change and needs for adaptation throughout the state: climate change is truly a state challenge. From the evidence assessed in this study, it appears that climate costs for the sectors studied without adaptation in New York State may approach $10 billion annually by midcentury. However, there also appears to be a wide range of adaptations that, if skillfully chosen and scheduled, can markedly reduce the impacts of climate change in excess of their costs. This is likely to be even more true when non-economic objectives, such as the environment and equity, are taken into account.

All sectors will have significant additional costs from climate change. The sectors that will require the most additional adaptations include transportation, the coastal zone, and water resources. Communications and agriculture are sectors in which costs could be large if there is no adaptation; but in these sectors, adaptation to climate is a regular part of investment, so that additional costs are likely to be moderate. This is also true to some extent of the energy sector. The ecosystem sector will see also significant impacts, but many of these costs estimates are preliminary and require further assessment. Finally, public health will be significantly impacted by climate change, but many of these impacts can be avoided with appropriate adaptations.

10.1. SECTOR RESULTS

Water Resources. Water supply and wastewater treatment systems will be impacted throughout the state. Inland supplies will see more droughts and floods, and wastewater treatment plants located in coastal areas and riverine flood plains will have high potential costs of impacts and adaptations. Adaptations are available that, as suggested in the case study for this sector, will have sizable benefits in relation to their costs.

Coastal Zones. Coastal areas In New York State have the potential to incur very high economic damages from a changing climate due to the enhanced coastal flooding as the result of sea level rise and continued development in residential and commercial zones, transportation infrastructure (treated separately in this study), and other facilities. Adaptation costs for coastal areas are expected to be significant, but relatively low as compared to the potential benefits.

Transportation. The transportation sector may have the highest climate change impacts in New York State among the sectors studied, and also the highest adaptation costs. There will be effects throughout the state, but the primary impacts and costs will be in coastal areas where a significant amount of transportation infrastructure is located at or below the current sea level. Much of this infrastructure floods already, and rising sea levels and storm surge will introduce unacceptable levels of flooding and service outages in the future. The costs of adaptation are likely to be very large and continuing.

Agriculture. For the agriculture sector, appropriate adaptation measures can be expected to offset declines in milk production and crop yields. Although the costs of such measures will not be insignificant, they are likely to be manageable, particularly for larger farms that produce higher value agricultural products. Smaller farms, with less available capital, may have more difficulty with adaptation and may require some form of adaptation assistance. Expansion of agricultural extension services and additional monitoring of new pests, weeds and diseases will be necessary in order to facilitate adaptation in the agricultural sector.

Ecosystems. Climate change will have substantial impacts on ecosystems in New York State. For revenue-generating aspects of the sector, including winter tourism and recreational fishing, climate change may impose significant economic costs. For other facets of the sector, such as forest-related ecosystems services, heritage value of alpine forests, and habitat for endangered species, economic costs associated with climate change are more difficult to quantify. Options for adaptation are currently limited within the ecosystems sector and costs of adaptation are only beginning to be explored. Development of effective adaptation strategies for the ecosystems sector is an important priority.

Energy. The energy sector, like communications, is one in which there could be large costs from climate change if ongoing improvements in system reliability are not implemented as part of regular and substantial reinvestment. However, it is expected that regular investments in system reliability will be made, so that the incremental costs of adaptation to climate change will be moderate. Even with regular reinvestments there may be increased costs from climate change. Moreover, the energy sector is subject to game-changing policy measures such as impacts on demand from a carbon tax (either directly or via cap and trade) and from the large investments in stability that could be undertaken to deal with the impacts of electromagnetic storms.

Communications. The communications sector is one in which there could be large costs from climate change if ongoing adaptations are not implemented as part of regular reinvestment in the sector or if storms are unexpectedly severe. However, it is expected that regular adaptations will be made, so that additional costs of adaptation for climate change will be relatively small.

Public Health. Public health will be impacted by climate change to the extent that costs could be large if ongoing adaptations to extreme events are not implemented. Costs could also be large if appropriate adaptations are not implemented in other sectors that directly affect public health, particularly water resources and energy. The costs associated with additional adaptations within the public health sector need further study.

10.2. SUMMARY

This study is an important starting point for assessing the costs of climate change impacts and adaptations in New York, although much further work needs to be done in order to provide detailed estimates of comprehensive costs and benefits associated with climate change. This work will have to deal with challenges such as the lack of climate-focused data sets and the fact that the feasibility of many potential adaptations has not been adequately analyzed. On the other hand, the basic conceptual approaches to future work have been identified, and even initial cost-benefit analyses of major impacts and corresponding adaptation options can help to illustrate the economic benefits of adaptation and thus to shape policy.

In terms of costs of adaptations, higher costs are projected for the Transportation sector, with its extensive capital infrastructure and less but still significant costs are projected for the Health, Water Resources, Ocean and Coastal Zones, Energy, and Communications sectors. Costs for adaptations in the Agriculture Sector are projected to be moderate, and costs for adaptations in the Ecosystems Sector require further assessment.

Net benefits comparing avoided impacts to costs of adaptation are most favorable for the Public Health and Ocean and Coastal Zones sectors, more moderate but still significant for the Water Resources, Agriculture, Energy, and Transportation sectors, and low for the Communications sector.

Planning for adaptation to climate change in New York State should continue to build on the State's significant climate change adaptation planning and implementation efforts to date, including further assessments of specific adaptation strategies. Benefits from adaptation are likely to be significant because there are many opportunities for development of resilience in all sectors and regions.

11 References

ACC, *see* Panel on Adapting to the Impacts of Climate Change

Agrawala, S. and S. Fankhauser, eds. 2008. *Economic Aspects of Adaptation to Climate Change: Costs, Benefits and Policy Instruments*. Paris: Organization for Economic Cooperation and Development.

AIR Worldwide Corporation. 2005. *The Coastline at Risk: Estimated Insured Value of Coastal Properties*. Boston MA: AIR Worldwide Corporation.

American Society of Civil Engineers Metropolitan Section (ASCE). 2009. "Against the Deluge: Storm Surge Barriers to Protect New York City." *Infrastructure Group Seminar*. Held at the Polytechnic Institute of New York University on March 30-31.

Apt, J., L.B. Lave, S. Talukdar, M.G. Morgan, and M. Ilic. 2004. "Electrical Blackouts: A Systemic Problem, Issues in Science and Technology." Summer 2004. Retrieved from http://www.issues.org/20.4/apt.html

Asian Development Bank. 2005. "Climate Proofing: A Risk-based Approach to Adaptation." *Pacific Studies Series*. Retrieved from http://www.adb.org/Documents/Reports/Climate-Proofing/chap6.pdf

Associated Press. 1986, February 25. "6 western states affected by flooding." *The New York Times*. Retrieved from http://www.nytimes.com/1986/02/25/us/around-the-nation-6-western-states-affected-by-flooding.html?scp=1&sq=6%20western%20states%20affected%20by%20flooding&st=cse

Associated Press. 2008, July 10. Like the Dollar, Value of American Life Has Dropped. *The New York Times*. Retrieved from http://www.nytimes.com/2008/07/10/business/worldbusiness/10iht-10life.14401415.html?scp=1&sq=Like%20the%20Dollar,%20Value%20of%20American%20Life%20Has%20Dropped&st=cse

Barber, L.M., J.J. Schleier III, and R.K.D. Peterson. 2010. "Economic cost analysis of West Nile virus outbreak, Sacramento County, California, USA 2005." *Emerging Infectious Disease* 16.3: 480-486.

Bell, M., R. Goldberg, C. Hogrefe, P.L. Kinney, K. Knowlton, B. Lynn, J. Rosenthal, C. Rosenzweig, and J.A. Patz . 2007. "Climate Change, Ambient Ozone, and Health in 50 US Cities." *Climatic Change:* 82, 61-76.

Bennet, E.M. and P. Balvanera. 2007. "The future of production systems in a globalized world." *Frontiers in Ecology and Environment.* 5(4): 191-198.

Berry, P. 2009. "Costing adaptation for natural ecosystems." In *Assessing the costs of adaptation to climate change: A review of the UNFCCC and other recent estimates*. M. Parry, N. Arnell, P. Berry, D. Dodman, S. Fankhauser, C. Hope, S. Kovats, R. Nicholls, D. Satterthwaite, R. Tiffin, and T. Wheeler, Contributors. London: International Institute for Environment and Development & Grantham Institute for Climate Change. 90-99.

Blake, E.S., E.N. Rappaport, C.W. Landsea. 2007. "The Deadliest, Costliest, and Most Intense United States Tropical Cyclones from 1851 to 2006 (And Other Frequently Requested Hurricane Facts)." *National Weather Service, National Hurricane Center*. Miami, Florida. Retrieved from http://www.nhc.noaa.gov/Deadliest_Costliest.shtml

Bowman, M.J., B. Colle, R. Flood, D. Hill, R.E. Wilson, F. Buonaiuto, P. Cheng and Y. Zheng. 2005. "Hydrologic Feasibility Of Storm Surge Barriers To Protect The Metropolitan New York – New Jersey Region, Final Report to New York Sea Grant." *HydroQual, Inc.* Retrieved from http://stormy.msrc.sunysb.edu/phase1/Phase%20I%20final%20Report%20Main%20Text.pdf

Brookhaven National Laboratory (BLN). 2001. "Technology Fact Sheet. Peconic River Remedial Alternatives: Wetland Restoration/Constructed Wetlands." Retrieved from http://www.bnl.gov/erd/Peconic/Factsheet/Wetlands.pdf

Buonaiuto, F., L. Patrick, E. Hartig, V. Gornitz, J. Stedinger, J. Tanski, J. Waldman. Forthcoming. "Coastal Zones." In *New York State ClimAID: Integrated Assessment for Effective Climate Change Adaptation Strategies in New York State.* C. Rosenzweig, W. Solecki, A. DeGaetano, S. Hassol, P. Grabhorn, M. O'Grady, Eds., Chapter 5.

Canadian Council of Professional Engineers. 2008. *Adapting to Climate Change: Canada's First National Engineering Vulnerability Assessment of Public Infrastructure.* Ottawa.

Center for Integrative Environmental Research. 2008. "Climate Change Impacts on Maryland and the Cost of Inaction." In *Maryland Climate Action Plan*. Maryland Commission on Climate Change, Chapter 3.

Chan, Sewell (2007, May 4) Con Ed Seeks to Raise Electricity Rates . *The New York Times*. Retrieved from http://empirezone.blogs.nytimes.com/tag/blackout-in-queens/

Chichilnisky, G., and G. Heal. 1998. "Economic returns from the biosphere." *Nature*. 391: 629-630.

Cline, W. 2007. *Global Warming and Agriculture: Impact Estimates by Country*. Washington, D.C.: Center for Global Development.

Connecticut Climate Change Infrastructure Workgroup of the Adaptation Subcommittee. 2010. *Climate Change Impacts on Connecticut Infrastructure: Draft 12/11/09.*
Connelly, N. 2010. Personal communication based on 2007 New York Statewide Angler Survey.

Connelly, N. and Brown. 2009a. *New York Statewide Angler Survey 2007. Report 1: Angler Effort and Expenditures.* New York State Department of Environmental Conservation, Bureau of Fisheries.

Connelly, N. and T. Brown. 2009b. *New York Statewide Angler Survey 2007. Report 1: Estimated Angler Effort and Expenditures in New York State Counties.* New York State Department of Environmental Conservation, Bureau of Fisheries.

Costanza, R., M. Wilson, A. Troy, A. Voinov, S. Liu, and J. Agostino. 2006. *The Value of New Jersey's Ecosystem Services and Natural Capital.* Rubenstein School of Environment and Natural Resources, University of Vermont: Gund Institute for Ecological Economics, Burlington, VT.

Dasgupta, Partha, Amartya Sen, and Stephen Marglin. 1972. *Guidelines for Project Evaluation,* United Nations Industrial Development 1972.

Dore, M. H., & Burton, I., 2000. *The Costs of Adaptation to Climate Change in Canada: A stratified Estimate by Sectors and Regions.* Canadian Comprehensive Auditing Foundation. Ottawa: Canadian Comprehensive Auditing Foundation.

Ebi, K. L., T.J. Teisberg, L.S. Kalkstein, L. Robinson, and R.F. Weiher. 2004. "Heat watch/warning systems save lives, estimated costs and benefits for Philadelphia." *American Meteorological Society.* 85(8): 1067-1073.

Eckstein, O. 1958. *Water-Resource Development.* Cambridge MA: Harvard University Press.

Economics of Climate Change Working Group (ClimateWorks Foundation, Global Environmental Facility, European Commission, McKinsey & Company, The Rockefeller Foundation, Standard Chartered Bank, and Swiss Re). 2009. Shaping Climate-Resilient Development: A Framework for Decision-making.

EcoWorld. (n.d.). *Gridpoint's Storage+.* Retrieved from EcoWorld Nature & Technology in Harmony: http://www.ecoworld.com/technology/gridpoints-storage.html

Fankhauser, S. 2010. "The costs of adaptation," *Wiley Interdisciplinary Reviews: Climate Change,* 1:1, January/February 2010: 23–30

Federal Emergency Management Agency. Floodsmart.gov, The Official Site of the National Flood Insurance Program, http://www.floodsmart.gov

Fischer, G., F.N. Tubiello, H. van Velthuizen, D.A. Wiberg. 2007. "Climate change impacts on irrigation water requirements: Effects of mitigation, 1990-2080." *Technological Forecasting and Social Change*. 74(7): 1083-1107.

Gittinger, J. Price. 1972. *Economic Analysis of Agricultural Projects*. Baltimore: The Johns Hopkins University Press.

Goodman, A.S., and M. Hastak.2006. *Infrastructure Planning Handbook: Planning, Engineering, and Economics*. ASCE Press.

Hammer, S. 2010. ClimAID Energy Sector Leader Interview. In D. Major, R. Leichenko & K. Johnson, eds. (Conference Call ed.). New York.

Hammer, S. and L. Parshall. Forthcoming. "Energy." In *New York State ClimAID: Integrated Assessment for Effective Climate Change Adaptation Strategies in New York State*. C. Rosenzweig, W. Solecki, A. DeGaetano, S. Hassol, P. Grabhorn, M. O'Grady, Eds., Chapter 8.

Hansler, G. and D.C. Major. 1999. "Climate change and the water supply systems of New York City and the Delaware Basin: Planning and action considerations for water managers." In *Proceedings of the Specialty Conference on Potential Consequences of Climate Variability and Change to Water Resources on the United States*. A. D. Briane, ed. American Water Resources Association. Herndon, VA. pp. 327-330.

Horton, R., D. Bader, A. DeGaetano, and C. Rosenzweig. Forthcoming. "Climate risks in New York state." In *New York State ClimAID: Integrated Assessment for Effective Climate Change Adaptation Strategies in New York State*, C. Rosenzweig, W. Solecki, A. DeGaetano, S. Hassol, P. Grabhorn, M. O'Grady, Eds., Chapter 1.

Jacob, K. 2010. ClimAID Transportation and Communication Sector Leader Interview. In D. Major, R. Leichenko & K. Johnson, eds. (Conference Call ed.). New York.

Jacob, K., G. Deodatis, J. Atlas, M. Whitcomb, M. Lopeman, O. Markogiannaki, Z. Kennett, and A. Morla. Forthcoming-a. "Transportation." In *New York State ClimAID: Integrated Assessment for Effective Climate Change Adaptation Strategies in New York State*. C. Rosenzweig, W. Solecki, A. DeGaetano, S. Hassol, P. Grabhorn, M. O'Grady, eds., Chapter 9.

Jacob, K., N. Maxemchuk, G. Deodatis, A. Morla, E. Schlossberg, I. Paung, and M. Lopeman. Forthcoming-b. "Telecommunications." In *New York State ClimAID: Integrated Assessment for Effective Climate Change Adaptation Strategies in New York State*. C. Rosenzweig, W. Solecki, A. DeGaetano, S. Hassol, P. Grabhorn, and M. O'Grady, eds., Chapter 10.

Jenkins, J. 2008. "Climate Change in the Adirondacks." *The Wild Center and The Wildlife Conservation Society.* Tupper Lake, NY. Available at: www.usclimateaction.org/userfiles/JenkinsBook.pdf

Kaiser Family Foundation, The. 2007. New York: Health Costs & Budgets. Retrieved February 2010, Data Sources: Health Expenditure Data, Health Expenditures by State of Residence, Centers for Medicare and Medicaid Services, Office of the Actuary, National Health Statistics Group. Retrieved from http://www.statehealthfacts.org/profileind.jsp?cat=5&rgn=34

Kalkstein, L.S. and J.S. Greene. 1997. "An evaluation of climate/mortality relationships in large U.S. cities and the possible impacts of a climate change." *Environmental Health Perspectives* 105(1): 84-93.

Kalkstein, L.S, and J.S. Greene. 2007. "An Analysis of potential heat-related mortality increases in U.S. cities under a business-as-usual climate change scenario." *Environment America*. 1-12.

Kinney, P., D. Shindell, E. Chae, and B. Winston. 2000. "Climate Change and Public Health: Impact Assessment for the NYC Metropolitan Region." *Metropolitan East Coast Assessment Report.* Retrieved from http://metroeast_climate.ciesin.columbia.edu/health.html

Kinney, PL. 2008. "Climate change, air quality, and human health." *Am J Prev Med.* 35(5): 459-467.

Kinney, P.L., M.S. O'Neill, M.L. Bell, and J. Schwartz. 2008. "Approaches for estimating effects of climate change on heat-related deaths: challenges and opportunities." *Environmental Science and Policy*. 11(1): 87-96, 2008.

Kinney, P. 2010. Public Health Sector Leader Interview. In D. Major, R. Leichenko & K. Johnson (Eds.) (Conference Call ed.). New York.

Kinney, P. L., P. Sheffield, R.S. Ostfeld, and J.L. Carr. Forthcoming. "Public Health." In *New York State ClimAID: Integrated Assessment for Effective Climate Change Adaptation Strategies in New York State*. C. Rosenzweig, W. Solecki, A. DeGaetano, S. Hassol, P. Grabhorn, M. O'Grady, Eds., Chapter 11.

Kirshen, P., M. Ruth and W. Anderson. 2006. "Climate's Long-Term Impacts on Urban Infrastructures and Services: The Case of Metro Boston", in M. Ruth, K. Donaghy and P.H. Kirshen (eds.), *Climate Change and Variability: Local Impacts and Responses*. Edward Elgar Publishers, Cheltenham, United Kingdom.

Kirshen, P. 2007. "Adaptation Options and Cost in Water Supply", a report to the UNFCCC Secretariat Financial and Technical Support Division, Retrieved from http://unfccc.int/files/cooperation_and_support/financial_mechanism/application/pdf/kirshen.pdf

Knowlton, K., J. Rosenthal, C. Hogrefe, B. Lynn, S. Gaffin, R. Goldberg, C. Rosenzweig, K. Civerolo, J-Y. Ku, and P.L. Kinney. 2004. Assessing ozone-related health impacts under a changing climate." *Environ. Health Perspect.*, 112:1557–1563.

Knowlton, K., B. Lynn, R. Goldberg, C. Rosenzweig, C. Hogrefe, J. Rosenthal, P. Kinney. 2007. "Projecting heat-related mortality impacts under a changing climate in the New York City Region. 2007. *American Journal of Public Health* 97(1): 2028-2034.

Kovacs, K.F., R.G. Haight, D.G. McCullough, R.J. Mercader, N.W. Siegert, and A.M. Liebhold. 2010. "Cost of potential emerald ash borer damage in U.S. communities, 2009-2019." *Ecological Economics.* 69(3): 569-578.

Krutilla, J.V., and O. Eckstein. 1958. *Multiple Purpose River Development*. Baltimore MD, Johns Hopkins University Press.

Leichenko, R., P. Vancura, and A. Thomas. Forthcoming. "Equity and Environmental Justice." In *New York State ClimAID: Integrated Assessment for Effective Climate Change Adaptation Strategies in New York State*, C. Rosenzweig, W. Solecki, A. DeGaetano, S. Hassol, P. Grabhorn, M. O'Grady, Eds., Chapter 3.

Lettenmaier, D.P., D.C. Major, L. Poff, and S. Running. 2008. "Water Resources." In *The Effects of Climate Change on Agriculture, Land Resources, Water Resources, and Biodiversity in the United States*. A Report by the U.S. Climate Change Science Program and the Subcommittee on Global Change Research. Washington, D.C. 121-150.

Lin S., M.I. Gomez, S.A. Hwang, E.M. Franko, and J.K. Bobier. 2004. "An evaluation of the asthma intervention of the New York State Healthy Neighborhoods Program." *J Asthma.* 41(5): 583-95.

Lin S., L. Ming, R.J. Walker, X. Liu, S. Hwang, and R. Chinary. 2009. "Extreme high temperatures and hospital admissions for respiratory and cardiovascular diseases." *Epidemiology* 20.5: 738-746.

Lin S., B. Fletcher, M. Luo, R. Chinery, and S. Hwang. 2010. *Health impacts in New York City during the Northeastern (US) Blackout of 2003.* Manuscript.

Lozano, P., S.D. Sullivan, D.H. Smith, and K.B. Weiss, 1999. "The economic burden of asthma in US children: Estimates from the National Medical Expenditure Survey." *Journal of Allergy and Clinical Immunology.* 104(5): 957 - 963.

Maass, A., M.M. Hufschmidt, R. Dorfman, H.A. Thomas, Jr., S.A. Marglin, and G.M. Fair. 1962. *Design of Water-Resource Systems.* Cambridge MA: Harvard University Press.

Major, D. C. 1977. *Multiobjective Water Resource Planning*. Washington, DC: American Geophysical Union, Water Resources Monograph 4.

Major, D.C., and M.C. O'Grady. 2010. "Adaptation Assessment Guidebook." In *Climate Change Adaptation in New York City: Building a Risk Management Response.* C. Rosenzweig and W. Solecki, eds. Annals of the New York Academy of Sciences, Volume 1185. New York, N.Y.

Major, D.C., and R.A. Goldberg. 2001. "Water Supply." In Rosenzweig, C. and W.D. Solecki, eds. *Climate Change and a Global City: The Potential Consequences of Climate Variability and Change - Metro East Coast.* New York: Columbia Earth Institute. Chapter 6.

Maloney, Congresswoman Carolyn B. 2001, August 2. Federal Study Shows Need for Seawall Repair on Roosevelt Island, Press Release. Retrieved from http://maloney.house.gov/index.php?option=com_content&task=view&id=661&Itemid=61

Marglin, S. A. 1967. *Public Investment Criteria*. Cambridge MA: MIT Press.

Margulis, S., A. Bucher, D. Corderi, U. Narain, H. Page, K. Pandey, T.T. Linh Phu, C. Bachofen, R. Mearns, B. Blankespoor, S. Dasgupta, S. Murray, E. Cushion, L. Gronnevet, L. Cretegny, P. Ghosh, B. Laplante, L. Leony, R. Schneider, P. Ward, and D. Wheeler. 2008. *The Economics of Adaptation to Climate Change: Methodology Report.* Washington DC: The World Bank.

McCarl, B. A. 2007. *Adaptation options for agriculture, forestry and fisheries*. A Report to the UNFCCC Secretariat Financial and Technical Support Division: UNFCCC.

McCulloch, M.M., D.L. Forbes, R.W. Shaw, and the CCAF A041 Scientific Team. 2002. *Coastal Impacts of Climate Change and Sea-Level Rise on Prince Edward Island: Synthesis Report*. Geological Survey of Canada, Open File 421.

Meltzer, M.I., D.T. Dennis, and K.A. Orloski. 1999. "The Cost Effectiveness of Vaccinating against Lyme Disease." *Emerging Infectious Diseases.* 5(3): 321-28.

Metropolitan Transportation Authority. 2007. August 8, 2007 Storm Report, Sep. 20.

Miller, K. A., and D. Yates, eds., 2005. *Climate Change and Water Resources: A Primer for Water Utilities.* Denver CO: American Water Resources Association Research Foundation.

Milly, P.C. D., J. Betancourt, M. Falkenmark, R.M. Hirsch, Z.W. Kundzewicz, D.P. Lettenmaier, and R.J. Stouffer. 2008. "Climate Change: Stationarity Is Dead: Whither Water Management?," *Science.* 1 February 2008: 573-574.

MKF Research. 2005. *Economic Impact of New York Grapes, Grape Juice and Wine*. MKF Research LLC, St. Helena, CA.

Morris, S. C., G. Goldstein, A. Singhi, and D. Hill. 1996. "Energy Demand and Supply in Metropolitan New York with Global Climate Change." In D. Hill, ed. *Annals of the New York Academy of Science: The Baked Apple? Metropolitan New York in the Greenhouse.* New York, New York Academy of Science.

Multihazard Mitigation Council. 2005a. *Natural Hazard Mitigation Saves: An Independent Study to Assess the Future Savings from Mitigation Activities, vol. 1: Findings, Conclusions, and Recommendations.* Retrieved from http://www.floods.org/PDF/MMC_Volume1_FindingsConclusionsRecommendations.pdf

Multihazard Mitigation Council, 2005b. *Natural Hazard Mitigation Saves: An Independent Study to Assess the Future Savings from Mitigation Activities, vol. 2: Study Documentation.* Washington DC: National Institute of Building Sciences.

National Research Council, Committee on Assessing and Valuing the Services of Aquatic and Related Terrestrial Ecosystems. 2004. *Valuing Ecosystem Services: Toward Better Environmental Decision-Making*. Washington, D.C.: The National Academies Press. Retrieved from http://www.nap.edu/openbook.php?record_id=11139&page=1

National Research Council, Committee on Climate change and U.S. Transportation. 2008. *Potential Impacts of climate change on U.S. transportation.* Transportation Research Board Special Report 290. Washington, DC: Transportation Research Board.

New York City Department of Environmental Protection. 2008a. *Climate Change Program: Assessment and Action Plan.* Retrieved from http://www.nyc.gov/html/dep/html/news/climate_change_report_05-08.shtml

New York City Department of Environmental Protection. 2008b. Request for Proposals for Contract: 826EE-GCC: Climate Change and Population Growth Effects on New York City Sewer and Wastewater Systems.

New York City Economic Development Corporation, Department of Information Technology and Telecommunications, & Department of Small Business Services. 2005. *Telecommunications and Economic Development in New York City: A Plan for Action*. New York City.

New York City Municipal Water Finance Authority. 2009, December 11. Water and Sewer System Second General Resolution Revenue Bonds, Adjustable Rate Fiscal 2010 Series CC (CUSIP #64972FK62).

New York City Panel on Climate Change. 2010. *Climate Change Adaptation in New York City: Building a Risk Management Response.* C. Rosenzweig and W. Solecki, Eds. Prepared for use by the New York City Climate Change Adaptation Task Force. Annals of the New York Academy of Sciences, Volume 1185. New York, N.Y., 354 pp.

New York Invasive Species Clearinghouse. 2010. New York Invasive Species Information. Retrieved from http://nyis.info/

New York State Department of Agriculture and Markets. 2007. Commissioner Announces $30 Million for Dairy Assistance. Retrieved from: http://www.agmkt.state.ny.us/AD/release.asp?ReleaseID=1604

New York State Department of Environmental Conservation. 1974. *Tidal Wetlands Inventory*.

New York State Department of Environmental Conservation. 2009. New York Statewide Angler Survey 2007. Retrieved from http://www.dec.ny.gov/outdoor/56020.html

New York State Department of Environmental Conservation (NYSDEC). 2010a. Lands and Waters. Retrieved from www.dec.ny.gov/61.html

New York State Department of Environmental Conservation (NYSDEC). 2010b. Status and Trends of Freshwater Wetlands in New York State. Retrieved from http://www.dec.ny.gov/lands/31835.html

New York State Department of Health. 2010, 2009, 2008, 2007, 2006, 2004, 2003, 2002, 2001. West Nile Virus update. West Nile Virus summary list. Retrieved February 2010, from http://www.health.state.ny.us/nysdoh/westnile/update/update.htm

Table 9.8 New York State Department of Health. The Burden of Cardiovascular Disease in New York: Mortality, Prevalence, Risk Factors, Costs, and Selected Populations [Data File]. Retrieved from http://www.nyhealth.gov/diseases/cardiovascular/heart_disease/docs/burden_of_cvd_in_nys.pdf

New York State Department of Health. 2009. *New York State Asthma Surveillance Summary Report 2009.* Retrieved from http://www.health.state.ny.us/statistics/ny_asthma/pdf/2009_asthma_surveillance_summary_report.pdf

New York State Department of Labor. 2008. Seasonally Adjusted Employment, New York State (data in thousands). Retrieved from http://www.labor.ny.gov/stats/lscesmaj.shtm

New York State Department of Public Service. 2007. In *the Matter of Consolidated Edison Company of New York, Inc. Case 07-E-0523 September 2007.* Prepared Testimony of Kin Eng, Utility Analyst 3, Office of Electric, Gas, and Water.

New York State Integrated Pest Management Program. 2010. *The New York State Agricultural IPM Program.* NYSIPM: Cornell University.

New York State Office of the Comptroller. 2010. *The Role of Agriculture in the New York State Economy.* Report 21-10. Available at: www.osc.state.ny.us/reports/other/agriculture21-2010.pdf

New York Times, The. 1994, October 22. "Texans striving to contain pipeline spills". Retrieved from http://www.nytimes.com/1994/10/22/us/texans-striving-to-contain-pipeline-spills.html?scp=1&sq=Texans+striving+to+contain+pipeline+spills&st=nyt

Niemi, E., M. Buckley, C. Neculae, and S. Reich. 2009. *An Overview of Potential Economic Costs to Washington of a Business-As-Usual Approach to Climate Change*. The Program on Climate Economics, Climate Leadership Initiative. Institute for a Sustainable Environment, University of Oregon.

Nordhaus, W. and J. Boyer. 2000. *Warming the World: Economic Models of Global Warming*. Cambridge, MA: MIT Press.

Nordhaus, Willliam. 2007, July 13. “Critical Assumptions in the Stern Review on Climate Change.” *Science 317*: 201-202.

Nordhaus William, 2009, “An Analysis of the Dismal Theorem,” http://cowles.econ.yale.edu/P/cd/d16b/d1686.pdf

North East Foresters Association. 2007. *The Economic Importance of Wood Flows from New York’s Forests.* Retrieved from http://www.nefainfo.org/

NYSERDA ClimAID Team. 2010. *Integrated Assessment for Effective Climate Change Adaptation Strategies in New York State.* C. Rosenzweig, W. Solecki, A. DeGaetano, S. Hassol, P. Grabhorn, M. O’Grady, eds. New York State Energy Research and Development Authority (NYSERDA), 17 Columbia Circle, Albany, New York, 12203.

Office of the Attorney General, Andrew Cuomo. 2007. RE: Case 06-E-0894 – Proceeding on Motion of the Commission to Investigate the Electric Power Outages in Consolidated Edison Company of New York, Inc.’s Long Island City Electric Network. State of New York; March 2, 2007.

Panel on Adapting to the Impacts of Climate Change, Board on Atmospheric Sciences and Climate, Division on Earth and Life Studies. 2010. National Research Council, *America’s Climate Choices: Adapting to the Impacts of Climate Change*, Washington, DC: The National Academies Press.

Paradis, A., J. Elkinton, K. Hayhoe, and J. Buonaccorsi. 2008. “Role of winter temperature and climate change on the survival and future range expansion of the hemlock woolly adelgid (Adelges tsugae) in eastern North America.” *Mitigation and Adaptation Strategies for Global Change.* 13:541-554.

Parry, M.L., C. Rosenzweig, A. Iglesias, M. Livermore, and G. Fischer. 2004. “Effects of climate change on global food production under SRES emissions and socio-economic scenarios.” *Global Environmental Change* 14. 1: 53-67.

Parry, M. L., O. F. Canziani, J. P. Palutikof and Co-authors. 2007. Technical Summary. *Climate Change 2007: Impacts, Adaptation and Vulnerability. Contribution of Working Group II to the Fourth Assessment Report of the Intergovernmental Panel on Climate Change*. M.L. Parry, O.F. Canziani, J.P. Palutikof, P.J. van der Linden and C.E. Hanson, eds. Cambridge University Press: Cambridge, UK. 23-78.

Parry, M., Arnell, N., Berry, P., Dodman, D., Fankhauser, S., Hope, C., Kovats, S., Nicholls, R., Satterthwaite, D., Tiffin, R., Wheeler, T. 2009. *Assessing the costs of adaptation to climate change: A review of the UNFCCC and other recent estimates*. London: International Institute for Environment and Development & Grantham Institute for Climate Change.

Pielke, Jr., R.A., Gratz, J., Landsea, C.W., Collins, D., Saunders, M.A., and Musulin, R. 2008. “Normalized Hurricane Damages in the United States: 1900-2005.” *Natural Hazards Review.* 9(1): 29-42.

Pimentel, D., R. Zuniga, and D. Morrison. 2005. “Update on the environmental and economic costs associated with alien-invasive species in the United States.” *Ecological Economics.* 52: 273-288.

Rich, J. 2008. “Winter nitrogen cycling in agroecosystems as affected by snow cover and cover crops.” M.S. Thesis. Cornell University. Ithaca, NY.

Rosenzweig, C. and W.D. Solecki, eds. 2001. *Climate Change and A Global City: The Potential Consequences of Climate Variability and Change–Metro East Coast.* Report for the U.S. Global Change Research Program, National Assessment of the Potential Consequences of Climate Variability and Change for the United States. New York: Columbia Earth Institute. p.224. Retrieved from http://metroeast_climate.ciesin.columbia.edu/

Rosenzweig, C., R. Horton, D.C. Major, V. Gornitz, and K. Jacob. 2007a. “Appendix 2: Climate Component, 8.8.07 MTA Task Force Report.” In *August 8, 2007, Storm Report*. Metropolitan Transportation Authority. p.73. Retrieved from http://www.mta.info/mta/pdf/storm_report_2007.pdf

Rosenzweig, C., D.C. Major, K. Demong, C. Stanton, R. Horton, and M. Stults. 2007b. “Managing climate change risks in New York City's water system: Assessment and adaptation planning.” *Mitig. Adapt. Strategies Global Change*, **12**, 1391-1409, doi:10.1007/s11027-006-9070-5.

Savonis, M. J., Virginia R. Burkett, and Joanne R. Potter. 2008. *Impacts of Climate Change and Variability on Transportation Systems and Infrastructure: Gulf Coast Study, Phase I.* Washington, DC, USA, U.S. Climate Change Science Program, Synthesis and Assessment Product 4.7, U.S. Department of Transportation.

Schneider, R. 2010. ClimAID Water Resources Sector Leader Interview. In D. Major, R. Leichenko & K. Johnson, eds. (Conference Call ed.). New York.

Schneider, R., A. McDonald, S. Shaw , S. Riha, L. Tryhorn, A. Frei, B. Montz. Forthcoming. "Water." In *New York State ClimAID: Integrated Assessment for Effective Climate Change Adaptation Strategies in New York State.* C. Rosenzweig, W. Solecki, A. DeGaetano, S. Hassol, P. Grabhorn, M. O'Grady, eds. Chapter 4.

Scott, D., J. Dawson, and B. Jones. 2008. "Climate change vulnerability of the US northeast winter recreation tourism sector." *Mitigation and Adaptation Strategies for Global Change*. 13(5-6), 577-596.

Sheffield, P. 2010. Personal communication regarding costs of air conditioning program in New York City.

Solecki, W., R. Leichenko, D. C. Major, M. Brady, D. Robinson, M. Barnes, M. Gerbush, and L. Patrick. 2011 (Forthcoming). *Economic Assessment of Climate Change Impacts in New Jersey*.

Stern, N. 2007. *The Economics of Climate Change: The Stern Review.* Cambridge, UK: Cambridge University Press.

Stern, N., 2009. *The Global Deal: Climate Change and the Creation of a New Era of Progress and Prosperity*. New York: Public Affairs.

St. Pierre, N.R., B. Cobanov, and G. Schnitkey. 2003. "Economic losses from heat stress by U.S. livestock industries." *J Dairy Sci* 86: (E Suppl): E52-E77.

Sussman, E. and D. C. Major, Coordinating Lead Authors, Climate Change Adaptation: Fostering Progress Through Law and Regulation," New York University Environmental Law Journal, *18*:1 (2010), 55-155.

The Center for Integrated Environmental Research . 2008. *Economic Impacts of Climate Change on New Jersey.* University of Maryland. College Park: The Center for Integrated Environmental Research.

The Nature Conservancy. n.d. *Coastal Resilience Long Island: Adapting Natural and Human Communities to Sea Level Rise and Coastal Hazards.* Retrieved from http://coastalresilience.org/

Titus, J. 2002. "Does Sea Level Rise Matter to Transportation Along the Atlantic Coast?" In *United States Department of Transportation Center for Climate Change and Environmental Forecasting, 2002, The Potential Impacts of Climate Change on Transportation*. Federal Research Partnership Workshop Summary and Discussion Papers.

Tol, R.S.J. 2002. "Estimates of the damage costs of climate change. Part 1. Benchmark Estimates." *Environmental and Resource Economics.* 21: 47-73.

Tourism Economics. 2009. *The Economic Impact of Tourism in New York State*. Oxford Economics, Oxford, UK and Wayne, PA. Retrieved from http://thebeat.iloveny.com/industry/wp-content/uploads/2009/04/statewideimpact.pdf

Turner, L.W., R.C. Warner and J.P. Chastain. 1997. *Micro-sprinkler and Fan Cooling for Dairy Cows: Practical Design Considerations; University of Kentucky.* Department of Agriculture. Cooperative Extension Service. AEN-75. Retrieved from http://www.ca.uky.edu/agc/pubs/aen/aen75/aen75.pdf

United Nations Development Programme (UNDP). 2007. *Fighting Climate Change: Human Solidarity in a Divided World*. Human Development Report 2007/2008. New York: Palgrave Macmillan.

United Nations Framework Convention on Climate Change. n.d. National Adaptation Programmes of Action (NAPAs). Retrieved from http://unfccc.int/national_reports/napa/items/2719.php

United Nations Framework Convention on Climate Change. 2007. *Investment and financial flows to address climate change*. Bonn: UNFCCC.

United States Army Corps of Engineers, New York District. n.d. Green Brook Sub Basin, NJ. Retrieved from http://www.nan.usace.army.mil/business/prjlinks/flooding/greenbk/

United States Army Corps of Engineers. 2010. Personal communication between Michael Morgan, U.S. Army Corps of Engineers, Ecosystem Restoration Team Leader, NY Region and Frank Buonaiuto on costs of recent restoration efforts on Long Island.

United States Census Bureau, American FactFinder. 2010a. *2007 Economic Census.* Sector 31: EC0731A1: Manufacturing: Geographic Area Series: Industry Statistics for the States, Metropolitan and Micropolitan Statistical Areas, Counties, and Places. Retrieved from http://factfinder.census.gov/servlet/IBQTable?_bm=y&-geo_id=04000US36&-fds_name=EC0700A1&-ds_name=EC0731A1&-_lang=en

United States Census Bureau, American FactFinder. 2010b. *Detailed Statistics, 2007 Economic Census*. Sector 62: EC0762A1: Health Care and Social Assistance: Geographic Area Series: Summary Statistics: 2007. Retrieved from http://factfinder.census.gov/servlet/IBQTable?_bm=y&-geo_id=04000US36&-fds_name=EC0700A1&-_skip=0&-ds_name=EC0762A1&-_lang=en

United States Congress. 1936. *Flood Control Act of 1936, Public Law 74-738.* 74th Congress, 2nd Session.

United States Department of Agriculture National Agricultural Statistics Service. 2009. *U.S. Census of Agriculture 2007.* Retrieved from http://www.agcensus.usda.gov/

United States Department of Agriculture National Agricultural Statistics Service, New York Office. 2009. *New York Annual Statistical Bulletin.* Retrieved from http://www.nass.usda.gov/Statistics_by_State/New_York/Publications/Annual_Statistical_Bulletin/2009/09-pages30-37%20-%20Fruit.pdf

United States Department of Agriculture National Agricultural Statistics Service. 2010. U.S. United States Department of Agriculture Economic Research Service, 2009. *New York State Fact Sheet.* Retrieved from http://www.ers.usda.gov/StateFacts/NY.htm

United States Department of Agriculture National Agricultural Statistics Service, New York Office. 2010. *New York Farm Land Protection Survey 2009.* Retrieved from http://www.nass.usda.gov/Statistics_by_State/New_York/Publications/Special_Surveys/Report-NY%20Farmland%20Protect%20Svy.pdf

United States Department of Commerce Bureau of Economic Analysis. 2009. *New York BEARFACTS.* 2010. Retrieved from http://www.bea.gov/regional/bearfacts/action.cfm

United States Department of Commerce Bureau of Economic Analysis. n.d. *Gross Domestic Product by State.* Retrieved from http://bea.gov/regional/gsp/

United States Department of Commerce Bureau of Economic Analysis, n.d. National Economic Accounts. Retrieved from: http://www.bea.gov/national/index.htm

United States Department of Commerce Bureau of Economic Analysis, 2010. *Regional Economic Accounts.* Retrieved from http://www.bea.gov/regional/

United States Environmental Protection Agency (USEPA). 2010. Guidelines for Preparing Economic Analysis, Frequently Asked Questions on Mortality Risk. http://yosemite.epa.gov/ee/epa/eed.nsf/pages/MortalityRiskValuation.html (updated 10/27/1010)

United States Environmental Protection Agency. 2004. *Value of Statistical Life Analysis and Environmental Policy: A White Paper*. prepared by C. Dockins, K. Maguire, N. Simon, M. Sullivan for the U.S. EPA, National Center for Environmental Economics; April 21, 2004 For presentation to Science Advisory Board - Environmental Economics Advisory Committee.

United States Environmental Protection Agency (USEPA). 2000. Guidelines for Preparing Economic Analyses, Office of the Administrator. EPA-240-R-00-003, December 2000.

United States Fish and Wildlife Service. 2006. *National Survey of Fishing, Hunting, and Wildlife-Associated Recreation.* Washington, D.C.

Weitzman. 2009. “Reactions to the Nordhaus Critique.” (Preliminary). Retrieved from http://www.economics.harvard.edu/faculty/weitzman/papers_weitzman.

Wharton School. 2003. “Lights Out: Lessons from The Blackout.” August 23. Wharton School of the University if Pennsylvania. Retrieved from http://knowledge.wharton.upenn.edu/article.cfm?articleid=838

Wheeler, T. and R. Tiffin. 2009. “Costs of adaptation in agriculture, forestry and fisheries.” In M. Parry et al. *Assessing the costs of adaptation to climate change: A review of the UNFCCC and other recent estimates*. London: International Institute for Environment and Development & Grantham Institute for Climate Change. 29-39.

Wilks, D.S. and D.W. Wolfe. 1998. “Optimal use and economic value of weather forecasts for lettuce irrigation in a humid climate.” *Agricultural and Forest Meteorology.* 89: 115-129.

Williams, P. J. and M. Wallis. 1995. "Permafrost and Climate Change: Geotechnical Implications." *Philosophical Transactions: Physical Sciences and Engineering.* 352(1699): 347-358.

World Bank. 2006. *Investment Framework for Clean Energy and Development*. Washington, D.C.: World Bank.

World Bank. 2010. *The Economics of Adaptation to Climate Change: A synthesis report;* Washington, D.C.: World Bank.

Wolfe, D. 2010. ClimAID Agriculture and Ecosystems Sector Leader Interview. In D. Major, R. Leichenko & K. Johnson, eds. (Conference Call ed.). New York.

Wolfe, D., and J. Comstock. Forthcoming-a. “Ecosystems.” In *New York State ClimAID: Integrated Assessment for Effective Climate Change Adaptation Strategies in New York State*, C. Rosenzweig, W. Solecki, A. DeGaetano, S. Hassol, P. Grabhorn, M. O’Grady, Eds., Chapter 6.

Wolfe, D., and J. Comstock. Forthcoming-b. “Agriculture.” In *New York State ClimAID: Integrated Assessment for Effective Climate Change Adaptation Strategies in New York State*, C. Rosenzweig, W. Solecki, A. DeGaetano, S. Hassol, P. Grabhorn, M. O’Grady, Eds., Chapter 7.

Yohe, G. and R. Leichenko. 2010. “Adopting a Risk-Based Approach,” in New York City Panel on Climate Change. 2010. *Climate Change Adaptation in New York City: Building a Risk Management Response.* C. Rosenzweig and W. Solecki, Eds. Prepared for use by the New York City Climate Change Adaptation Task Force. Annals of the New York Academy of Sciences, Volume 1185. New York, N.Y., 354 pp.